Lecture Notes in Computer Science 5810

Commenced Publication in 1973
Founding and Former Series Editors:
Gerhard Goos, Juris Hartmanis, and Jan van Leeuwen

Srečko Brlek Christophe Reutenauer
Xavier Provençal (Eds.)

Discrete Geometry for Computer Imagery

15th IAPR International Conference, DGCI 2009
Montréal, Canada, September 30 – October 2, 2009
Proceedings

Volume Editors

Srečko Brlek
Université du Québec à Montréal
Laboratoire de Combinatoire et d'Informatique Mathématique
CP 8888, Succ. Centre-ville, Montréal (QC) H3C 3P8, Canada
E-mail: brlek.srecko@uqam.ca

Christophe Reutenauer
Université du Québec à Montréal
Laboratoire de Combinatoire et d'Informatique Mathématique
CP 8888, Succ. Centre-ville, Montréal (QC) H3C 3P8, Canada
E-mail: reutenauer.christophe@uqam.ca

Xavier Provençal
Université de Montpellier II
Laboratoire d'Informatique de Robotique et de Microélectronique de Montpellier
161 Rue Ada, 34392 Montpellier, France
E-mail: provencal@lirmm.fr

Library of Congress Control Number: 2009934040

CR Subject Classification (1998): I.3.5, I.4.1, I.4, I.3, G.2

LNCS Sublibrary: SL 6 – Image Processing, Computer Vision, Pattern Recognition, and Graphics

ISSN 0302-9743
ISBN-10 3-642-04396-8 Springer Berlin Heidelberg New York
ISBN-13 978-3-642-04396-3 Springer Berlin Heidelberg New York

springer.com

Printed in Germany

Typesetting: Camera-ready by author, data conversion by Scientific Publishing Services, Chennai, India
Printed on acid-free paper SPIN: 12757718 06/3180 5 4 3 2 1 0

Preface

Held for the first time outside Europe, the 15th International Conference on Discrete Geometry for Computer Imagery took place in Montréal (Canada) from September 30 to October 2, 2009. This conference addressed a large international audience: 61 papers were submitted originating from 14 different countries.

Following a thorough reviewing process, remodeled for the previous conference, held in Lyon, 42 papers were accepted and scheduled for either oral (21) or poster presentation (21). All these papers appear in these proceedings, and whether a paper was presented orally or not was based on our appreciation of suitability rather than on ranking.

As discrete geometry is emerging as a theory from groundwork in automated representation and processing of digitized objects, we invited three distinguished speakers of international renown: Valérie Berthé from the LIRMM (Montpellier) gave an account on discrete planes from the point of view of combinatorics on words, with relations to number theory and in particular multidimensional continued fractions. Anders Kock, whose research is mostly in category theory, contributed to the development of what is known as synthetic differential geometry. The basis of the theory is the fact that classical differential calculus can be lifted in algebraic geometry where the limit process does not exist: this is achieved by enriching the affine line R with infinitesimals – nilpotent elements in this case – that are distinct from the infinitesimals in non-standard analysis. The research of Pierre Gauthier focuses on mathematical modeling of the climate. Modeling the atmosphere is necessary for numerical weather prediction (NWP), and the theoretical background for addressing the problem is based on fluid dynamics governed by the Navier-Stokes equations, thermodynamic relationships, and numerous other processes that influence its dynamics.

Finally, the "Laboratoire de Combinatoire et d'Informatique Mathématique (LaCIM)" was fortunate to be able to count on the conference secretary Lise Tourigny and the webmaster Jérôme Tremblay. Our warmest thanks for their invaluable contribution to the organization of the event.

July 2009

Srečko Brlek
Xavier Provençal
Christophe Reutenauer

Organization

DGCI 2009 was hosted and organized by the Laboratoire de Combinatoire et d'Informatique Mathématique (LaCIM) of the Université du Québec à Montréal.

Organizing Committee

Srečko Brlek	LaCIM, (General Chair)
Christophe Reutenauer	LaCIM, (General Co-chair)
Xavier Provençal	LIRMM, LAMA, FQRNT
Ariane Garon	LaCIM, (Local Arrangements)
Lise Tourigny	LaCIM, (Secretary)
Jérôme Tremblay	LaCIM, (Webmaster)

Steering Committee

Eric Andres	XLIM-SIC, Université de Poitiers, France
Gunilla Borgefors	Centre for Image Analysis, Uppsala University, Sweden
Achille Braquelaire	LABRI, Université Bordeaux 1, France
Jean-Marc Chassery	Gipsa-Lab, CNRS, Grenoble, France
David Coeurjolly	LIRIS, CNRS, Université Claude Bernard, Lyon, France
Ullrich Köthe	HCI, University of Heidelberg, Germany
Annick Montanvert	Gipsa-Lab, Université Pierre-Mendès France, Grenoble, France
Kálmán Palágyi	Department of Image Processing and Computer Graphics, University of Szeged, Hungary
Gabriella Sanniti di Baja	Istituto di Cibernetica Eduardo Caianiello, CNR, Italy

Program Committee

Barneva, Reneta
Couprie, Michel
Debled-Rennesson, Isabelle
Daurat, Alain
Fiorio, Christophe
Jonker, Pieter
Lachaud, Jacques-Olivier
Kiselman, Christer O.
Kropatsch, Walter G.
Malgouyres, Rémy
Pinzani, Renzo
Soille, Pierre
Thiel, Edouard
Veelaert, Peter

List of Reviewers

Andres, Eric
Barcucci, Elena
Barneva, Reneta
Batenburg, K. Joost
Becker, Jean-Marie
Berthé, Valérie
Bloch, Isabelle
Bodini, Olivier
Borgefors, Gunilla
Braquelaire, Achille
Brimkov, Valentin E.
Brlek, Srečko
Brun, Luc
Chassery, Jean-Marc
Coeurjolly, David
Couprie, Michel
Damiand, Guillaume
Daurat, Alain
Debled-Rennesson, Isabelle
Domenger, Jean-Philippe
Domenjoud, Eric
Eckhardt, Ulrich
Fernique, Thomas
Feschet, Fabien
Fiorio, Christophe
Frosini, Andrea
Gerard, Yan
Guedon, Jean-Pierre
Imiya, Atsushi
Jamet, Damien
Jean-Marie, Alain
Jonker, Pieter
Kenmochi, Yukiko
Kerautret, Bertrand
Kiselman, Christer O.
Klette, Reinhard
Koskas, Michel
Köthe, Ullrich
Kropatsch, Walter G.
Labelle, Gilbert
Lachaud, Jacques-Olivier
Lindblad, Joakim
Malgouyres, Rémy
Miguet, Serge
Montanvert, Annick
Monteil, Thierry
Normand, Nicolas
Nyul, Laszlo G.
Palagyi, Kalman
Peltier, Samuel
Pinzani, Renzo
Provençal, Xavier
Remy, Éric
Ronse, Christian
Sanniti di Baja, Gabriella
Sivignon, Isabelle
Soille, Pierre
Strand, Robin
Tajine, Mohamed
Talbot, Hugues
Thiel, Edouard
Tougne, Laure
Toutant, Jean-Luc
Veelaert, Peter
Vialard, Anne
de Vieilleville, François
Vuillon, Laurent

Sponsoring Institutions

DGCI 2009 was sponsored by the following institutions:

- The International Association for Pattern Recognition (IAPR)
- Centre de Recherches Mathématiques
- Université du Québec à Montréal
- Laboratoire de Combinatoire et d'Informatique Mathématique (LaCIM)
- Chaire de Recherche du Canada en Algèbre, Combinatoire et Informatique Mathématique

Table of Contents

Invited Papers

Discrete Shape Representation, Recognition and Analysis

Discrete and Combinatorial Tools for Image Segmentation and Analysis

Discrete and Combinatorial Topology

Models for Discrete Geometry

Geometric Transforms

Discrete Tomography

Arithmetic Discrete Planes Are Quasicrystals

Valérie Berthé

LIRMM, Université Montpellier II, 161 rue Ada, 34392 Montpellier Cedex 5 - France
berthe@lirmm.fr

Abstract. Arithmetic discrete planes can be considered as liftings in the space of quasicrystals and tilings of the plane generated by a cut and project construction. We first give an overview of methods and properties that can be deduced from this viewpoint. Substitution rules are known to be an efficient construction process for tilings. We then introduce a substitution rule acting on discrete planes, which maps faces of unit cubes to unions of faces, and we discuss some applications to discrete geometry.

Keywords: digital planes; arithmetic discrete planes; tilings; word combinatorics; quasicrystals; substitutions.

1 Introduction

Discrete planes are known since the work of Réveillès [Rev91], see also [AAS97], to be efficiently defined in arithmetic terms, which allows a precise and effective understanding of most of their properties, see e.g. [BCK07]. More precisely, let $\boldsymbol{v} \in \mathbb{R}^d$, $\mu, \omega \in \mathbb{R}$. The *arithmetic discrete (hyper)plane* $\mathfrak{P}(\boldsymbol{v}, \mu, \omega)$ is defined as the set of points $\boldsymbol{x} \in \mathbb{Z}^d$ satisfying $0 \leq \langle \boldsymbol{x}, \boldsymbol{v} \rangle + \mu < \omega$. We assume now for the sake of clarity that $d = 3$ but the results and methods mentioned in the present paper hold for any $d \geq 3$.

For the sake of simplicity $\boldsymbol{v} = (v_1, v_2, v_3)$ is assumed in all that follows to be a nonzero vector with nonnegative coordinates. We consider here integer as well as irrational parameters $\boldsymbol{v}, \mu, \omega$. We recall that the dimension of the lattice of the period vectors of an arithmetic discrete plane is equal to the dimension of the space d minus the dimension of the $\mathbb{Q}$-vector space generated by the coordinates of the normal vector $\boldsymbol{v}$.

Our aim is to focus on the properties of arithmetic discrete planes obtained when noticing that they can be described thanks to a so-called *cut and project construction* which consists in projecting a subset of a lattice onto a plane. Indeed, arithmetic discrete planes are obtained by selecting points of the lattice $\mathbb{Z}^3$: we take a slice of width ω of $\mathbb{Z}^3$ along the Euclidean hyperplane $P(\boldsymbol{v}, \mu)$ with equation $\langle \boldsymbol{x}, \boldsymbol{v} \rangle + \mu = 0$. We thus have performed the *cutting* part of the construction. The interval $[0, \omega)$ is called the *selection window*.

Consider now the *projection* part. Let π_0 be the orthogonal projection onto the hyperplane P_0 with equation $\langle \boldsymbol{x}, (1,1,1) \rangle = 0$. Recall that $\boldsymbol{v}$ is assumed to have nonnegative coordinates. By projecting by π_0 the arithmetic discrete plane $\mathfrak{P}(\boldsymbol{v}, \mu, \omega)$, one gets a set of points of P_0 which is a *Delone set*, that is,

S. Brlek, C. Reutenauer, and X. Provençal (Eds.): DGCI 2009, LNCS 5810, pp. 1–12, 2009.

it is *relatively dense* (there exists $R > 0$ such that any Euclidean ball of P_0 of radius R contains a point of this set) and *uniformly discrete* (there exists $r > 0$ such that any ball of radius r contains at most one point of this set). We have recovered via this construction a so-called *quasicrystal*, that is, a discrete structure which displays long-range order without having to be periodic. The object of the present paper is to stress it. For more details, see for instance [Sen95].

This selection and projection method is also widely used as an efficient method for constructing tilings. Recall that a *tiling* by translation of the plane by a set T of (proto)tiles is a union of translates of elements of T that covers the full plane, with any two tiles intersecting either on an empty set, or on a vertex or on an edge. For more details on tilings, see for instance [GS87].

Note that the choice of the projection π_0 for the definition of the cut and project set P_0 is noncanonical. We have chosen here a projection π_0 that maps in a one-to-one way our selected set of points to a *discrete set* endowed with an underlying lattice (see Section 2.1), whereas the orthogonal projection of this set onto the line orthogonal to P_0 is *dense* in the selection window $[0, \omega)$ when $\boldsymbol{v}$ has irrational entries. We will use this latter denseness result in Section 2.3. Such a choice for π_0 enters the framework of cut and project schemes. For precise definitions, see [BM2000].

This paper is organized as follows. Section 2 introduces a tiling and a two-dimensional coding associated with arithmetic discrete planes. We will show how to deduce properties of configurations and regularity properties such as repetitivity. The idea will be to associate with finite configurations occuring in the discrete plane subintervals of the selection window $[0, \omega)$. This allows us to handle simultaneously arithmetic discrete planes having the same normal vector. We then want to "compare" discrete planes having different normal vectors. We thus introduce in Section 3 a substitution rule acting on discrete planes, and whose action is described with respect to their normal vector $\boldsymbol{v}$. We conclude this survey of applications of tiling theory in the study of arithmetic discrete planes by focusing in Section 4 on so-called boundary words.

2 Discrete Planes Coded as Multidimensional Words

Let $\mathfrak{P}(\boldsymbol{v}, \mu, \omega)$ be an arithmetic discrete plane. We have obtained so far a discrete set of points $\pi_0(\mathfrak{P}(\boldsymbol{v}, \mu, \omega))$ of the plane P_0 which has a priori no specific algebraic structure. Neverthless, we have obtained this set via a selection process which involves the lattice $\mathbb{Z}^3$ and a projection π_0 which also displays some kind of regularity. We first show how to recover a lattice underlying the arithmetic discrete plane $\mathfrak{P}(\boldsymbol{v}, \mu, \omega)$. We assume that we are in the standard case $\omega = ||\boldsymbol{v}||_1$.

2.1 A Tiling by Lozenges

One way to recover this lattice is to associate with the quasicrystal $\pi_0(\mathfrak{P}(\boldsymbol{v}, \mu, \omega))$ a tiling of the plane P_0 obtained by connecting with edges points of the quasicrystal. One checks that one gets a tiling of the plane P_0 by three kinds of tiles, namely the three regular lozenges being the projections by π_0 of the three faces of the unit cube depicted on Fig. 1 below. We call them T_i, for $i = 1, 2, 3$. We will denote this tiling by $T(\boldsymbol{v}, \mu, \omega)$.

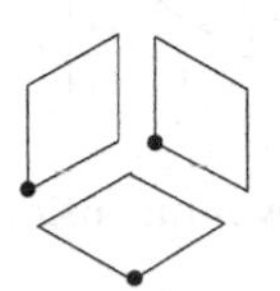

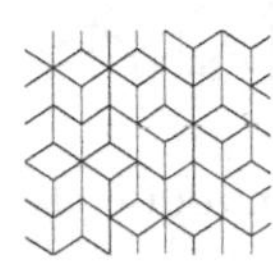

Fig. 1. Left: The three lozenge tiles T_1, T_2, T_3. Middle and Right: From the stepped plane $\mathfrak{P}(\boldsymbol{v}, \mu, ||\boldsymbol{v}||_1)$ to the tiling $T(\boldsymbol{v}, \mu, ||\boldsymbol{v}||_1)$.

We also associate with $\mathfrak{P}(\boldsymbol{v}, \mu, \omega)$ a surface $\mathcal{P}(\boldsymbol{v}, \mu, \omega)$ in $\mathbb{R}^3$ called *stepped plane* defined as the union of translates of faces of the unit cube whose vertices belong to $\mathfrak{P}(\boldsymbol{v}, \mu, \omega)$. The projection by π_0 of $\mathcal{P}(\boldsymbol{v}, \mu, \omega)$ yields the tiling $T(\boldsymbol{v}, \mu, \omega)$ (see Fig. 1).

Conversely, any tiling made of the three lozenge tiles T_i, for $i = 1, 2, 3$, admits a unique *lifting* as a surface in $\mathbb{R}^3$ up to translation by the vector $(1, 1, 1)$, with this lifting being equal to $\mathcal{P}(\boldsymbol{v}, \mu, \omega)$ if the tiling equals $T(\boldsymbol{v}, \mu, \omega)$. The idea of the proof is to associate with every vertex of the tiling a height function that is uniquely determined and whose definition is globally consistent. For more details, see [Thu89] and for a proof in this context, see [ABFJ07]. Tilings by the three tiles T_i are widely studied in the framework of dimers on the honeycomb graph (see [KO05]).

We now pick for each tile T_i a particular vertex, called its *distinguished vertex*. More precisely, let $(\boldsymbol{e}_1, \boldsymbol{e}_2, \boldsymbol{e}_3)$ stand for the canonical basis of $\mathbb{R}^3$. We consider the following faces of the unit cube: $F_1 = \{\lambda \boldsymbol{e}_2 + \mu \boldsymbol{e}_3 \mid 0 \leq \lambda, \mu \leq 1\}$, $F_2 = \{-\lambda \boldsymbol{e}_1 + \mu \boldsymbol{e}_3 \mid 0 \leq \lambda, \mu \leq 1\}$, $F_3 = \{-\lambda \boldsymbol{e}_1 - \mu \boldsymbol{e}_2 \mid 0 \leq \lambda, \mu \leq 1\}$. One has $T_i = \pi_0(F_i)$, for $i = 1, 2, 3$. For $\boldsymbol{x} \in \mathbb{Z}^3$, the distinguished vertex of $\boldsymbol{x} + F_i$ is then defined as $\boldsymbol{x}$, and similarly the distinguished vertex of the tile $\boldsymbol{y} + T_i$ of the tiling $T(\boldsymbol{v}, \mu, \omega)$ is defined as $\boldsymbol{y}$.

Note that one has a one-to-one correspondence between tiles $T_i + \boldsymbol{y}$ of the tiling $T(\boldsymbol{v}, \mu, \omega)$ and faces $F_i + \boldsymbol{x}$ in $\mathbb{R}^3$ of the stepped plane $(\boldsymbol{v}, \mu, \omega)$: indeed, for any tile $T_i + \boldsymbol{y}$ of the tiling $T(\boldsymbol{v}, \mu, \omega)$, there exists a unique $\boldsymbol{x}$ such that $\pi_0(\boldsymbol{x}) = \boldsymbol{y}$ and $0 \leq \langle \boldsymbol{x}, \boldsymbol{v} \rangle + \mu < ||\boldsymbol{v}||_1$.

Now one checks that the set of distinguished vertices of $\mathfrak{P}(\boldsymbol{v}, \mu, \omega)$ is a lattice (see [BV00]). We thus have found a lattice underlying the arithmetic discrete plane $\mathfrak{P}(\boldsymbol{v}, \mu, ||\boldsymbol{v}||_1)$ even if the coordinates of $\boldsymbol{v}$ are rationally independent, that is, even if the arithmetic discrete plane has no nonzero period vector. Since the set of distinguished vertices is a lattice that can be assimilated to $\mathbb{Z}^2$, we can code as a $\mathbb{Z}^2$-word over the alphabet $\{1, 2, 3\}$ any arithmetic discrete plane. See [BV00] for more details and Fig. 2 below for an illustration.

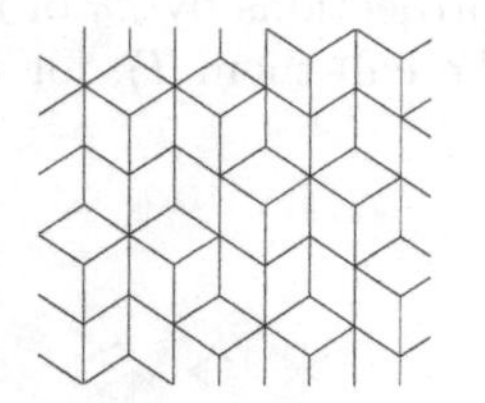
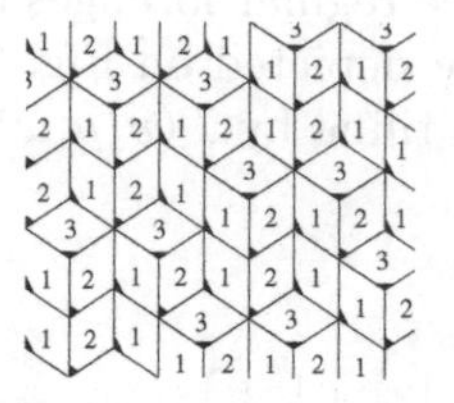

Fig. 2. From a discrete plane to its coding as a two-dimensional word

2.2 Configurations

We want now to localize with respect to the value $\langle \boldsymbol{x}, \boldsymbol{v} \rangle$ in the selection window $[0, \omega)$ the distinguished vertices of faces of a given type. Recall that we work in the standard case. One has $0 \leq \langle \boldsymbol{x}, \boldsymbol{v} \rangle + \mu < ||\boldsymbol{v}||_1 = v_1 + v_2 + v_3$.

Assume first that $0 \leq \langle \boldsymbol{x}, \boldsymbol{v} \rangle + \mu < v_1$. Then $\boldsymbol{x} + \boldsymbol{e}_2$, $\boldsymbol{x} + \boldsymbol{e}_3$, $\boldsymbol{x} + \boldsymbol{e}_2 + \boldsymbol{e}_3$ all belong to $\mathfrak{P}(\boldsymbol{v}, \mu, \omega)$. We thus deduce that the full face $F_1 + \boldsymbol{x}$ is included in $\mathcal{P}(\boldsymbol{v}, \mu, \omega)$.

Similarly, assume $v_1 \leq \langle \boldsymbol{x}, \boldsymbol{v} \rangle + \mu < v_1 + v_2$ (respectively $v_1 + v_2 \leq \langle \boldsymbol{x}, \boldsymbol{v} \rangle + \mu < v_1 + v_2 + v_3$). Then $\boldsymbol{x} - \boldsymbol{e}_1$, $\boldsymbol{x} + \boldsymbol{e}_3$, $\boldsymbol{x} - \boldsymbol{e}_1 + \boldsymbol{e}_3$ (respectively $\boldsymbol{x} - \boldsymbol{e}_1$, $\boldsymbol{x} - \boldsymbol{e}_2$, $\boldsymbol{x} - \boldsymbol{e}_1 - \boldsymbol{e}_2$) all belong to $\mathfrak{P}(\boldsymbol{v}, \mu, \omega)$. We thus deduce that the full face $F_2 + \boldsymbol{x}$ (respectively $F_3 + \boldsymbol{x}$) is included in $\mathcal{P}(\boldsymbol{v}, \mu, \omega)$.

In fact more can be said. One checks for $\boldsymbol{x} \in \mathbb{Z}^3$ that

$$\boldsymbol{x} + F_i \subset \mathcal{P}(\boldsymbol{v}, \mu, \omega) \text{ if and only if } \sum_{k<i} v_k \leq \langle \boldsymbol{x}, \boldsymbol{v} \rangle + \mu < \sum_{k \leq i} v_k. \quad (1)$$

Note that we have only sketched the proof of the "if " direction. For a full proof and more details, see [BV00].

In other words, our convention for the choice of a distinguished vertex of a face implies that a face of type i with distinguished vertex $\boldsymbol{x}$ is included in $\mathcal{P}(\boldsymbol{v}, \mu, \omega)$ if and only if $\langle \boldsymbol{x}, \boldsymbol{v} \rangle + \mu \in I_i$. Hence, we have cut the selection interval $[0, ||\boldsymbol{v}||_1)$ into three subintervals,

$$I_1 = [0, v_1), \ I_2 = [v_1, v_1 + v_2), \ I_3 = [v_1 + v_2, v_1 + v_2 + v_3),$$

each of them corresponding to the occurrences of the distinguished vertex of a particular type of face.

This holds not only for faces but also for finite unions of faces. A *configuration* of the tiling $T(\boldsymbol{v}, \mu, \omega)$ is an edge-connected finite union of lozenge tiles contained in the tiling. We asume that $\boldsymbol{0}$ is always a distinguished vertex of one of the faces of a configuration. We then consider occurrences of configurations up to translation. Note that liftings in $\mathcal{P}(\boldsymbol{v}, \mu, \omega)$ of configurations correspond to usual local configurations of discrete planes. By abuse of terminology, we also call them here configurations. The configuration C is said to occur at $\boldsymbol{y}$ in the tiling

$T(\boldsymbol{v},\mu,\omega)$ if $C+\boldsymbol{y}$ is included in it. We then can associate with the configuration C the set I_C defined as the closure of the set

$$\{\langle \boldsymbol{x},\boldsymbol{v}\rangle+\mu \mid \boldsymbol{y}=\pi_0(\boldsymbol{x}),\ \boldsymbol{x}\in\mathfrak{P}(\boldsymbol{v},\mu,\omega),\ C \text{ occurs in } T(\boldsymbol{v},\mu,\omega) \text{ at } \boldsymbol{y}\}.$$

One checks that the set I_C is an interval if the dimension of the $\mathbb{Q}$-vector space generated by the coordinates of $\boldsymbol{v}$ is at least 2. We use the denseness in the acceptance window $[0,\omega)$ of $(\langle \boldsymbol{x},\boldsymbol{v}\rangle)_{\boldsymbol{x}\in\mathbb{Z}^2}$. If $\boldsymbol{v}$ has integer coprime entries and μ is also an integer, I_C is a set of consecutive integers. We use in this latter case Bezout's lemma. For a proof, see [BV00]. It also uses the fact that I_C is described as an intersection of intervals which are preimages of the intervals I_i under the action of the translations $x\mapsto x+v_j$, for $i,j\in\{1,2,3\}$. We thus have here again divided the selection window $[0,\omega)$ into unions of intervals associated with configurations.

Let us illustrate it on one example. Consider the configuration $C=T_1\cup(T_1+\boldsymbol{e}_3)$. Configuration C occurs at $\boldsymbol{y}$ if and only if $\langle \boldsymbol{x},\boldsymbol{v}\rangle+\mu\in I_1$ and $\langle \boldsymbol{x}+\boldsymbol{e}_3,\boldsymbol{v}\rangle+\mu=\langle \boldsymbol{x},\boldsymbol{v}\rangle+v_3+\mu\in I_1$, that is, $\langle \boldsymbol{x},\boldsymbol{v}\rangle+\mu\in I_1\cap(I_1-v_3)$. Hence $I_C\neq\emptyset$ if and only if $v_1>v_3$. If $v_1>v_3$, then $I_C=[0,v_1-v_3)$.

Note that this approach still holds when one works with arithmetic discrete planes with general width $[0,\omega)$. One can similarly associate with a configuration C a set I_C defined here again as an intersection of intervals themselves intersected with $[0,\omega)$.

2.3 Applications

We thus have been able to associate with a configuration C an interval I_C of the selection window $[0,\omega)$. Let us discuss several properties that can be deduced from this correspondence.

A configuration C occurs if and only if I_C is nonempty. Let us come back to the example of the previous section. The lifting $F_1\cup(F_1+\boldsymbol{e}_3)$ of the configuration C occurs up to translation in $\mathcal{P}(\boldsymbol{v},\mu,\omega)$ if and only if $I_C=[0,v_1-v_3)$ is nonempty, that is, $v_1>v_3$. Note that this does not depend on the parameter μ.

We thus recover an effective way to check whether a configuration occurs in a given arithmetic discrete plane. Indeed, the set I_C can be effectively determined such as shown in Section 2.2. It thus remains to check whether the coresponding intersection is nonempty.

More generally, we deduce that two discrete planes with the same normal vector and same width have the same configurations. Indeed, intervals I_C do not depend on μ and only depend on $\boldsymbol{v}$. We furthermore can count the number of configurations of a given size and shape. Indeed we have to determine the bounds of the intervals I_C and then count them. We can also determine the frequencies of occurrence of configurations: they are given by the lengths of the intervals I_C. In this latter case, we use not only denseness results but equidistribution results. For more details, see e.g. [BV00] and [BFJP07]. Note that these methods hold for general widths ω.

These methods are classic in word combinatorics, symbolic dynamics, or tiling theory. Let us recall that standard and naive arithmetic discrete lines are coded thanks to the Freeman code as Sturmian words [Lot02, PF02].

Consider now repetitivity results. The *radius of a configuration* is defined as the minimal radius of a disk containing this configuration. Two configurations in the plane P_0 are said identical if they only differ by a translation vector. A tiling is said *repetitive* if for every configuration C of radius r there exists a positive number R such that every configuration of radius R contains C. This is a counterpart of the notion of uniform recurrence in word combinatorics and symbolic dynamics. In other words, configurations appear "with bounded gaps". Repetitive tilings can be considered as ordered structures.

Let C be a given configuration that occurs in the tiling T. We consider the interval associated with C. Repetitivity can then be deduced from the following fact. Given any interval I of $\mathbb{R}/\mathbb{Z}$, the sequence $(n\alpha)_{n\in\mathbb{N}}$ enters the interval I with bounded gaps, that is, there exists $N \in \mathbb{N}$ such that any sequence of N successive values of the sequence contains a value in I [Sla67]. We apply it to the interval I_C and work modulo ω in $\mathbb{R}/\omega\mathbb{Z}$. We thus deduce that the configuration C occurs in the tiling T with bounded gaps.

3 Substitutions

We have been so far able to describe properties of arithmetic discrete planes sharing the same normal vector $\boldsymbol{v}$. We now want to be able to relate two discrete planes with different normal vectors $\boldsymbol{v}$ and $\boldsymbol{v}'$. We focus on the case $\boldsymbol{v} = M\boldsymbol{v}'$, where M is a 3 by 3 square matrix with entries in $\mathbb{N}$ having determinant equal to 1 or -1 (such a matrix is called *unimodular*).

3.1 Multidimensional Continued Fractions

Our motivation relies on the fact that unimodular transformations are the basic steps when expanding vectors under the action of a unimodular multidimensional continued fraction algorithm, such as Jacobi-Perron or Brun algorithms (here we expand a normal vector $\boldsymbol{v}$ of a plane). It is well known that arithmetic discrete lines and their codings as Sturmian words are perfectly well described by Euclid's algorithm and by the continued fraction expansion of their slope [Lot02, PF02]. We want to generalize this to the higher-dimensional case.

It is also well-known that there exists no canonical multidimensional continued fraction algorithm. We focus here on unimodular algorithms such as described in [Bre81, Sch00]. Such multidimensional continued fraction algorithms produce sequences of matrices in $GL(d, \mathbb{Z})$ as follows.

Let $X \subset \mathbb{R}^d$. A d-dimensional continued fraction map over X is a map $T : X \to X$ such that $T(X) \subset X$ and, for any $\boldsymbol{x} \in X$, there is a matrix in $GL(d, \mathbb{Z})$ depending on $\boldsymbol{x}$ (we thus denote it by $M(\boldsymbol{x})$) satisfying: $\boldsymbol{x} = M(\boldsymbol{x}).T(\boldsymbol{x})$. The associated continued fraction algorithm consists in iteratively applying the map T on a vector $\boldsymbol{x} \in X$. This yields the following sequence of matrices, called the continued fraction expansion of $\boldsymbol{x}$: $(M(T^n(\boldsymbol{x})))_{n\in\mathbb{N}}$. If the matrices have nonnegative entries, the algorithm is said nonnegative.

Consider for instance the Jacobi-Perron algorithm. Its projective version is defined on the unit square $[0,1)\times[0,1)$ by:

$$(\alpha,\beta)\mapsto\left(\frac{\beta}{\alpha}-\left\lfloor\frac{\beta}{\alpha}\right\rfloor,\frac{1}{\alpha}-\left\lfloor\frac{1}{\alpha}\right\rfloor\right)=(\{\beta/\alpha\},\{1/\alpha\}).$$

Its linear version is defined on the positive cone $X=\{(a,b,c)\in\mathbb{R}^3|0\leq a,b<c\}$ by:

$$T(a,b,c)=(b-\lfloor b/a\rfloor a,c-\lfloor c/a\rfloor a,a).$$

We set $(a_0,b_0,c_0):=(a,b,c)$ and $(a_{n+1},b_{n+1},c_{n+1}):=T^n(a_n,b_n,c_n)$, for $n\in\mathbb{N}$. Let $B_{n+1}=\lfloor b_n/a_n\rfloor a_n$, $C_n=\lfloor c_n/a_n\rfloor$. One has

$$\begin{pmatrix}a_n\\b_n\\c_n\end{pmatrix}=\begin{pmatrix}0&0&1\\1&0&B_{n+1}\\0&1&C_{n+1}\end{pmatrix}\begin{pmatrix}a_{n+1}\\b_{n+1}\\c_{n+1}\end{pmatrix}.$$

The Jacobi-Perron algorithm is thus a unimodular nonnegative continued fraction algorithm. With the previous notation, $M(\boldsymbol{x})=M_{B_1,C_1}$ for $\boldsymbol{x}=(a_0,b_0,c_0)$, by setting $M_{B,C}:=\begin{pmatrix}0&0&1\\1&0&B\\0&1&C\end{pmatrix}$ for $B,C\in\mathbb{N}$.

Note that the sequence of admissible Jacobi-Perron digits $(B_n,C_n)_{n\geq1}$ satisfies $0\leq B_n\leq C_n$ and $B_n=C_n$ implies $B_{n+1}\geq1$ for all n.

The idea is now to consider the expansion of a given normal vector $\boldsymbol{v}\in X$. Let $\boldsymbol{v}^{(n)}$ stand for $T^n(\boldsymbol{v})$. One expands $\boldsymbol{v}$ as

$$\boldsymbol{v}=M_{B_1,C_1}\cdots M_{B_n,C_n}\boldsymbol{v}^{(n)}.$$

3.2 Substitution Rules

Let us come back to our geometric framework. We fix a matrix $M\in SL(3,\mathbb{N})$ (for instance $M=M_{B,C}$). We assume furthermore that we are in the standard case $\omega=||\boldsymbol{v}||_1$. We want to find an algorithmic way to go from $\mathcal{P}_{M\boldsymbol{v},\mu,||M\boldsymbol{v}||_1}$ to $\mathcal{P}_{\boldsymbol{v},\mu,||\boldsymbol{v}||_1}$.

The idea is to use the fact that

$$\langle\boldsymbol{x},M\,\boldsymbol{v}\rangle=\langle{}^tM\,\boldsymbol{x},\boldsymbol{v}\rangle\text{ for any }\boldsymbol{x}\in\mathbb{Z}^3.\tag{2}$$

Furthermore, for computational reasons, we change our convention for faces. We now define faces as follows: $E_1=\{\lambda e_2+\mu e_3\mid 0\leq\lambda,\mu\leq1\}$, $E_2=\{\lambda e_1+\mu e_3\mid 0\leq\lambda,\mu\leq1\}$, $E_3=\{\lambda e_1+\mu e_2\mid 0\leq\lambda,\mu\leq1\}$. We loose the notion of distinguished vertex but we obtain here a characterization of faces that are contained in $\mathcal{P}(\boldsymbol{v},\mu,||\boldsymbol{v}||_1)$ that will be here easier to handle than (1): the face $\boldsymbol{y}+E_j$ is a subset of $\mathcal{P}(\boldsymbol{v},\mu,||\boldsymbol{v}||_1)$ if and only if

$$0\leq\langle\boldsymbol{y},\boldsymbol{v}\rangle+\mu<v_j=\langle\boldsymbol{e}_j,\boldsymbol{v}\rangle.$$

Let us assume now that the face $\boldsymbol{y}+F_j$ is a subset of $\mathcal{P}(M\,\boldsymbol{v},\mu,||M\,\boldsymbol{v}||_1)$. One has

$$0 \leq \langle \boldsymbol{y}, M\,\boldsymbol{v}\rangle + \mu = \langle {}^tM\,\boldsymbol{y},\boldsymbol{v}\rangle + \mu < \langle \boldsymbol{e}_j, M\,\boldsymbol{v}\rangle = \langle {}^tM\,\boldsymbol{e}_j,\boldsymbol{v}\rangle.$$

The interval $[0,\langle {}^tM\,\boldsymbol{e}_j,\boldsymbol{v}\rangle)$ is nonempty according to the assumptions made on M. Since M has nonegative entries, we can divide the interval $[0,\langle {}^tM\,\boldsymbol{e}_j,\boldsymbol{v}\rangle)$ into subintervals of respective lengths $\langle \boldsymbol{e}_i,\boldsymbol{v}\rangle = v_i$, for $i=1,2,3$. We know that the point $\langle {}^tM\,\boldsymbol{y},\boldsymbol{v}\rangle + \mu$ belongs to one of these intervals. Hence there exist a vector $\boldsymbol{S}$ and $i \in \{1,2,3\}$ such that

$$0 \leq \langle {}^tM\,\boldsymbol{y},\boldsymbol{v}\rangle + \mu - \langle \boldsymbol{S},\boldsymbol{v}\rangle < v_i,$$

which implies that the face ${}^tM\boldsymbol{y} - \boldsymbol{S} + E_i$ is a subset of $\mathcal{P}(\boldsymbol{v},\mu,||\boldsymbol{v}_1||)$.

We thus have associated with a face $\boldsymbol{y}+E_j$ of $\mathcal{P}(M\boldsymbol{v},\mu,||M\boldsymbol{v}||_1)$ a face $\boldsymbol{x}+E_i$ where $\boldsymbol{x} = {}^tM\boldsymbol{y} - \boldsymbol{S}$ in $\mathcal{P}(\boldsymbol{v},\mu,||\boldsymbol{v}||_1)$, and the entry m_{ji} of M satisfies $m_{ji} \neq 0$.

One checks that this map is onto (we use the fact that M is an invertible matrix as a matrix acting on $\mathbb{Z}^3$) but not one-to-one. This is why we will work below (see Eq. (3)) with the inverse of this map.

3.3 Substitutions in Word Combinatorics

Let us formalize this map. In particular, we need to define more precisely the way we divide the interval $[0,\langle {}^tM\,\boldsymbol{e}_j,\boldsymbol{v}\rangle)$ into subintervals of respective lengths $\langle \boldsymbol{e}_i,\boldsymbol{v}\rangle = v_i$. In other words, we have to choose a way of tiling the larger interval by these smaller intervals. For that purpose, we first recall basic definitions from word combinatorics. For more details, see [Lot02, PF02].

We consider the set of finite words over the alphabet $\{1,2,3\}$. We denote this set as $\{1,2,3\}^*$. Endowed with the concatenation, it is a free monoid. A *substitution* is a morphism of the free monoid. A substitution σ thus replaces letters by words, and moreover, since it is a morphism, it satisfies $\sigma(uv) = \sigma(u)\sigma(v)$, for any u,v words in $\{1,2,3\}^*$. Consider for example the substitution σ defined by $\sigma(1)=12$, $\sigma(2)=13$, $\sigma(3)=1$. The *incidence matrix* M_σ of σ is defined as the 3 by 3 square matrix with entry (i,j) entry equal to the number of occurrences of the letter i in $\sigma(j)$. With our example, $M_\sigma = \begin{bmatrix} 1&1&1\\ 1&0&0\\ 0&1&0 \end{bmatrix}$.

A substitution is said *unimodular* if the determinant of its incidence matrix equals 1 or -1. Lastly, let $\boldsymbol{l} : \{1,2,3\}^* \to \mathbb{N}^3$ be the abelianization map $\boldsymbol{l}(w) = {}^t(|w|_1,|w|_2,|w|_3)$.

We now can introduce the notion of *generalized substitution*, such as introduced in [AI02]. Let σ be a unimodular substitution. The map $E_1^*(\sigma)$ maps faces on unions of faces. It is defined respectively on faces $\boldsymbol{x}+E_i$ and on sets of faces G,H as follows

$$E_1^*(\sigma)(\boldsymbol{x}+E_i) = \sum_{j\in\{1,2,3\}} \sum_{P,\,\sigma(j)=PiS} \left(M_\sigma^{-1}(\boldsymbol{x}+\boldsymbol{l}(S)) + E_j\right), \qquad (3)$$

$$E_1^*(G\cup H) = E_1^*(G) \cup E_1^*(H).$$

Let us explain the notation "P, $\sigma(j) = PiS$" used above. We consider letters j such that $\sigma(j)$ contains i and make the summation on all occurrences of the letter i in $\sigma(j)$. We thus decompose $\sigma(j)$ as PiS, where P, S are words in $\{1,2,3\}^*$ and we " label" each occurrence of i by the prefix P that precedes it. One of the main interests of this kind of geometric replacement map is that it maps stepped planes onto stepped planes:

Theorem 1. *[AI02, Fer06] Let σ be a unimodular substitution. Let $\boldsymbol{v} \in \mathbb{R}^d_+$ be a positive vector. The generalized substitution $E_1^*(\sigma)$ maps without overlaps the stepped plane $\mathcal{P}_{\boldsymbol{v},\mu,||\boldsymbol{v}||_1}$ onto $\mathcal{P}_{{}^tM_\sigma \boldsymbol{v},\mu,||{}^tM_\sigma \boldsymbol{v}||_1}$.*

The proof of the theorem is based on the ideas sketched above in Section 3.2. Note that if $\boldsymbol{y} = M_\sigma^{-1}(\boldsymbol{x} + \boldsymbol{l}(S))$, then $\boldsymbol{x} = M_\sigma \boldsymbol{y} - \boldsymbol{l}(S)$. We recover the expression obtained in Section 3.2.

3.4 Back to Discrete Planes

Let us see now how to use Theorem 1 in discrete geometry.

We first start with the question of the generation of discrete planes. We illustrate our strategy with the Jacobi-Perron algorithm. We also fix $\mu = 0$. We use the notation $\mathcal{P}(\boldsymbol{v})$ for $\mathcal{P}(\boldsymbol{v}, 0, ||\boldsymbol{v}||_1)$. We associate with the matrix $M_{B,C}$ the substitution $\sigma_{B,C}\colon 1 \mapsto 1,\ 2 \mapsto 1^B 3,\ 3 \mapsto 1^C 2$ having the transpose of $M_{B,C}$ as incidence matrix. Let $\boldsymbol{v} \in X$. By Theorem 1, one has

$$E_1^*(\sigma_{B,C})(\mathcal{P}(T(\boldsymbol{v})) = \mathcal{P}(\boldsymbol{v}).$$

Assume now we are given the sequence of Jacobi-Perron digits $(B_n, C_n)_{n\geq 1}$ produced by the Jacobi-Perron expansion of the vector $\boldsymbol{v}$. One has

$$E_1^*(\sigma_{(B_1,C_1)}) \circ E_1^*(\sigma_{(B_2,C_2)}) \cdots \circ E_1^*(\sigma_{(B_n,C_n)})(\mathcal{P}(\boldsymbol{v}^{(n)}) = \mathcal{P}(\boldsymbol{v}).$$

Note that the lower unit cube $\mathcal{U} := E_1 + E_2 + E_3$ belongs to every arithmetic discrete plane with parameter $\mu = 0$. We deduce that

$$E_1^*(\sigma_{(B_1,C_1)}) \circ \cdots \circ E_1^*(\sigma_{(B_n,C_n)})(\mathcal{U}) \subset \mathcal{P}(\boldsymbol{v}).$$

The question is now whether the patterns $E_1^*(\sigma_{(B_1,C_1)}) \circ \cdots \circ E_1^*(\sigma_{(B_n,C_n)})(\mathcal{U})$ generate the whole plane $\mathcal{P}(\boldsymbol{v})$, that is, whether

$$\lim_{n\to\infty} E_1^*(\sigma_{(B_1,C_1)}) \circ \cdots \circ E_1^*(\sigma_{(B_n,C_n)})(\mathcal{U}) = \mathcal{P}(\boldsymbol{v}).$$

If yes, then we get a simple algorithmic generation process of the arithmetic discrete plane $\mathcal{P}(\boldsymbol{v})$. This is illustrated in Fig. 3 below.

This question has been answered in the Jacobi-Perron case in [I093]. Indeed, it has been proved that if there exists no integer n such that for all k

$$B_{n+3k} = C_{n+3k},\ C_{n+3k+1} - B_{n+3k+1} \geq 1,\ B_{n+3k+1} \neq 0,\ B_{n+3k+2} = 0,$$

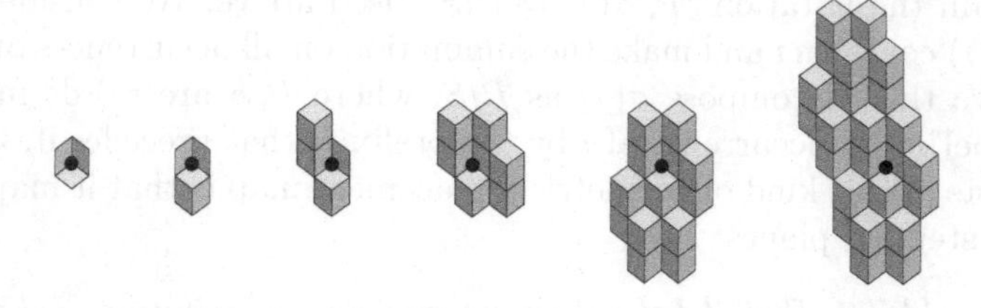

Fig. 3. Some iterations of a generalized substitution acting on the lower unit cube $\mathcal{U}$

the sequence of patterns $E_1^*(\sigma_{(B_1,C_1)}) \circ \cdots \circ E_1^*(\sigma_{(B_n,C_n)})(\mathcal{U})$ generates the whole plane $\mathcal{P}(\boldsymbol{v})$. Otherwise, it is not possible to generate the whole plane by starting only from $\mathcal{U}$, but there exists a finite set of faces $\mathcal{V}$ with $\mathcal{U} \subset \mathcal{V}$ such that the whole plane is generated starting from $\mathcal{V}$ instead of $\mathcal{U}$, that is, the sequence of patterns $E_1^*(\sigma_{(B_1,C_1)}) \circ \cdots \circ E_1^*(\sigma_{(B_n,C_n)})(\mathcal{V})$ generates the whole plane $\mathcal{P}(\boldsymbol{v})$.

Let us stress that we are not only able to substitute, i.e., to replace faces by unions of faces, but also to desubstitute, i.e., to perform the converse operation, by using the algebraic property $E_1^*(\sigma)^{-1} = E_1^*(\sigma^{-1})$ (where σ is considered as a morphism of the free group is an automorphism). For more details, see [BF09] in the case of Brun algorithm. This idea is also used in [Fer09], still for Brun algorithm, where algorithms are given for the digital plane recognition and digital plane generation problems.

Let us quote a further classical question in the study of discrete planes that can be handled under the formalism of generalized substitutions $E_1^*(\sigma)$. The question is to find the smallest width ω for which the plane $\mathcal{P}_{\boldsymbol{v},\mu,\omega}$ is connected (either edge connected or vertex connected). The case of rational parameters has been solved in [JT09a]. For the case of irrational parameters, see [DJT09]. The method used in both papers relies on the use of a particular unimodular multidimensional continued fraction algorithm (the so-called fully subtractive algorithm) and on the use of Eq. (2).

4 Conclusion

We have discussed here several ideas concerning the study of arithmetic discrete planes inspired by tiling theory and word combinatorics. We have focused on a multidimensional coding (via faces) of configurations and discrete planes. We can also associate with configurations of discrete planes a boundary word coding as a finite word their boundary. The use of word combinatorics and the study of boundary words has recently proven to be particularly useful in discrete geometry. Let us also quote in particular [BLPR08] which gives a nice characterization of digitally convex polyominoes in terms of the Lyndon decomposition of the word coding their boundary. See also [BKP09] which gives an efficient recognition algorithm of nonintersecting paths based on the use of suffix trees.

Boundary words also allow us to describe the topology of the configurations of discrete planes we generate via the method described in Section 3.4. Let us recall the following result of [Ei03]. Let σ be an invertible three-letter substitution (by invertible, we mean that σ considered as a morphism of the free group is an automorphism). The boundary word of $E_1^*(\sigma)(\mathcal{U})$ is produced by the mirror image of the inverse of σ. We thus can apply this result to get a description of the boundary of the generating patterns [BLPP09]. Note that these patterns can be described in terms of pseudo-squares and pseudo-hexagons such as introduced in [BN91] which gives a characterization of polyominoes which tile the plane. See also [BFP09] for corresponding efficient recognition algorithms and [BBGL09] for a study of a family of so-called double pseudo-squares.

Acknowledgements

We would like to thank Geneviève Paquin and Xavier Provençal for their very useful comments.

References

[AAS97] Andres, É., Acharya, R., Sibata, C.: The Discrete Analytical Hyperplanes. Graph. Models Image Process. 59, 302–309 (1997)

[AI02] Arnoux, P., Ito, S.: Pisot substitutions and Rauzy fractals. Bull. Bel. Math. Soc. Simon Stevin 8, 181–207 (2001)

[ABI02] Arnoux, P., Berthé, V., Ito, S.: Discrete planes, $\mathbb{Z}^2$-actions, Jacobi-Perron algorithm and substitutions. Ann. Inst. Fourier (Grenoble) 52, 1001–1045 (2002)

[ABFJ07] Arnoux, P., Berthé, V., Fernique, T., Jamet, D.: Functional stepped surfaces, flips and generalized substitutions. Theoret. Comput. Sci. 380, 251–267 (2007)

[BN91] Beauquier, D., Nivat, M.: On translating one polyomino to tile the plane. Discrete Comput. Geom. 6, 575–592 (1991)

[BF09] Berthé, V., Fernique., T.: Brun expansions of stepped surfaces (Preprint)

[BV00] Berthé, V., Vuillon, L.: Tilings and Rotations on the Torus: A Two-Dimensional Generalization of Sturmian Sequences. Discrete Math. 223, 27–53 (2000)

[BFJP07] Berthé, V., Fiorio, C., Jamet, D., Philippe, F.: On some applications of generalized functionality for arithmetic discrete planes. Image and Vision Computing 25, 1671–1684 (2007)

[BLPP09] Berthé, V., Lacasse, A., Paquin, G., Provençal, X.: Boundary words for arithmetic discrete planes generated by Jacobi-Perron algorithm (Preprint)

[BBGL09] Blondin Massé, A., Brlek, S., Garon, A., Labbé, S.: Christoffel and Fibonacci Tiles. In: Brlek, S., Reutenauer, C., Provençal, X. (eds.) DGCI 2009. LNCS, vol. 5810, pp. 68–79. Springer, Heidelberg (2009)

[BFP09] Brlek, S., Fédou, J.M., Provençal, X.: On the tiling by translation problem. Discrete Applied Mathematics 157, 464–475 (2009)

[BKP09] Brlek, S., Koskas, M., Provençal, X.: A linear time and space algorithm for detecting path intersection. In: Brlek, S., Reutenauer, C., Provençal, X. (eds.) DGCI 2009. LNCS, vol. 5810, pp. 398–409. Springer, Heidelberg (2009)

[BLPR08] Brlek, S., Lachaud, J.O., Provençal, X., Reutenauer, C.: Lyndon+Christoffel=Digitally Convex. Pattern Recognition 42, 2239–2246 (2009)

[Bre81] Brentjes, A.J.: Mathematical Centre Tracts. Multi-dimensional continued fraction algorithms 145, Matematisch Centrum, Amsterdam (1981)

[BCK07] Brimkov, V., Coeurjolly, D., Klette, R.: Digital Planarity - A Review. Discr. Appl. Math. 155, 468–495 (2007)

[BM2000] Baake, M., Moody, R.V., Robert, V. (eds.): Directions in mathematical quasicrystals. CRM Monograph Series, vol. 13. American Mathematical Society, Providence (2000)

[DJT09] Domenjoud, E., Jamet, D., Toutant, J.-L.: On the connecting thickness of arithmetical discrete planes. In: Brlek, S., Reutenauer, C., Provençal, X. (eds.) DGCI 2009. LNCS, vol. 5810, pp. 362–372. Springer, Heidelberg (2009)

[Ei03] Ei, H.: Some properties of invertible substitutions of rank d and higher dimensional substitutions. Osaka J. Math. 40, 543–562 (2003)

[Fer06] Fernique, T.: Multi-dimensional Sequences and Generalized Substitutions. Int. J. Fond. Comput. Sci. 17, 575–600 (2006)

[Fer09] Fernique, T.: Generation and recognition of digital planes using multi-dimensional continued fractions. Pattern Recognition 432, 2229–2238 (2009)

[GS87] Grunbaum, B., Shepard, G.: Tilings and patterns. Freeman, New-York (1987)

[I093] Ito, S., Ohtsuki, M.: Modified Jacobi-Perron algorithm and generating Markov partitions for special hyperbolic toral automorphisms. Tokyo J. Math. 16, 441–472 (1993)

[JT09a] Jamet, D., Toutant, J.-L.: Minimal arithmetic thickness connecting discrete planes. Discrete Applied Mathematics 157, 500–509 (2009)

[KO05] Kenyon, R., Okounkov, O.: What is a dimer? Notices Amer. Math. Soc. 52, 342–343 (2005)

[Lot02] Lothaire, N.: Algebraic combinatorics on words. Cambridge University Press, Cambridge (2002)

[PF02] Pytheas Fogg, N.: Substitutions in Dynamics, Arithmetics, and Combinatorics. In: Berthé, V., Ferenczi, S., Mauduit, C., Siegel, A. (eds.) Frontiers of Combining System. Lecture Notes in Mathematics, vol. 1794, Springer, Heidelberg (2002)

[Rev91] Reveillès, J.-P.: Calcul en Nombres Entiers et Algorithmique. Thèse d'état, Université Louis Pasteur, Strasbourg, France (1991)

[Sch00] Schweiger, F.: Multi-dimensional continued fractions. Oxford Science Publications, Oxford Univ. Press, Oxford (2000)

[Sen95] Senechal, M.: Quasicrystals and geometry. Cambridge University Press, Cambridge (1995)

[Sla67] Slater, N.B.: Gaps and steps for the sequence $n\theta$ mod 1. Proc. Cambridge Phil. Soc. 63, 1115–1123 (1967)

[Thu89] Thurston, W.P.: Groups, tilings and finite state automata. In: AMS Colloquium lectures. Lectures notes distributed in conjunction with the Colloquium Series (1989)

Affine Connections, and Midpoint Formation

Anders Kock

Department of Mathematical Sciences, University of Aarhus
Ny Munkegade 118, Building 1530, DK-8000 Aarhus C
kock@imf.au.dk

Preface

It is a striking fact that differential calculus exists not only in analysis (based on the real numbers $\mathbb{R}$), but also in algebraic geometry, where no limit processes are available. In algebraic geometry, one rather uses the idea of *nilpotent elements* in the "affine line" R; they act as infinitesimals. (Recall that an element x in a ring R is called *nilpotent* if $x^k = 0$ for suitable non-negative integer k.)

Synthetic differential geometry (SDG) is an axiomatic theory, based on such nilpotent infinitesimals. It can be proved, via topos theory, that the axiomatics covers both the differential-geometric notions of algebraic geometry and those of calculus.

I shall provide a glimpse of this synthetic method, by discussing its application to two particular types of differential-geometric structure, namely that of *affine connection* and of *midpoint formation*.

I shall not go much into the foundations of SDG, whose core is the so-called KL[1] axiom scheme. This is a very strong kind of axiomatics; in fact, a salient feature of it is: *it is inconsistent* – if you allow yourself the luxury of reasoning with so-called classical logic, i.e. use the "law of excluded middle", "proof by contradiction", etc. Rather, in SDG, one uses a weaker kind of logic, often called "constructive" or "intuitionist". Note the evident logical fact that there is a trade-off: with a *weaker* logic, *stronger* axiom systems become consistent. For the SDG axiomatics, it follows for instance that any function from the number line to itself is infinitely often differentiable (smooth); a very useful simplifying feature in differential geometry – but incompatible with the law of excluded middle, which allows you to construct the non-smooth function

$$f(x) = \begin{cases} 1 & \text{if } x = 0, \\ 0 & \text{if not} \end{cases} .$$

1 Nilpotents, and Neighbours

Nilpotent elements on the number line serve as *infinitesimals*[2], in a sense which is "forbidden" when the number line is $\mathbb{R}$. Nilpotent infinitesimals come in a

---

[1] For "Kock-Lawvere".

[2] They are not to be compared to the infinitesimals of non-standard analysis.

S. Brlek, C. Reutenauer, and X. Provençal (Eds.): DGCI 2009, LNCS 5810, pp. 13–21, 2009.

precise hierachy, since

$$x^k = 0 \quad \text{implies} \quad x^{k+1} = 0.$$

The method of SDG combines the "nilpotency" ideas from algebraic geometry, with category theory, and categorical logic: category theory has provided a sense by which reasoning in (constructive) naive set theory is *sound* for geometric reasoning. So the following is formulated in such naive set theoretic terms.

We plunge directly into the geometry of infinitesimals (in the "nilpotency" sense): let us denote by $D \subseteq R$ the set of $x \in R$ with $x^2 = 0$ (the "first order infinitesimals"), more generally, let $D_k \subseteq R$ be the set of kth order infinitesimals, meaning the set of $x \in R$ with $x^{k+1} = 0$. (So $D = D_1$.) The basic instance of the KL axiom scheme says that any map $D_k \to R$ extends uniquely to a polynomial map $R \to R$ of degree $\leq k$. Thus, given any map $f : R \to R$, the restriction of f to D_k extends uniquely to a polynomial map of degree $\leq k$, the kth *Taylor polynomial* of f at 0.

For x and y in R, we say that x and y are kth order *neighbours* if $x - y \in D_k$, and we write $x \sim_k y$. It is clear that $\sim_k$ is a reflexive and symmetric relation. It is not transitive. For instance, if $x \in D$ and $y \in D$, then $x + y \in D_2$, by binomial expansion of $(x+y)^3$; but we cannot conclude $x + y \in D$. So $x \sim_1 y$ and $y \sim_1 z$ imply $x \sim_2 z$, and similarly for higher k.

We now turn to the (first order) neighbour relations in the coordinate plane R^2. It is, in analogy with the 1-dimensional case, defined in terms of a subset $D(2) \subseteq R^2$; we put

$$D(2) = \{(x_1, x_2) \in R \times R \mid x_1^2 = 0, x_2^2 = 0, x_1 \cdot x_2 = 0\};$$

we define $\underline{x} \sim \underline{y}$ if $\underline{x} - \underline{y} \in D(2)$, where $\underline{x} = (x_1, x_2)$ and $\underline{y} = (y_1, y_2)$. So $D(2) \subseteq D \times D$. Similarly for $D(n) \subseteq R^n$, and the resulting first order neighbour relation on the higher "coordinate vector spaces" R^n.

The following is a consequence of the KL axiom scheme:

Theorem 1.1. *Any map* $f : R^n \to R^m$ *preserves the* k*th order neighbour relation,*

$$\underline{x} \sim_k \underline{y} \quad \textit{implies} \quad f(\underline{x}) \sim_k f(\underline{y}).$$

Proof sketch. for $n = 2$, $m = 1$, for the first order neighbour relation $\sim_1$. It suffices to see that $\underline{x} \sim_1 \underline{0}$ implies $f(\underline{x}) \sim_1 f(\underline{0})$, i.e to prove that $\underline{x} \in D(2)$ implies $f(\underline{x}) - f(\underline{0}) \in D$. Now from a suitable version of the KL axiom scheme follows that on $D(2)$, f agrees with a unique affine function $T_1 f : R^2 \to R$, so

$$f(\underline{x}) - f(\underline{0}) = a_1 x_1 + a_2 x_2.$$

Squaring the right hand side here yields 0, since not only $x_1 \in D$ and $x_2 \in D$, but also $x_1 \cdot x_2 = 0$. So $f(\underline{x}) - f(\underline{0}) \in D$.

From the Theorem follows that the relation $\sim_k$ on R^n is *coordinate free*, i.e. is a truly geometric notion: any re-coordinatization of R^n (by *any* map, not just by a linear or affine one) preserves the relation $\sim_k$.

For suitable definition of what an *open* subsets of R^n is, and for a suitable definition of "*n-dimensional manifold*" (something that locally can be coordinatized with open subsets of R^n), one concludes that on any manifold, there are canonical reflexive symmetric relations $\sim_k$: they may be defined in terms of a local coordinatization, but, by the Theorem, are independent of the coordinatization chosen.

Any map between manifolds preserves the relations $\sim_k$.

We shall mainly be interested in the *first* order neighbour relation $\sim_1$, which we shall also write just $\sim$. In Section 3, we study aspects of the second order neighbour relation $\sim_2$.

So for a manifold M, we have a subset $M_{(1)} \subseteq M \times M$, the "first neighbourhood of the diagonal", consisting of $(x, y) \in M \times M$ with $x \sim y$. It was in terms of this "scheme" $M_{(1)}$ that algebraic geometers in the 1950 gave nilpotent infinitesimals a rigourous role in geometry. Note that for $M = R^n$, we have $M_{(1)} \cong M \times D(n)$, by the map $(\underline{x}, \underline{y}) \mapsto (\underline{x}, \underline{x} - \underline{y})$.

Let us consider some notions from "infinitesimal geometry" which can be expressed in terms of the first order neighbour relation $\sim$ on an arbitrary manifold M. Given three points x, y, z in M. If $x \sim y$ and $x \sim z$ we call the triple (x, y, z) a *2-whisker at* x (sometimes: an *infinitesimal* 2-whisker, for emphasis); since $\sim$ is not transitive, we cannot in general conclude that $y \sim z$; if y happens to be $\sim z$, we call the triple (x, y, z) a *2-simplex* (sometimes an *infinitesimal* 2-simplex). Similarly for k-whiskers and k-simplices. A k-simplex is thus a $k + 1$-tuple of mutual neighbour points. The k-simplices form, as k ranges, a simplicial complex, which in fact contains the information of differential forms, and the de Rham complex of M, see [2], [6], [1], [7].

(When we say that $(x_0, x_1, \ldots, x_k)$ is a k-whisker, we mean to say that it is a k-whisker *at* x_0, i.e. that $x_0 \sim x_i$ for all $i = 1, \ldots, k$. On the other hand, in a simplex, none of the points have a special status.)

Given a k-whisker $(x_0, \ldots, x_k)$ in M. If U is an open subset of M containing x_0, it will also contain the other x_is, and if U is coordinatized by R^n, we may use coordinates to define the affine combination

$$\sum_{i=0}^{k} t_i \cdot x_i, \tag{1.1}$$

(where $\sum t_i = 1$; recall that this is the condition that a linear combination deserves the name of *affine* combination). The affine combination (1.1) can again be proved to belong to U, and thus it defines a point in M. The point thus obtained has in general *not* a good geometric significance, since it will depend on the coordinatization chosen. However (cf. [5], [7] 2.1), it does, if the whisker is a simplex:

Theorem 1.2. *Let $(x_0, \ldots, x_k)$ be a k-simplex in M. Then the affine combination (1.1) is independent of the coordinatization used to define it. All the points* *that arise in this way are mutual neigbours. And any map to another manifold M' preserves such combinations.*

Proof sketch. This is much in the spirit of the proof of Theorem 1.1: it suffices to see that any map $R^n \to R^m$ (not just a linear or affine one) preserves affine combinations of mutual neighbour points. This follows by considering a suitable first Taylor polynomial of f (expand from x_0), and using the following purely algebraic fact: If $\underline{x}_1, \ldots, \underline{x}_k$ are in $D(n)$, then any linear combination of them will again be in $D(n)$ *provided* the $\underline{x}_i$s are *mutual* neighbours.

Examples. If $x \sim y$ in a manifold (so they form a 1-simplex), we have the affine combinations "midpoint of x and y", and "reflection of x in y",

$$\tfrac{1}{2}x + \tfrac{1}{2}y \quad \text{and} \quad 2y - x,$$

respectively. If x, y, z form a 2-simplex, we may form the affine combination $u := y - x + z$; geometrically, it means completing the simplex into a parallelogram by adjoining the point u. Here is the relevant picture:

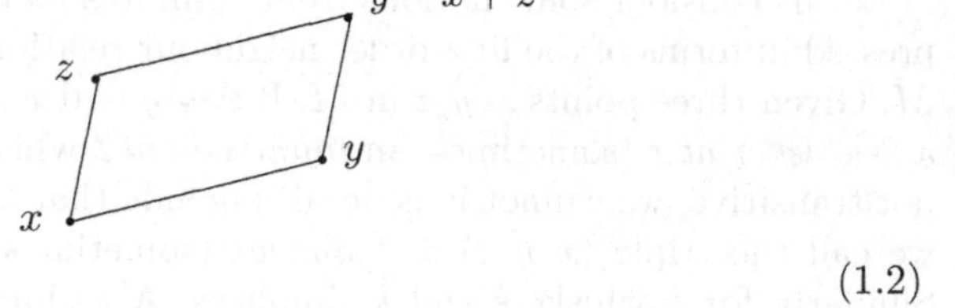

(1.2)

(All four points here are neighbours, not just those that are connected by lines in the figure.) The u thus constructed will be a neighbour of each of the three given points. Therefore, we may form the midpoint of x and u, and also we may form the midpoint of y and z; these two midpoints will agree, because they do so in R^n, from where the construction of affine combinations was imported.

Remark. If x, y, z and u are as above, and if x, y, and z belong to a subset $S \subseteq M$ given as a zero set of a function $f : M \to R$, then so does $u = y - x + z$; for, f preserves this affine combination.

2 Affine Connections

If x, y, z form a 2-whisker at x (so $x \sim y$ and $x \sim z$), we cannot canonically form a parallelogram as in (1.2); rather, parallelogram formation is an added *structure*:

Definition 2.1. *An* affine connection *on a manifold M is a law λ which to any 2-whisker x, y, z in M associates a point $u = \lambda(x, y, z) \in M$, subject to the conditions*

$$\lambda(x, x, z) = z, \quad \lambda(x, y, x) = y. \tag{2.1}$$

It can be verified, by working in a coordinatized situation, that several other laws follow; in a more abstract combinatorial situation than manifolds, these laws should probably be postulated. The laws are that for any 2-whisker (x, y, z)

$$\lambda(x, y, z) \sim y \text{ and } \lambda(x, y, z) \sim z \tag{2.2}$$

$$\lambda(y, x, \lambda(x, y, z)) = z \tag{2.3}$$

One will not in general have or require the "symmetry" condition

$$\lambda(x, y, z) = \lambda(x, z, y); \tag{2.4}$$

nor do we in general have, for 2-simplices x, y, z, that

$$\lambda(x, y, z) = y - x + z. \tag{2.5}$$

The laws (2.4) and (2.5) are in fact equivalent, and affine connections satisfying either are called *symmetric* or *torsion free*. We return to the torsion of an affine connection below.

If x, y, z, u are four points in M such that (x, y, z) is a 2-whisker at x, the statement $u = \lambda(x, y, z)$ can be rendered by a diagram

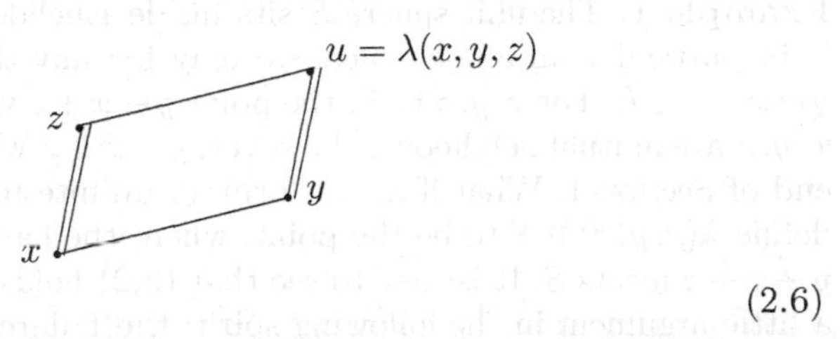

(2.6)

The figure[3] is meant to indicate that the data of λ provides a way of closing a whisker (x, y, z) into a *parallellogram* (one may say that λ provides a notion of *infinitesimal parallelogram*); but note that λ is not required to be symmetric in y and z, which is why we in the figure use different signatures for the line segments connecting x to y and to z, respectively.

Here, a line segment (whether single or double) indicates that the points connected by the line segment are neighbours.

If x, y, z, u are four points in M that come about in the way described, we say that the 4-tuple form a *λ-parallelogram* . The fact that we in the picture did not make the four line segments *oriented* contains some symmetry assertions, which can be proved by working in a coordinatized situation; namely that the 4-group $\mathbb{Z}_2 \times \mathbb{Z}_2$ acts on the set of λ-parallelograms; so for instance (u, z, y, x) is a λ-parallelogram, equivalently

$$\lambda(\lambda(x, y, z), z, y) = x.$$

[3] Note the difference between this figure and the figure (1.2), in which y and z are assumed to be neighbours, and where the parallelogram *canonically* may be formed.

On the other hand

$$\lambda(\lambda(x, y, z), y, z) \sim x, \tag{2.7}$$

but it will not in general be equal to x; its discrepancy from being x is an expression of the *torsion* of λ. Even when $y \sim z$ (so x, y, z form a simplex), the left hand side of (2.7) need not be x. Rather, we may define the *torsion* of λ to be the law b which to any 2-simplex x, y, z associates $\lambda(\lambda(x, y, z), y, z)$. Then $b(x, y, z) = x$ for all simplices iff λ is symmetric.

There is also a notion of *curvature* of λ: Let M be a manifold equipped with an affine connection λ. If x, y, z form an infinitesimal 2-simplex in M, $\lambda(x, y, -)$ takes any neighbour v of x into a neighbour of y, and to it, we can then apply $\lambda(y, z, -)$ to obtain a neigbour of z. But since $x \sim z$, we could also transport v directly by $\lambda(x, z, -)$. To say that λ is *flat* or *curvature-free* is to say that we get the same neighbour point of z as the result. An equivalent description of flatness of λ is to say that applying $\lambda(x, y, -)$, then $\lambda(y, z, -)$, then $\lambda(z, x, -)$ to any $v \sim x$ will return v as value. In this conception, the curvature r of λ is a law, which to any infinitesimal 2-simplex x, y, z provides a permutation of the set of neighbours of x. (In the terminology of [7], r is a group-bundle valued combinatorial 2-form.)

We give two examples of affine connections on the unit sphere S.

Example 1. The unit sphere S sits inside Euclidean 3-space, $S \subseteq E$. Since E is in particular an affine space, we may for any three points x, y, z in it form $y - x + z \in E$. For x, y, z in S, the point $y - x + z$ will in general be outside S; if x, y, z are mutual neighbours, however, $y - x + z$ will be in S, cf. Remark at the end of Section 1. What if x, y, z form an infinitesimal 2-whisker? Then we may define $\lambda(x, y, z) \in S$ to be the point, where the half line from the center of S to $y - x + z$ meets S. It is easy to see that (2.2) holds. The equation (2.3) requires a little argument in the following spirit: the failure of (2.3) to hold is quadratic in $x - y$, and therfore vanishes because $x - y \in D(3) \subset E = R^3$.

This affine connection is evidently symmetric in y and z, so is torsion free; it does, however, have curvature. It is the Riemann- or Levi-Civita connection on sphere.

Example 2. (This example does not work on the whole sphere, only away from the two poles.) Given x, y and z with $x \sim z$ ($x \sim y$ is presently not relevant). Since x and z are quite close, we can uniquely describe z in a rectangular coordinate system at x with coordinate axes pointing East and North. Now take $\lambda(x, y, z)$ to be that point near y, which in the East-North coordinate system at y has same coordinates as the ones obtained for z in the coordinate system that we considered at x.

The description of this affine connection is asymmetric in y and z, and it is indeed easy to calculate that it has torsion ([7], Section 2.4). It has no curvature. – One may think of this $\lambda(x, y, z)$ as parallel *transporting* z from x to y; so xy is an active aspect, z is a passive aspect.

Connections constructed in a similar way also occur in materials science: for a crystalline substance one may attach a coordinate system at each point, by using the crystalline structure to define directions (call them "East" and "North" and "Up", say). The torsion for a connection λ constructed from such coordinate systems is a measure for the imperfection of the crystal lattice (dislocations), – see [8], [4] and the references therein.

3 Second Order Notions; Midpoint Formation

The data of an affine connection on a manifold M is a (partially defined) ternary operation λ. We indicate in this Section how the data of a *symmetric* (= torsion free) affine connection may be encoded by a binary operation "midpoint formation" μ on pairs of *second order* neighbours in M.

Let $M_{(2)} \subseteq M \times M$ denote the set of pairs (x, u) of second order neighbours; $M_{(2)}$ is the "second neighbourhood of the diagonal", in analogy with the first neighbourhood $M_{(1)}$ described in Section 1. We have $M_{(1)} \subseteq M_{(2)}$. If λ is an affine connection on M, then for any 2-whisker x, y, z, we have that $x \sim_2 \lambda(x, y, z)$.

Recall that for $x \sim_1 y$ in M, we have canonically the affine combination $\frac{1}{2}x + \frac{1}{2}y$, the midpoint.

Definition 3.1. *A* midpoint formation structure *on M is a map $\mu : M_{(2)} \to M$, extending the canonical midpoint formation for pairs of first order neighbour points.*

Thus, $\mu(x, u)$ is defined whenever $x \sim_2 u$; and $\mu(x, u) = \frac{1}{2}x + \frac{1}{2}u$ whenever $x \sim_1 u$. It can be proved that such μ is automatically symmetric, $\mu(x, u) = \mu(u, x)$, and that $\mu(x, u) \sim_2 x$ and $\sim_2 u$.

Theorem 3.2. *There is a bijective correspondence between midpoint formation structures μ on M, and symmetric (= torsion free) affine connections λ on M.*

Proof sketch. Given μ, and given an infinitesimal 2-whisker (x, y, z). Since $x \sim_1 y$, we may form the affine combination $2y - x$ (reflection of x in y), and it is still a first order neigbour of x. Similarly for $2z - x$. So $(2y - x) \sim_2 (2z - x)$, and so we may form $\mu(2y - x, 2z - x)$, and we define

$$\lambda(x, y, z) := \mu(2y - x, 2z - x).$$

The relevant picture is here:

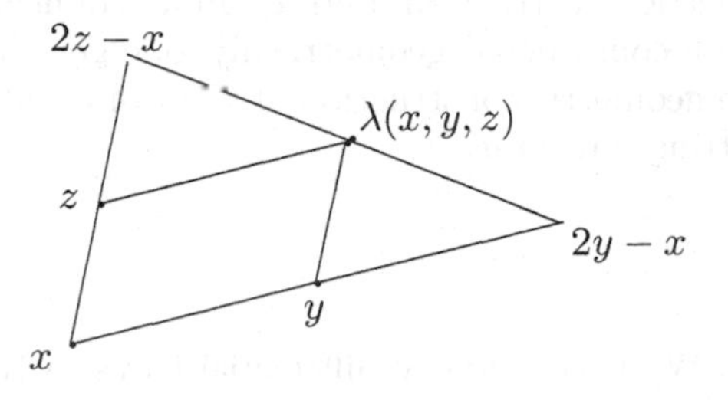

It is symmetric in y and z, by the symmetry of μ. Also, if $y = x$, we get

$$\lambda(x, x, z) = \mu(x, 2z - x) = \tfrac{1}{2}x + \tfrac{1}{2}(2z - x),$$

since $x \sim_1 2z - x$ and μ extends the canonical midpoint formation for first order neighbours. But this equals z, by evident equations for affine combinations. This proves the first equation in (2.1), and the second one then follows by symmetry.

The passage from a symmetric affine connection λ to midpoint formation μ is less evident. If $x \sim_2 u$ and if we have some y which "interpolates" in the sense that $x \sim_1 y \sim_1 u$, we may define $\mu(x, u)$ by

$$\mu(x, u) = \lambda(y, \tfrac{1}{2}x + \tfrac{1}{2}y, \tfrac{1}{2}y + \tfrac{1}{2}u),$$

(make a picture!); one can prove that this does not depend on the choice of the interpolating y.

Let us show that one gets the symmetric affine connection λ back from the midpoint formation μ to which it gives rise. Let $\tilde{\lambda}$ be the affine connection constructed from μ, so for a whisker x, y, z at x, use x as interpolating point between $2y - x$ and $2z - x$; so

$$\tilde{\lambda}(x, y, z) = \mu(2y - x, 2z - x) = \lambda(x, \tfrac{1}{2}x + \tfrac{1}{2}(2y - x), \tfrac{1}{2}x + \tfrac{1}{2}(2z - x)),$$

but $\frac{1}{2}x + \frac{1}{2}(2y - x) = y$ and $\frac{1}{2}x + \frac{1}{2}(2z - x) = z$, by purely affine calculations; so we get $\lambda(x, y, z)$ back.

Remark. The Theorem may also be seen as a manifestation of a simple algebraic fact: in R^n, a midpoint formation μ is given by $\mu(x, y) = \frac{1}{2}x + \frac{1}{2}y + M(x; y - x)$, where $x \sim_2 y$ and where $M(x; -)$ (for each x) is a function $D_2(n) \to R^n$, vanishing on $D_1(n)$. By a version of the KL axiom scheme, one sees that each such $M(x; -)$ extends uniquely to a quadratic R^n-valued form on R^n, and there is classically a bijective correspondence between such forms, and symmetric bilinear R^n-valued forms Γ on R^n; these Γs will serve as Christoffel symbols for the symmetric affine connection corresponding to μ, modulo a factor -8.

In [5], it is shown how a Riemannian metric geometrically gives rise to a midpoint formation (out of which, in turn, the Levi-Civita affine connection may be constructed, by the process given by the Theorem).

Problem: Since a midpoint formation structure μ gives rise to an affine connection λ by a geometric construction, and an affine connection λ gives rise to a curvature r, likewise constructed geometrically, one gets by concatenation of these constructions a geometric construction of r out of μ. Is there a more direct geometric way of getting r from μ?

References

1. Breen, L., Messing, W.: Combinatorial differential forms. Advances in Math. 164, 203–282 (2001)

2. Kock, A.: Synthetic Differential Geometry. London Math. Soc. Lecture Notes Series, vol. 51. Cambridge University Press, Cambridge (1981)*(Second edition, London Math. Soc. Lecture Notes Series, vol. 333. Cambridge University Press, Cambridge (2006))
3. Kock, A.: A combinatorial theory of connections. In: Gray, J. (ed.) Mathematical Applications of Category Theory, Proceedings 1983. AMS Contemporary Math., vol. 30, pp. 132–144 (1984)
4. Kock, A.: Introduction to Synthetic Differential Geometry, and a Synthetic Theory of Dislocations. In: Lawvere, F.W., Schanuel, S. (eds.) Categories in Continuum Physics, Proceedings Buffalo 1982. Springer Lecture Notes, vol. 1174. Springer, Heidelberg (1986)
5. Kock, A.: Geometric construction of the Levi-Civita parallelism. Theory and Applications of Categories 4, 195–207 (1998)
6. Kock, A.: Differential forms as infinitesimal cochains. Journ. Pure Appl. Alg. 154, 257–264 (2000)
7. Kock, A.: Synthetic Geometry of Manifolds. Cambridge Tracts in Mathematics 180 (2009)
8. Noll, W.: Materially uniform simple bodies with inhomogeneities. Arch. Rat. Mech. Anal. 27, 1–32 (1967)

Mathematics in Atmospheric Sciences: An Overview

Pierre Gauthier

Department of Earth and Atmospheric Sciences
Université du Québec à Montréal
pierre.gauthier@uqam.ca

Abstract. Several sectors of human activities rely on weather forecasts to plan in preparation of high impact weather events like snow storms, hurricanes or heat waves. Climate studies are needed to make decisions about the long-term development in agriculture, transport and land development. Observing and modeling the evolution of the atmosphere is needed to provide key *reliable* information for both weather prediction and climate scenarios. This paper gives an overview of the scientific research underlying the development and validation of numerical models of the atmosphere and the monitoring of the quality of the observations collected from several types of instruments. A particular emphasis will be given to data assimilation which establishes the bridge between numerical models and observations. The mathematical problems arising in atmospheric research are diverse as the problem is one of stochastic prediction for which errors in both the model and the observations need to be considered and estimated. Atmospheric predictability is concerned with the *chaotic* nature of the nonlinear equations that govern the atmosphere. Ensemble prediction is one area that has expanded significantly in the last decade. The interest stems from the necessity to evaluate more than just a forecast: it aims at giving an estimate of its accuracy as well. This brings up more questions than answers.

1 Introduction

Modeling and observing the atmosphere provide the basic information through which we improve our understanding of atmospheric processes and their interactions. At heart, the atmosphere is a fluid governed by the Navier-Stokes equations, thermodynamic relationships, and numerous other processes that influence its dynamics. Modeling the atmosphere is necessary for numerical weather prediction (NWP) and climate modeling. Weather prediction is concerned with an initial value problem : knowing the current state of the atmosphere, the primitive equations (e.g., Navier-Stokes, thermodynamics) are numerically integrated in time to obtain forecasts up to 10 days ahead. In climate studies, similar integrations are carried out but over longer periods of time from decades, to centuries and even millenia (paleoclimatology) and one is interested in the evolution of the average of the atmospheric state, filtering out transient processes. One has

S. Brlek, C. Reutenauer, and X. Provençal (Eds.): DGCI 2009, LNCS 5810, pp. 22–33, 2009.

to be concerned with both the first moment (the mean) and the second moments (variability) of the time series of the evolving climate.

Both problems (NWP and climate) are faced with similar challenges. The primitive equations must be discretized which implies that approximations need to be introduced to include the effect of unresolved processes. Numerical schemes are only approximating the "exact" solution and may not conserve some basic physical quantities (e.g., the total mass of the atmosphere). Discretization of the atmosphere is done horizontally over the globe or limited areas. The spatial resolution of the model must be sufficient to resolve the key physical processes that have an impact on the evolution of the atmosphere. Even for weather forecasts for a few days ahead, global models are needed as small changes in the weahter in distant regions may have an influence on what will happen here a few days later. This is often referred to as the *butterfly effect.* Local area models can only provide forecasts for shorter periods of time but can be run at higher spatial resolutions, which permit to resolve convection processes that are critical for accurate predictions of precipitations.

In both cases, computer efficiency of the numerical algorithms is also important and numerical models used for NWP is one of the main applications of high performance computing. NWP centres have some of the most powerful computers to be able to produce accurate weather forecasts within a limited time. More efficient algorithms and increasing the computing power enable to increase the complexity and accuracy of the models. In this presentation, an overview is presented of the current state of numerical modeling and some of the mathematical difficulties that are raised.

Numerical models include our knowledge about the physical laws that govern the motion of the atmosphere. They include also the limitations of our understanding and the results obtained with those models need to be validated by comparing their forecasts against observations. The global observing systems includes several types of instruments that provide measurements that are often only indirectly related to the atmospheric variables. Observations from satellite based instruments are measuring the electro-magnetic spectrum in the microwave, infrared and visible. This raises new problems associated with the propagation of electro-magnetic waves in highly heterogeneous media, particularly when clouds and aerosols are present. To make good use of the data collected from such instruments, it is important to be able to relate atmospheric variables to observations. Finally, acknowledging that both the numerical models and the observations have errors, the best estimate of the atmosphere is obtained by combining observations and models using the Bayesian framework of statistical estimation. The most important progress made over the last twenty years were in the development of data assimilation methods that are better suited for the assimilation of satellite data that are irregularly distributed in space and time.

This presentation aims at giving a sense of the mathematical problems that are raised when trying to build better numerical simulations of the evolving atmosphere.

2 Numerical Weather Prediction and Climate Modeling

The atmosphere being a fluid, it is governed by the Navier-Stokes equations. When expressed in a rotating referential, they can be written as

$$\frac{d\mathbf{v}}{dt} + 2\Omega \times \mathbf{v} = -\frac{1}{\rho}\nabla p + \mathbf{D}_{dissipation} + \mathbf{F}_{External} \quad (1)$$

where $\mathbf{v}$ is the wind vector, $\Omega = \frac{2\pi}{T_e} sin\varphi \ \mathbf{e}_r$ is the angular frequency associated with the rotation of the Earth, $T_e = 24$ hrs, φ is the latitude and $\mathbf{e}_r$ is a unit vector oriented along the radius of the sphere. The other elements appearing here are the pressure, p, the air density, ρ, while $\mathbf{D}$ and $\mathbf{F}$ stand respectively for the dissipation and external forces acting on the air parcel. Pressure is related to density and temperature by the gas law ($p = R\rho T$) and thermodynamic exchanges imply that

$$\frac{dln\theta}{dt} = \frac{1}{c_p}\frac{dQ}{dt} \quad (2)$$

where $\theta = T(\frac{1000}{c_p})^{R/c_p}$ is the potential temperature and $\frac{dQ}{dt}$ represents the heating rate associated with external sources (e.g., radiative heating, release of latent heat due to condensation). Finally, mass conservation yields another important constraint which is

$$\frac{\partial\rho}{\partial t} + \nabla \bullet \rho\mathbf{v} = 0. \quad (3)$$

In a nutshell, numerical modeling of the atmosphere consists in solving this system of equations by using numerical schemes. One therefore must discretize in space and time and devise numerical schemes that will insure numerical accuracy and stability. Moreover, it is also important to impose some dynamical constraints that will insure that the numerical solutions preserve the known invariants of the atmosphere as much as possible. The spatial discretization imposes a limit of what can be resolved by a numerical model. Operational global models of the atmosphere now operates at horizontal resolutions as low as 20 km while in the vertical the domain extends all the way to 60 km to include the stratosphere. All in all, this means that the grid of those models comprises $2000 \times 1000 \times 80 \sim 10^8$ points and one such grid is needed for each of the 4 basic dynamical fields (horizontal wind components, temperature and humidity). Typical synoptic weather systems extend over a 1000 km and are well resolved at those resolutions but there are numerous atmospheric processes (e.g., convection) that require more than that.

The dynamics of the atmosphere leads to several types of waves over a wide range of frequencies. Waves can be classified with respect to their restoring forces. At the larger scale, one finds the planetary waves with very long wavelengths (5000-10000 km) governed primarily by the temperature gradient between the equator and the poles, and the variation of the Coriolis force. These waves evolve slowly (order of days) and govern the motion of weather systems in the extra-Tropics. Gravity waves are associated with convergence (divergence) in the horizontal which creates vertical motion: those are external gravity waves. On the other hand, the atmosphere can be destabilized in the vertical due to heating at the surface: this is convection which is found within active cloud formations. The time scale associated with convection is of the order of hours. The Tropical regions are in a different regime as the Coriolis force is negligible there. Convection is very active and forced through different physical processes such as the heat exchange between the ocean and the atmosphere. Heat from the ocean is in a sense the fuel which permits to Tropical cyclones to intensify and become hurricanes (or Typhoons). Finally, the atmosphere being a gas, sound waves are also present and those are associated with very fast time scales.

Referring to [Durran, 1998], numerical schemes can become unstable if the time step is longer than the frequency of the waves. On the other hand, others are not unstable but will have the effect of damping any wave having a shorter wavelength. Different approaches can be taken: the problem can be cast in an Eulerian framework, a Lagrangian one or somewhere in between as in the semi-Lagrangian scheme [Robert, 1982]. Each has its own merits and weaknesses and must be looked at in the perspective of all the other elements that need to be included in the atmospheric models. The external forcing term of (1) includes many processes that are key to the way the atmosphere evolves. It includes the solar heating, the exchanges between the land and the atmosphere, and many other components. Finally, as stated earlier, an important aspect of this problem is its size. Algorithms must be able to take advantage of the computer architectures at hand. Nowadays, models are implemented on parallel architectures for which it is equally important to take into account the performance in terms of number of operations to do and the level of communication between the thousands of processors involved. The size of the problem demands such computing power and atmospheric numerical modeling is one of the important applications considered in high performance computing.

In numerical weather prediction, a model is run for a period up to ten days and the initial conditions play a key role: the weather that will occur over any specific region depends on the current conditions not only over that region but over the globe. Accurate forecasts in the 3 to 5-day range often critically depend of what is going in very distant regions on the globe. For example, weather forecasts over Canada depend of what is going on in the North Pacific, over Asia and in the Tropics near the South East of Asia.

Climate simulations are concerned with longer range predictions and only consider averages over long period of time. The objective is to describe the equilibrium state of the atmosphere due to the external forces exerted upon it. Climate models pay a lot of importance in having a model that, under certain conditions, preserves certain invariants of the basic equations. For instance, removing dissipation and external forces, total mass and total energy should be preserve. It is then important to test if the numerical schemes employed preserve such quantities. This is to make sure that if a climate simulation is indicating a warming of the atmosphere, that this is not the result of the numerical scheme.

A substantial part of the effort in model development is to try to validate the atmospheric model against known properties or directly against observations if available.

3 Observations and Model Validation

Numerical models are approximate and need to be validated by comparing forecasts against observations. The global observing system now includes a variety of observing systems from instruments on the ground, in ships and aircrafts. At the moment, most observations are collected from passive satellite instruments measuring the electro-magnetic spectrum in the microwave, infrared and visible bands. Active instruments like Lidars and radars are now onboard satellites and provide detailed information about the vertical structure of the atmosphere. This information is nevertheless only indirectly linked to the temperature, winds, humidity and other components of the atmosphere (e.g., chemical species, aerosols). A complex radiative transfer problem must be solved to be able to compare the observed radiances against those associated with the atmospheric state of a numerical model. Moreover, however abundant, observations do not provide neither the spatial or temporal coverage to uniquely characterize complete atmospheric fields, even when discretized. Some satellite instruments can cover the whole globe but it takes a number of hours. Moreover one must retrieves temperature out of radiances which only measure the integrated effect of temperature on the radiance emitted by the atmosphere. Methods based on observations alone compensate for the lack of spatial coverage by considering all observations available over a period of time (e.g., a month) to obtain an estimate of the mean temperature over that period. The *inversion problem* consists in reconstructing those atmospheric fields out of observations. Most inversion methods require a *background estimate* and one is trying to find how it can be changed to fit the observations. The background state (or *a priori* state) can be a reference standard atmospheric profile or a climatological mean. As will be discussed in the next section, a short-term forecast from a numerical weather prediction model is also a good *first-guess*.

Resolution is often lacking in observations to get quantitative measurements of atmospheric structures associated with particular physical processes. Convection in the Tropics is regulated by the diurnal cycle and numerical models still have difficulties to capture this. Measurements of precipitation are useful to validate the representation of convection. The Tropical Rainfall Measurement Mission (TRMM) had the objective of observing precipations in Tropical regions to better understand physical processes associated with Tropical cyclones. Instruments provided observations about the distribution of precipitations and the vertical structure of the distribution of water within clouds associated with such cyclones. However, the satellite being on a low orbit ($\sim$ 300 km) only flew over a given region at irregular intervals. Nevertheless, an average of the precipitation budget can be obtained by considering data over the whole hurricane season. This has been found to be very useful to understand the role of hurricanes on climate for instance.

Comparing observations with a model prediction must take into account the error that can occur in observing the atmosphere and modeling it. Knowledge of a prediction and an observation must be complemented by knowledge of their associated errors. Conversely, comparing a model against observations provides information about its accuracy. Moreover, observations from satellite instruments are not independent and error from measurements in different wavelengths but from the same instrument can be correlated. Analysis of the error can be used to estimate how much independent information we get from a large volume of data. Finally, it is important to say that the link between the atmosphere and the observed radiation from space is complex. Electromagnetic waves are travelling through a highly heterogeneous atmosphere which includes clouds. Detailed modeling of the diffusion, transmission and reflection of radiation is complex and is addressed by Monte-Carlo methods in which thousands of photons are emitted and their paths calculated. Some information about the average behavior can be obtained [Barker *et al.*, 2003].

Finally, geostationary satellites provide images of nearly a full hemisphere at frequent time intervals. Information about atmospheric winds can be retrieved by an analysis of these images. Clouds are detected in the visible and infrared and their motion can be related to the underlying atmospheric winds. The techniques employed involve pattern recognition to detect identifiable structures from one moment to the next. One of the difficulties however is to assign correctly the height of the clouds observed on the satellite image.

Although observations do provide a large volume of data about the atmosphere, they come from different sources and it is not obvious to piece those together to get a comprehensive representation of the atmosphere and its evolution in time.

4 Data Assimilation and Analyses

Considering that both the observations and the forecasts from a numerical model have errors, the current state of the atmosphere can be obtained by blending the two considering their respective accuracies here represented by the probability density functions (pdf) of observation and background error. Given that a set of observations $\mathbf{y} \in \mathbb{R}^m$ has been observed and an *a priori* knowledge $\mathbf{x}_b \in \mathbb{R}^n$, both having error characteristics defined by their respective pdf. In a Bayesian sense, the problem can be stated as

$$p(\mathbf{x}|\mathbf{y}) = \frac{p(\mathbf{y}|\mathbf{x})P(\mathbf{x})}{P(\mathbf{y})} \tag{4}$$

with $P(\mathbf{x})$ being the *a priori* pdf of having $\mathbf{x}$ as the true state, $\mathbf{x}_t$, $p(\mathbf{y}|\mathbf{x})$, the conditional probability of $\mathbf{y}$ being the true value *given* that $\mathbf{x} = \mathbf{x}_t$, and $p(\mathbf{x}|\mathbf{y})$ is the probability of $\mathbf{x}$ being the true value given that $\mathbf{y}$ has been observed. Finally, $P(\mathbf{y})$ is the marginal probability of observing $\mathbf{y}$.

The mode of $p(\mathbf{x}|\mathbf{y})$ is the most likely state given the information we have and it can be obtained by minimizing the objective function

$$J(\mathbf{x}) = -ln(p(\mathbf{x}|\mathbf{y}) \tag{5}$$

$$= -ln(p(\mathbf{y}|\mathbf{x}) - ln(P(\mathbf{x}) + C. \tag{6}$$

The complexity of the problem stems from its size as the number of observations m is of $O(10^6)$ while the dimension of the model space is $O(10^7)$. With such large dimensions, our knowledge of the probability distributions for the observation and background error can only be partial. Assuming these error distributions to be Gaussian, then

$$P(\mathbf{x}) \propto exp(-\frac{1}{2}(\mathbf{x} - \mathbf{x_b})^{\mathbf{T}}\mathbf{B}^{-1}(\mathbf{x} - \mathbf{x_b}))$$
$$p(\mathbf{y}|\mathbf{x}) \propto exp(-\frac{1}{2}(\mathbf{H}(\mathbf{x}) - \mathbf{y})^{\mathbf{T}}\mathbf{R}^{-1}(\mathbf{H}(\mathbf{x}) - \mathbf{y}))$$

in which case

$$J(\mathbf{x}) = \frac{1}{2}(\mathbf{x} - \mathbf{x_b})^{\mathbf{T}}\mathbf{B}^{-1}(\mathbf{x} - \mathbf{x_b})) + \frac{1}{2}(\mathbf{H}(\mathbf{x}) - \mathbf{y})^{\mathbf{T}}\mathbf{R}^{-1}(\mathbf{H}(\mathbf{x}) - \mathbf{y}). \tag{7}$$

This is the variational form of the data assimilation problem. When formulated at a given time, this corresponds to the 3D case (3D-Var) while if $\mathbf{x} = \mathbf{x}(t)$ is time dependent, this is extended to the 4D case (4D-Var). Over the last 10 years or so, 4D-Var has been implemented in several operational numerical weather prediction centres including Europe (ECMWF, [Rabier *et al.*, 2000]), Météo-France [Gauthier and Thépaut, 2001], the United Kingdom [Rawlins *et al.*, 2007] and Canada [Gauthier *et al.*, 2007]. It is used to assimilate more than a million data daily to produce meteorological analyses that define the initial conditions used to produce weather forecasts up to 10 days.

As they are our best estimate of the current state of the atmosphere, analyses are also useful to validate numerical weather forecasts by running historical cases. If available over long periods of time, they have proven to be extremely useful to validate climate simulations. However, observations of the atmosphere become scarce as we go back in time. The modern era includes large volume of data that became available only since the end of the 20th century ($\sim$ 1980). Renanalyses are obtained by assimilating past observations using the most recent data assimilation systems that benefited from the latest advances made in data assimilation methods and numerical modeling. As climate studies are concerned with periods of decades to centuries and longer, reanalyses over the complete 20th century would be needed to validate climate models. Because of the scarcity of obserations, the extension of reanalyses to the beginning of the 20th century presents a challenge. From the experience gained in using 4D-Var, it is known that because it embeds the dynamics of the atmosphere within the assimilation process, it has the ability to "propagate" information in a dynamically consistent way to regions that are poorly observed. In a recent study, [Uppala *et al.*, 2008] showed that with 4D-Var, it is possible to reconstruct the complete state of the atmosphere using only surface observations. Their experiments was to redo a portion of the reanalysis using only available surface observations. The analysis based on surface observations compared remarkbly well to the reference analysis that used all observations.

A data assimilation system requires that the error statistics for both the background and the observations be correct. The background error statistics are the results of the error growth in a short term weather forecast associated with the initial uncertainty in the initial conditions. Dynamical instabilities depend on the prevailing conditions and it is those that define the background error statistics. These uncertainties can be captured by realizing Monte-Carlo simulations which produce an ensemble of forecasts out of which partial information is obtained about the probability distribution of the background error. Due to the limited size of the ensemble and the large dimension of the phase space, this gives rise to a rank deficiency problem associated with the estimation of the background-error covariances. Observations on the other hand can be contaminated by systematic error (biases) associated with the instrument or the observation operator used to link the raw observations to atmospheric variables. When extended to the 4D case, the observation operator includes also several integrations of the numerical atmospheric model to either iteratively minimize the objective function of 4D-Var or to perform the Monte-Carlo simulations needed of ensemble-based assimilation methods.

5 Mathematical Challenges

Numerical modeling has been applied in atmospheric sciences, oceanography, hydrology and atmospheric chemistry. Although significant advances have been made in the different areas, their individual behavior is influenced by the other components. Oceans act as a huge heat reservoir that regulates the temperature of the atmosphere and it plays a key role in climate studies. The atmophere transport chemical constituents and aerosols which in turn can influence atmospheric heating by absorbing or reflecting atmospheric radiation. It is now necessary to have coupled systems to represent the Earth system as a whole. This raises a number of new problems that need to be addressed if such an endeavour is to be successful. The individual components often represent phenomena that act on very different time scales. The oceans evolve on longer time scales than the atmosphere and atmospheric chemistry includes fast and slow reacting species. To model a system with such a wide variety of time scales is not obvious. One could use the lowest time step that would be acceptable for all components but this would lead to very expensive models in terms of computing time. The same can be said about the spatial resolution that is needed. Several approaches have been considered involving adaptive grids, multigrid models, etc., ecah with its own aadvantages and disadvantages. Numerous model integrations need to be carried out over long periods of time to study different climate scenarios and their sensitivity to changes in the model inputs. Consequently, the algorithms need to be very efficient on parallel computer architecture.

Another modeling problem is the radiative transfer in presence of clouds. Incident radiation on heterogeneous surfaces is partly reflected in different directions, partly absorbed and then propagated through a highly heterogeneous medium [Barker *et al.*, 2003, Schertzer and Lovejoy, 1987]. In [Lovejoy et al., 2001], fractals are used to represent the heterogeneous nature of the distribution of particles in the ocean and used for the assimilation of satellite measurements of ocean color, which relates to the presence of phytoplankton in oceans.

Climate studies and numerical weather prediction use models which involve a lot of unknowns. Some of these are empirical parameters that have been *tuned* to bring the model to agree as much as possible with observations. Error bars can be associated with those parameters and this immediately raises the question: how will this influence a weather forecast or a climate simulation? Ensemble prediction is used to examine the sensitivity of numerical simulations to perturbations of these parameters within the bounds of what is known about the uncertaintly of the input parameters [Palmer and Hagedorn, 2006]. Finite ensembles of forecasts are generated by integrating the model with random perturbations of the parameters and the objective is to evaluate the time evolution of the probability density function (p.d.f.) of the atmospheric state. Given that the size of the sample is very small with respect to the number of degrees of freedom involved, this p.d.f. is only partially known. In [Candille and Talagrand, 2005], it is argued that the size of the ensemble puts a limit on what we can know about the p.d.f. in terms of both its reliability and resolution. These considerations are important for the design of ensemble prediction systems.

The use of observations in data assimilation is based on statistical estimation and requires a good knowledge of the forecast and observation error. Forecast error can be estimated from ensemble of forecasts produced by perturbing the initial conditions according to the analysis error. Since the forecast error covariances embed the error growth processes of associated with atmospheric instabilities, they are *flow-dependent*. These background-error statistics are used in the assimilation and should therefore reflect the changing nature of the meteorological situations. Casting the problem in the framework of the Kalman filter, the Ensemble Kalman filter (EnKF) estimates the analysis and forecast error covariances which are used in the assimilation of observations. This aims at correctly weighting the information gained from observations with respect to that of the background state [Evensen, 1994, Houtekamer *et al.*, 2005]. Forecast error can be attributed to error in the input to any given model but it can also be due to the error in the model itself. Somehow this must be taken into account but then, model error must be estimated through comparison with observations and it is not difficult to argue that there is simply not enough of data to properly estimate fully the characteristics of model error. All in all the difficulties associated with data assimilation methods are related to model error and nonlinearities of the dynamics which is not fully taken into account in ensemble methods. This is now investigated with particle filters [van Leeuwen, 2009] and Bayesian estimation. As for model errors, [Trémolet, 2006] presents the *weak-constraint* 4D-Var, which permits to use longer assimilation windows than is posssible with the strong constraint 4D-Var.

6 Conclusion

Complex mathematical and physical problems are associated with understanding the Earth system, and the atmosphere in particular. Given the numerous approximations that must be made, it is important to validate those models through comparison to observations. This paper presented an overview of different areas in atmospheric sciences that are faced with some mathematical challenges. These need to be translated as efficient algorithms that can tap into the computing power now available. There are vast amount of resources invested to build high performance computing centres to experiment with models at higher spatial resolutions, to study the impact of including complex physical processes, to see how the ocean, the atmosphere and land interact, and investigate how these processes impact the evolution of the future climate or the weather forecasts that are critical in many sectors of human activities.

These issues are important enough to justify also the investment made in deploying satellite instruments to observe the Earth system. In recent years, good progress has been made in the development of new data assimilation methods but the fact remains that less than 10% of the total volume of incoming satellite data can be assimilated. The presence of clouds is one important obstacle and the effort is now focusing on representing and includinging clouds in models of the atmospheric radiative transfer in the assimilation. There are questions associated with the estimation of the observations error statistics, taking into account that, for satellite instruments, data cannot be assumed to be independent as their error is spatially correlated. Another difficulty is that even though there is information in satellite imagery that the eye can see, the current assimilation methods have difficulties in following recognizable patterns in clouds and other constituents advected by the winds. Finally, one should not forget that the new instruments onboard satellite create extremely large volume of data that must be transmitted from the satellite to the ground, and then have to be processed and disseminated to operational numerical weather prediction centres all around the world. If this is not dealt with properly, this can be the bottle neck and the data would not be received in time to be assimilated.

The scientific community is constantly consulted for guidance on how to make the best use of these computing capabilities or the huge amounts of data collected by the global observing system. The research that needs to be done is often the result of exchange of ideas between different disciplines. The advent of quadri-dimensional data assimilation (4D-Var) is a nice example. It was the result from discussions between atmospheric dynamicists and mathematicians working on control theory. Dynamicists were interested to understand the genesis of weather systems that develop in just a few days. The problem was boiled down to minimizing a functional that had the initial conditions as the control variable. Results had been obtained with low dimensional systems (just a few hundreds of degrees of freedom) but the approach taken was to compute the gradient of this functional by finite differences which required a number of model integrations equal to the number of degrees of freedom. This was to compute *one* gradient and this had to be repeated for each iteration required by the descent algorithm. This was clearly a dead end unless an insane amount of computing power was available. The breakthrough came from control theory which stated that the same gradient could be obtained from a single *backward* integration of the adjoint model. This happened in the mid-1980's and now, 4D-Var is part of the operational suite of several NWP centres.

Some examples were given in the preceeding section in which observing and modeling the atmosphere (and the Earth system) give rise to complex problems. Following the work of [Lorenz, 1963], it was found that atmospheric processes being inherently nonlinear can lead to what is now known as *chaos* meaning that a small perturbation to the system can lead to dramatic changes in a finite time. This idea has sparked a lot of interest in mathematics which resulted in significant advances in research on dynamical systems [Guckenheimer and Holmes, 1983]. The chaotic nature of the atmosphere is reflected in studies on predictability which led to the developement of ensemble prediction and probabilistic forecasting. These applications are currently treated with severe approximations that leads to limitations in the information they provide. The question is to know what reliable information can we gain from the vast amount of observations and numerical results generated by these experiments. Analysing that much data is also a challenge.

References

[Barker *et al.*, 2003] Barker, H.W., Pincus, R., Morcrette, J.-J.: The Monte-Carlo Independent Column Approximation: Application within large-scale models. In: Proceedings GCSS/ARM Workshop on the Representation of Cloud System in Large-Scale Models, Kananaskis, Al, Canada, 10 pp. (May 2002), `http://www.met.utah.edu/skrueger/gcss-2002/Extended-Abstracts.pdf`

[Candille and Talagrand, 2005] Candille, G., Talagrand, O.: Evaluation of probabilistic prediction systems for a scalar variable. Quart. J.R. Metor. Soc. 131, 2131–2150 (2005)

[Durran, 1998] Durran, D.R.: Numerical methods for wave equations in geophysical fluids. Texts in Applied Mathematics, vol. 32, 463 pages. Springer, Heidelberg (1998)

[Evensen, 1994] Evensen, G.: Sequential data assimilation with a nonlinear quasi-geostrophic model using Monte Carlo methods to forecast error statistics. em J. Geophys. Res. 99(C5), 10143–10162 (1994)

[Gauthier and Thépaut, 2001] Gauthier, P., Thépaut, J.-N.: Impact of the digital filter as a weak constraint in the preoperational 4DVAR assimilation system of Météo-France. Mon. Wea. Rev. 129, 2089–2102 (2001)

[Gauthier *et al.*, 2007] Gauthier, P., Tanguay, M., Laroche, S., Pellerin, S., Morneau, J.: Extension of 3D-Var to 4D-Var: implementation of 4D-Var at the Meteorological Service of Canada. Mon. Wea. Rev. 135, 2339–2354 (2007)

[Guckenheimer and Holmes, 1983] Guckenheimer, J., Holmes, P.: Nonlinear Oscillations, Dynamical Systems and Bifurcations of vector fields, 453 pages. Springer, Heidelberg (1983)

[Houtekamer *et al.*, 2005] Houtekamer, P.L., Mitchell, H.L., Pellerin, G., Buehner, M., Charron, M., Spacek, L., Hansen, B.: Atmospheric Data Assimilation with an Ensemble Kalman Filter: Results with Real Observations. Mon. Wea. Rev. 133, 604–620 (2005)

[Lorenz, 1963] Lorenz, E.N.: Deterministic non-periodic flow. J. Atmos. Sci. 20, 130–141 (1963)

[Lovejoy et al., 2001] Lovejoy, S., Schertzer, D., Tessier, Y., Gaonach, H.: Multifractals and Resolution independent remote sensing algorithms: the example of ocean colour. Inter. J. Remote Sensing 22, 1191–1234 (2001)

[Palmer and Hagedorn, 2006] Palmer, T., Hagedorn, R.: Predictability of Weather and Climate, 695 pages. Cambridge University Press, Cambridge (2006)

[Rabier *et al.*, 2000] Rabier, F., Jrvinen, H., Klinker, E., Mahfouf, J.-F., Simmons, A.: The ECMWF operational implementation of four dimensional variational assimilation. Part I: experimental results with simplified physics. Quart. J.R. Meteor. Soc. 126, 1143–1170 (2000)

[Rawlins *et al.*, 2007] Rawlins, F., Ballard, S.P., Bovis, K.J., Clayton, A.M., Li, D., Inverarity, G.W., Lorenc, A.C., Payne, T.J.: The Met Office global four-dimensional variational data assimilation scheme. Quart. J.R. Meteor. Soc. 623(623), 347–362 (2007)

[Robert, 1982] Robert, A.: A semi-implicit and semi-Lagrangian numerical integration scheme for the primitive meteorological equations. J. Meteor. Soc. Japan 60, 319–325 (1982)

[Schertzer and Lovejoy, 1987] Schertzer, D., Lovejoy, S.: Physical modeling and Analysis of Rain and Clouds by Anisotropic Scaling of Multiplicative Processes. Journal of Geophysical Research D8(8), 9693–9714 (1987)

[Trémolet, 2006] Trémolet, Y.: Accounting for an imperfect model in 4D-Var. Quart. J.R. Meteorol. Soc. 132, 2483–2504 (2006)

[Uppala *et al.*, 2008] Ippala, S., Simmons, A., Dee, D., Källberg, P., Thépaut, J.N.: Atmospheric reanalyses and climate variations. In: Climate variability and Extremes during the past 100 years, 364 pages, pp. 103-118. Springer, Heidelberg (2008)

[van Leeuwen, 2009] van Leeuwen, P.J.: Particle filters in geophysical systems. To appear in Mon Wea. Rev. (2009)

On Three Constrained Versions of the Digital Circular Arc Recognition Problem

Tristan Roussillon[1,*], Laure Tougne[1], and Isabelle Sivignon[2]

[1] Université de Lyon,
Université Lyon 2, LIRIS, UMR5205, F-69676, France
{tristan.roussillon,laure.tougne}@liris.cnrs.fr
[2] Université de Lyon, CNRS
Université Lyon 1, LIRIS, UMR5205, F-69622, France
isabelle.sivignon@liris.cnrs.fr

Abstract. In this paper, the problem of digital circular arcs recognition is investigated in a new way. The main contribution is a simple and linear-time algorithm for solving three subproblems: online recognition of digital circular arcs coming from the digitization of a disk having either a given radius, a boundary that is incident to a given point, or a center that is on a given straight line. Solving these subproblems is interesting in itself, but also for the recognition of digital circular arcs. The proposed algorithm can be iteratively used for this problem. Moreover, since the algorithm is online, it provides a way to segment digital curves.

1 Introduction

This paper deals with three constrained versions of a well-known problem: the recognition of digital circular arcs (DCAs for short). Many authors have proposed a solution to the recognition of digital circles [2,3,5,6,7,9,11,12,14,16]. Some techniques are not adapted for DCAs, like [16], and only a few ones are online [2,9]. Even if a linear algorithm has been proposed for a long time [11], using a sophisticated machinery coming from linear programming [10], no solution is known to be truly fast and easy to implement. That's why further research on the topic is needed.

We opt for an original approach of the problem. Indeed, we study three constrained versions of the DCA recognition problem: (i) the case of disks of given radius, (ii) the case of disks whose boundary is incident to a given point, (iii) the case of disks whose center is on a given straight line. We show that deciding whether a part of a digital boundary is the digitization of one of these disks is done with a simple, online and linear-time algorithm.

Solving these constrained problems is interesting in itself. For instance, the proposed algorithm can be applied to the case of disks of infinite radius in order to provide a way for recognizing digital straight segments. Thus, our method is a nice tool to highlight what makes the problems studied in this paper be similar or different from the problem of digital straight segments recognition.

* Author supported by a grant from the DGA.

S. Brlek, C. Reutenauer, and X. Provençal (Eds.): DGCI 2009, LNCS 5810, pp. 34–45, 2009.

Moreover solving such constrained problems is also useful to solve the unconstrained problem. Indeed the proposed algorithm can be iteratively used for the recognition of DCAs. This new technique may be coarsely described as follows: if a new foreground (resp. background) point is located outside (resp. inside) the current smallest separating disk, then either the new point is on the new smallest separating disk, or the sets of foreground and background points are not circularly separable at all. In the aim of deciding between these two alternatives, the proposed algorithm can be used in the case (ii), *i.e.* when the boundary of the disks must be incident with a given point.

Section 2 is made up of formal definitions and a brief review of the literature. The main results are presented in Section 3. The main algorithm is described and proved in Section 3.3. We show how to use it for recognition of DCAs in Section 4.

2 Preliminaries

2.1 Digital Boundary and Digital Contour

A binary image I is viewed as a subset of points of $\mathbb{Z}^2$ that are located inside a rectangle of size $M \times N$. A digital object O is defined as a 4-connected subset of $\mathbb{Z}^2$ (Fig. 1.a). Its complementary set $\bar{O} = I \backslash O$, which is assumed to be connected, is the so-called background. The digital boundary B (resp. $\bar{B}$) of O (resp. $\bar{O}$) is defined as the 8-connected clockwise-oriented list of the digital points having at least one 4-neighbour in $\bar{O}$ (resp. O). (Fig. 1.b).

Let us assume that each digital point of O is considered as the center of a closed square of size 1×1. The topological border of the union of these squares defines the digital contour C of O (Fig. 1.c). C is a 4-connected clockwise-oriented list of points with half-integer coordinates (Fig. 1.c).

Each point of C is numbered according to its position in the list. The starting point, which is arbitrarily chosen, is denoted by C_0 and any arbitrary point of the list is denoted by C_k. A part (C_iC_j) of C is the list of points that are ordered increasingly from index i to j (Fig. 1.d).

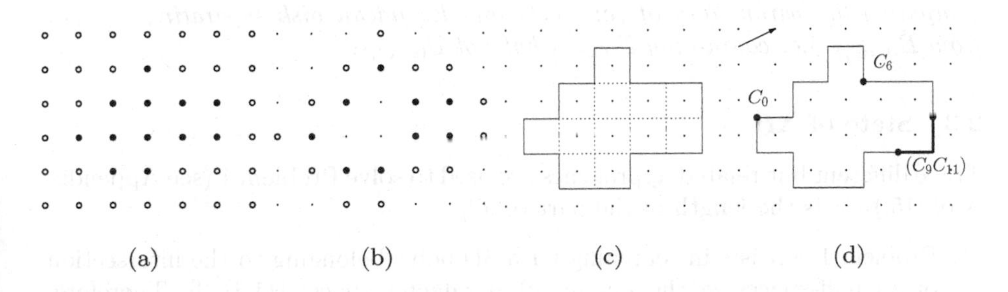

Fig. 1. (a) A digital object in black points and its complementary set in white points, (b) their digital boundaries and (c) their digital contour. (d) Notations.

An elementary part bounded by two consecutive points (C_kC_{k+1}) separates a point of B (on its right side) from a point of $\bar{B}$ (on its left side) (Fig. 2.a). Let us denote by $B_{(C_iC_j)}$ (resp. $\bar{B}_{(C_iC_j)}$) the list of digital points of B (resp. $\bar{B}$) that are located on the right (resp. left) side of each elementary part (C_kC_{k+1}) of (C_iC_j) with $i \leq k < j$.

2.2 Digital Circle and Circular Arc

Definition 1 (Digital circle (Fig. 2.b)). *A digital contour C is a digital circle iff there exists a Euclidean disk $\mathcal{D}(\omega, r)$ that contains B but no points of $\bar{B}$.*

Definition 2 is the analog of Definition 1 for parts of C.

Definition 2 (Circular arc (Fig. 2.c)). *A part (C_iC_j) of C (with $i < j$) is a circular arc iff there exists a Euclidean disk $\mathcal{D}(\omega, r)$ that contains $B_{(C_iC_j)}$ but no points of $\bar{B}_{(C_iC_j)}$.*

This definition is equivalent to the one of Kovalevsky [9].

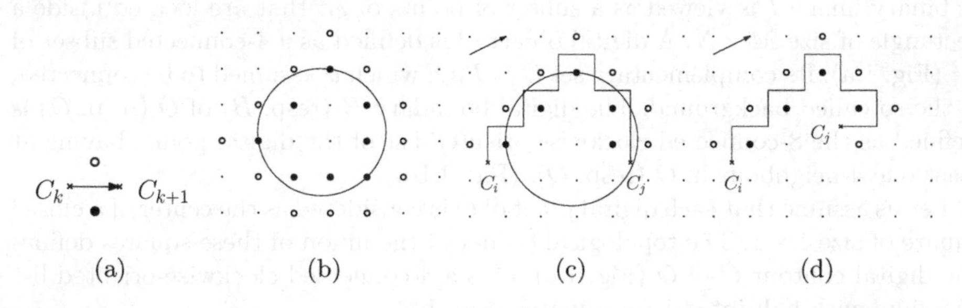

Fig. 2. (a) An elementary part. (b) A digital circle. (c) A circular arc. (d) A part that is not a circular arc.

Problem 1 (DCA recognition). *Given a part (C_iC_j) of C, the DCA recognition problem consists in deciding whether (C_iC_j) is a DCA or not, and if so, computing the parameters of (at least) one Euclidean disk separating $B_{(C_iC_j)}$ from $\bar{B}_{(C_iC_j)}$,* i.e. *containing $B_{(C_iC_j)}$ but not $\bar{B}_{(C_iC_j)}$.*

2.3 State of Art

Three different but related approaches are used to solve Problem 1 (see Appendix A of [15]) - n is the length of the part (C_iC_j):

1. Problem 1 consists in searching for a 3D point belonging to the intersection of $2n$ half-spaces in the (ω_x, ω_y, r) parameter space: [11,16,3]. Therefore, Megiddo's algorithm can be used [10] in order to derive an algorithm in $\mathcal{O}(n)$. An online version of this algorithm exists [1] but is difficult to implement.

2. If the parameter space is projected along the r-axis onto the (ω_x, ω_y)-plane, Problem 1 consists in searching for a 2D point belonging to the intersection of n^2 half-planes. This approach, which requires massive computation, has been widely used [6,9,12,2] and several optimizations have been proposed. Kovalevsky [9] removes some points during the computation but without improving the worst-case bound. Coeurjolly *et al.* [2] proposed a preprocessing stage using the arithmetic properties of digital curves so that the time complexity of their algorithm goes down from $\mathcal{O}(n^2 \log n)$ to $\mathcal{O}(n^{4/3} \log n)$. As noticed in [2], these algorithms may be made online with an incremental convex hull algorithm [13].
3. In the space that is dual to the parameter space, Problem 1 consists in searching for a plane separating two sets of n 3D points [11,5,14]. The quadratic algorithm of Kim [8] can be straightforwardly interpreted in this way. Using classical results about the computation of 3D convex hulls [13] and the computation of the vertical distance between two convex polyhedra [13], this approach leads to an algorithm whose time complexity is bounded by $\mathcal{O}(n \log n)$ [14]. Though, this algorithm is not online.

3 Main Results

In this section, we study three constrained versions of Problem 1.

3.1 Definitions

We define three classes of constrained disks that fulfill the following property: a constrained disk is uniquely defined by two points. The set of constrained disks fulfilling one of the three conditions of Definition 3 is called a *class of constrained disks*.

Definition 3 (Constrained disks). *A* constrained *disk is uniquely defined by two points because one of the three following conditions is fulfilled: (i) it has a given radius and an orientation is arbitrarily chosen (Fig. 3.a and Fig. 3.b), (ii) its boundary is incident with a third point (Fig. 3.c), (iii) its center is on a given straight line (Fig. 3.d).*

Problems 2, 3 and 4 are the analog of Problem 1 for a specific class of contrained disks:

Problem 2. *Computing the parameters of the set of Euclidean disks $\mathcal{D}(\omega, r = r_0)$ separating $B_{(C_iC_j)}$ from $\bar{B}_{(C_iC_j)}$, where r_0 is fixed and given as input.*

Problem 3. *Computing the parameters of the set of Euclidean disks $\mathcal{D}(\omega, r)$ separating $B_{(C_iC_j)}$ from $\bar{B}_{(C_iC_j)}$, such that $\mathcal{D}$ touches a fixed point P_0 given as input.*

Problem 4. *Computing the parameters of the set of Euclidean disks $\mathcal{D}(\omega, r)$ separating $B_{(C_iC_j)}$ from $\bar{B}_{(C_iC_j)}$, such that ω belongs to a fixed straight line $\mathcal{L}_0$ given as input.*

In Sections 3.2 and 3.3, we assume that a class of constrained disks is given.

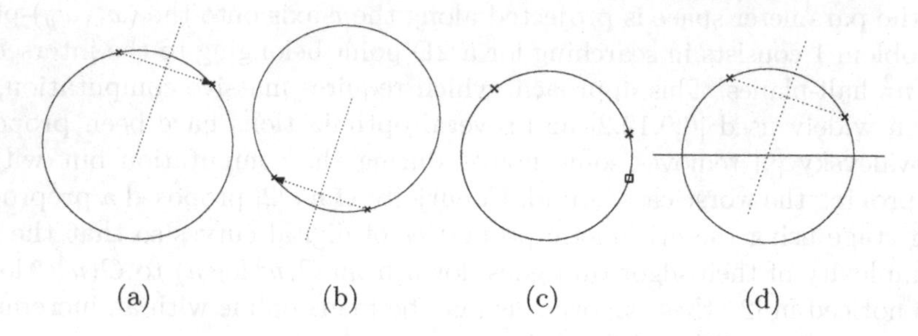

Fig. 3. One and only one disk is incident with the two points, labeled by a cross each, because a radius and an orientation have been chosen in (a) and (b), a third point depicted with a square has been given in (c) and the center has to be on the solid horizontal straight line in (d)

3.2 Circular Hulls and Points of Support

Definition 4 (Circular hull). *Let L be an ordered list of points. Its* inner *(resp.* outer*) circular hull is a list of some points of L such that, for each pair of consecutive points, all the points of L belong (resp. do not belong) to the constrained disk defined by the two points.*

Fig. 4 displays the inner and outer circular hulls of a list of points in the case of a given radius. Notice that the intrinsic order of the points determines a natural orientation so that one and only one disk of a given radius is defined by two points. If the radius is infinite, the circular hull of L is a part of the convex hull of L. As a consequence, the circular hull is easily computed with an online and linear-time algorithm in the style of the Graham's scan thanks to the intrinsic order of the points.

In order to solve Problems 2, 3 and 4, the constrained disks separating $B_{(C_iC_j)}$ from $\bar{B}_{(C_iC_j)}$ have to be computed. Some special points, called *points of support*, play a key role in the computation.

Definition 5 (Point of support). *A point of $B_{(C_iC_j)}$ or $\bar{B}_{(C_iC_j)}$ that is located on the boundary of a constrained disk separating $B_{(C_iC_j)}$ from $\bar{B}_{(C_iC_j)}$ is called* point of support.

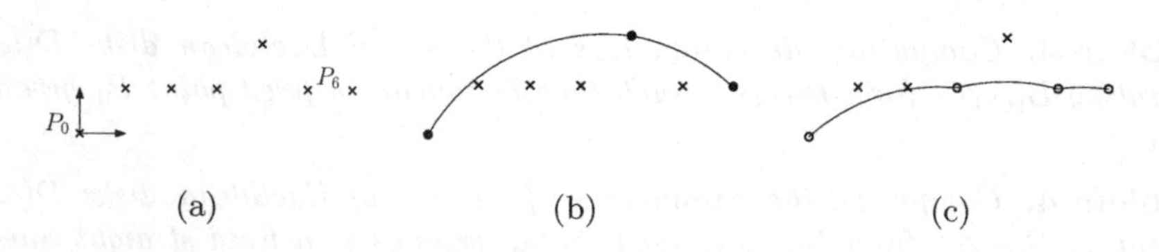

Fig. 4. Inner (b) and outer (c) circular hull of a list of points (a) when the radius of the disks is fixed ($r_0 = 4$)

The following propositions, which are related to the points of support, are proved in Appendix B of [15]:

Proposition 1. *$B_{(C_iC_j)}$ and $\bar{B}_{(C_iC_j)}$ are separable by a constrained disk iff $B_{(C_iC_j)}$ and $\bar{B}_{(C_iC_j)}$ contain at least one point of support.*

Proposition 2. *The points of support of $B_{(C_iC_j)}$ (resp. $\bar{B}_{(C_iC_j)}$) are consecutive points of the inner circular hull of $B_{(C_iC_j)}$ (resp. outer circular hull of $\bar{B}_{(C_iC_j)}$).*

Proposition 3. *The points of support of $B_{(C_iC_j)}$ and $\bar{B}_{(C_iC_j)}$ define the whole set of separating constrained disks.*

The first and last points of support of the inner circular hull of $B_{(C_iC_j)}$, respectively denoted by I_f and I_l, as well as the first and last points of support of the outer circular hull of $\bar{B}_{(C_iC_j)}$, respectively denoted by O_f and O_l, play a key role in the algorithm that checks the separability of $B_{(C_iC_j)}$ and $\bar{B}_{(C_iC_j)}$.

3.3 Separability

Algorithm 1 solves Problems 2, 3 and 4. The points of a part (C_iC_j) are processed one by one. Assume that the k first points have already been processed. When a new point C_{k+1} is taken into consideration, the inner and outer points defined by the elementary part (C_kC_{k+1}) (Fig. 2.b) are respectively added to the lists $B_{(C_iC_k)}$ (lines 4 and 5) and $\bar{B}_{(C_iC_k)}$ (lines 7 and 8).

Algorithm 1. Algorithm that solves Problems 2, 3 and 4

Input: A part (C_iC_j) (with $i < j$).
Output: The boolean value *areSeparable* and $OHull$, $IHull$, O_f, O_l, I_f, I_l.

```
   /* Initialization                                                    */
1  Initialization of IHull, I_f, I_l (resp. OHull, O_f, O_l) with the inner (resp. outer)
   point of (C_i C_{i+1});
2  areSeparable ← true; k ← i + 1;
   /* Scan                                                              */
3  while C_k < C_j and areSeparable do
4      N ← inner point of (C_k C_{k+1});
5      areSeparable ← Separability(N,OHull,IHull,O_f,O_l,I_f,I_l);
6      if areSeparable then
7          N ← outer point of (C_k C_{k+1});
8          areSeparable ← Separability(N,IHull,OHull,I_f,I_l,O_f,O_l);
9      k ← k + 1;
10 return areSeparable
```

Algorithm 1 calls Algorithm 2 on lines 5 and 8. It updates the inner and outer circular hulls as well as the points of support when a new point N is added.

Algorithm 2. Separability(N, $OHull$, $IHull$, O_f, O_l, I_f, I_l)

Input: $OHull$, $IHull$, O_f, O_l, I_f, I_l and a new point N
Output: a boolean value and updated $OHull$, $IHull$, O_f, O_l, I_f, I_l

```
1  if N is outside the constrained disk touching I_f and O_l then
2  |  return false;
3  else
4  |  if N is outside the constrained disk touching O_f and I_l then
   |  |  /* update of the inner circular hull                            */
5  |  |  while N is outside the constrained disk touching the last two points of
   |  |  IHull do
6  |  |  |  The last point of IHull is removed from IHull;
7  |  |  N is added to IHull;
   |  |  /* update of the points of support                              */
8  |  |  I_l ← N;
9  |  |  while N is outside the constrained disk touching the first two points of
   |  |  support of OHull do
10 |  |  |  O_f ← the point of OHull that is just after O_f;
11 |  return true;
```

Let us assume that N is an inner point. The case where N is an outer point is similar. If N does not belong to the constrained disk touching I_f and O_l (area 1 of Fig.5), $B_{(C_iC_{k+1})}$ and $\bar{B}_{(C_iC_k)}$ cannot be separated by a constrained disk and the Algorithm 2 returns false (lines 1 and 2).

If N belongs to the constrained disk touching I_f and O_l (areas 2 and 3 of Fig.5), $B_{(C_iC_{k+1})}$ and $\bar{B}_{(C_iC_k)}$ are still separable. If N does not belong to the disk touching O_f and I_l (area 2 of Fig.5), the inner circular hull is updated (lines 5-7) and the points of support are updated too (lines 8-10).

After this brief description of Algorithm 1 and 2, let us prove the following theorem:

Theorem 1. *Algorithm 1 correctly retrieves the set of constrained disks separating $B_{(C_iC_j)}$ from $\bar{B}_{(C_iC_j)}$ in linear time.*

Proof. Thanks to Proposition 3, we know that the whole set of separating constrained disks is given by the points of support of $B_{(C_iC_j)}$ and $\bar{B}_{(C_iC_j)}$. Therefore, showing that the algorithm properly retrieves the points of support of $B_{(C_iC_j)}$ and $\bar{B}_{(C_iC_j)}$ in linear time is sufficient to prove Theorem 1. Moreover, since Algorithm 1 completely depends on the correctness of Algorithm 2, we can focus on Algorithm 2. To prove that it properly updates the points of support, we have to show that a new point involves the removal of the p last points of support of $B_{(C_iC_j)}$ and the removal of the q first points of support of $\bar{B}_{(C_iC_j)}$. Because of the intrinsic order of the points, a new point cannot be in areas 4 and 5 of Fig. 5, but only in areas 1, 2 or 3. As a consequence, the points of support lying in the middle of the list of consecutive points of support of $B_{(C_iC_j)}$ (resp. $\bar{B}_{(C_iC_j)}$) cannot be removed if those lying at the front (resp. back) are not removed too.

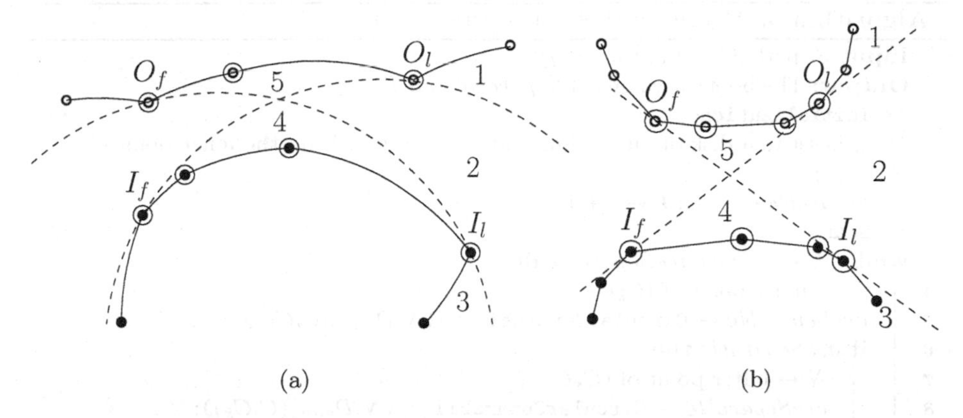

Fig. 5. The points of support are encircled. To help the reader to figure out why the role of the points of support is so important, here are two examples. In (a) the radius of the constrained disks is equal to 5, whereas in (b) the radius is infinite. The first and last points of support of each hull delineate 5 areas numbered from 1 to 5.

Algorithm 2 correctly computes p and q, because the points of the circular hulls are sequentially scanned respectively from front to back (lines 5-7) and back to front (lines 9 and 10).

Each point is added and removed at most once in the inner circular hull as well as in the list of points of support, implying that Algorithm 1 is linear-time. □

Our algorithm applies to Problems 2, 3 and 4: the only thing that changes is the implementation of the predicate: "is N outside the constrained disk touching P_1 and P_2 ?" (lines 1, 4, 5 and 9 in Algorithm 2). In addition, notice that in the three different implementations the computation may use integers only.

Fig. 6.a illustrates the outcome of Algorithm 1. A part of a digital ellipse has been scanned and iteratively decomposed into DCAs of radius 10.

4 Digital Circular Arc Recognition

In this section, we propose a method that iteratively solves Problem 3 (online recognition of a DCA coming from the digitization of a disk having a boundary that is incident with a given point) to solve the unconstrained problem, *i.e.* Problem 1.

The method computes the smallest disk that separates $B_{(C_iC_j)}$ from $\bar{B}_{(C_iC_j)}$. In the (ω_x, ω_y, r)-space (Section 2.3), the smallest separating disk corresponds to a 3D point that belongs to the intersection of $2n$ half-spaces and that is the closest to the paraboloid of equation $r^2 = {\omega_x}^2 + {\omega_y}^2$. Due to the convexity of the paraboloid, a classic result of convex programming [4] implies the following property: if a new inner (resp. outer) point is located outside (resp. inside) the current smallest separating disk, then either the new point is on the new smallest

Algorithm 3. Algorithm that solves Problem 1

```
Input: A part (C_iC_j) (with i < j).
Output: The boolean value areSeparable and D_min.
/* Initialization                                                    */
1  D_min is set to a disk of null radius and whose center is on the inner point of
   (C_iC_{i+1});
2  areSeparable ← true; k ← i + 1;
/* Scan                                                              */
3  while C_k < C_j and areSeparable do
4  |  N ← inner point of (C_kC_{k+1});
5  |  areSeparable ← CircularSeparability(N,D_min,(C_iC_k));
6  |  if areSeparable then
7  |  |  N ← outer point of (C_kC_{k+1});
8  |  |_ areSeparable ← CircularSeparability(N,D_min,(C_iC_k));
9  |_ k ← k + 1;
10 return areSeparable
```

Algorithm 4. CircularSeparability(N,$\mathcal{D}_{min}$,(C_iC_k))

```
Input: A new point N, the smallest separating disk D_min and the part (C_iC_k).
Output: A boolean value and updated D_min.
1  if N is inside D_min then
2  |_ return true;
3  else
   |  /* The boundary of the constrained disks must be incident with N
   |  */
4  |  Use Algorithm 1 to decide whether B_(C_iC_k) and B̄_(C_iC_k) are separable by a
   |  constrained disk whose boundary is incident with N;
5  |  if B_(C_iC_k) and B̄_(C_iC_k) are separable then       /* update of D_min */
6  |  |  D_min ← D_new, the smallest separating disk that is incident with N;
7  |  |_ return true;
8  |  else
9  |_ |_ return false;
```

separating disk, or the sets of inner and outer points are not circularly separable at all. In the aim of deciding between these two alternatives, Algorithm 1 can be used in the case where the disks have to be incident with a given point.

Similarly to Algorithm 1, the points of a part (C_iC_j) are processed one by one in Algorithm 3.

Algorithm 4 is called instead of Algorithm 2 when a new inner (resp. outer) point N is added to $B_{(C_iC_k)}$ (resp. $\bar{B}_{(C_iC_k)}$) (lines 5 and 8 of Algorithm 3).

Let us assume that N is an inner point. The case where N is an outer point is similar. If the inner point is located inside the current smallest separating disk, then Algorithm 4 returns true (line 2) because the current disk is still separating. Otherwise, the constrained disks are defined as touching the new point that

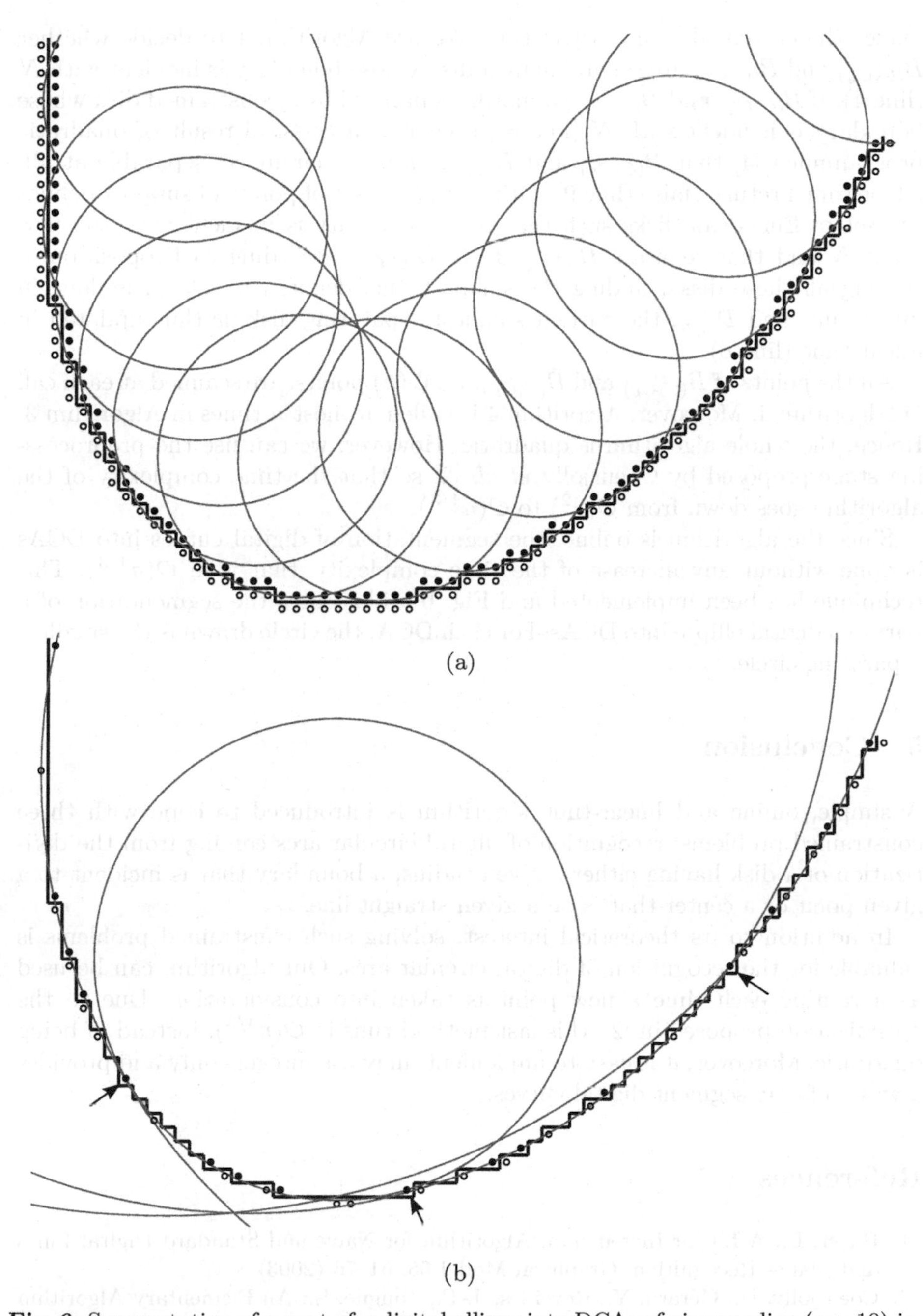

Fig. 6. Segmentation of a part of a digital ellipse into DCAs of given radius ($r = 10$) in (a) and of any radius in (b). The black polygonal line depicts the digital contour. The black and white points are those retained for the computation. In (b), the preprocessing stage proposed in [2] has discarded a great amount of black and white points. The arrows point to the end points of the DCAs.

makes the current disk not separating. We use Algorithm 1 to decide whether $B_{(C_iC_k)}$ and $\bar{B}_{(C_iC_k)}$ are separable by a disk whose boundary is incident with N (line 4). If $B_{(C_iC_k)}$ and $\bar{B}_{(C_iC_k)}$ cannot be separated by a constrained disk whose boundary is incident with N (line 5), then it is a classical result of quadratic programming [4] that $B_{(C_iC_k)}$ and $\bar{B}_{(C_iC_k)}$ are not circularly separable at all. Algorithm 4 returns false (line 9). Otherwise, the set of points of support defines the set of Euclidean disks such that one's boundary is incident with the new point N and that separate $B_{(C_iC_k)}$ from $\bar{B}_{(C_iC_k)}$ according to Proposition 3. Among all these disks, finding the smallest one, denoted by $\mathcal{D}_{new}$, is done in linear time and $\mathcal{D}_{min}$, the current smallest separating disk, is thus updated in linear time (line 6).

All the points of $B_{(C_iC_k)}$ and $\bar{B}_{(C_iC_k)}$, *i.e* $\mathcal{O}(n)$ points, are scanned at each call of Algorithm 4. Moreover, Algorithm 4 is called at most n times in Algorithm 3. Hence, the whole algorithm is quadratic. However, we can use the preprocessing stage proposed by Coeurjolly *et al.* [2] so that the time complexity of the algorithm goes down from $\mathcal{O}(n^2)$ to $\mathcal{O}(n^{4/3})$.

Since the algorithm is online, the segmentation of digital curves into DCAs is done without any increase of the time complexity, that is in $\mathcal{O}(n^{4/3})$. This technique has been implemented and Fig. 6.b illustrates the segmentation of a part of a digital ellipse into DCAs. For each DCA, the circle drawn is the smallest separating circle.

5 Conclusion

A simple, online and linear-time algorithm is introduced to cope with three constrained problems: recognition of digital circular arcs coming from the digitization of a disk having either a given radius, a boundary that is incident to a given point or a center that is on a given straight line.

In addition to its theoretical interest, solving such constrained problems is valuable for the recognition of digital circular arcs. Our algorithm can be used as a routine each time a new point is taken into consideration. Due to the optimization proposed in [2], this last method runs in $\mathcal{O}(n^{4/3})$, instead of being quadratic. Moreover, it is easy to implement, may use integers only and provides a way to fastly segment digital curves.

References

1. Buzer, L.: A Linear Incremental Algorithm for Naive and Standard Digital Lines and Planes Recognition. Graphical Model 65, 61–76 (2003)
2. Coeurjolly, D., Gérard, Y., Reveillès, J.-P., Tougne, L.: An Elementary Algorithm for Digital Arc Segmentation. Discrete Applied Mathematics 139(1-3), 31–50 (2004)
3. Damaschke, P.: The Linear Time Recognition of Digital Arcs. Pattern Recognition Letters 16, 543–548 (1995)
4. de Berg, M., van Kreveld, M., Overmars, M., Scharzkopf, O.: Computation geometry, algorithms and applications. Springer, Heidelberg (2000)

5. Efrat, A., Gotsman, C.: Subpixel Image Registration Using Circular Fiducials. International Journal of Computational Geometry & Applications 4(4), 403–422 (1994)
6. Fisk, S.: Separating Points Sets by Circles, and the Recognition of Digital Disks. IEEE Transactions on Pattern Analysis and Machine Intelligence 8, 554–556 (1986)
7. Kim, C.E.: Digital Disks. IEEE Transactions on Pattern Analysis and Machine Intelligence 6(3), 372–374 (1984)
8. Kim, C.E., Anderson, T.A.: Digital Disks and a Digital Compactness Measure. In: Annual ACM Symposium on Theory of Computing, pp. 117–124 (1984)
9. Kovalevsky, V.A.: New Definition and Fast Recognition of Digital Straight Segments and Arcs. In: Internation Conference on Pattern Analysis and Machine Intelligence, pp. 31–34 (1990)
10. Megiddo, N.: Linear Programming in Linear Time When the Dimension Is Fixed. SIAM Journal on Computing 31, 114–127 (1984)
11. O'Rourke, J., Kosaraju, S.R., Meggido, N.: Computing Circular Separability. Discrete and Computational Geometry 1, 105–113 (1986)
12. Pham, S.: Digital Circles With Non-Lattice Point Centers. The Visual Computer 9, 1–24 (1992)
13. Preparata, F.P., Shamos, M.I.: Computational geometry: an introduction. Springer, Heidelberg (1985)
14. Roussillon, T., Sivignon, I., Tougne, L.: Test of Circularity and Measure of Circularity for Digital Curves. In: International Conference on Image Processing and Computer Vision, pp. 518–524 (2008)
15. Roussillon, T., Sivignon, I., Tougne, L.: On-Line Recognition of Digital Arcs. Technical report, RR-LIRIS-2009-008 (2009)
16. Sauer, P.: On the Recognition of Digital Circles in Linear Time. Computational Geometry 2(5), 287–302 (1993)

Efficient Lattice Width Computation in Arbitrary Dimension*

Émilie Charrier[1,2,3], Lilian Buzer[1,2], and Fabien Feschet[4]

[1] Université Paris-Est, LABINFO-IGM, CNRS, UMR 8049
[2] ESIEE, 2, boulevard Blaise Pascal, Cité DESCARTES, BP 99
93162 Noisy le Grand Cedex, France
[3] DGA/D4S/MRIS
[4] Univ. Clermont 1, LAIC,
Campus des Cézeaux, 63172 Aubière Cedex, France
charriee@esiee.fr, buzerl@esiee.fr, feschet.research@gmail.com

Abstract. We provide an algorithm for the exact computation of the lattice width of an integral polygon K in linear-time with respect to the size of K. Moreover, we describe how this new algorithm can be extended to an arbitrary dimension thanks to a greedy approach avoiding complex geometric processings.

1 Introduction

Integer Linear Programming is a fundamental tool in optimization, in operation research, in economics... Moreover, it is interesting in itself since the problem is NP-Hard in the general case. Several works were done for the planar case [18,25,14] before Lenstra [21] proved that Integer Linear Programming can be solved in polynomial time when the dimension is fixed. Faster and faster algorithms are nowadays developed and available making the use of Integer Linear Programming reliable even for high dimensional problems. The approach of Lenstra uses the notion of *lattice width* for precise lattice definition to detect directions for which the polyhedron of solutions is thin. In polynomial time, the problem is then reduced to a feasibility question: given a polyhedron P, determine whether P contains an integer point. To solve it, Lenstra approximates the width of the polyhedron and gives a recursive solution solving problems of smaller dimension. The approximate lattice width is also used in the recent algorithms of Eisenbrand and Rote [9] and Eisenbrand and Laue [8] for the 2-variable problem.

Not surprisingly, following the arithmetical approach of Reveillès [23] [6], the lattice width is also a fundamental tool in digital geometry since it corresponds to the notion of width for digital objects [10]. Moreover, as an application of the

* Supported in part by the the French National Agency of Research under contract GEODIB ANR-06-BLAN-0225-02. A preliminary version of part of this paper appeared in the proceedings of the IWCIA 2006 conference.

S. Brlek, C. Reutenauer, and X. Provençal (Eds.): DGCI 2009, LNCS 5810, pp. 46–56, 2009.

lattice width computation, we mention the intrinsic characterization of linear structures [11]. Indeed, the lattice width can be computed for any digital set but it does not correspond to a direct measure of linearity. However, when combining the lattice width along a direction and along its orthogonal, it can be used as a linearity measure. The work in [11] is currently extended, by the third author of the present paper, to higher dimensions for detecting either linear or tubular structures. A preliminary algorithm dealing with the two-dimensional case was given in [10] with a geometrical interpretation. It has the advantage to be extensible to the incremental and to the dynamic case but it seems difficult to extend it to an arbitrary dimension. Thus, in the present paper, we propose a totally new algorithm which is efficient for any dimension and which runs in linear time for the two-dimensional case.

The paper is organized as follows. In Sect. 2, the main definitions and tools are presented. Then, we present the two-dimensional algorithm introduced in [10] in Sect. 3. As this method cannot be easily extensible to higher dimension, we introduce in Sect. 4 a new geometric method to estimate the lattice width in any dimension. This approach is based on the computation of a particular surrounding polytope which is used to bound the set of candidate vectors to define the lattice width. In Sect. 5, we describe geometrical methods to compute such a polytope. We first focus on the two-dimensional case and we provide two deterministic approaches. Then, since these geometric constructions might be difficult to extend in arbitrary dimension, we provide a greedy algorithm which runs in any dimension. Some conclusions and perspectives end the paper.

2 Definitions

In this section, we review some definitions from algorithmic number theory and we provide a precise formulation of the problem we solve. Definitions are taken from [2,9,26].

Let K be a set of n points of $\mathbb{Z}^d$. Moreover, we suppose that all numbers appearing in the points and in the vectors coordinates have their bit size bounded by $\log s$. The *width* of K along a direction $c \neq 0$ in $\mathbb{R}^d$ is defined as:

$$\omega_c(K) = \max\left\{c^T x \mid x \in K\right\} - \min\left\{c^T x \mid x \in K\right\} \tag{1}$$

Geometrically, if a set K has a width of l along the direction c then any integer point which lies on K also lies on a hyperplane of the form $c^T x = \lambda$ where λ corresponds to an integer value between $\min\{c^T x \mid x \in K\}$ and $\max\{c^T x \mid x \in K\}$. We say that K can be covered by these $\lfloor l \rfloor + 1$ parallel hyperplanes. It is straightforward to see that $\omega(K) = \omega(\text{conv}(K))$ where $\text{conv}(K)$ denotes the convex hull of K.

Let $\mathbb{Z}^{d*} = \mathbb{Z}^d \setminus \{0\}$ denote the set of integer vectors different from zero. The *lattice width* of K is defined as follows:

$$\omega(K) = \min_{c \in \mathbb{Z}^{d*}} \omega_c(K) \tag{2}$$

We notice that the lattice width is an integer value. We briefly recall some basic and important properties about inclusion and translation:

Lemma 1. *For any sets of points A and B, such that $conv(A) \subset conv(B)$ and for any vector $c \in \mathbb{Z}^{d*}$, we have $\omega_c(A) \leq \omega_c(B)$. Thus, it follows that $\omega(A) \leq \omega(B)$.*

Lemma 2. *Suppose that A' corresponds to the points of A translated in the same direction. By definition, we know that for any $c \in \mathbb{Z}^{d*}$, $\omega_c(A) = \omega_c(A')$ and so we have $\omega(A) = \omega(A')$.*

The problem we would like to solve is the following one:

Problem (Lattice Width)
Given a set of integer points $K \subset \mathbb{Z}^d$, find its lattice width $\omega(K)$ as well as all vectors $c \in \mathbb{Z}^d$ such that $\omega_c(K) = \omega(K)$.

It is known [21] that the lattice width of a convex set K is obtained for the shortest vector with respect to the *dual norm* whose unit ball is the polar set of the set $\frac{1}{2}(K + (-K))$. In the general case, computing the shortest vector is NP-hard. Thus, approximations of the solution can be computed via standard arguments [26,17,24], but it does not lead us to an easy exact algorithm in arbitrary dimension.

3 Planar Case

We recall in this part, the result obtained in [10] via connections with the notion of digital straightness and more precisely with the notion of arithmetical digital lines [23]. As this two-dimensional algorithm requires a convex polygon as input, we have to compute the convex hull H of K in $O(n)$ time ([16]).

The idea in [10] is based on the principle that the lattice width of H is necessarily reached for two *opposite* vertices of H. To define the notion of *opposite*, we rely on the notion of *supporting lines* well known in computational geometry [5]. A *supporting line* of H is a line D such that $D \cap K \neq \emptyset$ and H is contained entirely in one of the half-planes bounded by D. For each supporting line D, there exists at least one vertex v of H such that the parallel line D_v to D passing through v is such that H entirely lies in the strip bounded by D and D_v. If s denotes a vertex of H belonging to D then s and v are called *opposite* (see Fig. 1, left). Opposite pairs are also called *antipodal* pairs. Note that in general, a supporting line intersects H at only one point. The supporting line D intersecting H along an edge is called *principal* supporting line.

We now suppose H to be oriented counter clockwise. As in the classical Rotating Calipers algorithm of Toussaint [15], we can rotate the principal supporting lines D around the right vertex of $D \cap H$. D_v is also rotated around v to keep it parallel to D. This rotation can be pursued until D or D_v corresponds to another principal supporting line. Note that D and D_v are simply supporting lines during the rotation. At each position of the rotation s and v form an opposite pair of points which exactly define $\omega_c(H)$ where c is the normal direction to D. Hence, as depicted in Fig. 1 (right), s and v exactly define $\omega_c(H)$ for all D in a cone whose apex is v. The point r is such that the segment from v to r has exactly the same length than the opposite edge of H and the point u is either the next vertex of H after v or the point on the parallel of the line (st) such that the length of $[st]$ equals the length of $[vu]$.

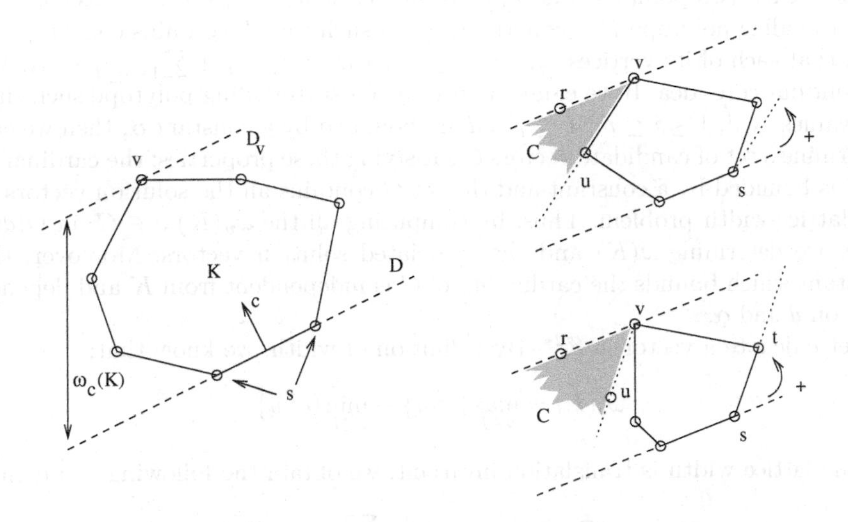

Fig. 1. (left) Supporting lines and width $\omega_c(K)$ (right) Cone of rotations

After one turn around H, we have constructed at most $2n$ opposite pairs and associated cones. Hence, the number of cones is $O(n)$. Moreover, the series of cones forms a partition of all possible directions of computation for the lattice width taking into account that $\omega_c(H) = \omega_{-c}(H)$. Hence, as previously announced, the computation of the lattice width is reduced to the computation of the minimal value of $\omega_c(H)$ for each cone.

In each cone C, the computation of the minimal value of $\omega_c(H)$ is the computation of the shortest vectors for the dual norm. They are thus located at the vertices of the border of the convex hull of integer points except v inside the cone [17]. This set is also known as Klein's sail [19,20,1]. Note that to allow the possibility to find all solution vectors, repetitions in the convex hull must be kept. To compute Klein's sail, we use an adapted version of the algorithm of Harvey [13] whose complexity is $O(\log s)$ arithmetic operations for each sail computation. To bound the complexity of the search, we could also rely on the general theorem of Cook et al [4] which says that there exists at most $O((\log s)^{d-1})$ vertices in dimension d, a result also shown by Harvey [13] with an explicit example of the worst case in two dimension. Thus, the complexity of the lattice width computation is $O(n + n \log s)$.

4 A New Algorithm

The geometric approach of the previous algorithm appears as a major drawback for an arbitrary dimension. Hence, we propose a new geometric method based on the possibility of bounding the set of candidate directions used to compute the lattice width. We introduce the new method in the d-dimensional case.

Let $(a_j)_{1\leq j\leq d}$ denote a sequence of d vectors in $\mathbb{Z}^d$ such that for any $j, 1 \leq j \leq d$, there exist two points u_j and v_j of K satisfying $a_j = u_j - v_j$. Let us consider a surrounding polytope $\Gamma = \text{conv}((\gamma_i)_{1\leq i\leq m})$ such that Γ contains conv(K) and such that each of its vertices γ_i, $1 \leq i \leq m$, satisfies $\gamma_i = p + \sum_{1\leq j\leq d} \alpha_{ij} a_j$. We present our core idea. If we can compute such a surrounding polytope such that the values $|\alpha_{ij}|, 1 \leq i \leq m, 1 \leq j \leq d$ are bounded by a constant α, then we can determine a set of candidate vectors C satisfying these properties: the cardinality of C is bounded by a constant and the set C contains all the solution vectors of our lattice width problem. Thus, by computing all the $\omega_u(K)$,$u \in C$, in $O(dn)$ time, we determine $\omega(K)$ and the associated solution vectors. Moreover, the constant which bounds the cardinality of C is independent from K and depends only on d and α.

Let c denote a vector in $\mathbb{Z}^{d*}$. By definition of width, we know that:

$$\omega_c(\Gamma) = \max_{u\in\Gamma}\{c^T u\} - \min_{u\in\Gamma}\{c^T u\}$$

As the lattice width is translation invariant, we obtain the following inequality:

$$\omega_c(\Gamma) \leq 2 \max_{1\leq i\leq m} \{|c^T \sum_{1\leq j\leq d} \alpha_{ij} a_j|\}$$

As the values $|\alpha_{ij}|, 1 \leq i \leq m, 1 \leq j \leq d$ are bounded by α, we obtain:

$$\omega_c(\Gamma) \leq 2\alpha d \sum_{1\leq j\leq d} |c^T a_j|$$

Let A denote the matrix whose rows correspond to the $a_j^T, 1 \leq j \leq d$, we can rewrite the previous expression:

$$\omega_c(\Gamma) \leq 2\alpha d||A^T c||_\infty \tag{3}$$

As conv(K) is included in conv(Γ), we immediately know:

$$\omega(K) \leq \omega(\Gamma) \leq \omega_c(\Gamma)$$

Thus, for a given vector c, we obtain an upper bound for $\omega(K)$. For solving our problem, it is sufficient to test only the vector u in $\mathbb{Z}^{d*}$ satisfying:

$$\omega_u(K) \leq 2\alpha d||A^T c||_\infty$$

Let us try to determine a lower bound for the term $\omega_u(K)$. For any j, $1 \leq j \leq d$, there exists two points u_j and v_j in K such that $a_k = u_k - v_k$. As for any

$j, 1 \leq j \leq d$, $\text{conv}(\{u_j, v_j\})$ is included in $\text{conv}(K)$, we have for any $u \in \mathbb{Z}^{d*}$, $\omega_u(\{u_j, v_j\}) \leq \omega_u(K)$. Thus, by definition of the lattice width along a direction $u \in \mathbb{Z}^{d*}$, we obtain:

$$|u^T a_i| \leq \omega_u(K) \text{ for } 1 \leq i \leq d$$

Then, it follows for any $u \in \mathbb{Z}^{d*}$:

$$||A^T u||_\infty \leq \omega_u(K)$$

Thus, we can conclude that it is sufficient to test only the vector u in $\mathbb{Z}^{d*}$ satisfying :

$$||A^T u||_\infty \leq 2\alpha d||A^T c||_\infty \tag{4}$$

As the right term is fixed for a given c, we can compute a vector c with some interesting properties. The more natural approach is to compute the shortest vector v in the lattice given by A and to choose c such that $A^T c = v$. Hence the upper bound becomes a bound independent on the direction c and we get:

$$\|v\|_\infty \leq \|A^T u\|_\infty \leq 2\alpha d\|v\|_\infty$$

It follows that the set of tested vectors is contained in a ball with a radius independent of the set K. Since there is a constant number of points in the ball, all points can be tested to extract the lattice width. Of course, an approximation of the shortest vectors can be used in place of v to avoid the difficulty of its computing in arbitrary dimension.

5 Surrounding Polytope Computation

Let K denote a set of n points in $\mathbb{Z}^d$. We look for a sequence of d vectors $(a_j)_{1 \leq j \leq d}$ such that for any $j, 1 \leq j \leq d$, there exists two points u_j and v_j of K satisfying $a_j = u_j - v_j$. From this sequence of vectors, we must be able to build a surrounding polytope $\Gamma = conv((\gamma_i)_{1 \leq i \leq m})$ such that Γ contains K and such that each $\gamma_i, 1 \leq i \leq m$ satisfies $\gamma_i = \sum_{j=1}^{d} \alpha_{ij} a_j$. An implicit goal is to obtain the smallest possible upper bound for the $|\alpha_{ij}|, 1 \leq i \leq m, 1 \leq j \leq d$ values in order to improve the performance of the algorithm. We show afterwards that, in the two-dimensional case, we can compute such a surrounding polygon in $O(n)$ time. The corresponding approaches become too difficult to extend to the three-dimensional case. So, we present a simpler and more efficient approach that can be used in any dimension.

5.1 Deterministic Approaches

Existing Approach. We first find in the literature the work of Fleischer et al. [12] that confirms that such a polytope exists. We recall this result in the two-dimensional case:

Theorem 1. *For any convex body P, let t denote a greatest area triangle contained in P. The convex body P is contained in the dilatation of t by an expansion factor of at most $9/4$.*

From [16], we know that the convex hull of the set of integer points in the plane can be computed in linear time. Thus, we can operate on $conv(K)$ without damaging the overall time complexity. Moreover, we know from [7] that there exists a greatest area triangle included in K whose vertices correspond to vertices of K. So, let T denote the dilatation of a maximal area triangle $t = ABC$ of K by a factor of 9/4. The triangle T corresponds to a solution of our surrounding polygon problem. Indeed, if we set the origin at the point A, its three vertices are of the form $\alpha_{i1}AB + \alpha_{i2}AC$ with $|\alpha_{i1}|$ and $|\alpha_{i2}|$ bounded by 1 for $1 \leq i \leq 3$. So, the first approach we propose consists in computing a maximum area triangle of K and then consider its dilatation with a factor 9/4. Boyce et al. propose an overall method to compute a greatest area k-gon which runs in $O(kn \log n + n \log^2 n)$ time and in $O(n \log n)$ time when $k = 3$ (see [3]). Dobkin et al. focus on the two-dimensional case and they show that the computation of a greatest area triangle runs in linear time performing at most $10n$ triangle area computations (see [7]). Their method consists in "walking" along the convex polygon K and determining for each vertex v two other vertices a and b such that the triangle abv has a maximal area. Their method runs in linear time because the three vertices move in clockwise order and they never have to backtrack during the traversal (see [7] for more details).

A Simpler Approach. We introduce a simpler approach to compute a surrounding polygon that only requires $2n$ distance computations. Moreover, it does not need the computation of the convex hull of K. This method consists in computing a surrounding parallelogram. Let A and B denote the leftmost and the rightmost point of K respectively according to the x-axis. Let N denote the normal vector of AB with positive y-coordinate. Let C and D denote the extremal point of K in the direction N and $-N$ respectively. We show afterwards that the set K is contained in a parallelogram of sides CD and $2AB$. This parallelogram corresponds to a solution of our problem because its vertices can be expressed as $\alpha_1 CD + \alpha_2 AB$ where $|\alpha_1|$ and $|\alpha_2|$ does not exceed 1 if the origin is well chosen.

Let $EFGH$ denote the parallelogram bounded by the two vertical straight lines passing through A and B and bounded by the two straight lines of direction vector AB passing trough C and D. By construction, the set K is included in the parallelogram $EFGH$, but the vertical sides of this parallelogram cannot be expressed using vectors rooted in points of K. As a result, we try to minimize δ such that $EFGH$ is contained in a parallelogram of side CD and of side δAB. We show that δ equals 2. Indeed, the points A and B may not be extremal according to the normal vector of CD and in this case, δ is strictly greater than 1 (see Fig. 2.a). The "worst case" happens when the points C and D coincide with opposite vertices of the parallelogram $EFGH$, (as in Fig. 2.b). In this case, we notice that half of the parallelogram $EFGH$ is included in a parallelogram of side AB and CD. As a consequence, it is sufficient to double the side AB of the parallelogram such that the new parallelogram becomes large enough.

These two methods run in linear time. Unfortunately, they are not easily extensible to the three-dimensional case (see [22]). As a result, we describe in the next section a greedy method to compute a surrounding tetrahedron.

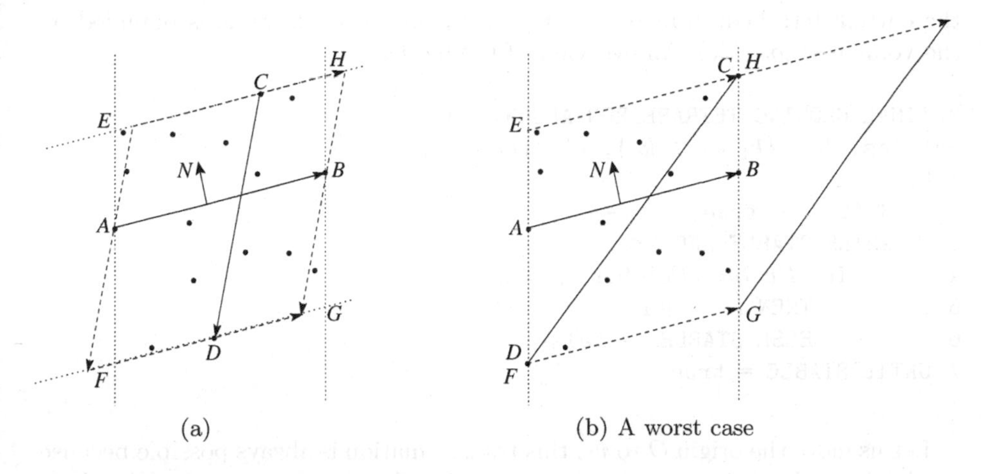

Fig. 2. Construction of the parallelograms

This method can be extended to higher dimension but we focus on the three-dimensional case for convenience of presentation.

5.2 A Greedy Approach

We introduce the following definition:

Definition 1. *Let K denote a three-dimensional set of points. A tetrahedron G whose vertices belong to K is a* maximal growing tetrahedron *relative to K if for each face f of G, the furthest point from f in the set K corresponds to a vertex of G.*

Notice that a maximal growing tetrahedron does not always correspond to a greatest volume tetrahedron. Such a tetrahedron can be stated using three vectors a_1, a_2 and a_3, each generated by two points of K. We show afterwards that we can compute a surrounding parallelepiped Γ of K such that each vertex γ_i of Γ satisfies $\gamma_i = \sum_{1 \leq j \leq 3} \alpha_{ij} a_j$ where the values $|\alpha_{ij}|$ are bounded by 1 when the origin is well chosen.

First, we compute a maximal growing tetrahedron G using a non-deterministic but efficient approach. Moreover, this method easily extends to the d-dimensional case contrary to the previous deterministic approaches. Let $(f_l)_{1 \leq l \leq 4}$ and $(v_l)_{1 \leq l \leq 4}$ denote the faces and the vertices of G respectively such that the vertex v_l is opposite to the face f_l for $1 \leq l \leq 4$. We first initialize G using four non-coplanar points of K. Our method consists in "pushing" the vertices of G among the points of K until G corresponds to a maximal growing tetrahedron. Repeatedly, for each face f_l of G, the algorithm tries to grow the tetrahedron G by moving the current vertex v_l to a point of K which is the furthest point from f_l. The method stops when no more growing can be done. This approach always finds a valid solution because at each step the volume of the current tetrahedron increases at least by one and this value is bounded by the volume of $conv(K)$. An overview of the method follows:

```
MAXIMAL GROWING TETRAHEDRON ALGORITHM:
Entries: K = {k1, k2, ..., kn},  G = (v1, v2, v3, v4)
1 DO
2     STABLE ← true;     l ← 1;
3     WHILE STABLE AND l ≤ 4
4         IF (FurthestPoint(K, f_l) = v_l)
5             THEN l ← l + 1
6             ELSE STABLE ← false;
7 UNTIL STABLE = true
```

Let us move the origin O to v_1, this transformation is always possible because the lattice width is invariant under translation. Let $(a_j)_{1\leq j\leq 3}$ denote the three vectors defined by $v_{j+1} - v_1$ where $(v_j)_{1\leq j\leq 4}$ denote the vertices of a maximal growing tetrahedron G of K. We show that the set K is contained in a parallelepiped Γ whose vertices $(\gamma_i)_{1\leq i\leq 8}$ are of the form $\gamma_i = \sum_{1\leq j\leq 3} \alpha_{ij} a_j$ where each $|\alpha_{ij}|$ is bounded by 1. As the maximal growing tetrahedron we use is non-flat by definition, any point k_l of K can be written as $k_l = O + \sum_{1\leq j\leq 3} \delta_{lj} a_j$. Let us show that we have $|\delta_{ij}| \leq 1$ for $1 \leq i \leq n$, $1 \leq j \leq d$. Indeed, suppose that there exists an index i and a index j such that a $|\delta_{ij}|$ is strictly greater than 1. It would contradict the fact that G corresponds to a maximal growing tetrahedron because the vertex v_{j+1} would not be extremal. As a result the set K is contained in a parallelepiped Γ whose vertices $(\gamma_i)_{1\leq i\leq 8}$ are of the form $\gamma_i = \sum_{1\leq j\leq 3} \alpha_{ij} a_j$ where each $|\alpha_{ij}|$ is bounded by 1. Fig. 3 and Fig. 4 show examples in the plane and in the three-dimensional space respectively: G corresponds to a maximal growing triangle (resp. a maximal growing tetrahedron) and Γ corresponds to a surrounding parallelogram (resp. a surrounding parallelepiped). This method can be extended to higher dimension.

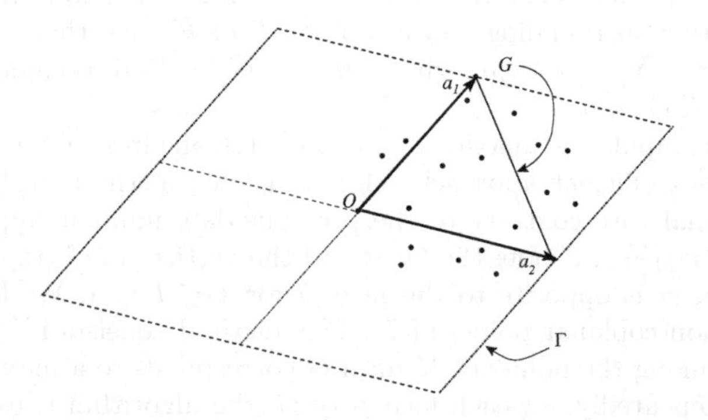

Fig. 3. Example of surrounding parallelogram

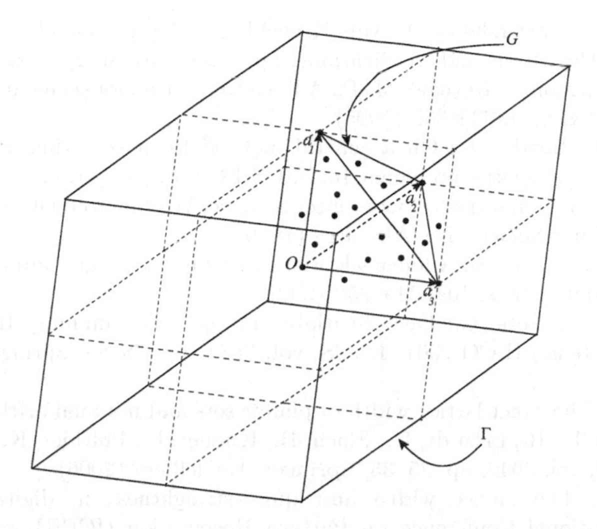

Fig. 4. Example of surrounding parallelepiped

6 Conclusion

We have described in this paper a new algorithm to compute the lattice width. It runs in linear time relative to the size of K, which is optimal. Moreover, its principle is directly extensible to an arbitrary dimension even if intermediate constructions become more complex in that case. Our greedy approach simplifies greatly the application of this algorithm to an arbitrary dimension since it avoids the computation of an inscribed $d-$simplex of greatest volume which is a problem in $O(n^4)$ at least in the three-dimensional case for the best known algorithm [22]. This problem is an interesting problem in itself and will be the subject of future research since it is known that the dilation constant of this d-simplex is independent of K and thus it guarantees the smallest possible space search in any case. We also plan to extend the definition of linearity given in [11] to an arbitrary dimension. Moreover, we intend to test our method in order to describe its numerical behaviour. The optimal complexity of our new algorithm is a key point in that construction.

References

1. Arnold, V.I.: Higher dimensional continued fractions. Regular and chaotic dynamics 3, 10–17 (1998)
2. Barvinok, A.: A Course in Convexity. Graduates Studies in Mathematics, vol. 54. Amer. Math. Soc, Providence (2002)
3. Boyce, J.E., Dobkin, D.P., Drysdale, R.L., Guibas, L.J.: Finding extremal polygons. In: STOC, pp. 282–289 (1982)
4. Cook, W., Hartman, M., Kannan, R., McDiarmid, C.: On integer points in polyhedra. Combinatorica 12, 27–37 (1992)

5. de Berg, M., Schwarzkopf, O., van Kreveld, M., Overmars, M.: Computational Geometry: Algorithms and Applications. Springer, Heidelberg (2000)
6. Debled-Rennesson, I., Reveillès, J.-P.: A linear algorithm for segmentation of digital curves. IJPRAI 9(4), 635–662 (1995)
7. Dobkin, D.P., Snyder, L.: On a general method for maximizing and minimizing among certain geometric problems. In: SFCS 1979: Proceedings of the 20th Annual Symposium on Foundations of Computer Science, Washington, DC, USA, pp. 9–17. IEEE Computer Society, Los Alamitos (1979)
8. Eisenbrand, F., Laue, S.: A linear algorithm for integer programming in the plane. Math. Program. Ser. A 102, 249–259 (2005)
9. Eisenbrand, F., Rote, G.: Fast 2-variable integer programming. In: Aardal, K., Gerards, B. (eds.) IPCO 2001. LNCS, vol. 2081, pp. 78–89. Springer, Heidelberg (2001)
10. Feschet, F.: The exact lattice width of planar sets and minimal arithmetical thickness. In: Reulke, R., Eckardt, U., Flach, B., Knauer, U., Polthier, K. (eds.) IWCIA 2006. LNCS, vol. 4040, pp. 25–33. Springer, Heidelberg (2006)
11. Feschet, F.: The lattice width and quasi-straightness in digital spaces. In: 19th International Conference on Pattern Recognition (ICPR), pp. 1–4. IEEE, Los Alamitos (2008)
12. Fleischer, R., Mehlhorn, K., Rote, G., Welzl, E., Yap, C.-K.: Simultaneous inner and outer approximation of shapes. Algorithmica 8(5&6), 365–389 (1992)
13. Harvey, W.: Computing two-dimensional Integer Hulls. SIAM Journal on Computing 28(6), 2285–2299 (1999)
14. Hirschberg, D., Wong, C.K.: A polynomial-time algorithm for the knapsack problem with two variables. J. Assoc. Comput. Mach. 23, 147–154 (1976)
15. Houle, M.E., Toussaint, G.T.: Computing the width of a set. IEEE Trans. on Pattern Analysis and Machine Intelligence 10(5), 761–765 (1988)
16. Hübler, A., Klette, R., Voss, K.: Determination of the convex hull of a finite set of planar points within linear time. Elektronische Informationsverarbeitung und Kybernetik 17(2/3), 121–139 (1981)
17. Kaib, M., Schnörr, C.-P.: The Generalized Gauss Reduction Algorithm. Journal of Algorithms 21(3), 565–578 (1996)
18. Kannan, R.: A polynomial algorithm for the two variable integer programming problem. J. Assoc. Comput. Mach. 27, 118–122 (1980)
19. Lachaud, G.: Klein polygons and geometric diagrams. Contemporary Math. 210, 365–372 (1998)
20. Lachaud, G.: Sails and klein polyhedra. Contemporary Math. 210, 373–385 (1998)
21. Lenstra, H.W.: Integer Programming with a Fixed Number of Variables. Math. Oper. Research 8, 535–548 (1983)
22. Lyashko, S.I., Rublev, B.V.: Minimal ellipsoids and maximal simplexes in 3D euclidean space. Cybernetics and Systems Analysis 39(6), 831–834 (2003)
23. Reveillès, J.-P.: Géométrie discrète, calcul en nombres entiers et algorithmique. Thèse d'etat, Université Louis Pasteur, Strasbourg, France (1991)
24. Rote, G.: Finding a shortest vector in a two-dimensional lattice modulo m. Theoretical Computer Science 172(1-2), 303–308 (1997)
25. Scarf, H.E.: Production sets with indivisibilities part i and part ii. Econometrica 49, 1–32, 395–423 (1981)
26. Schrijver, A.: Theory of Linear and Integer Programming. John Wiley and Sons, Chichester (1998)

Convergence of Binomial-Based Derivative Estimation for C^2 Noisy Discretized Curves

Henri-Alex Esbelin[1] and Rémy Malgouyres[2]

[1] Univ. Clermont 1, LAIC, IUT Dépt Informatique, BP 86, F-63172 Aubière, France
esbelin@laic.u-clermont1.fr

[2] Univ. Clermont 1, LAIC, IUT Dépt Informatique, BP 86, F-63172 Aubière, France
remy.malgouyres@laic.u-clermont1.fr

Abstract. We present new convergence results for the integer-only binomial masks method to estimate derivatives of digitized functions. The results work for C^2 functions and as a consequence we obtain a complete uniform convergence result for parametrized C^2 curves.

Introduction

In the framework of image and signal processing, as well as shape analysis, a common problem is to estimate derivatives of functions or tangents and curvatures of curves and surfaces, when only some (possibly noisy) sampling of the function or curve is available from acquisition. This problem has been investigated through finite difference methods, scale-space ([7],[4]), and discrete geometry ([5],[6]).

In [8], a new approach to derivative estimation from discretized data is proposed. As in scale-space approaches, this approach is based on simple computations of "convolutions". However, unlike scale-space methods, this approach is oriented towards integer-only models and algorithms, and is based on a discrete approach to analysis on $\mathbb{Z}$. Unlike existing approaches from discrete geometry, our approach is not based on discrete line reconstruction, which involves uneasy arithmetical calculations and complicated algorithms. Implementation of the binomial approach is straightforward.

As far as the speed of convergence is concerned, in [8] the method for tangents is proved uniform worst-case $O(h^{2/3})$ for C^3 functions, where h is the size of the pixel, while that of [6] is uniform $O(h^{1/3})$ and $O(h^{2/3})$ in average. Moreover, the estimator of [8] allows some (uniformly bounded) noise. Furthermore, the method of [8] allows to have a convergent estimation of higher order derivatives, and in particular a uniform $O(h^{4/9})$ estimation for the curvature of a generic curve.

In this paper, we prove a new convergence result which works for C^2 functions. Moreover, the upper bound for the error is the same as in [8] for small mask size and better for very large mask size.

To deal with parametrized curves in $\mathbb{Z}^2$, we introduced in [8] a new notion of pixel-length parametrization which solves the problem of correspondence between parametrizations of discrete and continuous curves, which arises from the

S. Brlek, C. Reutenauer, and X. Provençal (Eds.): DGCI 2009, LNCS 5810, pp. 57–66, 2009.

non isotropic character of $\mathbb{Z}^2$. However, the reparametrized curve is only C^2 so that [8] didn't contain a full uniform convergence result for parametrized curves. This problem is solved in this paper using our convergence results for C^2 functions, and we provide a complete uniform convergence result for parametrized curves.

The paper is organize as follows. First we find a definition of a uniform noise model and an error model, as well as statements of our main convergence results for real functions. For the sake of comparison, some results from [8] are quoted. Then we find a series of lemmas which outline the proofs. We also give hints as to how to reduce the computational complexity. Finally, we state similar results for parametrized curves. A couple of experiments are presented here. We plan to submit an extended version including all proofs and experiments soon.

1 Estimation of Derivatives for Real Function

Functions for which domain and range are 1−dimensional, without any assumption on the nature of these sets, are called *real functions.* We call *discrete function* a function from $\mathbb{Z}$ to $\mathbb{Z}$.

First we establish a relationship between a continuous function and its discretization. Let $f : R \longrightarrow R$ be a real function and let $\Gamma : \mathbb{Z} \longrightarrow \mathbb{Z}$ be a discrete function. Let h be the discretization step (i.e. the size of a pixel). We introduce a (possibly noisy) discretization of f:

Definition 1. *The function Γ is a discretization of f with error ϵ with discretization step h on the interval $[a; b]$ if for any integer i such that $a \leq ih \leq b$ we have:*

$$h\Gamma(i) = f(ih) + \epsilon_h(i)$$

We consider the following particular cases :

- Rounded case: $|\epsilon_h(i)| \leq \frac{1}{2}h$ which is equivalent to $\Gamma(i) = \left[\frac{f(ih)}{h}\right]$
- Floor case: $0 \leq \epsilon_h(i) \leq h$ which is equivalent to $\Gamma(i) = \lfloor \frac{f(ih)}{h} \rfloor$
- Uniform Noise case: $0 \leq |\epsilon_h(i)| \leq Kh^\alpha$ with $0 < \alpha \leq 1$ and K a positive constant. Note that the round case and the floor case are particular cases of uniform noise with $\alpha = 1$ (see Figure 1).

Definition 2. *The* discrete derivative *of a sequence u with mask size m, denoted by $\Delta_{2m-1} \star u$, is defined by:*

$$(\Delta_{2m-1} \star u)(n) = \frac{1}{2^{2m-1}} \left(\sum_{i=-m+1}^{i=m} \binom{2m-1}{m-1+i} (u(n+i) - u(n-1+i)) \right)$$

In order to show that the discrete derivative of a discretized function provides an estimate for the continuous derivative of the real function, we would like to evaluate the difference between $(\Delta_{2m-1} \star \Gamma)(n)$ and $f'(nh)$.

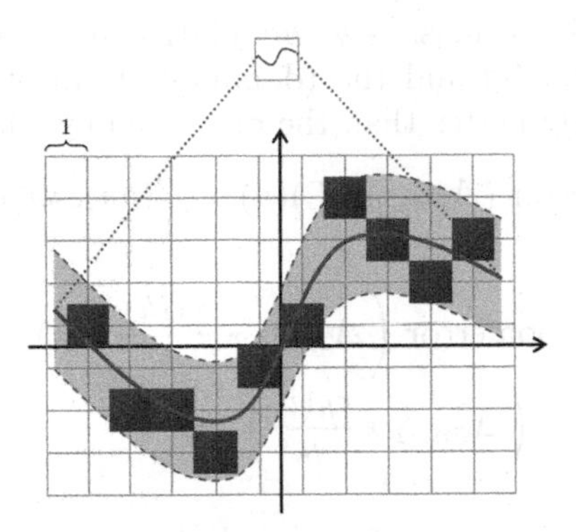

Fig. 1. The uniform noise model

Theorem 1 ([8]). *Suppose that* $f : R \longrightarrow R$ *is a* C^3 *function and* $f^{(3)}$ *is bounded,* $\alpha \in]0,1]$, $K \in \mathbb{R}_+^*$ *and* $h \in \mathbb{R}_+^*$. *Suppose* $\Gamma : Z \longrightarrow Z$ *is such that* $|h\Gamma(i) - f(hi)| \leq Kh^\alpha$ *(uniform noise case). Then for* $m = \lfloor h^{2(\alpha-3)/3} \rfloor$, *we have* $|(\Delta_{2m-1} \star \Gamma)(n) - f'(nh)| \in O(h^{2\alpha/3})$

The proof of this theorem is based on the following more general upper bound:

Theorem 2. *Under the assumptions of Theorem 1, for some constant* K' *and* m *sufficiently large,*

$$|\Delta_{2m-1} \star \Gamma(n) - f'(nh)| \leq \frac{h^2 m}{4} \|f^{(3)}\|_\infty + \frac{K' h^{\alpha-1}}{\sqrt{m}}$$

In this paper, we provide the following new result:

Theorem 3. *Suppose that* f *is a* C^2 *function and* $f^{(2)}$ *is bounded. Suppose* $\Gamma : Z \longrightarrow Z$ *is such that* $|h\Gamma(i) - f(hi)| \leq Kh^\alpha$ *(uniform noise case). Then if* $m = \lfloor h^{(\alpha-2)/1,01} \rfloor$ *we have*
$|(\Delta_{2m-1} \star \Gamma)(n) - f'(nh)| \in O(h^{(0,51(\alpha-0,01)/1,01})$.

This result is interesting mainly because it works for C^2 functions, which allows us to definitely solve the case of parametrized curves in Section 3. The proof of this theorem is based on the following new more general upper bound:

Theorem 4. *Under the assumptions of Theorem 3, for some constant* K' *and* m *sufficiently large,*

$$|\Delta_{2m-1} \star \Gamma(n) - f'(nh)| \leq h m^{0,51} \|f^{(2)}\|_\infty + \frac{K' h^{\alpha-1}}{\sqrt{m}}$$

To compare our new Theorem 4 with previous Theorem 2, besides the regularity assumption, note that

- For a sufficiently small mask size m relative to h (namely $m \leq \min(h^{2(\alpha-3)/3}, h^{(\alpha-2)/1.01})$), the upper bounds are the same because they have the same dominant terms.

– For a sufficiently large mask size m relative to h (namely $m \gg h^{-\frac{1}{0.49}}$) we have $hm^{0.51} \ll h^2 m$ and the (dominant term of the) upper bound of Theorem 4 is strictly better than the one given by Theorem 2.

In order to study the error $(\Delta_{2m-1} \star \Gamma)(n) - f'(nh)$, we decompose it into two errors:

– The real approximation error $\left(\Delta_{2m-1} \star \frac{f(ih)}{h}\right)(n) - f'(nh)$
– The input data error $\left(\Delta_{2m-1} \star \frac{\epsilon_h(i)}{h}\right)(n)$

1.1 Upper Bound for the Real Approximation Error

Here are the bounding results of this subsection. The second one is new:

Lemma 1. *Under the assumptions of Theorem 2*

$$\left|\left(\Delta_{2m-1} \star \frac{f(ih)}{h}\right)(n) - f'(nh)\right| \leq \frac{h^2 m}{4}\|f^{(3)}\|_\infty$$

Lemma 2. *Under the assumptions of Theorem 4*

$$\left|\left(\Delta_{2m-1} \star \frac{f(ih)}{h}\right)(n) - f'(nh)\right| \leq h\sqrt{m}\|f^{(2)}\|_\infty \left(2 + \left(\frac{3}{2}ln(m)\right)^{3/2}\right)$$

An important tool for our proofs are Bernstein polynomials defined by:

$$B_n(u_0, u_1, ..., u_n)(x) = \sum_{i=0}^{i=n} \binom{n}{i} u_i x^i (1-x)^{n-i}$$

We define the Bernstein approximation for a function ϕ with domain $[0; 1]$ by:

$$B_n(\phi)(x) = B_n\left(\phi(0), ..., \phi\left(\frac{i}{n}\right), ..., \phi(1)\right)(x)$$

We denote in a classical way $p_{n,i}(x) = \binom{n}{i} x^i(1-x)^{n-i}$ the polynomials of the Berntein basis.

We denote $\frac{f(ih)}{h} = f_i$ and when sequences $(u_{i,j,...})$ depend on several integer parameters, we denote it by $(u_{i,j,...})_i$ to specify that sums are made over different values of the parameter i, with other parameters fixed.

We make use of the following convergence theorems for Bernstein approximations:

Theorem 5 ([2]). *If ϕ is C^3 on $[0,1]$, then*

$$\left|(B_n\phi)'\left(\frac{1}{2}\right) - \phi'\left(\frac{1}{2}\right)\right| \leq \frac{\|\phi^{(3)}\|_\infty}{8n}$$

Now we have the new

Theorem 6. *If ϕ is C^2 on $[0,1]$, then*

$$\left|(B_{2m}\phi)'\left(\frac{1}{2}\right) - \phi'\left(\frac{1}{2}\right)\right| \leq \frac{\|\phi^{(2)}\|_\infty}{2\sqrt{m}}\left(2+\left(\frac{3}{2}ln(m)\right)^{3/2}\right)$$

1.2 Upper Bound for the Input Data Error

Lemma 3 ([8]). *For any bounded sequence u, we have:*

$$|(\Delta_{2m-1} \star u)(n)| \leq \frac{1}{4^{m-1}}\|u\|_\infty \binom{2m-1}{m-1}$$

2 Reducing the Complexity

Considering the mask size suggested in Theorem 3, the mask size is $O(h^{(\alpha-2)/1.01})$. Depending on the implementation, the computational complexity for computing one derivative value at discrete samples is then $O(h^{(\alpha-2)/1.01})$ if we precompute the binomial mask, which requires a space of $O(h^{(\alpha-2)/1.01})$. This complexity is not as good as the one induced by tangent estimators from discrete geometry which is in $O(h)$ in [5]. Indeed, some runtime tests for comparison processed in [3] show an important runtime for the method of [8] (without using the substantial complexity optimization suggested in Section 2.3 of [8]) for small values of h. As already noted there, it is possible to reduce the complexity of our estimator without changing the convergence speed. To do that, we show that values at extremities of the smoothing kernel H_n are negligible. This can be adapted to Theorem 3.

Theorem 7 ([8]). *Let $\beta \in \mathbb{N}^*$ and $m \in \mathbb{N}$. If $k = \frac{m}{2} - \sqrt{\frac{\beta m \ln(m)}{2}}$ then:*

$$\frac{1}{2^m}\sum_{j=0}^{k}\binom{m}{j} \leq \frac{1}{m^\beta} \tag{1}$$

Theorem 7 means that the sum of the k first and the k last coefficients of the smoothing kernel are negligible with respect to the whole kernel. The parameter β enables to define what negligible mean. If we take $\beta = \frac{0.51(\alpha-0.01)}{2-\alpha}$, then we have $\frac{1}{m^\beta} \leq h^{0.51(\alpha-0.01)/1.01}$ Using the result of Theorem 7, it is possible to reduce the size of the derivative kernel D_n recommended by Theorem 3 without affecting the proven convergence speed. Indeed, we compute in the computation of the derivative (Definition 2) only the terms of the sum between $n = -\sqrt{\frac{\beta m \ln(m)}{2}}$ and $n = +\sqrt{\frac{\beta m \ln(m)}{2}}$, which involves an $O(\sqrt{m \ln(m)})$ complexity. As a function of h, the complexity is then $O(h^{0.51(\alpha-0.01)/2.01}\sqrt{-ln(h)})$ (which is less than $O(h^{0.254(\alpha-0.01)})$).

3 Derivatives Estimation for Parametrized Curves

The proofs of this section are omitted for lack of space.

3.1 Tangent Estimation

We assume that a planar simple closed C^1−parametrized curve C (i.e. the parametrization is periodic and one-to-one on a period) is given together with a family of parametrized discrete curves (Σ_h) with Σ_h contained in a tube with radius $H(h)$ around C.

Here we estimate the tangent at a point of C by a binomial digital tangent at a point of Σ_h which is not too far. The goal of this section is to bound the error of this estimation and in particular to show that this error uniformly converges to 0 with h. Note that in [8], the results are valid for C^3 curves and don't work at points with horizontal of vertical tangents. We overcome these limitations, and also use weaker hypothesis on the discretization of C.

Definition 3. *The binomial discrete tangent at M_i is the real line going through M_i directed by the vector $(\Delta_{m+1} \star (x_i), \Delta_{m+1} \star (y_i))$, when this vector is nonzero.*

Theorem 8. *Let g be a C^2 parametrization of a simple closed curve C. Suppose that for all i we have $\|g(ih) - h\Sigma_h(i)\|_\infty \leq K h^{\alpha-1}$.*
Then for some constant K' and m sufficiently large,

$$\|\Delta_{2m-1} \star \Sigma_h(n) - g'(nh)\| \leq h m^{0,51} \|g^{(2)}\|_\infty + \frac{K' h^{\alpha-1}}{\sqrt{m}}$$

The proof of Theorem 8 is similar to that of Theorem 4, and we can derive a theorem similar to Theorem 3 for parametrized curves under the hypothesis of Theorem 8.

The hypothesis of Theorem 8 is stronger than the discretization being simply contained in a tube, because not only the discrete curve must be close to the continuous curve, but the parametrization of the discretization must be close to the parametrization of the continuous curve. This is the reason for introducing Pixel-Length parametrizations in the next section.

3.2 Pixel-Length Parametrization of a Curve

Definition 4. *A parametrization of a real curve γ is pixel-length if for all u et u' such that γ_x and γ_y are monotonic between u and u', we have*

$$\|\gamma(u) - \gamma(u')\|_1 = |u - u'|.$$

The idea is that for a curve with a pixel-length parametrization, the speed on the curve is the same as the speed of a discretization of the curve with each edgel taking the same time (see Figure 2).

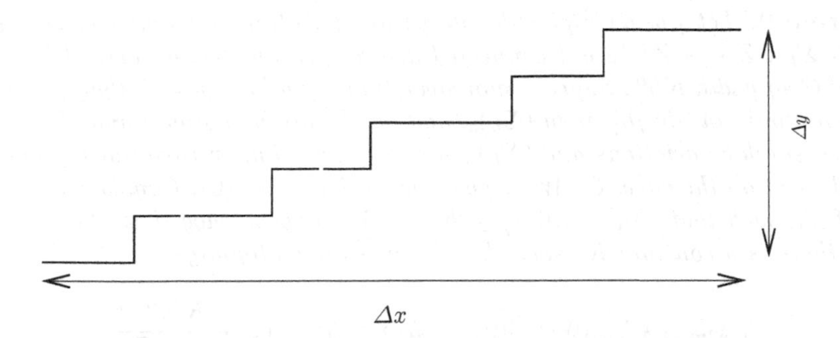

Fig. 2. On a monotonic parametrized discrete curve, the number of edgels between two points is equal to the norm $||.||_1$ of the difference between these two points

Lemma 4. *Let $g : R \longrightarrow R^2$ be a C^1 simple closed regular curve. Let us define σ_a by $\sigma_a(u) = \int_a^u \|g'(t)\|_1 dt$ and γ by $\gamma(u) = g[\sigma_a^{-1}(u)]$. Then γ is a C^1 pixel-length parametrization of g and $\|\gamma'(u)\|_1 = 1$.*
Moreover, suppose g is C^2 and generic in the sense that the set of values of u in a period of g for which either $g_x'(u) = 0$ or $g_y'(u) = 0$ is finite, and $g_x'(u) = 0$ implies $g_x''(u) = 0$ and $g_y'(u) = 0$ implies $g_y''(u) = 0$. Then γ is C^2.

Thanks to this lemma, we may now reduce the C^2 regular generic case to the C^2 pixel-length regular case.

3.3 Tangent Estimation for C^2 Pixel-Length Parametrization

Definition 5. *Two monotonic $\mathbb{R}-$ or $\mathbb{Z}-$valued functions are said to have* similar variations *if either they are both increasing or they are both decreasing.*

Lemma 5. *Let $\gamma : R \longrightarrow R^2$ be a pixel-length parametrization of a curve C, and let $\Sigma : Z \longrightarrow Z^2$ be a 4-connected discrete parametrized curve, lying in a tube of C with radius H. Suppose γ_x and Σ_x (resp. γ_y and Σ_y) are monotonous with similar variations. If $\|\gamma(0) - h\Sigma(0)\|_2 \leq D$, then for all i, we have $\|\gamma(ih) - h\Sigma(i)\|_2 \leq (H + D)\sqrt{2}$*

Lemma 6. *Let $\gamma : R \longrightarrow R^2$ be a pixel-length parametrization of a curve C, and let $\Sigma : Z \longrightarrow Z^2$ be a 4-connected discrete parametrized curve, lying in a tube of C of width H. Suppose moreover that for all $i < j$ such that γ_x and γ_y are monotonic in $[ih; jh]$, then Σ_x and Σ_y are monotonic and Σ_x and γ_x have similar variations and Σ_y and γ_y have similar variations.*

If $\|\gamma(0) - h\Sigma(0)\|_2 \leq D$, then for all i, we have

$$\|\gamma(ih) - h\Sigma(i)\|_2 \leq D2^{l/2} + (2h + H\sqrt{2})\frac{2^{l/2} - 1}{\sqrt{2} - 1} - 2h$$

where $l \geq 1$ is the number of points with horizontal or vertical tangents on the real curve between parameters 0 and ih.

Theorem 9. *Let γ be a C^2 pixel-length parametrization of a simple closed curve C. Let $\Sigma_h : Z \longrightarrow Z^2$ be a 4-connected discrete parametrized curve, lying in a tube of C of width Kh^α. Suppose moreover that for all $i < j$ such that γ_x and γ_y are monotonic in $[ih; jh]$, then $(\Sigma_h)_x$ and $(\Sigma_h)_y$ are monotonic and $(\Sigma_h)_x$ and γ_x have similar variations and $(\Sigma_h)_y$ and γ_y have similar variations. Consider a fixed point on the curve C. We suppose wlog that it is $\gamma(0)$. Consider any point M_0 of Σ_h such that $\|M_0 - \gamma(0)\|_2 \leq Kh^\alpha$. We suppose wlog that M_0 is $\Sigma_h(0)$. Then there is a constant K' such that for m sufficiently large,*

$$\|\Delta_{2m-1} \star \Sigma_h(0) - \gamma'(0)\| \leq hm^{0,51}\|\gamma^{(2)}\|_\infty + \frac{K'h^{\alpha-1}}{\sqrt{m}}$$

3.4 Tangent Estimation for a General C^2 Curve

Let C be a simple closed real curve C. If we assume that g is a C^2 generic parametrization of C, then Lemma 4 provides a C^2 pixel-length parametrization of C and Theorem 9 provides a convergent estimation of the derivative of the pixel-length parametrization, hence of the tangent of the curve. The method for solving the problem in a non generic case consists in a uniform C^2 approximation of C by a family of curves with a generic parametrization.

Lemma 7. *Let $\boldsymbol{g}$ be a C^2 parametrization of a real curve C. There exists a family $\boldsymbol{g}_n$ of C^2 generic parametrizations of real curves C_n such that $\boldsymbol{g}_n$ uniformly converges to $\boldsymbol{g}$ and $\boldsymbol{g}'_n$ converges uniformly to $\boldsymbol{g}'$.*

Theorem 10. *Let C be a simple closed curve C and $M_0 \in C$. Suppose that $\boldsymbol{g}$ is a regular C^2−parametrization of C and wlog $M_0 = \boldsymbol{g}(0)$. Let $\boldsymbol{\Sigma}_h : Z \longrightarrow Z^2$ be a 4-connected discrete parametrized curve, lying in a tube of C of width Kh^α. Suppose moreover that for all $i < j$ such that g_x and g_y are monotonic in $[ih; jh]$, then $(\Sigma_h)_x$ and $(\Sigma_h)_y$ are monotonic and $(\Sigma_h)_x$ and g_x have similar variations and $(\Sigma_h)_y$ and g_y have similar variations. Let $\boldsymbol{T}_0$ be a tangent vector to C in M_0 such that $\|\boldsymbol{T}_0\|_1 = 1$.*
Suppose that $\|\boldsymbol{\Sigma}_h(0) - M_0\|_2 \leq Kh^\alpha$ (up to a translation on the parameters of Σ, this is alway possible). Then there is constants K' and K" such that for m sufficiently large,

$$\|\Delta_{2m-1} \star \boldsymbol{\Sigma}_h(0) - \boldsymbol{T}_0\| \leq K'hm^{0,51} + \frac{K'h^{\alpha-1}}{\sqrt{m}}$$

Proof. (Sketch)

Following Lemma 7, let us introduce a family g_n. For a convenient choice of n, $\boldsymbol{\Sigma}_h$ lies in a tube of C with width $\frac{3K}{2}h^\alpha$. Following Lemma 4, we introduce γ_n as a C^2 pixel-length parametrization for each one. We use Theorem 9 to approximate $\Delta_{2m-1} \star \boldsymbol{\Sigma}_h(0)$ with $\gamma'_n(0)$ and a suitable n to approximate $\gamma'_n(0)$ with $\boldsymbol{T}_0$. □

4 Experiments

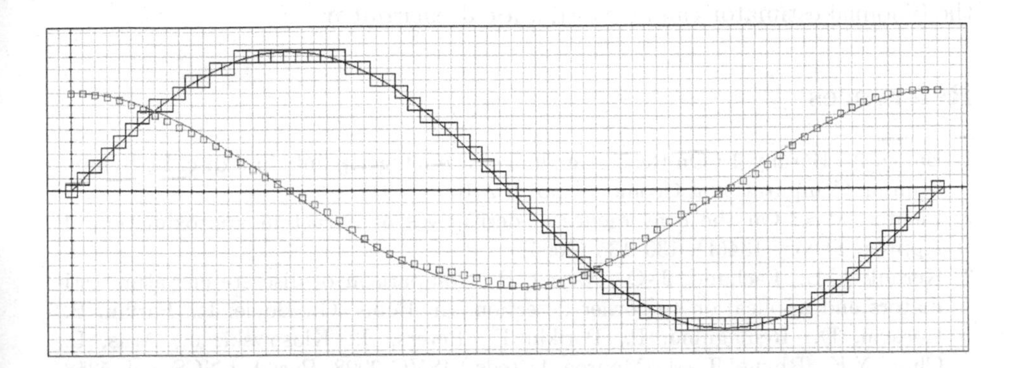

Fig. 3. Estimation of the derivative of the function $x \longmapsto \sin(2\pi x)/(2\pi)$ with $h = 0.014$ and $m = 31$

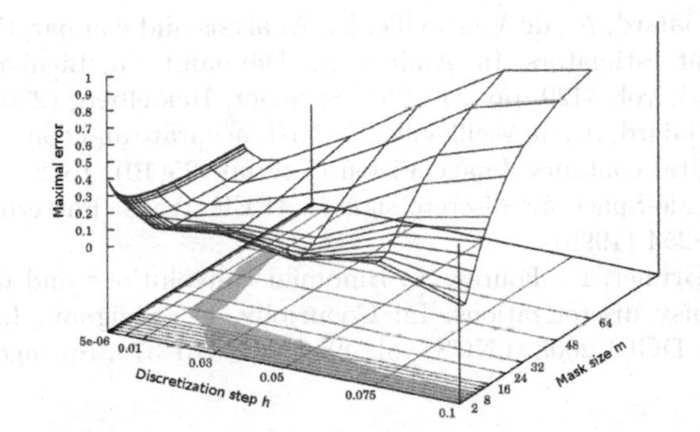

Fig. 4. Maximal error as a function of the discretization step and the mask size. The practical optimum remains to be theoretically determined as Theorem 1 and Theorem 3 make require different mask sizes.

5 Conclusion

We have provided convergence results for the binomial derivative estimator for C^2 real curves and parametrized curves. Note that the proofs in this papers are made for masks with odd size, but similar results can be obtained for masks with even size.

Note that the restriction to closed curve has been used here for convenience, but the proofs can be extended to non closed curves.

Note that when h tends to 0, the number of pixels of the boundary discretization of a convex shape is of the order of $h^{-2/3}$ (adapted from [1]). Hence the size of the required effective mask to ensure convergence (Section 2), which is $O(\sqrt{m \ln(m)})$. Hence under application of Theorem 3 and its generalization to parametrized curves, the mask size, is $O(\sqrt(h^{(\alpha-2)/1.01\ln(1/h)}))$, and so its real size tends to 0 as a proportion of the number of pixels involved. For this reason, the binomial estimator can be called a **local estimator**.

References

1. Balog, A., Bárány, I.: On the convex hull of the integer points in a disc. In: Symposium on Computational Geometry, pp. 162–165 (1991)
2. Floater, M.S.: On the convergence of derivatives of Bernstein approximation. J. Approx. Theory 134(1), 130–135 (2005)
3. Kerautret, B., Lachaud, J.-O., Naegel, B.: Comparison of discrete curvature estimators and application to corner detection. In: Bebis, G., Boyle, R., Parvin, B., Koracin, D., Remagnino, P., Porikli, F., Peters, J., Klosowski, J., Arns, L., Chun, Y.K., Rhyne, T.-M., Monroe, L. (eds.) ISVC 2008, Part I. LNCS, vol. 5358, pp. 710–719. Springer, Heidelberg (2008)
4. Kimmel, R., Sochen, N.A., Weickert, J.: Scale-Space and PDE methods in computer vision. In: Kimmel, R., Sochen, N.A., Weickert, J. (eds.) Scale-Space 2005. LNCS, vol. 3459. Springer, Heidelberg (2005)
5. Lachaud, J.-O., Vialard, A., de Vieilleville, F.: Analysis and comparative evaluation of discrete tangent estimators. In: Andrès, É., Damiand, G., Lienhardt, P. (eds.) DGCI 2005. LNCS, vol. 3429, pp. 240–251. Springer, Heidelberg (2005)
6. Lachaud, J.-O., Vialard, A., de Vieilleville, F.: Fast, accurate and convergent tangent estimation on digital contours. Image Vision Comput. 25(10), 1572–1587 (2007)
7. Lindeberg, T.: Scale-Space for discrete signals. IEEE Trans. Pattern Anal. Mach. Intell. 12(3), 234–254 (1990)
8. Malgouyres, R., Brunet, F., Fourey, S.: Binomial convolutions and derivatives estimation from noisy discretizations. In: Coeurjolly, D., Sivignon, I., Tougne, L., Dupont, F. (eds.) DGCI 2008. LNCS, vol. 4992, pp. 370–379. Springer, Heidelberg (2008)

Christoffel and Fibonacci Tiles*

Alexandre Blondin-Massé, Srečko Brlek**, Ariane Garon, and Sébastien Labbé

Laboratoire de Combinatoire et d'Informatique Mathématique,
Université du Québec à Montréal,
C.P. 8888 Succursale "Centre-Ville", Montréal (QC), Canada H3C 3P8
blondin_masse.alexandre@courrier.uqam.ca, brlek.srecko@uqam.ca,
garon.ariane@courrier.uqam.ca, labbe.sebastien@courrier.uqam.ca

Abstract. Among the polyominoes that tile the plane by translation, the so-called squares have been conjectured to tile the plane in at most two distinct ways (these are called double squares). In this paper, we study two families of tiles : one is directly linked to Christoffel words while the other stems from the Fibonacci sequence. We show that these polyominoes are double squares, revealing strong connections between discrete geometry and other areas by means of combinatorics on words.

Keywords: Polyomino, tiling, Christoffel word, Fibonacci sequence.

1 Introduction

In a discussion following the presentation of X. Provençal at the DGCI 2006 conference held in Szeged [1], E. Andres asked if there were a description of tesselations of the plane with tiles whose borders were discrete segments. That was the starting point of an investigation revealing interesting connections between discrete geometry, combinatorics on words, Lindenmayer systems, cristallography and number theory. The basic object of study is the ubiquitous *polyomino*, widely studied in the literature for having connections in numerous fields, whose list would be needless for our purpose. There are different types of polyominoes and here, by *polyomino* we mean a finite union of unit lattice squares (pixels) in the discrete plane whose boundary is a simple closed path. In particular, a polyomino is simply connected (without holes), and its boundary is simple (does not cross itself). Paths are conveniently encoded by words on the alphabet $\{a, \overline{a}, b, \overline{b}\}$, representing the elementary grid steps $\{\rightarrow, \leftarrow, \uparrow, \downarrow\}$. For instance, starting from S in counterclockwise way, the

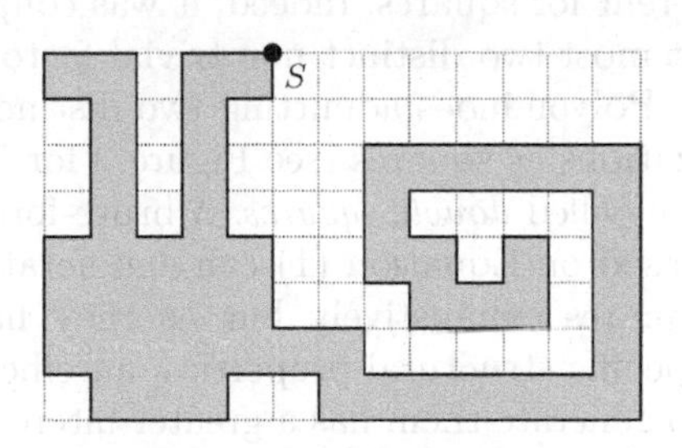

Fig. 1. A snail-shaped polyomino

* With support of the NSERC.
** Corresponding author.

S. Brlek, C. Reutenauer, and X. Provençal (Eds.): DGCI 2009, LNCS 5810, pp. 67–78, 2009.

boundary $b(P)$ of the polyomino P in Figure 1 is coded by the word

$$w = \overline{a}^2\overline{b}^4\overline{a}b^4\overline{a}^2\overline{b}a\overline{b}^3\overline{a}\overline{b}^3a\overline{b}a^3ba^2\overline{b}a^7b^6\overline{a}^6\overline{b}^3a\overline{b}a^3b^2\overline{a}\overline{b}\overline{a}b\overline{a}ba^4\overline{b}^4\overline{a}^5b\overline{a}^2b^2\overline{a}b^3ab.$$

Observe that we may consider the words as circular which avoids a fixed origin.

The *perimeter* of a polyomino P is the length of its boundary words and is of even length. The problem of deciding if a given polyomino tiles the plane by translation goes back to Wisjhoff and Van Leeuven [2] who coined the term *exact polyomino* for these. Beauquier and Nivat [3] gave a characterization stating that the boundary $b(P)$ of an exact polyomino P satisfies the following (not necessarily in a unique way) factorization

$$b(P) = A \cdot B \cdot C \cdot \widehat{A} \cdot \widehat{B} \cdot \widehat{C} \tag{1}$$

where at most one of the variables is empty, $\widehat{\cdot} = \overline{\cdot} \circ \widetilde{\cdot}$, $\widetilde{\cdot}$ is the usual reversal operation and $\overline{\cdot}$ is the morphism defined by $a \leftrightarrow \overline{a}$ and $b \leftrightarrow \overline{b}$ (see next section). Hereafter, this condition is referred as the BN-factorization.

For example, the polyomino in Figure 2 is exact, has semi-perimeter 14 and its boundary may be factorized as

$$ababa \cdot bba\overline{b}a \cdot bb\overline{a}b \cdot \overline{a}\overline{b}\overline{a}\overline{b}\overline{a} \cdot \overline{a}b\overline{a}\overline{b}\overline{b} \cdot \overline{b}a\overline{b}\overline{b}.$$

Polyominoes having a factorization with A, B and C nonempty are called *pseudo-hexagons*. If one of the variable is empty, they were called *pseudo-squares*, and from now on we call them *squares*. It has been shown in [4] that there exist polyominoes admitting an arbitrary number of distinct non trivial factorizations as pseudo-hexagons. The case is different for squares. Indeed, it was conjectured in [4] that a square polyomino has at most two distinct non trivial factorizations.

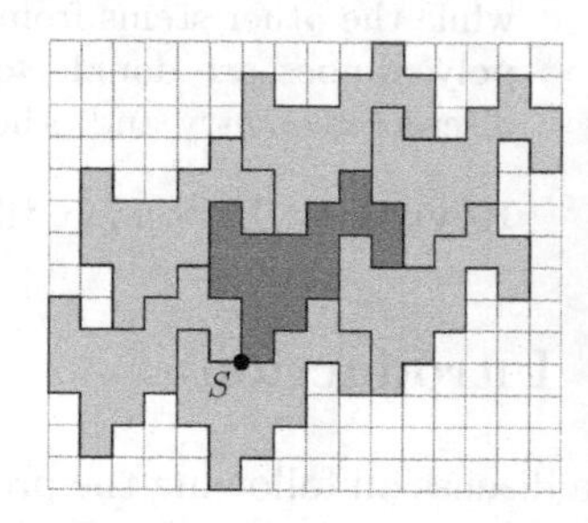

Fig. 2. A pseudo-hexagon

Polyominoes admitting two distinct factorizations as squares (see Figure 3 for instance) are called *double squares*. A brute-force search based on Equation (1) can enumerate double squares exhaustively but as they have very specific structural properties, an efficient way to generate them has a greater interest. Moreover, some conjectures (e.g. related to palindromes [4]) could be solved by a complete description of double squares. In this paper we use a combinatorial approach, relying on efficient techniques [1,5], to construct two classes of double square polyominoes. Those two families, interesting in themselves, are important for the zoology because they describe entirely the table of small double squares available in [4]. The first is composed of Christoffel tiles, those for which the boundary word is composed of crenellated versions of two digitized segments (answering partially to E. Andres' question), for which a characterization is provided (Theorem 2). The second is built on the Fibonacci recurrence: a special

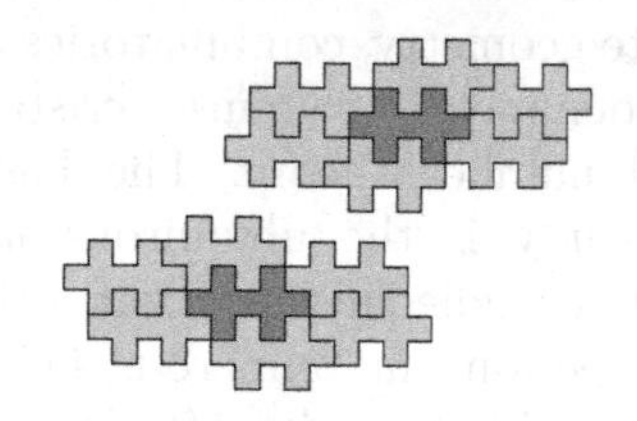

Fig. 3. A double square and its two square tilings

subclass of Fibonacci tiles is completely described (Theorem 3), and we describe four more derived classes of squares polyominoes. Finally, this study reveals new connections with Lindenmayer systems and number theory among others.

2 Preliminaries

The usual terminology and notation on words is from Lothaire [6]. An *alphabet* Σ is a finite set whose elements are called *letters*. A finite word w is a sequence of letters, that is, a function $w : \{1, 2, \ldots, n\} \to \Sigma$, where w_i is the i-th letter, $1 \le i \le n$. The *length* of w, denoted by $|w|$, is given by the integer n. The unique word of length 0 is denoted ε, and the set of all finite words over Σ is denoted Σ^*. The *free monoid* Σ^* is the set of all finite words over Σ, and $\Sigma^{\geq k}$ is the set of words of length at least k. The *reversal* $\widetilde{w}$ of $w = w_1 w_2 \cdots w_n$ is the word $\widetilde{w} = w_n w_{n-1} \cdots w_1$. Words p satisfying $p = \widetilde{p}$ are called *palindromes*. The set of all palindromes over Σ is denoted $\mathrm{Pal}(\Sigma^*)$. A word u is a *factor* of another word w if there exist $x, y \in \Sigma^*$ such that $w = xuy$. We denote by $|w|_u$ the number of times that u appears in w. Two words u and v are *conjugate* if there are words x and y such that $u = xy$ and $v = yx$. In that case, we write $u \equiv v$. Clearly, $\equiv$ is an equivalence relation. Given two alphabets Σ_1 and Σ_2, a *morphism* is a function $\varphi : \Sigma_1^* \to \Sigma_2^*$ compatible with concatenation, that is, $\varphi(uv) = \varphi(u)\varphi(v)$ for any $u, v \in \Sigma_1^*$. It is clear that a morphism is completely defined by its action on the letters of Σ_1.

Paths on the square lattice. The notation of this section is partially adapted from [5]. A path in the square lattice, identified as $\mathbb{Z} \times \mathbb{Z}$, is a polygonal path made of the elementary unit translations

$$a = (1, 0), \overline{a} = (-1, 0), b = (0, 1), \overline{b} = (0, -1).$$

A finite path w is therefore a word on the alphabet $\mathcal{F} = \{a, \overline{a}, b, \overline{b}\}$, also known as the Freeman chain code [7,8] (see [9] for further reading). Furthermore, we say that a path w is *closed* if it satisfies $|w|_a = |w|_{\overline{a}}$ and $|w|_b = |w|_{\overline{b}}$. A *simple path* is a word w such that none of its factor is a closed path. A *boundary word* is a closed path such that none of its proper factors is closed. Finally, a *polyomino* is a subset of $\mathbb{Z}^2$ contained in some boundary word. For instance, for the paths

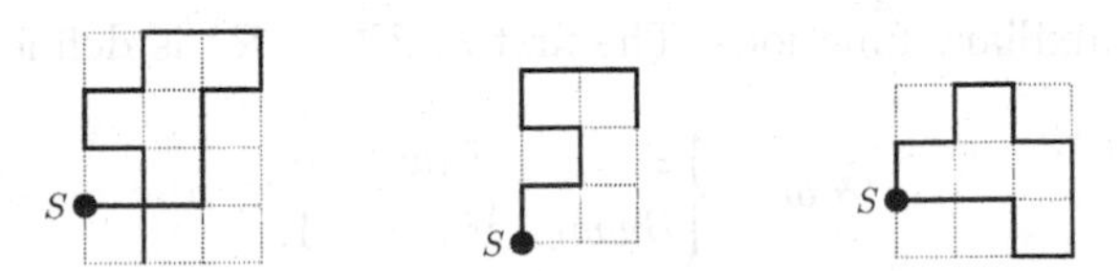

Fig. 4. Paths in the discrete plane: (a) arbitrary, (b) simple (c) closed

represented in Figure 4, the corresponding Freeman chain code is respectively

$$\text{(a) } p_1 = ab\overline{a}babaa\overline{b}\overline{a}\overline{b}\overline{b}\overline{a}\overline{b} \quad \text{(b) } p_2 = bab\overline{a}baa\overline{b}, \quad \text{(c) } p_3 = baba\overline{b}a\overline{b}\overline{b}\overline{a}b\overline{a}\overline{a}.$$

On the square grid, a path describes a sequence of basic movements in the left (L), right (R), forward (F) and backward (B) directions. Each pair of letters in $\mathcal{F}$ is associated to a movement on the alphabet $\mathcal{R} = \{\text{L}, \text{R}, \text{F}, \text{B}\}$ by a function $g : \mathcal{F}^2 \to \mathcal{R}$ defined by

$$g(u) = \begin{cases} \text{L} & \text{if } u \in V_{\text{L}} = \{ab, b\overline{a}, \overline{a}\overline{b}, \overline{b}a\}, \\ \text{R} & \text{if } u \in V_{\text{R}} = \{ba, a\overline{b}, \overline{b}\overline{a}, \overline{a}b\}, \\ \text{F} & \text{if } u \in V_{\text{F}} = \{aa, \overline{a}\overline{a}, \overline{b}\overline{b}, bb\}, \\ \text{B} & \text{if } u \in V_{\text{B}} = \{a\overline{a}, \overline{a}a, b\overline{b}, \overline{b}b\}, \end{cases}$$

that can be extended to a derivative function $\partial : \mathcal{F}^{\geq 1} \to \mathcal{R}^*$ by setting

$$\partial w = \begin{cases} \prod_{i=2}^{n} g(w_{i-1}w_i) & \text{if } |w| \geq 2, \\ \varepsilon & \text{if } |w| = 1. \end{cases}$$

where n is the length of the word w and the product is the concatenation. For example, if $p_2 = bab\overline{a}baa\overline{b}$ as defined above then $\partial p_2 = \text{RLLRRFR}$.

Note also that each path $w \in \mathcal{F}^{\geq 1}$ is completely determined, up to translation, by its initial step $\alpha \in \mathcal{F}$ and a word y on the alphabet $\mathcal{R} = \{\text{L}, \text{R}, \text{F}, \text{B}\}$. Thus, we use calculus notation and we introduce a function $\int_\alpha : \mathcal{R}^* \to \mathcal{F}^{\geq 1}$ defined recursively by

$$\int_\alpha y = \begin{cases} \alpha & \text{if } |y| = 0, \\ \alpha \int_\beta y' & \text{if } |y| \geq 1, \end{cases}$$

where $\beta \in \mathcal{F}$ is the letter such that $\alpha\beta \in V_x$ and $y = xy'$ with $x \in \mathcal{R}$. For example, if $y = \text{RLLRRFR}$, then $\int_b y = bab\overline{a}baa\overline{b} = p_2$. The next lemma gives some easily established statements and shows how both functions ∂ and $\int$ behave naturally.

Lemma 1. *Let $w, w' \in \mathcal{F}^*$, $y, y' \in \mathcal{R}^*$, $\alpha \in \mathcal{F}$ and $x \in \mathcal{R}$. Then*

(i) *$\int_{w_1} \partial w = w$ and $\partial \int_\alpha y = y$, where w_1 is the first letter of w.*
(ii) *$\partial(ww') = \partial w \cdot g(w_n w'_1) \cdot \partial w'$ and $\int_\alpha yxy' = \int_\alpha y \int_\beta y'$, where w_n is the last letter of w, w'_1 is the first letter of w' and β is the last letter of $\int_\alpha yx$.* □

Note that $|\partial w| = |w| - 1$ and $|\int_\alpha y| = |y| + 1$. In some situations, for example when w represents a circular word, this is not convenient. For that reason we introduce two auxiliary functions. The first $\mathring{\partial} : \mathcal{F}^* \to \mathcal{R}^*$ is defined by

$$\mathring{\partial}\, w = \begin{cases} \varepsilon & \text{if } |w| = 0, \\ \partial w w_1 & \text{if } |w| \geq 1, \end{cases}$$

where w_1 is the first letter of w. The second is $\oint_\alpha : \mathcal{R}^* \to \mathcal{F}^*$, defined as follows : if β is the last letter of $\int_\alpha y$, then $\oint_\alpha y$ is the word such that $\oint_\alpha y \cdot \beta = \int_\alpha y$. It is clear that $\oint_{w_1} \mathring{\partial}\, w = w$ and $\mathring{\partial} \oint_\alpha y = y$ for all $w \in \mathcal{F}^*$ and $y \in \mathcal{R}^*$.

In [5], the authors also introduced a valuation function Δ defined on $\mathcal{R}^*$ by $\Delta(y) = |y|_{\text{L}} - |y|_{\text{R}} + 2|y|_{\text{B}}$ as well as on $\mathcal{F}^{\geq 1}$ by setting $\Delta(w) = \Delta(\partial w)$. This valuation is nothing but the *winding number*.

Transformations. Some useful transformations on $\mathcal{F}^*$ are rotations by an angle $k\pi/2$ and reflections with respect to axes of angles $k\pi/4$, where $k \in \mathbb{N}$. The rotation of angle $\pi/2$ translates merely in $\mathcal{F}$ by the morphism:

$$\rho : a \mapsto b, b \mapsto \overline{a}, \overline{a} \mapsto \overline{b}, \overline{b} \mapsto a.$$

We denote the other rotations by ρ^2 and ρ^3 according to the usual notation. The rotation ρ^2 is also noted $\bar{\cdot}$ since it can be seen as the *complement* morphism defined by the relations $\overline{\overline{a}} = a$ and $\overline{\overline{b}} = b$. Similarly, for $k \in \{0, 1, 2, 3\}$, σ_k is the reflection defined by the axis passing through the origin and having an angle of $k\pi/4$ with the absciss. It may be seen as a morphism on $\mathcal{F}^*$ as well:

$$\sigma_0 : a \mapsto \overline{a}, \overline{a} \mapsto a, b \mapsto b, \overline{b} \mapsto \overline{b} \text{ and } \sigma_1 : a \mapsto b, b \mapsto a, \overline{a} \mapsto \overline{b}, \overline{b} \mapsto \overline{a}.$$

The two other reflections are $\sigma_2 = \sigma_0 \circ \rho^2$ and $\sigma_3 = \sigma_1 \circ \rho^2$. Another useful map is the antimorphism $\widehat{\cdot} = \bar{\cdot} \circ \tilde{\cdot}$ defined on $\mathcal{F}^*$. If $w \in \mathcal{F}^*$ is a path, then $\widehat{w}$ is the path run in the opposite direction. The effect of the operators $\widehat{\cdot}$, $\tilde{\cdot}$ and $\bar{\cdot}$ are illustrated in Figure 5.

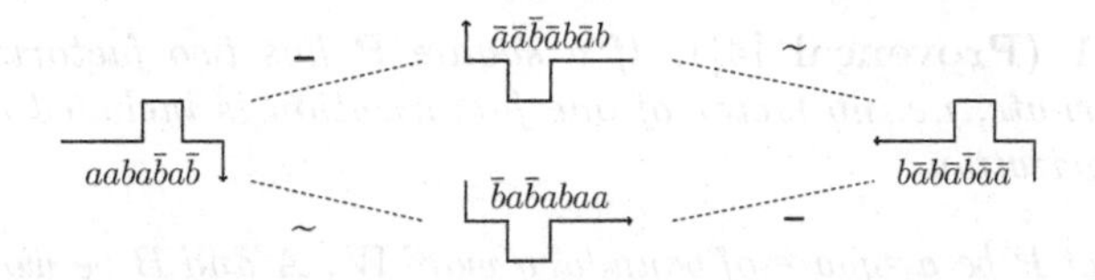

Fig. 5. Effect of the operators $\widehat{\cdot}$, $\tilde{\cdot}$ and $\bar{\cdot}$ on $\mathcal{F}^*$

On the alphabet $\mathcal{R}$, we define an involution

$$\imath : \mathrm{L} \mapsto \mathrm{R}, \mathrm{R} \mapsto \mathrm{L}, \mathrm{F} \mapsto \mathrm{F}, \mathrm{B} \mapsto \mathrm{B}.$$

This function $\imath$ extends to $\mathcal{R}^*$ as a morphism, so that the map $\widehat{\cdot}$ extends as well to $\widehat{\cdot} : \mathcal{R}^* \to \mathcal{R}^*$ by $\widehat{\cdot} = \imath \circ \tilde{\cdot}$. All these operations are closely related as shown in the lemmas hereafter. The proofs are left to the reader.

Lemma 2. *Let $w \in \mathcal{F}^*$, $y \in \mathcal{R}^*$ and $\alpha \in \mathcal{F}$. The following properties hold:*

(i) $\partial w = \partial \rho^i(w)$ *for all* $i \in \{1, 2, 3\}$,
(ii) $\imath(\partial w) = \partial \sigma_i(w)$ *for all* $i \in \{0, 1, 2, 3\}$,
(iii) $\partial \widehat{w} = \widehat{\partial w} = \partial \widetilde{w}$,
(iv) $\rho^i(\int_\alpha y) = \int_{\rho^i(\alpha)} y$ *for all* $i \in \{1, 2, 3\}$,
(v) $\sigma_i(\int_\alpha y) = \int_{\sigma_i(\alpha)} \imath(y)$ *for all* $i \in \{0, 1, 2, 3\}$,
(vi) $\widetilde{\int_\alpha y} = \int_\beta \widetilde{y}$ *where β is the last letter of* $\int_\alpha y$,
(vii) $\widehat{\int_\alpha y} = \int_{\overline{\beta}} \widehat{y}$ *where β is the last letter of* $\int_\alpha y$,
(viii) *If β is the last letter of $\int_\alpha y$, then $\beta = \rho^i(\alpha)$ where $i = \Delta(y)$.* □

For the rest of the paper, the words w of $\mathcal{F}^*$ and $\mathcal{R}^*$ satisfying $\widehat{w} = w$ are called *antipalindromes.*

Lemma 3. *Let $w \in \mathcal{F}^*$. Then the following statements are equivalent.*

(i) $\widehat{w} = \rho^2(w)$
(ii) *w is a palindrome*
(iii) *∂w is an antipalindrome.*

Finally, reflections on $\mathcal{F}^*$ are easily described on $\mathcal{R}^*$.

Lemma 4. *Let $w \in \mathcal{F}^*$. There exists $i \in \{0, 1, 2, 3\}$ such that $\widehat{w} = \sigma_i(w)$ if and only if ∂w is a palindrome.* □

Square Tilings. Let P be a polyomino having W as a boundary word and Q a square having $V = AB\widehat{A}\widehat{B}$ as a BN-factorization. Then *the product of P and Q*, denoted by $P \circ Q$, is the polyomino whose boundary word is given by $\gamma(W)$, where $\gamma : \mathcal{F}^* \to \mathcal{F}^*$ is the morphism defined by $\gamma(a) = A$, $\gamma(\overline{a}) = \widehat{A}$, $\gamma(b) = B$ and $\gamma(\overline{b}) = \widehat{B}$. In [4], a polyomino R is called *prime* if for all pair of polyominoes P and Q such that $R = P \circ Q$, we have either $R = P$ or $R = Q$.

Proposition 1 (Provençal [4]). *If a square P has two factorizations, then they must alternate, i.e. no factor of one factorization is included in a factor of the other factorization.* □

Lemma 5. *Let P be a square of boundary word W, A and B be words such that $W \equiv AB\widehat{A}\widehat{B}$. Then $A, B \in \mathrm{Pal}(\mathcal{F}^*)$ if and only if $W = w\overline{w}$ for some word w.*

Proof. If $W = w\overline{w}$ then every conjugate of W has this form. Therefore, if $W \equiv AB\widehat{A}\widehat{B}$, we have that $AB = \overline{\widehat{A}\widehat{B}} = \widetilde{A}\widetilde{B}$, showing that A and B are palindromes. Conversely, one shows that if A and B are palindromes, then $W \equiv AB\overline{AB}$. □

For more details on tiling by translation and square tilings see [1,10].

3 Christoffel Tiles

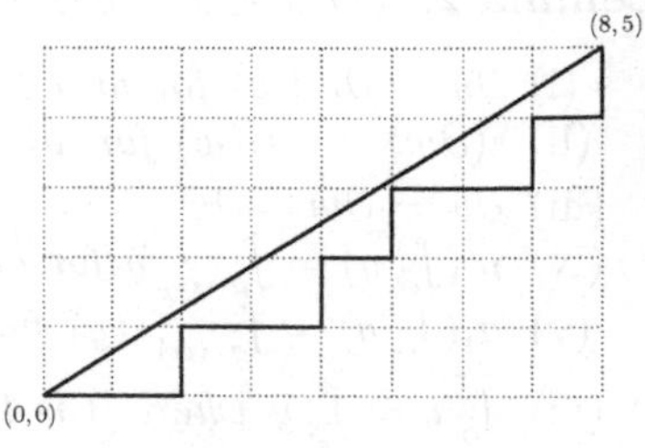

Recall that Christoffel words are finite Sturmian words, that is, they are obtained by discretizing a segment in the plane. Let $(p, q) \in \mathbb{N}^2$ with $\gcd(p, q) = 1$, and let S be the segment with endpoints $(0, 0)$ and (p, q). The word w is a *lower Christoffel word* if the path induced by w is under S and if they both delimit a polygon with no integral interior point. An *upper Christoffel word* is defined similarly. A *Christoffel word* is either a lower

Christoffel word or an upper Christoffel word. On the right is illustrated the lower one corresponding to

$$w = aabaababaabab.$$

It is well known that if w and w' are respectively the lower and upper Christoffel words associated to (p, q), then $w' = \widetilde{w}$. Moreover, we have $w = amb$ and $w' = bma$, where m is a palindrome and a, b are letters. We call *cutting word* the word m. They have been widely studied in the litterature (see e.g. [11], where they are also called *central words* . The next theorem gives a very useful characterization of Christoffel words.

Theorem 1 (Pirillo [12]). *A word m on the two-letter alphabet $\{a, b\} \subset \mathcal{F}$ is a cutting word if and only if amb and bma are conjugate.* □

Another useful result is the following.

Proposition 2 (Borel and Reutenauer [13]). *The lower and upper Christoffel words w and w' are conjugate by palindromes.* □

Let $\mathcal{B} = \{a, b\}$. Consider the morphism $\lambda : \mathcal{B}^* \to \mathcal{F}^*$ by $\lambda(a) = a\bar{b}ab$ and $\lambda(b) = ab$, which can be seen as a "crenellization" of the steps *east* and *north-east.* Two useful properties of λ are used through the rest of this section and are easy to establish.

Lemma 6. *Let $v, v' \in \mathcal{B}^*$. Then*

(i) *$b\lambda(v)$ is a palindrome if and only if v is a palindrome.*
(ii) *$\lambda(v) \equiv \lambda(v')$ if and only if $v \equiv v'$.* □

We call *crenellated tile* a polyomino whose boundary word is given by $\lambda(w)\overline{\lambda(w)}$ where $w = avb$ and $v \in \text{Pal}(\mathcal{B}^*)$. A crenellated tile is a square since

$$\lambda(w)\overline{\lambda(w)} = a\bar{b}ab\lambda(v)ab\bar{a}b\bar{a}\bar{b}\overline{\lambda(v)}\bar{a}\bar{b} \equiv \bar{b}a\bar{b} \cdot ab\lambda(v)a \cdot \widehat{\bar{b}a\bar{b}} \cdot \widehat{ab\lambda(v)a}.$$

We say that a crenellated tile obtained from a lower Christoffel word w is a *basic Christoffel tile* while a *Christoffel tile* is a polyomino isometric to a basic Christoffel tile under some rotations ρ and symmetries σ_i (see Figure 6).

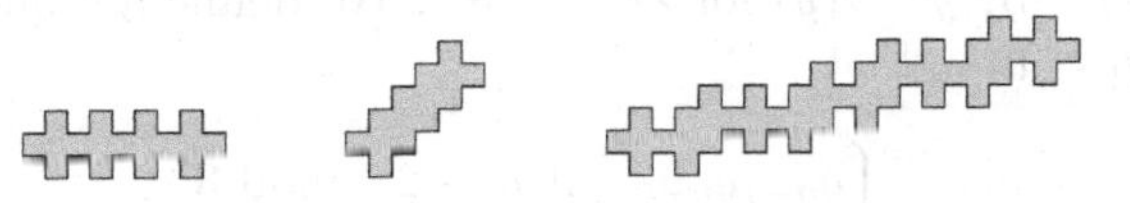

Fig. 6. Basic Christoffel tiles: (a) $w = aaaab$ (b) $w = abbbb$ and (c) $w = aabaababaabab$

Theorem 2. *Let P be a crenellated tile. Then P is a double square if and only if it comes from a Christoffel word.*

Proof. ($\Rightarrow$) Assume that P is a double square. Let $W = \lambda(w)\overline{\lambda(w)}$ be its boundary word, where $w = avb \in a\mathrm{Pal}(\mathcal{B}^*)b$. We know from Proposition 1 that the factorizations must alternate. Since P has the factorization

$$W = \overline{b}a\overline{b} \cdot ab\lambda(v)a \cdot \widehat{\overline{b}a\overline{b}} \cdot \widehat{ab\lambda(v)a},$$

it means that the second factorization is obtained from one of the conjugates $W' = a\overline{b}ab\lambda(v)ab\overline{a}b\overline{a}\overline{b}\overline{\lambda(v)}\overline{a}\overline{b}$ or $W'' = \overline{b}ab\lambda(v)ab\overline{a}b\overline{a}\overline{b}\overline{\lambda(v)}\overline{a}\overline{b}a$. Let V' and V'' be respectively the first half of W' and W''. Then, by Lemma 5, either V' or V'' is a product of two palindromes x and y. First, assume that the other factorization is obtained from V'. Then $V' = \lambda(avb) = a\overline{b}ab\lambda(v)ab = xy$. Taking the reversal on each side, we get $\widetilde{\lambda(avb)} = yx$, that is $\lambda(avb) \equiv \widetilde{\lambda(avb)}$. But

$$\widetilde{\lambda(avb)} = ba\widetilde{\lambda(v)}ba\overline{b}a = bab\widetilde{\lambda(v)}a\overline{b}a = bab\lambda(v)a\overline{b}a \equiv ab\lambda(v)a\overline{b}ab = \lambda(bva),$$

which means that $\lambda(avb) \equiv \lambda(bva)$. Thus, by Lemma 6, we deduce $avb \equiv bva$. Hence, by Theorem 1, we conclude that v is a cutting word so that $w = avb$ is a lower Christoffel word.

It remains to consider the case where the second factorization is obtained from V''. Hence, we could write $V'' = \overline{b}ab\lambda(v)ab\overline{a} = xy$. But such palindromes x and y cannot exist since $\overline{a}$ appears only at the end of V''.

($\Leftarrow$) Assume that w is a lower Christoffel word. We know from Proposition 2 that $w = amabm'b$ for some palindromes m and m'. Therefore,

$$\begin{aligned}\lambda(w)\overline{\lambda(w)} &= \lambda(amabm'b)\overline{\lambda(amabm'b)} \\ &= a\overline{b}ab\lambda(m)a\overline{b}a \cdot bab\lambda(m')ab \cdot \overline{a}b\overline{a}\overline{b}\overline{\lambda(m)}\overline{a}b\overline{a} \cdot \overline{b}\overline{a}\overline{b}\overline{\lambda(m')}\overline{a}\overline{b},\end{aligned}$$

showing that P admits a second square factorization. □

It can also be shown in view of Lemma 5, that for each Christoffel tile such that $W \equiv AB\widehat{A}\widehat{B}$, the factors A and B are palindromes which suggests that the conjecture of Provençal [4] is true.

4 Fibonacci Tiles

In this section, in order to simplify the notation, we overload the operator $\bar{\cdot}$ by defining it over $\mathcal{R}^*$ by $\overline{y} = \iota(y)$ for all $y \in \mathcal{R}^*$. We define a sequence $(q_n)_{n\in\mathbb{N}}$ in $\mathcal{R}^*$ by $q_0 = \varepsilon$, $q_1 = \mathrm{R}$ and

$$q_n = \begin{cases} q_{n-1}q_{n-2} & \text{if } n \equiv 2 \mod 3, \\ q_{n-1}\overline{q_{n-2}} & \text{if } n \equiv 0,1 \mod 3. \end{cases}$$

whenever $n \geq 2$. The first terms of $(q_n)_{n\in\mathbb{N}}$ are

$q_0 = \varepsilon$	$q_3 = \mathrm{RL}$	$q_6 = \mathrm{RLLRLLRR}$
$q_1 = \mathrm{R}$	$q_4 = \mathrm{RLL}$	$q_7 = \mathrm{RLLRLLRRLRRLR}$
$q_2 = \mathrm{R}$	$q_5 = \mathrm{RLLRL}$	$q_8 = \mathrm{RLLRLLRRLRRLRRLLRLLRR}$

Note that $|q_n| = F_n$ is the n-th Fibonacci number. Moreover, given $\alpha \in \mathcal{F}$, the path $\int_\alpha q_n$ presents strong symmetric properties, as shown by the next lemma.

Lemma 7. *Let $n \in \mathbb{N}$. Then $q_{3n+1} = p\alpha$, $q_{3n+2} = q\alpha$ and $q_{3n+3} = r\overline{\alpha}$ for some antipalindrome p, and some palindromes q, r and some letter $\alpha \in \{\text{L}, \text{R}\}$.*

Proof. By induction on n. For $n = 0$, we have indeed $q_1 = \varepsilon \cdot \text{R}$, $q_2 = \varepsilon \cdot \text{R}$ and $q_3 = \text{R} \cdot \text{L}$. Now, assume that $q_{3n+1} = p\alpha$, $q_{3n+2} = q\alpha$ and $q_{3n+3} = r\overline{\alpha}$ for some antipalindrome p, some palindromes q, r and some letter $\alpha \in \{\text{L}, \text{R}\}$. Then

$$\begin{aligned}
q_{3n+4} &= q_{3n+3}\overline{q_{3n+2}} = q_{3n+2}\overline{q_{3n+1}q_{3n+2}} = q\alpha\overline{p\alpha q} \cdot \overline{\alpha} \\
q_{3n+5} &= q_{3n+4}q_{3n+3} = q_{3n+3}\overline{q_{3n+2}}q_{3n+3} = r\overline{\alpha q\alpha}r \cdot \overline{\alpha} \\
q_{3n+6} &= q_{3n+5}\overline{q_{3n+4}} = q_{3n+4}q_{3n+3}\overline{q_{3n+4}} = q\alpha\overline{p\alpha q}\overline{\alpha}r\overline{\alpha}\overline{q\alpha}p\alpha q \cdot \alpha.
\end{aligned}$$

Since $q\alpha\overline{p\alpha q}$ is an antipalindrome and $r\overline{\alpha q\alpha}r$, $q\alpha\overline{p\alpha q}\overline{\alpha}r\overline{\alpha}\overline{q\alpha}p\alpha q$ are palindromes, the result follows. □

The proof of the following lemma is technical and can be done by induction. It is not included here due to lack of space.

Lemma 8. *Let $n \in \mathbb{N}$ and $\alpha \in \mathcal{F}$.*

(i) *The path $\int_\alpha q_n$ is simple.*
(ii) *The path $\oint_\alpha (q_{3n+1})^4$ is the boundary word of a polyomino.* □

A Fibonacci tile of order n is a polyomino having $\oint_\alpha (q_{3n+1})^4$ as a boundary word, where $n \in \mathbb{N}$. They are somehow related to the Fibonacci Fractals found in [14]. The first Fibonacci tiles are illustrated in Figure 7.

Fig. 7. Fibonacci tiles of order $n = 0, 1, 2, 3, 4$

Theorem 3. *Fibonacci tiles are double squares.*

Proof. We know from Lemma 7 that $q_{3n+1} = px$ for some antipalindrome p and some letter $x \in \{\text{L}, \text{R}\}$. If $x = \text{R}$, we consider the reversal of the path, i.e. $(\widehat{q_{3n+1}})^4$, so that we may suppose that $x = \text{L}$. Therefore, on the first hand, we obtain

$$\oint_\alpha (q_{3n+1})^4 = \int_\alpha p\text{L} \cdot p\text{L} \cdot \widehat{p}\text{L} \cdot \widehat{p} = \int_\alpha p \cdot \int_{\rho(\alpha)} p \cdot \widehat{\int_\alpha p} \cdot \widehat{\int_{\rho(\alpha)} p},$$

because $\Delta(p) = 0$. On the other hand, the conjugate $q'_{3n+1} = \overline{q_{3n-1}q_{3n}}$ of q_{3n+1} corresponds to another boundary word of the same tile. Using again Lemma 7, we can write $q_{3n} = r\text{L}$ and $q_{3n-1} = q\text{R}$, for some palindromes q and r. Therefore, $p\text{L} = q_{3n+1} = q_{3n}\overline{q_{3n-1}} = r\text{L}\overline{q}\text{L}$ so that $p = r\text{L}\overline{q}$. But p is an antipalindrome, which means that $q'_{3n+1} = \overline{q_{3n-1}q_{3n}} = \overline{q}\text{L}r\text{L} = \widetilde{p}\text{L} = \overline{p}\text{L}$. Hence, since $\overline{p}$ is an antipalindrome as well, we find

$$\oint_\alpha (q'_{3n+1})^4 = \int_\alpha \overline{p}\text{L} \cdot \overline{p}\text{L} \cdot \widehat{\overline{p}}\text{L} \cdot \widehat{\overline{p}} = \int_\alpha \overline{p} \cdot \int_{\rho(\alpha)} \overline{p} \cdot \widehat{\int_\alpha \overline{p}} \cdot \widehat{\int_{\rho(\alpha)} \overline{p}}. \qquad \square$$

As for Christoffel Tiles, Fibonacci Tiles also suggest that the conjecture of Provençal for double squares [4] is true as stated in the next result.

Corollary 1. *If $AB\widehat{A}\widehat{B}$ is a BN-factorisation of a Fibonacci tile, then A and B are palindromes.*

Proof. The conclusion follows from Lemmas 2 and 3 and Theorem 3. Indeed, since p is an antipalindrome, then $\int_\alpha p$ is a palindrome. The same argument applies for the second factorization. □

We end this section by presenting four families of double squares generalizing the Fibonacci tiles. Consider the sequence $(r_{d,m,n})_{(d,m,n)\in\mathbb{N}^3}$ satisfying the following recurrence, for $d \geq 2$,

$$r_{d,m,n} = \begin{cases} r_{d-1,n,m}\overline{r_{d-2,n,m}} & \text{if } d \equiv 0 \bmod 3 \\ r_{d-1,n,m}r_{d-2,n,m} & \text{if } d \equiv 1 \bmod 3 \\ r_{d-1,m,n}\overline{r_{d-2,m,n}} & \text{if } d \equiv 2 \bmod 3 \end{cases}$$

Using similar arguments as in the Fibonacci tiles case, one shows that both families obtained respectively with seed values

$$r_{0,m,n} = (\text{RLLR})^m\text{RLR}, \qquad r_{1,m,n} = (\text{RLLR})^n\text{R}, \tag{2a}$$
$$r_{0,m,n} = (\text{RL})^m\text{RLR}, \qquad r_{1,m,n} = (\text{RL})^n\text{RL} \tag{2b}$$

are such that $\oint_\alpha (r_{d,m,n}r_{d,n,m})^2$ is a boundary word whose associated polyomino is a double square (see Figure 8), where $\alpha \in \mathcal{F}$. Their level of fractality increases with d so that one could say that they are crenellated versions of the Fibonacci Tiles.

Similarly, let $(s_{d,m,n})_{(d,m,n)\in\mathbb{N}^3}$ be a sequence satisfying for $d \geq 2$ the recurrence

$$s_{d,m,n} = \begin{cases} s_{d-1,n,m}\overline{s_{d-2,n,m}} & \text{if } d \equiv 0, 2 \bmod 3, \\ s_{d-1,m,n}s_{d-2,m,n} & \text{if } d \equiv 1 \bmod 3. \end{cases}$$

Then the families obtained with seed values

$$s_{0,m,n} = (\text{RLLR})^m\text{RLR}, \qquad s_{1,m,n} = \text{RL}, \tag{3a}$$
$$s_{0,m,n} = (\text{RL})^m\text{RLR}, \qquad s_{1,m,n} = \text{R} \tag{3b}$$

yield double squares $\oint_\alpha (s_{d,m,n}s_{d,n,m})^2$ as well (see Figure 9). One may verify that $r_{d,0,0} = s_{d,0,0}$ for any $d \in \mathbb{N}$ if the seed values are respectively (2a) and (3b) or respectively (2b) and (3a).

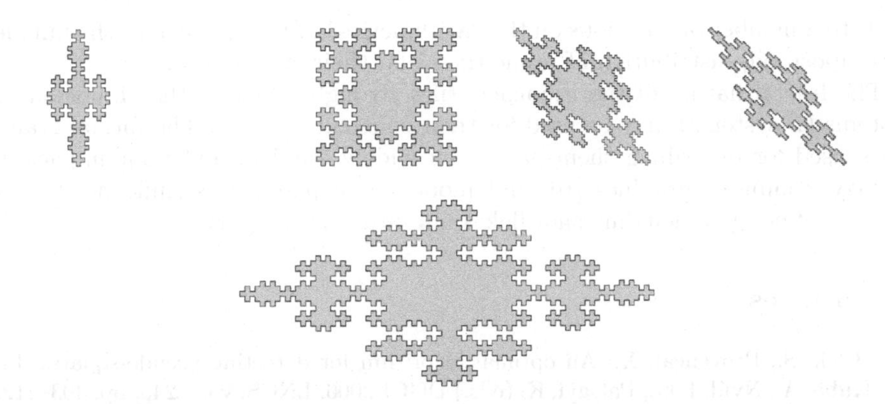

Fig. 8. *Upper row*: tile obtained from $r_{2,0,1}$ with seeds (2a); tiles obtained from $r_{3,m,0}$ with seeds (2b), for $m = 0, 1, 2$. *Lower row*: generalized Fibonacci tile obtained with parameters $d = 3$, $m = 1$ and $n = 0$ on sequence r with seed values (2a).

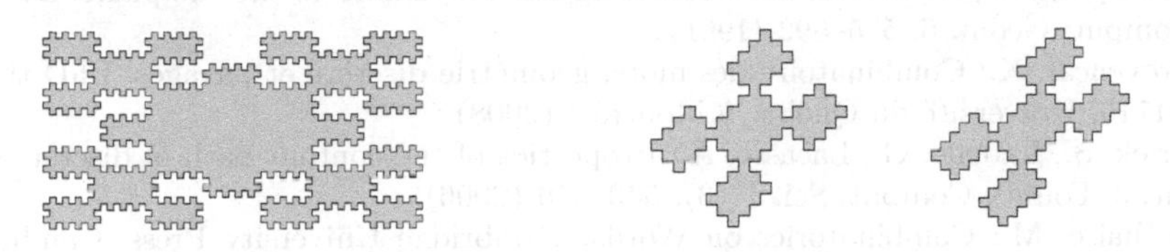

Fig. 9. Tile obtained from $s_{3,2,0}$ with seeds (3a); tiles obtained from $s_{2,0,n}$ with seeds (3b) for $n = 1, 2$

5 Concluding Remarks

The study of double squares suggests interesting and challenging problems. For instance, it is appealing to conjecture that a prime double square is either of Christoffel type or of Fibonacci type. However, that is not the case, as illustrated by Figure 10. This begs for a thorough study in order to exhibit a complete zoology of such tilings. On the other hand, there is a conjecture of [4] stating that if $AB\widehat{A}\widehat{B}$ is the BN-factorization of a prime double square, then A and B are palindromes, for which we have not been able to provide any counter-example. Another problem is to prove that Christoffel and Fibonacci tiles are prime, that is, they are not obtained by composition of smaller squares. This

Fig. 10. Three double squares not in the Christoffel and Fibonacci tiles families

leads to a number of questions on the "arithmetics" of tilings, such as the unique decomposition, distribution of prime tiles, and their enumeration.

The fractal nature of the Fibonacci tiles strongly suggests that Lindemayer systems (L-systems) may be used for their construction [15]. The formal grammars used for describing them have been widely studied, and their impact in biology, computer graphics [16] and modeling of plants is significant [17]. A number of designs including snowflakes fall into this category.

References

1. Brlek, S., Provençal, X.: An optimal algorithm for detecting pseudo-squares. In: Kuba, A., Nyúl, L.G., Palágyi, K. (eds.) DGCI 2006. LNCS, vol. 4245, pp. 403–412. Springer, Heidelberg (2006)
2. Wijshoff, H.A.G., van Leeuven, J.: Arbitrary versus periodic storage schemes and tesselations of the plane using one type of polyomino. Inform. Control 62, 1–25 (1984)
3. Beauquier, D., Nivat, M.: On translating one polyomino to tile the plane. Discrete Comput. Geom. 6, 575–592 (1991)
4. Provençal, X.: Combinatoire des mots, géométrie discrète et pavages. PhD thesis, D1715, Université du Québec à Montréal (2008)
5. Brlek, S., Labelle, G., Lacasse, A.: Properties of the contour path of discrete sets. Int. J. Found. Comput. Sci. 17(3), 543–556 (2006)
6. Lothaire, M.: Combinatorics on Words. Cambridge University Press, Cambridge (1997)
7. Freeman, H.: On the encoding of arbitrary geometric configurations. IRE Trans. Electronic Computer 10, 260–268 (1961)
8. Freeman, H.: Boundary encoding and processing. In: Lipkin, B., Rosenfeld, A. (eds.) Picture Processing and Psychopictorics, pp. 241–266. Academic Press, New York (1970)
9. Braquelaire, J.P., Vialard, A.: Euclidean paths: A new representation of boundary of discrete regions. Graphical Models and Image Processing 61, 16–43 (1999)
10. Brlek, S., Provençal, X., Fédou, J.M.: On the tiling by translation problem. Discrete Applied Mathematics 157(3), 464–475 (2009)
11. Berstel, J., Lauve, A., Reutenauer, C., Saliola, F.: Combinatorics on Words: Christoffel Words and Repetition in Words. CRM monograph series, vol. 27, 147 pages. American Mathematical Society, Providence (2008)
12. Pirillo, G.: A new characteristic property of the palindrome of standard sturmian word. Sém. Lothar. Combin. 43, 1–3 (1999)
13. Borel, J.P., Reutenauer, C.: On christoffel classes. RAIRO-Theoretical Informatics and Applications 40, 15–28 (2006)
14. Monnerot-Dumaine, A.: The Fibonacci fractal. Submitted (September 2008)
15. Rozenberg, G., Salomaa, A.: Mathematical Theory of L Systems. Academic Press, Inc., Orlando (1980)
16. Rozenberg, G., Salomaa, A. (eds.): Lindenmayer Systems: Impacts on Theoretical Computer Science, Computer Graphics, and Developmental Biology. Springer, Secaucus (2001)
17. Prusinkiewicz, P., Lindenmayer, A.: The algorithmic beauty of plants. Springer, New York (1990)

Optimal Partial Tiling of Manhattan Polyominoes*

Olivier Bodini and Jérémie Lumbroso

LIP6, UMR 7606, CALSCI departement, Université Paris 6 / UPMC,
104, avenue du président Kennedy,
F-75252 Paris cedex 05, France

Abstract. Finding an efficient optimal partial tiling algorithm is still an open problem. We have worked on a special case, the tiling of Manhattan polyominoes with dominoes, for which we give an algorithm linear in the number of columns. Some techniques are borrowed from traditional graph optimisation problems.

For our purpose, a *polyomino* is the (non necessarily connected) union of unit squares (for which we will consider the vertices to be in $\mathbb{Z}^2$) and a *domino* is a polyomino with two edge-adjacent unit squares.

To solve the domino tiling problem [6,9,10,11,12,13] for a polyomino P is equivalent to finding a perfect matching in the edge-adjacent graph G_p of P's unit squares. A specific point of view to study tiling problems has been introduced by Conway and Lagarias in [4]: they transformed the tiling problem into a combinatorial group theory problem. Thurston [13] built on this idea by introducing the notion of height, with which he devised a linear-time algorithm to tile a polyomino without holes with dominoes. Continuing these works, Thiant [12], and later Bodini and Fernique [3] respectively obtained an $O(n \log n)$ algorithm which finds a domino tiling for a polyomino whose number of holes is bound, and an $O(n \log^3 n)$ algorithm with no constraint as to the number of holes. All these advances involved the "exact tiling problem".

Naturally, all aforementioned exact tiling algorithms are useless when confronted with polyominoes that cannot be perfectly tiled with dominos. Such algorithms will output only that the polyominoes cannot be tiled—and this is inconvenient because a perfect tiling is not necessarily required: we might want partial tilings with at most x untiled squares.

Thus, by analogy with matching problems in graph theory and the notion of maximum matching, it seems interesting to study the corresponding notion of optimal partial tiling. No algorithm has been specifically designed with this purpose in mind; and whether the optimal partial tiling problem for polyominoes can be solved by a linear-time algorithm has been an open problem for the past 15 years. In this paper, we are studying the partial optimal domino tiling problem on the Manhattan class of polyominoes (see figure 1).

* Supported by ANR contract GAMMA, "Génération Aléatoire Modèles, Méthodes et Algorithmes", BLAN07-2 195422.

S. Brlek, C. Reutenauer, and X. Provençal (Eds.): DGCI 2009, LNCS 5810, pp. 79–91, 2009.

Fig. 1. A Manhattan polyomino

A *Manhattan polyomino* is a sequence of adjacent columns that all extend from the same base line ("*Manhattan*" refers to the fact that this polyomino looks like a skyline). We are exhibiting an algorithm that makes solving this problem much more efficient than solving the corresponding maximum matching problem. Indeed, the best known algorithm for maximum matchings in bipartite graphs is in $O(\sqrt{n}m)$ [7,8] (where n is the number of vertices and m the number of edges) and it yields an algorithm in $O(n^{3/2})$ for the partial tiling problem, whereas our algorithm is linear in the number of columns. The choice of the Manhattan class of polyominoes is justified as a generalisation of the seminal result of Bougé and Cosnard [2] which establishes that a trapezoidal polyomino (i.e. a monotic Manhattan polyomino, in which the columns are ordered by increasing height) is tilable by dominoes if and only if it is balanced. The partial tiling problem on more general classes seems to be out of reach at present.

This paper first succinctly recalls some definitions on tilings, polyominoes in general and Manhahattan polyominoes in particular; we list the notations that we will use throughout the paper (section 1). Follows a presentation of the main idea of reducing the partial tiling problem to a network flow problem (section 2); we then prove this reduction to be valid (section 3). A greedy algorithm to solve this specific network problem is given (section 4), which allows us to conclude (section 5).

1 Definitions

An *optimal partial tiling* by dominoes of a polyomino P is a set $\{D_1, \dots, D_k\}$ of vertical or horizontal dominoes placed in the discret plane such that:

(i) $\forall i \in \{1, \dots, k\}\, D_i \subset P$;
(ii) the dominoes $D_1, \dots, D_k$ have mutually disconnected interiors: for every i, j such that $1 \leqslant i \leqslant j \leqslant k$, we have $\mathring{D}_i \cap \mathring{D}_j = \emptyset$ where $\mathring{D}$ is the interior of D for the topology endowed with the euclidean distance on $\mathbb{R}^2$;
(iii) k is maximal: if k', $k' \geqslant k$, and $D'_1, \dots D'_k$ exists with the above two conditions, then $k = k'$.

A *column* is the union of unit squares placed one on top of the other. A *Manhattan polyomino* is the connected union of columns aligned on the same horizontal line.

We will identify a Manhattan polyomino with a tuple: the sequence of the heights of all the columns of the polyomino (ordered from left to right). For example, the polyomino in figure 1 is represented as $(4, 2, 4, 4, 1, 2, 2, 2, 4, 4)$. This representation is clearly more convenient in memory space than, say, the list of all unit squares of a polyomino.

It is convenient to define *inclusion*, and we will do so using the tuple representation: polyomino $P = (p_1, p_2, \ldots, p_n)$ is said to be included in $Q = (q_1, q_2, \ldots, q_n)$ if $\forall i, p_i \leqslant q_i$ (both polyominoes can be padded with zero-sized columns if necessary).

In addition, the unit squares of a polyomino have a chessboard-like coloration; specifically[1], the unit square with position (i, j) (i.e.: unit square $(i, j) + [0, 1[^2$) is *white* if $i + j$ is even and *black* otherwise. A column is said to be *white dominant* (or *black dominant*) if it has more white (or black) unit squares than black (or white) ones.

2 From Polyominoes to Flow Networks

In this section, without loss of generality, we consider the polyominoes are balanced (a polyomino is *balanced* if it contains the same count of white and black unit squares). Indeed, a polyomino can always be balanced by adding the necessary number of isolated unit squares of the lacking color.

Notations. Let $(h_1, \ldots, h_n)$ be a Manhattan polyomino P.

For $i \leqslant j$, we define $X(i, j)$ as the number $\min_{i \leqslant k \leqslant j} \{h_k\}$, the height of the smallest column contained in the subset $\{h_i, \ldots, h_j\}$.

We define $B(P)$ (and respectively $W(P)$) as the set of black unit squares, (or white unit squares) contained in P. Let I_P (and resp. J_P) be the set of indices of black dominant columns (and resp. white dominant columns). We define $s_1, \ldots, s_n$ as the elements of $I_P \cup J_P$ sorted in ascending order.

Let $G = (V, E)$ be a graph, and $S \subset V$. We define $\Gamma(S)$ as the subset of *neighbor vertices* of S, i.e. the vertices $y \in V$ such that there exists $x \in S$ with $(x, y) \in E$ or $(y, x) \in E$.

Construction of the flow network. For each polyomino P, we build a directed graph (*flow network*) that we call F_P:

- its vertex set is $\{s_1, \ldots, s_n\} \cup \{s, t\}$ where s and t are two additional vertices respectively called *source* and *sink* (be mindful of the fact that we use s_i both to refer to the vertices, and to the actual columns);
- its edge set E is defined by:

$$E = \{(s_i, s_{i+1})\, ; i \in \{1, \ldots, n-1\}\} \\ \cup \{(s_{i+1}, s_i)\, ; i \in \{1, \ldots, n-1\}\} \\ \cup \{(s, a); a \in I_P\} \cup \{(a, t); a \in J_P\},$$

[1] And regardless of the basis, as both colors are symmetrical in purpose (i.e.: it doesn't matter whether a given unit square is white or black, so long as the whole polyomino is colored in alternating colors).

and each arc of E is weighted by a function c called *capacity* function defined by:

$$c(e) = \begin{cases} \left\lceil \frac{X(s_i, s_{i+1})}{2} \right\rceil & \text{if } e = (s_i, s_{i+1}) \text{ or } e = (s_{i+1}, s_i) \\ 1 & \text{if } e = (s, a) \text{ with } a \in I_P \\ 1 & \text{if } e = (b, t) \text{ with } b \in J_P \end{cases}$$

where $x \mapsto \lceil x \rceil$ is the ceiling function (which returns the smallest integer not lesser than x).

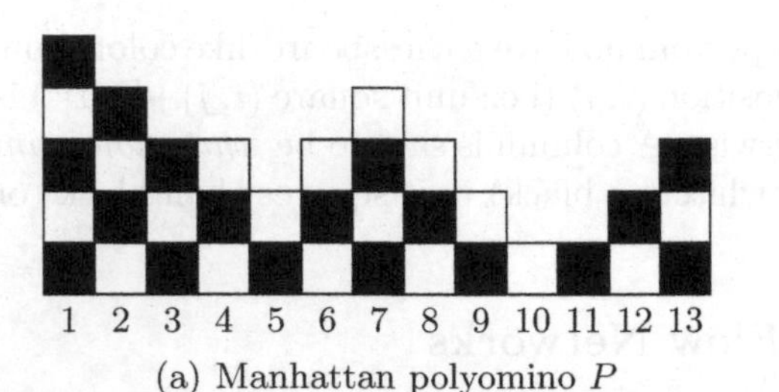

(a) Manhattan polyomino P

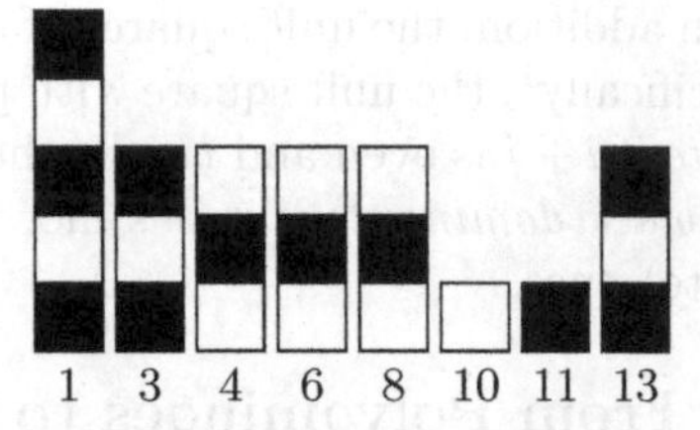

(b) The union $I_P \cup J_P$ of black and white dominant columns (with $s_1 = 1$, $s_2 = 3$ and so on)

Fig. 2. Construction of a $\{s_1, \ldots, s_n\}$ set

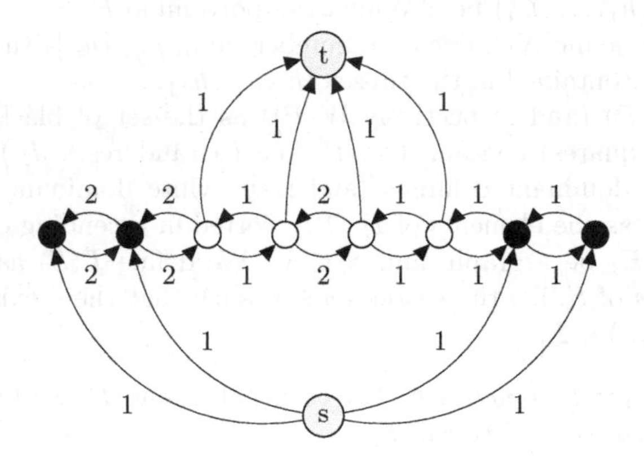

Fig. 3. Flow network associated with the polyomino of figure 2(a)

Example. Figures 2 and 3 illustrate the construction described above. The polyomino P in figure 2(a) contains four black dominant columns and four white dominant columns which have been extracted and are represented on figure 2(b).

This yields the set $I_P \cup J_P = \{s_1 = 1, s_2 = 3, \ldots, s_8 = 13\}$ from which we can build the flow network in figure 3.

For instance, the capacity of arc (s_1, s_2) is $c((s_1, s_2)) = 2$ because the smallest column between the indices $s_1 = 1$ and $s_2 = 3$ is the column 3 and its height is 3 (what this boils down to is $X(1, 3) = 3$).

3 Main Theorem: An Overview

As might have been evident in the previous sections, oddly-sized columns, whether white or black dominant, are what make domino tiling of Manhattan polyominoes an interesting problem: indeed, evenly-sized columns (of a polyomino comprising only such columns) can conveniently be tiled with half their size in dominoes.

3.1 Sketch of the Method

Let P be a polyomino. Our method can intuitively be summarized in several steps:

- we devise a "*planing*" transformation which partially tiles the polyomino P in a locally optimal way (i.e.: such that it is possible to obtain an optimal partial tiling for P by optimally tiling the squares which remain uncovered after the transformation); this transformation, as illustrated in figure 4(c), "*evens out*" a black dominant column and a white dominant column;
- we devise a way to construct a maximum flow problem that translates exactly how the planing transformation can be applied to the polyomino P;
- we prove that the solutions to the maximum flow problem constructed from P, and to the optimal partial tiling of P are closely linked (by a formula).

In truth, these steps are intertwined: the local optimality of the planing transformation is demonstrated using the maximum flow problem.

3.2 The Planing (or Leveling) Transformation

The planing transformation we just mentionned can be expressed as a map ϕ, which can only be applied to a polyomino P which has two *consecutive*[2] oddly-sized columns s_i, s_{i+1} of *different* dominant colors. This condition[3] can be summarized as

$$\exists i, (s_i \in I_P \text{ and } s_{i+1} \in J_P \quad \textbf{or} \quad s_i \in J_P \text{ and } s_{i+1} \in I_P).$$

Let $P = (h_1, \dots h_n)$ be a Manhattan polyomino, and let i be the smallest integer such that s_i, s_{i+1} are two oddly-sized columns of different dominant colors and $c(s_i, s_{i+1}) \neq 0$, then $P' = \phi(P)$ is a new polyomino defined by

$$P' = (h_1, \dots, h_{s_i-1}, a, \dots, a, h_{s_{i+1}+1}, \dots, h_n)$$

where $a = \min(h_{s_i} - 1, h_{s_i+1} - 2, \dots, h_{s_{i+1}-1} - 2, h_{s_{i+1}} - 1)$, i.e.: all columns from s_i to s_{i+1} are *leveled* to height a, which is at least one unit square smaller

[2] By this we mean that the two dominant columns are consecutive in the $s_1, s_2, \dots$ ordering (and this does not necessarily imply that they are adjacent in P); in figure 4(c), dominant columns s_1 and s_2 are consecutive (so are s_2 and s_3).

[3] Wherein it is obvious that the **or** is mutually exclusive.

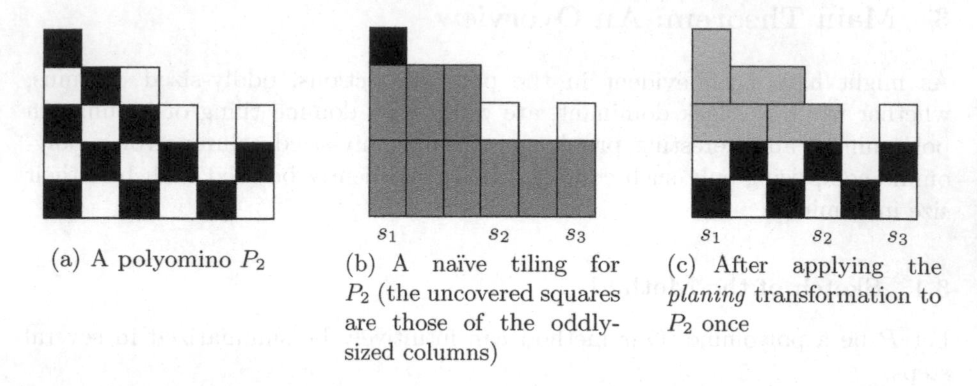

(a) A polyomino P_2

(b) A naïve tiling for P_2 (the uncovered squares are those of the oddly-sized columns)

(c) After applying the *planing* transformation to P_2 once

Fig. 4. Tilings of P_2, which has three oddly-sized columns

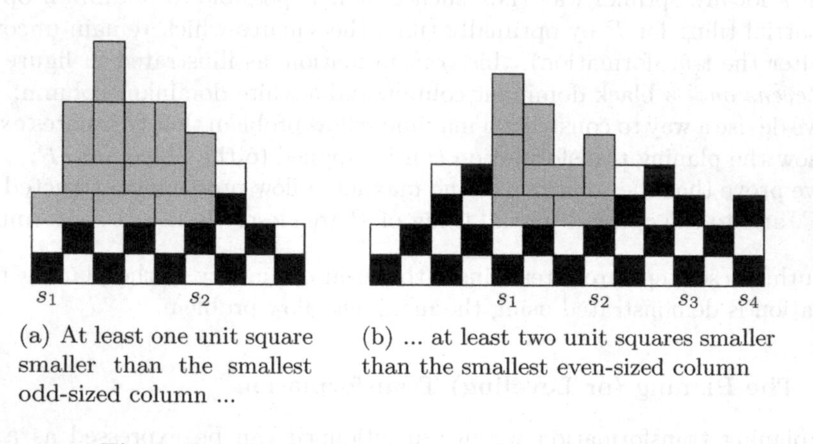

(a) At least one unit square smaller than the smallest odd-sized column ...

(b) ... at least two unit squares smaller than the smallest even-sized column

Fig. 5. Generic examples of the planing transformation

than the smallest of s_i and s_{i+1} and two unit squares smaller than the smallest even-sized column of P from s_i to s_{i+1} .

Since we have only decreased the heights of the columns of P to obtain P', $\phi(P) = P' \subset P$ is trivial (following the definition of inclusion we gave in section 1).

We will show that this transformation is locally optimal: again, this means that if optimal tilings of $\phi(P)$ have x non-covered unit squares, then so do optimal tilings of P (the number of non-covered unit squares is invariant with regards to the transformation).

3.3 Main Theorem

Having constructed a flow network F_P from a given Manhattan polyomino P, we now show that, if we call $d(P)$ the number of non-covered unit squares in an optimal partial tiling of P, then the value $v(P)$ of the maximum flow on F_P is such that:

$$|I_P| + |J_P| - 2v(P) = d(P).$$

The meaning of this equation is quite intuitive: it says that the number of uncovered squares $d(P)$, is exactly the number of single squares that come both from the black dominant columns I_P and from the white dominant columns J_P, from which we withdraw two squares for each of the $v(P)$ times we are able to apply the planing transformation (recall that this transformation evens out both a black dominant column and a white dominant column, hence the 2 factor).

Lemma 1. *Let P be a Manhattan polyomino. For all P', such that $P' = \phi^k(P)$ (the planing transformation applied k times to P), the following invariant holds:*

$$|I_P| + |J_P| - 2v(P) = |I_{P'}| + |J_{P'}| - 2v(P')$$

where $v(P)$ and $v(P')$ are the values of a maximum flow respectively in T_P and $T_{P'}$.

Proof (by induction on k). Let f be a maximum flow on $T_{P_{k-1}}$ which minimizes $\sum_e f(e)$ (i.e.: the flow takes the shortest path from s to t given a choice). Since it is possible to apply the planing transformation to P_{k-1}, then there must be two consecutive vertices s_i and s_{i+1} corresponding to oddly-sized columns of different dominant colors (given several such pairs, we consider that for which i is the smallest).

Let us now build T_{P_k}, the flow network associated with $P_k = \phi(P_{k-1})$: the vertices of T_{P_k} are the vertices of $T_{P_{k-1}}$ from which we remove $\{s_i, s_{i+1}\}$; the arcs of T_{P_K} are the trace of the arcs of the arcs of $T_{P_{k-1}}$ (*trace* is a non-standard term by which we mean all arcs that are incident neither to s_i nor to s_{i+1}) to which we add, should both vertices exist, the arcs (s_{i-1}, s_{i+2}) and (s_{i+2}, s_{i-1}) with capacity

$$\min\left(c\left((s_{i-1}, s_i)\right), c\left((s_i, s_{i+1})\right) - 1, c\left((s_{i+1}, s_{i+2})\right)\right).$$

Suppose s_i and s_{i+1} respectively represent a black and white dominant column (figure 6 illustrates the construction in this case); the other case is symmetrical. Then, because $\sum_e f(e)$ is minimal, maximum flow f routes a non-null amount through the path (s, s_i, s_{i+1}, t).

Thus the maximum flow f on $T_{P_{k-1}}$ induces a maximum flow f' on T_{P_k}, which is defined as the trace of f on T_{P_k} and updated, if necessary, by adding:

$$\begin{cases} f'((s_{i-1}, s_{i+2})) = f((s_{i-1}, s_i)) = f((s_{i+1}, s_{i+2})) \\ f'((s_{i+2}, s_{i-1})) = f((s_i, s_{i-1})) = f((s_{i+2}, s_{i+1})) \end{cases}.$$

Proving that f' is a maximum flow on T_{P_k} is straightforward (suppose there is a better flow, and reach a contradiction). We then immediately remark that $v(P_{k-1}) = v(P_k) + 1$; and because the planing transformation applied to P_{k-1}

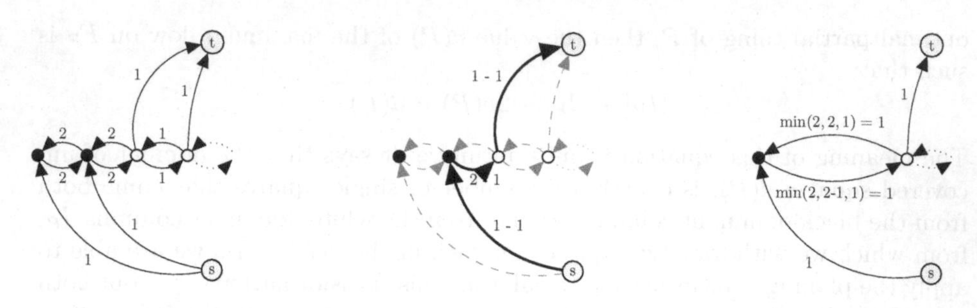

Fig. 6. Building T_{P_k} from $T_{P_{k-1}}$: a max flow for the network of figure 3 necessarily goes through the path (s, s_2, s_3, t). Here we show the impact of removing this path on the flow network's edges, vertices and capacities.

levels both a black dominant column and a white dominant column to obtain P_k, we have $|I_{P_k}| = |I_{P_{k-1}}| - 1$ and $|J_{P_k}| = |J_{P_{k-1}}| - 1$, thus:

$$|I_{P_{k-1}}| + |J_{P_{k-1}}| - 2v(P_{k-1}) = |I_{P_k}| + |J_{P_k}| - 2v(P_k),$$

from which, by induction on k, we derive the lemma's invariant. □

Lemma 2. *Let P be a Manhattan polyomino and $v(P)$ the value of a maximum flow in T_P (cf. section 2). There is a partial tiling of P in which exactly $|I_P| + |J_P| - 2v(P)$ squares are uncovered.*

Proof. We prove the property by induction on the size of P. Let f be a maximum flow in T_P. We consider whether there are two consecutive vertices s_i and s_{i+1} which correspond to oddly-sized columns of different dominant colors, such that the flow of f going from s_i to s_{i+1} is non-null.

Either: no such vertices s_i and s_{i+1} exist[4], in which case we have $v(P) = 0$ (this would mean f is the *zero flow*). Indeed, any path from s to t would use at least one transversal arc (s_i, s_{i+1}) because, by construction, a vertex s_i cannot be simultaneously linked to s and t. We remark that the planing transformation cannot be applied. If we partially tile P, as in figure 4(b), with vertical dominoes alone, leaving one square uncovered per oddly-sized column, we obtain a partial tiling which globally leaves $|I_P| + |J_P|$ squares uncovered—thus proving our property for this case.

Or: such vertices s_i and s_{i+1} exist. We can then apply the planing transformation ϕ, $P' = \phi(P)$. By induction, the property holds for P'. As we've seen, the (union of the) squares removed by the transformation is a domino-tileable Manhattan polyomino (figure 5); and by lemma 1,

$$|I_{P'}| + |J_{P'}| - 2v(P') = |I_P| + |J_P| - 2v(P).$$

□

[4] This case arises when there are no oddly-sized columns, or only oddly-sized columns of a given color.

Theorem 1. *Let P be a Manhattan polyomino, and $d(P)$, the number of uncovered unit squares in an optimal partial tiling of P. The construction outlined in the proof of lemma 2 is optimal, i.e.:*

$$|I_P| + |J_P| - 2v(P) = d(p).$$

Remark 1. We are now going to work on G_P which is the edge-adjacency graph of polyomino P: the vertices of G_P are the unit squares of P, and an edge connects two vertices of G_P if and only if the two corresponding squares share an edge in P. (By contrast, T_P is the flow network constructed in section 2.)

The idea behind the proof of theorem 1 is as follow: we isolate a subset $\mathcal{B}(P)$ of black squares of P, such that the (white) neighbors (in graph G_P) to squares of $\mathcal{B}(P)$ verify the relation

$$|\mathcal{B}(P)| - |\Gamma(\mathcal{B}(P))| = |I_P| + |J_P| - 2v(P).$$

We can then conclude, using Hall's theorem, that $d(P) \geqslant |\mathcal{B}(P)| - |\Gamma(\mathcal{B}(P))|$. To construct $\mathcal{B}(P)$, we use a minimal cut[5] of graph T_P, from which we deduce a list of "*bottlenecks*". These bottlenecks mark the boundary of zones containing squares of $\mathcal{B}(P)$.

Intuitively, once bottlenecks are planed (that is to say when we have tiled them following the template given by our planing transformation), they isolate zones which have either too many white or too many black squares.

3.4 Greedy Algorithm

Having proven that our optimal partial tiling problem can be reduced to a network flow problem, we will now present an algorithm to efficiently solve this specific brand of network flow problem (indeed, a general-purpose network flow algorithm would not take into account those properties of our network which are a consequence of the way it was constructed).

We will first introduce the notion of f-tractability for two vertices of our flow network. The pair (s_i, s_j) (for which we consider that $i < j$) is said to be f-tractable if:

- s_i and s_j correspond to two oddly-sized columns of *different* dominant columns[6];
- if s_i corresponds to a black-dominant column, then the arcs (s_i, s_{i+1}), ..., (s_{j-1}, s_j) are not saturated;

[5] Recall that a cut of graph $G = (V, E)$ is a partition of its vertices in two subsets X_1 and X_2; let E_C be the set of edges such that one vertex is in X_1 and the other in X_2: the value of the cut is the sum of the capacity of each edge in E_C; a minimal cut is a cut which minimizes this sum. **In the case of flow networks**, a cut separates the source s from the sink t.

[6] Recall that this is *almost* the condition under which the planing transformation can be applied: the planing transformation requires, in addition, that s_i and s_j be so that $j = i + 1$ —per the ordering which have defined in section 2.

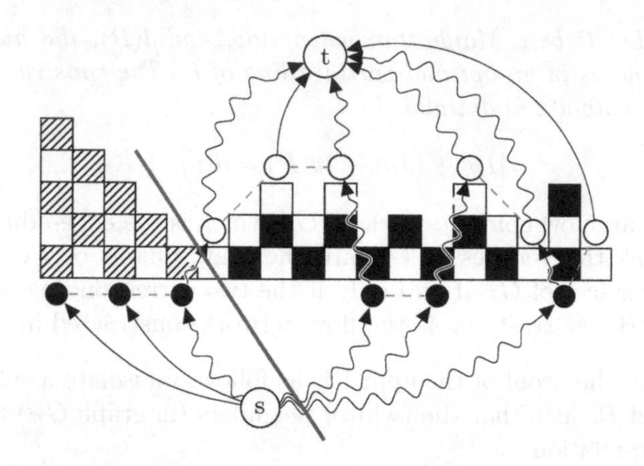

Fig. 7. A polyomino P, its flow network T_P (the transversal arcs are only hinted at, with dashed lines; the max flow is represented with snake-lines), and, in red, a minimal cut of T_P which separates P in two zones (each zone is connected here, but that may not be the case). The black striped squares on the left belong to $\mathcal{B}(P)$: in accordance with Hall's theorem, $|\mathcal{B}(P)|-|\Gamma(\mathcal{B}(P))| = 9-7 = 2$ is the number of uncovered squares in the optimal partial tiling.

– or else, if s_i corresponds to a white-dominant column, then the arcs (s_j, s_{j-1}), ..., (s_{i+1}, s_i) are not saturated;
– for every k, $i < k < j$, if $s_k \in I_P$ (resp. $s_k \in J_P$) then (s, s_k) (resp. (s_k, t)) is saturated.

This might seem like an elaborate notion, but in fact, it only translates whether, in terms of flow increase, we can apply our transformation.

We are going to prove that the following algorithm builds a maximum flow:

1. we begin with a null flow f;
2. while there exists two vertices s_i and s_j with $i < j$ which are f-tractable:
   (a) we take the first pair of vertices in lexicographic order which is f-tractable,
   (b) and we augment the flow f by 1 on the arcs belonging to the path $[s, s_i, s_{i+1}, \ldots, s_j, t]$ or $[s, s_j, s_{j-1}, \ldots, s_i, t]$ (depending on the dominant color of s_i);
3. we return the flow f.

Proof (Correctness of the greedy algorithm). Let (s_i, s_{i+1}) be the first pair of vertices in lexicographic order which, at the beginning of the algorithm, is f-tractable. We suppose here that s_i corresponds to a black dominant column (the other case is symetrical). We want to prove that there is a maximum flow f, which minimizes $\sum_{v \in T_P} f(v)$, such that none of the following values are null: $f((s, s_i)), f((s_i, s_{i+1})), f((s_{i+1}, t))$. If a maximum flow following this last condition does not exist, then one of those three arcs must be null and another one must be saturated, otherwise the flow f can be augmented. Now, by the minimality of $\sum_{v \in T_P} f(v)$, it follows that (s, s_i) is not saturated (otherwise we can remove 1 to f on every arc belonging to a minimal path from s to t using (s, s_i) with non-zero flow, and add 1 to $f((s, s_i)), f((s_i, s_{i+1})), f((s_{i+1}, t))$. We obtain a maximum flow which contradicts the minimality of $\sum_{v \in T_P} f(v)$.) The arc (s_i, s_{i+1}) cannot be saturated for the same reason. So, (s_{i+1}, t) is saturated. But in this case, as (s, s_i) and (s_i, s_{i+1}) are not saturated, we can remove 1 to f on every arc belonging to a non-zero flow path from s to t using (s_{i+1}, t), and add 1 to $f((s, s_i)), f((s_i, s_{i+1})), f((s_{i+1}, t))$. We again obtain a new maximum flow with total weigth at most $\sum_{v \in T_P} f(v)$. That proves what we want. Now, the next recursive steps of the algorithm are exactly proved by the same way but on transport networks obtained by convenient diminution of the arc capacities.

To briefly but correctly analyze the complexity of this algorithm, we have to describe the representation of the entry. If the Manhattan polyomino is given by the list of its column heights, it can be stored in $O(l \log(\max(h_i)))$ bits where l is the number of columns. We can consider two steps. The first one consists in obtaining the transport network T_P from P. This can be done in one pass and we make $O(l)$ elementary operations (min, division by 2). The second part consists in solving the transport networks problem. Using stacks, the algorithm can solve the problem making a unique pass on the vertices of T_P. So, its complexity is in $O(|I_P|)$. Thus, globally the algorithm works in $O(l)$ elementary operations. If we make the hypothesis that the number of columns is approximatively the average of the height of the columns. Our algorithm is clearly sublinear in n the number of unit squares. Moreover, if we consider the complexity according to the size of the input, this algorithm is linear.

4 Conclusion

In this paper, we have illustrated that particular optimal partial domino tiling problems can be solved with specific algorithms. We have not yet been able to evidence a linear algorithm for the general problem of partial domino tiling of polyominoes without holes. This appears to be an attainable goal, as [13] has devised such an algorithm for the exact domino tiling problem.

Currently, the best known complexity for the general partial domino tiling problem is $O(n \cdot \sqrt{n})$ using a *maximum matching algorithm in bipartite graphs*. To improve this bound, it seems irrelevant to obtain an analogue to the notion of height which is a cornerstone of the tiling algorithms. Nevertheless, it can be proven that two optimal partial tilings are mutually accessible using extended flips (which contain classical flips [1,11] and taquin-like moves). This fact involves in a sense a weak form of "lattice structure" over the set of optimal partial tilings. This is relevant and could be taken into account to optimize the algorithms.

References

1. Bodini, O., Latapy, M.: Generalized Tilings with Height Functions. Morfismos 7 3 (2003)
2. Bougé, L., Cosnard, M.: Recouvrement d'une pièce trapézoidale par des dominos, C. R. Académie des Sciences Paris 315, Série I, pp. 221–226 (1992)
3. Bodini, O., Fernique, T.: Planar Tilings. In: Grigoriev, D., Harrison, J., Hirsch, E.A. (eds.) CSR 2006. LNCS, vol. 3967, pp. 104–113. Springer, Heidelberg (2006)
4. Conway, J.H., Lagarias, J.C.: Tiling with polyominoes and combinatorial group theory. JCT Series A 53, pp. 183–208 (1990)
5. Ford, L.R., Fulkerson, D.R.: Maximal Flow through a Network. Canadian Journal of Mathematics, 399 (1956)
6. Fournier, J.C.: Tiling pictures of the plane with dominoes. Discrete Mathematics 165/166, 313–320 (1997)
7. Gabow, H.N., Tarjan, R.E.: Faster scaling algorithms for network problems. SIAM J. Comput. 18(5), 1013–1036 (1989)
8. Hopcroft, J.E., Karp, R.M.: An $n^{5/2}$ algorithm for maximum matchings in bipartite graphs. SIAM J. Comput. 2(4), 225–231 (1973)
9. Ito, K.: Domino tilings on planar regions. J.C.T. Series A 75, pp. 173–186 (1996)
10. Kenyon, R.: A note on tiling with integer-sided rectangles. JCT Series A 74, pp. 321–332 (1996)
11. Rémila, E.: The lattice structure of the set of domino tilings of a polygon. Theoretical Computer Science 322(2), 409–422 (2004)
12. Thiant, N.: An O (n log n)-algorithm for finding a domino tiling of a plane picture whose number of holes is bounded. Theoretical Computer Science 303(2-3), 353–374 (2003)
13. Thurston, W.P.: Conway's tiling groups. American Mathematics Monthly 95, 757–773 (1990)

A Annexe (Proof of Theorem 1)

So, we have proved that on the sequence of Manhattan polyominoes obtained by successive planing transformations, the function $|I_{\_}| + |J_{\_}| - 2v(\_)$ is invariant. We need another invariant to conclude that $|I_P|+|J_P|-2v(P) \leq d(P)$. It follows from a list of "bottlenecks" which prevent the matching of columns.

We denote by $P_{\text{red}} = (h_1^{\text{red}}, ..., h_n^{\text{red}})$ the Manhattan polyomino obtained from P after $v(P)$ planing transformations.

We can suppose that the least x such that (x, x') belongs to Y_P (the set of edges of F_P that crosses the minimal cut) corresponds to a black dominant column (if this is not the case, we can inverse the colors). Let $g(x, x')$ be the smallest $k \in \{x, \dots, x'\}$ such that $h_k = X(x, x')$. We denote by $\mathcal{G} = \{g(x, x'); (x, x') \in Y_P\}$ the *list of bottlenecks* of P and we sort the elements of $\mathcal{G}$ in ascending order. $\mathcal{G} = (g_1^0, \dots, g_k^0)$. Now, if i is odd (resp. even) and is the index of a white dominant column, we put $g_i = g_i^0 - 1$ (resp. $g_i = g_i^0 + 1$) otherwise we put $g_i = g_i^0$. With the hypothesis, we can observe that $(h_1, h_2, \dots, h_{g_1})$ has more black unit squares than white ones. We put

$$H_P = B\left((h_1, h_2, \ldots, h_{g_1})\right) \cup B\left((h_{g_2}, \ldots, h_{g_3})\right) \cup \ldots$$

and

$$K_P = W\left((h_1, h_2, \ldots, h_{g_1})\right) \cup W\left((h_{g_2}, \ldots, h_{g_3})\right) \cup \ldots$$

We have $2(|H_P| - |\Gamma(H_P)|) \leq d(P)$ and we want to prove that $2(|H_P| - |\Gamma(H_P)|) = |I_P| + |J_P| - 2v(P)$. In order to do that, let P' be the Manhattan polyomino obtained from P by a planing transformation as we have processed previously. Firstly, we are going to show that $|H_P| - |\Gamma(H_P)| = |H_{P'}| - |\Gamma(H_{P'})|$ where

$$H_{P'} = B\left((h'_1, h'_2, ..., h'_{g_1})\right) \cup B\left((h'_{g_2}, ..., h'_{g_3})\right) \cup ...$$

We put $\mathcal{B}(H_P)$ (resp. $\mathcal{W}(H_P)$) the number of odd black dominant columns (resp. white) in P which belongs to the columns of indices in $E = \{1, 2, ...g_1\} \cup \{g_2, ..., g_3\} \cup ....$ An easy observation allows us to see that

$$|H_P| - |\Gamma(H_P)| = |H_P| - |K_P| - \sum_{v \in Y_P} c(v)$$
$$= \mathcal{B}(H_P) - \mathcal{W}(H_P) - \sum_{v \in Y_P} c(v)$$

When we apply the planing transformation on P, we delete :

- a black dominant column in E and a white column in E^c and $\sum_{v \in Y_{P'}} c(v) = \sum_{v \in Y_P} c(v) - 1$,
- a black dominant and a white dominant column in E or a black dominant and a white column in E^c and $\sum_{v \in Y_{P'}} c(v) = \sum_{v \in Y_P} c(v)$.

This fact proves that $|H_{\_}| - |\Gamma(H_{\_})|$ is invariant on every sequence of Manhattan polyominos obtained by successive planing transformations.

Now, let us remark that

$$|H_{P_{\text{red}}}| - |\Gamma(H_{P_{\text{red}}})| = \mathcal{B}(H_{P_{\text{red}}}) = |I_{P_{\text{red}}}|$$

$(\mathcal{W}(H_{P_{\text{red}}}) = \sum_{v \in Y_{P_{\text{red}}}} c(v) = 0)$ and that $|I_{P_{\text{red}}}| = |I_P| - v(P)$. Indeed, we have made $v(P)$ planing transformations and each of them reduces by 1 the number of black dominant columns.

So, $|H_P| - |\Gamma(H_P)| = |H_{P_{\text{red}}}| - |\Gamma(H_{P_{\text{red}}})| = |I_P| - v(P)$.

Moreover, we have assumed initially that the polyomino is balanced. So, $|I_P| = |J_P|$. Consequently, we have proved that $2(|H_P| - |\Gamma(H_P)|) = |I_P| + |J_P| - 2v(P)$. Thus, we can conclude that $|I_P| + |J_P| - 2v(P) \leq d(P)$.

An Improved Coordinate System for Point Correspondences of 2D Articulated Shapes*

Adrian Ion, Yll Haxhimusa, and Walter G. Kropatsch

Pattern Recognition and Image Processing Group,
Faculty of Informatics, Vienna University of Technology, Austria
{ion,yll,krw}@prip.tuwien.ac.at

Abstract. To find corresponding points in different poses of the same articulated shape, a non rigid coordinate system is used. Each pixel of each shape is identified by a pair of distinct coordinates. The coordinates are used to address corresponding points. This paper proposes a solution to a discretization problem identified in a previous approach. The polar like coordinate system is computed in a space where the problem cannot occur, followed by mapping the computed coordinates to pixels.

Keywords: eccentricity transform, discrete shape, coordinate system.

1 Introduction

Shape as well as other perceptual properties such as depth, motion, speed and color are important for object detection and recognition. What makes shape a unique perceptual property is its sufficient complexity to make object detection and recognition possible [1]. In many cases object recognition is possible even without taking other perceptual properties into account. The animals in Fig. 1 can be recognized by humans without the need of other perceptual properties.

Object recognition has to be robust against scaling, rotation, and translation, as well as against some deformation (e.g. partial occlusion, occurrence of tunnels, missing parts etc). In order to recognize objects based on their shape we assume that shapes have enough 'complexity' to allow the discrimination between different objects. This 'complex' measure has to be small for intra-class shapes (eg. shapes of an elephant), and big for inter-class shapes (e.g. shapes of elephants vs deers). Usually, the measure is not computed on the shapes directly, but on the result of an image transformation. One class of image transforms that is applied to shapes, associates to each point of the shape a value that characterizes in some way its relation to the rest of the shape. This value can be the distance to other points of the shape (e.g. landmarks). Examples of such transforms include the distance transform [2, 3], the Poisson equation [4], the global geodesic function [5], and the eccentricity transform [6].

Having a measure for the shapes, matching methods can be employed to decide which shapes are 'similar'. Many matching methods use a similarity value to 'find

* Partially supported by the Austrian Science Fund under grants S9103-N13, P18716-N13.

S. Brlek, C. Reutenauer, and X. Provençal (Eds.): DGCI 2009, LNCS 5810, pp. 92–103, 2009.

Fig. 1. Shapes of elephants and deers (taken from MPEG7 CE Shape-1 Part B)

similar shapes' (e.g. [7–9]). Similarity values are well suited to find the class of a shape, like the elephant class, based on labeled examples.

Assuming that we already have a shape class (e.g. elephants), and we are interested in finding correspondence between parts/points in shapes of the same class (e.g. trunk of an elephant), similarity values cannot be used. Some matching methods produce correspondences of the used signature, usually border points/parts (e.g. [10–12]), but finding all point correspondences based on the obtained information is in most of the cases not straightforward or even impossible. In [7], a triangulation of the shape is used as a model, which could be used to find corresponding points, but an a priori known model is still needed. The task of finding correspondences of all points of the shape, is similar to the non-rigid registration problem used in the medical image processing community [13]. Differences include the usage of gray scale information to compute the deformation vs. the usage of a binary shape and, the registration of a whole image (in most cases) vs. the registration of a (in this approach) simply connected 2D shape. In the surface parametrization community [14] a coordinate system for shapes is defined, but articulation is not considered. In [15], for small variations, correspondences between points of 3D articulated shapes are found. Recently shape matching has also moved toward decomposition and part matching, e.g. [9], mainly due to occlusions, imperfect segmentation or feature detection.

The approach presented in this paper is similar in concept with the one in [16], but can handle cases where the method in [16] is not defined. We are interested in finding point correspondence of binary shapes of objects of the same class. It is mainly motivated by observations like: 'one might change his aspect, alter his pose, but the wristwatch is still located in the same place on the hand' (i.e. shapes under an articulation create a class). We map a coordinate system to an articulated shape, with the purpose of addressing the corresponding point (or a close one) in other instances of the same shape. Our 2D polar like coordinate system is created on the Euclidean eccentricity transform [6]. Each pixel of each shape will be uniquely identified by a pair of coordinates, a 'radius' and an 'angle' like coordinate. Discretization results in the problems encountered in [16], where for some pixels the coordinates cannot be computed as intended. In this paper we propose a solution to this problem, and identify each pixel by a distinct pair of coordinates.

The structure of the paper is as follows. In Section 2 we define the coordinate system and describe shortly the problem of finding correspondence of points. Section 3 recalls the eccentricity transform and its properties relevant for this paper. In Section 4 we recall the coordinate system directly mapped to the pixels of the shape [16]. Section 5 describes the proposed solution, with the experiments given in Section 5.2. Section 6 concludes the paper.

2 Coordinate System: Definition, Point Correspondences

Coordinate System: a system of *curvilinear coordinates* [17] is composed of intersecting surfaces. If all intersections are at angle $\pi/2$, then the coordinate system is called *orthogonal* (e.g. polar coordinate system). If not, a *skew* coordinate system is formed. Two classes of curves are needed for a planar system of curvilinear coordinates, one for each coordinate.

The problem of mapping a coordinate system to a shape $\mathcal{S}$ is equivalent with defining the following mapping:

Definition 1. (Coordinate map) *A 2D coordinate map, is an injective map* $coord(\mathbf{p}) : \mathcal{S} \rightarrow \mathbb{R} \times \mathbb{R}$ *that associates to each point a pair of coordinates, i.e.*

1. *(defined)* $\forall \mathbf{p} \in \mathcal{S}$, $coord(\mathbf{p})$ *is defined;*
2. *(distinct)* $\forall \mathbf{p}, \mathbf{q} \in \mathcal{S}$, $\mathbf{p} \neq \mathbf{q}$, *we have* $coord(\mathbf{p}) \neq coord(\mathbf{q})$;
3. *(single)* $\forall \mathbf{p} \in \mathcal{S}$, $coord(\mathbf{p})$ *returns a single pair of coordinates.*

Finding point correspondences: Consider the **continuous shapes** $\mathcal{S}_1$ and $\mathcal{S}_2$, and the continuous coordinate maps $coord_1$ and $coord_2$ corresponding to $\mathcal{S}_1$ respectively $\mathcal{S}_2$. Given a point $\mathbf{p} \in \mathcal{S}_1$ its corresponding point in $\mathcal{S}_2$ is $\mathbf{q} \in \mathcal{S}_2$ s.t. $coord_2(\mathbf{q}) = coord_1(\mathbf{p})$.

For $\mathcal{S}_1$ and $\mathcal{S}_2$ two **discrete shapes**, and the discrete maps $dcoord_1, dcoord_2$, a point $\mathbf{q} \in \mathcal{S}_2$ with the exact same coordinates as $\mathbf{p} \in \mathcal{S}_1$ might not exist. The corresponding point $\mathbf{q}$ is defined to be the one that has its coordinates 'closest' to $dcoord_1(\mathbf{p})$. As the local variation of the two coordinates might be very different and also depend on both coordinates of $\mathbf{q}$, just using the ℓ_2-norm $||dcoord_1(\mathbf{p}) - dcoord_2(\mathbf{q})||$ to find the 'closest' coordinates is not the best solution. For example, for polar coordinates the variation of the angular coordinate of two neighboring pixels having the same r can be as large as the variation of the 'radial' coordinate between e.g. 10 pixels (the modified ℓ_2-norm $||(\Delta r, r \cdot \Delta\theta)||$ cannot be used in our case, as the variation of one coordinate depends on both coordinate values).

We employ a two step scheme (Section 4 gives details): given a point $\mathbf{p} \in \mathcal{S}_1$ with $dcoord_1(\mathbf{p}) = (c_1, c_2)$:

1. use the **first coordinate** to find points $\mathbf{q} \in \mathcal{S}_2$ that have 'approximately' the same 'first' coordinate $dcoord_2(\mathbf{q}) = (c_1, c)$; and
2. use the **second coordinate** to select from the above the point that minimizes the difference in the second coordinate $|c - c_2|$.

3 The Eccentricity Transform

The definitions and properties follow [6, 18]. Let the shape $\mathcal{S}$ be a closed set in $\mathbb{R}^2$ (a 2D shape) and $\partial\mathcal{S}$ be its boundary[1]. A path π is the continuous mapping from the interval $[0,1]$ to $\mathcal{S}$. Let $\Pi(\mathbf{p}_1,\mathbf{p}_2)$ be the set of all paths within the set $\mathcal{S}$, between two points $\mathbf{p}_1,\mathbf{p}_2 \in \mathcal{S}$.

The *geodesic distance* $d(\mathbf{p}_1,\mathbf{p}_2)$ between $\mathbf{p}_1,\mathbf{p}_2$ is defined as the length $\lambda(\pi)$ of the shortest path $\pi \in \Pi(\mathbf{p}_1,\mathbf{p}_2)$, more formally:

$$d(\mathbf{p}_1,\mathbf{p}_2) = \min\{\lambda(\pi(\mathbf{p}_1,\mathbf{p}_2)) \mid \pi \in \Pi\} \tag{1}$$

$$\text{where } \lambda(\pi) = \int_0^1 |\dot{\pi}(t)|dt\,, \tag{2}$$

$\pi(t)$ is a parametrization of the path from $\mathbf{p}_1 = \pi(0)$ to $\mathbf{p}_2 = \pi(1)$, and $\dot{\pi}(t)$ is the differential of the arc length. A path $\pi \in \Pi(\mathbf{p}_1,\mathbf{p}_2)$ with $\lambda(\pi) = d(\mathbf{p}_1,\mathbf{p}_2)$ is called a *geodesic*.

The *eccentricity* of a point $\mathbf{p} \in \mathcal{S}$ is defined as:

$$ECC(\mathcal{S},\mathbf{p}) = \max\{d(\mathbf{p},\mathbf{q}) \mid \mathbf{q} \in \mathcal{S}\}. \tag{3}$$

The *eccentricity transform* $ECC(\mathcal{S})$ associates to each point of $\mathcal{S}$ the value $ECC(\mathcal{S},\mathbf{p})$ i.e. to each point $\mathbf{p}$ it assigns the length of the geodesic path(s) to the points farthest away.

In this paper, the class of 4-connected, planar, and simply connected discrete shapes $\mathcal{S}$ defined by points on the square grid $\mathbb{Z}^2$ are considered. Paths are contained in the area of $\mathbb{R}^2$ defined by the union of the support squares for the pixels of $\mathcal{S}$. The distance between any two pixels whose connecting segment is contained in $\mathcal{S}$ is computed using the ℓ^2-*norm*.

Fig. 2 shows two hand shapes (taken from the Kimia99 database [19]) and their eccentricity transform. Note the differences on the fingers of the shapes.

Fig. 2. Eccentricity transform of two shapes from the hand class. Geodesic center marked with '●'. (gray values are the eccentricity value modulo a constant.)

[1] It can be generalized to any continuous and discrete space.

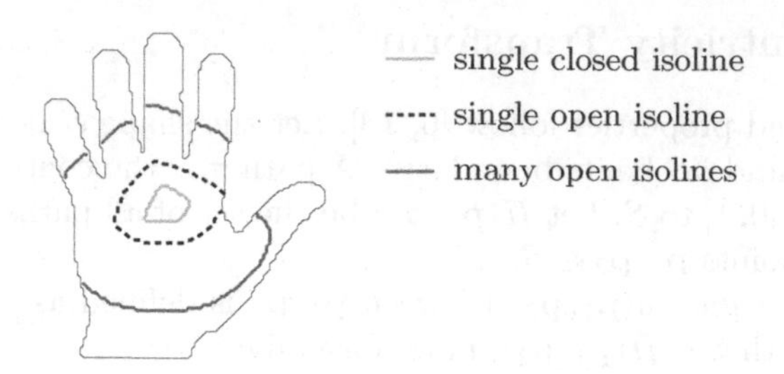

Fig. 3. Level sets made of different configurations of isolines

Terminology: An *eccentric point* of a point $\mathbf{p} \in \mathcal{S}$ is a point $\mathbf{e} \in \mathcal{S}$ s.t. $d(\mathbf{p}, \mathbf{e}) = ECC(\mathcal{S}, \mathbf{p})$. An *eccentric point* of a shape $\mathcal{S}$ is a point $\mathbf{e} \in \mathcal{S}$ that is eccentric for at least one point $\mathbf{p} \in \mathcal{S}$ i.e. $\exists \mathbf{p} \in \mathcal{S}$ such that $ECC(\mathcal{S}, \mathbf{p}) = d(\mathbf{p}, \mathbf{e})$.

The *geodesic center* $\mathcal{C} \subseteq \mathcal{S}$ is the set of points with the smallest eccentricity i.e. $\mathbf{c} \in \mathcal{C}$ iff $ECC(\mathcal{S}, \mathbf{c}) = min\{ECC(\mathcal{S}, \mathbf{p}) \mid \forall \mathbf{p} \in \mathcal{S}\}$.

Properties: If $\mathcal{S}$ is simply connected, the geodesic center $\mathcal{C}$ is a single point. Otherwise it can be a disconnected set of arbitrary size (e.g. for $\mathcal{S}$ = the points on a circle, all points are eccentric and they all make up the geodesic center). Also, for $\mathcal{S}$ simply connected and planar, $ECC(\mathcal{S})$ has a single local/global minimum which is $\mathcal{C}$ [20], and the shapes $\mathcal{P} = \{\mathbf{p} \in \mathcal{S} \mid ECC(\mathcal{S}, \mathbf{p}) \leqslant k\}$ for $k \in [\min(ECC(\mathcal{S})), \max(ECC(\mathcal{S}))]$ are geodesically convex in $\mathcal{S}$ (proof in [20]).

Definition 2. (Level Set [17]) *The level set of a function $f : \mathbb{R}^n \to \mathbb{R}$, corresponding to a value h, is the set of points $\mathbf{p} \in \mathbb{R}^n$ such that $f(\mathbf{p}) = h$.*

For $\mathcal{S} \in \mathbb{R}^2$, the level sets of $ECC(\mathcal{S})$ can be a closed curve or a set of disconnected open curves. The connected components of the level sets are called *isolines* (Fig. 3).

Due to using geodesic distances, the variation of ECC is bounded under articulated deformation to the width of the 'joints' [10]. The transform is robust with respect to salt & pepper noise, and the positions of eccentric points and geodesic center are stable [18].

Computation: In [6] different methods to compute $ECC(\mathcal{S})$ are presented.

4 The Pixel Mapped Coordinate System

This section recalls the method in [16] and shows the cases where it fails to produce results.

The proposed coordinate system is intuitively similar to the polar coordinate system, but forms a skew coordinate system. It is defined for simply connected, planar 2D shapes, and has two coordinates denoted by r and θ.

The origin of the coordinate system is taken to be the geodesic center of the shape $\mathcal{S}$ and has $r = 0$ (Fig. 2). r is a linear mapping from $ECC(\mathcal{S})$ and increases as ECC increases. The isolines of r correspond to the isolines of $ECC(\mathcal{S})$. Closed isolines of r have corresponding θ values of $[0, 360)$ mapped along the isoline. For each pixel of $\mathcal{S}$ the 'second' coordinate θ is mapped as described in the following. Note that θ is not really an angle, just denoted intuitively so.

The coordinate system is computed as follows:

1. extract 'discrete level sets' of the ECC of $\mathcal{S}$, made out of at least 8 connected components, using:
$$h \leqslant ECC < h+1,$$
for $h = \lfloor \min(ECC) \rfloor, \lfloor \min(ECC) \rfloor + 1, ..., \lfloor \max(ECC) \rfloor$ (Fig. 4a);
2. map the value
$$r = (h - \lfloor \min(ECC) \rfloor)/(\lfloor \max(ECC) \rfloor - \lfloor \min(ECC) \rfloor),$$
to each 'discrete level set';
3. decompose $\mathcal{S}$ into simply connected regions s.t.:
   - in each region there is at most one connected component of the 'discrete level sets' of $ECC(\mathcal{S})$;
   - all isolines in the same region are either closed or all are open;
   - the number of regions is minimal.

   This decomposition produces regions $\mathcal{R}$ where 'discrete level sets' correspond to all ECC values from some $e_1, e_1 + 1, ..., e_2$, with $\min(ECC(\mathcal{S})) \leqslant e_1 \leqslant e_2 \leqslant \max(ECC(\mathcal{S}))$, and for each e, $e_1 \leqslant e \leqslant e_2$, the pixels in $\mathcal{R}$ corresponding to e form a single at least 8 connected set;
   The region that contains the geodesic center and where all isolines are closed paths is called the *center region.*
   For example, the isolines of the level sets drawn in Fig. 3 belong to different regions.
4. use a precomputed geodesic, called the *zero path,* that connects the geodesic center with the border of the shape, to choose the pixel with $\theta = 0$ on each isoline of the *center region* i.e. the region where all isolines are closed (Fig. 4b). This geodesic can for example connect the geodesic center with a point having maximum eccentricity (Fig. 9), but could also use corresponding border points obtained by a matching method (e.g. [10]), or can be chosen to maximize the number of matched pixels between two given shapes $\mathcal{S}_1, \mathcal{S}_2$.
5. map angle values $[0, 360)$ to each closed isoline in the center region: $\theta = 0$ was computed before, the other values are given proportional to the arc length, along the same orientation (Fig. 4b);
6. for each region:
   - the 'start' and 'end' θ of the isolines $(\theta_{st}, \theta_{en})$ in the region are obtained by projecting the end points of the isoline with the smallest ECC to the isoline with the highest ECC in the adjacent region with θ already computed (Fig. 5a).
   - map angle values $[\theta_{st}, \theta_{en})$ to each isoline: θ_{st} and θ_{en} is given to the end points of each isoline, the other values are given proportional to the arc length (Fig. 5b);

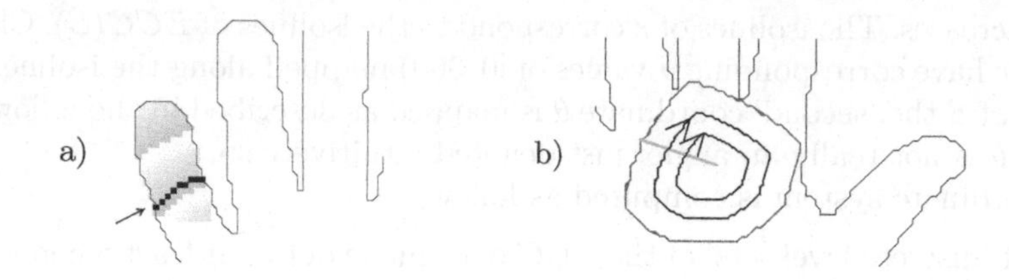

Fig. 4. Map the coordinate system: a) discrete isolines, b) zero θ in the center region

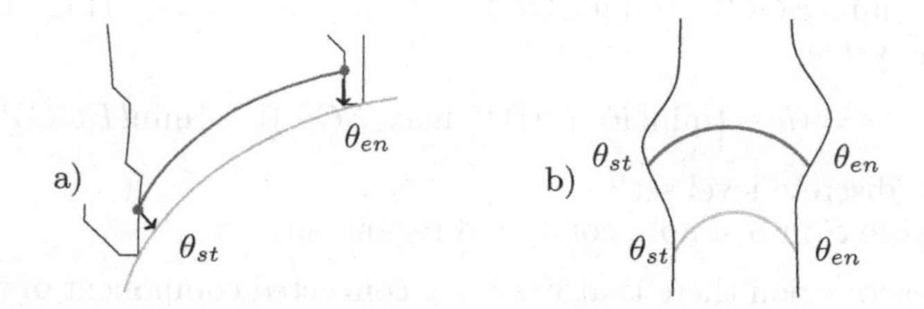

Fig. 5. Start and end theta θ_{st}, θ_{en} for isolines outside the center region: a) project from neighboring region, b) keep constant inside region

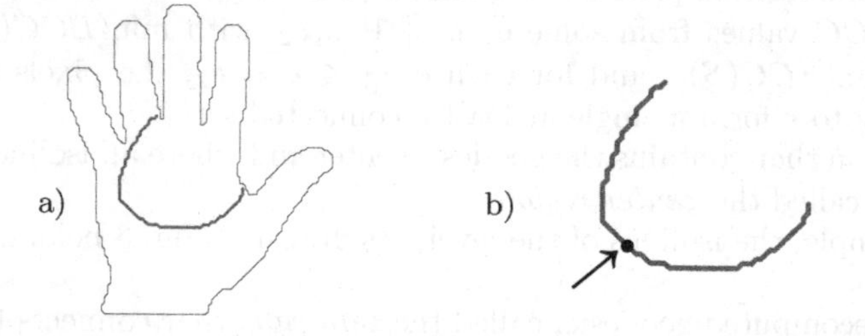

Fig. 6. Finding point correspondences: a) identify level set, b) find 'closest' point

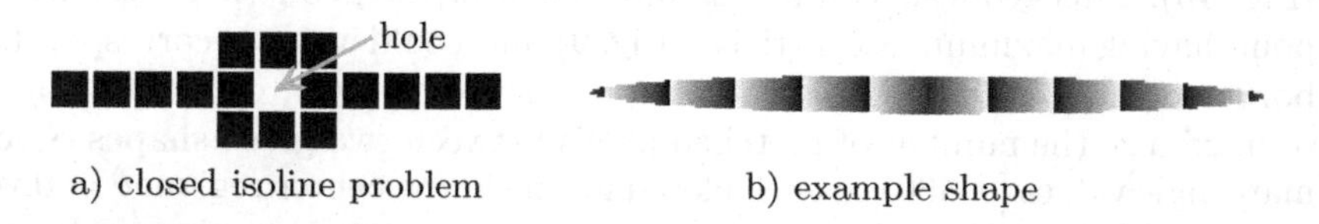

Fig. 7. Discretization problem

Finding point correspondences: To find the point $\mathbf{q} \in \mathcal{S}$ given the coordinates (c_r, c_θ) we proceed as discussed in Section 2 (Fig. 6):

1. $\mathcal{L}$ = 'discrete level set' corresponding to r;
2. find $\arg\min_{\mathbf{q} \in \mathcal{L}} \{|c'_\theta - c_\theta|\}$, where c'_θ is the θ value associated to $\mathbf{q}$.

r is chosen as the 'first coordinate' (see Section 2) because it has a constant variation and the obtained 'discrete level set' does not contain more than one pixel with the same θ. Due to its non-constant variation, choosing θ as the 'first coordinate' could produce many pixels with the same r ('thick' discrete level sets) or have r values for which no pixel was selected (disconnected level sets).

4.1 Problems due to Discretization

A reference point is a point $\mathbf{m} \in \partial\mathcal{S}$ s.t. $\mathbf{m} \in \pi(\mathbf{p}, \mathbf{e})$, where $\mathbf{e}$ is an eccentric point of $\mathbf{p}$, $\mathbf{m} \neq \mathbf{p}$, $\pi(\mathbf{p}, \mathbf{e})$ is a geodesic from $\mathbf{p}$ to $\mathbf{e}$, and $d(\mathbf{p}, \mathbf{m})$ is minimal. Reference points have the property that $\pi(\mathbf{p}, \mathbf{m})$ is a line segment and all points at the same distance to $\mathbf{e}$ that have $\mathbf{m}$ as a reference point, will lie on a circle centered at $\mathbf{m}$.

The isolines of $ECC(\mathcal{S})$ are made out of circle arcs, corresponding to different *reference points*. Depending on the position and distance to the reference points, these circle arcs can meet under very sharp angles. In the case of the center region, isolines are closed in the continuous domain. For the 'discrete level set' the pixels are at least 8 connected, but an 8 connected closed path, where each pixel would be passed through only once might not exist. Fig. 7 shows an example. In this case an ordering cannot be made between the pixels, and mapping the θ coordinate as mentioned before is not possible.

5 The Inter-pixel Coordinate System

Instead of extracting 'discrete level sets' made out of pixels, as mentioned in Section 4, we propose to extract 'inter-pixel' isolines. These isolines are polygonal lines made out of points on the line segments that connect the pixel centers. The inter-pixel isolines can be efficiently extracted using *marching squares*, the 2D version of the classical *marching cubes* [21]. The eccentricity of the points on these line segments is estimated by linear interpolation. We extract inter-pixel isolines of the ECC of $\mathcal{S}$ for the eccentricity values:

$$\min(ECC) + 0.5, \min(ECC) + 1.5, ...$$

Inside the center region the extracted inter-pixel isolines will be *non self-intersecting, convex polygonal lines. Outside the center region*, where level sets get disconnected, each inter-pixel isoline will be a *polygonal line with all angles on the side with smaller ECC, less or equal to 180 degrees.* Fig. 8 shows examples.

The inter-pixel isolines can be seen as connected paths where the ordering of the points is given and unique, and thus we can map the new coordinate system as follows:

1. extract inter-pixel isolines with marching squares;
2. map coordinate system as in Section 4 on the inter-pixel isolines extracted;
3. map the θ values from the isolines to the pixels (discussed below).

5.1 Mapping θ Values

The coordinate system has been mapped to the inter-pixel isolines. All vertices of the isolines correspond to points with the same r and, each vertex has a unique

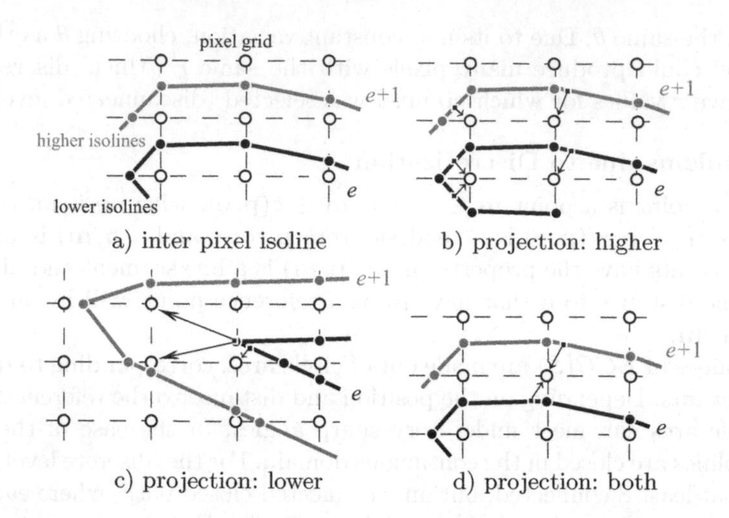

Fig. 8. Inter-pixel isolines. Projection of r, θ from the isoline to the pixels.

θ value. The next step is to map r and θ values to the pixels of the shape based on the r and θ values on the inter-pixel isolines.

We can identify all pixels located between inter-pixel isolines corresponding to two consecutive r values with one of the following relations (Fig. 8a):

$$e < ECC(\mathcal{S}) \leqslant e + 1 \tag{4}$$

$$e \leqslant ECC(\mathcal{S}) < e + 1 \tag{5}$$

where e, e+1 are the eccentricity values corresponding to two consecutive r values r_1, r_2. The choice between Equations 4 and 5 depends on the method chosen to compute the θ value for each pixel and is discussed below. All pixels satisfying the chosen equation (4 or 5) are considered to have the same r coordinate.

We extract inter-pixel isolines for eccentricity values with difference one. Thus, for each pixel the number of inter-pixel isolines crossing any of the adjacent line segments that connect pixel centers, is maximum four and at least one.

To compute the θ value for a pixel we consider the inter-pixel isolines that bound the pixel center and identify the following options (Fig. 8 illustrates the used notation):

1. **higher:** assign θ of the closest point(s) on the isoline with higher ECC value;
2. **lower:** assign θ of the closest point on the isoline with lower ECC value;
3. **both:** assign a function of θ of the closest points on both isolines.

Considering that the inter-pixel isolines are polygonal lines with angles less or equal to 180 degrees on the side with smaller ECC values, we can state the following about each option above:

higher: one pixel can get more than one θ value (Fig. 8b);
lower: more than one pixel can get the same θ value (Fig. 8c);

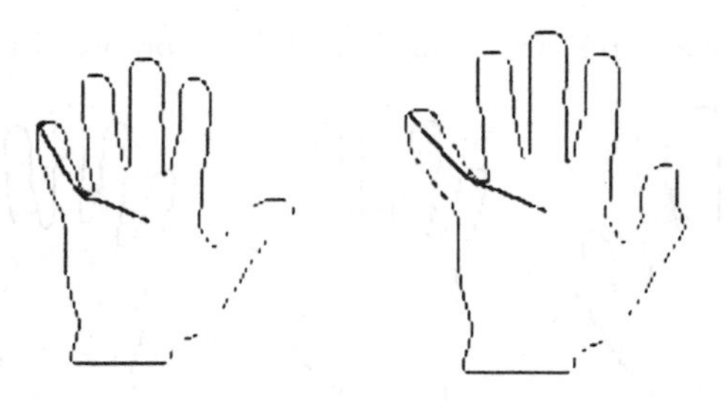

Fig. 9. Zero paths used in our experiments

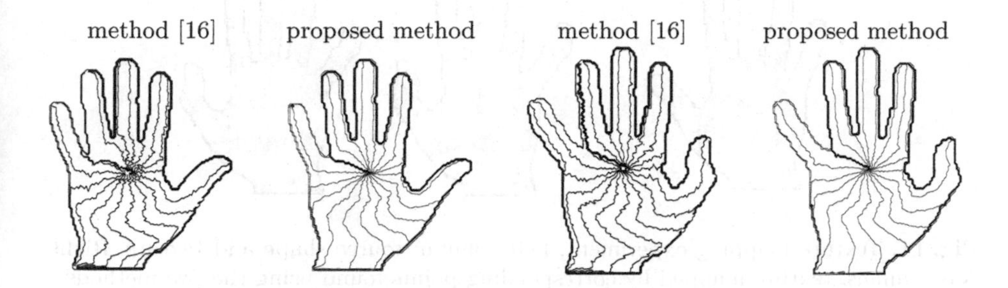

Fig. 10. Isolines of θ for the two shapes, with the two methods

both: could produce smoother results, but uniqueness of θ values is not guaranteed (Fig. 8d).

In addition, option 'higher' cannot compute θ values for the border pixels that have no higher eccentricity isolines, and option 'lower' cannot compute θ values for the geodesic center.

Going back to the definition of the coordinate system, the proposed method to find correspondences and the 3 requirements for the coordinate map (Section 2) we can state that properties 'defined' and 'distinct' are required to be able to find a corresponding point. We can relax the requirements and give up property 'single', and if a pixel gets more than one θ value, consider only the smallest one.

Following this reasoning, **option 'higher' is chosen**, as it assigns distinct θ values to all pixels having the same r.

5.2 Experiments

Fig. 9 shows the used zero path for the two hands in Fig. 2. In Fig. 10 the isolines of the θ coordinate mapped with the method in Section 4 and the proposed one (Section 5) is given. The θ values computed by the proposed method have a smoother profile, due to the inter-pixel isolines better approximating the continuous isolines.

A texture was laid on each hand - the *source*, and copied to the other one - the *destination*, by finding for each pixel $\mathbf{p}_d(r_d, \theta_d)$ of the *destination* the

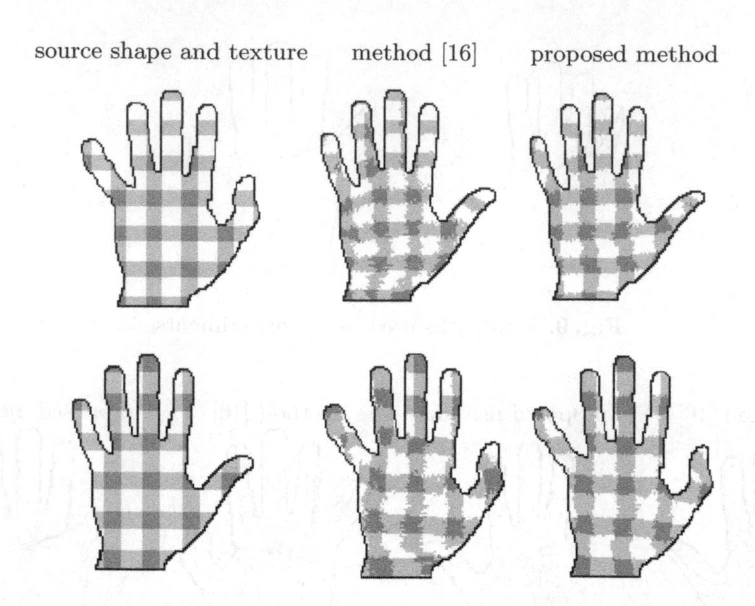

Fig. 11. Texture mapping experiment. Left column: source shape and texture. Right two columns: texture mapped by corresponding points found using the two methods.

corresponding pixel $\mathbf{p}_s(r_s, \theta_s)$ in the *source*. Results are shown in Fig. 11. Notice the smaller perturbation in the palm and the better mapping on the fingers.

Note that for these two shapes, the circle arcs making the isolines in the center region do not make any sharp angles and the problem mentioned in Section 4.1 does not appear. Thus, both methods could be applied.

6 Conclusion

This paper considered using a non-rigid coordinate system to find corresponding points in different poses of the same articulated shape. To each pixel distinct coordinates are associated. These coordinates are used to address corresponding points. A solution to a discretization problem identified in a previous approach was proposed. Instead of mapping the coordinate system directly to the pixels, the coordinate system is computed in a space where the problem cannot occur. The coordinates for the pixels are computed based on the computed coordinate system. Even smoother coordinates could be obtained by more refined mapping from the coordinate system in the transformed space to the pixels. Extension to non-simply connected 2D- and 3D shapes has to consider non-convex isolines, and connected components of the level sets merging at higher ECC values.

References

1. Pizlo, Z.: Shape: Its Unique Place in Visual Perception, 2nd edn. Springer, Heidelberg (2002)

2. Rosenfeld, A.: A note on 'geometric transforms' of digital sets. Pattern Recognition Letters 1(4), 223–225 (1983)
3. Borgefors, G.: Digital distance transforms in 2D, 3D, and 4D. In: Handbook of Pattern Recognition & Computer Vision, 3rd edn. World Scientific, Singapore (2005)
4. Gorelick, L., Galun, M., Sharon, E., Basri, R., Brandt, A.: Shape representation and classification using the poisson equation. In: IEEE CVPR, pp. 61–67 (2004)
5. Aouada, D., Dreisigmeyer, D.W., Krim, H.: Geometric modeling of rigid and non-rigid 3d shapes using the global geodesic function. In: NORDIA workshop in conjunction with IEEE CVPR, Anchorage, Alaska, USA. IEEE, Los Alamitos (2008)
6. Ion, A., Kropatsch, W.G., Andres, E.: Euclidean eccentricity transform by discrete arc paving. In: Coeurjolly, D., Sivignon, I., Tougne, L., Dupont, F. (eds.) DGCI 2008. LNCS, vol. 4992, pp. 213–224. Springer, Heidelberg (2008)
7. Felzenszwalb, P.F.: Representation and detection of deformable shapes. In: CVPR (2003)
8. Felzenszwalb, P.F., Schwartz, J.D.: Hierarchical matching of deformable shapes. In: CVPR (2007)
9. Ozcanli, O.C., Kimia, B.B.: Generic object recognition via shock patch fragments. In: BMVC 2007, Warwick Print, pp. 1030–1039 (2007)
10. Ling, H., Jacobs, D.W.: Shape classification using the inner-distance. IEEE TPAMI 29(2), 286–299 (2007)
11. Belongie, S., Malik, J., Puzicha, J.: Shape matching and object recognition using shape contexts. IEEE TPAMI 24(4), 509–522 (2002)
12. Siddiqi, K., Shokoufandeh, A., Dickinson, S., Zucker, S.W.: Shock graphs and shape matching. International Journal of Computer Vision 30, 1–24 (1999)
13. Crum, W.R., Hartkens, T., Hill, D.L.: Non-rigid image registration: theory and practice. The British Journal of Radiology 77 Spec. No. 2 (2004)
14. Brechbuhler, C., Gerig, G., Kubler, O.: Parametrization of closed surfaces for 3-D shape-description. CVIU 61(2), 154–170 (1995)
15. Kambhamettu, C., Goldgof, D.B.: Curvature-based approach to point correspondence recovery in conformal nonrigid motion. Computer Vision, Graphics and Image Processing: Image Understanding 60(1), 26–43 (1994)
16. Ion, A., Haxhimusa, Y., Kropatsch, W.G., López Mármol, S.B.: A coordinate system for articulated 2d shape point correspondences. In: Proceedings of 19th International Conference on Pattern Recognition (ICPR), IAPR. IEEE, Los Alamitos (2008)
17. Weisstein, E.W.: CRC Concise Encyclopedia of Mathematics, 2nd edn. Chapman & Hall/CRC, Boca Raton (2002)
18. Kropatsch, W.G., Ion, A., Haxhimusa, Y., Flanitzer, T.: The eccentricity transform (of a digital shape). In: Kuba, A., Nyúl, L.G., Palágyi, K. (eds.) DGCI 2006. LNCS, vol. 4245, pp. 437–448. Springer, Heidelberg (2006)
19. Sebastian, T.B., Klein, P.N., Kimia, B.B.: Recognition of shapes by editing their shock graphs. IEEE TPAMI 26(5), 550–571 (2004)
20. Ion, A.: The Eccentricity Transform of n-Dimensional Shapes with and without Boundary. PhD thesis, Vienna Univ. of Technology, Faculty of Informatics (2009)
21. Lorensen, W.E., Cline, H.E.: Marching cubes: A high resolution 3d surface construction algorithm. In: Stone, M.C. (ed.) Proceedings of the 14st Annual Conference on Computer Graphics and Interactive Techniques, SIGGRAPH 1987, pp. 163–169. ACM, New York (1987)

Two Linear-Time Algorithms for Computing the Minimum Length Polygon of a Digital Contour

Xavier Provençal[1,2,*] and Jacques-Olivier Lachaud[1,**]

[1] Laboratoire de Mathématiques, UMR 5127 CNRS, Université de Savoie, 73376 Le Bourget du Lac, France
[2] LIRMM, UMR 5506 CNRS, Université Montpellier II, 34392 Montpellier, France
provencal@lirmm.fr, jacques-olivier.lachaud@univ-savoie.fr

Abstract. The Minimum Length Polygon (MLP) is an interesting first order approximation of a digital contour. For instance, the convexity of the MLP is characteristic of the digital convexity of the shape, its perimeter is a good estimate of the perimeter of the digitized shape. We present here two novel equivalent definitions of MLP, one arithmetic, one combinatorial, and both definitions lead to two different linear time algorithms to compute them.

1 Introduction

The minimum length polygon (MLP) or minimum perimeter polygon has been proposed long ago for approaching the geometry of a digital contour [1,2]. One of its definitions is to be the polygon of minimum perimeter which stays in the band of 1 pixel-wide centered on the digital contour. It has many interesting properties such as: (i) it is reversible [1]; (ii) it is characteristic of the convexity of the digitized shape and it minimizes the number of inflexion points to represent the contour [2,3]; (iii) it is a good digital length estimator [4,5] and is proven to be multigrid convergent in $O(h)$ for digitization of convex shapes, where h is the grid step (reported in [6,7,8]); (iv) it is also a good tangent estimator; (v) it is the relative convex hull of the digital contour with respect to the outer pixels [2,9] and is therefore exactly the convex hull when the contour is digitally convex.

Several algorithms for computing the MLP have been published. We have already presented the variational definition of the MLP (length minimizer). It can thus be solved by a nonlinear programming method. The initial computation method of [1] was indeed an interative Newton-Raphson algorithm. Computational complexity is clearly not linear and the solution is not exact. We have also mentioned its set theoretic definition (intersection of relative convex sets). However, except for digital convex shapes, this definition does not lead to a specific algorithms. The MLP may also be seen as a solution to a shortest path query in some well chosen polygon. An adaptation of [10] to digital contour could be

* Author supported by a scholarship from FQRNT (Québec).
** Partially funded by ANR project FOGRIMMI (ANR-06-MDCA-008-06).

S. Brlek, C. Reutenauer, and X. Provençal (Eds.): DGCI 2009, LNCS 5810, pp. 104–117, 2009.

implemented in time linear with the size of the contour. It should however be noted that data-structures and algorithms involved are complex and difficult to implement. Klette *et al.* [11] (see also [4,6]) have also proposed an arithmetic algorithm to compute it, but as it is presented, it does not seem to compute the MLP in all cases.[1]

The MLP is in some sense characteristic of a digital contour. One may expect to find strong related arithmetic and combinatorial properties. This is precisely the purpose of this paper. Furthermore, we show that each of these definitions induces an optimal time integer-only algorithm for computing it. The combinatorial algorithm is particularly simple and elegant, while the arithmetic definition is essential for proving it defines the MLP. These two new definitions give a better understanding of what is the MLP in the digital world. Although other linear-time algorithms exist, the two proposed algorithms are simpler than existing ones. They are thus easier to implement and their constants are better.

The paper is organized as follows. First Section 2 recalls standard definitions. Section 3 presents how to split uniquely a digital contour into convex, concave and inflexion zones, the arithmetic definition of MLP follows then naturally. Section 4 is devoted to the combinatorial version of MLP. Section 5 illustrates our results and concludes. Due to space limitations, we are not able to provide the proofs that these definitions induce the MLP, we only give hints and discuss the algorithms to extract them.

2 Preliminaries

This section presents the generally standard definitions that we will used throughout the paper, in order to avoid any ambiguity. A *polyomino* is a set of digital squares in the plane such that its topological boundary is a Jordan curve. It is thus bounded. It is convenient to represent a polyomino as a subset of the digital plane $\mathbb{Z}^2$, which code the integer coordinates of the centroids of its squares, instead of representing it as a subset of the Euclidean plane $\mathbb{R}^2$. When seeing a polyomino as a subset of $\mathbb{R}^2$, we will say the *the body of the polyomino.* For instance, the Gauss digitization of a convex subset of the plane is a polyomino iff it is 4-connected. A subset of $\mathbb{Z}^2$, or *digital shape*, is a polyomino iff it is 4-connected and its complement is 4-connected.

In the following, a *digital contour* is the boundary of some polyomino, represented as a sequence of horizontal and vertical steps in the half-integer plane $(\mathbb{Z}+\frac{1}{2}) \times (\mathbb{Z}+\frac{1}{2})$. One can use for instance a Freeman chain to code it. Again, the *body* of a digital contour is its embedding in $\mathbb{R}^2$ as a polygonal curve. Now, since the body of a digital contour is a Jordan curve, it has one well-defined inner component in $\mathbb{R}^2$, whose closure is exactly the polyomino whose boundary is the digital contour. There is thus a one-to-one map from digital contours to polyominoes, denoted by I.

[1] Its edges seem restricted to digital straight segments such that the continued fraction of their slope has a complexity no greater than two.

Given a digital contour C, its *inner polygon* $\mathrm{L}_1(C)$ is the erosion of the body of $\mathrm{I}(C)$ by the open unit square centered on $(0,0)$. Its *outer polygon* $\mathrm{L}_2(C)$ is the dilation of the body of $\mathrm{I}(C)$ by the closed unit square centered on $(0,0)$. In the following, the inner and outer polygons play the role of inner and outer constraints for constructing the MLP. We only deal in this paper with simple digital contours (or grid continua in the terminology of [8]), therefore both $\mathrm{L}_1(C)$ and $\mathrm{L}_2(C)$ are assumed to be simple polygons.

The relative convex hull leads to a simple and elegant set-theoretic definition of the characteristic polygon, which is valid in arbitrary dimension. In the following, the notation $\overline{xy}$ stands for the straight line segment joining x and y, i.e. their convex hull.

Definition 1. *[9]. Let $U \subseteq \mathbb{R}^n$ be an arbitrary set. A set $C \subseteq U$ is said to be U-*convex *iff for every $x, y \in C$ with $\overline{xy} \subseteq U$ it holds that $\overline{xy} \subseteq C$.*

Let $V \subseteq U \subseteq \mathbb{R}^n$ be given. The intersection of all U-convex sets containing V will be termed convex hull of V relative to U, *and denoted by* $\mathrm{Conv}_U(V)$.

The *set-theoretic MLP* of a digital contour C is the convex hull of $\mathrm{L}_1(C)$ relative to $\mathrm{L}_2(C)$. It is easily proven that the MLP of a digitally convex contour is exactly the convex hull of its digital points.

The standard definition of the MLP of C [1,8,4] is the shortest Jordan curve whose digitization is (very close to) the polyomino of C. More precisely, letting $\mathcal{A}$ be the family of simply connected compact sets of $\mathbb{R}^2$, we define:

Definition 2. *The* minimum perimeter polygon, *MPP, of two polygons V, U with $V \subset U^\circ \subset \mathbb{R}^2$ is a subset P of $\mathbb{R}^2$ such that*

$$\mathrm{Per}(P) = \min_{A \in \mathcal{A},\ V \subseteq A,\ \partial A \subset U \setminus V^\circ} \mathrm{Per}(A), \tag{1}$$

where $\mathrm{Per}(A)$ *stands for the perimeter of A, more precisely the 1-dimensional Hausdorff measure of the boundary of A.*

In [7]-(Theorem 3), and in [9] it is shown that Equation (1) has a unique solution, which is a polygonal Jordan cuver whose convex vertices (resp. concave) belong to the vertices of the inner polygon (resp. the vertices of the outer polygon), and which is also the convex hull of V relative to U. The *variational MLP* of C is thus the MPP of $\mathrm{L}_1(C), \mathrm{L}_2(C)$.

3 Arithmetic MLP

3.1 Tangential Cover and Arithmetic Properties

We recall that the *tangential cover* of a digital contour is the sequence of its Maximal Digital Straight Segments (MDSS). In the following, the tangential cover is denoted by $(M_l)_{l=1..N}$, where M_l is the l-th MDSS of the contour. Let us denote by θ_l the slope direction (angle wrt x-axis) of M_l. All indices are taken modulo the number of MDSS N. Since the directions of two consecutive MDSS can differ of no greater than π, their variation of direction can always be casted in $]-\pi,\pi[$ without ambiguity. The angle variation $(\theta_l - \theta_{l+1}) \mod [-\pi,\pi[$ is denoted by $\Delta(\theta_l,\theta_{l+1})$. For clarity, we will also write $\theta_l > \theta_{l+1}$ when $\Delta(\theta_l,\theta_{l+1}) > 0$. We always consider the digital contour to turn clockwise around the polyomino. A couple of consecutive MDSS (M_l, M_{l+1}) is thus said to be a $\wedge$-*turn* (resp. $\vee$-*turn*) when $\Delta(\theta_l,\theta_{l+1})$ is negative (resp. positive). The symbol $\wedge$ stands for "convex" while the symbol $\vee$ stands for "concave".

We have the following theorem from [12], which relates convexity to MDSS directions. It also induces a linear time algorithm to check convexity.

Theorem 1. *[12] A digital contour is digitally convex iff every couple of consecutive MDSS of its tangential cover is made of* $\wedge$*-turns.*

For a given DSS M, its first and last upper leaning points are respectively denoted by $\mathrm{U}_f(M)$ and $\mathrm{U}_l(M)$, while its first and last lower leaning points are respectively denoted by $\mathrm{L}_f(M)$ and $\mathrm{L}_l(M)$. In the same paper, it is proven that the point $\mathrm{U}_l(M_l)$ is no further away than $\mathrm{U}_f(M_{l+1})$. The same property holds for lower leaning points.

We may now consider the succession of turns along a digital contour to cut it into parts. The above-mentioned ordering on leaning points between successive MDSS guarantees the consistency of the following definition.

$\cdots \; \boxed{\wedge \underline{M_i} \wedge \cdots \wedge \underline{M_j} \wedge} \; \underline{M_{j+1}} \; \boxed{\vee \underline{M_{j+2}} \vee \cdots \vee \underline{M_k} \vee} \; \underline{M_{k+1}} \; \boxed{\wedge \cdots}$

... convex zone inflexion concave zone inflexion ...

Definition 3. *A digital contour C is uniquely split by its tangential cover into a sequence of closed connected sets with a single point overlap as follows:*

1. *A* convex zone *or* $(\wedge,\wedge)$-zone *is defined by an inextensible sequence of consecutive* $\wedge$*-turns from* (M_{l_1}, M_{l_1+1}) *to* (M_{l_2-1}, M_{l_2})*. If it is a proper zone of C, it starts at* $\mathrm{U}_l(M_{l_1})$ *and ends at* $\mathrm{U}_f(M_{l_2})$*, otherwise it starts and ends on the first point.*
2. *A* concave zone *or* $(\vee,\vee)$-zone *is defined by an inextensible sequence of consecutive* $\vee$*-turns from* $(M_{l'_1}, M_{l'_1+1})$ *to* $(M_{l'_2-1}, M_{l'_2})$*. It starts at* $\mathrm{L}_l(M_{l'_1})$ *and ends at* $\mathrm{L}_f(M_{l'_2})$*.*
3. *A* convex inflexion zone *or* $(\wedge,\vee)$-zone *is defined by a* $\wedge$*-turn followed by a* $\vee$*-turn around MDSS* M_i*. It starts at* $\mathrm{U}_f(M_i)$ *and ends at* $\mathrm{L}_l(M_i)$*.*
4. *A* concave inflexion zone *or* $(\vee,\wedge)$-zone *is defined by a* $\vee$*-turn followed by a* $\wedge$*-turn around MDSS* $M_{i'}$*. It starts at* $\mathrm{L}_f(M_{i'})$ *and ends at* $\mathrm{U}_l(M_{i'})$*.*

Note that a convex or concave zone may be reduced to a single turn between two successive inflexions. In this case, the zone may or may not be a single contour point (point A in Fig. 1).

Since a MDSS is contained in a digital straight line, it is formed of exactly two kind of steps, with Freeman codes c and $(c+1) \mod 4$. These codes defines the *quadrant* of the MDSS. The *quadrant vector* is then the diagonal vector that is the sum of the two unit steps coded by the Freeman codes of the quadrant, rotated by $+\frac{\pi}{2}$.

We eventually associate pixels to contour points (C_i) as follows:

- the *inside pixel* $\mathrm{in}(C_i)$ of C_i is the pixel $C_i - \frac{\vec{v}}{2}$, where $\vec{v}$ is the quadrant vector of any MDSS containing it (or the last MDSS stricly containing it at a quadrant change).
- the *outside pixel* $\mathrm{out}(C_i)$ of C_i is the pixel $C_i + \frac{\vec{v}}{2}$, where $\vec{v}$ is the quadrant vector of any MDSS containing it (or the last MDSS stricly containing it at a quadrant change).

Fig. 1 illustrates these definitions. It is clear that inside pixels belong to $\partial \mathrm{L}_1(C)$ and outside pixels to $\partial \mathrm{L}_2(C)$.

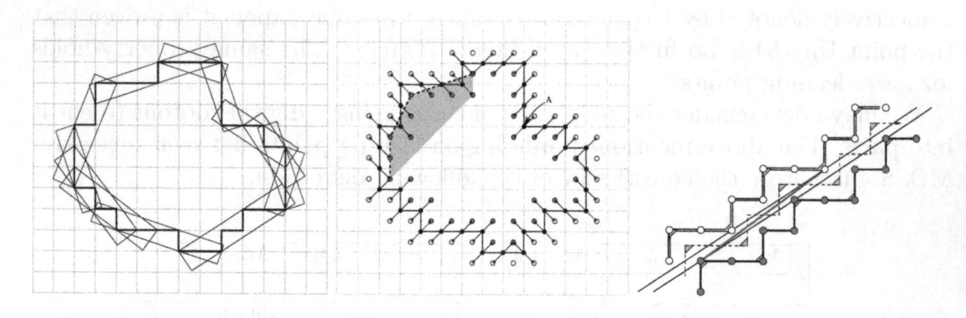

Fig. 1. Left: digital contour, tangential cover. Center: inside and outside pixels, edges of AMLP(C) in a convex part of C. Right : Geometry of a DSS of slope 2/3 along a digital contour. The digital contour is drawn in-between the corresponding inside and outside pixels. They clearly draw the same contour up to a translation. Upper and lower leaning points are denoted by grey right triangles. The thick red line connects the first upper leaning point on the inside contour to the last lower leaning point on the outside contour. The two straight lines define a straight band which separates inside from outside pixels. The thick red line is in this band, thus in $\mathrm{L}_2(C)$.

3.2 Definition of the Arithmetic MLP of C

The boundary word of any *hv*-convex polyomino admits a unique factorization as $Q_{0<1}0Q_{3<0}3Q_{2<3}2Q_{1<2}1$, where $Q_{a<b}$ is written over the two letters $\{a, b\}$ and ends by b. The words $Q_{a<b}$ are called its *quadrant words*. As a corollary, convex zones of a polyomino admits a unique factorization into a prefix of a conjugate of $Q_{0<1}0Q_{3<0}3Q_{2<3}2Q_{1<2}1$.

Definition 4. *Assume $C_{i,j}$ is a connected part of a contour, with only two kinds of steps. The* left enveloppe *of $C_{i,j}$ (resp.* right enveloppe *of $C_{i,j}$) is the sequence of edges of the convex hull of the inside (resp. outside) pixels of $C_{i,j}$, such that the first vertex is C_i, the last vertex is C_j and the edges turn clockwise (resp. counterclockwise) around the hull.*

We may now define a linear analog to a digital contour.

Definition 5. *The* arithmetic *MLP of a digital contour C is the polygon* AMLP(C) *defined by zones $C_{i,j}$ in C, according to its type:*

zone type of $C_{i,j}$	*associated part of* AMLP(C)
$(\wedge,\wedge)$ *convex*	*union of the left enveloppe of each quadrant word of $C_{i,j}$*
$(\vee,\vee)$ *concave*	*union of the right enveloppe of each quadrant word of $C_{i,j}$*
$(\wedge,\vee)$ *cvx-inflexion*	*segment joining the inside pixel of C_i to the outside pixel of C_j*
$(\vee,\wedge)$ *ccv-inflexion*	*segment joining the outside pixel of C_i to the inside pixel of C_j*

Lemma 1. AMLP(C) *is a closed polygonal line with vertices in $\mathbb{Z}^2$.*

Proof. By construction, all vertices of AMLP(C) are in $\mathbb{Z}^2$. Clearly, AMLP(C) is a polygonal line inside each zone of C. All that remains to check is that inflexion zones links the convex and concave zones correctly, and this is done using standard properties of MDSS. □

Let U be the closed unit square centered at $(0,0)$ and let $\oplus$ denotes the Minkowski sum of two sets. We recall that the set $C \oplus U$ is also the one pixel wide band $\mathrm{L}_2(C) \setminus \mathrm{L}_1(C)^\circ$.

Lemma 2. *In a $(\wedge,\wedge)$-zone $C_{i,j}$ of C, the corresponding edges of* AMLP(C) *form a simple polygonal line included in $\mathrm{L}_2(C)^\circ \setminus \mathrm{L}_1(C)^\circ$, and in $C_{i,j} \oplus U$.*

Proof. Consider a single quadrant word $Q_{a<b}$ that is part of a $(\wedge,\wedge)$-zone. For this part of C, the edges of AMLP(C) form the digital convex hull of the inside pixels of C so that there cannot be any intersection with $\mathrm{L}_1(C)^\circ$. On the other hand, again by the digital convex hull property, all the integer points in the AMLP(C), for this zone, must belong to the set of inner pixels of C. In particular, there are no points of $\mathrm{L}_2(C)$. □

The following lemmata are proven similarly to the previous one.

Lemma 3. *In a $(\vee,\vee)$-zone $C_{i',j'}$ of C, the corresponding edges of* AMLP(C) *form a simple polygonal line included in $\mathrm{L}_2(C) \setminus \mathrm{L}_1(C)$, and in $C_{i',j'} \oplus U$.*

Lemma 4. *In a $(\wedge,\vee)$ or $(\vee,\wedge)$ zone $C_{i'',j''}$ of C, the corresponding edges of* AMLP(C) *form a single straight segment included in $\mathrm{L}_2(C) \setminus \mathrm{L}_1(C)^\circ$, and in $C_{i'',j''} \oplus U$.*

Proof. We refer the reader to Fig. 1. The AMLP(C) is in this zone the thick red segment squeezed in the band that separates inside from outside pixels. The lemma follows. □

Theorem 2. AMLP(C) *is a simple polygon with boundary in $\mathrm{L}_2(C) \setminus \mathrm{L}_1(C)^\circ$.*

Proof. Lemmata 2 and 3 guarantee that the restriction of the edges of AMLP(C) to these parts is always a single polygonal line in $\mathrm{L}_2(C) \setminus \mathrm{L}_1(C)^\circ$. Furthermore these parts have always a pairwise empty intersection. Indeed, taking any two of these parts, they are defined by two pieces of C, say C_{l_1,l_2} and $C_{l'_1,l'_2}$. These pieces are separated by at least one inflexion zone, whose length is at least 1. Therefore $C_{l_1,l_2} \cap C_{l'_1,l'_2} = 0$. Moreover $(C_{l_1,l_2} \oplus U) \cap (C_{l'_1,l'_2} \oplus U)$ is either (i) empty, (ii) an integer point, (iii) a unit segment. If (i), the inflexion zone is large enough and the two parts of AMLP(C) may obviously not touch each other. If (ii) or (iii), the inflexion zone is two surfels long or one surfel long respectively. We notice furthermore that AMLP(C) ends on one side at an inside pixel and on the other side at an outside pixel. It is easy to check (there is one 1 surfel configuration and three 2 surfels configuration) that the two parts of AMLP(C) have an empty intersection.

Lemma 4 indicates that the edges of AMLP(C) in inflexion zones are also in $\mathrm{L}_2(C) \setminus \mathrm{L}_1(C)^\circ$. It is also clear that at a convex junction C_k, AMLP$(C) \cap (C_k \oplus U)$ is reduced to the point in(C_k), therefore with no other self-intersections. A symmetric result is obtained at a concave junction. Gathering everything, the so-formed polygonal line is simple. Jordan theorem concludes it has a finite inner component. AMLP(C) is thus a polygon with a boundary in $\mathrm{L}_2(C) \setminus \mathrm{L}_1(C)^\circ$. □

We finish by providing an algorithm to compute AMLP(C).

Algorithm 1. Computation of AMLP(C)

```
1 Compute (M_i)_{i=1..N} the tangential cover of C ;
2 Decompose (M_i)_{i=1..N} in (α,β)-zones; // WHERE α,β ∈ {∧,∨} (DEFINITION 3)
3 S ← ();
4 for each (α,β)-zone z of C do
5  └ add to the list S the part of AMLP(C) associate to z; // (DEFINITION 5)
6 return S
```

Theorem 3. *Algorithm 1 computes* AMLP(C) *in time linear with respect to the length of* C.

Proof. Let n be the number of points of C. Computation of the tangential cover on line 1 is performed in linear time according to [13]. The computation on line 1 is clearly proportional to N which is also $O(n)$. Finally, on line 1, there are two cases to consider. Given a convex or concave zone, the AMLP(C) is some convex hull of a simple polygonal line which is computed in time proportional to the length of the zone using [14] or [15]. Given an inflexion zone, the computation is reduced to a segment, thus a $O(1)$ operation. Total computation time is $O(n)$. □

3.3 AMLP(C) is the MLP of C

In order to show the equivalence between AMLP(C) and the convex hull of $\mathrm{L}_1(C)$ relatively to $\mathrm{L}_2(C)$, we need the following technical lemma.

Lemma 5. *Convex vertices of* AMLP(C) *are inside pixels of* C *(i.e.* $\in \partial\mathrm{L}_1(C)$*), concave vertices of* AMLP(C) *are outside pixels of* C *(i.e.* $\in \partial\mathrm{L}_2(C)$*).*

Proof. We already know that in a convex zone of C, vertices of AMLP(C) are by Definition 5 inside pixels with strictly decreasing edge directions. In a concave zone of C, vertices of AMLP(C) are outside pixels with strictly increasing edge directions. Around an inflexion zone of direction θ_i, one may show that $\theta_{i-1} > \theta_i$, where θ_{i-1} is the slope of the previous edge of AMLP(C), imposes a $(\wedge, \vee)$-zone, so the vertex is the inside pixel of an upper leaning point. The reasonning is similar when $\theta_{i-1} < \theta_i$. □

We can now prove that the polygon AMLP(C) is the minimum perimeter polygon of $\mathrm{L}_1(C)$, $\mathrm{L}_2(C)$, or the so-called MLP of C in the terminology of [4].

Theorem 4. *If C is a simple 4-connected digital contour, then* AMLP(C) *is the convex hull of* $\mathrm{L}_1(C)$ *relative to* $\mathrm{L}_2(C)$ *or, otherwise said,* AMLP(C) *is the intersection of every* $\mathrm{L}_2(C)$*-convex set containing* $\mathrm{L}_1(C)$.

Proof. We proceed in four steps :

1. AMLP(C) is a $\mathrm{L}_2(C)$-convex set containing $\mathrm{L}_1(C)$. Using Theorem 2 along with Lemma 5 one concludes that if there exist $x, y \in$ AMLP(C) such that $\overline{xy} \not\subset$ AMLP(C) but $\overline{xy} \subset \mathrm{L}_2(C)$ then $\overline{xy}$ splits $\mathrm{L}_2(C)^c$ in two disjoint components.
2. Every convex and every concave vertex of AMLP(C) belongs to every $\mathrm{L}_2(C)$-convex set containing $\mathrm{L}_1(C)$. For convex vertices, this is direct from Lemma 5. For a concave vertex v, using Lemma 3 one may find a pair of point $x, y \in \mathrm{L}_1(C)$ such that $v \in \overline{xy} \subset \mathrm{L}_2(C)$.
3. Every edge of AMLP(C) belongs to every $\mathrm{L}_2(C)$-convex set containing $\mathrm{L}_1(C)$. By lemmata 2, 3 and 4, no edge from AMLP(C) intersect $\mathrm{L}_2(C)^c$ so that the previous step of this proof allows to conclude.
4. Points (2) and (3) implies ∂AMLP(C) is included in every $\mathrm{L}_2(C)$-convex set containing $\mathrm{L}_1(C)$, which proves that AMLP(C) is included in the intersection of every $\mathrm{L}_2(C)$-convex set containing $\mathrm{L}_1(C)$. Being itself such a relative convex set (point (1)), it is necessarily the convex hull of $\mathrm{L}_1(C)$ relative to $\mathrm{L}_2(C)$. □

4 Combinatorial Definition

Based on the combinatorial characterization of digital convexity obtained in [15], we propose a new algorithmic definition of the minimum length polygon. Given an arbitrary ordered alphabet $\mathcal{A} = \{a_1, a_2, \ldots, a_n\}$ with the order $a_1 < a_2 < \cdots < a_n$, written $\mathcal{A} = \{a_1 < a_2 < \cdots < a_n\}$ for short, we extend this order to words over $\mathcal{A}$ using the lexicographic order. We note $|w|_a$ the number of occurences of the letter a in w and $|w| = \sum_{a \in \mathcal{A}} |w|_a$ is the length of w. Let $\mathcal{A}^n$ be the set of all words of length n over $\mathcal{A}$, in particular $\mathcal{A}^0 = \{\varepsilon\}$ where ε is called the the *empty word*. A word w is *non-empty* if $w \neq \varepsilon$. The i-th letter of a word w is $w[i]$ and we refer to factors of w like this : $w = w[1 : i-1] \cdot w[i : i+j] \cdot w[i+j+1 : n]$, where $\cdot$ is the concatenation and $|w| = n$. By reference to the Freeman coding, given a word $w \in \{0, 1, 2, 3\}^n$ the translation vector associated to w is $\overrightarrow{w} = (|w|_0 - |w|_2, |w|_1 - |w|_3)$.

Definition 6. *A non-empty word w over the ordered alphabet $\mathcal{A}$ is a* Lyndon word *if $w < v$ for any non-empty word v such that $w = uv$. We note $\mathbf{L}_\mathcal{A}$ the set of all Lyndon words over $\mathcal{A}$.*

Theorem 5 ([16] Theorem 5.1.1). *Any non-empty word w over $\mathcal{A}$ admits a unique factorization as a sequence of decreasing Lyndon words : $w = l_1^{n_1} l_2^{n_2} \cdots l_k^{n_k}$ with $l_1 > l_2 > \cdots > l_k$ where $n_i \geq 1$ and $l_i \in \mathbf{L}_\mathcal{A}$ for all $1 \leq i \leq k$.*

We define the function FLF, called *first Lyndon factor*, as $\text{FLF}(w, \mathcal{A}) = (l_1, n_1)$ where $w = l_1^{n_1} l_2^{n_2} \cdots l_k^{n_k}$ is the unique factorization of w in decreasing Lyndon words according to the ordered alphabet $\mathcal{A}$.

Introduced by Christoffel in [17], Christoffel words where reinvestigated by Borel and Laubie in [18]. Since then their impressive combinatorial structure has been studied by many, see [19] for a comprehensive self-contained survey. Here is one of the many equivalent definitions of Christoffel words.[2]

Definition 7. *A* Christoffel *word on the alphabet $\{a < b\}$ is the Freeman code of a the path joining two consecutive upper leaning points of a DSS with positive slope according to the convention that a codes an horizontal step and b codes a vertical one.*

Once again, referring to the Freeman code, one defines the *slope* of a word as $\rho(w) = |w|_b / |w|_a$ with the convention that $1/0 = \infty$. In the case of Christoffel words, unlike the general case, there is a direct link between lexicographic order and the slope : $u, v \in \mathbf{C}_{a<b} \implies (u < v \iff \rho(u) < \rho(v))$.

A convex polyomino being composed of only one convex zone the quadrant words $Q_{a<b}$ provide a natural decomposition of its border in four quadrant words. Our combinatorial view of convexity is based the following result which characterizes convex quadrant words.

Theorem 6 ([15]). *A hv-convex polyomino P is convex if and only if the factorization as decreasing Lyndon words of each quandrant words $Q_{a<b} = l_1^{n_1} l_2^{n_2} \cdots l_k^{n_k}$ is such that $l_i \in \mathbf{C}_{a<b}$ for all $1 \leq i \leq n_k$. Moreover, in the case where P is convex, for each quadrant, the edges of its convex hull coincide with the vectors $n_i \overrightarrow{l_i}$.*

4.1 Definition of the CMLP

We define the combinatorial minimum length polygon algorithmically using Algorithm 3 which simply computes vertices given by a list a edges determined by Algorithm 2. We suppose that the word w codes the boundary of a polyomino P starting from the point (x_0, y_0) the lowers point among the leftmost points of P (i.e. $x_0 = \min\{x | (x, y) \in P\}$ and $y_0 = \min\{y | (x_0, y) \in P\}$).

In order to illustrate how Algorithm 2 works, let us discuss the geometrical interpretation of the modifications performed to the alphabet A in Algorithm 2. First, notice that initially the alphabet is set as $A = \{3 < 0 < 1 < 2\}$ as we consider a convex zone with quadrant word $Q_{0<1}$. All through the algorithm, it

[2] These words are sometimes refered as *primitive lower Christoffel* words.

Algorithm 2. `nextEdge`

Input: (u, n, A) such that $u \in A^n$ and $A = \{a_1 < a_2 < a_3 < a_4\}$.
Output: (x, l, A) with $x \in \mathbb{Z}^2$, $l \in \mathbb{N}$.

```
1  (v, k, inC) ← FLF(u, A);
2  x = k v⃗; l = k|v|;
3  if v = a2 then
4  |   A ← {a4 < a1 < a2 < a3};
5  |   x ← x − v⃗;
6  else if not inC then
7  |   u[0] ← a3;
8  |   A ← {a4 < a3 < a2 < a1};
9  |   (x, l, A) ← nextEdge(u, n, A)
10 return (x, l, A)
```

shall always be the case that when analysing a convex quadrant word $Q_{a_2<a_3}$ the alphabet A is set as $\{a_1 < a_2 < a_3 < a_4\}$ so that the word a_2a_1 codes a quadrant change while a_3a_4 codes a change of convexity type.

A bijective map $\mu : \mathcal{A} \to \mathcal{A}$ over the letters $\mathcal{A}$ extends naturally to any word $w \in \mathcal{A}^n$ as $\mu(w) = \mu(w[1])\mu(w[2])\cdots\mu(w[n])$. Using the notation $\mu(\mathcal{A}) = \{\mu(a_1) < \mu(a_2) < \ldots\}$, clearly $\mu(w) \in \mathbf{L}_{\mathcal{A}} \iff w \in \mathbf{L}_{\mu(\mathcal{A})}$ and $\mu(w) \in \mathbf{C}_{a<b} \iff w \in \mathbf{C}_{\mu(a)<\mu(b)}$. Algorithm 2 uses this fact so that instead of applying transformations to the whole word w, only the order relation over the four letter alphabet is changed. Let w by the contour word of C over $\mathcal{A} = \{3 < 0 < 1 < 2\}$ and define $r : \mathcal{A} \to \mathcal{A}$ as $r(3) = 2$, $r(0) = 3$, $r(1) = 0$ and $r(2) = 1$. One verifies that the contour coded by $r(w)$ corresponds to a rotation by $\pi/2$. This explains line 4 of Algorithm 2 which is called when a quadrant change occurs.

Similarly, let $A = \{a_1 < a_2 < a_3 < a_4\}$ and define the *bar* operator $\overline{()}$ as $\overline{a_1} = a_4$, $\overline{a_2} = a_3$, $\overline{a_3} = a_2$ and $\overline{a_4} = a_1$. Consider a quadrant word $Q_{a<b}$, one verifies that $\overline{Q_{a<b}}$ corresponds to a reflexion by the line $y = x$ if $\{a, b\} \in \{\{0, 1\}, \{2, 3\}\}$ or by the line $y = -x$ if $\{a, b\} \in \{\{0, 3\}, \{1, 2\}\}$. Roughly speaking, this transformation turns this part of the contour *inside out*, so that computing the left enveloppe of $Q_{a<b}$ is equivalent to the computation of the right enveloppe of $\overline{Q_{a<b}}$. This explains line 8 of Algorithm 2.

The modification of the first letter of w at line 7 is due to the fact that w code the inter-pixel path. Since the condition at line 6 detects a change in the convexity type, the inner pixel adjacent to the step coded by $w[1]$ must now be consider as an outside pixel. This is done by switching the value of the first letter of w from a_2 to a_3.

Definition 8. *The* combinatorial MLP *of a digital contour C, noted* CMLP(C), *is obtained by joining consecutives vertices given as output of Algorithm 3.*

We suppose that the lower pixel among the leftmost pixels of the polyomino P is centered at $(0, 0)$. This ensures that point $v_0 = (0, 0)$ is a vertex of $CMLP(C)$. Moreover, in order to close the polygonal path computed, the word 10 is added

Algorithm 3. Computation of vertices from Algorithm 2.

Input: $w \in \{0,1,2,3\}^n$ the boundary word of P.
Output: $(x_0, x_1, x_2, \dots)$ a list of vertices that form the CMLP of P.

```
1 x ← (0,0); i ← 0; A ← {3 < 0 < 1 < 2};
2 w ← w · 10; n ← n + 2 ;
3 while w ≠ ε do
4     (v, l, A) = nextEdge(w, n, A);
5     x_{i+1} ← x_i + v; i ← i + 1 ;
6     w ← w[l + 1 : n]; n ← n − l ;
7 return (x_0, x_1, x_2, ..., x_i)
```

at the end of w so that an extra vertex located at $(0,0)$ is added at the end of the list closing the polygonal line.

Proposition 1. *The AMLP(C) and CMLP(C) are the same.*

Proof. By Theorem 6, for any convex part that stays in the same quadrant, the edges computed by Algorithm 2 are the same as those of the convex hull. Quadrant changes are detected by the condition at line 3 of Algorithm 2. Note that in such case, the first letter is ignored since the inter-pixel path is always one step longer than the corresponding part of the border of $L_1(C)$.

On a concave part of C, the reversal of the alphabet at line 8 of Algorithm 2 reverses the perspective and allow to use exactly the same algorithms to compute the convex hull of the outside pixels of C. Finally, it remains to check that after having performed the operations on lines 7 and 8 of Algorithm 2, the edge computed by FLF(w, A) is the same as the segment of the AMLP(C) corresponding to this inflexion zone. □

4.2 An Adapted Implementation of FLF

In order to compute function FLF one may use Duval's algorithm [20] which computes the pair $(u,k) = \text{FLF}(w, \mathcal{A})$ for any word w with a time complexity of $O(k|u|)$. This optimal algorithm is attributed to [21] in [22]. In Algorithm 2 we compute $(u,k) = \text{FLF}(w, \mathcal{A})$ but, in the case where u is not a Christoffel word, we do not care about this specific output (u,k). Based on this idea, we propose a modified version of Duval's algorithm. Algorithm 4 has the ability to determine dynamically if the word being read might lead to a Christoffel word and otherwise the computation immediately stops. Note that by removing lines identified by numbers from Algorithm 4, one obtains exactly Duval's algorithm. The extra variables p and q have the following interpretation: p is the biggest index smaller than j such that $w[1:p]$ is a Christoffel word, while q is the smallest index bigger or equal to j such that $w[1:q]$ might be a Christoffel word. The way p and q are updated is based on the following result.

Proposition 2. *Given a Christoffel word w, define $\chi_w = \{wz \in \mathbf{C}_{a<b} \,|\, z \neq \varepsilon\}$. For any $x \in \chi_w$, there exist $k \geq 1$ and v such that $x = w^k vz$ for some z where*

Algorithm 4. Duval++

```
    Input: (w, n, {a1 < a2 < a3 < a4}) where w ∈ {a2, a3} · {a1, a2, a3, a4}^(n−1).
    Output: (u, k, b) where b is a boolean such that b ⟺ (u, k) = FLF(w).
    i ← 1; j ← 2;
1   p ← 1; q ← 2;
    while j ≤ n and w[i] ≤ w[j] do
        if w[i] = w[j] then
2           if j = q then
3               q ← q + p;
            i ← i + 1;
        else
4           if j ≠ q or w[j] ≠ a3 then
5               return (ε, 0, false);
6           tmp ← p; p ← q; q ← 2q − tmp;
            i ← 1;
        j ← j + 1;
    return (w[1 : j − i], ⌊(j−1)/(j−i)⌋, true);
```

$w = uv$ is the unique factorization of w as two Christoffel words[3] *if $|w| \geq 2$ and $v = b$ if $|w| = 1$. Moreover, $w^l v \in \mathbf{C}_{a<b}$ for any $l \geq 1$.*

Proof. This is mainly due to a result of [18,23] stating that any $w = uv \in \mathbf{C}_{a<b}$ is obtained in a unique way as $(u, v) = H_1 \circ H_2 \circ \cdots \circ H_n(a, b)$ where each H_i is either the function $(x, y) \to (xy, y)$ or $(x, y) \to (x, xy)$. □

Proposition 3. *Algorithm 3 computes* CMLP(C) *in time proportional to the length of C.*

Proof. It suffices to show that the time complexity of Algorithm 2 is bounded by $O(l)$ where (v, l) is its output. First, suppose there is no recursive call. In this case, time complexity is simply given by Duval's algorithm.

On the other hand, suppose that there is a recursive call. This means that the call to FLF has output $(\varepsilon, 0, \texttt{false})$. Let m be the number of letters read in this first call to Algorithm 4. One may show that in such case, in the recursive call the computation of FLF will provide an output of the form $(v, k, \texttt{true})$ where $k|v| \geq m/2$. □

5 Concluding Remarks

We have presented two different definitions for the MPL of a digital contour, one based on an arithmetic approach, the other based on a combinatorial one. Both are shown to be the unique MLP of the contour and linear time algorithm

[3] The factorization of a Christoffel word as a product of two Christoffel word is equivalent to the *splitting formula* of DSS.

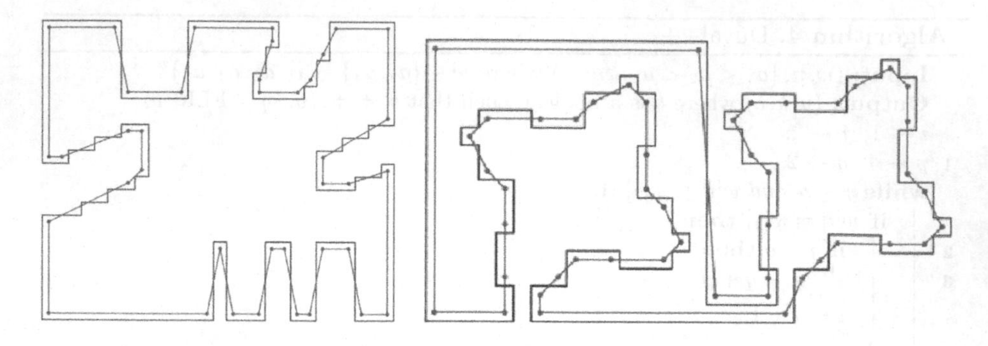

Fig. 2. Examples of CMLP

to compute them are provided. Even though we did not showed it, our notion of MLP has no problem dealing with one pixel wide areas, i.e. holes or bars of one pixel wide, as illustrated in Fig. 2.

Finally, note that one could modify the given algorithms in order to remove aligned points. This is done in a time proportional to the number of vertices of the MLP so that the overall linearity of the algorithms is not affected.

Acknowledgment. The authors would like to thank François de Vieilleville for helpful discussions about MLP.

References

1. Montanari, U.: A note on minimal length polygonal approximation to a digitized contour. Communications of the ACM 13(1), 41–47 (1970)
2. Sklansky, J., Chazin, R.L., Hansen, B.J.: Minimum perimeter polygons of digitized silhouettes. IEEE Trans. Computers 21(3), 260–268 (1972)
3. Hobby, J.D.: Polygonal approximations that minimize the number of inflections. In: SODA 1993: Proceedings of the fourth annual ACM-SIAM Symposium on Discrete algorithms, pp. 93–102. Society for Industrial and Applied Mathematics, Philadelphia (1993)
4. Klette, R., Yip, B.: The length of digital curves. Machine Graphics Vision 9(3), 673–703 (2000) (Also research report CITR-TR-54, University of Auckland, NZ. 1999)
5. Coeurjolly, D., Klette, R.: A comparative evaluation of length estimators of digital curves. IEEE Trans. Pattern Analysis and Machine Intelligence 26(2), 252–258 (2004)
6. Klette, R., Rosenfeld, A.: Digital Geometry - Geometric Methods for Digital Picture Analysis. Morgan Kaufmann, San Francisco (2004)
7. Sloboda, F., Stoer, J.: On piecewise linear approximation of planar Jordan curves. J. Comput. Appl. Math. 55(3), 369–383 (1994)
8. Sloboda, F., Zaťko, B.: On one-dimensional grid continua in R^2. Technical report, Institute of Control Theory and Robotics, Slovak Academy of Sciences, Bratislava, Slovakia (1996)

9. Sloboda, F., Zaťko, B.: On approximation of jordan surfaces in 3D. In: Bertrand, G., Imiya, A., Klette, R. (eds.) Digital and Image Geometry. LNCS, vol. 2243, pp. 365–386. Springer, Heidelberg (2002)
10. Guibas, L.J., Hershberger, J.: Optimal shortest path queries in a simple polygon. In: SCG 1987: Proceedings of the third annual symposium on Computational geometry, pp. 50–63. ACM, New York (1987)
11. Klette, R., Kovalevsky, V., Yip, B.: On length estimation of digital curves. In: Vision geometry VIII, SPIE Proceedings, vol. 3811, pp. 117–128 (1999)
12. Doerksen-Reiter, H., Debled-Rennesson, I.: Convex and concave parts of digital curves. In: Klette, R., Kozera, R., Noakes, L., Weickert, J. (eds.) Geometric Properties for Incomplete Data. Computational Imaging and Vision, vol. 31, pp. 145–160. Springer, Heidelberg (2006)
13. Lachaud, J.O., Vialard, A., de Vieilleville, F.: Fast, accurate and convergent tangent estimation on digital contours. Image and Vision Computing 25(10), 1572–1587 (2007); Discrete Geometry for Computer Imagery 2005
14. Melkman, A.A.: On-line construction of the convex hull of a simple polyline. Inf. Process. Lett. 25(1), 11–12 (1987)
15. Brlek, S., Lachaud, J.O., Provençal, X., Reutenauer, C.: Lyndon + christoffel = digitally convex. Pattern Recognition 42(10), 2239–2246 (2009); Selected papers from the 14th IAPR International Conference on Discrete Geometry for Computer Imagery 2008
16. Lothaire, M.: Combinatorics on words. Cambridge Mathematical Library. Cambridge University Press, Cambridge (1997)
17. Christoffel, E.B.: Observatio arithmetica. Annali di Mathematica 6, 145–152 (1875)
18. Borel, J.P., Laubie, F.: Quelques mots sur la droite projective réelle. J. Théor. Nombres Bordeaux 5(1), 23–51 (1993)
19. Berstel, J., Lauve, A., Reutenauer, C., Saliola, F.V.: Combinatorics on words of CRM Monograph Series, vol. 27. American Mathematical Society, Providence (2009); Christoffel words and repetitions in words
20. Duval, J.P.: Factorizing words over an ordered alphabet. J. Algorithms 4(4), 363–381 (1983)
21. Fredricksen, H., Maiorana, J.: Necklaces of beads in k colors and k-ary de Bruijn sequences. Discrete Math. 23(3), 207–210 (1978)
22. Lothaire, M.: Applied combinatorics on words. Encyclopedia of Mathematics and its Applications, vol. 105. Cambridge University Press, Cambridge (2005)
23. Berstel, J., de Luca, A.: Sturmian words, Lyndon words and trees. Theoret. Comput. Sci. 178(1-2), 171–203 (1997)

Multiscale Discrete Geometry*

Mouhammad Said[1,2], Jacques-Olivier Lachaud[1], and Fabien Feschet[2]

[1] Laboratoire de Mathématiques, UMR 5127 CNRS, Université de Savoie, 73376 Le Bourget du Lac, France
mouhammadsaiid@hotmail.com, jacques-olivier.lachaud@univ-savoie.fr

[2] LAIC, Univ. Clermont-Ferrand, IUT, Campus des Cézeaux, 63172 Aubière Cedex, France
feschet@laic.u-clermont1.fr

Abstract. This paper presents a first step in analyzing how digital shapes behave with respect to multiresolution. We first present an analysis of the covering of a standard digital straight line by a multi-resolution grid. We then study the multi-resolution of Digital Straight Segments (DSS): we provide a sublinear algorithm computing the exact characteristics of a DSS whenever it is a subset of a known standard line. We finally deduce an algorithm for computing a multiscale representation of a digital shape, based only on a DSS decomposition of its boundary.

Keywords: multiscale geometry, digital contours, standard lines, digital straight segment recognition, Stern-Brocot tree, multi-resolution.

1 Introduction

Multiscale analysis is a classical tool in image processing [15,11]. It has been introduced for addressing the fact that derivatives or geometric features are sensitive to the notion of scale. They do not intrinsically exist but they are often present only at a finite number of scales. To compute this representation, the original signal is embedded in a parameterized family of derived signals, obtained by convolutions of a parameterized family of Gaussian kernels, whose variance is the scale. At any scale, smooth versions of the derivatives are computed. Combined together, they provide a multiscale representation of the feature geometry, thus providing information about its significance for later interpretation or post-processing. Originally defined in the continuous space, these approaches have been extended to the discrete case [13] to better handle both the finite support and the finite resolution of digital images.

The use of Gaussian convolutions does not process well binary data such as digital curves. So, another approach must be followed. We can consider the approach of Vacavant *et al.* [14] as a multiscale approach in the sense that geometric objects are represented by a multi-resolution set of rectilinear tiles. However, the set of tiles is not intrinsic and is computed from the digital objects as the result of an optimization process. Indeed, they determine the minimal

* Part of this work was funded by ANR project GeoDIB (ANR-06-BLAN-0225-03).

S. Brlek, C. Reutenauer, and X. Provençal (Eds.): DGCI 2009, LNCS 5810, pp. 118–131, 2009.

number of rectangles whose union covers the considered object. In this way, they define geometric primitives such as lines and use them to analyze thick digital objects. Its potential is yet unclear for multiscale analysis of digital object features, and the constructed objects are not analytically defined.

We believe that, in the context of digital geometry, geometric primitives such as lines, circles or polynomials are of a great importance. Pieces of digital lines are excellent tangent estimators [7,12], circular arcs estimate curvature [1]. It is thus fundamental to keep them in the multiscale analysis of digital boundaries. Our point of view is therefore similar to the one of Figueiredo [8] who studied the behavior of 8-connected lines when changing the resolution of the grid. He was the first to characterize when the image of such a digital line in various resolution grids is another digital line.

In the present paper, we pursue on this idea and we provide in Section 2 an analytic description of how 4-connected digital straight lines, called *digital standard lines* (DSL), behave when the resolution of the grid is changed by an arbitrary factor. We prove first that their subsampling is also a standard line, whose parameters can all be analytically defined (Theorems 1 and 2).

The multiresolution of a digital straight segment (DSS) is addressed in Section 3. However, a full analytic approach seems out of reach at the moment. This is why we take an algorithmic approach. We present a new algorithm which computes the exact characteristics of a DSS that is a subset of a known DSL, given the endpoints of the DSS. Not surprisingly, it is based on continued fractions. Its worst-case computational complexity is $\Theta(\sum_{i=0}^{k} u_i)$, where $[u_0; u_1, \ldots, u_k]$ is the continued fraction of the slope of the input DSL. The expectation of this sum for fractions $\frac{a}{b}$ with $a+b=n$ is experimentally lower than $\log^2 n$, and this sum is upper bounded by n. This is to compare with classical DSS recognition algorithms (e.g., see [4]), whose complexities are at best $\Theta(n)$.

Section 4 applies the preceding results to the multiscale computation of a digital shape. We show there that if the digital shape was previously polygonalized as a sequence of M DSS, its exact multiresolution is computed in time linear with M times the average of the sums of the partial cofficients of its output DSS slopes. In most cases, this is clearly sublinear, and at worst, linear in the size of the contour. This paper is therefore a first step towards the multiscale computation of the tangential cover [6,12], a fundamental tool for representing and analyzing digital curves.

2 Covering of a Standard Line by a Lower Resolution Grid

A first step for defining multiscale discrete geometry is to find the covering of a digital straight line by a lower resolution grid. We shall prove here that this covering is indeed a digital straight line with computable characteristics (Theorem 1) and that this line is standard (Theorem 2). The proof is essentially technical and is based on the same principle as the proof of Figueiredo [8].

Let $D(a,b,\mu)$ be a standard digital line such that $0 < a < b$ and $\gcd(a,b)=1$. Let us consider the subgroup $S(h,v)=(Xh,Yv)$ of $\mathbb{Z}^2$ (where X, Y, h, v all are integers). Obviously the fondamental domain $[0,h)\times[0,v)$ of $S(h,v)$ and its translations by the vectors $X(h,0)+Y(0,v)$ induce a tiling of $\mathbb{Z}^2$ where each tile contains exactly one point of $S(h,v)$ (for which reason we will refer indifferently to the tile itself or to the point of $S(h,v)$ it contains). We are interested in the set of tiles of $S(h,v)$ that intersect $D(a,b,\mu)$. The set of tiles is a discrete line in the subgroup $S(h,v)$. We denote this line by Δ and call it the *covering line* of $D(a,b,\mu)$ in $S(h,v)$.

The tiling generated by $S(h,v)$ on $\mathbb{Z}^2$ induces a new coordinate system where coordinates (X,Y) are related to the canonical coordinates of $\mathbb{Z}^2$ by the following obvious relations where $\left[\frac{x}{h}\right]$ is the quotient of the euclidean division of x by h and $\left\{\frac{x}{h}\right\}$ is the remainder of this division:

$$X=\left[\frac{x}{h}\right] \qquad Y=\left[\frac{y}{v}\right] \qquad x=hX+\left\{\frac{x}{h}\right\} \qquad y=vY+\left\{\frac{y}{v}\right\}.$$

By definition, $D(a,b,\mu)$ can be written as

$$\mu \le ax+by < \mu+a+b \tag{1}$$

Hence the equation of Δ in the coordinate system related to S writes:

$$\mu \le a\left(hX+\left\{\frac{x}{h}\right\}\right)+b\left(vY+\left\{\frac{y}{v}\right\}\right) < \mu+a+b \tag{2}$$

In order to simplify (2) let us introduce

$$m_x=\left\{\frac{x}{h}\right\} \qquad m_y=\left\{\frac{y}{v}\right\} \tag{3}$$

Since m_x and m_y vary when x steps through $\mathbb{Z}$, equation (2) becomes:

$$\mu-\max_{x\in\mathbb{Z}}(am_x+bm_y) \le ahX+bvY < \mu+a+b-\min_{x\in\mathbb{Z}}(am_x+bm_y) \tag{4}$$

Now to fully determine this equation that defines the covering of the digital line by the tiling, the exact range of am_x+bm_y, i.e, the values of $\min_{x\in\mathbb{Z}}(am_x+bm_y)$ and $\max_{x\in\mathbb{Z}}(am_x+bm_y)$, need to be calculated.

By definition of m_x and m_y, we have $0\le m_x\le h-1$ and $0\le m_y\le v-1$. Using these bounds in (4) suggests for Δ an equation of the form:

$$\mu-a(h-1)-b(v-1) \le ahX+bvY < \mu+a+b. \tag{5}$$

But since $ahX+bvY$ can only assume values that are multiples of $g=\gcd(ah,bv)$, the bounds of equation (4) can be refined. We further denote $\frac{ah+bv}{g}$ by p.

However since m_x and m_y are linked through (2) and (3), the precise bounds of am_x+bm_y when x steps through $\mathbb{Z}$, need to be determined. In fact, we show in what follows, that even though am_x+bm_y does not reach the absolute bounds 0 and $a(h-1)+b(v-1)$ in some cases, equation (20) (given on page 123) always holds.

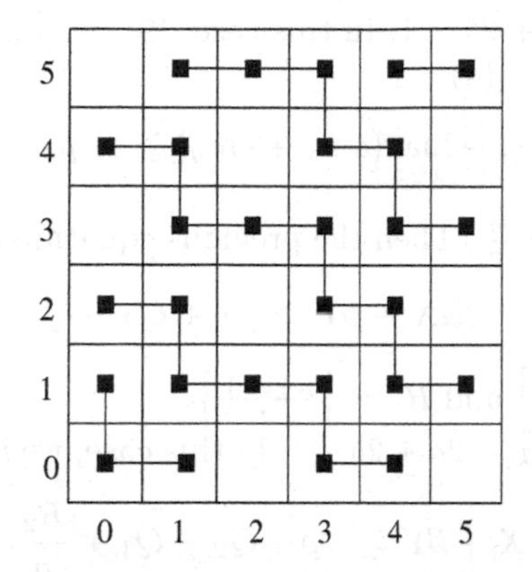

Fig. 1. Determination of the range of $2m_x+3m_y$. The black squares represent the family of standard digital lines defined by $5-6t \leq 2m_x+3m_y < 10-6t$ restricted to $[0,6) \times [0,6)$ and hence the possible values of (m_x, m_y).

Inverting (3) yields $x = kh + m_x, k \in \mathbb{Z}$ and $y = lv + m_y, l \in \mathbb{Z}$. Hence, after insertion into (1):

$$\mu - akh - blv \leq am_x + bm_y < \mu + a + b - akh - blv \tag{6}$$

The term $akh + blv$ has values which are multiples of g. Equation (6) can thus be rewritten as

$$\mu - tg \leq am_x + bm_y < \mu + a + b - tg, t \in \mathbb{Z} \tag{7}$$

Equation (7) describes a family of digital lines of direction (a, b) parameterized by $t \in \mathbb{Z}$, which we denote by D_t. The intersection of D_t with the domain of (m_x, m_y), $[0, h) \times [0, v)$ defines the possible pairs (m_x, m_y) and therefore the range of possible values for the sum $am_x + bm_y$ (see Figure 1).

Let us denote with (n_x, n_y) the pair of $D_t \cap [0, h) \times [0, v)$ that yields the maximum value for the sum $am_x + bm_y$. Let us also denote by d the difference between $am_x + bm_y$ evaluated at $(h-1, v-1)$ and at (n_x, n_y):

$$d = a(h-1) + b(v-1) - (an_x + bn_y) \tag{8}$$

Determining the maximum value of t verifying (7) for the point (n_x, n_y) provides a more precise expression of the lower bound of d. (8) becomes

$$d = (p+t)g - (\mu + 2a + 2b - 1).$$

To find the maximum value, we take $d = 0$ which implies that $(n_x, n_y) = (h - 1, v - 1)$, and

$$t = -p + Q_2 + \frac{R_2}{g}, \tag{9}$$

where $Q_2 = \left[\frac{\mu+2a+2b-1}{g}\right]$ and $R_2 = \left\{\frac{\mu+2a+2b-1}{g}\right\}$. Now, we must distinguish two possible cases for R_2.

1. If g divides $\mu + 2a + 2b - 1$. In this case, (9) becomes $t = -p + Q_2$ and we insert this equality in (4).

$$ahX + bvY \geq \mu - \max_{x \in \mathbb{Z}}(am_x + bm_y) \geq (-p + Q_2)\,g - a - b + 1$$

Let $\alpha = \frac{ah}{g}$ and $\beta = \frac{bv}{g}$. Then the previous equation can be rewritten as:

$$\alpha X + \beta Y \geq -p + Q_2 - Q_1 \tag{10}$$

where $Q_1 = \left[\frac{a+b-1}{g}\right]$ and $R_1 = \left\{\frac{a+b-1}{g}\right\}$.

2. If g does not divide $\mu + 2a + 2b - 1$. In this case, we insert (9) directly in (4).

$$\alpha X + \beta Y \geq -p + Q_2 - Q_1 + \frac{R_2}{g} - \frac{R_1}{g}$$

We calculate the difference between the division of R_2 by g and the division of R_1 by g, by comparing R_1 and R_2:

$$\text{If } R_2 \leq R_1, \text{ then } \alpha X + \beta Y \geq -p + Q_2 - Q_1 \tag{11}$$

$$\text{If } R_2 > R_1, \text{ then } \alpha X + \beta Y \geq -p + Q_2 - Q_1 + 1 \tag{12}$$

A similar reasoning can be applied with the minimum value of $am_x + bm_y$ throughout $D_t \cap [0, h) \times [0, v)$ obtained at point (n'_x, n'_y). Let us also denote by d' the difference between $am_x + bm_y$ evaluated at 0 and at (n'_x, n'_y):

$$d' = 0 - (an'_x + bn'_y). \tag{13}$$

Determining the maximum value of t verifying (7) for the point (n'_x, n'_y) provides a more precise expression of the upper bound of d'. (13) becomes

$$d' = 0 - (\mu - tg) = (\mu + 2a + 2b - 1) + tg - (2\mu + 2a + 2b - 1)$$

We take $d' = 0$ to find the minimum value, implying $(n'_x, n'_y) = (0, 0)$, then:

$$t = -Q_2 - \frac{R_2}{g} + \left(\frac{2\mu + 2a + 2b - 1}{g}\right) \tag{14}$$

At this point, we must distinguish two possible cases according to R_2.

1. If g divides $\mu + 2a + 2b - 1$. In this case, (14) becomes $t = -Q_2 + \left(\frac{2\mu+2a+2b-1}{g}\right)$ and we insert this equality in (4):

$$ahX + bvY < \mu + a + b - (\mu - tg) < -Q_2 g + 2\mu + 3a + 3b - 1$$

and,

$$\alpha X + \beta Y < -Q_2 + Q_3 + \frac{R_3}{g}, \tag{15}$$

where $Q_3 = \left[\frac{2\mu+3a+3b-1}{g}\right]$ and $R_3 = \left\{\frac{2\mu+3a+3b-1}{g}\right\}$.
We must further distinguish two possible cases for R_3.

(a) If g divides $2\mu + 3a + 3b - 1$. In this case Equation (15) becomes

$$\alpha X + \beta Y < -Q_2 + Q_3. \quad (16)$$

(b) If g does not divide $2\mu + 3a + 3b - 1$. In this case Equation (15) becomes

$$\alpha X + \beta Y < -Q_2 + Q_3 + 1. \quad (17)$$

2. If g does not divide $\mu + 2a + 2b - 1$. In this case we insert (14) directly in (4):

$$ahX + bvY < \mu + a + b - (\mu - tg) < -Q_2 g - Q_1 + 2\mu + 3a + 3b - 1.$$

Then the previous equation can be rewritten as:

$$\alpha X + \beta Y < Q_3 - Q_2 + \frac{R_3}{g} - \frac{R_2}{g}.$$

We calculate the difference between the division of R_3 by g and the division of R_2 by g, by comparing R_2 and R_3:

$$\text{If } R_3 \leq R_2, \text{ then } \alpha X + \beta Y < -Q_2 + Q_3 \quad (18)$$
$$\text{If } R_3 > R_2, \text{ then } \alpha X + \beta Y < -Q_2 + Q_3 + 1 \quad (19)$$

More precisely, eleven equations involving $\alpha X + \beta Y$ can be obtained by combining the lower bound (equations (10), (11), (12)) and the upper bound (equations (16), (17), (18), (19)) of $\alpha X + \beta Y$. Among the eleven possible cases, four cases are impossible. For the remaining cases we have also conditions which yield the same result. Therefore those equations can be formulated as a single expression given below.

Theorem 1. *The digital straight line Δ of $S(h, v)$ covering the standard digital line $D(a, b, \mu)$ of $\mathbb{Z}^2$ is defined by:*

$$-p + Q_2 - Q_1 + SI \leq \alpha X + \beta Y < Q_3 - Q_2 + SS \quad (20)$$

Where $\alpha = \frac{ah}{g}$, $\beta = \frac{bv}{g}$, $g = \gcd(ah, bv)$, $p = \alpha + \beta$ *and*

$$SI = \begin{cases} 0 \text{ if } & R_2 \leq R_1 \\ 1 & \textit{otherwise} \end{cases} \qquad SS = \begin{cases} 0 \text{ if } & R_3 \leq R_2 \\ 1 & \textit{otherwise} \end{cases}$$

Figure 2 illustrates the covering of $D(7, 9, 6)$ in the subsampling $(6, 6)$.

Theorem 2. *The covering line Δ is standard.*

In order to prove that the difference between the upper and lower bounds of (20) is equal to $\alpha + \beta$, we first need the following proposition and lemma.

Proposition 1.
$$\frac{R_2 - R_1 - R}{g} = \begin{cases} 0 \text{ if } & R_1 \leq R_2 \\ -1 & \textit{otherwise.} \end{cases} \quad \textit{and} \quad \frac{R_3 - R_2 - R}{g} = \begin{cases} 0 \text{ if } & R_2 \leq R_3 \\ -1 & \textit{otherwise.} \end{cases}$$

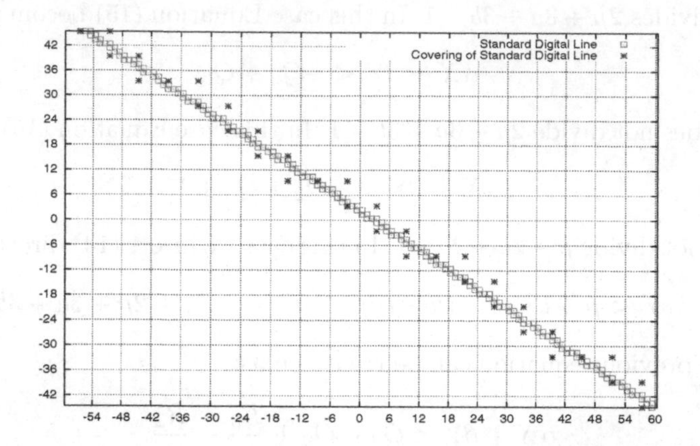

Fig. 2. The covering line of $D(7,9,6)$ by $S(6,6) : -12 \leq 7X + 9Y < 4$

Proof. Since $0 \leq R_1 < g$, $0 \leq R_2 < g$ and $0 \leq R < g$, then $0 \leq \frac{R_1}{g} < 1$, $0 \leq \frac{R_2}{g} < 1$ and $0 \leq \frac{R}{g} < 1$. Thus we can write $-2 < \frac{R_2-R_1-R}{g} < 1$. To specify the exact integer value of $\frac{R_2-R_1-R}{g}$, we must compare R_1 and R_2.

- If $R_1 \leq R_2$, then $R_2 = (\mu + 2a + 2b - 1) \bmod g = (\mu + a + b) \bmod g + (a + b - 1) \bmod g = R + R_1$. Therefore $R_2 - R_1 - R = 0$ and $\frac{R_2-R_1-R}{g} = 0$.
- If $R_1 > R_2$, then $R_2 = (\mu + 2a + 2b - 1) \bmod g = (\mu + a + b) \bmod g + (a + b - 1) \bmod g - g = R + R_1 - g$. Therefore $R_2 - R_1 - R = -g$ and $\frac{R_2-R_1-R}{g} = -1$.

The proof of the second condition is analogous to the first one. □

Lemma 1.

$$Q_2 - Q_1 = Q + \begin{cases} 0 \text{ if } & R_1 \leq R_2 \\ 1 & \textit{otherwise.} \end{cases} \quad \textit{and} \quad Q_3 - Q_2 = Q + \begin{cases} 0 \text{ if } & R_2 \leq R_3 \\ 1 & \textit{otherwise.} \end{cases}$$

Proof. We only provide the proof of the first result and the second result has a similar proof, $\mu + a + b = (\mu + 2a + 2b - 1) - (a + b - 1)$, then by definition of the Euclidean division, we can write $\mu + a + b = gQ + R$, $a + b - 1 = gQ_1 + R_1$, and $\mu + 2a + 2b - 1 = gQ_2 + R_2$. So, the first equality can be rewritten as $gQ + R = (gQ_2 + R_2) - (gQ_1 + R_1)$. Then,

$$g(Q - Q_2 + Q_1) = R_2 - R_1 - R$$

To reduce this equality, we must compare R_1 and R_2

1. If $R_1 \leq R_2$, then by Proposition 1, $\frac{R_2-R_1-R}{g} = 0$ which implies that $Q = Q_2 - Q_1$,
2. If $R_1 > R_2$, then by Proposition 1, $\frac{R_2-R_1-R}{g} = -1$ which implies that $Q = Q_2 - Q_1 - 1$.

□

Proof (Theorem 2). We want to prove that the difference between the upper and lower bounds of (20) is equal to $\alpha + \beta$. There are 5 cases to consider. From Proposition 1 and Lemma 1, we get:

- If $R_3 < R_2 < R_1$, then $SI = 0$, $SS = 0$, $Q_2 - Q_1 = Q + 1$ and $Q_3 - Q_2 = Q + 1$.
- If $R_2 < R_1$ and $R_2 < R_3$, then $SI = 0$, $SS = 1$, $Q_2 - Q_1 = Q + 1$ and $Q_3 - Q_2 = Q$.
- If $R_1 < R_2 < R_3$, then $SI = 1$, $SS = 1$, $Q_2 - Q_1 = Q$ and $Q_3 - Q_2 = Q$.
- If $R_1 < R_2$ and $R_3 < R_2$, then $SI = 1$, $SS = 0$, $Q_2 - Q_1 = Q$ and $Q_3 - Q_2 = Q + 1$.
- If $R_1 = R_2 = R_3$, then $SI = 0$, $SS = 0$, $Q_2 - Q_1 = Q$ and $Q_3 - Q_2 = Q$.

It is easy to see for all cases that, $(Q_3 - Q_2 + SS) - (-p + Q_2 - Q_1 + SI) = p$.

□

3 A Fast DSS Recognition Algorithm

We focus now our attention on the multiresolution of digital straight segments (DSS), which are finite connected parts of digital lines. In the first subsection, we recall first some links between DSS, patterns, continued fraction of the slope, and Stern-Brocot tree representation of fractions. Let S be some DSS included in a line D. We then obtain easily Proposition 3, which states that the slope of the (h, v)-multiresolution S' of S is some reduced fraction of the slope of the (h, v)-multiresolution D' of D. However, the full analytic description of S' seems out of reach at the moment. This is why we propose an algorithmic approach in the second subsection. All the characteristics of S' can be determined in time sublinear in the number of points of S' (Algorithm 2 and Proposition 4).

3.1 DSS, Patterns, Irreducible Fractions and Continued Fractions

Given a standard line (a, b, μ), we call *pattern* of characteristics (a, b) the succession of Freeman moves between any two consecutive upper leaning points. The Freeman moves defined between any two consecutive lower leaning points is the previous word read from back to front and is called the *reversed pattern.* As noted by several authors (e.g. see [10], or the work of Berstel reported in [2]), the pattern of any slope can be constructed from the continued fraction of the slope. We recall that a *simple continued fraction* is an expression:

$$z = \frac{a}{b} = [u_0, u_1, u_2, ..., u_i, ..., u_n] = u_0 + \frac{1}{u_1 + \frac{1}{\ldots + \frac{1}{u_{n-1} + \frac{1}{u_n}}}},$$

where n is the *depth* of the fraction, and u_0, u_1, etc, are all integers and called the *partial quotients.* We call *k-th convergent* the simple continued fraction formed of the k first partial quotients: $z_k = \frac{p_k}{q_k} = [u_0, u_1, u_2, ..., u_k]$.

The role of partial quotients can be visualized with a structure called *Stern-Brocot tree* (see [9] for a complete definition) which is a hierarchy containing all the positive irreductible rational fractions. The idea under its construction is to begin with the two fractions $\frac{0}{1}$ and $\frac{1}{0}$ and to repeat the insertion of the median of these two fractions as follows: insert the median $\frac{m+m'}{n+n'}$ between $\frac{m}{n}$ and $\frac{m'}{n'}$. The sequence of partial quotients defines the sequence of right and left moves down the tree. Many works deal with the relations between irreducible rational fractions and digital lines (see [5] for characterization with Farey series, and [16] for a link with decomposition into continuous fractions). In [3], Debled and Réveillès first introduced the link between this tree and the recognition of digital line. Recognizing a piece of digital line is like going down the Stern-Brocot tree up to the directional vector of the line. To sum up, the classical online DSS recognition algorithm **DR95** [3] (reported in [10]) updates the DSS slope when adding a point that is just exterior to the current line (weak exterior points). The slope evolution is analytically given by next property.

Proposition 2. *[2] The slope evolution in **DR95** depends on the parity of the depth of its slope, the type of weakly exterior point added to the right. This is summed up in the table below, where the slope is $[0, u_1, ..., u_k]$, $k = 2i$ even or $k = 2i + 1$ odd, δ pattern(s) and δ' reversed pattern(s):*

	Even k	*Odd k*
Upper weakly exterior	$[0, u_1, ..., u_{2i}, \delta]$	$[0, u_1, ..., u_{2i+1} - 1, 1, \delta]$
Lower weakly exterior	$[0, u_1, ..., u_{2i} - 1, 1, \delta']$	$[0, u_1, ..., u_{2i+1}, \delta']$

A slope evolution during recognition is then clearly equivalent to descending in the Stern-Brocot tree, to the left or to the right. Since S' is a piece of a DSL of slope $\frac{a'}{b'}$, we therefore obtain:

Proposition 3. *If $D'(a', b', \mu')$ is the covering of some digital line D by a tiling $S(h, v)$, then any segment $S \subset D$ induces a segment $S' \subset D'$ by the same tiling, such that the slope of S' is $\frac{a'}{b'}$ or one of its ancestors in the Stern-Brocot tree.*

3.2 Fast DSS Recognition When DSL Container Is Known

We provide a general algorithmic solution to the multiresolution of a DSS that is sublinear in the number of its points (Algorithm 2). We exploit the fact that in our case, we *know* that it is a piece of a standard digital line of known characteristics (with Theorems 1 and 2). Therefore most of the points tested in **DR95** are useless since they do not lead to any slope evolution. Proposition 3 tells us that we must find the slope of S' in the ancestors of the slope of D'. We must also determine the shift to origin of S'.

Starting from the initial correct quadrant, Algorithm 2 determines progressively the positions of the weakly exterior points. They are related to the Bézout coefficient of the current DSS slope $\frac{p_k}{q_k}$ (line 1). Since we update at each step the continued fraction of the slope, these coefficients are given in $O(1)$ time by the formula

$$\text{Bezout}(p, q, k) = \begin{cases} (q_k - q_{k-1}, p_k - p_{k-1}), \text{when } k \text{ is even} \\ (q_{k-1}, p_{k-1}), \text{when } k \text{ is odd} \end{cases} .$$

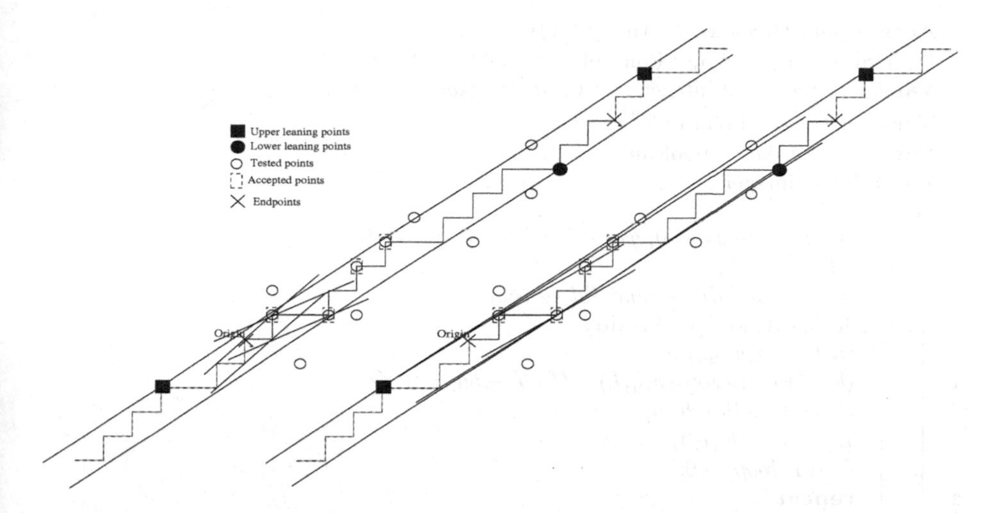

Fig. 3. A digital straight line $D(13, 17, -5)$ with an odd slope. Computes the characteristics (a, b, μ) of a DSS S that is the subset of D between the origin and the point $(12, 9)$. The intermediate slopes are drawn with solid lines on the left and on the right, tested points are circled.

Action UpdateSlope(**In** uw, δ, **InOut** k, u, p, q) ;
uw : boolean /* *True iff upper weak leaning point* */ ;
δ : integer /* *number of (reversed) patterns* */ ;
k : integer /* *depth of slope continued fraction* */ ;
u, p, q : array of integers /* *slope cont. fraction* */ ;
begin
  if *(uw = true and k is odd) or (uw = false and k is even)* **then**
    $u_k \leftarrow u_k - 1$, $p_k \leftarrow p_k - p_{k-1}$, $q_k \leftarrow q_k - q_{k-1}$;
    $u_{k+1} \leftarrow 1$, $p_{k+1} \leftarrow p_k + p_{k-1}$, $q_{k+1} \leftarrow q_k + q_{k-1}$;
    $k \leftarrow k + 1$;
  if $\delta = 1$ **then**
    $u_k \leftarrow u_k + 1$, $p_k \leftarrow p_k + p_{k-1}$, $q_k \leftarrow q_k + q_{k-1}$;
  else
    $u_{k+1} \leftarrow \delta$, $p_{k+1} \leftarrow \delta p_k + p_{k-1}$, $q_{k+1} \leftarrow \delta q_k + q_{k-1}$;
    $k \leftarrow k + 1$;
end

Algorithm 1. Updates in $O(1)$ the slope of a DSS according to the addition of an upper leaning point (uw is $true$) or lower leaning point (uw is $false$), to the number of patterns or reversed patterns δ, and to the current continued fraction of the slope

The algorithm then checks in sequence upper and lower weakly exterior points so as to find the first in the DSL (repeat block at line 2). The number of iteration gives the number of pattern repetitions. Once such a point is found, Proposition 2 indicates how to update the slope, depending if it is a slope increase (line 3) or

```
Action SmartDSS( In D, In P, Q, Out S ) ;
D : DSL (α, β, μ'), P, Q : Point of Z²S : DSS (a, b, μ) ;
Var u, p, q : array of integers  /* Cont. fraction a/b = [u_0, ..., u_k] = p_k/q_k */ ;
Var U, L, U', L' : Point of Z² ;
Var ulu, lul, inside : boolean ;
Var k, loop : integer ;
begin
    k ← 0, u_0 ← 0, p_0 ← 0, q_0 ← 1, p_{-1} ← 1, q_{-1} ← 0 ;
    U ← P, L ← P ;
    inside ← true, ulu ← true, lul ← true ;
    while inside and p_k ≠ α do
        (a, b) ← (p_k, q_k) ;
1       (b', a') ← Bézout(p, q, k)  /* ab' − ba' = 1 */ ;
        U' ← U + (b − b', a − a');
        L' ← L + (b', a');
        δ ← 1, loop ← 0;
2       repeat
            U' ← U' + (b, a), L' ← L' + (b, a);
            if U'_y ≤ Q_y and U' ∈ D then loop ← 1 ;
            else if L'_x ≤ Q_x and L' ∈ D then loop ← 2 ;
            else δ ← δ + 1 ;
        until loop ≠ 0 or U'_y ≥ Q_y or L'_x ≥ Q_x ;
3       if loop = 1 then
            /* Increase slope with weak upper leaning point U' */ ;
            UpdateSlope(true, δ, k, u, p, q ) ;
4           L ← L' − (b', a') ;
            if not lul then L ← L − (b, a);
            ulu ← true, lul ← false ;
5       if loop = 2 then
            /* Decrease slope with weak lower leaning point L' */ ;
            UpdateSlope(false, δ, k, u, p, q) ;
6           U ← U' − (b − b', a − a') ;
            if not ulu then U ← U − (b, a);
            ulu ← false, lul ← true ;
        else
            inside ← false;
    a ← p_k, b ← q_k, μ ← aU_x − bU_y ;
end
```

Algorithm 2. Computes the characteristics (a, b, μ) of a DSS that is some subset of a DSL D, given a starting point P and an ending point Q $(P, Q \in D)$. The computation time is linear with the sum $\sum_{i=0}^{k} u_k$ with $\frac{a}{b} = [u_0, \ldots, u_k]$.

decrease (line 5). This is effectively implemented in $O(1)$ time with Algorithm 1. It remains to update consistently the new first lower leaning point (line 4) or upper leaning point (line 6). We have to be a little careful whether the current DSS is a pattern (ULU) or a reversed pattern (LUL): it is just to adapt δ to be the number of patterns or the number of reversed patterns. The algorithm stops either when point Q has been reached or when DSL slope has been reached.

An execution of this algorithm is illustrated on Figure 3. 11 points have been tested compared with a DSS length of 23. We show below that the worst-case time complexity of Algorithm 2 is related to the partial quotients of the DSS slope and not to the slope itself.

Proposition 4. *Let S be a DSS of slope $\frac{a}{b} = [u_0, u_1, ..., u_k]$, $T(n)$ the number of points on S tested by Algorithm 2 to recognize $\frac{a}{b}$ with $\sum_{i=0}^{k} u_i = n$, then $T(n) \leq 2n$ (it only depends on the sum of u_i).*

Proof. We prove by induction on n that $T(n) \leq 2n$. The initial state $T(0) = 0$ is obvious since P and the quadrant is known. Assume that $T(n) \leq 2n$ and we shall prove that $T(n+1) \leq 2n+2$.

As $T(n) = 2n$, then $\frac{a}{b}$ is some $[u_0, u_1, ..., u_k]$. According to Stern-Brocot tree and Proposition 2, there are only two possible evolutions for the slope, either $[u_0, u_1, ..., u_k, 1]$ or $[u_0, u_1, ..., u_k - 1, 1, 1]$. The computation of the next slope on $n+1$ depends of the parity of the slope depth and the type of weakly exterior point added.

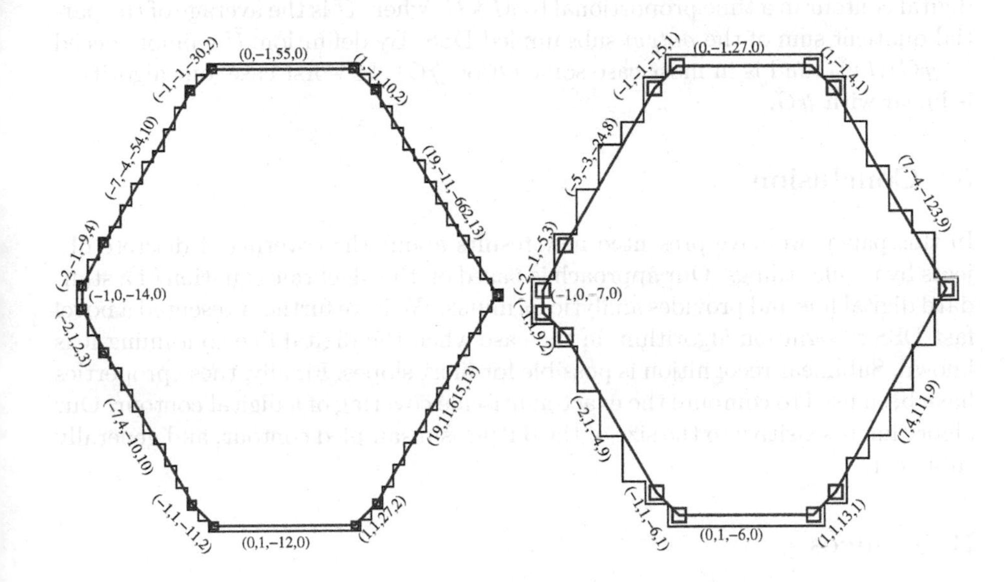

Fig. 4. (a) Multiscale computation of the boundary of a digital shape according to several tiling (h, v). (b) $(h, v) = (1, 1)$ and original DSL is $(4, 7, -6)$. (c) $(h, v) = (2, 2)$, DSL is $(4, 7, -8)$ and extracted DSS is $(4, 7, -8)$. (d) $(h, v) = (3, 4)$, DSL is $(16, 21, -32)$ and extracted DSS is $(3, 4, -6)$.

Assume first that the slope has even depth $k = 2i$. If this point is upper weakly exterior, then the new slope increases to $[u_0, u_1, ..., u_{2i}, 1]$ and $T(n+1) = T(n)+1$. In the other case, we test the lower weakly exterior point, then the slope decreases to $[u_0, u_1, ..., u_{2i} - 1, 1, 1]$ and $T(n+1) = T(n) + 2$. Therefore we conclude that $T(n+1) \leq T(n) + 2 \leq 2n + 2$.

The reasonning is similar in the other case. We note that the loop at line 2 agglomerates a sequence of δ patterns, instead of doing it one by one as in this proof. □

4 Multiscale Covering of a Digital Contour

The preceding results allow us to compute an exact multiscale representation of a digital contour C with $\#C$ points, if this contour has been beforehand decomposed into a sequence of M digital straight segments. The process is illustrated on Figure 4. The 4-connected inner pixel contour of a digital shape is first extracted and decomposed greedily into digital straight segments. Taking for instance DSL $(19, -11, -662)$ associated to a DSS of length 32, we compute two (h, v) coverings of this segment. The $(1, 1)$-covering requires 13 tests and the $(2, 2)$ covering requires 9 tests. A multiresolution algorithm tracing the original DSS would have made about 32 tests, whichever the chosen covering.

Our technique allows us to compute the exact multiscale representation of a digital contour in a time proportional to $M \times \overline{U}$, where $\overline{U}$ is the average of the partial quotient sum of the *output* subsampled DSS. By definition, $\overline{U}$ cannot exceed $O(\#C/(hv))$, and is in most case some $O(\log \#C)$. In worst case, our algorithm is linear with $\#C$.

5 Conclusion

In this paper, we have presented new results about the covering of discrete objects by regular tilings. Our approach is based on the algebraic equation of a standard digital line and provides analytic formulas. We have further presented a novel fast DSS recognition algorithm, in the case when the digital line containing it is known. Sublinear recognition is possible for most slopes. Finally, these properties have been used to compute the exact multiscale covering of a digital contour. Our algorithm is sensitive to the size of the output subsampled contour, and generally sublinear.

References

1. Coeurjolly, D., Grard, Y., Reveills, J.-P., Tougne, L.: An elementary algorithm for digital arc segmentation. Discrete Applied Mathematics 139, 31–50 (2004)
2. de Vieilleville, F., Lachaud, J.-O.: Revisiting digital straight segment recognition. In: Kuba, A., Nyúl, L.G., Palágyi, K. (eds.) DGCI 2006. LNCS, vol. 4245, pp. 355–366. Springer, Heidelberg (2006)
3. Debled-Rennesson, I.: Etude et reconnaissance des droites et plans discrets. PhD thesis, Université Louis Pasteur, Strasbourg (1995)
4. Debled-Rennesson, I., Reveillès, J.-P.: A linear algorithm for segmentation of discrete curves. International Journal of Pattern Recognition and Artificial Intelligence 9, 635–662 (1995)
5. Dorst, L., Smeulders, A.W.M.: Discrete representation of straight lines. IEEE transactions Pattern Analysis Machine Intelligence 6, 450–463 (1984)

6. Feschet, F.: Canonical representations of discrete curves. Pattern Anal. Appl. 8(1-2), 84–94 (2005)
7. Feschet, F., Tougne, L.: Optimal time computation of the tangent of a discrete curve: Application to the curvature. In: Bertrand, G., Couprie, M., Perroton, L. (eds.) DGCI 1999. LNCS, vol. 1568, pp. 31–40. Springer, Heidelberg (1999)
8. Figueiredo, O.: Advances in discrete geometry applied to the extraction of planes and surfaces from 3D volumes. PhD thesis, EPFL, Lausanne (1994)
9. Hardy, G.H., Wright, E.M.: An introduction to the Theory of Numbers. Oxford Society (1989)
10. Klette, R., Rosenfeld, A.: Digital Geometry - Geometric Methods for Digital Picture Analysis. Morgan Kaufmann, San Francisco (2004)
11. Koenderink, J.J.: The structure of images. Biol. Cyb. 50, 363–370 (1984)
12. Lachaud, J.-O., Vialard, A., de Vieilleville, F.: Fast, accurate and convergent tangent estimation on digital contours. Image Vision Comput. 25(10), 1572–1587 (2007)
13. Lindeberg, T.: Discrete derivative approximations with scale-space properties: A basis for low-level features extraction. Journal of Mathematical Imaging and Vision 3(4), 349–376 (1993)
14. Vacavant, A., Coeurjolly, D., Tougne, L.: Dynamic reconstruction of complex planar objects on irregular isothetic grids. In: Bebis, G., Boyle, R., Parvin, B., Koracin, D., Remagnino, P., Nefian, A., Meenakshisundaram, G., Pascucci, V., Zara, J., Molineros, J., Theisel, H., Malzbender, T. (eds.) ISVC 2006. LNCS, vol. 4292, pp. 205–214. Springer, Heidelberg (2006)
15. Witkin, A.P.: Scale-space filtering. In: Proc. 8th Int. Joint Conf. Art. Intell., Larlsruhe, Germany, pp. 1019–1022 (1983)
16. Yaacoub, J.: Enveloppes convexes de réseaux et applications au traitement d'images. PhD thesis, Université Louis Pasteur, Strasbourg (1997)

Vanishing Point Detection with an Intersection Point Neighborhood*

Frank Schmitt and Lutz Priese

Institute for Computational Visualistics, University of Koblenz-Landau, Germany
{fschmitt,priese}@uni-koblenz.de
http://www.uni-koblenz-landau.de/koblenz/fb4/institute/icv/agpriese

Abstract. A new technique to automatically detect the vanishing points in digital images is presented. The proposed method borrows several ideas from various papers on vanishing point detection and segmentation in sparse images and recombines them with a new intersection point neighborhood on $\mathbb{Z}^2$.

Keywords: Image analysis, measurable lines, intersection points, vanishing points.

1 Introduction

There exists a large literature on automatic detection of vanishing points in digital images by an analysis of intersection points of straight lines, using rather different techniques. There are three main problems: i) finding straight lines, ii) construction of an accumulator space of intersection points, and iii) interpretation of the values in the intersection point accumulator space.

We add new ideas to i) in 5.1, where we add a wildness map to reduce noise in the Hough transformation, to ii) in 4.1, by using a finite part of $\mathbb{Z}^2$ with a new intersection point neighborhood as a substitute for an "accumulator" of intersection points, and to iii) in 4.2, where we cluster intersection points to candidates for vanishing points with the AGS algorithm of [1]. The introduction of the intersection point neighborhood is our main contribution. This neighborhood defines a proximity that is proportional to the amount of theoretically possible intersection points of straight lines in an image. This amount depends on the way one measures straight lines. The intersection point neighborhood technique does not require calibrated cameras, does not impose restrictions on the geometry and number of vanishing points, but improves vanishing point detection.

2 Notations

We regard an *image* I as a mapping $I : Loc \to Val$ that maps coordinates $p \in Loc$ to values $I(p)$ in Val. Usually $Loc = [0, N-1] \times [0, M-1]$ in 2-dimensional images

* This work was supported by the DFG under grant PR161/12-1 and PA599/7-1.

S. Brlek, C. Reutenauer, and X. Provençal (Eds.): DGCI 2009, LNCS 5810, pp. 132–143, 2009.

and $Val = [0, 2^n - 1]$ for gray value images or $Val = [0, 2^n - 1]^3$ for color images, where $[N, M]$ denotes the interval of integers between N and M. However, we also allow $Loc \subseteq \mathbb{Z}^2$. I is a *binary* image if $|Val| = 2$. Often the values 0 and 255 are used in binary images.

A *Hough transformation* of I is a function $h_I : H \to \mathbb{N}$. H is called the *accumulator*, an element $b \in H$ a *bin*. Usually H is of some higher dimension and a bin is a formal description of an object or a set of objects in I. The value $h_I(b)$ tells how often the object described by b appears in the image I. It is often helpful to regard h_I as an image with H as its location and values from $\mathbb{N}$. The Hough transformation is frequently used to detect straight lines in a binary image I. For this, a straight line is represented in the *Hesse normal form* and the origin $(0, 0)$ of $\mathbb{R}^2$ is thought to be in the middle of Loc_I. The Hesse normal form describes a straight line l by two parameters: α, the *angle* between the normal of the line and the x-axis, and d, the *distance* between line and image origin. In this representation the accumulator H is 2-dimensional, one coordinate for α and another for d, and a bin (α, d) corresponds to the straight line $l = \{(x, y) \in \mathbb{R}^2 | x \cdot cos\, \alpha + y \cdot sin\, \alpha = d\}$.

For any image I we denote by I^e the binary image of all edges in I that one may compute by transforming I into a greyscale image and applying a Canny edge detector. I^l denotes the binary image of all straight lines in I. An edge point in I^e and a line point in I^l is set to the value 255. By I^i we denote a list of measured intersection points from pairs of straight lines in I^l. I^i is used to compute candidates for the vanishing points.

For a set $M \subseteq \mathbb{R}^n$ of n-dimensional vectors the mean μ_M and standard deviation σ_M are the vectors of the means μ^i_M and standard deviations σ^i_M of all i-th coordinates in M.

3 Previous Work on Vanishing Point Detection

We mention only a few papers from the large literature on vanishing point detection that have influenced our approach.

One may regard the list I^i of measured intersection points as an infinite image $I^i : \mathbb{Z}^2 \to \mathbb{N}$, where $I^i(p)$ tells us how many pairs of straight lines from I intersect in $p \in \mathbb{Z}^2$. Seo et al present in [2] a technique where the infinite image I^i is transformed into three finite images I^i_1, I^i_2, and I^i_3. I^i_1 is I^i restricted to the locations Loc_I of I. Let Q_i be one of four infinite parts of $\mathbb{R}^2$ between two neighbored infinite diagonals through the origin (0,0) in $\mathbb{R}^2$ with Q_1 left , Q_2 right, Q_3 above, and Q_4 below (0,0). Further, set Q'_i to be the infinite part of I^i inside Q_i but outside *Loc*. Seo now uses a reciprocal transformation to put Q'_1 and Q'_2 into a single finite image I^i_2 and Q'_3 and Q'_4 into I^i_3. However, the transformation seems to be ad hoc without a theoretical justification. Also, a cluster of intersection points in I^i near a corner of *Loc* may lead to three smaller clusters in I^i_1, I^i_2 and I^i_3 and will complicate the cluster analysis.

Almansa et al use in [3] the idea to build a histogram of I^l. The bins of the histogram are areas in $\mathbb{R}^2$. Positions inside the image location *Loc* are represented by a circle c in $\mathbb{R}^2$ with center in (0,0) and diameter equaling the image diagonal. All histogram bins inside c are of the same size. Outside the circle the bins are areas of the shape of a segment of a circle from radius r_1 to r_2 around (0,0) and from angle ω_1 to ω_2. The values of $r_1, r_2, \omega_1, \omega_2$ are chosen in such a way that all bins have the same probability to contain a straight line in $\mathbb{R}^2$ that intersects c. This is a nice idea. However, the distribution of the discrete straight lines in digital images is very different from that idealistic distribution in $\mathbb{R}^2$ and depends on the way one measures discrete straight lines. Also, a histogram of the intersection points in I^i seems to be more adequate than of straight lines in I^l.

In architectural environments with parallel buildings on a plane there are normally two or three vanishing points. Rother [4] tries to use triples of three intersection points in the list I^i as candidates for vanishing points and searches triples fulfilling geometric constraints. However, if there are n straight lines in I^l one gets up to $O(n^2)$ intersection points and thus $O(n^5)$ possible triples of intersection points, leading to an algorithm with a running time that is $O(n^5)$, which is not acceptable in practice.

Seo [2] constructs a histogram of the angles of all found straight lines. The histogram is separated into three intervals. It is argued that intersection points of two straight lines from different intervals cannot become a vanishing point. Thus, only the intersections of straight lines of the same interval are regarded in I^i. This should simplify I^i substantially.

4 Some Theoretical Considerations

4.1 Intersection Point Neighborhood

Measurable Straight Lines. We follow the ideas in [2], [3] and [4]. A first consideration is that the size of the bins for some accumulator for straight lines or intersection points should depend on the number of *measurable* straight lines that intersect an image I. Suppose, one uses an algorithm where the straight lines are detected with a Hough transformation using the Hesse normal form. The accumulator H is usually set to be $H = [0, A-1] \times [0, D-1]$ of size $A \cdot D$ where $[0, A-1]$ represents the discrete angles and $[0, D-1]$ represents the discrete distances of the Hesse normal form. Usually, not all straight lines (α, d) represented in H will intersect with the coordinates $Loc(I)$ of the image. Remove those "false" bins of straight lines not intersecting with the image from H and call the resulting set H_l the *admissible accumulator of size* (A, D). A bin in H_l is called a *measurable line* (for I). If one prefers a technique that does not use a Hough transformation the definition of a measurable straight line must be adopted. In any case, a measurable line must be

- straight,
- intersecting with the image,
- detectable.

For simplicity, we continue our argumentation with a Hough transformation and an admissible accumulator H_l.

(a) Clipping of size 4000x3000 (b) Clipping of size 1600x1200

Fig. 1. Φ, measurable lines per pixel

Frequency Functions. Let $\Phi(p)$ be the number of measurable lines that run through $p \in \mathbb{Z}^2$. We call $\Phi : \mathbb{Z}^2 \rightarrow \mathbb{N}$ the *line frequency* function. As $\Phi(p)$ tells how many measurable lines run through p exactly $\Psi(p) := (\Phi^2(p) - \Phi(p))/2$ many pairs of those lines may intersect in p. Ψ is called the *intersection frequency* function.

Figure 1 presents a part of Φ for an admissible accumulator H_l of size (500, 500) for an image of location size 800 × 600. Φ is restricted to an area in $\mathbb{N}^2$ of size 4000 × 3000 in (a) respectively of size 1600 × 1200 in (b), where in both cases Loc_I is in the upper left corner. The values of Φ are transformed into the interval [0,255] to present Φ as a greyscale image. We have calculated the mapping $\Phi : \mathbb{Z}^2 \rightarrow \mathbb{N}$ iteratively. Initially $\Phi(p)$ is set to 0 for all $p \in \mathbb{Z}^2$. For each measurable line $l \in H_l$ we virtually draw l into the image Φ by a standard technique for drawing lines in discrete images. Whenever our virtual drawing touches a coordinate $p \in \mathbb{Z}^2$ Φ is accumulated by 1 in p.

It turns out that the frequency function is not constant inside Loc_I, however the line frequencies inside Loc_I are very similar. Outside Loc_I the line frequencies differ heavily. Let $\hat{\Phi}$ denote the mean value of $\Phi(p)$ inside Loc_I. We approximate with $\hat{\Psi} := (\hat{\Phi}^2 - \hat{\Phi})/2$ the mean intersection frequency inside Loc_I.

Intersection Point Neighborhoods. We want to define for any position $p \in \mathbb{Z}^2$ a neighborhood $N(p) \subseteq \mathbb{Z}^2$ in such a way that the aggregated intersection frequencies $\sum_{p' \in N(p)} \Psi(p')$ are independent of p. For the sake of simplicity, a *neighborhood* $N(p)$ *of* p shall become a circle

$$c_r(p) := \{p' \in \mathbb{Z}^2 | d_E(p, p') < r\}$$

of radius r around p. d_E is the Euclidean distance in $\mathbb{R}^2$. For $p \in Loc_I$ we choose some fixed radius $\hat{r}$. The question is how to choose a radius r_p for a point p outside Loc_I. We assume for a moment that the intersection frequency function is constant inside the circle $c_{r_p}(p)$. As the area of a circle of radius r is proportional to r^2 one should choose r_p in such a way that $r_p^2 \cdot \Psi(p) = \hat{r}^2 \cdot \hat{\Psi}$.

Thus,

$$\frac{r_p^2}{\hat{r}^2} = \frac{\hat{\Psi}}{\Psi(p)} = \frac{\hat{\Phi}^2 - \hat{\Phi}}{\Phi^2(p) - \Phi(p)} \approx \frac{\hat{\Phi}^2}{\Phi^2(p)} \text{ and } r_p \approx \hat{r} \cdot \frac{\hat{\Phi}}{\Phi(p)}.$$

We therefore choose the neighborhood of p outside Loc_I as

$$N(p) := c_{r_p}(p) \text{ with } r_p := \hat{r} \cdot \frac{\hat{\Phi}}{\Phi(p)}.$$

Those neighborhoods replace the reciprocal transformation in [2] and the bins in [3]. Although they form no topological space they define a concept of proximity and connectedness: two positions $p, p' \in \mathbb{Z}^2$ are *close* if $p \in N(p')$ and $p' \in N(p)$ and a set $S \subseteq \mathbb{Z}^2$ is *connected* if for any two points $p, p' \in S$ there exist points $p_1, ..., p_n \in S$ s.t. $p \in N(p_1)$, $p' \in N(p_n)$ and $p_{i+1} \in N(p_i)$ for $1 \leq i < n$.

4.2 Cluster Analysis with the Intersection Point Neighborhood

Suppose we have computed the image I^l of all straight lines in I. Then $I^i : \mathbb{Z}^2 \to \mathbb{N}$ is the image of all intersection points, where $I^i(p) = n$ means that n pairs of lines in I^l are intersecting in $p \in Z^2$. We also call n the *multiplicity* of intersection points at p. One may regard I^i as a list or as a very sparse image with the value 0 almost everywhere. A next step is to find clusters in I^i and to take the centers of gravity of those clusters as candidates for vanishing points. Of course, clusters must not be build according to the Euclidean distance in $\mathbb{Z}^2$ but according to the proximity given by the intersection point neighborhood.

As our neighborhood does not define a metric space we must carefully choose a cluster anaysis technique. Therefore we apply the AGS (**A**utomatic **G**rouping of **S**emantics) algorithm that was introduced by Priese, Schmitt, Hering in [1] for an automatic grouping of locations of a similar semantics. The advantage of the AGS is that it can find groupings in spaces with a concept of a neighborhood but without a topology. The AGS follows ideas of the CSC (Color Structure Code) segmentation technique of Priese and Rehrmann [5] and works as follows:

One starts with $\mathcal{N} := [N(p)|p \in I^i]$, a list of the neighborhoods $N(p)$ of all intersection points p in I^i, as an initial grouping. AGS will merge overlapping groups G, G' in $\mathcal{N}$ if they are similar enough as shown in the following pseudo code. Similarity is measured by the overlap rate $O(G, G') := |G \cap G'|/min(|G|, |G'|)$ and some threshold t_{ov}.

```
  H := N;
* G:= empty list;
  for 0 ≤ i < |H| do G := H[i];
      for 0 ≤ j < |H|, i ≠ j do
          if G = H[j]
              then remove H[j] from H
              else if O(G, H[j]) > t_ov then G := G ∪ H[j]
  end for;
  insert G into G
```

```
end for;
if H ≠ G
    then H := G; goto line *
    else end.
```

Each computed group G in $\mathcal{G}$ consists of connected intersection points with a multiplicity. The *center* c_G of such a group is the weighted mean μ_G of all intersection points in G according to their multiplicity. The center represents the coordinate in $\mathbb{R}^2$ of a possible vanishing point. The *size* $|G|$ of G is the sum of the multiplicities of all points in G. It tells how many intersection points have contributed to c_G. The *weight* w_G of G is the number of measured lines that have produced the intersection points in G. There are two extreme cases: G may result from a single line that intersects with n more or less parallel lines, leading to $w_G = n + 1$ and $|G| = n$, or from n' lines all intersecting with each other within G, leading to $w_G = n'$ and $|G| = (n'^2 - n')/2$. Thus, in any case we have

$$w_G - 1 \leq |G| \leq \frac{w_G^2 - w_G}{2}.$$

5 An Application

We present some details how we detect vanishing points with the intersection point neighborhood. In a current project we have to find the position and direction of a camera from the camera image and a 3d model of the environment. For this, we must detect the facades of buildings. If there is no perspective projection distortion, as in Figure 2a, we can do so without a knowledge of vanishing points. However, in images as in Figure 2b the vanishing points become important. Thus, we apply some simple geometrical considerations inspired by [4].

Basically, we expect to see two to three vanishing points, namely an upper, a left and a right vanishing point where either the left or right one might be missing (see, e.g., figure 2c). Thus, we need two or three groups of measured lines, where the lines of one group lead to one vanishing point.

We transform an image I into a greyscale image and apply a Canny edge detector to get the edge image I^e. Now, a Hough transformation is applied to I^e to get the line image I^l. However, even in such rather simple architectural environments standard implementations of a Hough transformation do not lead to satisfactory line images I^l. We thus have been forced to turn our attention to improve the Hough transformation, a step that is more or less independent from our application scenario.

5.1 Noise Reduction in the Hough Transformation

We apply two filters on the Hough accumulator to improve I^l. The first one from [6] removes accumulator maxima resulting from edge points which are not part of longer lines in the image (in outdoor images, such edge points typically occur, e.g., in trees) by increasing accumulator values in the middle of

(a) No perspective distortion, three vanishing points: top, center, infinity

(b) Perspective distortion, three vanishing points: top, left, right

(c) Perspective distortion, two vanishing points: top, right

Fig. 2. Vanishing points in facades

the typical butterfly formed structures resulting from true lines while lessening other accumulator values. When no such filter is applied, the numerous edges in strongly textured structures often yield artificial straight lines in the Hough transformation. The second filter, developed by us, thins out straight lines closely neighbored in the accumulator.

Unfortunately, even this is insufficient as often Moir lines in window shutters or straight structures in roofs introduce many "false" lines. We therefore have developed a "wildness map" which characterizes "wild" parts of an image and masks them in I^e. The remaining straight lines are the *measured lines.*

Wildness at an image position $p \in Loc_I$ is measured by calculating two values in a window w_p of size 11×11 around p:

- the standard deviation σ of all values in w_p, and
- the sum of the differences between a pixel $p' \in w_p$ and its four direct neighbor pixels is averaged across all $p' \in w_p$ and gives $\mu_D(w_p)$.

The value of the wildness map at p is calculated as $c_1 \cdot \mu_D(w_p) - c_2 \cdot \sigma$ with constants $c_1, c_2 \in \mathbb{R}$ assuring that both values are in the same number range. Figure 3 shows the effect of removing wild structures in the image (a) of figure 2. Lines found by the Hough transform are shown in white, edge points in I^l in grey.

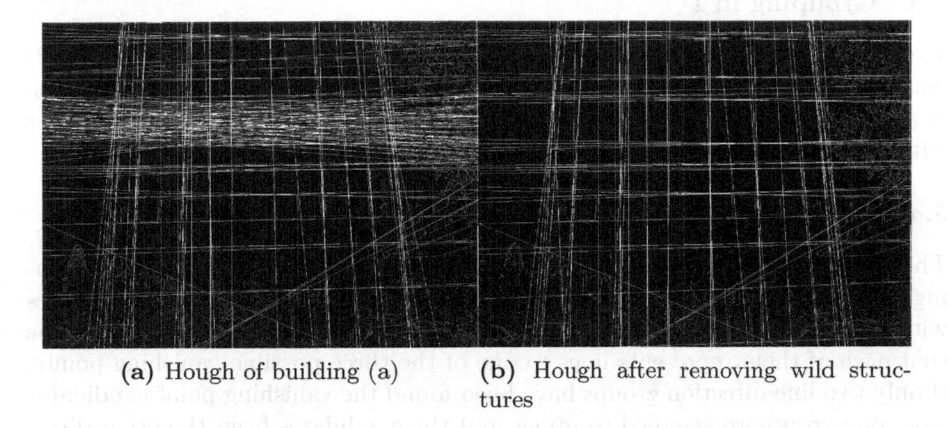

(a) Hough of building (a) **(b)** Hough after removing wild structures

Fig. 3. Removing wild textures for better line extraction

5.2 Construction of I^i

The previous step is rather independent from our application but the following steps follow our application scenario. We expect to find two or three vanishing points with certain geometrical restrictions. Thus, we first follow the idea of Seo [2] and group all measured lines according to their angle α. For this we use two independent k-harmonic-means clustering [7] runs, one time with two clusters and the other time with three clusters. Only when the intra-cluster variance drops substantially the result from the clustering in three groups is used. Otherwise we use only two groups and the group of vertical lines must have a width less than 30°.

For each pair of measured lines in one group we compute their intersection point and store it in a list I^i. It remains to handle vanishing points in infinity that may result from parallel lines. We expect two vanishing points on the horizon line and a third orthogonal to and above that line. The maximal distance Δ_{max} of a measurable intersection point from Loc_I depends on the size of Loc_I and the admissible accumulator H_l and is easily computed. Two measured lines $l_1, l_2 \in I^l$ with parameters α_i, d_i are parallel if $\alpha_1 = \alpha_2$ holds. In this case we regard the middle line $l_{1,2}$ between l_1 and l_2 with the parameters α_1 and $(d_1 + d_2)/2$. If $l_{1,2}$ is roughly a vertical line we chose as an point the point on $l_{1,2}$ in distance Δ_{max} above $loc(I)$. In case of a horizontal line we choose two intersection points in distance Δ_{max} on $l_{1,2}$ right and left from $loc(I)$. Those intersection points are the substitute of the intersection of the two parallel lines in infinity and they are added to I^i. If one just changes the angle of one of the two parallel lines by the smallest angle possible in H_l the resulting intersection point would be in a distance less than Δ_{max}.

An intersection point with a multilicity n will thus occur exactly n times in our list I^i.

5.3 Grouping in I^i

We now cluster the intersection points in I^i according to their neighborhoods with the AGS algorithm. The threshold t_{ov} is set to 0.6. The output of this step is a list $\mathcal{C} = [(c_G, |G|, w_G)|G \in \mathcal{G}]$ of all centers c_G of groups G in $\mathcal{G}$ together with the size and weight of G.

5.4 Construction of Vanishing Points

The list $\mathcal{C}$ contains weighted candidates for vanishing points. Each of the vanishing point candidates in $\mathcal{C}$ results from intersections of lines in one group of lines with similar directions. If three line direction groups have been found we assume that each of them represents lines to one of the three possible vanishing points. If only two line direction groups have been found the vanishing point candidates from one group are assigned to upper and the candidates from the other direction group are split into left and right. In our application a bird's eye view and an against the horizon line rotated camera are quite unlikely. This helps to use some geometrical restrictions. We firstly remove all vanishing point candidates c_G in $\mathcal{C}$ which are located inside the upper third of Loc_I or with $|G| < 3$, i.e., that result from less than three intersection points.

Often, inside Loc_I many intersection points are computed between lines pointing in very different directions. This intersections shall not contribute to a vanishing point. The introduction of two or three groups for different line angles helps to avoid this problem but will not always solve it. However, outside Loc_I this problem almost never arises. We therefore try to solve this problem by two actions:

Firstly, we use a different reference radius $\hat{r}$ inside and outside Loc_I where the value outside is multiplied by 1.5. In the following evaluation $\hat{r} = 6$ is used outside and $\hat{r} = 4$ inside Loc_I.

Secondly, we regard the situation that a single measured line crosses a bundle of almost parallel measured lines, resulting in many neighbored intersection points. To prevent that they form a vanishing point we compare the weight w_G with the size $|G|$ for any group G with its center c_G inside Loc_I. $|G|$ becomes close to $max_G := (w_G^2 - w_G)/2$, the maximal possible number of intersection points in G, if most lines of G accumulate into a neighbored set of intersection points. The size $|G|$ becomes rather small if one or two lines intersect a bundle of almost parallel lines. Thus, we also drop c_G if $|G| < 0.5 \cdot max_G$ holds.

Now, we calculate for each triple $(c_{G^l}, c_{G^r}, c_{G^u})$ of left, right and upper vanishing point candidates the size $S := \sqrt{2 \cdot |G^l|} \cdot \sqrt{2 \cdot |G^r|} \cdot \sqrt{2 \cdot |G^u|}$ and search for the triple with the biggest size fulfilling the following conditions:

- the angle between x-axis and the horizon line l_h between left and right vanishing point must be below 5°,

- l_h must be below the image center,
- the difference between the mean angle of lines to the upper vanishing point and the angle of the normal to l_h must be below 7.5°,
- the left vanishing point must be left of the image center and the right must be right of the image center.
- let d_l be the distance from image center to left vanishing point, d_r the distance from image center to right vanishing point and *diag* the length of the image diagonal.
  - At least one of d_l and d_r must be bigger than $\frac{diag}{2}$
  - If $d_l > 2 \cdot diag$ than d_r must be smaller than *diag* and vice versa

6 Evaluation

We tested our algorithm on 70 images taken at the campus of our university for which the "true" vanishing points have been annotated manually. This has been done by another group, not by the authors, by manually annotating several straight lines in an image that run to the same vanishing point. However, one has to note that in many images the location of those "true" vanishing points can only be estimated vaguely. Due to imperfect parallelism in 3d, lens distortions in the camera, and limited angular accurateness, straight lines that should intersect in one vanishing point in fact show intersections scattering in a rather large area.

Nevertheless, we attempt a quantitative evaluation where we compare each calculated vanishing point $c = (c_x, c_y)$ against a true vanishing point $t = (t_x, t_y)$. Such a measure of similarity can not be the Euclidean distance. Suppose we have an imgae $I : [0, 799] \times [0, 599] \to [0, 255]^3$ with a horizon line on $y = 400$. A computed vanishing point $(-20000, 390)$ far outside the image location is obviously similar to a true vanishing point $(-18000, 400)$ in spite of their rather large Euclidean distance. If the true vanishing point is $t = (350, 400)$ a computed vanishing point $c_1 = (380, 400)$ is better than $c_2 = (350, 420)$, although $d_E(t, c_2) < d_E(t, c_1)$. It turns out that the 3d angle between t and c measured from some height on a tower on the origin reflects very well the quality of the computed vanishing points: the lower this angle the better. In our images with a resolution of 1024×768 pixels the height of 40 to 90 pixels of this tower gives the best angles for an evaluation. As a suitable similarity measure for c and t we therefore calculate the angle between $\tilde{c} = (c_x, c_y, 40)$ and $\tilde{t} = (t_x, t_y, 40)$. This is $\alpha_{\tilde{c},\tilde{t}} = acos\left(\frac{\tilde{c} \cdot \tilde{t}}{|\tilde{c}| \cdot |\tilde{t}|}\right)$.

We say t is *successfully detected* if a vanishing point c is computed with $\alpha_{\tilde{c},\tilde{t}} \leq 3°$, *unsuccessfully detected* if $\alpha_{\tilde{c},\tilde{t}} > 3°$, and *undetected* if no vanishing point for t is computed at all. Figure 4 shows the distribution of $\alpha_{\tilde{c},\tilde{t}}$ across all detected vanishing points.

Table 1 gives the number of successfully detected, unsuccessfully detected, and undetected annotated vanishing points as well as the mean and standard deviation σ of $\alpha_{c,t}$ in degrees for all successfully detected vanishing points. Our method has detected 154 of 189 annotated vanishing points, resulting in a success rate of

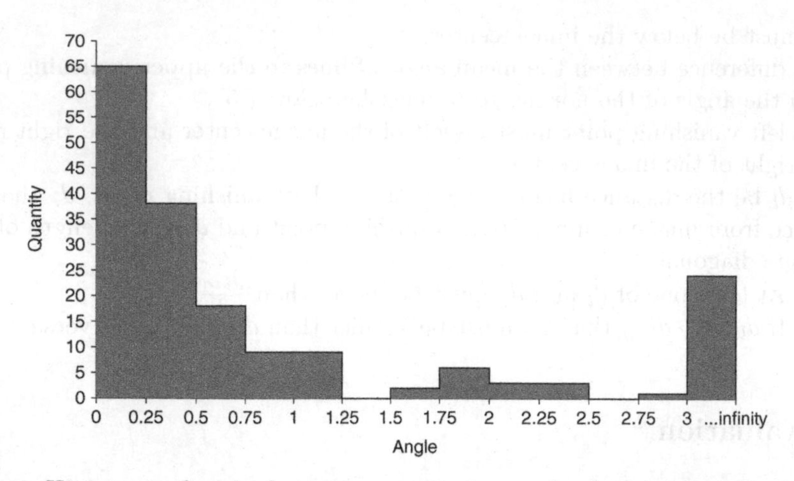

Fig. 4. Histogram of angles between vector to true and calculated vanishing point

Table 1. Evaluation of vanishing point detection

	total number	successfully detected	unsuccessfully detected	undetected	mean of $\alpha_{c,t}$ for successful t	σ of $\alpha_{c,t}$ for successful t
left	64	49	9	6	0.62	0.66
right	55	37	13	5	0.57	0.59
upper	70	68	2	0	0.45	0.51
total	189	154	24	11	0.54	0.58

81.48%. In most cases we find very small angles but a few rather large ones increase the mean. This is why σ becomes 0.58 with a smaller mean of just 0.54.

Only 7 out of the 189 calculated vanishing points do not correspond to a true vanishing point at all. This gives an error rate of 3.79%.

A visual inspection of the calculated vanishing points shows even less errors as the annotated "true" vanishing points are sometimes questionable. Further, in most of the images where vanishing points are undetected or unsuccessfully detected the Hough transform has not produced reasonable straight lines to operate with.

7 Conclusion

We have presented a new neighborhood on line intersection points and its application in detection of vanishing points in architectural environments. Our algorithm combines ideas from several algorithms in literature in order to achieve reliable results in reasonable time. In the future, we want to apply our new algorithm for facade detection and matching between a 3d model and camera images. However, we believe that the intersection point neighborhood can improve vanishing point detection in very different application scenarios.

We would like to encourage the community to use our improved algorithms for Hough and vanishing points detection in further research. The algorithms are freely available under `http://www.uni-koblenz-landau.de/koblenz/fb4/institute/icv/agpriese/downloads/vpoints`

Acknowledgment

We would like to thank the unknown reviewers for their valuable comments.

References

1. Priese, L., Schmitt, F., Hering, N.: Grouping of semantically similar image positions. In: Salberg, A.-B., Hardeberg, J.Y., Jenssen, R. (eds.) SCIA 2009. LNCS, vol. 5575, pp. 726–734. Springer, Heidelberg (2009)
2. Seo, K.S., Lee, J.H., Choi, H.M.: An efficient detection of vanishing points using inverted coordinates image space. Pattern Recogn. Lett. 27(2), 102–108 (2006)
3. Almansa, A., Desolneux, A., Vamech, S.: Vanishing point detection without any a priori information. IEEE Trans. Pattern Anal. Mach. Intell. 25(4), 502–507 (2003)
4. Rother, C.: A new approach for vanishing point detection in architectural environments. In: Proc. 11th British Machine Vision Conference, pp. 382–391 (2000)
5. Rehrmann, V., Priese, L.: Fast and robust segmentation of natural color scenes. In: Chin, R., Pong, T.-C. (eds.) ACCV 1998. LNCS, vol. 1351, pp. 598–606. Springer, Heidelberg (1997)
6. Leavers, V.F., Boyce, J.F.: The radon transform and its application to shape parametrization in machine vision. Image Vision Comput. 5(2), 161–166 (1987)
7. Zhang, B., Hsu, M., Dayal, U.: K-harmonic means - a spatial clustering algorithm with boosting. In: Roddick, J., Hornsby, K.S. (eds.) TSDM 2000. LNCS (LNAI), vol. 2007, pp. 31–45. Springer, Heidelberg (2001)

Ellipse Detection with Elemental Subsets

Peter Veelaert

University College Ghent, Engineering Sciences - Ghent University Association,
Schoonmeersstraat 52, B9000 Ghent, Belgium
peter.veelaert@hogent.be

Abstract. We propose a simple method for fitting ellipses to data sets. The method first computes the fitting cost of small samples, called elemental subsets. We then prove that the global fitting cost can be easily derived from the fitting cost of the samples. Since fitting costs are computed from small samples, the technique can be incorporated in many ellipse detection and recognition algorithms, and in particular, in algorithms that make use of incremental fitting. Some of the theoretical results are formulated in the more general setting of implicit curve fitting.

1 Introduction

Due to the ubiquitous presence of round objects in natural scenes and the need for robust recognition algorithms, ellipse detection and fitting remains an important research problem in image processing as well as metrology. Many recently proposed detection algorithms are multistage processes that combine several techniques. Zhang and Liu developed a method for robust, real-time ellipse detection [8]. They propose a global multistage algorithm paying attention to several important issues in shape recognition: robust edge detection, removal of spurious points, an efficient implementation of a real-time ellipse Hough transform, and proven accuracy. A somewhat different approach is followed in a recent paper by Qiao and Ong. They use the concepts of model-based distance and angular connectivity in an iterative algorithm for ellipse fitting [6].

Also in the field of digital geometry ellipses and conics have received some attention. Coeurjolly et al have proposed a circle recognition algorithm which constructs the polygon in which the center of the circle must lie, and verifies whether this polygon is empty or not [1]. Earlier, some work had been done on the coding of conic sections by Zunic and Sladoje [9]. These few, isolated results stand in stark contrast to the amount of work done in this field on the recognition of digital straight lines and digital planes.

In this paper we focus on one particular aspect of ellipse detection: the selection of small subsets, called elemental subsets, from an edge map to estimate the ellipse parameters. The results are useful in several ways. When developing new algorithms for real-time ellipse detection (as in [8]), incremental tracking (as in [6]) or fitting (as in [7]), a better understanding of the quality of the point samples will facilitate and improve the detection process. In fact, the concept of elemental subsets has already proved to be useful for developing fast, incremental fitting algorithms for the segmentation of edge maps into lines and parabola [4, 5, 2]. In this work we provide the theoretical framework for incremental fitting with ellipses.

S. Brlek, C. Reutenauer, and X. Provençal (Eds.): DGCI 2009, LNCS 5810, pp. 144–155, 2009.

Although elemental subsets have been introduced in previous work [4], in this paper we introduce signed elemental subsets and the related concept of elemental thickness. Furthermore we examine the relation between elemental thickness and algebraic thickness. The focus is on ellipse recognition, but some of the theoretical results are formulated in a more general setting: fitting implicit functions to discrete point sets.

2 Algebraic and Elemental Thickness

2.1 Algebraic Thickness

Suppose we are given a finite set $S \subset \mathbb{Z}^2$, and a family of curves C. How can we determine whether there is a curve in C that will pass "'close"' to all the points of S? The answer to this question depends in a crucial way on the definition of closeness, a notion which can be expressed in various ways. For example, for digital straight lines the distance between a line $y = ax + b$ and a grid point $p_i = (x_i, y_i)$ is often defined as $|y_i - ax_i - b|$. It is the Euclidean distance measured along a vertical line crossing the point p_i. Alternatively, closeness could also be measured as $|(y_i - ax_i - b)/a|$, which coincides with the shortest Euclidean distance between the point (x_i, y_i) and the line, but with the drawback that the parameter a now occurs in a non-linear way.

A notion related to closeness is the concept of supporting curves. Suppose we have found a line $y = ax + b$ such that all points of S satisfy $|y_i - ax_i - b| \leq \epsilon$, where ϵ has been chosen as small as possible. Then all the points lie between the straight lines $y - ax - b = \epsilon$ and $y - ax - b = -\epsilon$. These two parallel lines encompass (or support) the set S as tightly as possible.

In this work we shall use the **algebraic distance** to measure how close a set of points lies to a curve. Let C be a family of curves implicitly defined as $f(x, y; a_1, \ldots, a_n) = 0$, where f is a real polynomial. The n parameters $a_1, \ldots, a_n$ parametrize the family. The algebraic distance between the point (x_i, y_i) and the curve is simply defined as

$$f(x_i, y_i; a_1, \ldots, a_n).$$

Although the algebraic distance differs from the Euclidean distance, in particular where the curvature of a curve is high, the algebraic distance has the advantage of simplicity. Furthermore, our aim is to find approximating curves that pass at a given distance between points, which can be accomplished relatively easily with algebraic distances.

Definition 1. Let S be a finite subset of $\mathbb{Z}^2$ and let C a family of curves defined by $f(x, y; a_1, \ldots, a_n) = 0$, where f is a real polynomial in x, y and the parameters $a_1, \ldots, a_n$. The **algebraic thickness** τ_a of S is defined as

$$\tau_a(S) = \min_{a_1,\ldots,a_n} \left(\max_{(x,y)\in S} |f(x, y; a_1, \ldots, a_n)| \right). \quad (1)$$

Thus the algebraic thickness $\tau_a(S)$ is the smallest real number for which there exists a curve $f(x, y; a_1, \ldots, a_n) = 0$ in C such that all the points of S lie within algebraic distance $\tau_a(S)$ of the curve. The curve for which the minimum is achieved is called the **algebraic best fit**.

2.2 Elemental Thickness and Supporting Systems

Supporting lines are known from digital straight lines. Suppose we have a set S of grid points and a line $y = ax + b$ such that all the points of S satisfy $|y_i - ax_i - b| \leq \tau_a(S)$, where $\tau_a(S)$ denotes the algebraic thickness of S as defined for the family of straight lines. Then it is a well-known property that there are at least three points that lie on the two supporting lines $y_i - ax_i - b = \pm\tau_a(S)$ [4]. Moreover, $\tau_a(S)$ can be found by examining all 3-point subsets of S and computing the thickness for each of them separately. The thickness $\tau_a(S)$ is simply the maximum of all the 3-point thicknesses. Thus elemental subsets provide a powerful tool for finding the best algebraic linear fit to a set of points. It suffices to compute the algebraic thickness of all 3-points subsets. The subset with the largest thickness yields the supporting lines and the best fit for the entire set. Because of its proven usefulness for lines we will extend the concept of supporting curves to implicitly defined curves of general type.

Definition 2. Let C a family of curves be defined by n parameters a_i, and let S be a finite subset of $\mathbb{Z}^2$ containing at least $n+1$ points. An **elemental subset** of S is a subset of $n+1$ distinct points of S. A **signed elemental subset** is an elemental subset S where either the sign + or - has been assigned to each point.

For example, $\{p_1^+, p_2^+, p_3^-, p_4^-, p_5^+\}$ could be a signed elemental subset for a family parametrized with 4 parameters. The aim of the signs is to determine at which side of a curve the points lie. In this case the points p_1, p_2, p_5 all lie on the same side while p_3, p_4 lie on the opposite side. We also introduce a sign vector σ, which in this example is equal to $\sigma = (\sigma_i) = (+1, +1, -1, -1, +1)$.

Definition 3. Let C a family of curves be defined by n parameters. A **supporting system** $\{U_\sigma, C^0, C^{+\epsilon}, C^{-\epsilon}\}$ consists of the following parts:

- a signed elemental subset U_σ;
- a real curve C^0 of the form $f(x, y, a_1, a_2, \ldots) = 0$ in C and a value for ϵ such that the points $p_i = (x_i, y_i)$ of U_σ satisfy the $n+1$ equations

$$f(x_i, y_i, a_1, a_2, \ldots) = \sigma_i \epsilon, \quad \text{for } i = 1, \ldots, n+1. \tag{2}$$

- the two real curves $C^{+\epsilon}$ and $C^{-\epsilon}$ defined by $f(x, y, a_1, a_2, \ldots) = \epsilon$ and $f(x, y, a_1, a_2, \ldots) = -\epsilon$, respectively.

If such a system exists, the value for ϵ is called the **elemental thickness** of the system, and it will be denoted as $\tau_e(\{U_\sigma, C^0, C^{+\epsilon}, C^{-\epsilon}\})$. The two curves $C^{+\epsilon}$ and $C^{-\epsilon}$ are called **supporting curves** for U_σ. The curve C^0 is called an **elemental best fit** for U_σ. The sign σ_i determines at which side the point (x_i, y_i) lies from C^0.

Definition 3 actually gives us a means for computing supporting systems for an elemental subset. Given the signs σ_i it is sufficient to solve (2), for the unknowns a_j and ϵ. Supporting systems can therefore be computed for a wide variety of curve families. Furthermore, according to Definition 3 the elemental best fit always belongs to the family of curves C. For the two supporting curves $C^{+\epsilon}$ and $C^{-\epsilon}$, however, this may not be true, depending on the way in which the family of curves has been defined.

Example. We consider the family of lemniscates with equations of the form

$$-2a^2((-p+x)^2 - (-q+y)^2) + ((-p+x)^2 + (-q+y)^2)^2 = 0.$$

As shown in Fig. 1, the signed elemental subset $\{p_1 = (34, 14)^-, p_2 = (38, 15)^+, p_3 = (29, 8)^-, p_4 = (18, 17)^+\}$ has two supporting curves. The inner curve is not connected and consists of two separate components. The best fit is a lemniscate (the figure eight), which passes between the points of the signed elemental subset. The best fit was computed by solving the system $-2a^2((-p+x)^2-(-q+y)^2)+((-p+x)^2+(-q+y)^2)^2 = \sigma_i\epsilon, i = 1, \ldots, 4$, with $\sigma_i = -1, 1, -1, 1$, for a, p, q and ϵ. In this family the supporting curves are not lemniscates themselves, but their points lie at a constant algebraic distance (ϵ or $-\epsilon$) of a lemniscate.

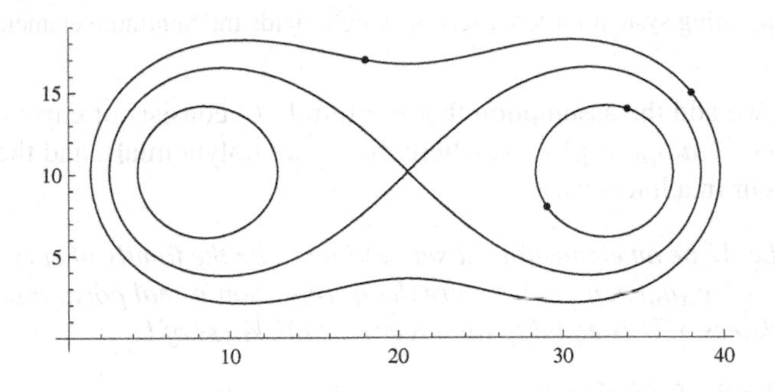

Fig. 1. A supporting system for lemniscates

From the elemental thicknesses of the supporting systems we derive the elemental thickness of an elemental subset.

Definition 4. Let U be an elemental subset with $n + 1$ points and let C be a family curves defined by n parameters. The **elemental thickness** of U is defined as $\tau_e(U) = \min_\sigma \tau_e(\{U_\sigma, C^0, C^{+\epsilon}, C^{-\epsilon}\})$, where the minimum is taken over all the 2^{n+1} possible ways in which signs can be assigned to the points of U and over all the supporting systems of each U_σ.

Fig. 2 shows another supporting system for the same points as in Fig. 1. Three points lie on $C^{-\epsilon}$ and one point on $C^{+\epsilon}$. The supporting curves enclose the set U much more tightly than in Fig. 1. In fact, there is no other supporting system with smaller thickness, which means that the ϵ of the supporting system in Fig. 2 is also the elemental thickness of U as defined in Definition 4.

2.3 Relation between Algebraic and Elemental Thickness

Up to now we have introduced two distinct forms of thickness. Algebraic thickness is defined by an optimization problem, while elemental thickness can be found by solving systems of equations. Both concepts will coincide, however, if we impose additional constraints on how a family of curves is parametrized. While all the previous definitions hold even when some of the parameters a_i occur in a non-linear way, in the

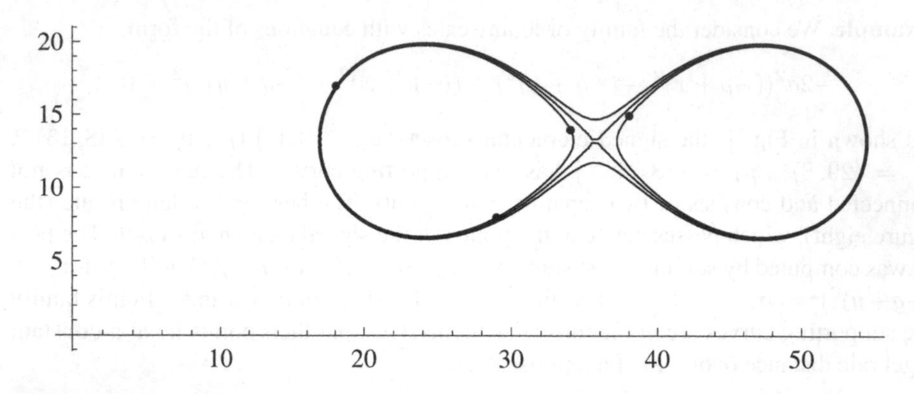

Fig. 2. The supporting system for lemniscates, which yields the minimum elemental thickness

next theorem, we add the assumption that the family C consists of curves of the form $a_1 g_1(x, y) + \cdots + a_n g_n(x, y) = 1$, where the g_i are polynomials, and the parameters a_j must all occur in a linear way.

Theorem 1. *Let U be an elemental subset, and let C be the family of curves defined as $a_1 g_1(x, y) + \cdots + a_n g_n(x, y) = 1$, where the g_i represent n real polynomials. Then the elemental thickness of U is equal to the algebraic thickness of U.*

Proof. Consider the following linear programming problem:
Minimize the objective function ϵ subject to the $2n + 2$ constraints

$$\begin{aligned} a_1 g_1(x_i, y_i) + \cdots + a_n g_n(x_i, y_i) - 1 &\leq \epsilon, \quad i = 1, \ldots, n+1 \\ \epsilon \leq a_1 g_1(x_i, y_i) + \cdots + a_n g_n(x_i, y_i) - 1, &\qquad i = 1, \ldots, n+1. \end{aligned} \tag{3}$$

This minimal value for ϵ is equal to the algebraic thickness of the elemental subset U. In fact, the above linear programming problem clarifies how the algebraic thickness $\tau_a(U)$ of Definition 1 can be computed for families with linear parameters.

Furthermore, since the $2n + 2$ inequalities define a polyhedron while the objective function is a linear function, the minimum will be attained in at least one of the vertices of the polyhedron. This vertex satisfies equations of the form

$$a_1 g_1(x_i, y_i) + \cdots + a_n g_n(x_i, y_i) - 1 = \sigma_i \epsilon, \tag{4}$$

where the signs $\sigma_i \in \{-1, 1\}$ must take the values that correspond to this particular vertex. It follows that the solution of (3) can be found by evaluating ϵ for all systems of the form (4) for different choices for σ_i, and taking the minimum, which boils down to computing the minimum of ϵ over all supporting systems for U, or in other words, computing the elemental thickness of U.

The restriction that the family must be parametrized in a linear way may seem to exclude some interesting curves. We will see, however, that at least for ellipses, the family of ellipses can be enlarged to the linear class of axis-aligned conics in such a way that the result of Theorem 1 can be extended to ellipses.

2.4 Closed Formula for Algebraic Thickness

The computation of thickness can be further simplified. We will derive a closed analytical expression for algebraic thickness. To this end we need the following lemma.

Lemma 1. *Consider the following system of inequalities in the unknowns a_i:*

$$|a_1 g_1(x_i, y_i) + \cdots + a_n g_n(x_i, y_i) - 1| \leq \epsilon, \text{ for } i = 1, \ldots, n+1. \quad (5)$$

Let

$$M = \begin{pmatrix} g_1(x_1, y_1) & \cdots g_n(x_1, y_1) & -1 \\ \cdots & & \\ g_1(x_{n+1}, y_{n+1}) & \cdots g_n(x_{n+1}, y_{n+1}) & -1 \end{pmatrix}$$

denote the coefficient matrix of (5), and let D_i denote the cofactor of the element on the last column and i-th row of M. If at least one of the cofactors is non zero, then (5) has a solution if and only if ϵ satisfies $|\sum_i D_i| / \sum_i |D_i| \leq \epsilon$. When all the cofactors are zero then the system has a solution for $\epsilon = 0$.

Proof. Assume all the cofactors of the last column are zero. If we then replace the inequalities by equalities and ϵ by zero in (5), we obtain a homogeneous system. Since all cofactors D_i are zero, the determinant of M is zero, and therefore the homogeneous system has a non-trivial solution. Furthermore, this solution is also a solution of the original system (5) with $\epsilon = 0$. Next, assume that at least one of the cofactors is non-zero and that (5) has a solution. By multiplying the i-th inequality by the cofactor D_i, and after adding the left and the right sides of the inequalities separately, we find that $\sum_i |D_i||a_1 g_1(x_i, y_i) + \cdots + a_n g_n(x_i, y_i) - 1| \leq \epsilon \sum_i |D_i|$, and therefore also $|\sum_i D_i(a_1 g_1(x_i, y_i) + \cdots + a_n g_n(x_i, y_i) - 1)| \leq \epsilon \sum_i |D_i|$. However, the terms $\sum_i D_i g_j(x_i, y_i)$ vanish for all j, because each term represents the determinant of a matrix with two identical columns. Hence, $|\sum_i D_i(-1)| = \sum_i |D_i| \leq \epsilon \sum_i |D_i|$, which concludes the first part of the proof.

The converse is proved by contra position. Assume that (5) does not have a solution. Then for each vector $(\delta_1, \ldots, \delta_{n+1})$ that satisfies $|\delta_i| \leq \epsilon$, the system of $n+1$ linear equations in n unknowns

$$a_1 g_1(x_i, y_i) + \cdots + a_n g_n(x_i, y_i) - (1 + \delta_i) = 0, \text{ for } i = 1, \ldots, n+1 \quad (6)$$

does not have a solution either. Therefore the homogeneous system with $n+1$ equations

$$a_1 g_1(x_i, y_i) + \cdots + a_n g_n(x_i, y_i) - (1 + \delta_i) b = 0, \text{ for } i = 1, \ldots, n+1 \quad (7)$$

in the $n+1$ unknowns $a_1, \ldots, a_n, b$ cannot have a non-trivial solution. Since, if such a solution would exist, it would have the form $(a_1, \ldots, a_n, b)$, with $b \neq 0$, because the columns of the matrix $m_{ij} = g_j(x_i, y_i)$ are independent. Then $(a_1/b, \ldots, a_n/b, 1)$ would be a solution of (7), and $(a_1/b, \ldots, a_n/b)$ a solution of (6). Since we assume that no such solution exists, it follows that the determinant of the system matrix of (7) cannot vanish, that is, $|\sum_i D_i(1 + \delta_i)| > 0$, for δ_i satisfying $|\delta_i| \leq \epsilon$. Since, $\sum_i D_i \delta_i$ can take any value in the closed interval $[\epsilon \sum_i |D_i|, \epsilon \sum_i |D_i|]$, it follows that $|\sum_i D_i| > \sum_i |D_i|$.

A closed formula for the algebraic thickness follows easily now.

Theorem 2. *Let C be the family of curves defined by $a_1 g_1(x, y) + \cdots + a_n g_n(x, y) = 1$. Let U be an elemental subset of $n+1$ points (x_i, y_i), and let M and its cofactors D_i be defined as in Lemma 1. If the determinant of M is non-zero, then the algebraic thickness of U is equal to $|\sum_i D_i| / \sum_i |D_i|$. If the determinant of M is zero, then the algebraic thickness is also zero.*

Proof. Reconsider the linear programming problem (3), which corresponds to finding the algebraic thickness. This minimization problem can be reformulated as follows. What is the minimum value for ϵ such that the polyhedron defined by the constraints in (3) is still non-empty? By Lemma 1 this minimal value for ϵ is given by $|\sum_i D_i| / \sum_i |D_i|$, when $\det M \neq 0$.

2.5 Algebraic Thickness of Non-elemental Sets

The algebraic thickness of a larger set can be computed from the algebraic thickness of its elemental subsets.

Theorem 3. *Let C be the family of curves defined by $a_1 g_1(x, y) + \cdots + a_n g_n(x, y) = 1$. The algebraic thickness of a set S is equal to the maximum of the algebraic thicknesses of all its elemental subsets.*

Proof. By definition the algebraic thickness is equal to the minimum value for ϵ of the following optimization problem:
Minimize the objective function ϵ subject to the constraints

$$|a_1 g_1(x_i, y_i) + \cdots + a_n g_n(x_i, y_i) - 1| \leq \epsilon, \ (x_i, y_i) \in S. \tag{8}$$

Let ϵ be the minimal value so that we can still find values for $a_1, \ldots, a_n$ that satisfy the above inequalities. Given ϵ the inequalities define a convex set in the parameter space $\mathbb{R}^n$. According to Helly's theorem this convex set is non-empty provided the intersection of every $n+1$ of its subsets is non-empty [3]. Or in other words, for the constraints to be satisfied, it suffices that they are satisfied for the points (x_i, y_i) of each elemental subset U of S separately. It follows that the minimal value for ϵ is the maximum of $\tau_a(U)$, where U ranges over all the elemental subsets of S.

3 Axis-Aligned Ellipses

We will apply the previous results to conics, and in particular ellipses. An ellipse has an equation of the form

$$\frac{(x-p)^2}{r^2} + \frac{(y-q)^2}{s^2} = 1.$$

We will denote this family of curves as $C_{ellipse}$. Since this family is characterized by 4 parameters, each elemental subset U has 5 points, i.e., $U = \{(x_1, y_1), \ldots, (x_5, y_5)\}$. In

$C_{ellipse}$ a signed elemental subset is part of a supporting system if the following system has a solution for p, q, r, s and $\epsilon_{ellipse}$:

$$\begin{cases} \frac{(x_1-p)^2}{r^2} + \frac{(y_1-q)^2}{s^2} - 1 = \sigma_1 \epsilon_{ellipse} \\ \dots \\ \frac{(x_5-p)^2}{r^2} + \frac{(y_5-q)^2}{s^2} - 1 = \sigma_5 \epsilon_{ellipse} \end{cases} \tag{9}$$

Although this system is non-linear it can easily be handled by equation solvers to determine the two supporting curves and the best elemental fit. Within the family of ellipses not every signed elemental subset will give rise to a supporting system, however. The existence of a supporting system can be determined by extending the family of ellipses to the family conics C_{conic} which consists of curves of the form

$$ax_i^2 + by_i^2 + dx_i + ey_i - 1 = 0.$$

This family includes parabola, hyperbola, and ellipses. Since the term xy is missing, all conics are axis-aligned. The family of conics is also characterized by 4 parameters and has therefore elemental subsets with 5 points.

In C_{conic} a signed elemental subset is part of a supporting system provided the following system can be solved:

$$\begin{cases} ax_1^2 + by_1^2 + dx_1 + ey_1 - \sigma_1 \epsilon_{conic} = 1 \\ \dots \\ ax_5^2 + by_5^2 + dx_5 + ey_5 - \sigma_5 \epsilon_{conic} = 1 \end{cases} \tag{10}$$

Therefore the signed elemental subset is part of a supporting system when the determinant of the following matrix is non-zero:

$$M_{conic} = \begin{pmatrix} x_1^2 \; y_1^2 \; x_1 \; y_1 \; \sigma_1 \\ \dots \\ x_5^2 \; y_5^2 \; x_5 \; y_5 \; \sigma_5 \end{pmatrix} \tag{11}$$

When a signed elemental subset gives rise to a supporting system in C_{conic} then the solution we find for ϵ_{conic} yields the elemental thickness $\tau_e(U_\sigma, C^0, C^{+\epsilon}, C^{-\epsilon})$. The elemental thickness of U itself, was defined as $\tau_e(U) = \min_\sigma \tau_e(U_\sigma, C^0, C^{+\epsilon}, C^{-\epsilon})$ where the minimum is taken over all supporting systems of U.

3.1 Relation between Elliptical and Conic Fits

By Theorems 1 and 2 the elemental thickness of U in C_{conic} is equal to

$$\tau_{e,conic}(U) = \tau_{a,conic}(U) = \frac{|D_1 + D_2 + D_3 + D_4 + D_5|}{|D_1| + |D_2| + |D_3| + |D_4| + |D_5|}, \tag{12}$$

where D_i denotes the cofactor of the element on the last column and i-th row of M_{conic}.

In general, we have $\tau_{e,ellipse}(U) \neq \tau_{e,conic}(U)$. Nevertheless, when they exist, the solutions of (9) and (10) are closely related. A solution of (9) can always be rewritten as a solution for (10). Conversely, suppose we find a solution for (10). Then this conic will be an ellipse if the parameters a and b have the same sign, and are both non-vanishing. In that case the subset also has a supporting system in the family of ellipses. Since these parameters are always unique, we have the following result: A signed elemental subset has at most one supporting system within the family of ellipses. We note that this is not true in general. For example, for lemniscates a single signed elemental set may be part of more than one supporting system.

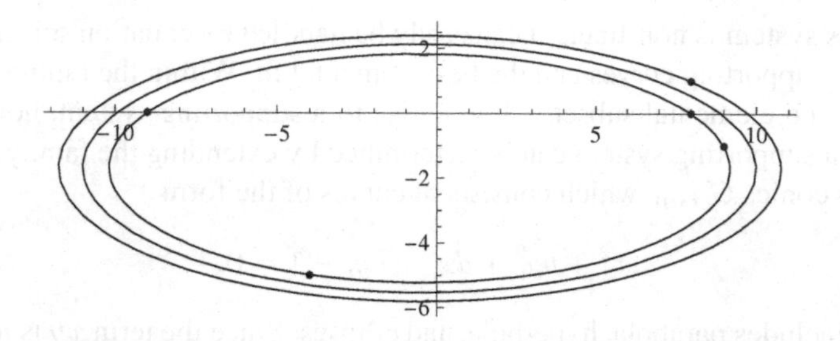

Fig. 3. A supporting system of 5 points in the family of aligned conics

Fig. 3 shows five points of an elemental subset. The best fit and the two supporting curves $C^{+\epsilon}$, $C^{-\epsilon}$ have been computed using (10) for all 32 possible sign vectors. Fig. 3 shows one particular supporting system where the signs have been attributed so that 4 points lie on the inner ellipse $C^{-\epsilon}$ and one point lies on the outer ellipse C^{ϵ}. The supporting system with the smallest value for ϵ_{conic} is the one shown in Fig. 4. For some choices of the signs σ_i we obtain a hyperbola as best fit in C_{conic}, as shown in Fig. 5. This means that when the signs are attributed as in Fig. 5, we will not find a real solution when we use (9) to compute a supporting system in $C_{ellipse}$.

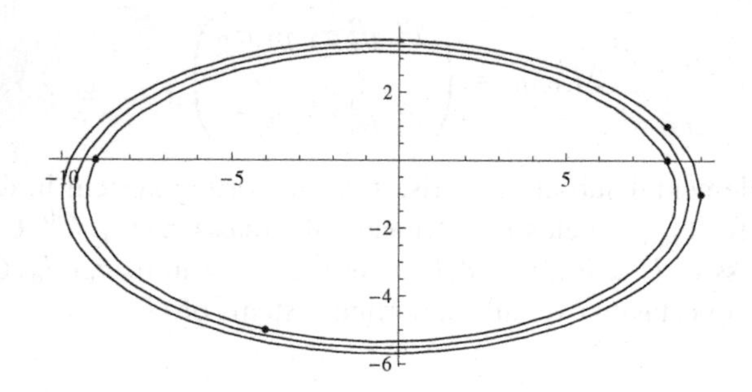

Fig. 4. The supporting system that yields the minimum elemental thickness

The elemental thickness $\tau_{e,conic}$ of an elemental subset in C_{conic} is not the same as its elemental thickness $\tau_{e,ellipse}$ in $C_{ellipse}$. The supporting curves are the same however. Let

$$ax^2 + by^2 + dx + ey = 1 \tag{13}$$

be the best fit found for a signed elemental subset U_σ in the family C_{conic}, and let

$$ax^2 + by^2 + dx + ey = 1 \pm \epsilon_{conic} \tag{14}$$

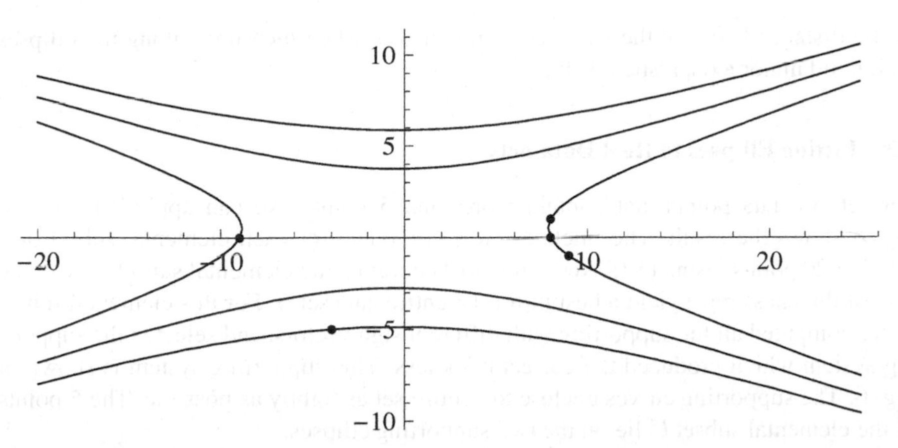

Fig. 5. A supporting system consisting of hyperbola

be its two supporting curves. We can rewrite (13) as $\frac{(x-p)^2}{r^2} + \frac{(y-q)^2}{s^2} = 1$ and (14) as $\frac{(x-p)^2}{r^2} + \frac{(y-q)^2}{s^2} = 1 \pm \epsilon_{ellipse}$ where $p = \frac{-d}{2a}$, $q = \frac{-e}{2b}$, $r^2 = \frac{4ab+bd^2+ae^2}{4a^2b}$, $s^2 = \frac{4ab+bd^2+ae^2}{4ab^2}$, and $\epsilon_{ellipse} = \frac{4ab\epsilon_{conic}}{4ab+bd^2+ae^2}$. The three curves defined by the rewritten expressions clearly yield a supporting system for U_σ in $C_{ellipse}$.

3.2 Real Thickness

The elemental thickness and the algebraic thickness can be easily calculated, but since they are based on algebraic measures rather than geometrical ones, they differ from the thickness as we would define it using Euclidean distance functions. In particular, the elemental thickness $\tau_{e,conic}(U)$ is not translation invariant. That is, if we translate all the points of U by a translation vector to obtain a second set V, then the thickness of V will be not the same as the thickness of U.

On the other hand, the elemental thickness $\tau_{e,ellipse}(U)$ is translation invariant. In fact, $\tau_{e,ellipse}(U)$ is invariant for any transformation of the form

$$\begin{pmatrix} x' \\ y' \\ 1 \end{pmatrix} = \begin{pmatrix} \alpha_{11} & 0 & \alpha_{13} \\ 0 & \alpha_{22} & \alpha_{13} \\ 0 & 0 & 1 \end{pmatrix} \begin{pmatrix} x \\ y \\ 1 \end{pmatrix}.$$

For example, after scaling by a factor 2, the thickness $\tau_{e,ellipse}(U)$ of the transformed set will still be that of the original elemental set, although the transformed points will lie apart twice as far. Hence neither $\tau_{e,ellipse}(U)$ or $\tau_{e,conic}(U)$ are good measures for the real distance between the supporting curves of an elemental set. Neither $\tau_{e,conic}(U)$ nor $\tau_{e,ellipse}(U)$ behave as normal thickness functions, because the first is not invariant for translations, while the second is invariant even for scaling.

A simple calculation shows, however, that the Euclidean distance between the two supporting ellipses can be easily be derived from the two supporting curves. We find

$$\tau_{major} = r(\sqrt{1+\tau_{e,ellipse}} - \sqrt{1-\tau_{e,ellipse}})$$
$$\tau_{minor} = s(\sqrt{1+\tau_{e,ellipse}} - \sqrt{1-\tau_{e,ellipse}})$$

as the distance between the two supporting curves when measured along the ellipses major and minor axis, respectively.

3.3 Fitting Ellipses to Real Data Sets

For sets of data points that contain more than 5 points, we can apply Theorem 3. Fig. 6 shows the result. The thickness was computed for each elemental subset of a set S of 20 points, using (12). According to Theorem 3 the elemental subset U with the largest thickness must yield a best fit for the entire data set S. For this elemental subset U we computed all the supporting with different sign vectors, and selected the supporting system which produced the correct thickness. This supporting system is shown in Fig. 6. The supporting curves enclose the entire set as tightly as possible. The 5 points of the elemental subset U lie on the two supporting ellipses.

Although this approach produces a correct result, its major drawback is that we must compute the thickness of a large collection of elemental subsets (all 5-point subsets of S).

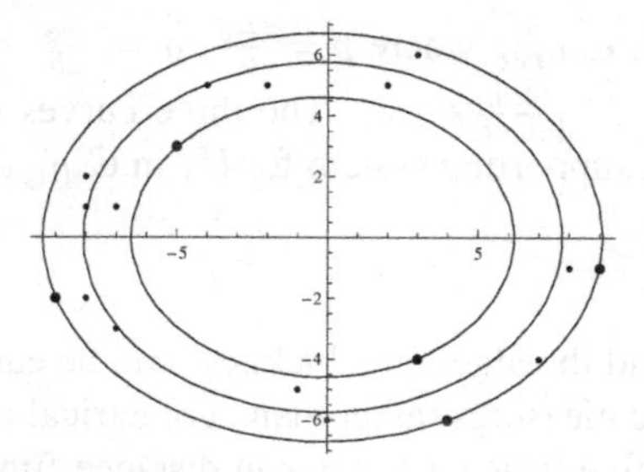

Fig. 6. Best fit and supporting system for non-elemental sets

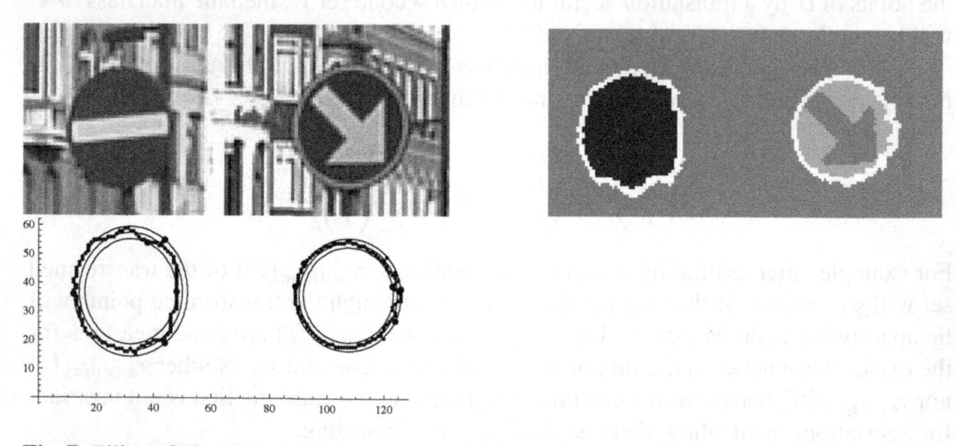

Fig. 7. Ellipse fitting of road sign contours, which were extracted by colour segmentation. The ellipses were fitted by a RANSAC process. The fitting cost was computed for randomly chosen elemental subsets from the edge map, typically between 1000 and 10000 subsets. To be robust against outliers, the RANSAC process stopped when 90% of the edge points of a closed contour were in the region between $C^{+\epsilon}$, $C^{-\epsilon}$.

Fig. 7 shows a typical application where ellipses were fitted to road sign contours. Computing all elemental thicknesses becomes impractical here because there are too many data points. In [5] it has been shown, however, that a small subcollection of elemental sets is sufficient to obtain a reliable supporting system for parabola. The same technique can be used to estimate a reliable best fit for ellipses. In paricular, only a relatively small number of elemental subsets were used to obtain the fits in Fig. 7.

4 Conclusion and Future Work

We have introduced the two important concepts of elemental and algebraic thickness, a distinction that as far as we know, has not been made before. Although their values coincide for linearly defined families of curves, they are computed in fundamentally different ways. Furthermore, the most interesting question for future research is whether we can find a relation between algebraic and elemental thickness for more generally defined families of implicit curves.

References

1. Coeurjolly, D., Gerard, Y., Reveilles, J.-P., Tougne, L.: An elementary algorithm for digital arc segmentation. Discrete Applied Mathematics 139, 31–50 (2004)
2. Deboeverie, F., Veelaert, P., Philips, W.: Face recognition using parabola edge map. In: Blanc-Talon, J., Bourennane, S., Philips, W., Popescu, D., Scheunders, P. (eds.) ACIVS 2008. LNCS, vol. 5259, pp. 994–1005. Springer, Heidelberg (2008)
3. Stoer, J., Witzgall, C.: Convexity and Optimization in Finite Dimensions I. Springer, Berlin (1970)
4. Veelaert, P.: Constructive fitting and extraction of geometric primitives. CVGIP: Graphical Models and Image Processing 59, 233–251 (1997)
5. Veelaert, P., Teelen, K.: Fast polynomial segmentation of digitized curves. In: Kuba, A., Nyúl, L.G., Palágyi, K. (eds.) DGCI 2006. LNCS, vol. 4245, pp. 482–493. Springer, Heidelberg (2006)
6. Qiao, Y., Ong, S.H.: Arc-based evaluation and detection of ellipses. Pattern Recognition 40, 1990–2003 (2007)
7. Umbach, D., Jones, K.: A few methods for fitting circles to data. IEEE Trans. on Instrumentation and Measurement 52, 1881–1885 (2003)
8. Zhang, S.-C., Liu, Z.-Q.: A robust, real-time ellipse detector. Pattern Recognition 38, 273–287 (2005)
9. Zunic, J., Sladoje, N.: Efficiency of characterizing ellipses and ellipsoids by discrete moments. IEEE Trans. on Pattern Analysis and Machine Intelligence 22, 407–414 (2000)

Multi-Label Simple Points Definition for 3D Images Digital Deformable Model*

Alexandre Dupas[1], Guillaume Damiand[2], and Jacques-Olivier Lachaud[3]

[1] Université de Poitiers, CNRS, SIC-XLIM, UMR6172, F-86962, France
dupas@sic.univ-poitiers.fr
[2] Université de Lyon, CNRS, LIRIS, UMR5205, F-69622, France
guillaume.damiand@liris.cnrs.fr
[3] Université de Savoie, CNRS, LAMA, UMR5127, F-73376, France
jacques-olivier.lachaud@univ-savoie.fr

Abstract. The main contribution of this paper is the definition of multi-label simple points that ensures that the partition topology remains invariant during a deformable partition process. The definition is based on simple intervoxel properties and is easy to implement. A deformation process is carried out with a greedy energy minimization algorithm. A discrete area estimator is used to approach at best standard regularizers classically used in continuous energy minimizing methods. The effectiveness of our approach is shown on several 3D image segmentations.

Keywords: Simple Point, Deformable Model, Multi-Label Image.

1 Introduction

Segmentation is a crucial step in any image analysis process. Over the past twenty years, energy-minimizing techniques have shown a great potential for segmentation. They combine in a single framework two terms, one expressing the fit to data, the other describing shape priors and acting as a regularizer. Furthermore, as noted by many authors, the parameter balancing the two terms acts as a scale factor, providing a very natural multiscale analysis of images. Deformable models [14], Mumford-Shah approximation [19], geometric or geodesic active contours and other levelset variants [5,18,6,24], are classical variational formulation (i.e. continuous) of such techniques. Our objective is to propose a novel energy-minimizing model for segmenting 3D images into regions, a kind of deformable digital partition with the following specific features.

(i) It is a purely digital formulation of energy minimization, which can be solved by combinatorial algorithms. We use a simple greedy algorithm.
(ii) The standard area regularizer is mimicked in this digital setting by a discrete geometric estimator.
(iii) It encodes both region structures and the geometry of their interfaces. It may thus incorporate any kind of fit to data energy, region-based like quadratic deviation [19,7] or contour-based like strong gradients [14].

* Partially supported by the ANR program ANR-06-MDCA-008/FOGRIMMI.

S. Brlek, C. Reutenauer, and X. Provençal (Eds.): DGCI 2009, LNCS 5810, pp. 156–167, 2009.

(iv) We propose a new method in order to guarantee that the topology of the whole partition is preserved during the deformation process.

Point (i) is interesting from a fundamental point of view. Continuous variational problems induce partial differential equations which are solved iteratively. They are most often bound to get stuck in local minimas, except in specific cases [9,7,1]. To our knowledge, none of them are able to find the optimal image partition if more than two regions are expected. In discrete settings, the optimal solution to the two label partitioning is computable [11]. For more regions, optimization algorithms can guarantee to be no further away than two times the optimal value [4], and scale-sets within pyramids present solutions that are experimentally very close to the optimal solution [12,21]. However, the regularization/shape prior term of these discrete methods is most often reduced to the number of surfels of the region boundaries, a very poor area estimator.

Point (ii) addresses this problem. We indeed propose an original regularization term which uses a discrete geometric estimator for computing the area of each surfel. Its principle is to extract maximal digital straight segments to estimate the surfel normal, area being a byproduct [17]. We get therefore a digital equivalent of continuous active surfaces minimizing their area, which is also an 3D extension of discrete deformable boundaries [16].

Point (iii) is important to get a versatile segmentation tool. According to the image characteristics, it is well known that contour or region based approaches are more or less adapted. From a minimization point of view, region-based energies are generally more "convex", thus easier to optimize [7,24]. Our partition model allows to mix energies defined on regions and energies defined on boundaries. To our knowledge, very few explicit or implicit variational or deformable models can do that in 3D, except perhaps the work of Pons and Boissonnat [20], but they may not model energies depending on the inclusion between regions.

In this paper we focus on the last point which is mandatory for such deformable model. Point (iv) is important in several specific image applications where the topology of anatomic components is a prior information, like atlas matching. This is even truer in 3D images, where anatomic components are intertwined in a deterministic way. Preserving the topology of a two label partition in a discrete setting is generally done by computing and locating simple points [3]. Similar tools are used in level set techniques to control topology changes [13,22]. For a multi-label partition, a few authors have proposed an equivalent to simple points in a discrete setting [23,2]. However, they are computationally too costly to be used to drive the evolution of a digital partition. We propose a new definition of simple points in multi-label partitions, that we call *ML-Simple points* (ML for Multi-Label). ML-Simpleness is stronger than simpleness, therefore deforming ML-Simple points preserves the partition topology. ML-Simpleness is easy to decide thanks to our intervoxel encoding. ML-Simpleness is sometimes a bit too restrictive and may forbid valid evolution. But our experiments show that it was not a problem in our context.

The paper is organized as follows. Section 2 recalls standard notions of digital geometry used later on. Section 3 presents the definition of ML-simpleness and proves that it implies simpleness. The ML-simpleness test derives from its definition. Section 4 describes a preliminary digital deformable partition model that uses ML-Simple points to ensure that the topology is preserved and Sect. 5 shows some first experiments.

2 Preliminaries Notions

A *voxel* is an element of the discrete space $\mathbb{Z}^3$. In the following, the symbol I designates a *3D image*, which is a couple (I_d, I_f), where I_d is a set of voxels (the image domain) and I_f is a map from I_d to a set of colors or to a set of grey levels (the image values). Each voxel v is associated with a label $l(v)$, a value in a given finite set L.

We use the classical notion of *α-adjacency*, with $\alpha \in \{6, 18, 26\}$. $N^*_\alpha(v)$ is the set of voxels α-adjacent to v, and $N_\alpha(v) = N^*_\alpha(v) \cup \{v\}$. An *$\alpha$-path* between two voxels v_1 and v_2 is a sequence of voxels from v_1 to v_2 such that each pair of consecutive voxels is α-adjacent. A set of voxels C is *α-connected* iff there is an α-path between any pair of voxels of C, with all its voxels in C.

The relation induced by being 6-connected and having the same label is an equivalence relation over the image domain. The equivalence classes are the *regions* of the image. We consider an infinite region r_0 that surrounds the image (i.e. $r_0 = \mathbb{Z}^3 \setminus I_d$). The complement set of a region X in I_d is denoted by $\bar{X}$.

In order to describe the boundaries of the regions within an image, we use the classical notion of *intervoxel* [15]. In the intervoxel framework, the discrete space is considered as a subdivision of the space in unit elements: *voxels* are unit cubes, *surfels* are unit squares between voxels, *linels* are unit segments between surfels, and *pointels* are the points between linels.

For a voxel v, we denote by surfels(v) the set of six surfels between v and all its 6-neighbors. For a surfel s, we denote by linels(s) the set of four linels between s and its adjacent surfels, and for a linel l we denote by pointels(l) the set of two pointels between l and its adjacent linels. We use also the notation linels(v) to denote the set of twelve linels around v. We say that a pointel p and a linel l (resp. a linel l and a surfel s, a surfel s and a voxel v) are *incident* if $p \in \text{pointels}(l)$ (resp. $l \in \text{linels}(s)$, $s \in \text{surfels}(v)$).

We denote by SF the set of boundary surfels of I_d, i.e. $SF = \{\text{surfel } s | s$ separates two voxels with different labels$\}$. Note that all the surfels incident to a voxel of the infinite region belong to SF since the label of the infinite region is by convention distinct from any other label. Given a voxel v, $sf(v) = \text{surfels}(v) \cap SF$ is the set of boundary surfels incident to the voxel v.

The *degree of a linel* $d(l)$ is the number of boundary surfels incident to l. Note that $d(l)$ is 0, 2, 3 or 4, but never 1. We denote by $d(l, v)$ the degree of l restricted to boundary surfels incident to v ($sf(v)$).

We recall now notations and definition from [3]. The set of α-connected components of a set of voxels X is called $C_\alpha(X)$. The geodesic neighborhood of v in X of order k is the set $N^k_\alpha(v, X)$ defined recursively by: $N^1_\alpha(v, X) = N^*_\alpha(v, X) \cap X$, and $N^k_\alpha(v, X) = \bigcup\{N_\alpha(Y) \cap N^*_{26}(v) \cap X,\ Y \in N^{k-1}_\alpha(v, X)\}$.

In other words, $N^k_\alpha(v, X)$ is the set of voxels x belonging to $N^*_{26}(v) \cap X$ such that it exists an α-path π from v to x of length at most k, all the voxels of π belonging to $N^*_{26}(v) \cap X$.

In this paper, we use only the couple of neighborhood $(6, 18)$ (6 for object and 18 for background). In this framework, we obtain the 6-geodesic neighborhood $G_6(x, X) = N^3_6(x, X)$ and the 18-geodesic neighborhood $G_{18}(x, X) = N^2_{18}(x, X)$.

From these notations, Bertrand [3] defines the notion of simple points in a $(6, 18)$-connectivity as given in Definition 1.

Definition 1 (Simple Points [3])
A voxel v is simple for a set X if $\#C_6\left[G_6(v, X)\right] = \#C_{18}\left[G_{18}(v, \bar{X})\right] = 1$, where $\#C_k[Y]$ denotes the number of k-connected components of a set Y.

3 Multi-Label Simple Points

Our goal is to modify an image partition by preserving the topology of described objects. Given a voxel v in some region R, we want to remove v from R by modifying the label of v. The simple point definition is the main tool to control topology change. However, (1) we deal with multi-label images and not binary images; (2) we want to preserve the topology of regions but also the relations between surfaces (*surface inclusion or intersection*).

For these two reasons, we cannot directly use simple points. To our knowledge, there is no variant of simple points that follows these two constraints. Thus, we propose a new variant of simple point in multi-label images. Our main idea is to preserve the linels incident to the considered voxel in order to avoid removal or creation of surface intersections.

3.1 Definition of Multi-Label Simple Points

Definition 2 gives the definition of *multi-label simple points* (called *ML-Simple points*) which are points preserving both the topology of regions and the surface relations:

Definition 2 (ML-Simple Points). *A voxel v is ML-Simple if:*

1. $\forall l \in \text{linels}(v), d(l) \in \{0, 2\}$;
2. *The body of $sf(v)$ is homeomorphic to a 2-disk;*
3. $\forall l \in \text{linels}(v), d(l, v) = 0 \Rightarrow d(l) = 0$.

Intuitively, the three conditions of Definition 2 allow:

1. to avoid cases of several regions around voxel v: this condition avoids $d(l) > 2$ which is the case when more than 2 regions touch linel l.
2. to preserve the topology of the surface: if the set of surfels incident to v and separating two voxels with different labels is not homeomorphic to a disk, the removal of voxel v from its region induces a topological modification on the surface (cases A, D, H and J in Fig. 1);

3. to preserve the relations between surfaces: if a linel l is such that $d(l, v) = 0$ and $d(l) > 0$, the removal of voxel v from its region forces the surface to touch another surface. This creates a new contact between two surfaces that were previously not adjacent (in Fig. 1, linels such that $d(l, v) = 0$ are drawn in bold; for these linels, $d(l)$ must be equal to 0).

Figure 1 displays the possible cases depending on the number of surfels of $sf(v)$ (from 0 to 6) for configurations satisfying condition (1). All other cases may be obtained from these cases by rotations and symmetries. In Fig. 1, cases A, D, H and J show non ML-Simple points since condition (2) is violated. In case A, $sf(v)$ is empty and thus is not homeomorphic to a disk; in case D, $sf(v)$ is composed of two distinct components; in case H, $sf(v)$ is homeomorphic to an annulus, and in case J, $sf(v)$ is homeomorphic to a sphere. Note that case I displays an ML-Simple point because there is no linel such that $d(l, v) = 0$ thus condition (3) holds. For all other cases, points are ML-Simple if $d(l) = 0$ for each bold linel l.

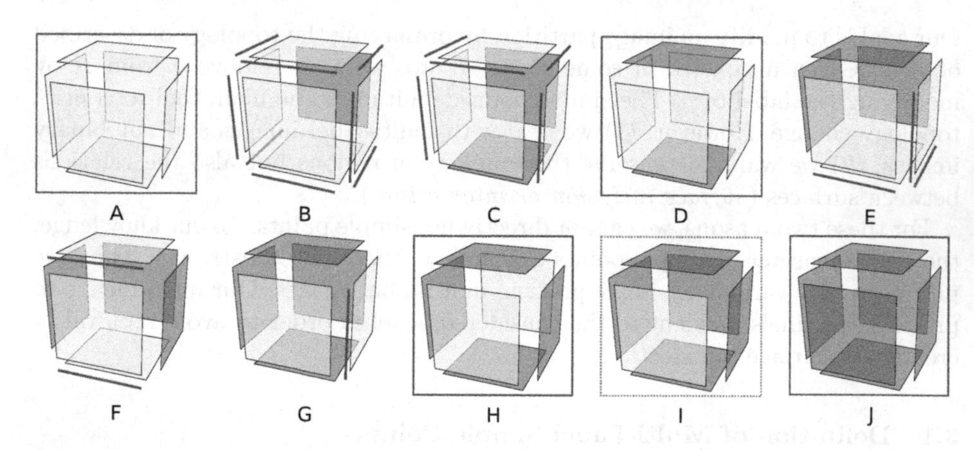

Fig. 1. All the possible cases depending on the number of surfels of $sf(v)$ (drawn in dark) incident to the considered voxel. All the configurations can be obtained from these 10 cases by rotation and symmetries. (A) Case of inner voxel with 0 boundary surfel. (B) 1 surfel. (C) and (D) 2 surfels. (E) and (F) 3 surfels. (G) and (H) 4 surfels. (I) 5 surfels. (J) Case of isolated voxel with 6 surfels. Bold linels are those with $d(l, v) = 0$. Plain frames highlight non ML-Simple point configurations. The dashed frame highlights the case where the voxel is always ML-Simple.

3.2 Multi-Label Simple Points Are Simple Points

To prove that the topology of regions is preserved when removing a ML-Simple point, we show that ML-Simple points are simple points. Therefore, we prove that ML-Simple points occur only in binary neighborhood, which is required to prove that ML-Simple points are simple points since simple points are only defined for binary images.

Lemma 1. *Let v be an ML-Simple point, the number of distinct labels in $N_{18}(v)$ is two.*

Idea of the proof. The principle of the proof is to study the neighboorhod of v, and to show that all the voxels in $N_{18}(v)$ have either the same label as v, or have all the same label w, with $w \neq l(v)$. This is proved by contradiction, assuming this is not the case and then showing that each possible configuration contradicts one condition of Definition 2. □

Note that v is an ML-Simple point does not imply that the number of distinct labels in $N_{26}(v)$ is two. However, Lemma 1 ensures that for each ML-Simple point v, removing v from its region R can be done in a unique way by setting $l(v) = w$ with w the second label of $N_{18}(v)$. We naturally call this operation a *swap*, and denote it by swap(v), since swap(swap(v)) is the identity.

Now we show that the topologies of both the region R containing v and its complementary $\bar{R}$ are preserved by proving Proposition 1 which links ML-Simple points to simple points.

Proposition 1. *If $v \in R$ is an ML-Simple point, then v is a simple point for R.*

Proof. We prove the contrapositive of Proposition 1, i.e. if v is not a simple point for R, then v is not an ML-Simple point. Let $n_1 = \#C_6\left[G_6(v, R)\right]$ and $n_2 = \#C_{18}\left[G_{18}(v, \bar{R})\right]$. Voxel x is not simple in the four following cases: (1) $n_1 = 0$, (2) $n_2 = 0$, (3) $n_1 \geq 2$, (4) $n_2 \geq 2$. We prove that, in each case, voxel v is not an ML-Simple point.

1. $n_1 = 0$. There is no 6-connected component of voxels belonging to R in $G_6(v, R)$: v is an isolated point. In this case, $sf(v)$ contains all the surfels incident to v, and thus is homeomorphic to a sphere (case J in Fig. 1) which contradicts condition (2) of Definition 2.
2. $n_2 = 0$. There is no 18-connected component of voxels belonging to $\bar{R}$ in $G_{18}(v, \bar{R})$: v is inside the region (i.e. all 18-neighbor voxels have the same label than v, case A in Fig. 1). In this case, $sf(v)$ is empty, which also contradicts condition (2).
3. $n_1 \geq 2$: there are at least two 6-connected components of voxels belonging to R in $G_6(v, R)$. If there are two 18-adjacent voxels v_1 and v_2 in two different connected components, then the voxel $v_3 \neq v$ 6-adjacent to v_1 and to v_2 belongs to $\bar{R}$ (otherwise there is only one connected component) and thus the linel l incident to v, v_1 and v_2 is such that $d(l, v) = 0$ (because the two voxels v_1 and v_2 belong to R thus there is no surfel between these voxels and v) and $d(l) = 2$ (there are two surfels, one between v_3 and v_1 and one between v_3 and v_2), which contradicts condition (3) of Definition 2.

   If there is no two voxels v_1 and v_2 in two different connected components and which are 18-adjacent too, the connected components are separated by v (case H in Fig. 1). In this case, $sf(v)$ is not homeomorphic to a disk (it is a annulus), which contradicts condition (2).
4. $n_2 \geq 2$: there are at least two 18-connected components of voxels belonging to $\bar{R}$ in $G_{18}(v, \bar{R})$. If there are two voxels $v_1, v_2 \in N_6(v)$ in two different connected components, then v_1 and v_2 are not 18-adjacent (otherwise there

is only one connected component). Hence the two surfels of $sf(v)$ between v_1 and v and between v_2 and v are not adjacent (case D in Fig. 1). Thus, $sf(v)$ is not homeomorphic to a disk, which contradicts condition (2) of Definition 2.

If there is no two voxels of $N_6(v)$ in two different connected components, it means that one of them (say v_1) belongs to $N_{18}(v) \setminus N_6(v)$. The two 6-neighbors of v_1 in $N_6(v)$ belong to R (otherwise we are in the case of the previous paragraph). Hence the linel l incident to v_1 and v is such that $d(l, v) = 0$ (because the two 6-neighbors of v_1 in $N_6(x)$ belong to R thus there is no surfel between these surfels and v). But there are two surfels between v_1 and its two 6-neighbors in $N_6(v)$ and $d(l) = 2$. This contradicts condition (3). □

Now we prove that the topology of the image is preserved (i.e. the topology of each region of the image is preserved and the surface relations are preserved). This proof is necessary since we deal with multiple regions and Proposition 1 proves only that the topology is preserved for binary cases.

Proposition 2. *If v is an ML-Simple point, the topology of the image partition is unchanged by swapping v.*

Idea of the proof. The proof is made in two steps. First, we show that the topology of each region is preserved by using the link with simple points (Proposition 1). Second, we prove that the topology of surfaces is preserved (inclusion and adjacency between surfaces). With these two facts, we can prove that for each region, its Betti numbers[1] are preserved. □

3.3 Detection of Multi-Label Simple Points

To efficiently retrieve intervoxel information, we use an intervoxel matrix to encode the borders of the regions in the 3D image. This matrix stores the state (on or off) of each intervoxel cell, determined by the three following rules:

- a surfel s is *on* iff $s \in SF$ (i.e. s is between 2 voxels with different labels);
- a linel l is *on* iff l is incident to > 2 on surfels;
- a pointel p is *on* iff p is incident to 1 or > 2 on linels.

Using this intervoxel matrix, Algo. 1 determines if voxel v is ML-Simple. We prove that our algorithm returns true iff v is an ML-Simple point.

Proof. For each linel l incident to v, l is on implies $d(l) > 2$ which contradicts condition (1) of Definition 2: the algorithm returns false. The second test, if both surfels incident to l and v are off, corresponds to case $d(l, v) = 0$. In this case, if at least one surfel incident to l and not to v is on, we have $d(l) > 0$ which contradicts condition (1): the algorithm returns false. The last test verifies

[1] In 3D, the first 3 Betti numbers count the number of connected components, the number of tunnels and the number of cavities.

Algorithm 1. Detection of ML-Simple points

```
Data: The intervoxel matrix;
      A voxel v.
Result: true iff v is an ML-Simple point.
foreach l ∈ linels(v) do
    if l is on then return false;
    if both surfels incident to l and v are off then
        if at least one surfel incident to l and not to v is on then
            return false;
if configuration of surfels is A, D, H, J then return false;
return true;
```

condition (2). We test if $sf(v)$ is homeomorphic to a disk by testing all the cases where this condition is not satisfied (cases A, D, H, J in Fig. 1) and returns false in such a case. After all these tests, the three conditions of Definition 2 are satisfied, thus v is ML-Simple. The algorithm returns true accordingly. □

The complexity of Algo. 1 is $O(1)$. There are 12 linels in linels(v), and each test (if a cell is on or off) is an atomic operation. Checking if the configuration is A, D, H or J is easily achieved in constant time for each case, just by probing some particular surfels.

4 Deformable Model Process

We define a digital deformable partition model, whose geometry is encoded with an intervoxel matrix. The elementary deformations are swaps of ML-Simple voxels. Proposition 2 ensures the preservation of the topology of the partition. The deformation is governed by an energy-minimizing process. The energy used here is a preliminary and simple version allowing to show the feasibility of a deformable partition model based on ML-Simple voxels swaps. The energy of a partition is the sum of the energies of each digital surface S between pairs of regions (r_1, r_2). The total energy for S is defined as $E(S) = \omega_r E_r(S) + \omega_s E_s(S)$, where E_r and E_s are respectively the region and surface energies. Parameters ω_r and ω_s are the weights defining the relative significance of the corresponding terms.

Energy E_r is an energy describing the quality of the fit of regions to image data: $E_r(S) = MSE(r_1) + MSE(r_2)$ where $MSE(x)$ is the *Mean Squared Error* of region x. $E_r(S)$ decreases when the region becomes more homogeneous.

Energy E_s is based on a discrete area estimator proposed in [17]. The area estimation is not the same whether it is computed on one side or the other of the surface. Thus, the energy of S is defined as $E_s(S) = \sum_{s \in S} \text{area}_{r_1}(s) + \text{area}_{r_2}(s)$ where $\text{area}_{r_1}(s)$ and $\text{area}_{r_2}(s)$ are the area estimators for a surfel s considering respectively the side of region r_1 and the side of region r_2. E_s decreases when the surface becomes smoother.

The principle of the deformation process of one surface S is the following. The initial energy is computed. For each surfel s of S, the process tries two swaps, one for each voxel incident to s: v_1 and v_2. If this voxel is an ML-Simple point we temporary swap it and compute the energy associated to the move. The move of minimum energy is selected. If its energy is lower than the initial energy, we apply the deformation and set the new initial energy. In the case of an energy minimization process, the deformation algorithm is executed on every border faces represented in the intervoxel matrix. The process iterates until no deformations occur (i.e. a local minimum energy is reached). Since we process surfels, and since a swap is made if and only if the global energy strictly decreases, deformations will stop at some point whatever the input configuration.

5 Experiments

In this section we present three experiments on the deformation process and the energies used in the paper.

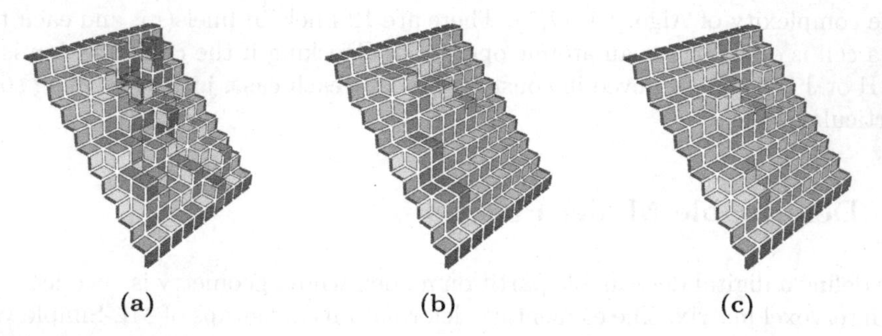

Fig. 2. (a) A noisy discrete plane surface. (b) Minimization of the number of surfels. (c) Minimization of the area based on the discrete estimator.

The first experiment highlights the interest of the discrete area estimator for regularization. We experiment a deformation process on a noisy plane surface in order to retrieve the optimal discrete plane. Two energies are compared: one using the number of surfels and one using the discrete area estimator. Figure 2a shows the noisy discrete area ($|S| = 240$, $E_s(S) = 338.0$) corresponding to an inclined plane: the optimal result is ($|S| = 180$, $E_s(S) = 259.4$). Figure 2b is the result of the energy minimization process using the energy based on the number of surfels ($|S| = 190$, $E_s(S) = 266.0$). Figure 2c shows the result of the energy minimization process using the energy based on the discrete area estimator ($|S| = 192$, $E_s(S) = 262.2$). The discrete area estimator gives a better visual and estimated area than the energy based on the number of surfels. Remark that, in both cases, a local minimum is reached.

The second experiment shows a segmentation of a 3D medical image with a poor initialization, in a way similar to continuous deformable partition models

[24]. Starting with a topologically correct segmentation of the image, the deformation process is used to retrieve shapes in the image while keeping topological information. The algorithm is applied on a simulated MRI brain image obtained from [8]. According to a-priori knowledge the image is composed of a sphere with two surrounding shells. Figure 3a shows a slice of the original image, the initial partition on the same slice is presented Fig. 3b and the resulting segmentation is shown in Fig. 3c). The algorithm ensures that the topology of the last segmentation is the same as the topology of the image partitionned by three surfaces. The resulting partition is not satisfactory due to the expression of the energies which needs to be adressed in future works.

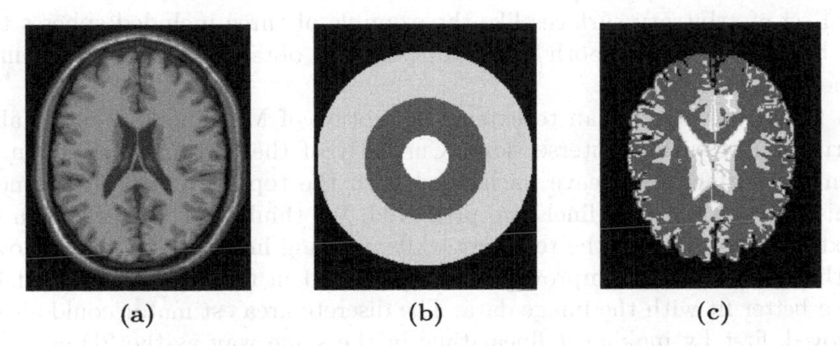

Fig. 3. (a) Slice of a simulated MRI brain image. (b) Initial partition with three overlapping spheres. (c) Resulting segmentation after deformation.

The last experiment consists in optimizing an initial segmentation containing several regions in order to enhance the result. The first partition, produced by a split and merge algorithm, contains block shaped regions. The deformation slightly modifies surfaces of the image to get a better partition according to the criterion. Figure 4a and Fig. 4b present a slice of the partition before and after the deformation processes. Borders of regions match more accurately image data. Figure 4c and Fig. 4d present the surface of the central dark region region before and after the deformation processes.

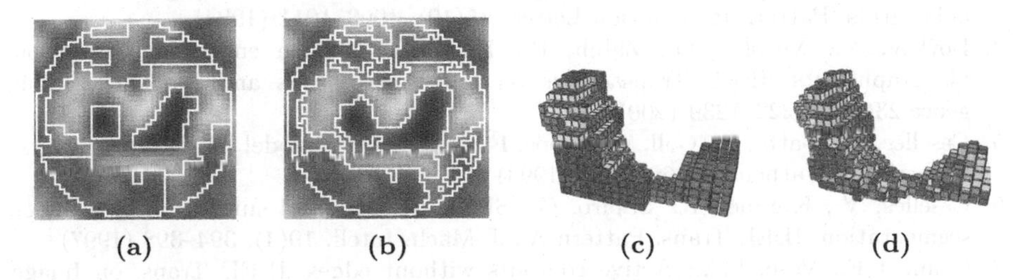

Fig. 4. (a) Slice of the initial segmentation. (b) Same slice after deformation. (c) Surface of the stripped region. (d) Same surface after deformation.

6 Conclusion

The main contributions of this work are: (i) The definition of ML-Simple points: a voxel is ML-Simple if its removal preserves the topology of the partition. The ML-Simple test algorithm is local, short and easy to implement. (ii) Our method is generic: regions and surfaces information can be mixed to define energies specialized for various applications. (iii) Our work deals with arbitrary multi-label image partitions: we can deform any number of surfaces while preserving their topology. The overall computational complexity depends on the number of surfels of the partition, not on its topological complexity. These interests have been illustrated in several preliminaries experiments. We may either deform an initial set of arbitrary surfaces like the example of three included spheres that fit a brain image, or smooth an initial partition obtained from a preliminary segmentation.

In future works, we plan to extend the notion of ML-Simple point to allow deformation of surface intersections. Currently, if there is an intersection between two surfaces, we leave unchanged both the topology and the geometry of this intersection since linels are preserved. We think that the definition can be extended to preserve the topology while allowing intersections to be moved. Another prospect is to improve the energies used in the deformable model to have a better fit with the image data. The discrete area estimator could also be improved, first by making it linear-time in the same way as the 2D case, secondly by making it dynamic to avoid global recomputation. This would allow the processing of big 3D images. Another research track is to find an area estimator with less local minima, in a way similar to [10] in 2D.

References

1. Ardon, R., Cohen, L.D.: Fast constrained surface extraction by minimal paths. International Journal on Computer Vision 69(1), 127–136 (2006)
2. Bazin, P.-L., Ellingsen, L.M., Pham, D.L.: Digital homeomorphisms in deformable registration. In: Karssemeijer, N., Lelieveldt, B. (eds.) IPMI 2007. LNCS, vol. 4584, pp. 211–222. Springer, Heidelberg (2007)
3. Bertrand, G.: Simple points, topological numbers and geodesic neighborhoods in cubic grids. Pattern Recognition Letters 15(10), 1003–1011 (1994)
4. Boykov, Y., Veksler, O., Zabih, R.: Fast approximate energy minimization via graph cuts. IEEE Transactions on Pattern Analysis and Machine Intelligence 23(11), 1222–1239 (2001)
5. Caselles, V., Catte, F., Coll, T., Dibos, F.: A geometric model for active contours. Numerische Mathematik 66, 1–31 (1993)
6. Caselles, V., Kimmel, R., Sapiro, G., Sbert, C.: Minimal surfaces based object segmentation. IEEE Trans. Pattern Anal. Mach. Intell. 19(4), 394–398 (1997)
7. Chan, T.F., Vese, L.A.: Active contours without edges. IEEE Trans. on Image Processing 10(2), 266–277 (2001)
8. Cocosco, C.A., Kollokian, V., Kwan, R.K.-S., Evans, A.C.: Brainweb: Online interface to a 3d MRI simulated brain database. In: Proc. of 3-rd Int. Conference on Functional Mapping of the Human Brain, Copenhagen, Denmark (May 1997)

9. Cohen, L.D., Kimmel, R.: Global minimum for active contour models: a minimal path approach. Int. Journal of Computer Vision 24(1), 57–78 (1997)
10. de Vieilleville, F., Lachaud, J.-O.: Toward a digital deformable model simulating open active contours. In: Brlek, S., Reutenauer, C., Provençal, X. (eds.) DGCI 2009. LNCS, vol. 5810, pp. 156–167. Springer, Heidelberg (2009)
11. Greig, D., Porteous, B., Seheult, A.: Exact maximum a posteriori estimation for binary images. Journal of the Royal Statistical Society (B) 51(2), 271–279 (1989)
12. Guigues, L., Cocquerez, J.-P., Le Men, H.: Scale-sets image analysis. International Journal on Computer Vision 68(3), 289–317 (2006)
13. Han, X., Xu, C., Prince, J.L.: A topology preserving level set method for geometric deformable models. IEEE Trans. on Pattern Analysis and Machine Intelligence 25(6), 755–768 (2003)
14. Kass, M., Witkin, A., Terzopoulos, D.: Snakes: Active contour models. International Journal of Computer Vision 1(4), 321–331 (1988)
15. Kovalevsky, V.A.: Finite topology as applied to image analysis 46, 141–161 (1989)
16. Lachaud, J.-O., Vialard, A.: Discrete deformable boundaries for the segmentation of multidimensional images. In: Arcelli, C., Cordella, L.P., Sanniti di Baja, G. (eds.) IWVF 2001. LNCS, vol. 2059, pp. 542–551. Springer, Heidelberg (2001)
17. Lachaud, J.-O., Vialard, A.: Geometric measures on arbitrary dimensional digital surfaces. In: Nyström, I., Sanniti di Baja, G., Svensson, S. (eds.) DGCI 2003. LNCS, vol. 2886, pp. 434–443. Springer, Heidelberg (2003)
18. Malladi, R., Sethian, J.A., Vemuri, B.C.: Shape Modelling with Front Propagation: A Level Set Approach. IEEE Trans. on Pattern Analysis and Machine Intelligence 17(2), 158–174 (1995)
19. Mumford, D., Shah, J.: Optimal approximations by piecewise smooth functions and associated variational problems. Comm. Pure Appl. Math. 42, 577–684 (1989)
20. Pons, J.-P., Boissonnat, J.-D.: Delaunay deformable models: Topology-adaptive meshes based on the restricted Delaunay triangulation. In: Proc. IEEE Conference on Computer Vision and Pattern Recognition, pp. 1–8 (2007)
21. Pruvot, J.H., Brun, L.: Scale set representation for image segmentation. In: Escolano, F., Vento, M. (eds.) GbRPR. LNCS, vol. 4538, pp. 126–137. Springer, Heidelberg (2007)
22. Ségonne, F.: Active contours under topology control - genus preserving level sets. Int. Journal of Computer Vision 79, 107–117 (2008)
23. Ségonne, F., Pons, J.-P., Grimson, W.E.L., Fischl, B.: Active contours under topology control genus preserving level sets. In: Int. Workshop Computer Vision for Biomedical Image Applications, pp. 135–145 (2005)
24. Vese, L.A., Chan, T.F.: A multiphase level set framework for image segmentation using the Mumford and Shah model. Int. J. Comput. Vis. 50(3), 271–293 (2002)

Marching Triangle Polygonization for Efficient Surface Reconstruction from Its Distance Transform

Marc Fournier, Jean-Michel Dischler, and Dominique Bechmann

Image Sciences, Computer Sciences and Remote Sensing Laboratory
LSIIT – UMR 7005 – CNRS – Louis Pasteur University, Strasbourg, France
{fournier,dischler,bechmann}@lsiit.u-strasbg.fr

Abstract. In this paper we propose a new polygonization method based on the classic Marching Triangle algorithm. It is an improved and efficient version of the basic algorithm which produces a complete mesh without any cracks. Our method is useful in the surface reconstruction process of scanned objects. It works over the scalar field distance transform of the object to produce the resulting triangle mesh. First we improve the original algorithm in finding new potential vertices in the mesh growing process. Second we modify the Delaunay sphere test on the new triangles. Third we consider new triangles configuration to obtain a more complete mesh. Finally we introduce an edge processing sequence to improve the overall Marching Triangle algorithm. We use a relevant error metric tool to compare results and show our new method is more accurate than Marching Cube which is the most widely used triangulation algorithm in the surface reconstruction process of scanned objects.

Keywords: Marching Triangle, Scalar field distance transform, Polygonization algorithm, Surface reconstruction, 3D scanned objects, Triangle mesh surface.

1 Introduction

In the past decade a lot of improvements have been made in the field of 3D scanners to acquire and to digitize real world objects. The advancement in computer technologies have made possible the design of 3D scanners to respond to the increasing needs in many fields such as digitizing precious cultural heritage [1]. The most common output data structure produced by 3D scanners is range images. From these raw data, many operations are needed in order to produce the final mesh which represents the real object geometry. All these operations can be achieved in the scalar field distance transform (SFDT) domain which is often used because it produces good results for each step of the reconstruction procedure.

To perform surface reconstruction in the SFDT domain, one needs to convert the explicit range images, or other initial triangle meshes dataset, by computing their SFDT implicit representation which is defined over a regular 3D grid created inside a mesh bounding box. The cubic grid cells are called voxels and for each voxel, the closest point on the mesh surface is found and the shortest distance to the mesh is saved in the voxel. For a given surface $S \subset \Re^3$ this volume representation consist of a

S. Brlek, C. Reutenauer, and X. Provençal (Eds.): DGCI 2009, LNCS 5810, pp. 168–179, 2009.

scalar value function $f:\Re^3 \rightarrow \Re$ such as the zero-set $f(x,y,z)=0$ defines the surface and in that case $[x,y,z] \in S$. To obtain a unique volumetric description for a given surface, this distance field is also signed according to surface normal vectors.

The surface reconstruction process begins with a mesh registration procedure [2] which is performed to express all range images in the same coordinates system. The first operation in the SFDT domain is mesh fusion [3] to integrate all initial range images into a unique representation, followed by mesh repair [4] to fill holes in the model, and then mesh smoothing [5] to remove acquisition noise introduced by the scanner and finally mesh simplification [6] to produce a more compact model without loss of details. Since the SFDT is an implicit representation, at the end of the reconstruction process a polygonization algorithm such as Marching Cube [7] or Marching Triangle [8] is needed to produce the final explicit mesh which describes the scanned object surface geometry.

In this paper we introduce a new polygonization method to triangulate the resulting mesh of the surface reconstruction process in the SFDT domain. We propose a set of improvements based on the Marching Triangle algorithm to obtain an efficient method which overcomes some crack problems in the basic version to produce a higher quality resulting mesh. Besides improving several steps of the original Marching Triangle algorithm, we propose an edge processing sequence to obtain better results from the overall algorithm. The remaining parts of the paper are organized as follows: Section 2 overviews related work. Section 3 presents our improvements to the original Marching Triangle algorithm steps. Section 4 describes our global edge processing sequence. And in Section 5, before concluding, we show and compare our triangulation method results with previously introduced algorithms.

2 Related Work

The most widely used algorithm to triangulate a SFDT is Marching Cube [7]. It is a volume-based approach which is very suitable to triangulate implicit representations such as SFDT in the surface reconstruction process of scanned objects. The original Marching Cube method has some ambiguities and many other algorithms, based on the original one, have been proposed to improve the resulting mesh quality. For example the original algorithm has 14 cube triangulation configurations which lead to face ambiguities resolved in [9], and [10] in which the 14 basic configurations are expended in 32 different cases. A remaining cube ambiguity was then solved in [11] to guaranty the resulting mesh topology.

Methods derived from the basic Marching Cube algorithm have been proposed to also resolve the ambiguities. Cubical Marching Squares [12] which opens the cubes into six squares and Marching Tetrahedron [13] which divides the cubes into tetrahedrons are two examples of cube configurations which resolve original ambiguities. Other algorithms have been introduced to improve the resulting mesh quality. The Extended Marching Cube [14] provides a sensitivity feature to better recover sharp edges and Dual Contouring [15] focus on preserving the resulting mesh topology. More recently, the VFDT model [16] which is a vector extension of the SFDT has been proposed to improve the implicit field accuracy and the Marching Cube algorithm has been adapted to this new representation in order to produce a higher quality resulting mesh.

Another well known algorithm to triangulate a SFDT is Marching Triangle [8]. It is a surface-based approach built on Delaunay triangulation definition. Starting from a seed triangle, a region-growing process enables the creation of triangles following the SFDT isosurface. This surface tracking process has been designed to triangulate SFDT of scanned objects and it has also been applied to continuous implicit surfaces describing virtual objects with a set of equations such as parametric ones. The basic Marching Triangle algorithm leaves some part of the model un-triangulated creating cracks in the resulting mesh as shown in Fig. 1.

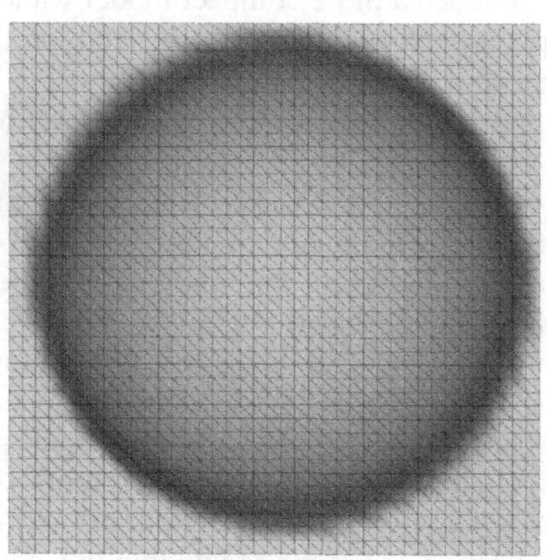

a) Original model

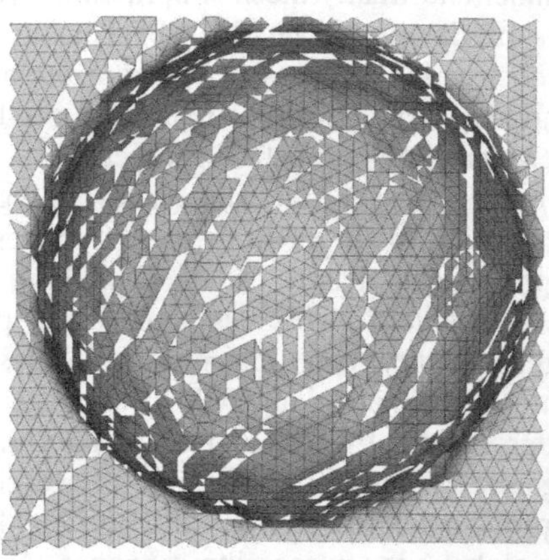

b) Marching Triangle result

Fig. 1. Basic Marching Triangle algorithm result on the Half-Sphere model

In Fig. 1a) a Half-Sphere model lying on a plane is used and its SFDT is computed. Fig. 1b) shows the basic Marching Triangle algorithm result which contains cracks in the triangulated resulting mesh. Theses cracks occur because of different reasons which will be detailed in next section.

Beside the classic Marching Triangle, other methods based on a region-growing and surface tracking process have been proposed such as Gopi algorithm [17] which works over a set of unorganized points and Hartmann algorithm [18] which works on continuous implicit surfaces. In this paper we focus on improving the steps of the original Marching Triangle to directly triangulate SFDT of scanned objects. The methods designed to work over point clouds are also suitable for SFDT as described in the Ball-Pivoting method [19] in which a first pass generate points from the SFDT before applying the triangulation on the set of generated points. This Ball-Pivoting method is similar to Marching Triangle, it also uses the Delaunay sphere test but the constraint is applied on the sphere radius instead of the sphere center as for the original Marching Triangle algorithm. In this paper we also modify the Delaunay sphere for the triangulating test which contributes to reduce the resulting mesh cracks.

Previous methods [20, 21, 22] applied to continuous implicit surfaces have been introduced to overcome the basic Marching Triangle algorithm cracks problem. These method triangulations are adaptive to local implicit curvature to obtain variable size triangles. In this paper we also modify the basic algorithm with a variable projection distance to obtain a better adaptive result over SFDT. Some of these methods [20, 21] operate in two passes. They first triangulate a resulting mesh according to the basic algorithm which contains cracks and then they introduce a crack filling algorithm to complete the resulting mesh. The other method [22] introduces a different region-growing algorithm compared to the original Marching Triangle. It is based on a hexagonal triangulation expansion pattern which is able to resolve cracks and to produce a complete resulting mesh in a single pass. In this paper we use the original Marching Triangle algorithm procedure and we propose improvements to several steps of the method. Our new triangulation works over SFDT in a single pass, as for the original algorithm, and produce a complete mesh without cracks. We do not need a second specific post-processing crack filling pass but instead we introduce an iterative process on our single pass to obtain a complete mesh.

Other algorithms have been proposed to improve the basic Marching Triangle triangulation over discrete implicit surfaces and to achieve specific goals. A topology preserving method [23] introduces a normal consistency constraint which guaranties the resulting mesh topology. An edge constrained method [24] detects discontinuities in the implicit surface and constrains triangle edges to match and better preserve these sharp features in the resulting mesh. In this paper we have a more global approach to produce a higher quality mesh by improving the entire original Marching Triangle algorithm. Each step of the algorithm is revised and in addition we introduce an edge processing sequence to obtain a more globally accurate resulting mesh including a better sharp features preserving compared to the original Marching Triangle algorithm.

3 Marching Triangle Improved Algorithm

In this section we describe our new Marching Triangle improved algorithm. We refer to the original algorithm [8] procedure numbering to identify the steps to improve. In Subsection 3.1 we improve steps 1 and 2 of the original algorithm in finding a new vertex. In Subsection 3.2 we improve step 4 in testing a new triangle and in Subsection 3.3 we improve step 6 in considering new triangles.

3.1 Variable Projection Distance and Vertex Interpolation

The first and second steps of Marching Triangle algorithm are illustrated in Fig. 2. The first step is the estimation of a new vertex position P' with a projection perpendicular to the mid-point P of the boundary edge C in the plane of the model boundary triangle ABC by a constant distance PP'. The second step is the evaluation of the nearest point to P' which is on the implicit surface. That new potential vertex is V_2 in Fig. 2 example.

The constant distance projection contributes to produce cracks in the mesh. In high curvature regions the projection can be further away from the isosurface and the new vertex estimation may cause either a test failure resulting in a crack beginning or a bad approximation of the surface local geometry. To correct this problem we propose a variable projection distance which is equal to √3/2 times the length of the boundary edge. This projection distance corresponds to the height of the equilateral triangle composed of the boundary edge. This improvement contributes to obtain more equilateral triangles which are less deformed and which sizes adapt gradually to local geometry curvature.

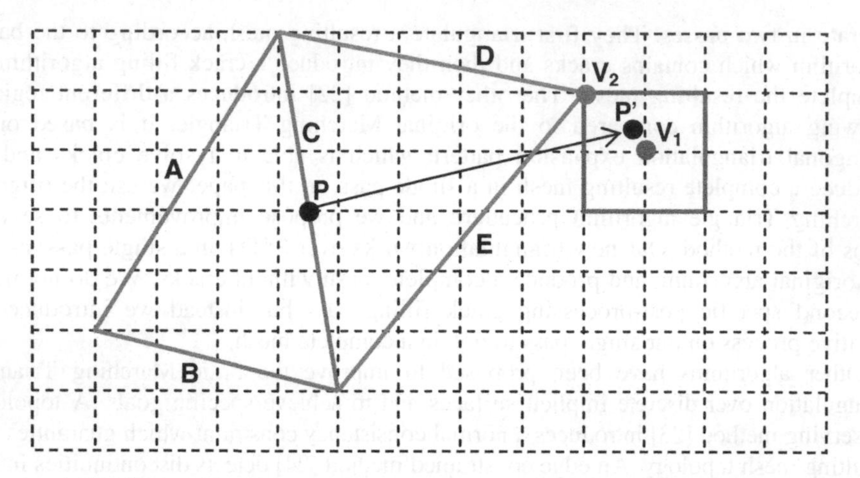

Fig. 2. Marching Triangle projection and nearest vertex on the isosurface

From the projection point P' on Fig. 2 example, we suppose the nearest voxel which is considered on the isosurface is V_2. While working with a discrete SFDT a threshold distance comparison is needed to find this nearest voxel. To consider that a voxel is on the isosurface it must contain a distance smaller than half the grid resolution. This approximation introduces a significant error in the vertex position according to the underlying implicit surface and this error can also contributes to a crack in the mesh if the new triangle test fails. To correct this problem we propose a linear vertex position interpolation between the nearest voxel found and its closest neighbor with opposite distance sign. The interpolation finds the new vertex position which corresponds to a distance equal to zero between the two distances of opposite sign. This interpolation uses the implicit surface definition to obtain a more accurate approximation of the isosurface.

3.2 Modified Delaunay Sphere Test

Step 4 of the original algorithm suggests testing the new potential triangle according to the Delaunay sphere which is circumscribed to the new triangle as shown in Fig. 3a) in which the current boundary edge E_b forms a new triangle with the new estimated vertex A. In Fig. 3a) example the test would fail because there are parts of the six numbered triangles inside the sphere. According to the original algorithm, further tests would be made with edge E_b and vertices B, C and D to consider these three new triangles. These tests would also fail because the Delaunay sphere would still contain parts of neighbor triangles. This test limits the original algorithm performances and contributes to produce cracks in the mesh. The Delaunay test sphere was designed to triangulate a set of unorganized points which is not exactly the same situation in the case of a region-growing algorithm over a SFDT.

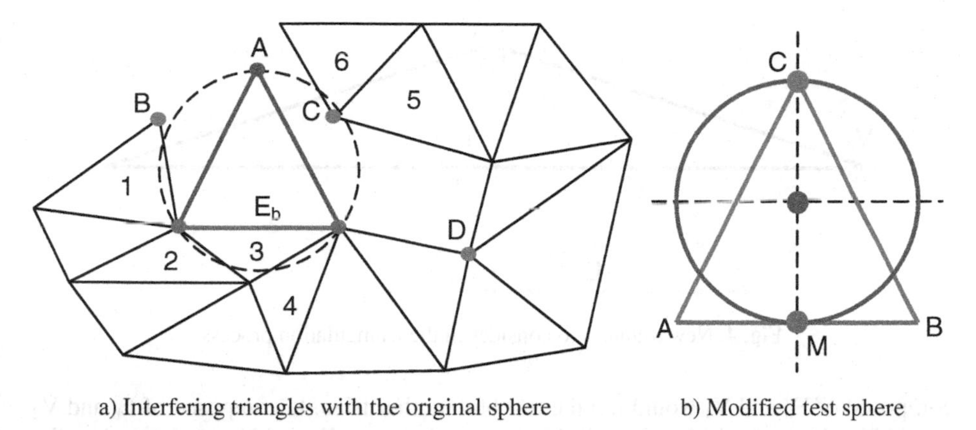

a) Interfering triangles with the original sphere b) Modified test sphere

Fig. 3. Interfering triangles with the original Delaunay sphere and modified test sphere

We propose the modified sphere shown in Fig. 3b) for new triangles testing. The modified sphere passes through the mid-point M of the current boundary edge and the new estimated vertex C. Its diameter is equal to the distance between these two points and its center is the mid-point between M and C. The modified sphere is smaller than the original one and it allows obtaining more successful tests which improve the algorithm and reduce the cracks in the resulting mesh. Since some parts of the new triangle are outside the sphere, we also need to test if there is no intersection between the new triangle and other triangles of the mesh. We test all triangles which have parts inside the original Delaunay sphere with the new triangle according to Moller intersection test [25] which is fast and efficient.

3.3 New Triangles to Consider

Step 6 of the original algorithm suggests considering two new potential triangles illustrated in Fig. 3a) if the first new triangle test fails at step 4. These new triangles are composed of the current boundary edge E_b and vertices B or D which are the previous and the next vertices along the boundary from the current boundary edge E_b. If the sphere test fails for these two new triangles, other new triangles should be tested to improve the algorithm. The original algorithm was upgraded in [26] with a seventh step which suggests to test another new triangle composed of the current boundary edge E_b and vertex C in reference of Fig. 3a). Vertex C is the nearest boundary vertex of overlapping triangles number 5 and 6 in the sphere test. This new potential triangle contributes to reduce the cracks in the mesh compared to the original algorithm but if the test fails with this triangle other triangles should be tested to better improve the algorithm.

We propose to test not only the nearest but all boundary vertices of overlapping triangles if they exist. In some particular cases as the one shown in Fig. 4 the sphere test would fail with the nearest boundary vertex leaving a part of the mesh un-triangulated and a new triangle could be added to the mesh if another boundary vertex of the overlapping triangle was considered. In Fig. 4 E_b is the current boundary edge and V_1 is the triangle nearest vertex from E_b. The new potential triangle

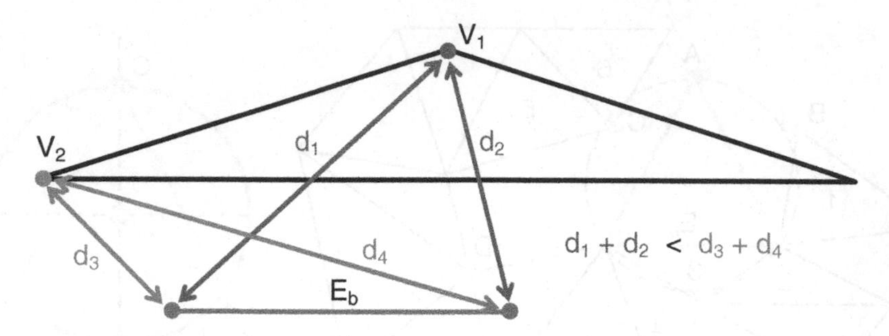

Fig. 4. New triangles to consider in the triangulation process

composed of E_b and V_1 would fail the test but another triangle composed of E_b and V_2 would be a better candidate even if the distance between E_b and V_2 is greater than the distance between E_b and V_1.

4 Edge Processing Sequence and Iterative Process

The original Marching Triangle algorithm does not specify any boundary edges processing sequence. It is only defined as a single pass into the edge list to process all boundary edges with the procedure steps including the estimation of a new potential triangle and its sphere test to determine if it will be added or not to the mesh. According to the implemented data structure to triangulate the SFDT and the method to add new edges in the edge list, the edge processing sequence can be different from one implementation to another. The resulting mesh from the Marching Triangle depends on the edge processing sequence and it can be different if the sequence is changed as shown in Fig. 5.

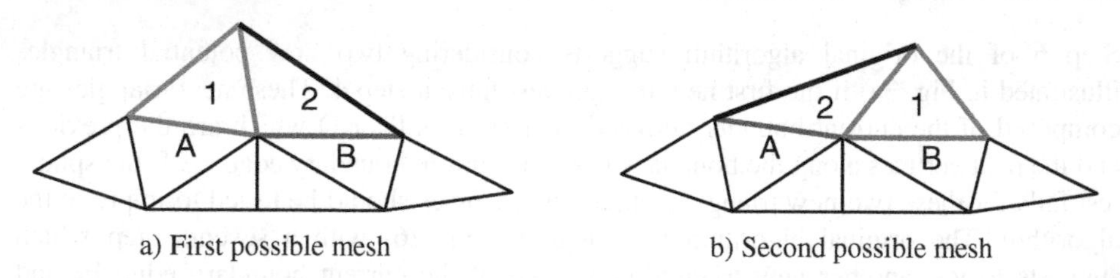

Fig. 5. Modified mesh according to the edge processing sequence

In Fig. 5 example we assume the current mesh is composed of the six bottom triangles, therefore edges A and B are boundary edges. In Fig. 5a) edge A is considered first and triangle 1 is added then edge B is processed and triangle 2 is added next. In Fig. 5b) edge B is considered first and triangle 1 is added first then edge A is tested and triangle 2 is also added. In Fig. 5 simple example both results are different just because of an edge permutation. The results would have been more different if for example after adding triangle 1 in Fig. 5a) the new boundary edges of that triangle were processed before edge B and if that same processing sequence was applied in Fig. 5b) to the new boundary edges of triangle 1.

We tested different edge sequences and combinations on the original algorithm and we selected the one which optimizes the result in terms of minimizing the cracks in the mesh. We propose the following procedure to improve the resulting mesh quality. Starting from a boundary edge, an arbitrary direction is selected and the next edge to consider is the neighbor boundary edge always in that same direction around the contour of the current mesh. Newly added boundary edges are not considered immediately, they will only be considered on the next turn around. This procedure is illustrated in Fig. 6 with E_C corresponding to the current boundary edge to be processed and E_N the next boundary edge to consider according to the chosen arrow direction D.

In Fig. 6a) New triangle and Fig. 6b) Previous triangle cases the next boundary edge to consider is straight forward. In Fig. 6c) Next triangle case if the test is successful and the new triangle is added then the next boundary edge to consider jumps over edge A which is lo longer a boundary edge. The new added boundary edge B will be considered on the next turn around only. In Fig. 6d) Overlapping triangle case a bridge is created in the mesh if the new triangle is added. In that case the next edge to consider is either E_{N1} or E_{N2} depending on the chosen direction D_1 or D_2 respectively. In that special case the inside contour in the region P is triangulated in priority immediately after the new triangle is added to the mesh and before pursuing with the outside contour. Starting from the new inside boundary edge A an arbitrary direction is selected and the previously described procedure is applied to the inside contour until no more triangle can be added.

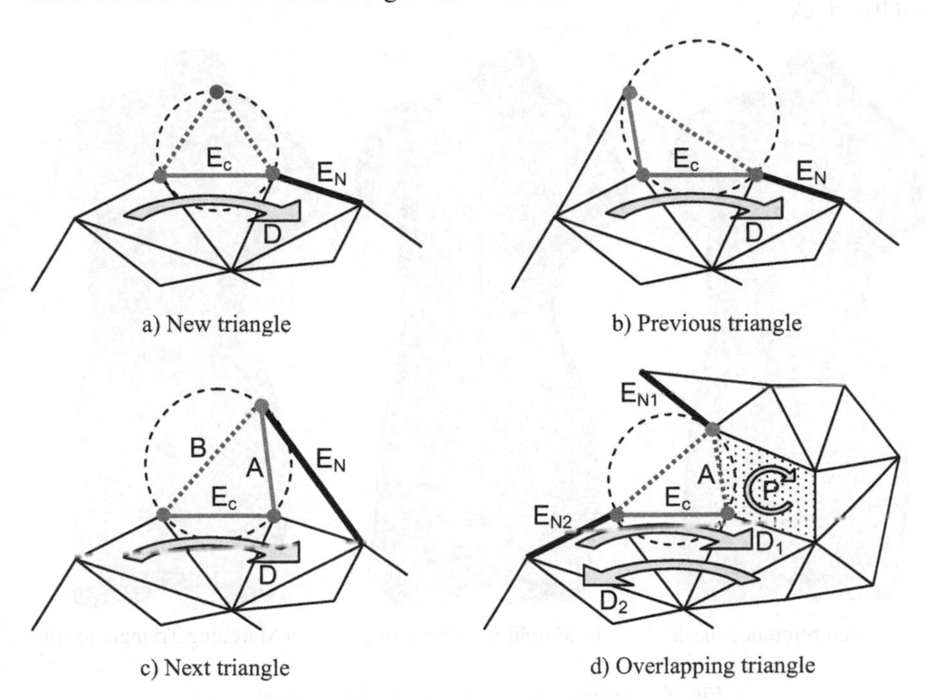

Fig. 6. Next edge to consider according to the edge processing sequence

The original Marching Triangle is defined as a single pass into the edge list. Our proposed procedure is iterative and the boundary edges are tested more than once, until a complete loop is made around the mesh contour without adding any triangle. A boundary edge can be tested without adding any triangle to it at the first pass and at the second pass the test could be successful depending on the local mesh neighbourhood configuration which can be different from one pass to another. Our iterative procedure continues as long as new triangles are added. Compared to the original algorithm our procedure contributes to add more triangles and to reduce the cracks in the final resulting mesh.

5 Results

In this section we evaluate and compare our improved algorithm results with Marching Cube triangulation using the vertex to surface error metric defined in [5]. We used the Venus model reference mesh shown in Fig. 7a) to compute its SFDT with an appropriate grid resolution according to the model level of details. Then we triangulated the SFDT with Marching Cube and our improved Marching Triangle algorithms and these results are shown in Fig. 7. We compared these two results to the reference mesh using the vertex to surface error metric and Table 1 shows these error values along with the number of triangles and the triangulation computing times for both results. The timing measures were made using a Pentium 4 CPU computer with a 3.03GHz clock.

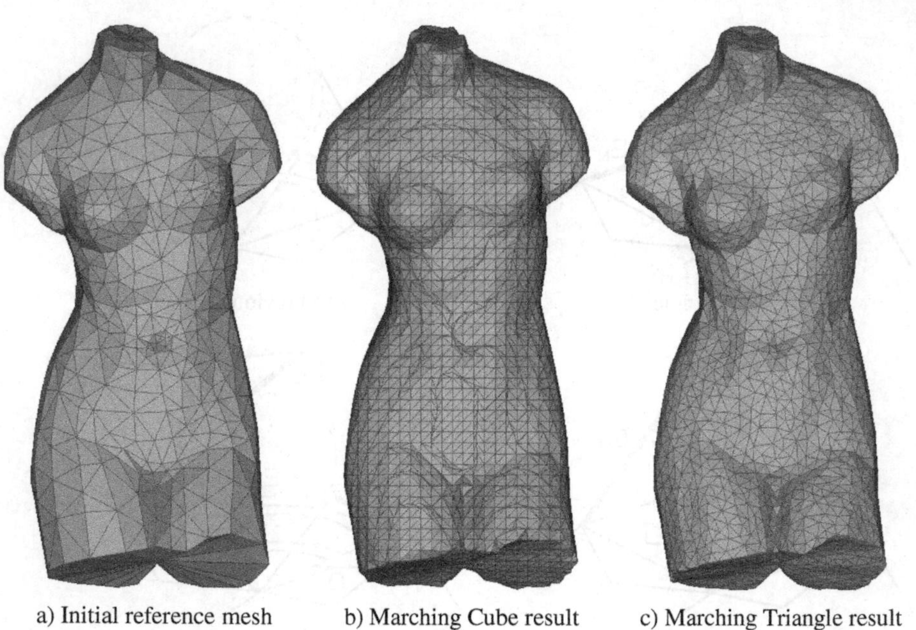

Fig. 7. Triangulation results on the Venus model

In Fig. 7 both results are of good quality but we see that the triangles of Marching Cube result are more dependent on the voxels size. Fig. 7b) result also shows small degenerated triangles forming elevation lines depending on the grid resolution which is the classic signature of Marching Cube algorithm. Marching Triangle result in Fig. 7c) shows a more homogeneous triangulation and sharp edges such as the one at the bottom are better preserved compared to Marching Cube result. Table 1 also shows that Marching Triangle result is of better quality according to the error metric. Our result also contains fewer triangles, thus optimizing the model quality, storage and memory space with almost one third less triangles than Marching Cube result. The drawback of our algorithm is the computation time which is almost the double compared to Marching Cube. The triangulation time is still reasonable for this model but it could make a difference for example in real time applications on large models.

Table 1. Triangulation parameters for the Venus model

Parameters	Marching Cube	Marching Triangle	Difference
Nb Triangle	7 465	5 152	31.0%
Error	7.58×10^{-3}	6.88×10^{-3}	9.23%
Time (ms)	26.7	48.5	81.6%

6 Conclusion and Future Work

In this paper we designed a new Marching Triangle algorithm to improve the triangulation result of such method over the SFDT of scanned objects in their surface reconstruction process. Our contribution focused on improving several steps of the original algorithm to overcome the crack forming problem. We proposed a variable projection distance and a vertex position interpolation to provide a better isosurface approximation. We introduced a modified sphere for testing potential triangles geometry consistency before adding them to the resulting mesh. We also proposed testing new potential triangles which can lead to more complete results in particular cases. We structured an edge processing sequence which is more efficient during the triangulation process. We compared our algorithm to Marching Cube triangulation and demonstrated its relevancy based on an error metric measure.

Future work will include optimizing the processing time of every step of our new algorithm since it is a drawback compared to Marching Cube performances. We will mainly focus on the triangle geometry consistency testing step since it is the most time consuming one of the overall algorithm. We will also work on adapting our new algorithm to other representations such as continuous implicit surfaces, point clouds and 3D volumetric datasets for medical and related applications. Adapting our algorithm to these representations will provide a useful tool to a wider range of applications in computer graphics. Surface-based triangulation algorithms such as Marching Triangle are more complex to design to obtain efficient results but in general their resulting meshes are of higher quality compared to volume-based methods which are usually simpler to implement.

References

1. Rocchini, C., Cignoni, P., Montani, C., Pingi, P., Scopigno, R.: A Low Cost 3D Scanner Based on Structured Light. In: EUROGRAPHICS 2001, Manchester, UK, vol. 20(3) (2001)
2. Besl, P.J., McKay, N.D.: A Method of Registration of 3D Shapes. IEEE Transactions on Pattern Analysis and Machine Intelligence 14(2), 239–256 (1992)
3. Hilton, A., Stoddart, A.J., Illingworth, J., Windeatt, T.: Implicit Surface-Based Geometric Fusion. Computer Vision and Image Understanding 69(3), 273–291 (1998)
4. Davis, J., Marschner, S., Garr, M., Levoy, M.: Filling Holes in Complex Surfaces Using Volumetric Diffusion. In: First International Symposium on 3D Data Processing, Visualization, and Transmission, Padua, Italy, pp. 428–438 (2002)
5. Fournier, M., Dischler, J.-M., Bechmann, D.: 3D Distance Transform Adaptive Filtering for Smoothing and Denoising Triangle Meshes. In: International Conference on Computer Graphics and Interactive Techniques, Kuala Lumpur, Malaysia, pp. 407–416 (2006)
6. Nooruddin, F.S., Turk, G.: Simplification and Repair of Polygonal Models Using Volumetric Techniques. IEEE Visualization and Computer Graphics 9(2), 191–205 (2003)
7. Lorensen, W.E., Cline, H.E.: Marching Cubes: A High Resolution 3D Surface Reconstruction Algorithm. In: ACM SIGGRAPH 1987, Anaheim, USA, Computer Graphics, pp. 163–169 (1987)
8. Hilton, A., Stoddart, A.J., Illingworth, J., Windeatt, T.: Marching Triangles: Range Image Fusion for Complex Object Modelling. In: International Conference on Image Processing, Lausanne, Switzerland, vol. 1 (1996)
9. Nielson, G.M., Hamann, B.: The Asymptotic Decider: Resolving the Ambiguity in Marching Cubes. In: IEEE Conference on Visualization 1991, San Diego, USA, pp. 83–91 (1991)
10. Chernyaev, E.V.: Marching Cubes 33: Construction of Topologically Correct Isosurfaces. Tech. Report, No. CN/95-17, CERN, Geneva, Switzerland (1995)
11. Lewiner, T., Lopes, H., Vieira, A.W., Tavares, G.: Efficient Implementation of Marching Cubes Cases with Topological Guarantees. Journal of Graphics Tools 8(2) (2003)
12. Chien-Chang, H., Fu-Che, W., Bing-Yu, C., Yung-Yu, C., Ming, O.: Cubical Marching Squares: Adaptive Feature Preserving Surface Extraction from Volume Data. In: EUROGRAPHICS 2005, Dublin, Ireland, Computer Graphics Forums, vol. 24(3) (2005)
13. Chan, S.L., Purisima, E.O.: A New Tetrahedral Tesselation Scheme for Isosurface Generation. Computers and Graphics 22(1), 83–90 (1998)
14. Kobbelt, L.P., Botsch, M., Schwanecke, U., Seidel, H.P.: Feature Sensitive Surface Extraction from Volume Data. In: ACM SIGGRAPH 2001, Los Angeles, USA. Computer Graphics (2001)
15. Zhang, N., Hong, W., Kaufman, A.: Dual Contouring with Topology-Preserving Simplification Using Enhanced Cell Representation. In: IEEE Visualization, Austin, USA, pp. 505–512 (2004)
16. Fournier, M., Dischler, J.-M., Bechmann, D.: A New Vector Field Distance Transform and its Application to Mesh Processing from 3D Scanned Data. The Visual Computer Journal 23(9-11), 915–924 (2007)
17. Gopi, M., Krishnan, S., Silva, C.T.: Surface Reconstruction Based on Lower Dimensional Localized Delaunay Triangulation. Computer Graphics Forum 19(3) (2000)
18. Hartmann, E.: A Marching Method for the Triangulation of Surfaces. The Visual Computer 14(3), 95–108 (1998)

19. Bernardini, F., Mittleman, J., Rushmeier, H., Silva, C., Taubin, G.: The Ball-Pivoting Algorithm for Surface Reconstruction. IEEE Transactions on Visualization and Computer Graphics 5(4), 349–359 (1999)
20. Akkouche, S., Galin, E.: Adaptive Implicit Surface Polygonization Using Marching Triangles. Computer Graphics Forum 20(2), 67–80 (2001)
21. Karkanis, T., Stewart, A.J.: Curvature-Dependent Triangulation of Implicit Surfaces. IEEE Computer Graphics and Applications 21(2), 60–69 (2001)
22. Araujo, B.R., Jorge, J.A.P.: Curvature Dependent Polygonization of Implicit Surfaces. In: Brazilian Symposium on Computer Graphics and Image Processing, Curitiba, Brazil (2004)
23. Xi, Y., Duan, Y.: A Region-Growing Based Iso-Surface Extraction Algorithm. In: IEEE International Conference on Computer-Aided Design and Computer Graphics, Beijing, China, pp. 120–125 (2007)
24. McCormick, N.H., Fisher, R.B.: Edge-Constrained Marching Triangles. In: International Symposium on 3D Data Processing Visualization and Transmission, Padova, Italy (2002)
25. Moller, T.: A Fast Triangle-Triangle Intersection Test. Journal of Graphics Tools 2(2), 25–30 (1997)
26. Hilton, A., Illingworth, J.: Marching Triangles: Delaunay Implicit Surface Triangulation. Tech. Rep. EPSRC-GR/K04569, Centre for Vision Speech and Signal Processing, University of Surrey, Guildford, UK, 1-12 (1997)

Multivariate Watershed Segmentation of Compositional Data*

Michael Hanselmann[1], Ullrich Köthe[1], Bernhard Y. Renard[1], Marc Kirchner[1], Ron M.A. Heeren[2], and Fred A. Hamprecht[1,]**

[1] Heidelberg Collaboratory for Image Processing, University of Heidelberg, Germany
[2] FOM-Institute for Atomic and Molecular Physics, Amsterdam, The Netherlands
fred.hamprecht@iwr.uni-heidelberg.de

Abstract. Watershed segmentation of spectral images is typically achieved by first transforming the high-dimensional input data into a scalar boundary indicator map which is used to derive the watersheds. We propose to combine a Random Forest classifier with the watershed transform and introduce three novel methods to obtain scalar boundary indicator maps from class probability maps. We further introduce the multivariate watershed as a generalization of the classic watershed approach.

1 Introduction

The watershed transform [1] is a region-based segmentation algorithm for gray-scale images and is a popular method in image segmentation [2]. Figuratively speaking, the gray-valued boundary indicator image is considered as a height map which is flooded with water. Whenever two water basins that originate from different local minima meet, a watershed is constructed. Various definitions for the continuous and discrete case have been given, most of which operate on scalar-valued input. Typically the gradient is used as a boundary indicator, featuring high values at border locations and low values in homogeneous areas.

High-dimensional data is typically transformed into a scalar boundary map, such that the conventional watershed can be applied. A direct generalization for color images is the color gradient. Alternatively, the watershed transform is individually performed on each color band and the obtained segmentation results are integrated into a single segmentation map [3]. Some authors have suggested to use color models that feature an intensity channel [4,3]. However, most of these approaches are based on the assumption that color channels are highly correlated. This is typically not the case for more complex data, e.g. in remote sensing [5], the analysis of medical data [6] and data acquired in imaging mass spectrometry experiments [7]. In fluorescence microscopy, experimentalists often deliberately select dyes that highlight different, uncorrelated structures [8].

To analyze such kind of data, authors have used (weighted) channel-wise gradients [9,10,11] or the metric-based gradient [11]. Noyel [11] proposed the sum

* We thank Erika R. Amstalden and Kristine Glunde for acquiring and providing data.
** Corresponding author.

S. Brlek, C. Reutenauer, and X. Provençal (Eds.): DGCI 2009, LNCS 5810, pp. 180–192, 2009.

or supremum of channel-wise morphological gradients that are computed as difference between channel dilation and erosion. Malpica [12] and Karvelis [13] construct the Jacobian matrix J of partial derivatives (in direction of the two spatial axes) in each pixel and calculate the Eigenvalues of $J'J$. The difference between the two Eigenvalues is used as a boundary indicator. Zhang [14] uses the spectral angle between a pixel's underlying spectrum and a reference spectrum to obtain a scalar boundary indicator. Angulo proposed a stochastic watershed algorithm [15] which was later extended to multispectral images [16]. Soille [17] combines a histogram-based clustering with shape priors and pixel-wise gradients which are used as input for the classic watershed transform. Authors have also suggested dimensionality reduction [11,16,6] to reduce the effect of noise.

In many segmentation scenarios, more than two classes are of interest (e.g. background, healthy cells, diseased cells, ...). Often, prior knowledge on the composition of the data exists which can be used to assign class probabilities to pixels. This way, the dimensionality of the data is significantly reduced. Such data, where the probability vector in each pixel sums to one, is termed *compositional* [18]. In this paper, we propose three novel methods to obtain a scalar boundary indicator based on such compositional data, and we introduce the multivariate watershed which is a multi-class generalization of the classic watershed transform. We quantitatively compare the presented methods to three algorithms that previously showed good performance [11,12]: Noyel's [11] sum/supremum of channel-wise morphological gradients and Malpica's [12] tensor approach.

2 Watershed Segmentation of Compositional Data

The discrete and sequential watershed methods on scalar-valued input can be grouped into watershed by immersion [19] and watershed by topographic distance [20]. In this paper we argue in terms of the latter, however our results also carry over to the former setting. Watershed by topographic distance is based on gradient descent. An efficient algorithm proceeds as follows: for each pixel we store a reference to the neighboring pixel with the maximum slope

$$LS(u) = \max_{v \in \text{neighbors}(u) \cup \{u\}} \frac{f_u - f_v}{d(u, v)} \tag{1}$$

where f_u is the gray value at pixel u and $d(u, v)$ is the distance between pixels u and v. Minima pixels point towards themselves and a distinct label is assigned to each local minimum. Starting from the minima, the constructed paths are traversed in reverse order and all pixels are labeled with the corresponding label. In this implementation, the resulting watersheds lie *between* pixels (inter-pixel boundaries) and the algorithm has linear complexity in the number of pixels.

In the following, we work with spectral data in two spatial dimensions. Note, however, that the methods presented below are also applicable to other high-dimensional data. Let $S = \{(\mathbf{x}^1, y^1), ..., (\mathbf{x}^K, y^K)\}$ be a set of available M-dimensional training samples, i.e. K spectra $\mathbf{x}^i$ with M channels and corresponding class labels $y^i \in L_1, ..., L_D$, e.g. "cancerous", "healthy tissue" and

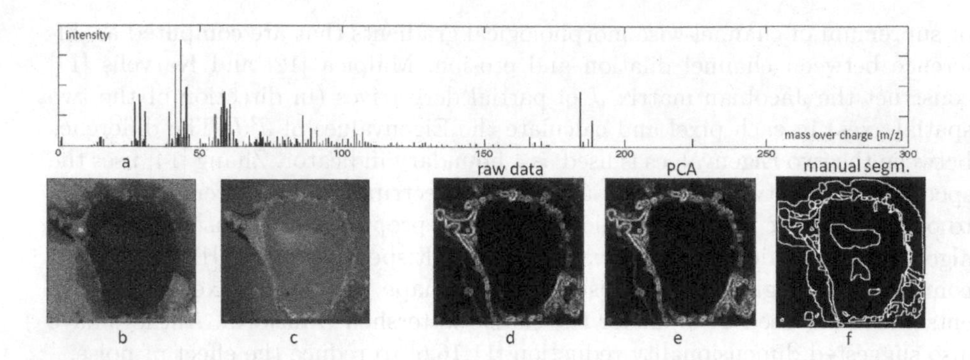

Fig. 1. An excerpt of a typical mass spectrum (a) and 2 channels of the IMS dataset (b,c). Only parts of the images are correlated. A boundary map is obtained with the sum of channel-wise morphological gradients on the raw data (d) or the first few principal components (e), revealing only few boundaries of the manual segmentation (f).

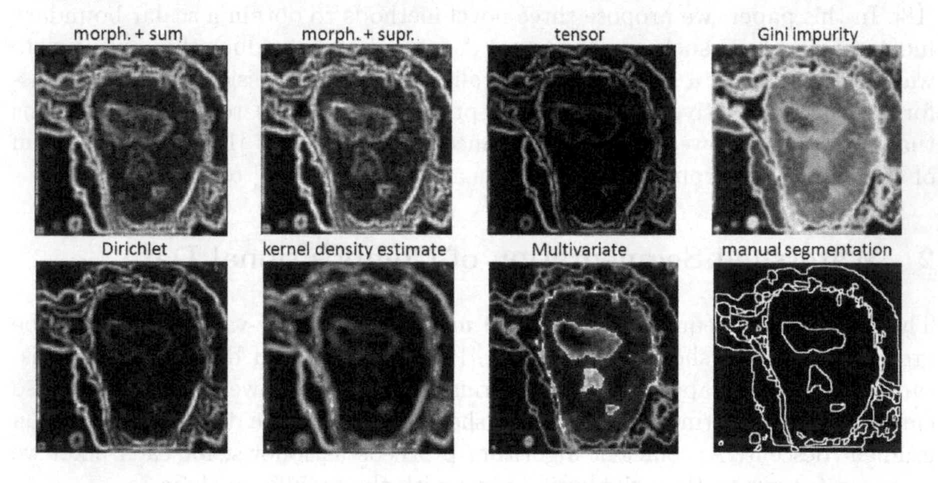

Fig. 2. The probability map-based boundary indicators obtained with Noyel's sum/supremum of channel-wise morphological gradients, Malpica's tensor and the methods introduced here. The result of an approximate manual labeling is also shown.

"blood vessels" in medical applications or "faulty" or "intact" in quality control. The Random Forest classifier [21] is a state-of-the-art supervised learning method that features high classification performance [22], allows fast training, can cope with a large number of input variables and is relatively robust to its hyperparameters. We train a Random Forest on S and classify the whole data set comprising N data points to obtain the posterior probability for each of the D classes in each pixel. The resulting data set has D dimensions in each pixel and is compositional since each probability vector sums to one.

We use data from an imaging mass spectrometry (IMS) [7] experiment of human breast cancer grown in mice [23,24] to illustrate the performance of the different methods. IMS allows the detailed analysis of spatial distributions of molecules in organic samples. In each pixel a mass spectrum was acquired that comprises several thousand mostly uncorrelated channels (see Fig. 1a for an example). Chemical staining of an adjacent slice was used to identify 5 different tissue classes, to label parts of the data and obtain an approximate manual segmentation. Random Forest classification hence yields a compositional data set with 5 dimensions in each pixel.

Noyel's sum of morphological gradients [11] was used to calculate boundary indicator maps directly from the raw input and the first principal components of the raw input (Fig. 1d+e) as well as from the probability maps (Fig. 2). Visual inspection suggests that the latter contains more information. Therefore, we used probability maps as input for all following boundary map computations.

3 Multivariate Gradients

We next present three novel methods that create scalar boundary maps from such compositional input data as well as the multivariate watershed which is a generalization of the classic watershed definition.

Gini impurity. The underlying idea for the Gini impurity watershed is that the class impurity in the classification results can be used to identify borders between different regions. Gini impurity [21] essentially is a measure of vector sparseness. For probability distribution f_u at pixel location u it is defined by

$$GI(f_u) = 1 - \sum_{j=1}^{D} f_u(j)^2 \qquad (2)$$

where $f_u(j)$ is the j-th value in f_u. The minimum degree of impurity is obtained if one of the $f_u(j)$ equals one. We calculate the Gini impurity index for each pixel and perform the conventional watershed segmentation on the obtained scalar boundary map. Slight smoothing of the probability maps with a channel-wise Gaussian filter (zero mean, unit variance) prior to calculation of the impurity indices preserves the sum-constraint and ensures that the border between two adjacent points with (different) pure components has indeed higher impurity.

Dirichlet boundary indicator. The observed probability vectors at pixel u and its neighbors $O(u)$ can be interpreted as realizations of a (single) Dirichlet distribution [25] (which can be seen as the multivariate generalization of the beta distribution). Its D-dimensional realizations sum to one as do the class probabilities for each pixel. The Dirichlet distribution is parameterized by a vector $\alpha = (\alpha_1, ..., \alpha_D)$ where $\alpha_j, j = 1, ..., D$ is positive, i.e. $f \sim Dir$ with

$$f(x_1, ..., x_D | \alpha_1, ..., \alpha_D) = \frac{\Gamma\left(\sum_{j=1}^{D} \alpha_j\right)}{\prod_{j=1}^{D} \Gamma(\alpha_j)} \prod_{j=1}^{D} x_j^{\alpha_j - 1} \qquad (3)$$

for all $x_j > 0$ with $\sum_j^D x_j = 1$ and Γ is the Gamma function. For given observations $O(u)$ its optimal parameters $\hat{\alpha}(u)$ can be estimated by maximizing the log-likelihood of the data [25]. This maximum likelihood estimation is performed in a neighborhood of each pixel. The shape of the distribution is determined by the parameters $\hat{\alpha}_j$. At pixel locations within homogeneous regions of the spectral image their sum is high and we obtain highly peaked distributions with low variances. In contrast, in the vicinity of borders the sum of the $\hat{\alpha}_j$ is low and the distributions are broad. Thus, we propose to use the inverted precision – defined as 1 divided by the sum of all $\hat{\alpha}_j(u)$ – as boundary indicator at pixel location u. The resulting boundary indicator map is used as input for the classic watershed.

Kernel density estimate. In this method, in each pixel a kernel density estimation [26] is performed. Here we use Gaussian kernels k, but other choices are also possible. The density KDE at a pixel u is calculated from

$$KDE(u) = \frac{c}{N\sigma_{spat}^2 \sigma_{prob}^D} \sum_{i=1}^{N} k_{spat}\left(\left\|\frac{g_u - g_{u^i}}{\sigma_{spat}}\right\|^2\right) k_{prob}\left(\left\|\frac{f_u - f_{u^i}}{\sigma_{prob}}\right\|^2\right) \quad (4)$$

where u^i is the i-th pixel in the data set, g_u is the two-dimensional vector of spatial components corresponding to pixel u, and c is a normalization constant. Contributions are weighted by the distance in space (term with k_{spat}) as well as by distance in composition (term with k_{prob}). σ_{spat} and σ_{prob} are the corresponding (Gaussian) kernel bandwidths. The inverse density is used as boundary indicator map for the classic watershed. The density estimation formula in eq. 4 is well known from the literature. It constitutes a link of the watershed algorithm with the mean shift procedure [27] and bilateral filtering [28].

Multivariate watershed. In the classic watershed algorithm, only one type of basin is known. For compositional data, we generalize this by defining one basin type per class, i.e. basin types $B_1, ..., B_D$. Each pixel is assigned to the class whose posterior probability dominates. In absence of prior knowledge, dominance is established by a simple "winner takes all" rule (corresponding to a fair arbitrator); but certain classes *can* be favored if desired by introducing bias (corresponding to a bribed arbitrator). We first consider the two-class case (cf. Fig. 3): for pixel u the probabilities for classes 1 and 2 are equal to $f_u(1)$ and $f_u(2) = 1 - f_u(1)$. Maximum likelihood estimation (MLE) for the assignment of pixels to basin types corresponds to introducing a threshold $T = 0.5$ where u is assigned to basin type 1, i.e. $w_T(u) = B_1$, if $f_u(1) \leq T$, and to B_2 otherwise. By moving T, different risks can be assigned to the two classes. The basin assignment is then obtained in a risk-weighted maximum a-posteriori decision.

When compositional data of higher dimension is analyzed, the points live on a simplex. In the $3D$ case, the MLE corresponds to using the threshold $T = [1/3; 1/3; 1/3]$ and the perpendiculars to the lines connecting the corners of the simplex to assign points (see Fig. 7). Moving T on the simplex again controls the emphasis of the different classes. If the threshold point is close to one corner of the simplex, the corresponding class will be less influential in the

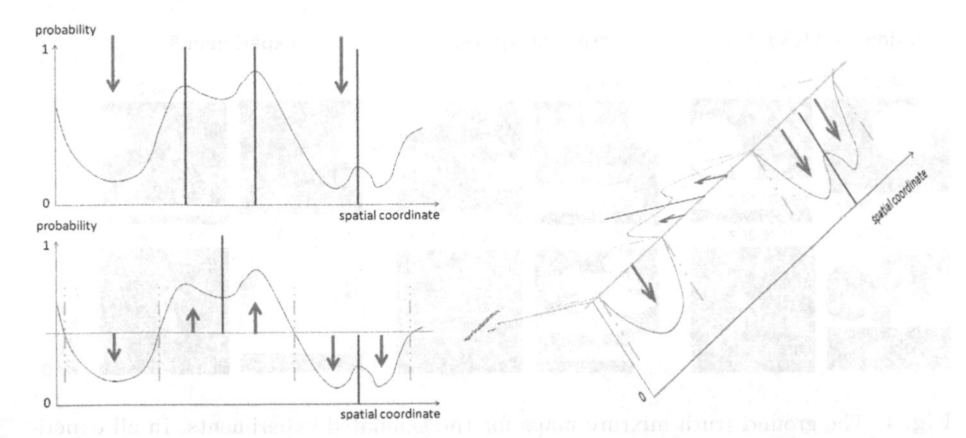

Fig. 3. Visualization of the multivariate watershed for $D = 2$. The classic approach (top left) only has one basin type. Figuratively, rain falls from above and fills the basins. Whenever two basins meet, a watershed is constructed (vertical lines). In the multivariate case (bottom left), a threshold is introduced that controls the assignment of each point to one of the two basin types. Figuratively, the rain now falls from a higher dimension onto a surface that has been folded at the threshold value (right). New watersheds emerge for the parts above the threshold and between basins of different type. Classic watershed emerges as a special case in which the threshold has been set to the maximum intensity.

segmentation. In the D-dimensional case, T has D components and the set of pixels that are assigned to basin B_k is given by $Z_k := \{u \mid w_T(u) = B_k\}$, i.e.

$$Z_k = \left\{ u \,\middle|\, 1 - \frac{f_u(k)}{f_u(k) + f_u(j)} < 1 - \frac{T(k)}{T(k) + T(j)} \forall j \in \{1, ..., D\} \setminus \{k\} \right\} . \quad (5)$$

For each of the D sets of points, the Euclidean distance between each point in the set and the respective basin (i.e. corner of the simplex) is calculated. On the resulting scalar map, a watershed-like segmentation is performed. It differs from the conventional watershed transform in the following way. The basin assignments define areas of influence for each basin type. Water is never allowed to cross borders between different influence zones. This can be included in the conventional watershed algorithm by changing its distance function $d(u, v)$ such that points that have been assigned to different basin types have infinite distance: $d(u, v) = \infty \; \forall v \in neighbors(u) : w_T(v) \neq w_T(u)$. The threshold T is a parameter which can be set by the user. Its default value is $[1/D; ...; 1/D] \in \mathbb{R}^D$.

4 Data and Experiments

We used the topographic distance version of the watershed algorithm and a neighborhood system with 8-connectivity. For the real data studied at the end of this section, no exact ground truth is available; for a quantitative evaluation, we hence resort to simulated data which is described next.

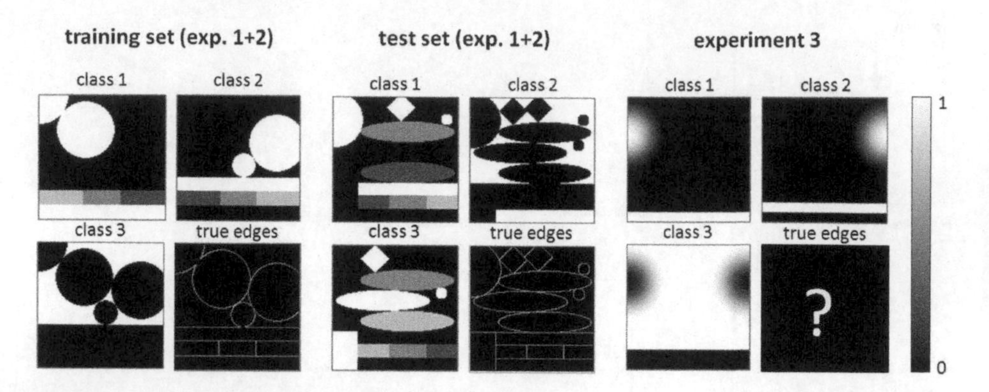

Fig. 4. The ground truth mixture maps for the simulated experiments. In all experiments we used three ground truth spectra and mixed them according to the mixture maps shown above. White areas correspond to a pure concentration, black indicates that this class is absent at the respective location. The correct boundaries - if unambiguous - are also given.

Simulation. First, three spectra from a real imaging mass spectrometry measurement were taken and defined as "pure" spectra (see Fig. 1a for an example). Then, different mixture maps were generated that contain pure areas as well as impure ones (Fig. 4). The "observed" data was created by mixing the pure spectra according to the mixture maps. IMS measurements typically consist of ion counts. A Poisson noise model was used to simulate instabilities in the data acquisition process. Samples from the noisy data were used to train a Random Forest with 250 trees and 100 samples per class that were selected as described below. After training, the whole dataset was classified and a set of probability maps was obtained. A total of three experiments was performed: in the first experiment, the mixture map also contained impure regions, but training was performed with samples from pure regions only (3 classes). For experiment 2 the same setting as in experiment 1 was used, but each mixture area was considered an individual class and the classifier was trained with samples from each of them (6 classes). Finally, the last experiment demonstrates the influence of the threshold of the multivariate watershed on the obtained segmentation.

Real data. We also applied the methods to the IMS data set described at the end of section 2 (250 trees, 100 samples/class). This spectral image comprises a total of 4000 mass over charge channels, two of which are shown in Fig. 1.

Postprocessing. The segmentation maps obtained with the different watershed methods typically contain oversegmentation. We use watershed dynamics [29,30] to amend this problem. For each edge a dynamics value is calculated and edges that correspond to a dynamics value below a given threshold are removed, i.e. the respective basins are merged.

Evaluation criteria. The true edges as well as the watersheds obtained with the topographic distance watershed algorithm lie between pixels of the grid. We use inter-pixel edges represented on a half-integer grid to quantitatively compare the estimated edge map E_{est} with the ground truth edge map E_{gt} with the Baddeley distance [31] which is calculated for all pixels u:

$$dist_{Bad}(E_{gt}, E_{est}) = \left[\frac{1}{N} \sum_{u} |DT(u, E_{gt}) - DT(u, E_{est})|^q \right]^{1/q} . \quad (6)$$

$DT(u, E)$ is the closest distance between pixel u and any of the edges in the edge map E [32]. In our study, q was set to 2 and we only considered distances up to 5 pixels, i.e. pixels that were more than 5 pixels away from any edge in both of the edge maps were ignored. The Baddeley distance penalizes both over- and undersegmentation. Since the optimal dynamics threshold value is different for the various watershed algorithms under consideration, for each method the best dynamics threshold with respect to the Baddeley distance was chosen.

Besides using the Baddeley distance, we quantify the segmentation quality by means of sensitivity and specificity. The former measures which percentage of the true edges are identified by a method, the latter how many background points are wrongly classified as edges. Since the edges in the obtained segmentation maps E_{est_i} can be displaced from their positions in the ground truth edge map E_{gt} by a few pixels, we first match them to the ground truth edges. To this aim, we employ the stable marriage algorithm [33] that uniquely assigns each pixel in E_{est_i} to a close pixel in E_{gt} as long as the maximum distance is below a given threshold (here 2 pixels). Edge pixels in E_{gt} without a "partner" in E_{est_i} are considered false negatives (FN, indicating undersegmentation), edge pixels in E_{est_i} without a partner in E_{gt} are false positives (FP, oversegmentation), pairs are considered true positives (TP) and the rest are true negatives (TN). We then calculate $sensitivity = \frac{TP}{TP+FN}$ and $specificity = \frac{TN}{FP+TN}$. An estimate for the test error is obtained by testing the watershed methods on a independent test image set where we keep the parameter settings. Training and test set differ in geometry and class mixtures (cf. Fig. 4).

In case of the kernel density estimation watershed, σ_{spat} and σ_{prob} have to be specified. We calculated the segmentations for a variety of different parameter settings out of a given range and the best settings in the training set were used in the comparison. In experiment 1, the best choice was $\sigma_{spat} = 2.0$ and $\sigma_{prob} = 1.0$, in experiment 2 $\sigma_{spat} = \sigma_{prob} = 1.0$. Similarly, different structuring elements can be used for the calculation of the morphological gradient. We experimented with discs of varying sizes and found a radius of 1 to be most adequate.

5 Discussion

We now discuss the outcome of the three experiments described in section 4.

Experiment 1. Regarding the training set, the results of Malpica's tensor and especially the Dirichlet approach are very close to the ground truth (cf. Fig. 5 and Tab. 1). The Dirichlet boundary indicator reliably finds edges between different

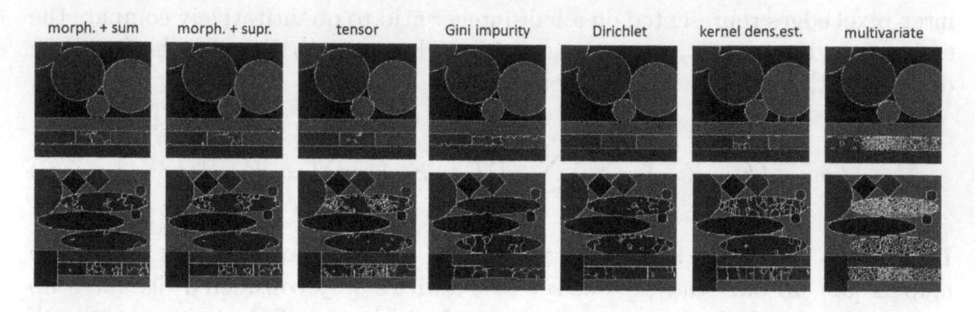

Fig. 5. Experiment 1: Segmentation results on training (top) and test set (bottom) after training of the classifier with samples from pure mixture regions only

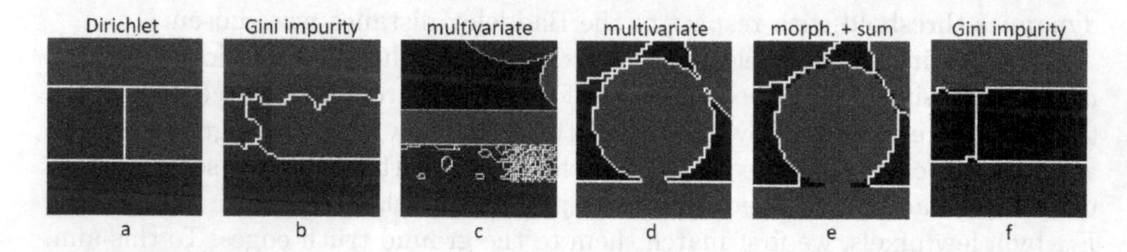

Fig. 6. Experiment 1: Zoomed-in areas of the segmentation results. See text for details.

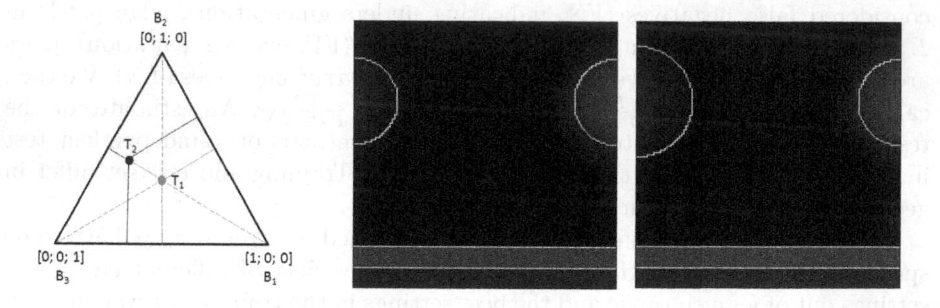

Fig. 7. Experiment 3: The influence of the threshold selection on the segmentation result of the multivariate watershed. Changing the threshold $T_1 = [1/3; 1/3; 1/3]$ to $T_2 = [1/7; 3/7; 3/7]$ in the simplex (left) leads to an accentuation of class 1 (red). The boundaries between classes 2 and 3 remain unchanged.

pure mixtures and impure mixtures and results in straight boundary lines (cf. Fig. 6a). The optimal dynamics threshold for the kernel density-based method leads to oversegmentation in the lower right part of the image, but the remaining part of the image is well segmented. The Gini impurity watershed and the multivariate watershed accurately identify boundaries between pure mixture areas, but have some problems with boundaries separating impure regions (cf. Fig. 6b). This task is indeed difficult for most of the methods since the classifier output

Table 1. The results obtained on the test set show that there is no clear winner. However, the methods introduced here compete well with existing ones.

method	threshold	Experiment 1 distance	sensitivity	specificity	Experiment 2 distance	sensitivity	specificity
morph./sum	0.0502	2.4958	0.9396	0.9864	0.2193	0.9408	0.9995
morph./supr.	0.0752	2.4835	0.9423	0.9855	0.3150	0.9253	0.9997
tensor	0.0019	3.1687	0.9710	0.9702	0.2000	0.9648	0.9996
Gini impurity	0.0576	**2.3467**	0.8734	**0.9906**	0.3755	0.9199	0.9999
Dirichlet	0.0033	2.5908	0.9311	0.9882	0.2955	0.9210	0.9999
kernel dens.e.	0.0013	3.2484	0.9206	0.9757	0.4054	0.9001	0.9998
multivariate	0.0118	4.7705	**0.9965**	0.8917	**0.0**	**1.0**	**1.0**

shows a relatively low gradient in these areas. The Gini impurity watershed itself cannot detect edges between pure and highly impure regions since highly impure mixtures have a high Gini boundary indicator and are therefore interpreted as boundaries instead of regions. However, in the postprocessing step, some of these boundaries are removed by merging basins and the real edges can be recovered. In the multivariate watershed the two regions in the lower left of Fig. 6c are both assigned to the same basin type (class) and thus, the boundary pixels are reduced to jumps in the distance function used to construct the boundary indicator maps. The oversegmentation of the 50% to 50% mixture area results from the fact that depending on the Poisson noise, the pixels are randomly assigned to classes 1 and 2. Here, some smoothing prior to the watershed segmentation could improve results. In contrast to all other methods including the summation of morphological gradients, the multivariate watershed is able to reconstruct the contours in areas of narrow bends (cf. Fig. 6d+e).

On the test set, the morphological gradients, the Dirichlet boundary indicator and the Gini impurity perform best. The latter leads to the best distance value, partly because it is least oversegmented. However, some of the boundaries are less precise, but displaced by a few pixels.

Experiment 2. One can argue that the 3 mixture areas (with 25%, 50% and 75% contributions) show clear spatial extent and, thus, constitute classes in their own right. To account for this effect, the classifier was trained with samples from 6 classes. Table 1 shows that in this scenario both the Gini impurity and the multivariate watershed achieve very good results. For example, the edges between pure and impure areas are now much better identified by the Gini impurity watershed (cf. Fig. 6f). The multivariate watershed even results in a perfect reconstruction of all boundaries and thus in the best sensitivity and specificity values. The kernel density and Dirichlet boundary indicators again compete well with the existing methods.

Experiment 3. Fig. 7 shows the effect that the threshold of the multivariate watershed has on the segmentation result. The threshold was varied so that the

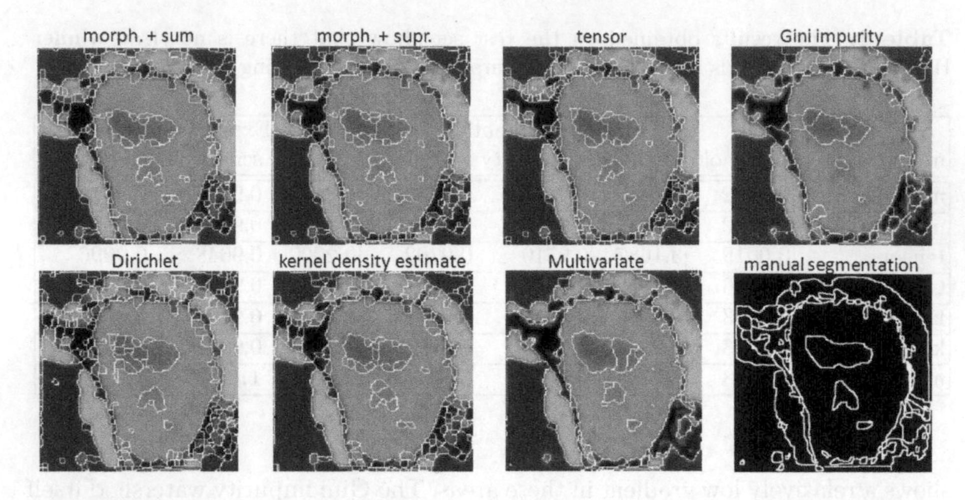

Fig. 8. Segmentation results on the real data (manually selected parameters)

proportions of classes 2 and 3 did not change, but the first class was emphasized with respect to the other classes. It can be seen from the two segmentation results that the border between classes 2 and 3 remains unchanged whereas the border between classes 1 and 3 is shifted in favor of class 1. By setting the threshold the user can control the class weights. Thus, the multivariate watershed provides the user with more control over the segmentation result than other methods like the watershed based on morphological gradients.

Real data. Fig. 8 shows the results of the different methods after postprocessing with watershed dynamics. The optimal parameters have been tuned manually. Judging from the approximate manual segmentation, all approaches lead to reasonable results. However, some oversegmentation remains that cannot be removed with the concept of dynamics.

6 Conclusion

We have introduced four watershed-based methods for segmenting multivariate compositional data: the Gini impurity watershed, the Dirichlet boundary indicator watershed, the kernel density estimate based watershed and the multivariate watershed. The former three approaches use novel techniques to obtain a scalar boundary indicator map which is used as input for the classic watershed transform. The latter one generalizes the classic definition of the watershed to multispectral compositional data. In our experiments on simulated data, no overall best performing method could be identified. However, the methods introduced in this paper have been shown to compete well with existing methods and are superior in some scenarios.

References

1. Digabel, H., Lantuéjoul, C.: Iterative algorithms. In: Proc. 2nd Europ. Symp. Quant. Anal. of Microstr. in Material Science, Biology and Medicine, pp. 85–99 (1978)
2. Roerdink, J.B.T.M., Meijster, A.: The watershed transform: Definitions, algorithms and parallelization strategies. Fundamenta Informaticae 41, 187–228 (2001)
3. Lezoray, O.: Supervised automatic histogram clustering and watershed segmentation. Image Anal. Stereol. 22, 113–120 (2003)
4. Chang, C.Y., Shie, W.S., Wang, J.H.: Color image segmentation via fuzzy feature tuning and feature adjustment. In: Conf. on Syst., Man & Cybernetics, vol. 4, p. 6 (2002)
5. Huguet, A.B., de Andrade, M.C., Carceroni, R.L., Araujo, A.A.: Color-based watershed segmentation of low-altitude aerial images. In: 17. Brazilian Symp. on Comp. Graphics and Image Proc., pp. 138–145 (2004)
6. Noyel, G., Angulo, J., Jeulin, D., Balvay, D., Cuenod, C.-A.: Filtering, segmentation and region classification by hyperspectral mathematical morphology of DCE-MRI series for angiogenesis imaging. In: Intern. Symp. on Biomedical Imaging: From Nano to Macro, pp. 1517–1520 (2008)
7. McDonnell, L.A., Heeren, R.M.A.: Imaging mass spectrometry. Mass Spectrometry Reviews 26, 606–643 (2006)
8. Moffat, J., et al.: A lentiviral RNAi library for human and mouse genes applied to an arrayed viral high-content screen. Cell 124(6), 1283–1298 (2006)
9. Scheunders, P.: Multivalued image segmentation based on first fundamental form. In: Proc. of the Intern. Conf. on Image Anal. and Processing, pp. 185–190 (2001)
10. Li, P., Xiao, X.: Evaluation of multiscale morphological segmentation of multispectral imagery for land cover classification. In: Proc. of the IEEE Geoscience and Remote Sensing Symposium, vol. 4, pp. 2676–2679 (2004)
11. Noyel, G., Angulo, J., Jeulin, D.: Morphological segmentation of hyperspectral images. Image Anal. Stereol. 26, 101–109 (2007)
12. Malpica, N., Ortuno, J.E., Santos, A.: A multichannel watershed-based algorithm for supervised texture segmentation. Patt. Rec. Letters 24, 1545–1554 (2003)
13. Karvelis, P.S., et al.: Region based segmentation and classification of multispectral chromosome images. IEEE Trans. on Medical Imaging 27, 697–708 (2008)
14. Zhang, Y., Feng, X., Le., X.: Segmentation on multispectral remote sensing image using watershed transformation. Congr. on Image & Sign. Proc. 4, 773–777 (2008)
15. Angulo, J., Jeulin, D.: Stochastic watershed segmentation. In: Intern. Symp. on Mathematical Morphology, vol. 8, pp. 265–276 (2007)
16. Noyel, G., Angulo, J., Jeulin, D.: Random germs and stochastic watershed for unsupervised multispectral image segmentation. In: Apolloni, B., Howlett, R.J., Jain, L. (eds.) KES 2007, Part III. LNCS (LNAI), vol. 4694, pp. 17–24. Springer, Heidelberg (2007)
17. Soille, P.: Morphological Image Analysis. Springer, Heidelberg (1999)
18. Aitchison, J.: The Statistical Analysis of Compositional Data. Monographs on Statistics and Applied Probability. Chapman and Hall, Boca Raton (1986)
19. Vincent, L., Soille, P.: Watersheds in digital spaces: an efficient algorithm based on immersion simulations. IEEE Trans. on PAMI 13, 583–598 (1991)
20. Meyer, F.: Topographic distance and watersheds lines. Sig. Proc. 38, 113–125 (1994)
21. Breiman, L.: Random forests. Machine Learning 45, 5–32 (2001)

22. Ulintz, P.J., Zhu, J., Qin, Z.S., Andrews, P.C.: Improved classification of mass spectrometry database search results using newer machine learning approaches. Molecular and Cellular Proteomics 5, 497–509 (2006)
23. Hanselmann, M., et al.: Concise representation of mass spectrometry images by probabilistic latent semantic analysis. Anal. Chem. 80(24), 9649–9658 (2008)
24. Hanselmann, M., et al.: Toward Digital Staining using Imaging Mass Spectrometry and Random Forests. J. of Prot. Res. (2009) DOI: 10.1021/pr900253y
25. Casella, G., Berger, R.L.: Statistical Inference. Duxbury Advanced Series (2002)
26. Godtliebsen, F., Marron, J.S., Chaudhuri, P.: Significance in scale space for bivariate density estimation. Journal of Comp. and Graph. Stat. 11, 1–21 (2002)
27. Comaniciu, D., Meer, P.: Mean shift: A robust approach toward feature space analysis. Trans. on Patt. Anal. and Mach. Intelligence 24(5), 603–619 (2002)
28. Tomasi, C., Manduchi, R.: Bilateral filtering for gray and color images. Proc. of the Conf. on Computer Vision, 836–846 (1998)
29. Grimaud, M.: New measure of contrast: the dynamics. Image Algebra and Morphological Image Processing III, 292–305 (1992)
30. Brun, L., Mokhtari, M., Meyer, F.: Hierarchical watersheds within the combinatorial pyramid framework. In: Andrès, É., Damiand, G., Lienhardt, P. (eds.) DGCI 2005. LNCS, vol. 3429, pp. 34–44. Springer, Heidelberg (2005)
31. Baddeley, A.J.: An error metric for binary images. Robust Comp. Vis., 59–78 (1992)
32. Rosenfeld, A., Pfaltz, J.: Sequential operations in digital picture processing. Robust Computer Vision 13(4), 471–494 (1966)
33. Gale, D., Shapley, L.S.: College admissions and the stability of marriage. American Mathematical Monthly 69, 9–14 (1962)

Pixel Approximation Errors in Common Watershed Algorithms

Hans Meine[1], Peer Stelldinger[1], and Ullrich Köthe[2]

[1] Cognitive Systems Laboratory, University of Hamburg, Germany
[2] Heidelberg Collaboratory for Image Processing, University of Heidelberg, Germany

Abstract. The exact, subpixel watershed algorithm delivers very accurate watershed boundaries based on a spline interpolation, but is slow and only works in 2D. On the other hand, there are very fast pixel watershed algorithms, but they produce errors not only in certain exotic cases, but also in real-world images and even in the most simple scenarios. In this work, we examine closely the source of these errors and propose a new algorithm that is fast, approximates the exact watersheds (with pixel resolution), and can be extended to 3D.

1 Introduction

Watershed algorithms are among the most important approaches to image segmentation. This is in large part due to the fact that they are based on a sound mathematical definition of segmentation boundaries: watershed segmentations are equivalent to Voronoi tessellations of a suitable boundary indicator function (e.g. the image gradient magnitude) with respect to the topographic distance [1]. That is, catchment basins (regions) are defined as the set of points which are closer (in topographic distance) to a particular minimum than to any other minimum, and watersheds form the boundaries between the basins.

The *exact watershed algorithm* [2,3] detects such boundaries with an accuracy that is only limited by adjustable numerical tolerances and the quality of the input image. The results are very precise and achieve theoretically predicted limits under realistic image acquisition models [4]. Unfortunately, the exact watershed algorithm is also rather time consuming, and its computational principle is restricted to two dimensions.

In contrast, the popular pixel-based watershed algorithms are much faster, and can be applied in any dimension, but these advantages come at the cost of much lower accuracy: pixel-based algorithms only produce inter-pixel boundaries or 8-connected pixel boundaries, i.e. they round boundary locations to certain fixed coordinates defined by the grid. Furthermore, loss of accuracy is not only observed in terms of geometry (i.e. displacement of boundaries from their true position), but also in terms of topology: it is not uncommon that some catchment basins are missing or are split into several regions. Furthermore, there are systematic errors that lead to watersheds snapping into "preferred" directions.

In this paper, we are going to compare pixel-accurate results to the corresponding exact watersheds and, for the first time, investigate systematically the

S. Brlek, C. Reutenauer, and X. Provençal (Eds.): DGCI 2009, LNCS 5810, pp. 193–202, 2009.

errors caused by common watershed discretization schemes. This investigation reveals a number of interesting findings:

- Correct detection of minima (i.e. catchment basin seeds) in discrete images is much more difficult than commonly believed.
- Grid-based flowline discretization suffers from accumulated quantization errors leading to systematic errors in the boundaries, i.e. "preferred directions".

We propose a compromise between the exact and pixel-accurate algorithms, which we call the *fast flowline algorithm.* While our new algorithm only computes the assignment of each pixel to a catchment basin (instead of a subpixel-accurate boundary), it does so by tracing the *exact, subpixel-accurate flowline* starting at each pixel center. In this way, the topological inconsistencies described above can be avoided, and the approximation effect indeed reduces to boundary rounding. This principle is applicable in higher dimensions as well, and the boundary can be refined by adaptive subdivision (i.e. use of a finer grid near the boundary) if desired. The algorithm's speed is significantly increased by early flowline termination. This optimization resembles the UNION-FIND variety of the watershed transform, but instead of tracing flowlines only to an immediate neighbor, we follow them for as long as is necessary for a correct decision about basin assignment.

2 The Exact Watershed Transform

The exact watershed transform assumes that a continuous boundary indicator image is given, i.e. an image that takes high values near boundaries and low values within regions, and which is defined over a compact subset D of $\mathbb{R}^2$. The latter assumption is not a significant restriction for image analysis, because band-limited images of standard quality can be accurately interpolated to the real domain, e.g. by spline interpolation. In addition, we assume that the boundary indicator function fulfills a Morse condition in the domain of interest, i.e. all points with zero gradient are isolated, so that the image has only extrema and saddles, but no plateaus and horizontal ridges (this is a sufficient, but not a necessary condition which we pose for convenience of exposition). In contrast to discrete watersheds, where the plateau problem is central [1], it is not very significant in interpolated images because band-limited functions can only violate the Morse condition in degenerate cases. Moreover, since the space of Morse functions is dense in the space of continuous functions, an infinitesimal change of the pixel values (e.g. by adding noise with standard deviation of a fraction of a gray level, and/or application of an infinite impulse response filter) would turn a non-Morse function into a Morse function.[1]

Due to the above assumptions, every point (with the exception of extrema and saddles) is crossed by a unique flowline which runs along the local gradient

[1] We always imply that pixel values are stored in a floating point representation during analysis.

direction, and all flowlines converge to and diverge from a critical point. With reflective boundary conditions, this even applies at the image border. The *catchment basin* of a minimum is now the set of points whose flowlines converge at this minimum, and watersheds are those points that do not belong to any catchment basin. This definition is the basis for the flowline (or "tobogganing" [5]) type of watershed algorithms.

Equivalently, one can base the watershed definition on the *topographic distance* [1] between two points $\boldsymbol{x}$ and $\boldsymbol{y}$

$$d_t(\boldsymbol{x}, \boldsymbol{y}) = \min_C \int_C |\nabla f(\boldsymbol{z})| \, d\boldsymbol{z} \tag{1}$$

where f is the boundary indicator function, and the minimum is taken over all paths C connecting $\boldsymbol{x}$ and $\boldsymbol{y}$. When $\boldsymbol{x}$ and $\boldsymbol{y}$ are on the same flowline, that flowline is necessarily a path of smallest distance, and the topographic distance reduces to $d_t(\boldsymbol{x}, \boldsymbol{y}) = |f(\boldsymbol{x}) - f(\boldsymbol{y})|$. In particular, when $\boldsymbol{x}$ is a minimum point, $d_t(\boldsymbol{x}, \boldsymbol{y}) + f(\boldsymbol{x}) = f(\boldsymbol{y})$ holds for all points $\boldsymbol{y}$ whose flowline converges to that minimum.

Suppose for a moment that all minima of f have the same value $f_{\min}$ (when this is not the case, one may change f in an infinitely small neighborhood around each minimum to fulfill the condition). Let $\{\boldsymbol{x}_i\}$ be the set of minima. Then we can define the Voronoi tessellation of f with respect to its minima as

$$r_i = \{\boldsymbol{y} \in D \,|\, \forall j \neq i : \, d_t(\boldsymbol{x}_i, \boldsymbol{y}) < d_t(\boldsymbol{x}_j, \boldsymbol{y})\} \tag{2}$$

where r_i denotes the Voronoi region ("catchment basin") around minimum i. This is the same definition as for standard Voronoi tessellation, except that the Euclidean distance has been replaced with the topographic distance. Watersheds are now precisely the points in D whose distance to at least two minima is equal. When the restriction that all minima have the same value is dropped, we get the same tessellation as before by replacing (2) with

$$r_i = \{\boldsymbol{y} \in D \,|\, \forall j \neq i : \, d_t(\boldsymbol{x}_i, \boldsymbol{y}) + f(\boldsymbol{x}_i) < d_t(\boldsymbol{x}_j, \boldsymbol{y}) + f(\boldsymbol{x}_j)\} \tag{3}$$

This definition is the basis for the flooding variety of watershed algorithms [6].

Yet another way to define a watershed segmentation is via the watersheds themselves: as Maxwell noted, watersheds are precisely the flowlines that converge to saddle points of the boundary indicator. In the continuous domain, all three definitions are equivalent. However, in practice, the last definition is the best choice when geometric and topological accuracy is of major concern. This so called the *exact watershed algorithm* [2,3] can be described as follows:

1. Use Newton-Raphson iterations to find all saddle points of f.
2. For every saddle $\boldsymbol{s}$ (optionally only for those that fulfill some application specific predicate, such as sufficient edge strength), solve the inverse flowline PDE

$$\frac{\partial \boldsymbol{x}(t)}{\partial t} = \nabla f(\boldsymbol{x}(t)) \tag{4}$$

   starting in the two directions of positive curvature until converging to a maximum (e.g. using the second-order Runge-Kutta method).

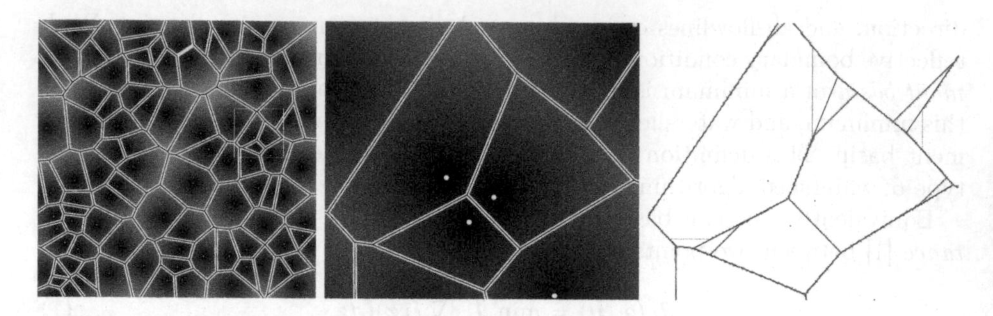

Fig. 1. Comparison of the results of the exact watershed algorithm (*black, thin*) with the known ground truth (*yellow, thick*) of a Euclidean distance transform – *middle:* close-up of ROI – *right:* pixel-based results with same colors as in Fig. 3

3. Connect all such flowline pairs into edges of a graph whose nodes consist of maxima of f.
4. Determine the cyclic order of edges around nodes (taking their tangential convergence into account) in order to reconstruct the graph's faces (i.e. the catchment basins) by contour traversal [3].

Experiments show that this algorithm produces very accurate watershed segmentations that are close to the ground truth (if available), cf. Fig. 1.

3 Error Analysis of Common Watershed Algorithms

Discrete watershed algorithms only determine an assignment of pixels (i.e. points at integer coordinates) to catchment basins. Since there are only a finite number of these points, algorithms of the flowline or flooding variety are most common. Often, these algorithms are defined without reference to continuous watershed definitions. In contrast, we interpret them as discretizations of the corresponding continuous definitions in order to better understand their properties and differences. Discretization then boils down to two questions: i) which discrete approximations of flowlines (or inverse flowlines for flooding) are used, and ii) how are minima detected?

The latter is usually done by detecting minima in a 4- or 8-neighborhood; however, this misses some real minima, and many false minima are detected. For instance, Fig. 2 shows the minima detected on digital images generated by sampling two test functions with known ground truth minima. Every false minimum gives rise to a false catchment basin.

The first function exhibits a cross-like arrangement of four valleys and is given by $10(\mathrm{s}(\frac{u}{8}) + \mathrm{s}(\frac{v}{8})) + 164\mathrm{s}(\frac{u}{4})\mathrm{s}(\frac{v}{4})$ with $u = \sin\frac{\pi}{8}x + \cos\frac{\pi}{8}y$, $v = \cos\frac{\pi}{8}x - \sin\frac{\pi}{8}y$ and $s(t) = 1 - \frac{\sin \pi t}{\pi t}$. The plot shows the window $-8 \leq x, y \leq 8$. Sampling at integer coordinates results in an oversampling factor of 2.8 with regard to the bandlimit of the function. There is only one minimum in the origin, which has

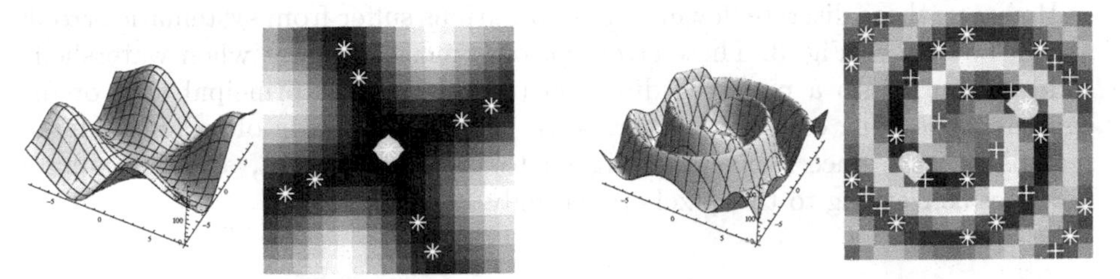

: ground-truth minima; : minima of spline-interpolated image (order 5)
+/✱: minima in 4/8-neighborhood, respectively (ranges shown: $x, y \in [-8 \ldots 8]$)

Fig. 2. Prototypical spurious results of discrete minima detection. The ground truth is only reproduced by subpixel-accurate minima detection in a spline-interpolated image.

been correctly found by all methods. Due to the discretization, the pixel-accurate methods detect 8 wrong minima, inside the slowly decreasing valleys.

The second function shows a spiral having two minima at (2.1904, −3.7276) and (−2.1904, 3.7276), two maxima at (−2.1904, −3.7276) and (2.1904, 3.7276) and a saddle point at the origin. There are two spiral-shaped catchment basins. Here, the sampling preserves 99.98% of the energy of the function and is thus effectively band-limited.

Furthermore, algorithms also differ in how they approximate flowlines with respect to their location and direction:

1. True flowlines may be rounded to grid-based paths, usually connecting pixel centers according to 4- or 8-connectivity or (at best) using adaptive support points on grid lines. Pixel paths have the obvious speed advantage that each path can be terminated after only one step (i.e. in a neighboring pixel) when basins are constructed incrementally.
2. The flowline direction may locally be derived from different gradient approximations (e.g. using a Gaussian or Sobel kernel) or even from simple forward differences (i.e. looking for the smallest neighbor [5]).

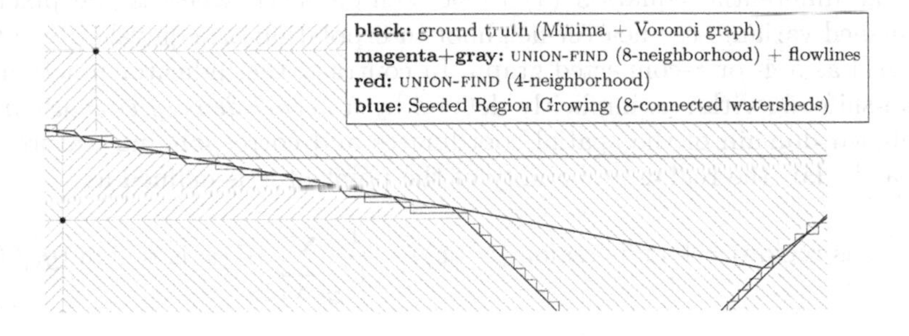

black: ground truth (Minima + Voronoi graph)
magenta+gray: UNION-FIND (8-neighborhood) + flowlines
red: UNION-FIND (4-neighborhood)
blue: Seeded Region Growing (8-connected watersheds)

Fig. 3. Systematic errors due to discrete flowline directions (cf. Fig. 1)

However, these discrete flowline approximations suffer from systematic errors, as can be seen in Fig. 3. These errors become most apparent when watersheds suddenly turn into a preferred direction (parallel to the principal axes or diagonals, depending on the neighborhood). This phenomenon often affects large regions because once a pixel is assigned to the wrong basin, all approximate flowlines converging to this pixel are wrongly assigned as well.

4 Subpixel Flowline Algorithms

As shown in the previous section, common grid-based flowline discretization schemes lead to discretization errors. These can only be fixed by lifting the restriction that flowline sampling points coincide with grid points. Interestingly, an algorithm realizing this idea already exists, although its close relationship to the watershed algorithm is not immediately obvious: the *mean-shift algorithm.*

4.1 Mean-Shift as a Watershed Algorithm

The mean-shift algorithm [7] is a popular segmentation algorithm whose results are similar to watersheds, but with markedly less oversegmentation. We show in the sequel that the latter is not a consequence of a different algorithmic approach, but stems solely from the differences in the boundary indicator definition. Indeed, one trivial difference is that mean-shift does not use a boundary indicator, but a region indicator (i.e. a measure of homogeneity rather than edgeness). Accordingly, when searching for region centers (catchment basins), one looks for local maxima of the region indicator instead of local minima of the boundary indicator, as in the watershed algorithm. That is, one traces flowlines upwards instead of downwards. But this difference is easily eliminated by inversion of the indicator function.

A more interesting difference is found in the way the indicator function itself is defined. In the watershed context, we simply assume that the values of the indicator function are given at the grid points. The subpixel watershed algorithm then interpolates these values into a smooth function over $\mathbb{R}^2$ (in order for the flowline differential equation (4) to be well-defined), whereas the pixel-based watershed variants use nearest-neighbor interpolation or, equivalently, interpret the grid as a 4- or 8-connected graph. In contrast, the indicator function of the mean-shift algorithm is implicitly defined by a *kernel density estimation* in the combined domain of the spatial coordinates and corresponding feature values. In particular, the region probability of the point at $\boldsymbol{x}$ is defined as

$$\mathbb{P}\left(\boldsymbol{x} \text{ is region}\right) = \mathbb{P}\left(\boldsymbol{x} \text{ is region } | \mathcal{N}\left(\boldsymbol{x}\right)\right) = \frac{1}{C} \sum_{\boldsymbol{y} \in \mathcal{N}(\boldsymbol{x})} \mathbb{P}\left(\boldsymbol{x} \text{ is region } | \boldsymbol{y}, f\left(\boldsymbol{y}\right)\right)$$

where $\mathcal{N}(\boldsymbol{x})$ denotes some neighborhood of $\boldsymbol{x}$, C is a normalization constant, and $f(\boldsymbol{y})$ is the feature vector at point $\boldsymbol{y}$ (e.g. its RGB color). The probability

Fig. 4. Example image and illustration of $\mathbb{P}(\boldsymbol{x} \text{ is boundary}) = 1 - \mathbb{P}(\boldsymbol{x} \text{ is region})$, i.e. the boundary indicator implicitly used by the mean-shift algorithm (parameters used: $\sigma_{\text{spatial}} = 2$, $\sigma_{\text{rgb}} = 30$)

$\mathbb{P}(\boldsymbol{x} \text{ is region} \,|\boldsymbol{y}, f(\boldsymbol{y}))$ is expanded as a product of spatial similarity ("nearness") between $\boldsymbol{x}$ and $\boldsymbol{y}$, and feature similarity between the values at $\boldsymbol{x}$ and $\boldsymbol{y}$:

$$\mathbb{P}(\boldsymbol{x} \text{ is region} \,|\boldsymbol{y}, f(\boldsymbol{y})) = \mathbb{P}(|\boldsymbol{x} - \boldsymbol{y}|)\, \mathbb{P}(|f(\boldsymbol{x}) - f(\boldsymbol{y})|) \tag{5}$$

In other words, $\boldsymbol{x}$ is considered part of a region when its neighboring points have similar feature vectors, which is clearly a measure of feature homogeneity. In the framework of kernel density estimation, the probabilities in (5) are expressed by means of kernel functions which give high values when their argument (i.e. the distance in the spatial or feature domain) is small. A popular kernel choice is the *Gaussian kernel* $\mathbb{P}_\sigma(d) = \frac{1}{C(\sigma)} \exp\left(-\frac{d^2}{2\sigma^2}\right)$ where $C(\sigma)$ is a normalization constant depending on σ and the dimension of the space where d is defined. In effect, the spatial probability defines a Gaussian window around $\boldsymbol{x}$, and the feature probability measures the degree of similarity between $\boldsymbol{x}$ and the points in this window[2]. The width of the spatial kernel is usually set to a few pixel distances, whereas the width of the feature kernel equals the noise standard deviation in the data. It should be noted that the coordinate $\boldsymbol{x}$ is *not* restricted to lie on the grid, even if all points in $\mathcal{N}(\boldsymbol{x})$ must be on the grid. Therefore, $\mathbb{P}(\boldsymbol{x} \text{ is region})$ is defined over all of $\mathbb{R}^2$, similarly to the spline-interpolated boundary indicator we are using in the exact watershed algorithm. In fact, spline interpolation is a special case of kernel-based interpolation, where the kernel is a cardinal spline function instead of a Gaussian.

The similarity between watersheds and mean-shift is not readily visible because the kernel density estimates $\mathbb{P}(\boldsymbol{x} \text{ is region})$ are never explicitly computed in the algorithm, for reasons of efficiency. However, there is no theoretical obstacle for doing so, and Fig. 4 (right) shows an example for an inverted region indicator. This image very much resembles the typical boundary indicators that are used in watershed algorithms.

[2] In practice, one truncates the spatial-domain Gaussian $\mathbb{P}(|\boldsymbol{x} - \boldsymbol{y}|)$ at some maximal distance d_0 in order to avoid infinite window sizes.

Instead of computing $\mathbb{P}(\boldsymbol{x} \text{ is region})$, the mean-shift algorithm directly computes an incremental flowline. Let $\boldsymbol{z} = (\boldsymbol{x} \quad f(\boldsymbol{x}))^\top$ denote a point in the combined domain, i.e. the coordinate vector and feature vector stacked on top of each other, and $\boldsymbol{z}^{(t)}$ such a vector at time step t. Then the flowline is computed via iterations[3]

$$\boldsymbol{z}^{(t+1)} = \boldsymbol{z}^{(t)} + \frac{\sum_{\boldsymbol{z}' \in \mathcal{N}(\boldsymbol{z}^{(t)})} \boldsymbol{z}' \, \nabla \mathbb{P}\left(\boldsymbol{z}^{(t)} \text{ is region} \,|\boldsymbol{z}'\right)}{\sum_{\boldsymbol{z}' \in \mathcal{N}(\boldsymbol{z}^{(t)})} \mathbb{P}\left(\boldsymbol{z}^{(t)} \text{ is region} \,|\boldsymbol{z}'\right)} \tag{6}$$

These iterations converge toward a local maximum of the probability, and all points $\boldsymbol{x}$ whose flowlines converge at the same maximum are considered as one region. This procedure immediately turns into a flowline-type watershed algorithm when the probability is inverted and flowlines are traced downwards to minima. Equation (6) is simply a first order approximation of the flowline equation (4) with $f(\boldsymbol{x})$ replaced by $\mathbb{P}(\boldsymbol{z} \text{ is region } |\mathcal{N}(\boldsymbol{z}))$. Since $\boldsymbol{z}^{(t)}$ is not restricted to the grid for $t > 0$, the mean-shift procedure provides more accurate flowline approximations than purely grid-based algorithms. Still, it is not quite as accurate as the Runge-Kutta flowline tracing employed in the exact watershed algorithm, because it lacks adaptive step length control. This means that mean-shift iterations occasionally converge to a maximum far from the original point.

4.2 The Fast Flowline Algorithm

Our new watershed algorithm is based on three considerations:

1. Since the computation of exact watersheds is expensive, we only determine an assignment of pixels to catchment basins, i.e. an inter-pixel boundary like in pixel-based watershed algorithms. (Boundary accuracy can be improved by refining the grid, possibly only near boundaries.)
2. As in the mean-shift algorithm, we compute flowlines and their points of convergence with subpixel accuracy using the flowline equation (4).
3. Since the computation of complete flowlines for all pixels would be even more expensive than the computation of exact watersheds, we introduce an early stopping criterion that provides correct region assignments after only a few tracing steps.

The only missing ingredient of the new algorithm is the early stopping criterion. The underlying idea is very simple: stop flowline tracing as soon as the flowline crosses a grid line whose end-points are already assigned to the same catchment basin. In order for this criterion to lead to *early* stopping, we have to process the pixels in a clever order: pixels close to minima must be investigated first, so that areas of already assigned pixels spread around each minimum. This is exactly the same idea as in the usual region-growing algorithm, with one important exception: in our approach, the initial seeds need *not* be true minima of the

[3] Notice that $\mathbb{P}(\boldsymbol{x} \text{ is region } |\mathcal{N}(\boldsymbol{x})) = \mathbb{P}(\boldsymbol{z} \text{ is region } |\mathcal{N}(\boldsymbol{z}))$ since $f(\boldsymbol{x})$ is a deterministic function of $\boldsymbol{x}$. The introduction of $\boldsymbol{z}$ just serves notational convenience.

boundary indicator, because seeds only serve to speed up flowline computations. Early flowline termination will simply succeed more often when the initial seeds are close to true minima. But the choice of seeds (be it good or poor) will have no influence on the actual region assignments, because those assignments only depend on the point of convergence of the flowlines.

Given a boundary indicator function f, we define the *fast flowline algorithm* as follows:

1. Find minima of f (e.g. in a 4-neighborhood) and put them into a priority queue which assigns highest priority to the point with minimal $f(\boldsymbol{x})$.
2. While the queue is not empty:
   (a) Pop the next point from the queue and trace its flowline by Runge-Kutta iterations with adaptive step size control according to the flowline equation
   $$\frac{\partial \boldsymbol{x}(t)}{\partial t} = -\nabla f(\boldsymbol{x}(t))$$
   Stop the iteration if either
      i. $\boldsymbol{x}(t \to \infty)$ converges at a minimum. When this minimum was already labeled, assign this label to the starting point $\boldsymbol{x}(0)$. Otherwise, create a new label and use that for both $\boldsymbol{x}(0)$ and the minimum.
      ii. If the path defined by $\boldsymbol{x}(t)$ crosses a grid line (i.e. the connection between two neighboring pixels) whose end points have already been assigned to the same label. Assign this label to $\boldsymbol{x}(0)$.

   (b) Put the neighbors of the just labelled point into the queue.

Notice that early termination will fail (i.e. will assign the starting point to the wrong region) if the flowline makes a sharp turn *after* the termination point and crosses the grid line again in opposite direction. Fortunately, the smoothness of real boundary indicators ensures that such errors occur very rarely.

4.3 Experimental Results

In order to aid the comparison of the performance of different watershed algorithms, we decided to use two sets of test images: first, we computed Gaussian gradient magnitude images (with $\sigma \in \{1, 1.5, 2, 3\}$) from a set of images with various real-world image analysis tasks. Here, the ground truth is unknown, and we use the exact watershed transform (with a spline interpolation of order 5) as the gold standard to compare against. Second, we generated images containing the Euclidean distance transform of randomly placed, isolated minimum points (cf. Fig. 1 on page 196) with varying density, on integer and sub-pixel positions. For the latter images, we could compute the correct catchment basins (i.e. the Voronoi tessellation w.r.t. the minima) for error analysis.

For every label image generated with one of the watershed algorithm variants, we performed a pixel-wise comparison of the labels with the reference image (after an optimal mapping of the labels using the stable marriage algorithm). The following table summarizes the results of more than 2000 single experiments:

	real-world images		distance transforms	
	mean err.	std.dev.	mean err.	std.dev.
subpixel watersheds	(gold standard)		0.23%	0.24%
Seeded Region Growing	12.8%	2.2%	1.25%	0.93%
UNION-FIND (4-neighborhood)	15.7%	3.9%	1.6%	1.0%
UNION-FIND (8-neighborhood)	10.3%	3.0%	1.8%	0.9%
flowline watersheds	0.25%	0.5%	0.18%	0.22%
fast flowline watersheds	0.28%	0.23%	0.19%	0.23%

5 Conclusions

Our experimental results show that subpixel-accurate algorithms do not suffer from the systematic errors of common grid-based watershed implementations: their errors are roughly an order of magnitude lower. However, they are also an order of magnitude slower (the "flowline watersheds" algorithm *without* early flowline termination even two orders of magnitude). Our current implementation of the fast flowline algorithm does not perform much faster than the subpixel watershed algorithm, but there is substantial potential for optimization, mainly because the time consuming detection of critical points is avoided completely.

A major advantage of the fast flowline algorithm is that it can be applied to 3D and higher dimensions, whereas the exact watershed algorithm is restricted to 2D. Our algorithm is therefore the first 3D watershed algorithm that avoids the systematic errors of grid-based approaches. Furthermore, our algorithm allows to further increase the segmentation accuracy by adaptive grid subdivision near boundaries. Our future work will explore these directions. At the same time, we will search for a faster (but equally accurate) alternative to the Runge-Kutta method for flowlines tracing.

References

1. Roerdink, J.B.T.M., Meijster, A.: The watershed transform: Definitions, algorithms, and parallellization strategies. In: Goutsias, J., Heijmans, H. (eds.) Mathematical Morphology, vol. 41, pp. 187–228. IOS Press, Amsterdam (2000)
2. Meine, H., Köthe, U.: Image segmentation with the exact watershed transform. In: Villanueva, J. (ed.) Proc. Vis., Imaging, and Image Processing, pp. 400–405 (2005)
3. Meine, H.: The GeoMap Representation: On Topologically Correct Sub-pixel Image Analysis. PhD thesis, Dept. of Informatics, Univ. of Hamburg (2009)
4. Meine, H., Köthe, U., Stelldinger, P.: A topological sampling theorem for robust boundary reconstruction and image segmentation. Discrete Applied Mathematics 157, 524–541 (2008) DGCI special issue
5. Yao, X., Hung, Y.P.: Fast image segmentation by sliding in the derivative terrain. In: Casasent, D.P. (ed.) Intelligent Robots and Computer Vision X: Algorithms and Techniques. SPIE Conference Series, vol. 1607, pp. 369–379 (1991)
6. Vincent, L., Soille, P.: Watersheds in digital spaces: an efficient algorithm based on immersion simulations. IEEE T-PAMI 13, 583–598 (1991)
7. Comaniciu, D., Meer, P.: Mean shift: A robust approach toward feature space analysis. IEEE T-PAMI 24, 603–619 (2002)

Digital Deformable Model Simulating Active Contours*

François de Vieilleville[1] and Jacques-Olivier Lachaud[2]

Laboratoire de Mathématiques, UMR CNRS 5127
Université de Savoie, 73776 Le-Bourget-du-Lac, France
francois.de-vieilleville@univ-savoie.fr,
jacques-olivier.lachaud@univ-savoie.fr

Abstract. Deformable models are continuous energy-minimizing techniques that have been successfully applied to image segmentation and tracking since twenty years. This paper defines a novel purely digital deformable model (DDM), whose internal energy is based on the minimum length polygon (MLP). We prove that our combinatorial regularization term has "convex" properties: any local descent on the energy leads to a global optimum. Similarly to the continuous case where the optimum is a straight segment, our DDM stops on a digital straight segment. The DDM shares also the same behaviour as its continuous counterpart on images.

1 Introduction

Since the pioneering works of Kass *et al.* [14], deformable models or active contour models have been extremely studied over the past 20 years. They have many applications in image segmentation, tracking and computer vision. They combine in a single framework two terms, one expressing the fit to data (image energy), the other describing shape priors and acting as a regularizer (internal energy). Furthermore, as noted by many authors, the parameter balancing the two terms acts as scale factor, providing a very natural multiscale analysis of images. Deformable models [15], Mumford-Shah approximation [20], geometric or geodesic active contours and other levelset variants [18,5], are classical variational formulation (*i.e.* continuous) of such techniques. Energy-minimization for image segmentation can also be expressed in a discrete setting: structural split and merge [8], weighted graph with cut optimization [3], irregular and combinatorial pyramids [12], Markov fields and stochastic processes [10].

Continuous variational problems induce partial differential equations which are solved iteratively. They are most often bound to get stuck in local minima, except in specific cases [7,6,1], sometimes with a loss of generality regarding the energies. On the other hand, in several discrete settings, optimal solutions can sometimes be found or better approached [3,12]. However, it should be noted that the first-order regularization term — the length — is generally very coarsely

* Partially funded by ANR project FOGRIMMI (ANR-06-MDCA-008-06).

S. Brlek, C. Reutenauer, and X. Provençal (Eds.): DGCI 2009, LNCS 5810, pp. 203–216, 2009.

approximated as the number of digital steps. Furthermore, as far as we know, there is no discrete setting that approaches an optimum for gradient-based image energies. Noticing this limitation, Boykov and Kolmogorov [2] have proposed to enrich the neighborhood graph to get finer area estimators — in a way similar in spirit to chamfer distances — but their approach is for now limited to an 8-neighborhood, and seems costly in practice. The work presented here shares the same objective, but it is based on discrete geometry tools.

There are several attemps to define digital analogs of continuous deformable models, migrating processes [9], discrete deformable boundaries, digital curvature flow [13]. However, none of them are able to mimick correctly continuous models. They do not solve the same variational problem, even asymptotically with finer and finer grids.

We present here a digital deformable model which incorporates in a digital way the length regularization term. The length estimation is based on an adaptation of the minimum length polygon (e.g., see [17]). It is defined as the shortest euclidean path within a one pixel wide band centered on an open simple digital 4-connected path. As in the case of active contour, this length constraint energy is a good regularizer since it defines a kind of convex energy. Optimization heuristics lead to the global optimum. Moreover our deformable model has a behaviour on images which is very close to the one of the original continuous active contour (with first-order regularization).

The paper is organized as follows. We first introduce some remarks and properties on minimum length polygons (Section 2). We then define our digital model and proves the above-mentioned properties (Section 3). Some experiments illustrate the use of our DDM on synthetic images, in particular the impact of the internal energy is shown (Section 4). Eventually, we conclude on the benefits of the proposed approach and propose some future works.

2 MLP Algorithm and Its Properties

A 4-connected path in the discrete plane is a sequence of digital points, such that any two consecutive points are 4-neighbors. We restrict our study to simple paths, i.e. paths such that all points are pairwise distinct. Such paths are conveniently described with a Freeman word. The points of a path may be embedded as unit squares, forming a grid continuum in the terminology of Sloboda and Zaťko [24]. We call $\mathcal{C}$ the set of simple digital paths, and $\mathcal{C}(A, B)$ the subset of $\mathcal{C}$ composed of the paths starting at a point A and ending at a point B.

Choosing an orientation (counterclockwise in the remaining of the paper) an inner and outer boundary can be defined. As illustrated on Fig. 1, the inner boundary is the frontier to the left (denoted by ∂I) and the outer boundary is the frontier to the right (denoted by ∂O). We are interested in computing the shortest Euclidean path starting and ending at the centroids of the two extremities of the open digital path while staying inside ∂I and ∂O (see Fig. 1). We may notice that the inner and outer boundary may not be simple 4-connected path. This shortest path is very close to the minimum length polygon (MLP),

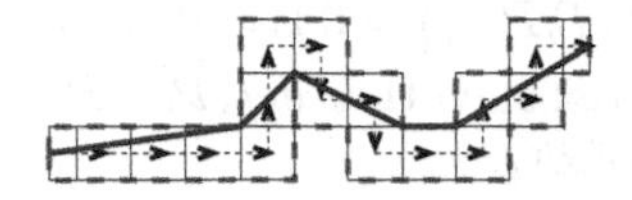

Fig. 1. Portion of a digital contour whose Freeman code word is EEEENNESESEENENE. In terms of concave and convex turns, given the CCW orientation, the contour is read as $(000+0--+--+0+-+-)$. Its inner and outer boundary are drawn with dashed lines. These 4-connected paths are not necessarily simple. The thick line is the shortest Euclidean path starting and ending at its endpoints.

except that we have two constrained points. We will say therefore say that we compute the *constrained* MLP, CMLP for short.

Minimum length polygons have a long history [19,22,17] . It was shown that MLPs are unique even for arbitrary constraint polygons. Furthermore, the MLP vertices are either positive turns of ∂I or negative turns of ∂O. Moreover there exists an application between the set of negative turns of ∂I and the set of positive turns of ∂O. This application just translate the negative vertices of ∂I by one of the following vectors according to the quadrant of the turn: $(1,1)$, $(-1,1)$, $(-1,-1)$, $(1,-1)$. As a consequence, the digital MLP can be computed from ∂I. In the case of elements of $\mathcal{C}$, the preceding rule applies except for the first and last vertices of the MLP, which are imposed.

Klette and Yip [17] (reported also in [16]) have proposed a linear time algorithm which seems to fail at finding the proper MLP (see Appendix A).

A correct linear time algorithm could be achieved for instance with computational geometry shortest path algorithms (e.g. see [11]), but would be tough to implement. Since our main objective is related to digital deformable models and not to MLP computation, we have slightly adapted the algorithm of Klette and Yip to compute a CMLP, in a time that may not be optimal in some cases. Certainly an optimal MLP would be an interesting work in itself, and may be a perspective to this work. Let us first introduce a property regarding the elements of $\mathcal{C}$ with a $CMLP$ of minimal length.

Lemma 1. *There exists $C \in \mathcal{C}(A,B)$ such that $CMLP(C) = [AB]$. Furthermore, for any $C' \in \mathcal{C}(A,B)$, the length of $CMLP(C')$ is no smaller than the length of the segment $[AB]$.*

It is enough to remark that the grid-intersection digitization of the Euclidean segment $[AB]$ is in $\mathcal{C}(A,B)$. Of course there exists other 4-connected paths whose $CMLP$ equals $[AB]$.

In the remaining of the paper, we call C_i the points associated to the path C and P_j the vertices of the $CMLP$. Let us now consider the constrained minimum length polygon of C, denoted by $CMLP(C)$, which is described as a list of pairwise distinct vertices $P_0P_1\dots P_nP_{n+1}$ with $A = P_0$ and $B = P_{n+1}$. Assuming that $n \geq 1$ and that there exists $1 \leq i \leq n$, such that $det(\overrightarrow{P_{i-1}P_i}, \overrightarrow{P_iP_{i+1}}) \neq 0$, $CMLP(C)$ is not the Euclidean segment $[AB]$. According to Sloboda [23], a vertex such that $det(\overrightarrow{P_{i-1}P_i}, \overrightarrow{P_iP_{i+1}}) > 0$ is called *convex* and is always a positive point of ∂I, similarly, a vertex such that $det(\overrightarrow{P_{i-1}P_i}, \overrightarrow{P_iP_{i+1}}) < 0$ is *concave* and is always a negative point of ∂O.

As explained earlier on we are interested in the deformation of constrained open digital contours. Let us now elaborate on our digital deformable model and the associated energy based on the CMLP.

3 A Digital Deformable Model

3.1 Digital Formulation of a Deformable Model

In [14] Kass *et al.* proposed the active contour model, which describes a curve whose evolution is monitored by a minimization scheme balancing two terms, an internal regularizing energy and an external energy expressing the fit to data. Let v be a parametrization of the curve embodying the active contour geometry:

$$v = [0,1] \to \mathbb{R}^2$$

$$v(s) = \begin{pmatrix} x(s) \\ y(s) \end{pmatrix}$$

with its energy function being defined as:

$$E_{DM}(v) = \int_0^1 (E_{int}(v(s), v'(s), v''(s)) + E_{image}(v(s)))\, ds.$$

This formulation allows to define various internal energies based on the position of the curve, its first and second derivative. The internal energy is most commonly found as follow, with α and β being parameters that drive the behaviour of the model.

$$E_{int}(v(s), v'(s), v''(s)) = \left(\alpha(s)\, |v'(s)|^2 + \beta(s)\, |v''(s)|^2\right)/2$$

Considering that the sought curve is a minimum for E_{DM}, its Euler-Lagrange equations induce differential equations whose discretization leads to the usual formulation of active contours.

We propose here a digital vision of the same minimization problem. The geometry of our digital analog is a digital curve, a 4-connected path Γ. Moreover, we do not consider curvature regularization in this paper, that is we consider β equal to zero. As noticed by many authors (see for instance the geometric active contours of [4]), the internal energy is in fact the length of the curve. Our digital analog to E_{int} is therefore defined as the length of the CMLP of Γ, that is the sum of the Euclidean length of its edges. Thus our internal energy becomes:

$$E^D_{int}(\Gamma) = \alpha \mathcal{L}_2(CMLP(\Gamma)).$$ [1]

[1] $\mathcal{L}_2(CMLP(\Gamma))$ is to be read as the sum of the Euclidean lengths of the edges of the polygon defined by the CMLP of Γ.

The image energy term is defined as a sum over the digital path and is positive everywhere:

$$E^D_{image}(\Gamma) = \sum_{c\in\Gamma} (\max(||\nabla I||) - ||\nabla I(c)||) \,.$$

Our digital combinatorial minimization problem becomes:

$$\Gamma^* = \arg \min_{\Gamma\in\mathcal{C}(A,B)} E^D_{DM}(\Gamma), \text{ where } E^D_{DM}(\Gamma) = E^D_{int}(\Gamma) + E^D_{image}(\Gamma).$$

Let us remark that in the case of the defined energies, geodesic active contours provide a way of efficiently finding a global minimum for this problem [7]. However our model allows more general energies such as region energy terms which do not fit in the previously cited approach.

3.2 Features and Elementary Deformations

Once an energy is defined for the space of admissible shapes, a set of elementary transitions must be provided. Enough transitions or deformations must be defined so that this space is connected.

We propose several elementary modifications for digital contours, which we call 'flips', 'bumps' and 'flats', see Fig. 2 for an illustration of some elementary deformations. A digital contour is a 4-connected path and is equivalently defined as an ordered list of turns, either toward the interior (denoted "+") or the exterior (denoted "−") or straight ahead moves (denoted "0"). Since they are elements of $\mathcal{C}$, no contour has three consecutive "+" or "−". All proposed deformations preserve this property. We say that an operation is valid if and only if the digital contour is still simple after the operation. Of course all operations are reversible.

- Outside corners, that is $(-+-)$, $(-+0)$,$(0+-)$, $(0+0)$ can be respectively "flipped" into $(0-0)$, $(0-+)$,$(+-0)$, $(+-+)$. This operation is denoted by $Flip^+$.
- Inside corners, that is $(0-0)$, $(0-+)$,$(+-0)$, $(+-+)$ can be respectively "flipped" into $(-+-)$, $(-+0)$,$(0+-)$, $(0+0)$. This operation is denoted by $Flip^-$, it is the inverse of $Flip^+$.
- Inside or outside bumps, that is $(--)$ or $(++)$ can be flatened to $()$, this is a valid operation. This operation is denoted by $Flat$.
- Flat inside parts, that is $(--)$, (-0), (00) can be respectively bumped inside to $(0--0)$, $(0--+)$,$(+--+)$. This operation is denoted by $Bump^-$.
- Flat outside parts, that is $(++)$, $(+0)$, (00) can be respectively bumped outside to $(0++0)$, $(0++-)$,$(-++-)$. This operation is denoted by $Bump^+$.

Remark that some deformations may turn an element of $\mathcal{C}$ into a self-crossing 4-connected path, it is of course not a valid operation in such cases. We can now define operations on specific contour points. They are related to the convexity of their corresponding vertex on the CMLP.

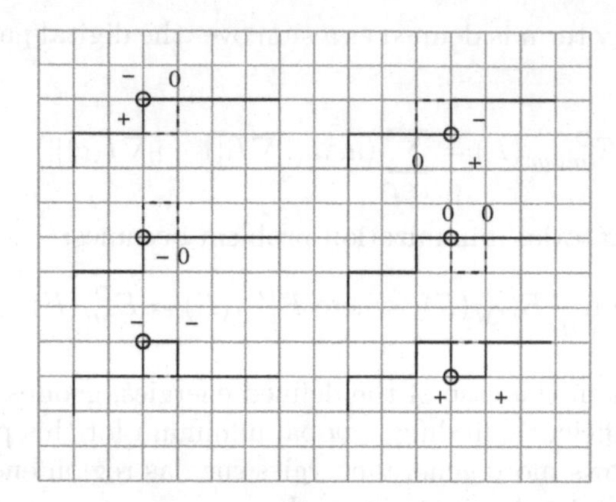

Fig. 2. Example of some local deformations used by the digital deformable model, solid line for Γ and dashed line for Γ after the local deformation, the points where the deformation is applied are circled. From left to right and top to bottom : flip of an inside corner, flip of an outside corner, inside bump of a flat part, outside bump of a flat part, flat of an inside bump and flat of an outside bump.

Definition 1. *Given a convex vertex of the CMLP, it is associated either to one outside corner on ∂I or to one outside bump on ∂I. We call* convex deformation *the $Flip^+$ of this outside corner or the Flat of this outside bump.*

Definition 2. *Given a concave vertex of the CMLP, it is associated either to one inside corner on ∂O or to one inside bump on ∂O. We call* concave deformation *the $Flip^-$ of this inside corner or the Flat of this inside bump.*

As we have explained earlier on there exists at least one element of $\mathcal{C}(A, B)$ with minimal energy. Our aim is now to show that despite the fact that some deformations are not valid, there always exists at least one valid convex deformation on a convex vertex or a concave deformation on a concave vertex. For this purpose, we introduce three lemmas:

Lemma 2. *On the digital plane, every simple 4-connected path composed of only two Freeman letters and containing a positive or negative corner, say C_i, cannot also visit the nearby digital point Flip(C_i).*

Proof. Let us consider a positive vertex given by the Freeman code NE with its corresponding points $P_0(x, y)$, $P_1(x, y+1)$ and $P_2(x+1, y+1)$. The corresponding flipped point $Flip(P_1)$ is $(x + 1, y)$. In order the path to visit $Flip(P_1)$ *after* visiting P_2, it must contain a S letter in the Freeman code word since its y coordinate is lower than that of P_2. If it were to visit it *before* P_0 then it should contain a W letter since its x coordinate is greater than that of P_0. □

Lemma 3. *On the digital plane every simple 4-connected path C defined with 3 or more Freeman letters is such that whenever $a \ldots \bar{a}$ is a factor of C then the associated turns of $a \ldots \bar{a}$ contain either $(+0^*+)$ or (-0^*-).*

Proof. It is enough to consider the associated points to the a letter and $\bar{a}$ letter, without loss of generality, let $a = E$ and $\bar{a} = O$, $P_{a_1} = (x,y)$ and $P_{a_2} = (x+1,y)$, $P_{\bar{a}_1} = (x',y')$ and $P_{\bar{a}_2} = (x'-1,y')$. C being simple we have necessarily $(x+1,y) \neq (x'-1,y')$ and $(x,y) \neq (x'-1,y')$ and $(x,y) \neq (x',y')$.

Considering the minimal bounding box containing the factor $a \ldots \bar{a}$, its height is greater than one, therefore its right side intersects $a \ldots \bar{a}$ with a N^m or S^m move, with $m \geq 1$, that is an EN^mO move or a ES^mO move. □

Lemma 4. *On the digital plane, if C is a simple 4-connected path containing an outside or inside corner, that is three points C_{i-1}, C_i and C_{i+1} such that $sign(C_i)$ is positive or negative, and visiting also the point $Flip(C_i)$, then it contains either -0^*- or $+0^*+$ between C_i and $Flip(C_i)$.*

Proof. Without loss of generality, we choose a NE corner, that is an outside corner. Let $C_{i-1} = (x,y), C_i = (x,y+1), C_{i+1} = (x+1,y+1)$ be the points embodying it on the digital plane. Since C is simple and has to visit the point $Flip(C_i) = (x+1,y) = C_j$, with $j > i$, let us consider two cases:

- either $C_{j-1} = (x+2,y)$ and $C_{j+1} = (x+1,y-1)$, that is a OS corner;
- or $C_{j-1} = (x+1,y-1)$ and $C_{j+1} = (x+2,y)$, that is a NE corner.

In the first case Lemma 3 shows that there exists either -0^*- or $+0^*+$ between E and O.

In the second case we see that $C_{i+1} = (x+1,y+1)$ and $C_{j-1} = (x+1,y-1)$ entails that there is at least two S letters between the two NE corners. Thus by Lemma 3 there exists either -0^*- or $+0^*+$ between S and N. □

We may now prove that any digital contour whose CMLP is not a segment is "deformable".

Proposition 1. *Let $C \in \mathcal{C}(A,B)$ being finite , if $CMLP(C) \neq [AB]$ there exists at least one valid convex deformation on a convex vertex or one valid concave deformation on a concave vertex of $CMLP(C)$.*

Proof. We prove it by contradiction. Let us consider that $CMLP(C) \neq [AB]$ and that there do not exist any valid convex deformation on a convex vertex or a valid concave deformation on a concave vertex of $CMLP(C)$.

First, all deformations on $CMLP(C)$ are necessarily flips on outside or inside turns since the flat deformation is always in $\mathcal{C}(A,B)$ and is by definition always a valid convex deformation on a convex vertex or a valid concave deformation on a concave vertex.

C is thus constituted of inside turns, outside turns and straight parts, and at least one of the inside or outside turn is associated to a CMLP vertex. Since by hypothesis those corners are not allowed to be flipped, the contour visits also the points associated to each corner flip.

In the case where C is only constituted of two Freeman letters, Lemma 2 concludes, since convex deformations can be applied to convex vertices and concave deformation can be applied to concave vertices.

Let us now consider that C has at least 3 Freeman letters and, without loss of generality, the case of a positive vertex of the CMLP. By definition it is associated to an outside corner in C, let C_i be the corresponding point associated to the positive turn. From Lemma 4, C contains at least $(+0^m+)$ with $m \geq 1$ between C_i and $Flip(C_i)$. Since bumps would entail valid deformations, we have further that $m \geq 1$. As a result, the $(+0^m+)$ quadrant change would entail at least one more outside corner. Since they are points of quadrant changes, a vertex of the CMLP is associated to it. Applying this reasoning recursively leads to an infinite number of points on C. As we are on a finite simple 4-connected path, this raises a contradiction. Similarly, the same reasoning applies on an inside vertex. □

3.3 Two Greedy Deterministic Optimization Schemes

We define two simple deterministic greedy optimization scheme. Algorithm 1 selects the deformation which brings the lowest energy from all the valid deformations that can be applied to the digital deformable model (DDM for short). Algorithm 2, even greedier, picks the first deformation that diminishes the energy of the DDM.

Function Greedy1(**In** Γ, **Out**) : boolean ;
Input: Γ: a Digital Deformable Model
Output: Γ': when returning *true*, elementary deformation of Γ with
$E^D_{DM}(\Gamma') < E^D_{DM}(\Gamma)$, otherwise $\Gamma' = \Gamma$ and it is a local minimum.
Data: Q : Queue of (Deformation, double) ;
$E_0 \leftarrow E^D_{DM}(\Gamma)$;
foreach *valid Deformation d on* Γ **do**
  Γ.applyDeformation(d);
  Q.push_back($d, E^D_{DM}(\Gamma)$);
  Γ.revertLastDeformation();
end
(d, E_1) = SelectDeformationWithLowestEnergy (Q);
$\Gamma' \leftarrow \Gamma$;
if $E_1 < E_0$ **then** Γ'.applyDeformation(d);
return $E_1 < E_0$;

Algorithm 1. Greedy1 algorithm: extracts the deformation that brings the lowest energy among all possibles ones

3.4 Properties

We will now show a remarkable property of our discrete internal energy. It is similar to a convex energy in the continuous case, in the sense that *any* local descent leads to one of the *optimal* solutions.

The following theorem proves that we always reach a configuration of minimal energy when choosing at each step an arbitrary valid deformation which decreases the energy. This induces that both Greedy1 and Greedy2 algorithms lead to one of the optimal solution.

Function Greedy2(**In** Γ, **Out**) : boolean ;
Input: Γ: a Digital Deformable Model
Output: Γ': when returning *true*, elementary deformation of Γ with $E^D_{DM}(\Gamma') < E^D_{DM}(\Gamma)$, otherwise $\Gamma' = \Gamma$ and it is a local minimum.
$E_0 \leftarrow E^D_{DM}(\Gamma)$;
foreach *valid Deformation d on* Γ **do**
  Γ.applyDeformation(d);
  if $E^D_{DM}(\Gamma) < E_0$ **then** $\Gamma' \leftarrow \Gamma$;
  return *true*;
  Γ.revertLastDeformation();
end
$\Gamma' \leftarrow \Gamma$;
return *false*;

Algorithm 2. Greedy2 algorithm : extracts the first valid deformation that decreases the energy

Theorem 1. *Let* $C \in \mathcal{C}(A, B)$ *being finite, choosing at each step any one of the valid convex or concave deformation leads to* $CMLP(C) = [AB]$ *after a finite number of steps.*

Proof. There are 2 cases, either:

- $CMLP(C) = [AB]$, and we have reached the global minimum of energy, by definition of a shortest path.
- $CMLP(C) \neq [AB]$ and in this case we have $CMLP(C) = P_0P_1 \ldots P_nP_{n+1}$ with $A = P_0$ and $B = P_{n+1}$ with $n \geq 1$ and there exist at least one vertex of the CMLP which is convex or concave, that is there exists i such that either $det(\overrightarrow{P_{i-1}P_i}, \overrightarrow{P_iP_{i+1}}) > 0$ or $det(\overrightarrow{P_{i-1}P_i}, \overrightarrow{P_iP_{i+1}}) < 0$.

In the second case, let us first suppose that P_i is associated to an outside or inside bump on C. Without loss of generality we consider the convex case, that is $det(\overrightarrow{P_{i-1}P_i}, \overrightarrow{P_iP_{i+1}}) > 0$ and an outside bump. By construction, $[P_{i-1}P_i]$ and $[P_iP_{i+1}]$ intersect the sides of the unit square centered on P_i on two distinct points I_1 and I_2. The triangle $I_1P_iI_2$ is within the unit square centered on P_i and of non zero area. The length of $[I_1I_2]$ is obviously shorter than that of $[P_{i-1}P_i] + [P_iP_{i+1}]$. As a result, the flat operation on the outside bump entails that the transformed CMLP has a length no greater than $P_0 \ldots P_{i-1}I_1I_2P_{i+1} \ldots P_{n+1}$. The energy has then strictly diminished.

Let us now suppose that P_i is associated to an outside or inside corner on C. From Prop. 1, if the corner associated to P_i cannot be flipped, then there exists at least one inside or outside corner on C associated to a convex or concave vertex of the CMLP. Without loss of generality, let it be an outside corner associated to a convex vertex of the CMLP. By construction, $[P_{i-1}P_i]$ and $[P_iP_{i+1}]$ intersect the sides of the unit square centered on P_i on two points I_1 and I_2. The triangle $I_1P_iI_2$ is within the unit square centered on P_i and of non zero area. Following the same reasoning as earlier, the transformed CMLP brings a lower energy. □

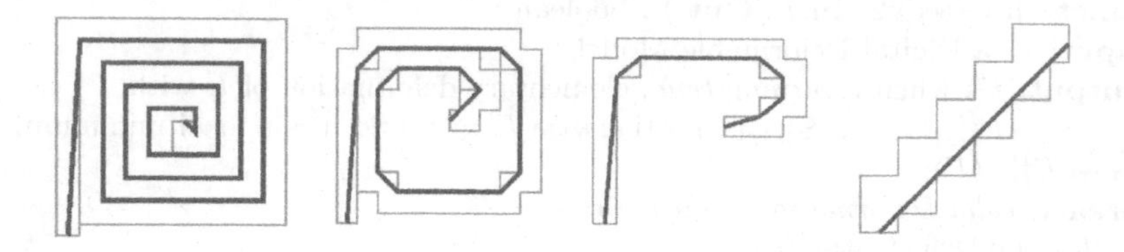

Fig. 3. Example of the minimization process on a spiral using the Greedy1 algorithm. Only the length penalisation term is used in this process. In green the inner and outer boundary are drawn, in red the CMLP is drawn. From left to right: Initialisation of the DDM. Result of the minimisation energy process after 50 iterations. Result of the minimisation energy process after 100 iterations. End of the process after 139 iterations, which is also a global minimum.

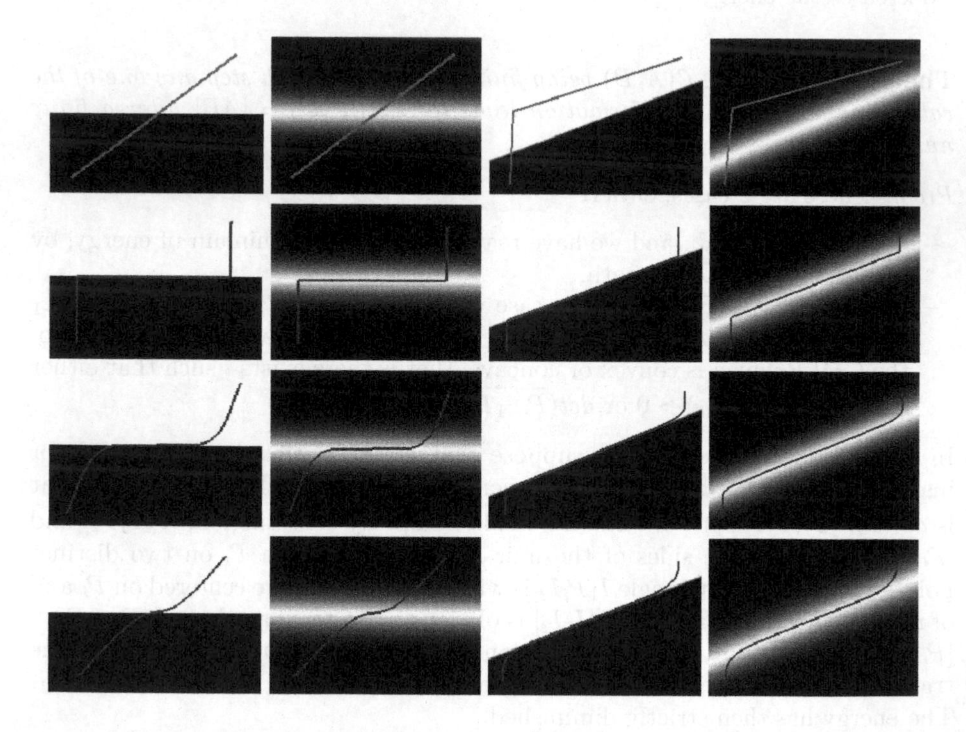

Fig. 4. Example of the minimization process using the Greedy1 algorithm. The gradient is computed with the Canny-Deriche method with scale coefficient 0.2. The input image represents a half-plane. (First row) Initialisation of the DDM. (Second row) Results of the minimisation process, the α coefficient used is equal to 0. (Third row) Results with $\alpha = 200$. (Fifth row) Results with $\alpha = 300$.

Note that it is possible to force the minimization process to choose the grid-intersection digitization of $[AB]$ by selecting the path with lowest energy and maximizing the number of positive turns. This slight modification reduces the number of optimal solutions to exactly one, as in the continuous case.

4 Experimental Evaluation

We here show some preliminary results of our DDM. Figure 3 shows that the internal energy, i.e. the regularization term, is indeed a kind of convex energy: the spiral is unwinded to the digital straight segment by Greedy1 algorithm. Figure 4 illustrates the balance between internal energy and gradient-based image energy. Again, the observed behaviour matches the expected bahaviour of classical snakes: the lower the length penalisation term, the more the DDM sticks

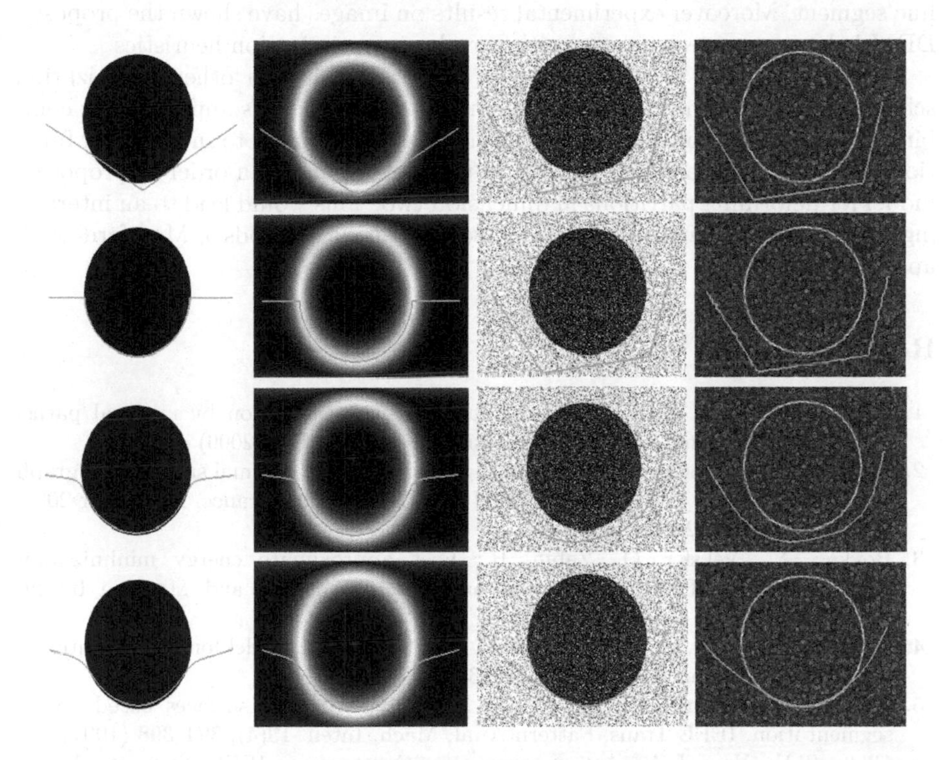

Fig. 5. Example of the minimization process using the Greedy2 algorithm. In order to test our DDM in extreme cases, the gradient is computed with the Canny-Deriche method with scale coefficient 0.2 for the ellipse image without noise, and coefficient 2.0 for the noisy ellipse image. The gradient field is thus very perturbated in the latter case. (First row) Initialisation of the DDM. (Second row) Results of the minimization process, the α coefficient is equal to 0. (Third row) Results with $\alpha = 800$ for the circle without noise and $\alpha = 4000$ for the noisy circle. (Fourth row) Results with $\alpha = 1600$ for the circle without noise and $\alpha = 8000$ for the noisy circle.

to points having strong gradients. Finally, we tested our model on images degradated with a strong Gaussian noise, as displayed on Fig. 5. In an extremely noisy environment, a very strong regularizing term brings effectively the DDM to escape local image energy minima so as to fit more consistent gradient information that is further away (bottom right of figure).

5 Conclusion

In this paper we have proposed a Digital Deformable Model based using the Minimum Length Polygon as length penalisation energy. We have given theoretical results that back up our claim that it is a digital analog to the continuous variational formulation of standard active contours. The regularizing term has indeed "convex-like" properties, and the optimum is the digitization of straight line segment. Moreover experimental results on images have shown the proposed DDM behave as expected, even with very basic optimization heuristics.

Future works will focus on the use of this DDM with other optimization scheme, such as stochastic processes (simulated annealing is appealing) or combinatorial optimization algorithms (our model seems related to submodular functions [21]). We plan also to embed it in topological maps, in order to propose a model for deforming partitions in dimension two. This would lead to an interesting digital analog to multiphase levelset segmentation methods or Mumford-Shah approximation.

References

1. Ardon, R., Cohen, L.D.: Fast constrained surface extraction by minimal paths. International Journal on Computer Vision 69(1), 127–136 (2006)
2. Boykov, Y., Kolmogorov, V.: Computing geodesics and minimal surfaces via graph cuts. In: Proc. Int. Cof. Comput. Vis (ICCV 2003), Nice, France, November 2003, vol. 1, pp. 26–33 (2003)
3. Boykov, Y., Veksler, O., Zabih, R.: Fast approximate energy minimization via graph cuts. IEEE Transactions on Pattern Analysis and Machine Intelligence 23(11), 1222–1239 (2001)
4. Caselles, V., Catte, F., Coll, T., Dibos, F.: A geometric model for active contours. Numerische Mathematik 66, 1–31 (1993)
5. Caselles, V., Kimmel, R., Sapiro, G., Sbert, C.: Minimal surfaces based object segmentation. IEEE Trans. Pattern Anal. Mach. Intell. 19(4), 394–398 (1997)
6. Chan, T.F., Vese, L.A.: Active contours without edges. IEEE Trans. on Image Processing 10(2), 266–277 (2001)
7. Cohen, L.D., Kimmel, R.: Global minimum for active contour models: a minimal path approach. Int. Journal of Computer Vision 24(1), 57–78 (1997)
8. Dupas, A., Damiand, G.: First results for 3D image segmentation with topological map. In: Coeurjolly, D., Sivignon, I., Tougne, L., Dupont, F. (eds.) DGCI 2008. LNCS, vol. 4992, pp. 507–518. Springer, Heidelberg (2008)

9. Fejes, S., Rosenfeld, A.: Discrete active models and applications. Pattern Recognition 30(5), 817–835 (1997)
10. Geman, S., Geman, D.: Stochastic relaxation, Gibbs distributions, and the Bayesian restoration of images, pp. 564–584. Morgan Kaufmann Publishers Inc., San Francisco (1987)
11. Guibas, L.J., Hershberger, J.: Optimal shortest path queries in a simple polygon. In: SCG 1987: Proceedings of the third annual symposium on Computational geometry, pp. 50–63. ACM, New York (1987)
12. Guigues, L., Le Men, H., Cocquerez, J.-P.: Scale-sets image analysis. In: Int. Conf. on Image Processing (ICIP 2003), vol. 2, pp. 45–48. IEEE, Los Alamitos (2003)
13. Imiya, A., Saito, M., Tatara, K., Nakamura, K.: Digital curvature flow and its applications for skeletonization. Journal of Mathematical Imaging and Vision 18, 55–68 (2003)
14. Kass, M., Witkin, A., Terzopoulos, D.: Snakes: Active contour models. International Journal of Computer Vision, 321–331 (1988)
15. Kass, M., Witkin, A., Terzopoulos, D.: Snakes: Active contour models. International Journal of Computer Vision 1(4), 321–331 (1988)
16. Klette, R., Rosenfeld, A.: Digital Geometry - Geometric Methods for Digital Picture Analysis. Morgan Kaufmann, San Francisco (2004)
17. Klette, R., Yip, B.: The length of digital curves. Machine Graphics Vision 9(3), 673–703 (2000) (Also research report CITR-TR-54, University of Auckland, NZ. 1999)
18. Malladi, R., Sethian, J.A., Vemuri, B.C.: Shape Modelling with Front Propagation: A Level Set Approach. IEEE Trans. on Pattern Analysis and Machine Intelligence 17(2), 158–174 (1995)
19. Montanari, U.: A note on minimal length polygonal approximation to a digitized contour. Commun. ACM 13(1), 41–47 (1970)
20. Mumford, D., Shah, J.: Optimal approximations by piecewise smooth functions and associated variational problems. Comm. Pure Appl. Math. 42, 577–684 (1989)
21. Schrijver, A.: Combinatorial optimization: polyhedra and efficiency. Springer, Heidelberg (2004)
22. Sloboda, F., Stoer, J.: On piecewise linear approximation of planar jordan curves. J. Comput. Appl. Math. 55(3), 369–383 (1994)
23. Sloboda, F., Stoer, J.: On piecewise linear approximation of planar jordan curves. Journal of Computational and Applied Mathematics 55, 369–383 (1994)
24. Sloboda, F., Zaťko, B.: On one-dimensional grid continua in r2. Technical report, Institute of Control Theory and Robotics, Slovak Academy of Sciences, Bratislava, Slovakia (1996)

A Detailed Steps of Klette and Yip Algorithm

Klette and Yip [17] (reported also in [16]) have proposed a linear time algorithm which fails sometimes at finding the proper MLP, as shown on Fig. 6. This is explained in where several steps of the algorithm are reproduced on Table 1 and match those illustrated on the previous figure.

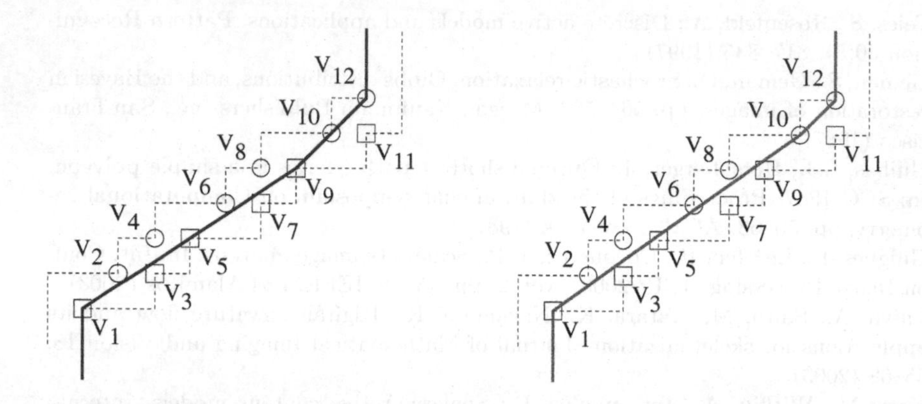

Fig. 6. Portion of a boundary where the algorithm of [17] fails at electing the correct Minimum Length Polygon vertices. The positive points are circled and the negative points are boxed. (Left) The MLP elected by Klette's algorithm, vertices are (v_1, v_5, v_9, v_{12}). (Right) the correct portion of MLP, whose vertices are (v_1, v_{10}, v_{12}).

Table 1. Several steps of the algorithm of Fig. 6 are reproduced here. The input list is $(v_1, v_2, v_3, v_4, v_5, v_6, v_7, v_8, v_9, v_{10}, v_{11})$. Where v_1 is the first MLP vertex considered.

$k=1, a=1, b=1, i=2, p1=v1$ 3) $j=2$ 5) $D(p_k, v_b, v_j) = D(p_1, v_1, v_2) = 0$ 5.1) $D(p_k, v_a, v_j) = D(p_1, v_1, v_2) = 0$ $v_2.sign > 0, b = 2$ 5bis) $i = 3$
$k=1, a=1, b=2, i=3, p1=v1$ 3) $j=3$ 5) $D(p_k, v_b, v_j) = D(p_1, v_2, v_3) = 0$ 5.1) $D(p_k, v_a, v_j) = D(p_1, v_1, v_3) = 0$ $v_3.sign < 0, a = 3$ 5bis) $i = 4$
$k=1, a=3, b=2, i=4, p1=v1$ 3) $j=4$ 5) $D(p_k, v_b, v_j) = D(p_1, v_2, v_4) = 0$ 5.1) $D(p_k, v_a, v_j) = D(p_1, v_3, v_4) > 0$ $v_4.sign > 0, b = 4$ 5bis) $i = 5$
$k=1, a=3, b=4, i=5, p1=v1$ 3) $j=5$ 5) $D(p_k, v_b, v_j) = D(p_1, v_4, v_5) < 0$ 5.1) $D(p_k, v_a, v_j) = D(p_1, v_3, v_5) > 0$ $v_5.sign < 0, a = 5$ 5bis) $i = 6$
$k=1, a=5, b=4, i=6, p1=v1$ 3) $j=6$ 5) $D(p_k, v_b, v_j) = D(p_1, v_4, v_6) < 0$ 5.1) $D(p_k, v_a, v_j) = D(p_1, v_5, v_6) > 0$ $v_6.sign > 0, b = 6$ 5bis) $i = 7$
$k=1, a=5, b=6, i=7, p1=v1$ 3) $j=7$ 5) $D(p_k, v_b, v_j) = D(p_1, v_6, v_7) < 0$ 5.1) $D(p_k, v_a, v_j) = D(p_1, v_5, v_7) < 0$ $k = 2, p_2 = v_5, i = 5, b = 5$ // v_5 should NOT be elected as an mlp vertex 5bis) $i = 6$
$k=2, a=5, b=5, i=6, p2=v5$ 3) $j=7$ 5) $D(p_k, v_b, v_j) = D(p_1, v_6, v_7) < 0$ 5.1) $D(p_k, v_a, v_j) = D(p_1, v_5, v_7) < 0$ $k = 2, p_2 = v_5, i = 5, b = 5$ 5bis) $i = 6$

Topology-Preserving Thinning in 2-D Pseudomanifolds

Nicolas Passat[1], Michel Couprie[2], Loïc Mazo[1,2], and Gilles Bertrand[2]

[1] Université de Strasbourg, LSIIT, UMR CNRS 7005, Strasbourg, France
[2] Université Paris-Est, Laboratoire d'Informatique Gaspard-Monge, Équipe A3SI, ESIEE Paris, France
{passat,loic.mazo}@unistra.fr, {m.couprie,g.bertrand}@esiee.fr

Abstract. Preserving topological properties of objects during thinning procedures is an important issue in the field of image analysis. In the case of 2-D digital images (*i.e.* images defined on $\mathbb{Z}^2$) such procedures are usually based on the notion of simple point. By opposition to the case of spaces of higher dimensions (*i.e.* $\mathbb{Z}^n$, $n \geq 3$), it was proved in the 80's that the exclusive use of simple points in $\mathbb{Z}^2$ was indeed sufficient to develop thinning procedures providing an output that is minimal with respect to the topological characteristics of the object. Based on the recently introduced notion of *minimal simple set* (generalising the notion of simple point), we establish new properties related to topology-preserving thinning in 2-D spaces which extend, in particular, this classical result to more general spaces (the 2-D pseudomanifolds) and objects (the 2-D cubical complexes).

Keywords: Topology preservation, simple points, simple sets, cubical complexes, collapse, confluence, pseudomanifolds.

1 Introduction

Topological properties are fundamental in many applications of image analysis, in particular in cases where the retrieval and/or the preservation of topology of real complex structures is required. In this context, numerous methods have been developed to process discrete 2-D and 3-D binary images, essentially to perform skeletonisation, homotopic thinning, or segmentation.

Such methods are generally based on the notion of *simple point* [8]. Intuitively, a point (or pixel) of a discrete object X is said to be simple if it can be removed from X without altering its topology.

Let us consider an object X, *i.e.* a set of points in $\mathbb{Z}^n$, and a subset Y of X called constraint set. A very common topology-preserving thinning scheme [4] consists of repeating the following steps until stability:

- choose (according to some given priority function) a point x in $X \setminus Y$ that is simple for X;
- remove x from X.

The result of such a procedure, called homotopic skeleton of X constrained by Y, is a subset Z of X, which (*i*) is topologically equivalent to X, (*ii*) includes Y and (*iii*) has no simple point outside of Y. We show an illustration in Fig. 1, notice in particular that the constraint set is useful to preserve some geometrical characteristics of the object.

S. Brlek, C. Reutenauer, and X. Provençal (Eds.): DGCI 2009, LNCS 5810, pp. 217–228, 2009.

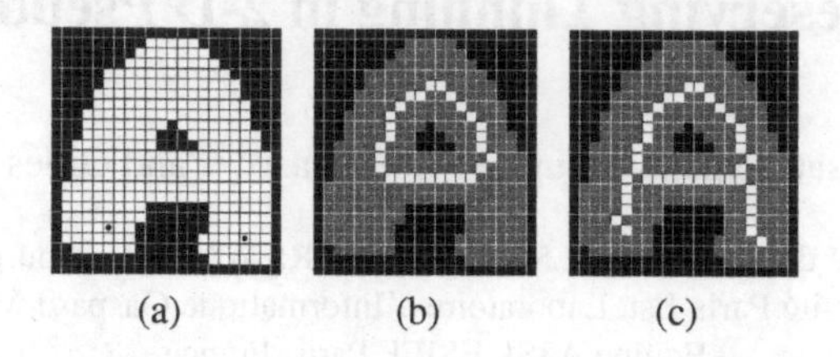

Fig. 1. (a) An object X (in white) and a subset Y of X (two pixels marked by black dots). (b) A homotopic skeleton of X (empty constraint set). (c) A homotopic skeleton of X constrained by Y.

The following question is fundamental with regard to the behaviour of sequential thinning procedures:

(1) Is Z always a minimal result, in the sense that it does not strictly include a subset Z' having the same properties (*i*), (*ii*) and (*iii*)?

If we consider the 3-D case, the answer to this question is no. For example, if X is a cuboid and $Y = \emptyset$, then, depending on the order of the point removals, the result Z of the above procedure might not be composed of a single point. As pointed out recently [11], there exist various kinds of configurations in which a 3-D topology-preserving thinning algorithm can be "blocked" before reaching a minimal result.

In the discrete plane $\mathbb{Z}^2$, question (1) was answered positively by Ronse in the 80's, after a partial answer was given in the early 70's by Rosenfeld. In 1970, in the same article where he introduced the notion of simple point [15], Rosenfeld proved that any finite subset of $\mathbb{Z}^2$ that is connected and has no holes, could be reduced to a single point by iterative removal of simple points, in any order. In 1986, Ronse introduced the notion of strong deletability in $\mathbb{Z}^2$ [14]. It is, to the best of our knowledge, the first attempt to generalise the notion of simple points to a more general notion of simple sets.

According to Def. 2.5 of [14], and skipping formal details, a subset S of $X \subseteq \mathbb{Z}^2$ is *strongly deletable from X* if (*i*) each connected component of X includes exactly one connected component of $X \setminus S$, and (*ii*) each connected component of $\overline{X} \cup S$ includes exactly one connected component of $\overline{X}$, where $\overline{X}$ denotes the complementary of X.

In the same article, Ronse proposed several results related to strongly deletable sets, which can be summarised as follows.

Theorem 1 (From [14], Lem. 3.1, 3.2, Prop. 3.3). *Let $X \subseteq \mathbb{Z}^2$. Let $S \subseteq X$. If S is strongly deletable from X, then:*

(i) there exists $x \in S$ such that x is a simple point for X;
(ii) for all $x \in S$ such that x is a simple point for X, $S \setminus \{x\}$ is strongly deletable for $X \setminus \{x\}$.

Consequently, if $Y \subseteq X \subseteq \mathbb{Z}^2$ and Y is topologically equivalent to X (more precisely, if $X \setminus Y$ is strongly deletable from X), then Y may be obtained from X by iterative removal of simple points, in any arbitrary order.

To summarise, question (1) received a positive answer in $\mathbb{Z}^2$ and a negative one in $\mathbb{Z}^3$ (and also for higher dimensions). Still, there are spaces for which this question remained open until now: the case of two-dimensional structures in n-dimensional spaces, $n \geq 3$. Such structures are often used in practice, *e.g.* to represent thin objects or (parts of) boundary of objects in 3-D image analysis and in finite element modelling.

The main outcome of this article is a theorem (Th. 26) that states a property analogous to Th. 1, holding in a large family of 2-D digital spaces, namely the pseudomanifolds.

This study is developed in the framework of cubical complexes [9], in which we can retrieve and generalise the concepts of digital topology in $\mathbb{Z}^n$. The definition of simple sets that we use here is based on the operation of collapse, a topology-preserving transformation known in algebraic topology.

The proof of Th. 26 is based on a property of collapse, that we call a confluence property (Th. 21), which is introduced and proved in this article.

Th. 26 is also closely related to the notion of minimal simple set [11,13], as we derive it using the following property: if X is a strict subset of a pseudomanifold, then any minimal simple subset of X is a simple point (Prop. 25).

Thanks to a correspondence between the notion of minimal simple set used here and the one of simple point [7], we retrieve as particular cases of Th. 26 the results of Rosenfeld and Ronse discussed before. However, the techniques of proof used in this article are essentially different from the ones used by these authors, and the generalisation of their results is not trivial.

This article is self-contained, however some of the proofs cannot be included due to space limitation (they can however be found in the following research report [12]).

2 Background Notions

In this section, we provide basic definitions and properties related to the notions of cubical complexes, collapse and simple sets (the last two ones enabling to modify a complex without altering its topology), see also [9,13].

2.1 Cubical Complexes

If T is a subset of S, we write $T \subseteq S$. Let $\mathbb{Z}$ be the set of integers. Let $k, \ell \in \mathbb{Z}$, we denote by $[k, \ell]$ the set $\{i \in \mathbb{Z} \mid k \leq i \leq \ell\}$.

We consider the families of sets $\mathbb{F}^1_0$, $\mathbb{F}^1_1$, such that $\mathbb{F}^1_0 = \{\{a\} \mid a \in \mathbb{Z}\}$ and $\mathbb{F}^1_1 = \{\{a, a+1\} \mid a \in \mathbb{Z}\}$. A subset f of $\mathbb{Z}^n$ ($n \geq 2$) that is the Cartesian product of m elements of $\mathbb{F}^1_1$ and $n - m$ elements of $\mathbb{F}^1_0$ is called a *face* or an *m-face* of $\mathbb{Z}^n$, m is the *dimension of* f, we write $\dim(f) = m$ (see Fig. 2(a,b)).

Let $n \geq 2$, we denote by $\mathbb{F}^n$ the set composed of all faces of $\mathbb{Z}^n$.

An *m-face* of $\mathbb{Z}^n$ is called a *point* if $m = 0$, a *(unit) edge* if $m = 1$, a *(unit) square* if $m = 2$.

Let f be a face in $\mathbb{F}^n$. We set $\hat{f} = \{g \in \mathbb{F}^n \mid g \subseteq f\}$. Any $g \in \hat{f}$ is a *face of* f (or *of* $\hat{f}$).

If X is a set of faces of $\mathbb{F}^n$, we write $X^- = \bigcup_{f \in X} \hat{f}$, and we say that X^- is the *closure of* X.

A set X of faces of $\mathbb{F}^n$ is a *cell* or an *m-cell* if there exists an m-face $f \in X$, such that $X = \hat{f}$. The *boundary of a cell* $\hat{f}$ is the set $\hat{f}^* = \hat{f} \setminus \{f\}$ (see Fig. 2).

A finite set X of faces of $\mathbb{F}^n$ is a *complex (in $\mathbb{F}^n$)* if for any $f \in X$, we have $\hat{f} \subseteq X$.

Let S, X be two sets of faces of $\mathbb{F}^n$. If X is a complex and $X \subseteq S$, we write $X \preceq S$. Furthermore, if S is also a complex, then we say that X is a *subcomplex of* S.

Let $X \subseteq \mathbb{F}^n$. A face $f \in X$ is a *facet of* X if there is no $g \in X$ such that $f \in \hat{g}^*$, in other words, if f is maximal for inclusion. A facet of X that is an m-face is also called an *m-facet of* X. We denote by X^+ the set composed of all facets of X (see Fig. 3).

If X is a complex, observe that in general, X^+ is not a complex, and that $(X^+)^- = X$. More generally, for any subset Y of $\mathbb{F}^n$, $(Y^+)^- = Y^-$.

Let $X \subseteq \mathbb{F}^n$, $X \neq \emptyset$. The *dimension of* X is the number $\dim(X) = \max\{\dim(f) \mid f \in X\}$, and we set $\dim(\emptyset) = -1$. We say that X is *pure* if for each $f \in X^+$, we have $\dim(f) = \dim(X)$. Let m be an integer. We say that X is an m-complex if X is a complex and $\dim(X) = m$. If X is an m-complex with $m \leq 1$, then we also say that X *is a graph* (see [5]).

Let $Y \preceq X \preceq \mathbb{F}^n$. If $Y^+ \subseteq X^+$, we say that Y is a *principal subcomplex of* X and we write $Y \sqsubseteq X$ (see Fig. 4).

Let $X \subseteq \mathbb{F}^n$. A sequence $\pi = \langle f_i \rangle_{i=0}^{\ell}$ ($\ell \geq 0$) of faces in X is a *path in X (from f_0 to f_ℓ)* if for each $i \in [0, \ell - 1]$, either f_i is a face of f_{i+1} or f_{i+1} is a face of f_i; the integer ℓ is the *length of* π. The path π is said to be *closed* whenever $f_0 = f_\ell$, it is a *trivial path* whenever $\ell = 0$.

Let $X \subseteq \mathbb{F}^n$. A path in X made of 0- and 1-faces is called a *1-path*. A 1-path from a 0-face x to a 0-face y (with possibly $x = y$), is said to be *elementary* if its 1-faces are all distinct. A non-trivial elementary closed path is called a *cycle*.

Let $X \subseteq \mathbb{F}^n$. We say that X is *connected* if, for any pair of faces (f, g) in X, there is a path in X from f to g. It is easily shown that, if X is a complex, then X is connected if and only if there exists an elementary path from x to y in X whenever x and y are 0-faces in X.

Let $X \subseteq \mathbb{F}^n$, and let Y be a non-empty subset of X, we say that Y is a *connected component of* X if Y is connected and if Y is maximal for these two properties (*i.e.*, if we have $Z = Y$ whenever $Y \subseteq Z \subseteq X$ and Z is connected). We will sometimes write *component* as a shortcut for connected component. The number of components of X is denoted by $|C(X)|$. Notice that $|C(\emptyset)| = 0$.

2.2 Collapse and Simple Sets

Let X be a complex in $\mathbb{F}^n$ and let $f \in X$. If there exists a face $g \in \hat{f}^*$ such that f is the only face of X that strictly includes g, then g is said to be *free for* X, and the pair (f, g) is said to be a *free pair for* X. Notice that, if (f, g) is a free pair for X, then we have necessarily $f \in X^+$ and $\dim(g) = \dim(f) - 1$.

Let X be a complex, and let (f, g) be a free pair for X. Let $m = \dim(f)$. The complex $X \setminus \{f, g\}$ is an *elementary collapse of* X, or an *elementary m-collapse of* X.

Let X, Y be two complexes. We say that X *collapses onto* Y, and we write $X \searrow Y$, if $Y = X$ or if there exists a *collapse sequence from X to Y*, *i.e.*, a sequence of complexes $\langle X_i \rangle_{i=0}^{\ell}$ ($\ell \geq 1$) such that $X_0 = X$, $X_\ell = Y$, and X_i is an elementary collapse of X_{i-1}, for each $i \in [1, \ell]$ (see Fig. 5). Let $J = \langle (f_i, g_i) \rangle_{i=1}^{\ell}$ be the sequence of pairs of faces of X such that $X_i = X_{i-1} \setminus \{f_i, g_i\}$, for any $i \in [1, \ell]$. We also call the sequence J a *collapse* *sequence (from X to Y)*. If X collapses onto Y and Y is a complex made of a single point, we say that *X is collapsible*.

Let $Y, X \subseteq \mathbb{F}^n$. We say that *X is an extension of Y* [1] if $Y \subseteq X$ and each connected component of X includes exactly one connected component of Y.

Proposition 2 (proved in [12]). *Let $Y \preceq X \preceq \mathbb{F}^n$. If $X \searrow Y$, then X is an extension of Y. In consequence, collapse preserves the number of connected components.*

Let $X \preceq \mathbb{F}^n$, the complex that is the closure of the set of all free faces for X, is called the *boundary of X*, and is denoted by $Bd(X)$. We denote by $Bd_1(X)$ the complex that is the closure of the set of all free 1-faces for X (see Fig. 6). Of course, we have $Bd_1(X) \preceq Bd(X)$.

Proposition 3 (proved in [12]). *Let $Y \preceq X \preceq \mathbb{F}^n$, let α be a set of facets of X that are not in Y,* i.e., *$\alpha \subseteq X^+ \setminus Y$. If $Bd(\alpha^-) \subseteq Y$, then X does not collapse onto Y.*

Proposition 4 (proved in [12]). *Let $Z \preceq X \preceq \mathbb{F}^n$ be two complexes such that $X \searrow Z$. Let $J = \langle (f_i, g_i) \rangle_{i=1}^{\ell}$ be a collapse sequence from X to Z. Suppose that there exists $Y \preceq X$ such that $Z \preceq Y$ and for any $i \in [1, \ell]$, either $\{f_i, g_i\} \subseteq Y$ or $\{f_i, g_i\} \subseteq X \setminus Y$. Then, $X \searrow Y$ and $Y \searrow Z$.*

Let $J = \langle (f_i, g_i) \rangle_{i=1}^{\ell}$ be a collapse sequence. This collapse sequence is said to be *decreasing* if for any $i \in [1, \ell - 1]$, we have $\dim(f_i) \geq \dim(f_{i+1})$.

Proposition 5 ([16]). *Let $Y \preceq X \preceq \mathbb{F}^n$. If X collapses onto Y, then there exists a decreasing collapse sequence from X to Y.*

Let X, Y be two complexes. Let Z be such that $X \cap Y \preceq Z \preceq Y$, and let $f, g \in Z \setminus X$. The pair (f, g) is a free pair for $X \cup Z$ if and only if (f, g) is a free pair for Z. Thus, by induction, we have the following property.

Proposition 6 ([2]). *Let $X, Y \preceq \mathbb{F}^n$. The complex $X \cup Y$ collapses onto X if and only if the complex Y collapses onto $X \cap Y$.*

The operation of detachment allows us to remove a subcomplex from a complex while guaranteeing that the result is still a complex (see Fig. 7).

Definition 7 ([2]). *Let $Y \preceq X \preceq \mathbb{F}^n$. We set $X \oslash Y = (X \setminus Y)^-$. The set $X \oslash Y$ is a complex that is called the* detachment of Y from X.

Intuitively a cell $\hat{f}$ or a subcomplex Y of a complex X is simple if its removal from X "does not modify the topology of X". Let us now recall a definition of simplicity [2] based on the collapse operation, which can be seen as a discrete counterpart of the one given by Kong [7].

Definition 8 ([2]). *Let $Y \preceq X \preceq \mathbb{F}^n$. We say that Y is* simple for X *if X collapses onto $X \oslash Y$.*

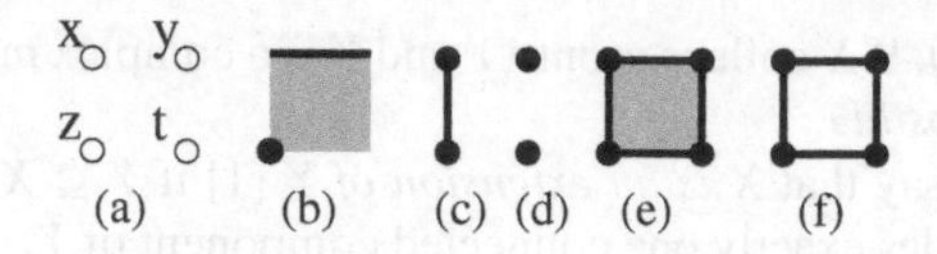

Fig. 2. (a) Four points of $\mathbb{Z}^2$: $x = (0, 1)$; $y = (1, 1)$; $z = (0, 0)$; $t = (1, 0)$. (b) A representation of the set of faces $\{f_0, f_1, f_2\}$ in $\mathbb{F}^2$, where $f_0 = \{z\}$ (0-face), $f_1 = \{x, y\}$ (1-face), and $f_2 = \{x, y, z, t\}$ (2-face). (c) A 1-cell $\hat{c}$. (d) A 2-cell $\hat{d}$. (e) The boundary $\hat{c}^*$ of $\hat{c}$. (f) The boundary $\hat{d}^*$ of $\hat{d}$.

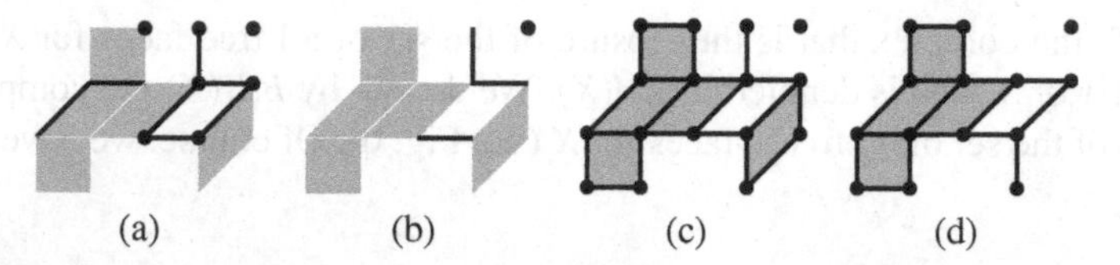

Fig. 3. (a) A set X of 0-, 1- and 2-faces in $\mathbb{F}^3$, which is not a complex. (b) The set X^+, composed of the facets of X. (c) The set X^-, *i.e.* the closure of X, which is a complex. (d) A subcomplex of X^-.

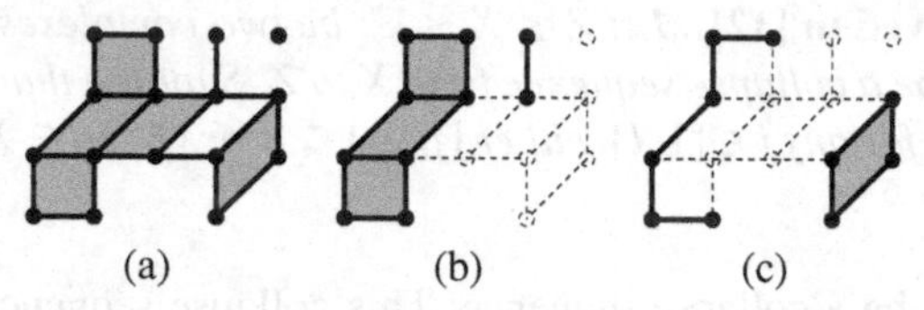

Fig. 4. (a) A complex X. (b) A subset Y of X, which is a principal subcomplex of X (*i.e.*, $Y \sqsubseteq X$). (c) A subset Z of X, which is a subcomplex of X but not a principal subcomplex of X.

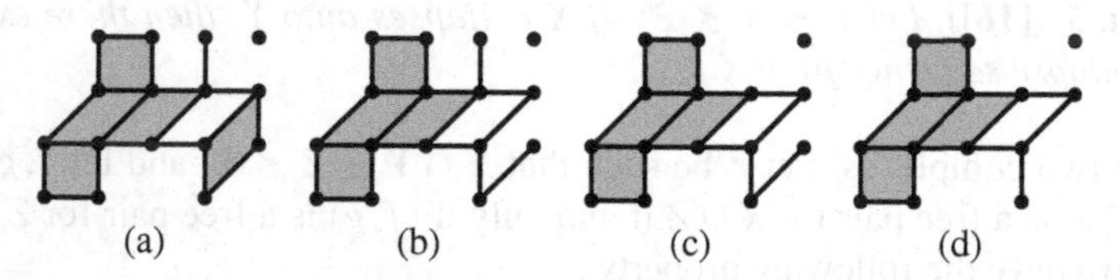

Fig. 5. (a) A complex X. (d) A subcomplex Y of X. (a,b,c,d) A collapse sequence from X to Y.

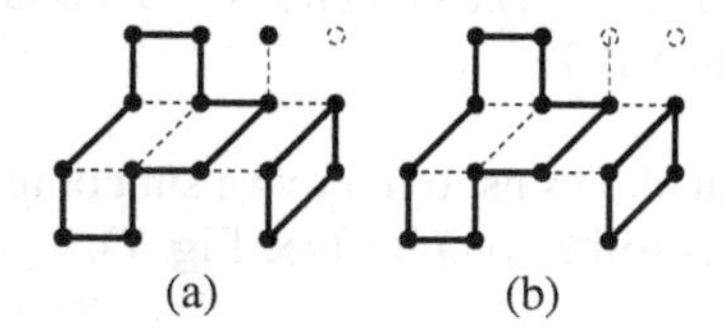

Fig. 6. (a) $Bd(X)$, where X is the complex of Fig. 5(a). (b) $Bd_1(X)$.

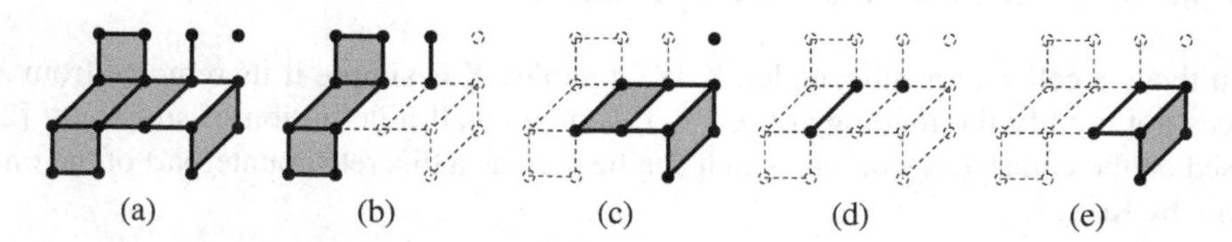

Fig. 7. (a) A complex X. (b) A subcomplex Y of X that is simple for X. (c) The detachment of Y from X. (d) The attachment of Y to X. (e) A subcomplex Z of X that is *not* simple for X.

If $\hat{f}$ is a simple cell, we will also say that f is simple.

The notion of attachment, as introduced by Kong [6,7], leads to a local characterisation of simple sets (Prop. 9).

Let $Y \preceq X \preceq \mathbb{F}^n$. The *attachment* of Y for X is the complex defined by $Att(Y,X) = Y \cap (X \oslash Y)$ (see Fig. 7). Remark that any facet f of X such that $Att(\hat{f}, X) \neq \hat{f}^*$ includes a free face for X.

Proposition 9 **([2]).** *Let $Y \preceq X \preceq \mathbb{F}^n$. The complex Y is simple for X if and only if Y collapses onto $Att(Y,X)$.*

Remark 10. *If $Y = \emptyset$, or if $Y \preceq X$ contains no facet of X, then Y is obviously a simple set for X, as we have $X \oslash Y = X$. More generally, it can be proved [13] that the detachment of a subcomplex Y from X is equal to the detachment of the maximal principal subcomplex Z of X included in Y. Without loss of generality, the study of the simple sets Y of a complex X can then be restricted to those verifying $Y \sqsubseteq X$ and $Y \neq \emptyset$. From now on, we will always implicitly consider that a simple set verifies these hypotheses.*

3 Confluence Properties in Cubical Complexes

Consider three complexes A, B, C. If A collapses onto C and A collapses onto B, then we know that A, B and C "have the same topology". If furthermore we have $C \preceq B \preceq A$, it is tempting to conjecture that B collapses onto C. We call this a confluence property. For example, this property implies that any complex in $\mathbb{F}^2$ obtained by a collapse sequence from a full rectangle indeed collapses onto a point.

Quite surprisingly, such a property does not hold in $\mathbb{F}^3$ (and more generally in $\mathbb{F}^n$, $n \geq 3$). A classical counter-example to this assertion is Bing's house ([3], see also [11]). A realisation of Bing's house as a 2-complex can be obtained by collapse from a full cuboid, and has no free face: it is thus a counter-example for the above conjecture, with A: a cuboid, B: Bing's house, and C: a point in B.

As we will show in this article, in the two-dimensional discrete plane $\mathbb{F}^2$ and more generally in the class of discrete spaces called pseudomanifolds, a confluence property indeed holds (Th. 21).

In this section, we establish confluence properties that are essentially 1-dimensional, a step for proving the more general confluence properties of Section 5.

A *tree* is a graph that is collapsible. It may be easily proved that a graph is a tree if and only if it is connected and does not contain any cycle (see [5]).

Let $X \preceq \mathbb{F}^n$ be a complex. The set of all i-faces of X, with $i \in [0,n]$, is denoted by $F_i(X)$. We denote by $|F_i(X)|$ the number of i-faces of X, $i \in [0,n]$. The *Euler characteristic of* X, written $\chi(X)$, is defined by $\chi(X) = \sum_{i=0}^{n}(-1)^i|F_i(X)|$. The Euler characteristic is a well-known topological invariant; in particular, it can be easily seen that collapse preserves it.

Let $X, Y \preceq \mathbb{F}^n$. A fundamental and well-known property of the Euler characteristic, deriving from the so-called inclusion-exclusion principle in set theory, is the following: $\chi(X \cup Y) = \chi(X) + \chi(Y) - \chi(X \cap Y)$.

The following property generalises a classical characterisation of trees: a graph X is a tree if and only if X is connected and $\chi(X) = 1$.

Proposition 11 (proved in [12]). *Let X, Y be such that $Y \preceq X \preceq \mathbb{F}^n$, and* $\dim(X \setminus Y) \leq 1$. *Then, X collapses onto Y if and only if X is an extension of Y and $\chi(Y) = \chi(X)$.*

From Props. 2, 11, and the fact that collapse preserves the Euler characteristic, we derive straightforwardly the following two propositions.

Proposition 12. *Let A, B, C be such that $C \preceq B \preceq A \preceq \mathbb{F}^n$ and such that* $\dim(B \setminus C) \leq 1$. *If A collapses onto C and A collapses onto B, then B collapses onto C.*

Proposition 13. *Let A, B, C be such that $C \preceq B \preceq A \preceq \mathbb{F}^n$ and such that* $\dim(A \setminus B) \leq 1$. *If A collapses onto C and B collapses onto C, then A collapses onto B.*

4 Pseudomanifolds

Intuitively, a (2-D) manifold [10] is a 2-D (finite or infinite) space which is locally "like" the 2-D Euclidean space (spheres and tori are, for instance, manifolds).

The notion of (2-D) pseudomanifold is less restrictive since it authorises several pieces of surface to be adjacent in a singular point (as two cones sharing the same apex, for instance). Note that any manifold is a pseudomanifold, but the converse is not true. Some examples of pseudomanifolds are provided in Fig. 8.

In the framework of cubical complexes, a 2-D pseudomanifold can be defined as follows. We denote by $\mathbb{F}_2^n$ the set composed of all m-faces of $\mathbb{Z}^n$, with $m \in [0, 2]$. We say that π is a *2-path (in X)* if π is a path in X composed of 1- and 2-faces.

Definition 14. *Let $M \subseteq \mathbb{F}_2^n$ be such that* $\dim(M) = 2$. *We say that M is a* (2-D) pseudomanifold *if the following four conditions hold:*

(i) for any $f \in M$, we have $\hat{f} \subseteq M$;
(ii) M is pure;
(iii) for any pair of 2-faces (f, g) in M, there is a 2-path in M from f to g;
(iv) any 1-face of M is included in exactly two 2-faces of M.

Notice that, in particular, $\mathbb{F}_2^2 = \mathbb{F}^2$ (namely the discrete plane) is a pseudomanifold. Notice also that, if M is a finite pseudomanifold, then M is a pure 2-complex that cannot be collapsed, since M has no free face by definition.

In the sequel, we focus on complexes that are strict subsets of a pseudomanifold.

Proposition 15 (proved in [12]). *Let $M \subseteq \mathbb{F}_2^n$ be a pseudomanifold, and let $X \preceq M$. Then, $Bd(Bd_1(X)) = Bd(Bd(X)) = \emptyset$.*

Proposition 16 (proved in [12]). *Let $M \subseteq \mathbb{F}_2^n$ be a pseudomanifold, let $B \preceq M$ such that* $\dim(B) = 2$ *and $B \neq M$, let f be a 2-face of B, and let g be a 2-face in $M \setminus B$. If π is a 2-path from f to g in M, then π necessarily contains a 1-face of $Bd(B)$.*

Prop. 17 follows easily from Prop. 16.

Proposition 17. *Let $M \subseteq \mathbb{F}_2^n$ be a pseudomanifold, let $B \preceq M$. If* $\dim(B) = 2$ *and $B \neq M$, then there exists at least one pair (f, g) that is free for B, with* $\dim(f) = 2$.

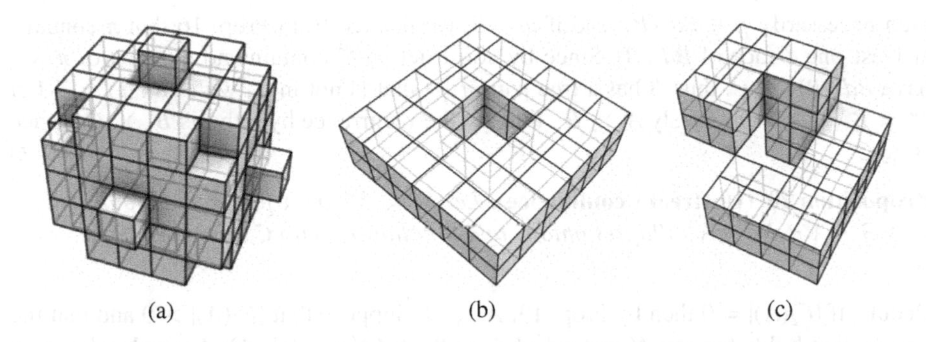

Fig. 8. 2-D pseudomanifolds. (a) A topological sphere. (b) A topological torus. (c) A pinched torus. (a) and (b) are (pseudo)manifolds, (c) is a pseudomanifold but not a manifold.

5 Confluence Properties in Pseudomanifolds

Our goal in this section is to establish confluence properties, similar to Props. 12 and 13, in the case where A, B and C are complexes that are subsets of a pseudomanifold.

It is tempting to try to generalise Prop. 11 to this case, for confluence properties would immediately follow from such a result. But in fact, the backward implication of Prop. 11 does not hold in the general case (that is, when $\dim(X \setminus Y)$ is not constrained), even if X and Y are complexes that are subsets of a pseudomanifold.

A counter-example is given by M: a pinched torus (see Fig. 8(c)), $A \preceq M$: a topological disk (*e.g.*, a square and all the faces included in it), $X = M \oslash A$, and $Y = Bd(X) = X \cap A$ (a topological circle). It is easily checked that $\chi(M) = \chi(X \cup A) = 1$, $\chi(A) = 1$ and $\chi(Y) = 0$, and since $\chi(X \cup A) = \chi(X) + \chi(A) - \chi(X \cap A)$ we deduce $\chi(X) = 0 = \chi(Y)$. We have also $Y \preceq X$ and $|C(Y)| = |C(X)| = 1$, thus X is an extension of Y. However, by construction, X has no free face outside Y, thus X does not collapse onto Y.

A similar counter-example could be built from a sphere M, which is a manifold, and a ring A (a closed ribbon that is a pure 2-complex). In this case X is made of two topological discs and $X \cap A$ is made of two topological circles. We have $\chi(M) = \chi(X) = 2$ and $\chi(A) = \chi(X \cap A) = 0$.

Nevertheless, we have the following property.

Proposition 18 (proved in [12]). *Let $M \subseteq \mathbb{F}_2^n$ be a pseudomanifold, and let $X \preceq M$, $X \neq M$. The complex X is collapsible if and only if $|C(X)| = \chi(X) = 1$.*

Proposition 19 (Downstream confluence). *Let $M \subseteq \mathbb{F}_2^n$ be a pseudomanifold, and let $C \preceq B \preceq A \preceq M$. If A collapses onto C and A collapses onto B, then B collapses onto C.*

Proof. If $|F_2(B)| = |F_2(C)|$ then by Prop. 12, $B \searrow C$. Suppose that $|F_2(B)| > |F_2(C)|$. Suppose that the proposition holds for any B' instead of B, with $|F_2(C)| \leq |F_2(B')| < |F_2(B)|$. Let q be a 2-face of B not in C. Since $A \searrow C$ and by Prop. 5, there exists a sequence of 2-collapse operations starting from A and that removes q. Let S be the set of faces removed by this sequence, clearly there exists a 2-path π in S, from a 1-face p that belongs to $Bd_1(A)$, to q. Let h be the 2-face just following p in π. If $h \in B$

then necessarily $p \in Bd_1(B)$; and if $h \notin B$, we deduce from Prop. 16 that π contains at least one 1-face of $Bd_1(B)$. Since by construction C contains no element of π, we have $Bd_1(B)^+ \not\subseteq C$, thus B has a free pair (f, g) that is not in C, with $\dim(f) = 2$. Let $B' = B \setminus \{f, g\}$. Obviously $A \searrow B'$, thus by the recurrence hypothesis $B' \searrow C$, hence $B \searrow C$. □

Proposition 20 (Upstream confluence). *Let $M \subseteq \mathbb{F}_2^n$ be a pseudomanifold, and let $C \preceq B \preceq A \preceq M$. If A collapses onto C and B collapses onto C, then A collapses onto B.*

Proof. If $|F_2(A)| = 0$ then by Prop. 13, $A \searrow B$. Suppose that $|F_2(A)| > 0$ and that the proposition holds for any A' instead of A, with $|F_2(A')| < |F_2(A)|$. Consider the set α of 1-faces that are free for A and not in C, *i.e.*, $\alpha = F_1(Bd(A) \setminus C)$. If $\alpha = \emptyset$, then the hypothesis $A \searrow C$ implies that $|F_2(A)| = |F_2(C)| = |F_2(B)|$, and the result follows from Prop. 13. We now suppose that $\alpha \neq \emptyset$. By Prop. 15, no face in $Bd(A)$ is free for $Bd(A)$, hence no face in α^- is free for $\alpha^- \cup C$. Thus, all the faces in α cannot be facets of B, for otherwise by Prop. 3, B could not collapse onto C. From this, we deduce that there exists a 1-face g in α such that either $g \in Bd(B)$ or $g \notin B$. Let f be the 2-face of A that includes g.
Case 1: $g \in Bd(B)$. Thus, (f, g) is a free pair for both A and B. Let $A' = A \setminus \{f, g\}$ and $B' = B \setminus \{f, g\}$. We have $C \preceq B' \preceq A'$, $A' \searrow C$ (by Prop. 19) and $B' \searrow C$ (also by Prop. 19), thus by the recurrence hypothesis $A' \searrow B'$. It can be seen that any sequence of collapse operations from A' to B' is also a sequence of collapse operations from A to B.
Case 2: $g \notin B$. Thus, (f, g) is a free pair for A that is not in B, let $A' = A \setminus \{f, g\}$. We have $C \preceq B \preceq A'$, $A' \searrow C$ (by Prop. 19) and $B \searrow C$, thus by the recurrence hypothesis $A' \searrow B$ hence $A \searrow B$. □

The following theorem summarises Props. 19 and 20.

Theorem 21 (Confluences). *Let $M \subseteq \mathbb{F}_2^n$ be a pseudomanifold, and let $C \preceq B \preceq A \preceq M$ be such that A collapses onto C. Then, A collapses onto B if and only if B collapses onto C.*

6 Minimal Simple Sets in Pseudomanifolds

Informally, a minimal simple set [11,13] is a simple set which does not strictly include any other simple set. In this section, we first establish the equivalence between the notions of simple cell and minimal simple set in pseudomanifolds (Prop. 25). Then we demonstrate that, in such spaces, any simple set can be fully detached, while preserving topology, by iterative detachment of simple cells, in any possible order (Th. 26).

Definition 22 ([13]). *Let $X \preceq \mathbb{F}^n$ and $S \sqsubseteq X$. The subcomplex S is a* minimal simple set (for X) *if S is a simple set for X and S is minimal with respect to the relation $\sqsubseteq$ (i.e. $Z = S$ whenever $Z \sqsubseteq S$ and Z is a simple set for X).*

Proposition 23 ([12]). *Let $M \subseteq \mathbb{F}_2^n$ be a pseudomanifold, and let $S \sqsubseteq X \preceq M$ such that S is a minimal simple set for X. Then, S is connected.*

This property of minimal simple sets indeed holds in more general conditions, see [13].

Proposition 24 (proved in [12]). *Let $M \subseteq \mathbb{F}_2^n$ be a pseudomanifold, let $X \preceq M$ be a connected 2-complex, let $S \sqsubseteq X$ be a simple subcomplex of X, and let f be a facet of S such that $Att(\hat{f}, X)$ is not empty and not connected. Then, $X \oslash \hat{f}$ is an extension of $Att(\hat{f}, X)$.*

Proposition 25. *Let $M \subseteq \mathbb{F}_2^n$ be a pseudomanifold, and let $S \sqsubseteq X \preceq M$ such that S is a minimal simple set for X. Then, S is necessarily a 1-cell or a 2-cell.*

Proof. Suppose that S is not reduced to one cell. Then, each facet of S must be non-simple for X. No facet f of S is such that $Att(\hat{f}, X) = \emptyset$. If S contains a 1-facet, then let f be such a facet. If S is a pure 2-complex, then at least one 2-face of S must include a free face for X, otherwise X could not collapse onto $X \oslash S$, let us assume that f is such a 2-face. Let $A = Att(\hat{f}, X)$. In both cases ($\dim(f) = 1$ or $\dim(f) = 2$), we know that A is disconnected. From now, we suppose that $\dim(f) = 2$ (the case where $\dim(f) = 1$ is similar and simpler).

From Prop. 23, S is connected and from Props. 9 and 2, $Att(S, X)$ is connected. Without loss of generality, we assume that X is connected (otherwise we replace X by the component of X that includes S). By Prop. 24, each component of $X \oslash \hat{f}$ includes exactly one component of A. Let X_1 be the component of $X \oslash \hat{f}$ that includes $Att(S, X)$ (and thus also $X \oslash S$), and let A_1 be the component of A that is in X_1. Let g and h be the two 1-faces of $\hat{f}^* \setminus A$ that include each a 0-face of A_1. Obviously (f, g) is a free pair for X, let $X' = X \setminus \{f, g\}$. Remark that h is a facet of X'. We have $X \searrow X'$ and $X \searrow X \oslash S$, by Prop. 19 we deduce $X' \searrow X \oslash S$.

Let $J = \langle (f_i, g_i) \rangle_{i=1}^{\ell}$ be a collapse sequence from X' to $X \oslash S$. Let $t \in [1, \ell]$ be such that $f_t = h$. It can be seen that $g_t \notin X_1$ (otherwise the result of the collapse operation would be disconnected, for by construction any path in X' from $X \oslash S$ to the remaining face in h would contain h), and of course $f_t \notin X_1$. Furthermore, any other pair of J is either in X_1 or in $X' \setminus X_1$, since the only facet of $X' \setminus X_1$ that includes a face of X_1 is f_t. Thus by Prop. 4, $X' \searrow X_1$, hence $X \searrow X_1$, *i.e.*, $X \oslash X_1$ is a simple set for X. Remark that by construction, we have $X \oslash X_1 \sqsubseteq S$. Thus, the minimality of S implies that $S = X \oslash X_1$, hence $Att(S, X) = A_1$.

It is plain that $\hat{f} \searrow A_1$, thus by Prop. 6 we have $X_1 \cup \hat{f} \searrow X_1$; and since $X \searrow X_1$, by Prop. 20 we deduce that $X \searrow X_1 \cup \hat{f}$, *i.e.*, $X \oslash (X_1 \cup \hat{f})$ is a simple set for X. This contradicts the minimality of S, since $X \oslash (X_1 \cup \hat{f}) \sqsubset S$. □

From Props. 25 and 19, we derive straightforwardly our main theorem.

Theorem 26. *Let $M \subseteq \mathbb{F}_2^n$ be a pseudomanifold, and let $S \sqsubseteq X \preceq M$ such that S is a simple set for X. Then:*

(i) there is a facet of X in S which is simple for X; and
(ii) for any cell $\hat{f}$ in S which is simple for X, $S \oslash \hat{f}$ is a simple set for $X \oslash \hat{f}$.

7 Conclusion

In this article we have established, in the case of digital 2-D pseudomanifolds, a confluence property of the collapse operation (Th. 21). From this result, we have proved that in pseudomanifolds, any minimal simple set is a simple cell (Prop. 25). This led us to the property stating that any simple set can be removed by iterative removal of simple cells in any order (Th. 26).

It is indeed possible to retrieve Ronse's theorem (Th. 1) from the results presented above, based on the equivalence between $\mathbb{Z}^2$ equipped with a $(8,4)$-adjacency framework and the set of pure 2-complexes in $\mathbb{F}^2$ [7]. To this aim, it is necessary to prove that any subcomplex $S \sqsubseteq X$ (where X is a pure 2-complex in $\mathbb{F}^2$) that is strongly deletable for X, is also simple for X in the sense of Def. 8 (the converse also holds). The Jordan's theorem is needed for this proof, which is not in the scope of the present article.

References

1. Bertrand, G.: On topological watersheds. Journal of Mathematical Imaging and Vision 22 (2–3), 217–230 (2005)
2. Bertrand, G.: On critical kernels. Comptes Rendus de l'Académie des Sciences, Série Mathématiques 1(345), 363–367 (2007)
3. Bing, R.H.: Some aspects of the topology of 3-manifolds related to the Poincaré conjecture. Lectures on Modern Mathematics II, pp. 93–128 (1964)
4. Davies, E.R., Plummer, A.P.N.: Thinning algorithms: A critique and a new methodology. Pattern Recognition 14(1–6), 53–63 (1981)
5. Giblin, P.: Graphs, surfaces and homology. Chapman and Hall, Boca Raton (1981)
6. Yung Kong, T.: On topology preservation in 2-D and 3-D thinning. International Journal of Pattern Recognition and Artificial Intelligence 9(5), 813–844 (1995)
7. Yung Kong, T.: Topology-preserving deletion of 1's from 2-, 3- and 4-dimensional binary images. In: Ahronovitz, E. (ed.) DGCI 1997. LNCS, vol. 1347, pp. 3–18. Springer, Heidelberg (1997)
8. Kong, T.Y., Rosenfeld, A.: Digital topology: Introduction and survey. Computer Vision, Graphics, and Image Processing 48(3), 357–393 (1989)
9. Kovalesky, V.A.: Finite topology as applied to image analysis. Computer Vision, Graphics, and Image Processing 46(2), 141–161 (1989)
10. Maunder, C.R.F.: Algebraic topology. Dover, New York (1996)
11. Passat, N., Couprie, M., Bertrand, G.: Minimal simple pairs in the 3-D cubic grid. Journal of Mathematical Imaging and Vision 32(3), 239–249 (2008)
12. Passat, N., Couprie, M., Mazo, L., Bertrand, G.: Topological properties of thinning in 2-D pseudomanifolds. Technical Report IGM2008-5, Université de Marne-la-Vallée (2008)
13. Passat, N., Mazo, L.: An introduction to simple sets. Technical Report LSIIT-2008-1, Université de Strasbourg (2008)
14. Ronse, C.: A topological characterization of thinning. Theoretical Computer Science 43(1), 31–41 (1986)
15. Rosenfeld, A.: Connectivity in digital pictures. Journal of the Association for Computer Machinery 17(1), 146–160 (1970)
16. Zeeman, E.C.: Seminar on Combinatorial Topology. In: IHES (1963)

Discrete Versions of Stokes' Theorem Based on Families of Weights on Hypercubes

Gilbert Labelle and Annie Lacasse

Laboratoire de Combinatoire et d'Informatique Mathématique, UQAM
CP 8888, Succ. Centre-ville, Montréal (QC) Canada H3C3P8
Laboratoire d'Informatique, de Robotique et de Microélectronique de Montpellier,
161 rue Ada, 34392 Montpellier Cedex 5 - France
labelle.gilbert@uqam.ca, annie.lacasse@lirmm.fr

Abstract. This paper generalizes to higher dimensions some algorithms that we developed in [1,2,3] using a discrete version of Green's theorem. More precisely, we present discrete versions of Stokes' theorem and Poincaré lemma based on families of weights on hypercubes. Our approach is similar to that of Mansfield and Hydon [4] where they start with a concept of difference forms to develop their discrete version of Stokes' theorem. Various applications are also given.

Keywords: Discrete Stokes' theorem, Poincaré lemma, hypercube.

1 Introduction

Typically, physical phenomenon are described by differential equations involving differential operators such as *divergence*, *curl* and *gradient*. Numerical approximation of these equations is a rich subject. Indeed, the use of simplicial or (hyper)cubical complexes is well-established in numerical analysis and in computer imagery. Such complexes often provide good approximations to geometrical objects occuring in diverse fields of applications (see Desbrun and al. [5] for various examples illustrating this point). In particular, discretizations of differential forms occuring in the classical Green's and Stokes' theorems allows one to reduce the (approximate) computation of parameters associated with a geometrical object to computation associated with the boundary of the object (see [4,5,6,7]).

In the present paper, we focus our attention on hypercubic configurations and systematically replace differential forms by weight functions on hypercubes. We extend the classical difference and summation operators (Δ, Σ), of the finite difference calculus, to weight functions on hypercubic configurations in order to formulate our discrete versions of Stokes' theorem. For example, using these operators, the total volume, coordinates of the center of gravity, moment of inertia and other statistics of an hypercubic configuration can be computed using suitable weight functions on its boundary (see Section 4 below).

Although our approach have connections with the works [4,6] on cubical complexes and cohomology in $\mathbb{Z}^n$, we think that it can bridge a gap between differential calculus and discrete geometry since no differential forms or operations

S. Brlek, C. Reutenauer, and X. Provençal (Eds.): DGCI 2009, LNCS 5810, pp. 229–239, 2009.

of standard exterior algebra calculus are involved. Moreover, our results are easily implemented on computer algebra systems such as Maple or Mathematica since families of weight functions on hypercubes, difference and summation operators are readily programmable. Finally, weight functions can take values in an arbitrary commutative ring. In particular, taking values in the boolean ring $B = \{0, 1, \text{and}, \text{xor}\}$, our results provide algorithms to decide, for example, whether a given hypercube (or pixel, in the bi-dimensional case) belongs to an hypercubic configuration (or polyomino) using a suitable weight on its boundary (see [1,2,3], for similar examples in the case of polyominoes).

For completeness, in Section 2 we state the classical Stokes' theorem and recall some elementary concepts on differential forms. Section 3 describes discrete geometrical objects such as *hypercubes, hypercubic configurations* together with exterior *difference* and *summation* operators on weight functions which are used to formulate our discrete versions of Stokes' theorem. Section 4 illustrates the theory through examples. Due to the lack of space, most of the proofs (see [3]) are omitted in this extended abstract.

2 Classical Differential Forms and Stokes' Theorem

We start our analysis with the following version of Stokes' theorem [8]

Theorem 1. *Let ω be a differential $(k-1)$-form and K a k-surface in $\mathbb{R}^n$ with positively oriented boundary ∂K. Then,*

$$\int_K d\omega = \int_{\partial K} \omega,$$

where d is the exterior differential operator.

In this statement, we suppose implicitly that ω is sufficiently differentiable and K sufficiently smooth. Moreover, a differential $(k-1)$-form ω can be written in the standard form

$$\omega = \sum_{1 \le j_1 < \cdots < j_{k-1} \le n} \omega_{j_1,\ldots,j_{k-1}}(x_1,\ldots,x_n)\, dx_{j_1} \wedge \cdots \wedge dx_{j_{k-1}}$$

and its exterior derivative, $d\omega$ is a k-form described by

$$d\omega = \sum_{1 \le j_1 < \cdots < j_{k-1} \le n} \left(d\omega_{j_1,\ldots,j_{k-1}}(x_1,\ldots,x_n)\right) \wedge dx_{j_1} \wedge \cdots \wedge dx_{j_{k-1}}$$

where $df(x_1,\ldots,x_n) = \frac{\partial f}{\partial x_1}\, dx_1 + \cdots + \frac{\partial f}{\partial x_n}\, dx_n$. Making use of the general formulas $dx_j \wedge dx_i = -dx_i \wedge dx_j$, we obtain,

$$d\omega = \sum_{1 \le i_1 < \cdots < i_k \le n} (d\omega)_{i_1,\ldots,i_k}(x_1,\ldots,x_n)\, dx_{i_1} \wedge \cdots \wedge dx_{i_k},$$

where component $(d\omega)_{i_1,\ldots,i_k}$ is the function of $(x_1,\ldots,x_n)$, given by

$$(d\omega)_{i_1,\ldots,i_k} = \sum_{\nu=1}^{k} (-1)^{\nu-1} \frac{\partial \omega_{i_1,\ldots,\widehat{i_\nu},\ldots,i_k}}{\partial x_{i_\nu}},$$

where $\widehat{i_\nu}$ means that index i_ν is omitted. One of the fundamental properties of the exterior differential operator is $d^2 = 0$, meaning that for every differential form ω, we have $d(d\omega) = 0$. We now recall some basic notions and facts about differential forms. A differential form ω satisfying $d\omega = 0$ is said to be *closed.* Moreover, if there exists another form η, such that $\omega = d\eta$, then we say that ω is *exact.* The following result is immediate: *If ω is exact then ω is closed.* Poincaré lemma states that, under some conditions, the converse is also true. In this work, the following version of Poincaré lemma [8] will be sufficient:

Lemma 1. *Let ω be a k-form with $k \geq 1$. If ω is closed and defined on a starshaped open set of $\mathbb{R}^n$, then ω is exact. Hence, $\omega = d\eta \iff d\omega = 0$.*

In particular, if ω is defined everywhere on $\mathbb{R}^n$ then the notions of closedness and exactness coincide and no cohomology issues occur.

3 Discretization of Stokes' Theorem

A first step in our discretization is to notice that K may have *corners* which can be seen as limiting cases of *continuously differentiable round corners.* This leads to consider the case where K is a k-dimensional hypercube in $\mathbb{R}^n$. In this context, Mansfield and Hydon [4] have replaced differential forms

$$\sum_{1\leq i_1<\cdots<i_k\leq n} \omega_{i_1,\ldots,i_k}(x_1,\ldots,x_n)\, dx_{i_1}\wedge\cdots\wedge dx_{i_k},$$

where $(x_1,\ldots,x_n)\in\mathbb{R}^n$, by *difference* forms

$$\sum_{1\leq i_1<\cdots<i_k\leq n} \omega_{i_1,\ldots,i_k}(x_1,\ldots,x_n)\, \Delta_{i_1}\wedge\cdots\wedge \Delta_{i_k},$$

where $(x_1,\ldots,x_n)\in\mathbb{Z}^n$ instead of $\mathbb{R}^n$ and $\Delta_i = \Delta_{x_i}$ for $i = 1,\ldots,n$ denote the finite partial difference operators with respect to the i-th component, that is $\Delta_i u(x_1,\ldots,x_n) = u(x_1,\ldots,x_i+1,\ldots,x_n) - u(x_1,\ldots,x_i,\ldots,x_n)$ for every function $u = u(x_1,\ldots,x_n)$.

We propose a variant of this approach in which the difference forms are systematically replaced by *families of weights* defined on hypercubes. First, consider the set

$\mathcal{H}_{k,n} = \{H \mid H \text{ is a unit hypercube} \subseteq \mathbb{R}^n \text{ of dim } k \text{ with vertices in } \mathbb{Z}^n\}$. An *hypercube* $H\in\mathcal{H}_{k,n}$ can be seen as a *generalized pixel* and is denoted

$$\begin{aligned} H &= \mathrm{Pix}_{i_1,\ldots,i_k}(\alpha_1,\ldots,\alpha_n) \\ &= \left\{ (x_1,\ldots,x_n) \,\middle|\, \begin{array}{ll} x_j = \alpha_j, & \text{if } j\notin\{i_1,\ldots,i_k\} \\ \alpha_j \leq x_j \leq \alpha_j + 1, & \text{if } j\in\{i_1,\ldots,i_k\} \end{array} \right\}, \end{aligned}$$

where $(\alpha_1,\ldots,\alpha_n) \in \mathbb{Z}^n$ and $\{i_1 < \cdots < i_k\} \subseteq \{1,\ldots,n\}$ denotes the ordered set of indices of coordinates varying in the hypercube. We say that the hypercube H comes from the point $(\alpha_1,\ldots,\alpha_n)$ according to the directions $i_1,\ldots,i_k$. The point $(\alpha_1,\ldots,\alpha_n)$ is also called *principal corner* of the hypercube and the two orientations of H are denoted H and $-H$. Using this convention, an *hypercubic configuration* denoted $\mathbf{P}$, of dimension k in $\mathbb{R}^n$, with vertices in $\mathbb{Z}^n$, is defined as a finite $\mathbb{Z}$-linear combination of hypercubes $\in \mathcal{H}_{k,n}$ and $\mathbb{Z}\mathcal{H}_{k,n}$ denotes the set of hypercubic configurations of dimension k in $\mathbb{R}^n$. Changing sign corresponds to changing orientation. Obviously, if all the coefficients are equal to 1, an hypercubic configuration is interpreted as a union of hypercubes. In the general case, an hypercubic configuration can be interpreted as a union of hypercubes with several multiplicities and orientations. Since the boundary of a region is a natural way to describe objects, we define the *boundary* operator ∂ as follows.

Definition 1. *The* boundary ∂H *of an hypercube* $H = \mathrm{Pix}_{i_1,\ldots,i_k}(\alpha_1,\ldots,\alpha_n)$ *is the hypercubic configuration*

$$\partial H = \partial\,\mathrm{Pix}_{i_1,\ldots,i_k}(\alpha_1,\ldots,\alpha_n) = \sum_{\nu=1}^{k}(-1)^{\nu-1}\Delta_{i_\nu}\,\mathrm{Pix}_{i_1,\ldots,\widehat{i_\nu},\ldots,i_k}(\alpha_1,\ldots,\alpha_n).$$

More generally, by linearity, the boundary *of an hypercubic configuration* $\mathbf{P} = \sum_j n_j H_j, n_j \in \mathbb{Z}$ *is defined by* $\partial\mathbf{P} = \sum_j n_j \partial H_j$.

For example, let $n = k = 2$ and $\mathbf{P} = \mathrm{Pix}_{1,2}(1,1) + \mathrm{Pix}_{1,2}(2,1) + \mathrm{Pix}_{1,2}(1,2)$, then, after simplifications,

$$\begin{aligned}\partial\mathbf{P} = {} & \mathrm{Pix}_1(1,1) + \mathrm{Pix}_1(2,1) + \mathrm{Pix}_2(3,1) - \mathrm{Pix}_1(2,2) + \mathrm{Pix}_2(2,2)\\ & - \mathrm{Pix}_1(1,3) - \mathrm{Pix}_2(1,2) - \mathrm{Pix}_2(1,1),\end{aligned}$$

which corresponds to Figure 1 (a). This figure can be reinterpreted as a conventional oriented path as in Figure 1 (b).

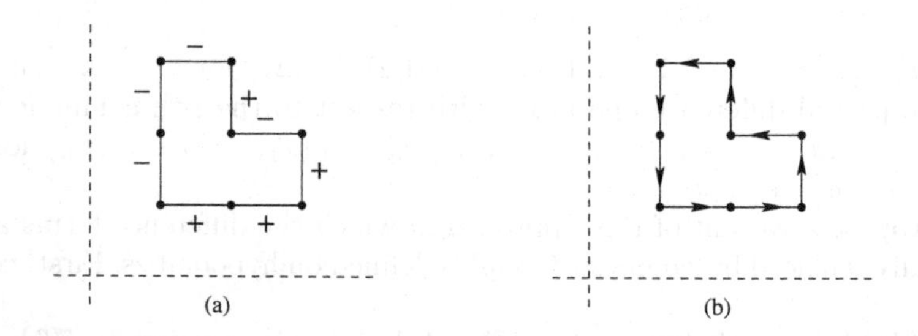

Fig. 1. Boundary: (a) linear combination of segments (b) oriented path

Since every differential k-form ω gives rise to a weight function w defined on hypercubes of dimension k by $w(H) = \int_H \omega$, this allows us to replace integrals by weight functions on hypercubes in the following way.

Definition 2. *An arbitrary* $w : \mathcal{H}_{k,n} \longrightarrow \mathbb{R}$ *is called a* weight function *on* k*-dimensional hypercubes in* $\mathbb{R}^n$.

Note that such a weight function is equivalent to a family $(w_{j_1,\ldots,j_k})_{1\leq j_1<\cdots<j_k\leq n}$ made of $\binom{n}{k}$ functions $w_{j_1,\ldots,j_k} : \mathbb{Z}^n \longrightarrow \mathbb{R}$ defined by

$$(\alpha_1,\ldots,\alpha_n) \mapsto w_{j_1,\ldots,j_k}(\alpha_1,\ldots,\alpha_n) = w(\mathrm{Pix}_{j_1,\ldots,j_k}(\alpha_1,\ldots,\alpha_n)).$$

More generally, we can take functions $w_{j_1,\ldots,j_k} : \mathbb{Z}^n \longrightarrow \mathbb{A}$, where $\mathbb{A}$ is an arbitrary commutative ring or a vector space. For example, $\mathbb{A} = \mathbb{K}[q_1,\ldots,q_n]$ where $\mathbb{K}$ is a field or $\mathbb{A} = \mathbb{Z}/(2)$.

Definition 3. *The* exterior finite difference *operator, denoted* Δ*, acts on a weight function* $w : \mathcal{H}_{k,n} \longrightarrow \mathbb{R}$ *by producing a new weight function, denoted* Δw*, defined on the* $(k+1)$*-dimensional hypercubes in* $\mathcal{H}_{k+1,n}$ *by*

$$\Delta w = ((\Delta w)_{i_1,\ldots,i_{k+1}})_{1\leq i_1<\cdots<i_{k+1}\leq n},$$

where

$$(\Delta w)_{i_1,\ldots,i_{k+1}}(\alpha_1,\ldots,\alpha_n) = \sum_{\nu=1}^{k+1}(-1)^{\nu-1}\Delta_{i_\nu} w_{i_1,\ldots,\widehat{i_\nu},\ldots,i_{k+1}}(\alpha_1,\ldots,\alpha_n).$$

Proposition 1. *The exterior difference operator* Δ *satisfies* $\Delta^2 = 0$. ■

Definition 4. *Let* w *be a weight function on hypercubes. We say that* w *is* closed *if* $\Delta w = 0$ *and* exact *if* $\exists\, \rho$ *such that* $w = \Delta\rho$.

By linearity, we can extend w and Δw to hypercubic configurations.

Theorem 2. (Discrete version of Stokes' theorem) *For every* k*-dimensional hypercubic configuration* $\mathbf{P}$ *in* $\mathbb{R}^n$ *and every weight function* w *defined on* $(k-1)$*-dimensional hypercubic configurations in* $\mathbb{R}^n$, $(\Delta w)(\mathbf{P}) = w(\partial\mathbf{P})$. ■

As in [1,2,3], it is more useful to write $(\Delta w)(\mathbf{P})$ in the form $W(\mathbf{P})$, where W is a *given* weight function and $\mathbf{P}$ is an arbitrary hypercubic configuration. Then, the right member $w(\partial\mathbf{P})$ can be described as an algorithm to compute $W(\mathbf{P})$ using another weight function w, applied to the boundary $\partial\mathbf{P}$.
By analogy with the finite difference calculus, we can use the suggestive notation $w = \Sigma W$. Then $(\Delta w)(\mathbf{P}) = w(\partial\mathbf{P})$ can be rewritten as,

$$W(\mathbf{P}) = (\Sigma W)(\partial\mathbf{P}). \tag{1}$$

Before defining the *exterior summation* operator Σ we need to introduce the *partial summation* operator Σ_i: *for* $i = 1,\ldots,n$, Σ_i *applied on a function* f *produces a function* $\Sigma_i f : \mathbb{Z}^n \longrightarrow \mathbb{R}$ *defined by*

$$\Sigma_i f(\alpha_1,\ldots,\alpha_n) = \begin{cases} \sum_{\nu_i=0}^{\alpha_i-1} f(\alpha_1,\ldots,\nu_i,\ldots,\alpha_n) & \text{if } \alpha_i > 0,\\ 0 & \text{if } \alpha_i = 0,\\ -\sum_{\nu_i=\alpha_i}^{-1} f(\alpha_1,\ldots,\nu_i,\ldots,\alpha_n) & \text{if } \alpha_i < 0. \end{cases}$$

Lemma 2. *The operators Δ_i and Σ_i are related as follows*

$$\Delta_i \Sigma_i = \mathrm{id} \quad \textit{and} \quad \Sigma_i \Delta_i = \mathrm{id} - \mathrm{eval}_{i,0},$$

where $\mathrm{eval}_{i,0}$ *is the evaluation operator of the i-th component at the origin, defined by* $\mathrm{eval}_{i,0} f(\alpha_1, \dots, \alpha_i, \dots, \alpha_n) = f(\alpha_1, \dots, 0, \dots, \alpha_n)$. *Moreover, if* $i \neq j$, *then* $\Delta_j \Sigma_i = \Sigma_i \Delta_j$.

The vector space of weight functions on k-dimensional hypercubes in $\mathbb{R}^n$ is denoted by $\mathbb{W}_{k,n} = \{w | w : \mathcal{H}_{k,n} \longrightarrow \mathbb{R}\}$. We are now ready to define the exterior summation operator Σ.

Definition 5. *The* exterior finite summation *operator* $\Sigma : \mathbb{W}_{k,n} \longrightarrow \mathbb{W}_{k-1,n}$ *is defined by*

$$(\Sigma W)_{j_1,\dots,j_{k-1}}(\alpha_1, \dots, \alpha_n)$$
$$= (-1)^{k-1} \left(\sum_{j_{k-1}<i} \Sigma_i W_{j_1,\dots,j_{k-1},i}(\alpha_1, \dots, \alpha_i, 0, \dots, 0) \right).$$

In order to have (1), we made the assumption that W was exact. In general, this condition is difficult to check. Fortunately, in our context of everywhere defined weight functions on hypercubes, exactness is equivalent to closedness, which is a much easier condition to check. This is precisely the content of the following discrete version of Poincaré lemma. It is noteworthy that our proof is non inductive in contrast with the corresponding proof given in [4] in the context of difference forms.

Theorem 3. *Let W be a weight function defined on hypercubes. Then W is closed if and only if W is exact.*

Proof. Obviously, $W = \Delta w \Longrightarrow \Delta W = \Delta^2 w = 0$. To establish the converse, we assume that $\Delta W = 0$ and show that W is exact. It suffices to see that $\Delta \Sigma W + \Sigma \Delta W = W$, since, taking $w = \Sigma W$, we will find $\Delta w = \Delta \Sigma W = \Delta \Sigma W + 0 = \Delta \Sigma W + \Sigma 0 = \Delta \Sigma W + \Sigma \Delta W = W$.
On one hand, we have,

$$\begin{aligned}
&(\Delta \Sigma W)_{i_1,\dots,i_k}(\alpha_1, \dots, \alpha_n) \\
&\quad = \sum_{1 \le \nu \le k} (-1)^{\nu-1} \Delta_{i_\nu} (\Sigma W)_{i_1,\dots,\widehat{i_\nu},\dots,i_k}(\alpha_1, \dots, \alpha_n) \\
&\quad = \sum_{1 \le \nu \le k, i_k < i \le n} (-1)^{\nu-1+k-1} \Delta_{i_\nu} \Sigma_i W_{i_1,\dots,\widehat{i_\nu},\dots,i_k,i}(\alpha_1, \dots, \alpha_i, 0, \dots, 0) \\
&\qquad + \sum_{\nu=k, i=i_k} (-1)^{2k-2} \Delta_{i_k} \Sigma_i W_{i_1,\dots,i_{k-1},\widehat{i_k},i}(\alpha_1, \dots, \alpha_i, 0, \dots, 0) \\
&\quad = S + \Delta_{i_k} \Sigma_{i_k} W_{i_1,\dots,i_k}(\alpha_1, \dots, \alpha_{i_k}, 0, \dots, 0), \quad \text{say} \\
&\quad = S + W_{i_1,\dots,i_k}(\alpha_1, \dots, \alpha_{i_k}, 0, \dots, 0).
\end{aligned}$$

On the other hand,

$$\begin{aligned}(\Sigma\Delta W)_{i_1,\dots,i_k}(\alpha_1,\dots,\alpha_n)\\ &= \sum_{i_k<i\le n, 1\le\nu\le k}(-1)^{k+\nu-1}\Sigma_i\Delta_{i_\nu}W_{i_1,\dots,\widehat{i_\nu},\dots,i_k,i}(\alpha_1,\dots,\alpha_i,0,\dots,0)\\ &\quad+\sum_{i_k<i\le n}(-1)^{2k}\Sigma_i\Delta_i W_{i_1,\dots,i_k,\widehat{i}}(\alpha_1,\dots,\alpha_i,0,\dots,0)\\ &= -S+\sum_{i_k<i\le n}\Sigma_i\Delta_i W_{i_1,\dots,i_k}(\alpha_1,\dots,\alpha_i,0,\dots,0).\end{aligned}$$

Hence, using Lemma 2,

$$\begin{aligned}&(\Delta\Sigma W)_{i_1,\dots,i_k}(\alpha_1,\dots,\alpha_n)+(\Sigma\Delta W)_{i_1,\dots,i_k}(\alpha_1,\dots,\alpha_n)\\ &= W_{i_1,\dots,i_k}(\alpha_1,\dots,\alpha_{i_k},0,\dots,0)+\sum_{i_k<i\le n}\Sigma_i\Delta_i W_{i_1,\dots,i_k}(\alpha_1,\dots,\alpha_i,0,\dots,0)\\ &= W_{i_1,\dots,i_k}(\alpha_1,\dots,\alpha_{i_k},0,\dots,0)\\ &\quad+\sum_{i_k<i\le n}(W_{i_1,\dots,i_k}(\alpha_1,\dots,\alpha_i,0,\dots,0)-W_{i_1,\dots,i_k}(\alpha_1,\dots,\alpha_{i-1},0,\dots,0))\\ &= W_{i_1,\dots,i_k}(\alpha_1,\dots,\alpha_n).\end{aligned}$$

■

We have the following useful variant of discrete Stokes' theorem.

Theorem 4. (Variant of the discrete Stokes' theorem) *If W is closed, then for every hypercubic configuration $\mathbf{P}$, we have $W(\mathbf{P}) = (\Sigma W)(\partial\mathbf{P})$.* ■

Note that the weight ΣW is only one of an infinite number of possible weights for the right-hand side of the equality in Theorem 4.

Corollary 1. *Every weight function $\Lambda : \mathcal{H}_{k-1,n} \longrightarrow \mathbb{R}$ satisfying*

$$W(\mathbf{P}) = \Lambda(\partial\mathbf{P})$$

is of the form $\Lambda = \Sigma W + \Theta$, where Θ is an exact weight function, that is of the form $\Theta = \Delta\Omega$, where $\Omega : \mathcal{H}_{k-2,n} \longrightarrow \mathbb{R}$. ■

4 Examples and Applications

Theorem 4 gives rise to many applications for the reduction of the computation of a weight function on hypercubic configurations to the computation of another weight function on the boundary of the configurations.

Our discrete version of Stokes' theorem generalizes the classical one in the particular case of hypercubic configurations. Indeed, let ω be a $(k-1)$- form in $\mathbb{R}^n$ and define the weight function $w : \mathcal{H}_{k-1,n} \longrightarrow \mathbb{R}$ by $w(H) = \int_H \omega$ for any

$(k-1)$-dimensional hypercube H. Then, it can be checked that $\Delta w : \mathcal{H}_{k,n} \longrightarrow \mathbb{R}$ satisfies $(\Delta w)(K) = \int_K d\omega$. Hence, $(\Delta w)(K) = w(\partial K)$ is equivalent to the classical Stokes' theorem for hypercubes. But when a weight function is not written in the form $w(H) = \int_H \omega$ or have values in an arbitrary commutative ring, Theorems 2 and 4 can be directly applied using no integration.

4.1 The Special Case $k = n$

Consider an arbitrary weight function $W \in \mathbb{W}_{n,n}$. Such a weight function is always closed since it reduces to a single function $W_{1,\ldots,n}$ defined on n-dimensional hypercubes in $\mathbb{R}^n$ and ΔW is an empty sum. Writing $\Phi = W_{1,\ldots,n}$, then without further hypotheses on Φ, Theorem 4 reduces to the following:

Proposition 2. *Let $\Phi : \mathcal{H}_{n,n} \longrightarrow \mathbb{R}$. Then for any n-dimensional hypercubic configuration $\mathbf{P}$ in $\mathbb{R}^n$, we have $\Phi(\mathbf{P}) = (\Sigma\Phi)(\Delta\mathbf{P})$. Explicitly,*

$$(\Sigma\Phi)_{j_1,\ldots,j_{n-1}}(\alpha_1,\ldots,\alpha_n) = \begin{cases} (-1)^{n-1}\Sigma_n\Phi(\alpha_1,\ldots,\alpha_n) & \text{if } (j_1,j_2,\ldots,j_{n-1}) = (1,2,\ldots,n-1), \\ 0 & \text{otherwise.} \end{cases}$$

Take $n = 3$ and consider, for example, the computation of the moment of inertia, $\mathrm{I}(\mathbf{P})$, relative to the x_1-axis, of a 3-dimensional cubical configuration $\mathbf{P} = \sum_{k=1}^{N} H_k$, which is a union of N distinct unit cubes each having mass 1. Taking $\Phi(\alpha_1,\alpha_2,\alpha_3) = \mathrm{I}(\mathrm{Pix}_{123}(\alpha_1,\alpha_2,\alpha_3)) = (\alpha_2+1/2)^2 + (\alpha_3+1/2)^2 + 1/6$, we find, using Proposition 2, that $\Sigma\Phi : \mathcal{H}_{2,3} \rightarrow \mathbb{R}$ is given by

$$(\Sigma\Phi)_{12}(\alpha_1,\alpha_2,\alpha_3) = \Sigma_3\Phi(\alpha_1,\alpha_2,\alpha_3) = \alpha_3(\alpha_2^2+\alpha_2+1/3\alpha_3^2+1/3),$$
$$(\Sigma\Phi)_{13}(\alpha_1,\alpha_2,\alpha_3) = 0, \quad (\Sigma\Phi)_{23}(\alpha_1,\alpha_2,\alpha_3) = 0.$$

Hence $\mathrm{I}(\mathbf{P}) = (\Sigma\Phi)(\partial\mathbf{P})$. This is often much shorter to compute since the number of squares in $\partial\mathbf{P}$ is generally $O(N^{2/3})$ compared to the number N of cubes in $\mathbf{P}$.

The moment of inertia of $\mathbf{P}$ relative to any axis (oblique or not) can be computed in a similar way. For the center of gravity $g(\mathbf{P}) = (\bar{x}_1,\bar{x}_2,\bar{x}_3)$ of the cubical configuration $\mathbf{P}$, one can take the vector-valued function $\Phi : \mathcal{H}_{3,3} \rightarrow \mathbb{R}^3$, defined by

$$\Phi(\alpha_1,\alpha_2,\alpha_3) = g(\mathrm{Pix}_{123}(\alpha_1,\alpha_2,\alpha_3)) = (\alpha_1+1/2,\alpha_2+1/2,\alpha_3+1/2).$$

In this case, $\Sigma\Phi : \mathcal{H}_{2,3} \rightarrow \mathbb{R}^3$ is given by

$$(\Sigma\Phi)_{12}(\alpha_1,\alpha_2,\alpha_3) = ((\alpha_1+1/2)\alpha_3,(\alpha_2+1/2)\alpha_3,1/2\alpha_3^2),$$
$$(\Sigma\Phi)_{13}(\alpha_1,\alpha_2,\alpha_3) = \mathbf{0}, \quad (\Sigma\Phi)_{23}(\alpha_1,\alpha_2,\alpha_3) = \mathbf{0}.$$

and we have $g(\mathbf{P}) = 1/N(\Sigma\Phi)(\partial\mathbf{P})$.

For the special case $k = n = 2$, taking $\Phi(\mathbf{P}) = \iint_{\mathbf{P}} f(x,y)\,dx\,dy$ and using the graphical convention of Figure 1, Proposition 2 generalizes the vertical algorithm

(V-algo) given in [1,2,3]: *for a lattice polyomino* $\mathbf{P} \subseteq \mathbb{R}^2$ *given by its contour* $(x_0, y_0), (x_1, y_1), \ldots, (x_n, y_n) = (x_0, y_0)$, *then*

$$\iint_{\mathbf{P}} f(x,y)\, dx\, dy = \sum_{\rightarrow} \Phi_r(x_i, y_i) + \sum_{\uparrow} \Phi_u(x_i, y_i) + \sum_{\leftarrow} \Phi_l(x_i, y_i) + \sum_{\downarrow} \Phi_d(x_i, y_i),$$

where functions Φ_r, Φ_u, Φ_l, Φ_d *are given by :*

$$\Phi_r = 0, \quad \Phi_d = \int_0^1 f_1(x, y+t)\, dt, \quad \Phi_l = 0, \quad \Phi_u = -\int_0^1 f_1(x, y-t)\, dt,$$

where $f_1(x,y) = \int^x f(u,y)\, du$. By Corollary 1, the general form for Λ such that $\iint_{\mathbf{P}} f(x,y)\, dx\, dy = \Lambda(\partial H)$, is given by $\Lambda = \Sigma\Phi + \Theta$, where $\Sigma\Phi$ is given (in the context of [1,2,3]) by the right-hand side of the above equation and $\Theta = \Delta\Omega$, with an arbitrary $\Omega : \mathcal{H}_{0,2} \longrightarrow \mathbb{R}$, that is, $\Omega : \mathbb{Z} \times \mathbb{Z} \longrightarrow \mathbb{R}$. Therefore, writing $\Omega = \Omega(\alpha_1, \alpha_2)$, then $\Theta : \mathcal{H}_{1,2} \longrightarrow \mathbb{R}$, where

$$\Theta_1 = (\Delta\Omega)_1 = (-1)^{1-1}\Delta_1\Omega(\alpha_1, \alpha_2) = \Omega(\alpha_1 + 1, \alpha_2) - \Omega(\alpha_1, \alpha_2),$$
$$\Theta_2 = (\Delta\Omega)_2 = (-1)^{1-1}\Delta_2\Omega(\alpha_1, \alpha_2) = \Omega(\alpha_1, \alpha_2 + 1) - \Omega(\alpha_1, \alpha_2).$$

This corresponds to Corollary 8 of [2] using the convention of Figure 1. For an example involving weight values in the ring $\mathbb{Z}[q_1, \ldots, q_n]$, consider the weight function $\Phi = W_{1,\ldots,n}$, where $\Phi(\alpha_1, \ldots, \alpha_n) = q_1^{\alpha_1} \ldots q_n^{\alpha_n}$. Then,

$$(\Sigma\Phi)_{1,\ldots,n-1}(\alpha_1, \ldots, \alpha_n) = (-1)^{n-1} q_1^{\alpha_1} \ldots q_{n-1}^{\alpha_{n-1}} [\alpha_n]_{q_n},$$

where $[\alpha]_q = 1 + q + \cdots + q^{\alpha-1}$ is the q-analogue of α. This generalizes the concept of q-area introduced in [1,2,3].

4.2 The Case $k < n$

We focus now on some specific families of weights in the context of Theorem 4 for $k < n$. Consider weights of the form

$$W_{i_1,\ldots,i_k}(\alpha_1, \ldots, \alpha_n) = c_{i_1,\ldots,i_k},$$

depending only on the directions $i_1, \ldots, i_k$ of the hypercubes. The condition $\Delta W = 0$ is satisfied and after some computations

$$(\Sigma W)_{j_1,\ldots,j_{k-1}}(\alpha_1, \ldots, \alpha_n) = (-1)^{k-1} \sum_{j_{k-1} < i} c_{j_1,\ldots,j_{k-1},i}\,\alpha_i.$$

For example, if $c_{i_1,\ldots,i_k} = 1$ for all $i_1, \ldots, i_k$ (= volume of an hypercube), then $W(\mathbf{P})$ is the signed volume of the hypercubic configuration $\mathbf{P}$. In other words, if $\mathbf{P} = \sum_{i=1}^m n_i H_i$ then $W(\mathbf{P}) = \sum_{i=1}^m n_i$. We then obtain,

$$(\Sigma W)_{j_1,\ldots,j_{k-1}}(\alpha_1, \ldots, \alpha_n) = (-1)^{k-1}(\alpha_{j_{k-1}+1} + \cdots + \alpha_n).$$

This generalizes the *H-algo* for the area of a polyomino given in [1,2,3].

Consider now weights of the form

$$W_{i_1,\ldots,i_k}(\alpha_1,\ldots,\alpha_n) = f(\alpha_1,\ldots,\alpha_n)$$

depending only on the principal corner of the hypercubes. We have two cases to consider:

(i) If $0 \le k \le n-2$, then it can be shown that $\Delta W = 0$ if and only if

$$f(\alpha_1,\ldots,\alpha_n) = \begin{cases} \varphi(\alpha_1+\cdots+\alpha_n) & \text{if } k \text{ odd}, \\ \text{a constant } c & \text{if } k \text{ even}, \end{cases}$$

where $\varphi : \mathbb{Z} \to \mathbb{R}$ is an arbitrary function. It turns out that

$$(\Sigma W)_{j_1,\ldots,j_{k-1}}(\alpha_1,\ldots,\alpha_n) = \begin{cases} F(\alpha_1+\cdots+\alpha_n) - F(\alpha_1+\cdots+\alpha_{j_{k-1}}) & \text{if } k \text{ odd} \le n-2, \\ c\cdot(\alpha_{j_{k-1}+1}+\cdots+\alpha_n) & \text{if } k \text{ even} \le n-2, \end{cases}$$

where $F(x) = \sum_x \varphi(x)$, $1 \le j_1 < \cdots < j_{k-1} \le n$.

(ii) If $k = n-1$ and $f(\alpha_1,\ldots,\alpha_n)$ is a polynomial, then it can be shown that $\Delta W = 0$ if and only if

$$f(\alpha_1,\ldots,\alpha_n) = (I + \Delta_2 - \Delta_3 + \cdots + (-1)^{n-1}\Delta_n)^{\alpha_1}\Psi(\alpha_2,\ldots,\alpha_n),$$

where $\Psi : \mathbb{Z}^{n-1} \to \mathbb{R}$ is an arbitrary polynomial.

5 Concluding Remarks

The results of this paper can be used to study many other statistics about hypercubic configurations in relation to parameters associated to their boundary. This is particularly true when the weight functions are polynomials written in the form

$$W_{i_1,\ldots,i_k}(\alpha_1,\ldots,\alpha_n) = \sum_{p_1,\ldots,p_n \ge 0} c^{i_1,\ldots,i_k}_{p_1,\ldots,p_n} \binom{\alpha_1}{p_1}\cdots\binom{\alpha_n}{p_n},$$

where $\binom{\alpha}{p} = \alpha(\alpha-1)\ldots(\alpha-p+1)/p!$. Since $\Delta_i\binom{\alpha_i}{p_i} = \binom{\alpha_i}{p_i-1}$, we have

$$\Delta W = 0 \Longleftrightarrow \sum_{\nu=1}^{k+1}(-1)^{\nu-1} c^{j_1,\ldots,\widehat{j_\nu},\ldots,j_{k+1}}_{p_1,\ldots,p_{j_\nu}+1,\ldots,p_n} = 0$$

for every $j_1 < \cdots < j_{k+1}$. Moreover, since $\sum_i \binom{\alpha_i}{p_i} = \binom{\alpha_i}{p_i+1}$, we have

$$(\Sigma W)_{j_1,\ldots,j_{k-1}}(\alpha_1,\ldots,\alpha_n) = \sum_{p_1,\ldots,p_n \ge 0} \gamma^{j_1,\ldots,j_{k-1}}_{p_1,\ldots,p_n} \binom{\alpha_1}{p_1}\cdots\binom{\alpha_n}{p_n}$$

where $\gamma_{0,\dots,0}^{j_1,\dots,j_{k-1}} = 0$ and for $(p_1,\dots,p_n) \neq (0,\dots,0)$,

$$\gamma_{p_1,\dots,p_n}^{(j_1,\dots,j_{k-1})} = (-1)^{k-1} \sum_{j_{k-1}<i} c_{p_1,\dots,p_{\nu_0}-1,0,\dots,0}^{(j_1,\dots,j_{k-1},i)}$$

where $\nu_0 = \min\{\nu | p_\nu \neq 0\}$.

Acknowledgements. The authors wish to thank the anonymous referees for the careful reading of the paper and their valuable comments.

References

1. Brlek, S., Labelle, G., Lacasse, A.: Incremental algorithms based on discrete Green theorem. In: Nyström, I., Sanniti di Baja, G., Svensson, S. (eds.) DGCI 2003. LNCS, vol. 2886, pp. 277–287. Springer, Heidelberg (2003)
2. Brlek, S., Labelle, G., Lacasse, A.: The discrete Green theorem and some applications in discrete geometry. Theoret. Comput. Sci. 346(2), 200–225 (2005)
3. Lacasse, A.: Contributions à l'analyse de figures discrètes en dimension quelconque. PhD thesis, Université du Québec à Montréal (September 2008)
4. Mansfield, E.L., Hydon, P.E.: Difference forms. Found. Comp. Math. (2007), http://personal.maths.surrey.ac.uk/st/P.Hydon/
5. Desbrun, M., Kanso, E., Tong, Y.: Discrete differential forms for computational modeling. In: Discrete differential geometry. Oberwolfach Semin, vol. 38, pp. 287–324. Birkhäuser, Basel (2008)
6. Kaczynski, T., Mischaikow, K., Mrozek, M.: Computing homology. Homology Homotopy Appl 5(2), 233–256 (2003); (electronic) Algebraic topological methods in computer science (Stanford, CA, 2001)
7. González-Díaz, R., Real, P.: On the cohomology of 3D digital images. Discrete Appl. Math. 147(2-3), 245–263 (2005)
8. Spivak, M.: Calculus on manifolds. A modern approach to classical theorems of advanced calculus. W. A. Benjamin, Inc., New York (1965)

Distances on Lozenge Tilings

Olivier Bodini[1], Thomas Fernique[2], and Éric Rémila[3]

[1] LIP6 - CNRS & Univ. Paris 6,
4 place Jussieu 75005 Paris - France
olivier.bodini@lip6.fr
[2] LIF - CNRS & Univ. de Provence,
39 rue Joliot-Curie 13453 Marseille - France
thomas.fernique@lif.univ-mrs.fr
[3] LIP - CNRS & Univ. de Lyon 1,
46 allée d'Italie 69364 Lyon - France
eric.remila@ens-lyon.fr

Abstract. In this paper, a structural property of the set of lozenge tilings of a $2n$-gon is highlighted. We introduce a simple combinatorial value called *Hamming-distance*, which is a lower bound for the the number of *flips* – a local transformation on tilings – necessary to link two tilings. We prove that the flip-distance between two tilings is equal to the Hamming-distance for $n \leq 4$. We also show, by providing a pair of so-called *deficient* tilings, that this does not hold for $n \geq 6$. We finally discuss the $n = 5$ case, which remains open.

1 Introduction

Lozenge tilings are widely used by physicists as a model for quasicrystals [9], following the celebrated Penrose tilings and its pentagonal symmetry. A basic operation acting over such tilings is the *flip*: whenever three lozenge tile a hexagon, rotating this hexagon by 180 yields a new lozenge tiling. This leads to define the *tiling space* of a fixed domain as the graph whose vertices are all the possible tilings of this domain, with two tilings being connected by an edge if and only if one can transform one into the other by a performing a flip. Combinatorial properties of tiling spaces are not trivial, especially for $n \geq 4$. The connectivity of these spaces has been proved in [10] in the case of finite and simply connected domains. Nevertheless, connectivity turns out to not hold for infinite tiling (*e.g.*, the Penrose tiling), even by allowing infinite sequences of flips [4].

In this paper, we consider the case of finite domains, more precisely $2n$-gons. Tilings are thus flip-connected. It is however generally unclear what is the flip-distance between two given tilings, *i.e.*, the minimal number of flips necessary to link both tilings. We give a natural lower bound, called the *Hamming-distance*, which relies on elementary geometrical considerations (namely de Bruijn lines), and we adress the question, whether this bound is tight or not.

S. Brlek, C. Reutenauer, and X. Provençal (Eds.): DGCI 2009, LNCS 5810, pp. 240–251, 2009.

The first result of this paper (section 4) is that this bound is tight for octogonal tilings ($n = 4$ case). This extends a previously known similar result for hexagonal tilings [11] ($n = 3$ case). However, the lack of a distributive lattice structure on the octogonal tiling space makes the proof more difficult and the result more surprising.

The second result of this paper (section 5) is that this bound is not tight for $n \geq 6$. Indeed, there exist pairs of tilings (not that much) such that their Hamming-distance is strictly lower than their flip-distance (we explicitly provide such a pair in the $n = 6$ case).

The $n = 5$ case remains open, because of a huge case-study that we did not manage to complete. We only here sketch (section 6) the way the proof should be carried out.

Let us stress that our results, although they looks like the ones obtained in [8], are different. Indeed, in [8], it is proven that, in any unitary $2n$-gon, there exists a fixed special tiling $\mathcal{T}_0$ such that the flip-distance and the Hamming-distance between any other tiling $\mathcal{T}$ and $\mathcal{T}_0$ are equal. But this situation cannot be extended to any arbitrary pairs of tilings of general (non-unitary) $2n$-gons.

2 $2n$-Gons, Tilings and de Bruijn Lines

For $n > 1$, let $V = (\boldsymbol{v}_1, \ldots, \boldsymbol{v}_n)$ be pairwise non-colinear vectors of $\mathbb{R}^2$ and $(m_1, \ldots, m_n)$ be positive integers. We call $(m_1, \ldots, m_n)$*-gon* the subset $Z(V, M)$ of $\mathbb{R}^2$ defined by:

$$Z(V,M) = \left\{ \sum_{k=1}^{n} \lambda_k \ \boldsymbol{v}_k \mid \lambda_k \in [-m_k, m_k] \right\}.$$

This is called, for short, a $2n$-gon. It is said to be *regular* if $\boldsymbol{v}_k = (\cos \frac{k\pi}{n}, \sin \frac{k\pi}{n})$ for $k = 1, \ldots, n$. We here consider only regular $2n$-gons. It is said to be *unitary* if $m_k = 1$ for $k = 1, \ldots, n$.

For $1 \leq i < j \leq n$, we denote by T_{ij} the *lozenge prototile* $T_{ij} = \{\lambda \boldsymbol{v}_i + \mu \boldsymbol{v}_j, -1 \leq \lambda, \mu \leq 1\}$. A *lozenge tiling* of a $2n$-gon is a set of translated copies of lozenge prototiles with pairwise disjoint interiors and whose union is the whole $2n$-gon. Let $\mathcal{T}$ be a lozenge tiling, the *vertices* (resp. *edges*) of $\mathcal{T}$ are the vertices (resp. edges) of the tiles which belong to $\mathcal{T}$.

The combinatorial structure of tilings of a $(m_1, \ldots, m_n)$-gon depends only on $(m_1, \ldots, m_n)$, and not on the v_i's. This important property, which does not hold in dimension 3 or higher [7], ensures that we can w.l.o.g. restrict our study to regular $2n$-gons.

Height functions for tilings have been introduced by Thurston [12]. Here, for each integer $k \in \{1, ..., n\}$ and each tiling $\mathcal{T}$ of $Z(V, M)$, we define the *k-located height function* $h_{\mathcal{T},k}$ as the function from vertices to $\mathbb{Z}$ such that, for any edge $[x, x + \boldsymbol{v}_i]$ of $\mathcal{T}$:

$$h_{\mathcal{T},k}(x + \boldsymbol{v}_i) = \begin{cases} h_{\mathcal{T},k}(x) + 1 & \text{if } i = k, \\ h_{\mathcal{T},k}(x) & \text{otherwise.} \end{cases}$$

We use the normalized k-located height function such that, for each vertex x, $h_{T,k}(x) \geq 0$ and there exists a vertex x_0 with $h_{T,k}(x_0) = 0$. The existence of height function and uniqueness of normalized height functions is well known [3].

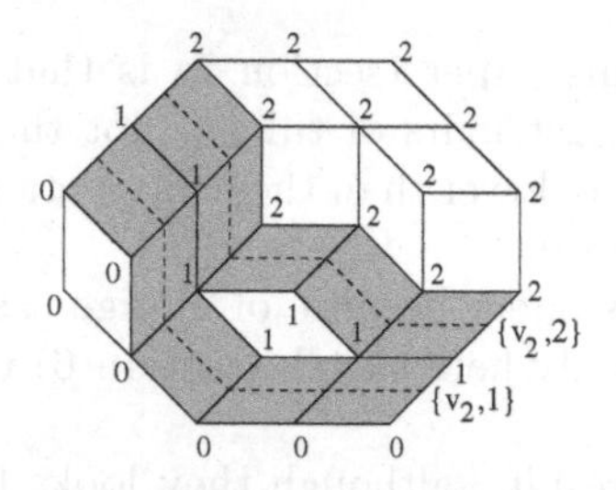

Fig. 1. The 2-located height function and two de Bruijn lines

Following [6], we define the *de Bruijn line* $\mathcal{S}_{i,j}$ of T as the set of tiles whose normalized i-located function is $j-1$ on one edge, and j on the opposite one. See Figure 1.

A de Bruijn line $\mathcal{S}_{i,j}$ is said of *type* i. It is interesting to note that two distinct de Bruijn lines of the same type do not intersect, while two de Bruijn lines of different types share a single tile. Conversely, each tile is the intersection of exactly two de Bruijn lines of different types.

Each de Bruijn line $\mathcal{S}_{i,j}$ disconnects $\mathcal{T}$

- $\triangle(\mathcal{S}_{i,j})$ is the set of tiles for which the i-located function is at least j on any vertex;
- $\nabla(\mathcal{S}_{i,j})$ is the set of tiles for which the i-located function is at most $j-1$ on any vertex.

Three de Bruijn lines of pairwise different types i, j, k define a sub-tiling of the zonotope, called *pseudo-triangle of type* ijk. A *minimal* (for inclusion) pseudo-triangle is reduced to three tiles.

3 Hamming-Distance and Flip-Distance

We introduce in this section two distance over tilings.

3.1 Flip-Distance

Two tiles are *adjacent* if they share an edge. Assume that three tiles of a tiling T are pairwise adjacent (*i.e.*, form a minimal pseudo-triangle). In this case, one can replace in a unique way these three tiles by three other tiles of the same type, to obtain another tiling T' of the same domain as $\mathcal{T}$. This operation is called a *flip* (see. Fig.3). The *tiling space* of $Z(V, M)$ is the undirected graph whose vertices are tilings of $Z(V, M)$, with two of them being connected by an edge if and only if they differ by a flip.

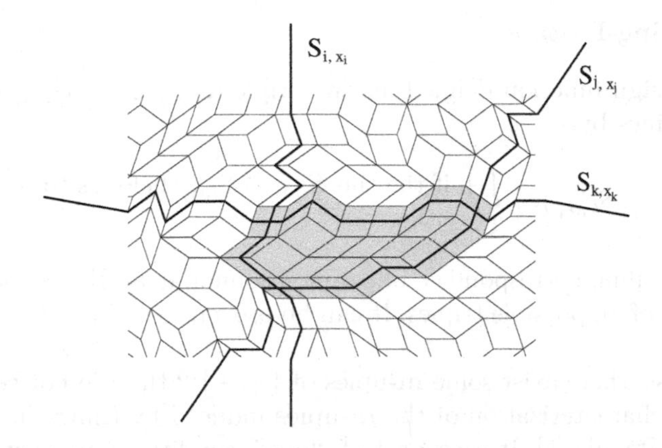

Fig. 2. In gray, a pseudo-triangle

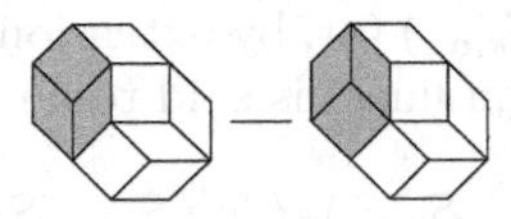

Fig. 3. Two neighbor tilings. One can pass from one to the other one by a single flip

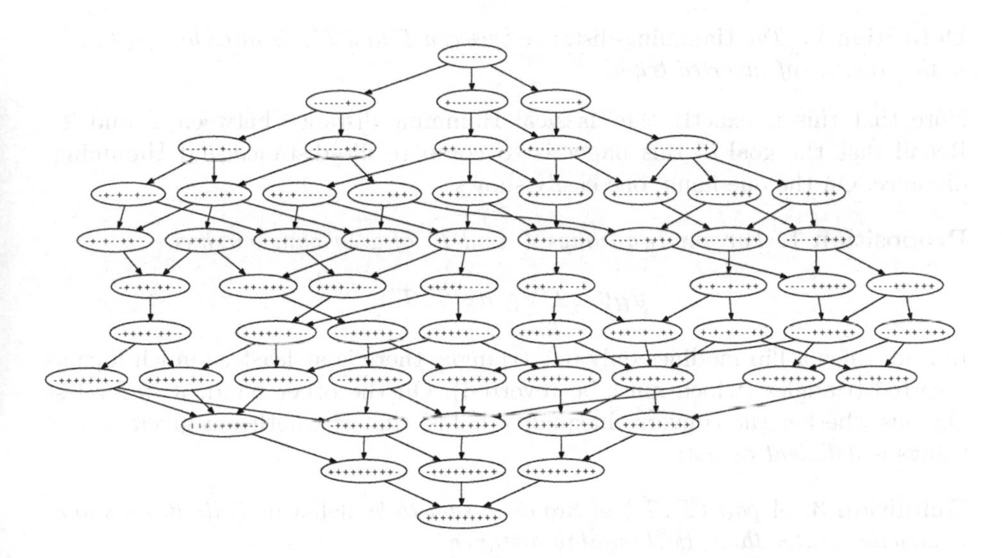

Fig. 4. The tiling space of the unitary decagon

Definition 1. *The* flip-distance *between two tilings $\mathcal{T}_1$ and $\mathcal{T}_2$ of $Z(V, M)$, denoted by $d_F(\mathcal{T}_1, \mathcal{T}_2)$, is the length of the shortest path connecting $\mathcal{T}_1$ to $\mathcal{T}_2$ in the tiling space. It is a finite value because the tiling space is connected [10].*

The figure 4 illustrates the topology of such a graph.

3.2 Hamming-Distance

Let $\mathbb{T}$ be the sign function defined on any triple $(\mathcal{S}_{i,\alpha_i}, \mathcal{S}_{j,\alpha_j}, \mathcal{S}_{k,\alpha_k})$, $i < j < k$, of de Bruijn lines by:

$$\mathbb{T}(\mathcal{S}_{i,\alpha_i}, \mathcal{S}_{j,\alpha_j}, \mathcal{S}_{k,\alpha_k}) = \begin{cases} + \text{ if the tile } \mathcal{S}_{i,\alpha_i} \cap \mathcal{S}_{j,\alpha_j} \text{ belongs to } \triangle(\mathcal{S}_{k,\alpha_k}), \\ - \text{ otherwise.} \end{cases}$$

Thus, to each tiling corresponds a one dimensionnal array $\mathbb{T}$ of $+$ or $-$, indexed on the set $\mathcal{L}_{\mathbf{m}}$ of all possible triples. It can be proven that $\mathbb{T}$ totally characterizes the tiling.

Nevertheless, there exist some $\mathbf{m}$-uples of $\{-,+\}^{\mathbf{m}}$ that do not correspond to any tiling. A characterization of the $\mathbf{m}$-uples induced by tilings has been given by Chavanon-Rémila [5]. It uses a set of "local" conditions, in a sense that each of them involve finitely many entries of the array.

Let $\mathcal{T}$ and $\mathcal{T}'$ be two tilings. Let respectively $\mathbb{T}$ and $\mathbb{T}'$ be the corresponding arrays. A triple $(\mathcal{S}_{i,\alpha_i}, \mathcal{S}_{j,\alpha_j}, \mathcal{S}_{k,\alpha_k})$ (or, by extension, the pseudo-triangle defined by the corresponding de Bruijn lines) is said to be *inverted* when

$$\mathbb{T}(\mathcal{S}_{i,\alpha_i}, \mathcal{S}_{j,\alpha_j}, \mathcal{S}_{k,\alpha_k}) \neq \mathbb{T}'(\mathcal{S}_{i,\alpha_i}, \mathcal{S}_{j,\alpha_j}, \mathcal{S}_{k,\alpha_k}).$$

This leads to define:

Definition 2. *The* Hamming-distance *between $\mathcal{T}$ and $\mathcal{T}'$, denoted by $d_H(\mathcal{T}, \mathcal{T}')$, is the number of inverted triples.*

Note that this is exactly the classical Hamming-distance between $\mathbb{T}$ and $\mathbb{T}'$. Recall that the goal of this paper is to compare flip-distance and Hamming-distance. On the one hand, one easily shows:

Proposition 1. *For any two tilings $\mathcal{T}$ and $\mathcal{T}'$ of a $2n$-gon, one has:*

$$d_H(\mathcal{T}, \mathcal{T}') \leq d_F(\mathcal{T}, \mathcal{T}').$$

Indeed, since a flip modifies only one triangle, there is at least as much flips as inverted triangles (which must be reverted). On the other hand, it is far less obvious whether the converse inequality holds, that is whether a given pair of tilings is *deficient* or not:

Definition 3. *A pair $(\mathcal{T}_1, \mathcal{T}_2)$ of tilings is said to be* deficient *if its flip-distance is strictly greater than its Hamming-distance.*

4 The Octogonal Case

In this section, we show that there is no deficient pair of tilings of a 8-gon:

Proposition 2 ([11]). *For hexagons (i. e. $n = 3$), the Hamming-distance between two lozenge tilings is equal to the flip-distance between them.*

This result, which also holds for any polygon, is strongly related to the structure of distributive lattice of the space of lozenge tilings, (for $n = 3$). It also can be interpreted in terms of "stepped surface" [1]: the Hamming-distance is exactly the volume of the solid inbetween the stepped surfaces defined by the tilings. It is possible to crush this gap by deforming the stepped surfaces decreasing the current volume. But for the general $2n$-gons one cannot use the same type of arguments, since the lattice structure and the volume interpretation are both lost.

Distances between the tilings of the unitary 8-gon. It is well-known that there exists only 8 tilings of the unitary 8-gon centered in Ω. These 8 tilings are isometrically equivalent. They can be obtain of acting the dihedral group of order 16 (which is isomorph to the group of isometry that preserves the octagon) from one of them. More precisely, the flip which is in general a local move can be seen here as the global map $s \circ \rho$ or $s \circ \rho^{-1}$ where ρ is the rotation of angle $2\pi/2n$ centered in Ω and s is the central symmetry of center Ω. The tiling space of these tilings is a cycle (of length 8) which is also the orbit of a tiling under $s \circ \rho$. We can easily remark that the Hamming-distance between two unitary 8-gon tilings is equal to their flip-distance. Up to isometry, there only exists 4 types of pair of tilings which corresponds to important configurations related to the next lemma 1.

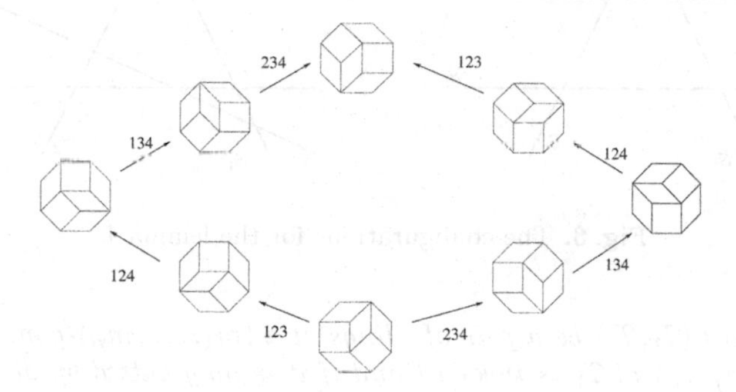

Fig. 5. The tiling space of the unitary octagon

Distances between general 8-gon tilings. The previous result, for unitary 8-gons, turns out to be true for general 8-gons. Let $(\mathcal{T}_1, \mathcal{T}_2)$ be a pair of tilings. Let us introduce another array $\mathbb{B}$ as follows:

- $\mathbb{B}(\mathcal{S}_{i,\alpha_i}, \mathcal{S}_{j,\alpha_j}, \mathcal{S}_{k,\alpha_k}) = 1$ if the pseudo-triangle $(\mathcal{S}_{i,\alpha_i}, \mathcal{S}_{j,\alpha_j}, \mathcal{S}_{k,\alpha_k})$ is inverted (*i.e.*, $\mathbb{T}(\mathcal{S}_{i,\alpha_i}, \mathcal{S}_{j,\alpha_j}, \mathcal{S}_{k,\alpha_k}) \neq \mathbb{T}'(\mathcal{S}_{i,\alpha_i}, \mathcal{S}_{j,\alpha_j}, \mathcal{S}_{k,\alpha_k})$.),
- $\mathbb{B}(\mathcal{S}_{i,\alpha_i}, \mathcal{S}_{j,\alpha_j}, \mathcal{S}_{k,\alpha_k}) = 0$ otherwise.

Some consistence conditions (lemmas 1 and 2) in $\mathbb{B}$ are similar (and local) to those appearing in the characterization of $\mathbb{T}$.

Lemma 1. *Let $(\mathcal{T}_1, \mathcal{T}_2)$ be a pair of tilings of a $(m_1, \ldots, m_n)$-gon. Let us consider four de Bruijn lines, say S_1, S_2, S_3, S_4. We assume that their relative positions in $\mathcal{T}_1$ are such that (see respectively Fig.6a and 6b):*

1. *S_3 and S_4 have the same type and S_4 cuts the pseudo-triangle (S_1, S_2, S_3);*
2. *the pseudo-triangle (S_2, S_3, S_4) is included in the pseudo-triangle (S_1, S_2, S_3).*

Then:

1. *$(\mathbb{B}(S_1, S_2, S_3),\ \mathbb{B}(S_1, S_2, S_4))$ belongs to $\{(0,0), (0,1), (1,1)\}$;*
2. *$(\mathbb{B}(S_1, S_3, S_4),\ \mathbb{B}(S_1, S_2, S_4),\ \mathbb{B}(S_1, S_2, S_3),\ \mathbb{B}(S_2, S_3, S_4))$ belongs either to $\{(0,0,0,0), (0,0,0,1), (0,0,1,1), (0,1,1,1)\}$, or to the set obtained by exchanging 0's and 1's (in other words, it is a monotonic sequence).*

Proof. First, note that we have to keep the order between the de Bruijn lines of a same type in both tilings. Thus, it suffices to consider only four lines of different types, what is equivalent to consider the case of the unitary 8-gon. Since there is only eight tilings of the unitary 8-gon, the claim can be check by an exhaustive case-study.

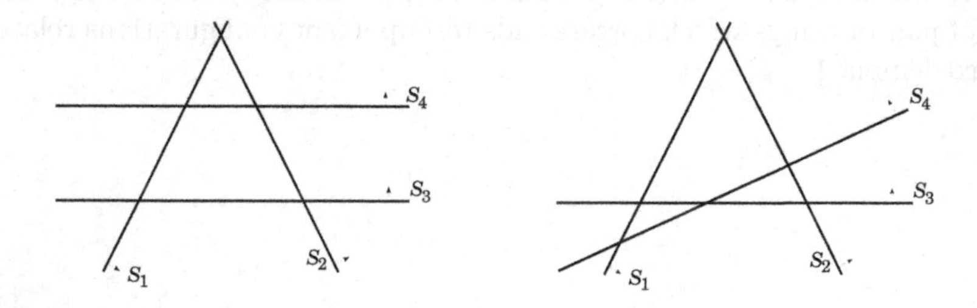

Fig. 6. The configurations for the lemma 1

Lemma 2. *Let $(\mathcal{T}_1, \mathcal{T}_2)$ be a pair of tilings of a $(m_1, \ldots, m_n)$-gon. If a pseudo-triangle (S_i, S_j, S_k) of $\mathcal{T}_1$ is inverted and if it is only cutted by de Bruijn lines of type i, j or k, then every pseudo-triangle (S'_i, S'_j, S'_k) included in (S_i, S_j, S_k) is inverted. In particular, (S_i, S_j, S_k) contains an inverted minimal sub-pseudo-triangle.*

Proof. The fact that every pseudo-triangle (S'_i, S'_j, S'_k) included in (S_i, S_j, S_k) is inverted follows from lemma 1, by induction on the number of de Bruijn lines.

Theorem 1. *Let $\mathcal{T}_1$ and $\mathcal{T}_2$ be two tilings of a $(m_1, \ldots, m_4)$-gon. Then, their Hamming-distance is equal to their flip-distance.*

Proof. We first prove that for every pair of distinct tilings $(\mathcal{T}_1, \mathcal{T}_2)$ of the 8-gon, $\mathcal{T}_1$ contains an inverted minimal pseudo-triangle (a closer flip is feasible on it).

Considering all these previous lemmas, an induction on the number m_4 of de Bruijn lines of type 4 can be done.

For initialization, if $m_4 = 0$, the 8-gon is actually a hexagon for which the result is previously known. Suppose that $m_4 = 1$ and when we remove the de Bruijn line S of type 4, $\mathcal{T}_1 \backslash S$ are identical to $\mathcal{T}_2 \backslash S$. In this case, the positions of the de Bruijn line S in $\mathcal{T}_1$ and $\mathcal{T}_2$ mark the boundary of a stepped surface U in $\mathcal{T}_1 \backslash S$ which is a hexagonal tiling (see figure 7).

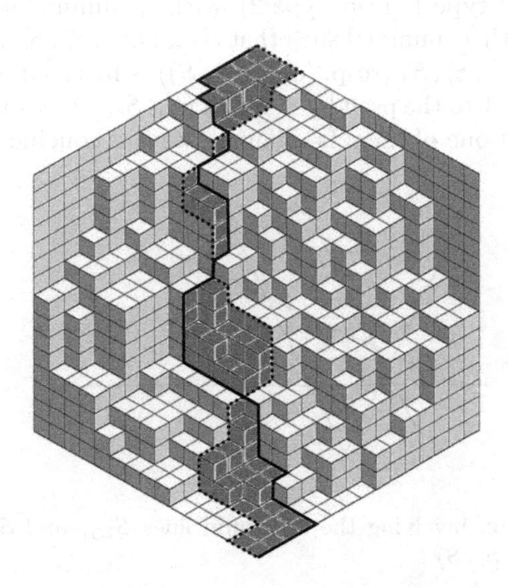

Fig. 7. The bold (resp. dotted) line indicates the position of the de Bruijn line S in $\mathcal{T}_1$ (resp. in $\mathcal{T}_2$). The stepped surface U is in dark gray inbetween the two de Bruijn lines.

The flip-distance and the Hamming-distance between $\mathcal{T}_1$ and $\mathcal{T}_2$ are exactly the number of tiles in U. Indeed, one easily proves that, in any case, a flip can be done, corresponding to a local transformation on the position of the de Bruijn line S in $\mathcal{T}_2$ (i.e. surrounding a unique tile by the other side).

Now, for the other cases, we can remove a de Bruijn line $S = \mathcal{S}_{4,\alpha}$, of type 4, in both tilings $\mathcal{T}_1$, $\mathcal{T}_2$ in such a way that $\mathcal{T}_1 \backslash S$ are distinct to $\mathcal{T}_2 \backslash S$ (this is always possible when $m_4 > 1$). Henceforward, we are going to only work on the tiling $\mathcal{T}_1$. By hypothesis of induction, there exists an inverted minimal pseudo-triangle (S_a, S_b, S_c), in the tiling $\mathcal{T}_1$ obtained by removing of S and sticking the two remaining parts of the initial tiling.

After this, let us replace the removed de Bruijn line S. The only tricky case arises when S cuts the pseudo-triangle (S_a, S_b, S_c) and the types of de Bruijn lines S_a, S_b and S_c are respectively 1, 2 and 3. Moreover, the minimal sub-pseudo-triangle of (S_a, S_b, S_c) (which is (S_a, S_b, S) or (S_b, S_c, S)) is not inverted (in any other configuration, the existence of an inverted minimal pseudo-triangle is trivial).

We can consider without loss of generality that $\mathbb{T}(S_a, S_b, S_c) = +$ and that (S_a, S_b, S) is the non-inverted minimal sub-pseudo-triangle of (S_a, S_b, S_c). The 3 other cases are in fact isometrically equivalent.

Since (S_a, S_b, S) is not inverted, we have by lemma 1.2 applying on the pseudo-lines $\{S_a, S_b, S_c, S\}$ that the pseudo-triangle (S_a, S_c, S) is inverted. If this pseudo-triangle is minimal, then the result ensues. Otherwise, (S_a, S_c, S) can only be cut by de Bruijn lines of type 1 or 2, because of the minimality of the pseudo-triangle (S_a, S_b, S_c) in $\mathcal{T}_1 \backslash S$. Let $\mathcal{S}_{1,j_1}$ (resp. $\mathcal{S}_{2,j_2}$) be (if there exists) the de Bruijn line of type 1 (resp. type 2), with j_1 minimal such that $\mathcal{S}_{1,j_1}$ cuts (S_a, S_c, S) (resp. with j_2 minimal such that $\mathcal{S}_{2,j_2}$ cuts (S_a, S_c, S)) (see fig.8). The pseudo-triangle $(\mathcal{S}_{1,j_1}, S_c, S)$ (resp. $(\mathcal{S}_{2,j_2}, S_c, S)$) is inverted. Indeed this follows of applying lemma 1.1 to the pseudo-triangles (S_a, S_c, S) (resp. (S_b, S_c, S)) which are both inverted. If one of them is minimal, we can conclude.

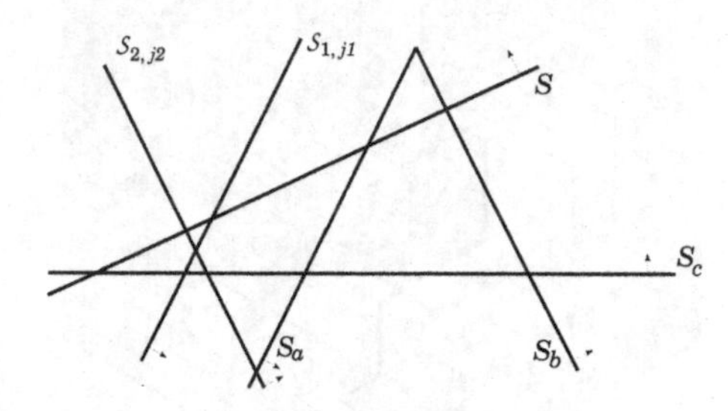

Fig. 8. A configuration involving the de Bruijn lines $\mathcal{S}_{1,j_1}$ and $\mathcal{S}_{2,j_2}$ such that their intersection is in (S_a, S_c, S)

Otherwise, the tile $\mathcal{S}_{1,j_1} \cap \mathcal{S}_{2,j_2}$ belongs to $\triangle(S_c) \cap \triangledown(S)$ and necessarily at least one of the pseudo-triangles $(\mathcal{S}_{1,j_1}, \mathcal{S}_{2,j_2}, S_c)$ and $(\mathcal{S}_{1,j_1}, \mathcal{S}_{2,j_2}, S)$ is inverted (by applying lemma 1.2 on $\{S_c, S, \mathcal{S}_{1,j_1}, \mathcal{S}_{2,j_2}\}$ with $(\mathcal{S}_{1,j_1}, S_c, S)$ inverted). But $(\mathcal{S}_{1,j_1}, \mathcal{S}_{2,j_2}, S_c)$ (resp. $(\mathcal{S}_{1,j_1}, \mathcal{S}_{2,j_2}, S)$) can only be cut by de Bruijn lines of type 1 (resp. type 2). Because of the minimality of j_1 and j_2, it includes by lemma 2 an inverted minimal sub-pseudo-triangle of type 123 (resp. type 124). Thus, we always have a inverted minimal pseudo-triangle in $\mathcal{T}_1$.

Now, we can prove the theorem. Assume that $(\mathcal{T}_1, \mathcal{T}_2)$ is a deficient pair of tiling of a $(m_1, \ldots, m_4)$-gon, with their flip-distance being minimal among such pairs. In particular, $\mathcal{T}_1$ could not contain any inverted minimal pseudo-triangle, because such a pseudo-triangle would correspond be flippable, thus yielding a tiling $\mathcal{T}_1'$ which would contradict the minimality of the flip-distance between $\mathcal{T}_1$ and $\mathcal{T}_2$. This is however impossible, according to what we proved above.

Corollary 1. *A $(m_1, \ldots, m_4)$-gon tiling (resp. $4 \to 2$ tiling of the plane $\mathbb{R}^2$) is uniquely determined by the value of $\mathbb{T}$ for its minimal pseudo-triangles.*

Proof. Consider $\mathcal{T}_1$ and $\mathcal{T}_2$ be a pair of tiling of a $(m_1, \ldots, m_4)$-gon with the same set M of minimal pseudo-triangle. If $\mathbb{T}_1(v) = \mathbb{T}_2(v)$ for every $v \in M$, then $\mathcal{T}_1$ has no inverted minimal pseudo-triangle. So, by theorem 1, $\mathcal{T}_1 = \mathcal{T}_2$. For the $4 \to 2$ tilings of the plane, it suffices to take the limit when (m_1, m_2, m_3, m_4) tends to infinity (no matter how).

5 The Dodecagonal Case (and beyond)

It turns out that there exist deficient pairs of tilings of the unitary 12-gon (and thus of any $2n$-gon, for $n \geq 6$). To shows this, we achieved (by using a computer) an exhaustive search of all the deficient pair of tilings of the unitary 12-gon. More precisely, this shows that there exist, up to isometry, only two deficient pairs of tilings. For these two pairs the Hamming-distance is equal to 16, while the flip-distance is equal to 18. The first pair (Fig. 9) yields, by symmetry, 12 different deficient pairs of tilings. The second pair (Fig.10) yields, by symmetry, 4 different deficient pairs of tilings. These 16 pairs of tilings are the only deficient ones, among the $(908)^2$ possible pairs of tilings.

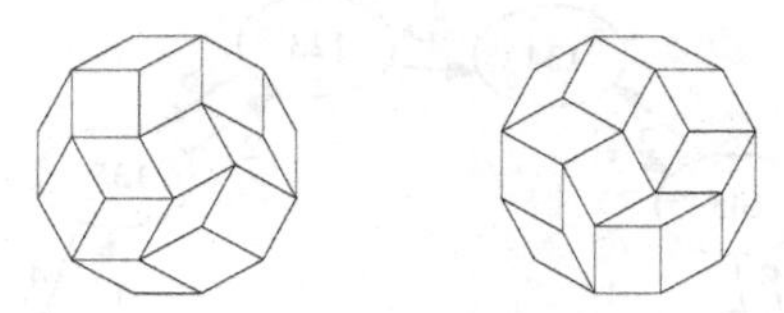

Fig. 9. The first deficient pair of tilings of the unitary 12-gon

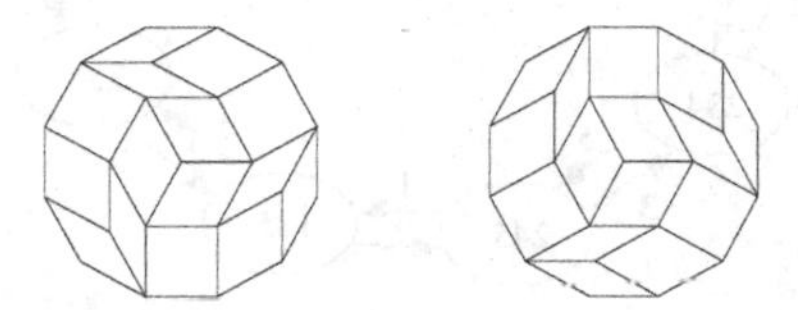

Fig. 10. The second deficient pair of tilings of the unitary 12-gon

6 Towards the Decagonal Case

The two previous sections solved the $n \leq 4$ and the $n \geq 6$ cases. We here discuss the $n = 5$ case, where we conjecture:

Conjecture 1. *The Hamming-distance between any two tilings of a* 10*-gon is equal to their flip-distance.*

The problem for proving (or disproving!) this conjecture is mainly technical. Indeed, a huge and tedious case-study need to be achieved. Let us just here state two proven lemmas, which should be the cornerstones of a complete proof.

Lemma 3 (Harp lemma). *Consider an inverted pseudo-triangle P of type ijk. Assume that it is cutted only by a set $\mathcal{S}$ of de Bruijn lines whose types are not in $\{i, j, k\}$. Then, the configuration formed by $\mathcal{S}$ and the three de Bruijn lines defining P contains an inverted minimal pseudo-triangle.*

Lemma 4 (10-cycle lemma). *Assume that there exist tuples* $(m_1, \ldots, m_5)$ *such that the* $(m_1, \ldots, m_5)$*-gon has deficient pairs of tilings. Let us consider a minimal such tuple. Consider a pair of tilings with minimal flip-distance among the deficient pairs of the* $(m_1, \ldots, m_5)$*-gon. In particular, the inverted pseudo-triangles of this pair cannot be minimal,* i.e., *flippable, because flipping such a pseudo-triangle would yield a deficient pair with a smaller flip-distance. Then, there is an infinite sequence* $(P_n)_n$ *of inverted pseudo-triangles such that:*

- P_n *is cutted by a de Bruijn line, say* S_n *;*
- P_{n+1} *is defined by* S_n *and two of the three lines defining* P_n*, with the types of these two lines being characterized by the types of* S_n *and* P_n *(Fig. 11).*

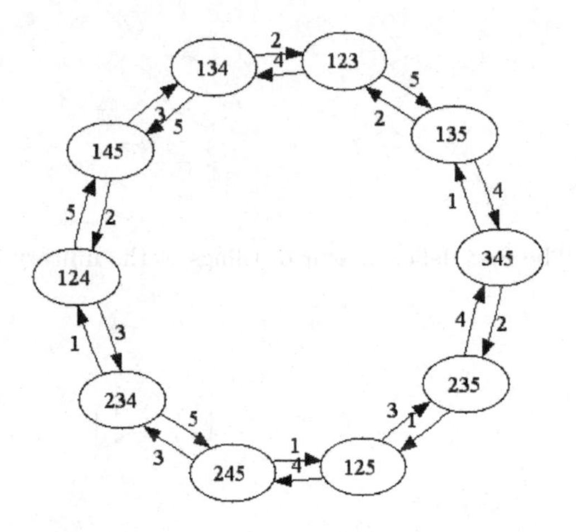

Fig. 11. The vertices of this 10-cycle are the possible types of the P_n's. Each edge links a P_n to P_{n+1}, with its label being the type of the de Bruijn line S_n which cuts P_n.

To fill the gap between the previous lemma and the above conjecture, it would remain to show that a pseudo-triangle cannot appear twice in the sequence (P_n), and thus that this sequence cannot be infinite. This would prove that decagons cannot have deficient pairs of tiling. However, proving (or disproving) that a pseudo-triangle cannot appear twice requires a huge case-study that has not yet been achieved (except for the unitary decagon, where an exhaustive computer-assisted search has shown that there is no deficient pair). This is a work in progress.

7 Conclusion

To conclude this paper, let us mention that the problem here adressed is only a part of the general study of combinatorics of lozenge tilings spaces, where a lot of questions remain so far open (*e.g.*, the size or the diameter of the set of tilings of a given domain, as well as the way to perform random sampling on this set *etc.*). More specifically, without going into details which are beyond the scope of this paper, our question is motivated by the problem of the growth of quasicrystals (seen as lozenge tilings), which can be roughly stated as follows: "given a random tiling, can we transform it into a certain specific tiling (quasicrystal) by performing only some restricted types of flips?"

References

1. Arnoux, P., Berthé, V., Fernique, T., Jamet, D.: Functional stepped surfaces, flips and generalized substitutions. Theor. Comput. Sci. 380, 251–265 (2007)
2. Björner, A., Las Vergnas, M., Sturmfels, B., White, N., Ziegler, G.M.: Oriented matroids. Encyclopedia of Mathematics, vol. 46. Cambridge University Press, Cambridge (1993)
3. Bodini, O., Latapy, M.: Generalized Tilings with height functions. Morfismos 7, 47–68 (2003)
4. Bodini, O., Fernique, T., Rémila, É.: Characterizations of flip-accessibility for rhombus tilings of the whole plane. Info. & Comput. 206, 1065–1073 (2008)
5. Chavanon, F., Rémila, É.: Rhombus tilings: decomposition and space structure. Disc. & Comput. Geom. 35, 329–358 (2006)
6. De Bruijn, N.G.: Dualization of multigrids. J. Phys. Fr. C3 47, 9–18 (1986)
7. Desoutter, V., Destainville, N.: Flip dynamics in three-dimensional random tilings. J. Phys. A: Math. Gen. 38, 17–45 (2005)
8. Felsner, S., Weil, H.: A theorem on higher Bruhat orders. Disc. & Comput. Geom. 23, 121–127 (2003)
9. Grnbaum, B., Shephard, G.C.: Tilings and patterns. W. H. Freeman, New York (1986)
10. Kenyon, R.: Tiling a polygon with parallelograms. Algorithmica 9, 382–397 (1993)
11. Rémila, É.: On the lattice structure of the set of tilings of a simply connected figure with dominoes. Theor. Comput. Sci. 322, 409–422 (2004)
12. Thurston, W.P.: Conway's tiling group. American Math. Month. 97, 757–773 (1990)

Jordan Curve Theorems with Respect to Certain Pretopologies on $\mathbb{Z}^2$

Josef Šlapal

Brno University of Technology, Department of Mathematics,
616 69 Brno, Czech Republic
slapal@fme.vutbr.cz
http://at.yorku.ca/h/a/a/a/10.htm

Abstract. We discuss four quotient pretopologies of a certain basic topology on $\mathbb{Z}^2$. Three of them are even topologies and include the well-known Khalimsky and Marcus-Wyse topologies. Some known Jordan curves in the basic topology are used to prove Jordan curve theorems that identify Jordan curves among simple closed ones in each of the four quotient pretopologies.

1 Introduction

When studying geometric and topological properties of (two-dimensional) digital images, we need the digital plane $\mathbb{Z}^2$ to be provided with a convenient structure. Here, the convenience means that such a structure satisfies some analogues of basic geometric and topological properties of the Euclidean topology on $\mathbb{R}^2$. For example, it is usually required that an analogue of the Jordan curve theorem be valid. (Recall that the classical Jordan curve theorem states that any simple closed curve in the Euclidean plane separates this plane into exactly two components). In the classical approach to this problem (see e.g. [11] and [12]), graph theoretic tools were used for structuring $\mathbb{Z}^2$, namely the well-known binary relations of 4-adjacency and 8-adjacency. Unfortunately, neither 4-adjacency nor 8-adjacency itself allows an analogue of the Jordan curve theorem - cf. [8]. To eliminate this deficiency, a combination of the two binary relations has to be used. Despite this inconvenience, the graph-theoretic approach was used to solve many problems of digital image processing and to create useful graphic software. In [5], a new, purely topological approach to the problem has been proposed which utilizes a convenient topology on $\mathbb{Z}^2$, called Khalimsky topology (cf. [4]), for structuring the digital plane. At present, this topology is one of the most important concepts of the theory called digital topology. It has been studied and used by many authors, e.g., [2] and [6]-[9]. The possibility of structuring $\mathbb{Z}^2$ by using closure operators more general than the Kuratowski ones is discussed in [13] and [14].

In [15], a new, convenient topology on $\mathbb{Z}^2$ has been introduced and studied. It was shown there that this topology has some advantages over the Khalimsky one. The new topology was further investigated in [16] where it was shown that

S. Brlek, C. Reutenauer, and X. Provençal (Eds.): DGCI 2009, LNCS 5810, pp. 252–262, 2009.

its quotient topologies include the Khalimsky topology as well as two other convenient topologies on $\mathbb{Z}^2$. A certain quotient pretopology of the new topology was also studied in [16]. In the present note we continue the study from [16] and prove a Jordan curve theorem for each of the three quotient topologies and also for the quotient pretopology from [16].

2 Preliminaries

For the topological terminology used we refer to [3]. Throughout the note, all topologies dealt with are thought of as being (given by) Kuratowski closure operators. Thus, a *topology* on a set X is a map p: $\exp X \to \exp X$ (where exp X denotes the power set of X) fulfilling the following four axioms:

(i) $p\emptyset = \emptyset$,
(ii) $A \subseteq pA$ for all $A \subseteq X$,
(iii) $p(A \cup B) = pA \cup pB$ for all $A, B \subseteq X$,
(iv) $ppA = pA$ for all $A \subseteq X$.

The pair (X, p) is then called a *topological space.* If p satisfies (i)-(iii) but not necessarily (iv), then it is called a *pretopology* and (X, p) is called a *pretopological space.* The topological concepts used for pretopologies in this note are obtained by natural extensions of these concepts defined for topologies. In the literature, pretopologies are also known as *Cech closure spaces* - cf. [1].

A map $f : (X, p) \to (Y, q)$ between topological (or pretopological, respectively) spaces (X, p) and (Y, q) is said to be *continuous* if $f(pA) \subseteq q(f(A))$ whenever $A \subseteq X$ - cf. [3] (or [1], respectively).

As usual, given a topological space (X, p) and a surjection $e : X \to Y$, a topology (pretopology) q on Y is called the *quotient topology* (*quotient pretopology*) of p generated by e if q is the finest topology (pretopology) on Y for which $e : (X, p) \to (Y, q)$ is continuous. Here, as usual, given topologies (pretopologies) q and r on X, q is said to be *finer* than r (and r *coarser* than q) if $qA \subseteq rA$ for every $A \subseteq X$.

We will need the following lemma resulting from [14], Corollary 1.5:

Lemma 1. *Let (X, p) be a pretopological space, $e : X \to Y$ be a surjection and let q be the quotient pretopology of p on $\mathbb{Z}^2$ generated by e. Let e have the property that $e^{-1}(\{y\})$ is connected in (X, p) for every point $y \in Y$ and let $B \subseteq Y$ be a subset. Then B is connected in (Y, q) if and only if $e^{-1}(B)$ is connected in (X, p).*

Let us note that, for a topological space (X, p) and a quotient topology of p, the statement of the previous Lemma need not be true. It is true if, for example, B is closed or open in (Y, q).

Recall that the connectedness graph of a pretopology p on X is the undirected simple graph whose vertex set is X and in which a pair of vertices x, y is adjacent (i.e., joined by an edge) if and only if $x \neq y$ and $\{x, y\}$ is a connected subset of (X, p). A pretopology p on X is called an *Alexandroff pretopology* if

$pA = \bigcup_{x \in A} p\{x\}$ for every $A \subseteq X$ and it is called a *T_0-pretopology* if $x \in p\{y\}$ and $y \in p\{x\}$ imply $x = y$ whenever $x, y \in X$. So, if p is an Alexandroff pretopology, then it is given by determining closures of all points of $\mathbb{Z}^2$. Thus, p is then also given by its connectedness graph provided that every edge of the graph is adjacent to a point which is known to be closed or to a point which is known to be open (in which case p is T_0). Therefore, in all the (connectedness) graphs displayed, the closed points are ringed while the mixed ones (i.e., the points that are neither closed nor open) are boxed. Thus, the points that are neither ringed nor boxed are open (note that no points may be both closed and open in a connected space with more than one point).

By a *(discrete) simple closed curve* in a pretopological space (X, p) we mean, in accordance with [16], a nonempty, finite and connected subset $C \subseteq X$ such that, for each point $x \in C$, there are exactly two points of C adjacent to x in the connectedness graph of p. A simple closed curve C in (X, p) is said to be a *(discrete) Jordan curve* if it separates (X, p) into precisely two components (i.e., if the subspace $X - C$ of (X, p) consists of precisely two components). In the remaining part of the paper, pretopologies on $\mathbb{Z}^2$ only will be dealt with.

Let $z = (x, y) \in \mathbb{Z}^2$ be a point. We put

$$H_2(z) = \{(x + k, y);\ k \in \{-1, 0, 1\}\},$$
$$V_2(z) = \{(x, y + l);\ l \in \{-1, 0, 1\}\},$$
$$D_4(z) = H_2(z) \cup \{(x - 1, y - 1), (x + 1, y - 1)\},$$
$$U_4(z) = H_2(z) \cup \{(x - 1, y + 1), (x + 1, y + 1)\},$$
$$L_4(z) = V_2(z) \cup \{(x - 1, y - 1), (x - 1, y + 1)\},$$
$$R_4(z) = V_2(z) \cup \{(x + 1, y - 1), (x + 1, y + 1)\}.$$

Next, we put

$$A_4(z) = H_2(z) \cup V_2(z),$$
$$A_8(z) = H_2(z) \cup L_4(z) \cup R_4(z) (= V_2(z) \cup D_4(z) \cup U_4(z)), \text{ and}$$
$$A_4'(z) = \{z\} \cup (A_8(z) - A_4(z)).$$

Thus, each of the sets introduced above consists of the point z and some other points whose number equals the index of the symbol denoting this set. In the literature, the points of $A_4(z)$ and $A_8(z)$ different from z are said to be *4-adjacent* and *8-adjacent* to z, respectively. It is natural to call the points, different from z, of $H_2(z)$, $V_2(z)$, $D_4(z)$, $U_4(z)$, $L_4(z)$, $R_4(z)$ and $A_4'(z)$ *horizontally 2-adjacent, vertically 2-adjacent, down 4-adjacent, up 4-adjacent, left 4-adjacent, right 4-adjacent* and *diagonally 4-adjacent* to z, respectively. Clearly, each of these adjacencies implies 8-adjacency.

3 Topology w and Some of Its Quotients

In this section, we repeat some results from [16] (without proofs) that will be needed in the sequel.

We denote by w the Alexandroff topology on $\mathbb{Z}^2$ given as follows:
For any point $z = (x, y) \in \mathbb{Z}^2$,

$$w\{z\} = \begin{cases} A_8(z) \text{ if } x = 4k,\ y = 4l,\ k, l \in \mathbb{Z}, \\ A'_4(z) \text{ if } x = 2 + 4k,\ y = 2 + 4l,\ k, l \in \mathbb{Z}, \\ D_4(z) \text{ if } x = 2 + 4k,\ y = 1 + 4l,\ k, l \in \mathbb{Z}, \\ U_4(z) \text{ if } x = 2 + 4k,\ y = 3 + 4l,\ k, l \in \mathbb{Z}, \\ L_4(z) \text{ if } x = 1 + 4k,\ y = 2 + 4l,\ k, l \in \mathbb{Z}, \\ R_4(z) \text{ if } x = 3 + 4k,\ y = 2 + 4l,\ k, l \in \mathbb{Z}, \\ H_2(z) \text{ if } x = 2 + 4k,\ y = 4l,\ k, l \in \mathbb{Z}, \\ V_2(z) \text{ if } x = 4k,\ y = 2 + 4l,\ k, l \in \mathbb{Z}, \\ \{z\} \text{ otherwise.} \end{cases}$$

Clearly, w is connected and T_0. A portion of the connectedness graph of w is shown in the following figure:

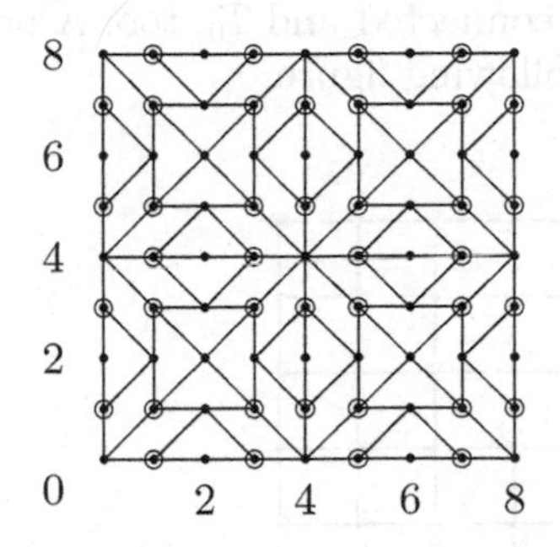

Recall [5] that the Khalimsky topology on $\mathbb{Z}^2$ is the Alexandroff topology t given as follows:
For any $z = (x, y) \in \mathbb{Z}^2$,

$$t\{z\} = \begin{cases} A_8(z) \text{ if } x, y \text{ are even,} \\ H_2(z) \text{ if } x \text{ is even and } y \text{ is odd,} \\ V_2(z) \text{ if } x \text{ is odd and } y \text{ is even,} \\ \{z\} \text{ otherwise.} \end{cases}$$

The Khalimsky topology is connected and T_0; a portion of its connectedness graph is shown in the following figure.

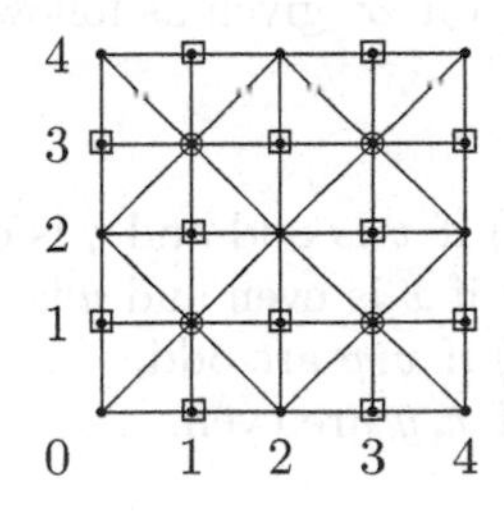

Theorem 1. *The Khalimsky topology is the quotient topology of w generated by the surjection $f : \mathbb{Z}^2 \to \mathbb{Z}^2$ given as follows:*

$$f(x,y) = \begin{cases} (2k,2l) \text{ if } (x,y) = (4k,4l), \ k,l \in \mathbb{Z}, \\ (2k,2l+1) \text{ if } (x,y) \in A_4(4k,4l+2), \ k,l \in \mathbb{Z}, \\ (2k+1,2l) \text{ if } (x,y) \in A_4(4k+2,4l), \ k,l \in \mathbb{Z}, \\ (2k+1,2l+1) \text{ if } (x,y) \in A_4'(4k+2,4l+2), \ k,l \in \mathbb{Z}. \end{cases}$$

Another well-known topology on $\mathbb{Z}^2$ is the Marcus-Wyse one (cf. [10]), i.e., the Alexandroff topology s on $\mathbb{Z}^2$ given as follows:

For any $z = (x,y) \in \mathbb{Z}^2$,

$$s\{z\} = \begin{cases} A_4(z) \text{ if } x+y \text{ is odd}, \\ \{z\} \text{ otherwise}. \end{cases}$$

The Marcus-Wyse topology is connected and T_0, too. A portion of its connectedness graph is shown in the following figure.

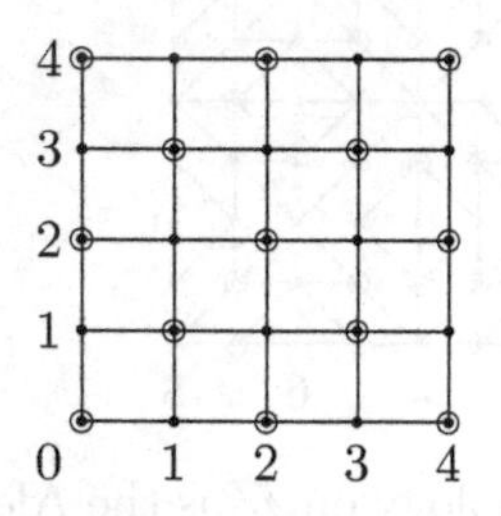

Theorem 2. *The Marcus-Wyse topology is the quotient topology of w generated by the surjection $g : \mathbb{Z}^2 \to \mathbb{Z}^2$ given as follows:*

$$g(x,y) = \begin{cases} (k+l, l-k) \text{ if } (x,y) \in A_8(4k,4l), \ k,l \in \mathbb{Z}, \\ (k+l+1, l-k) \text{ if } (x,y) = (4k+2,4l+2) \text{ for some } k,l \in \mathbb{Z} \\ \quad \text{with } k+l \text{ odd or } (x,y) \in A_{12}(4k+2,4l+2) \text{ for some} \\ \quad k,l \in \mathbb{Z} \text{ with } k+l \text{ even}. \end{cases}$$

Let v be the Alexandroff topology on $\mathbb{Z}^2$ given as follows:

For any $z = (x,y) \in \mathbb{Z}^2$,

$$v\{z\} = \begin{cases} H_2(z) \text{ if } x \text{ is odd and } y \text{ is even}, \\ V_2(z) \text{ if } x \text{ is even and } y \text{ is odd}, \\ A_4'(z) \text{ if } x,y \text{ are odd}, \\ \{z\} \text{ if } x,y \text{ are even}. \end{cases}$$

Evidently, v is connected and T_0. A a portion of its connectedness graph is shown in the following figure:

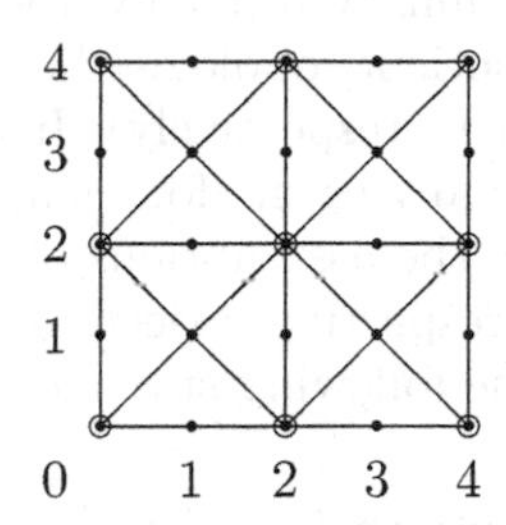

Theorem 3. *v is the quotient topology of w generated by the surjection $h : \mathbb{Z}^2 \to \mathbb{Z}^2$ given as follows:*

$$h(x,y) = \begin{cases} (2k, 2l) \text{ if } (x,y) \in A_8(4k, 4l),\ k,l \in \mathbb{Z}, \\ (2k, 2l+1) \text{ if } (x,y) \in H_2(4k, 4l+2),\ k,l \in \mathbb{Z}, \\ (2k+1, 2l) \text{ if } (x,y) \in V_2(4k+2, 4l),\ k,l \in \mathbb{Z}, \\ (2k+1, 2l+1) \text{ if } (x,y) = (4k+2, 4l+2),\ k,l \in \mathbb{Z}. \end{cases}$$

Let u be the Alexandroff pretopology on $\mathbb{Z}^2$ defined as follows:
For any point $z = (x,y) \in \mathbb{Z}^2$,

$$u\{z\} = \begin{cases} A_4(z) \text{ if both } x \text{ and } y \text{ are odd or } (x,y) = (4k+2l, 2l+2),\ k,l \in \mathbb{Z}, \\ A_8(z) \text{ if } (x,y) = (4k+2l, 2l),\ k,l \in \mathbb{Z}, \\ \{z\} \text{ otherwise.} \end{cases}$$

Then u is connected and T_0 with a portion of its connectedness graph shown in the following figure:

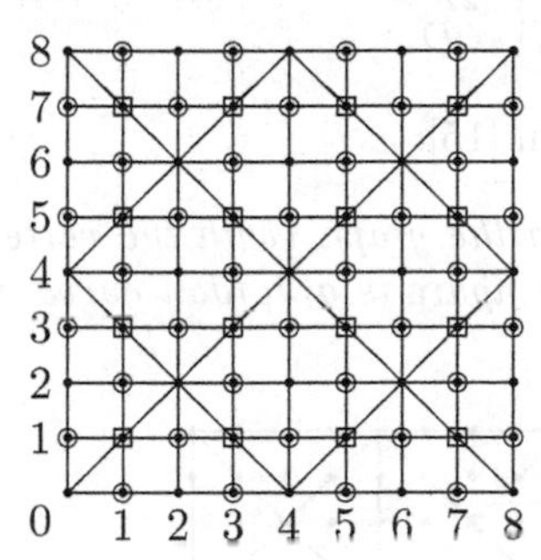

Theorem 4. *u is the quotient pretopology of w generated by the surjection $d : \mathbb{Z}^2 \to \mathbb{Z}^2$ given as follows:*

$$d(x,y) = \begin{cases} (2k+2l+1, 2l-2k+1) \text{ if } (x,y) \in A_4(4k, 4l+2),\ k,l \in \mathbb{Z}, \\ (2k+2l+1, 2l-2k-1) \text{ if } (x,y) \in A_4(4k+2, 4l),\ k,l \in \mathbb{Z}, \\ (\frac{x+y}{2}, \frac{y-x}{2}) \text{ if } x,y \text{ are odd or } (x,y) = (4k+2l, 2l),\ k,l \in \mathbb{Z}. \end{cases}$$

4 Jordan Curve Theorems

It can easily be seen that the Khalimsky topology, the Marcus-Wyse topology and the topology v are not only quotient topologies but also quotient pretopologies of w (generated by f, g and h, respectively). It is also evident that the inverse image of a singleton under any of the four mappings f, g, h and d is a connected subset of $(\mathbb{Z}^2, w)$. Thus, the assumptions of Lemma 1 are satisfied if $p = w$, $q \in \{t, s, v, u\}$ and e is the respective surjection f, g, h or d. The same is true also for the assumptions of the following statement:

Proposition 1. *Let $(\mathbb{Z}^2, p)$ be a pretopological space and let q be the quotient pretopology of p on $\mathbb{Z}^2$ generated by a surjection $e : \mathbb{Z}^2 \to \mathbb{Z}^2$. Let e have the property that $e^{-1}(\{y\})$ is connected for every point $y \in \mathbb{Z}^2$ and let $D \subseteq \mathbb{Z}^2$ be a simple closed curve in $(\mathbb{Z}^2, q)$. Then D is a Jordan curve in $(\mathbb{Z}^2, q)$ if the following two conditions are fulfilled:*

(1) There is a Jordan curve C in $(\mathbb{Z}^2, p)$ such that $e(C) = D$.
(2) $C_i - e^{-1}(D)$ is nonempty and connected in $(\mathbb{Z}^2, p)$ for $i = 1, 2$ where C_1 and C_2 are the two components of $\mathbb{Z}^2 - C$.

Proof. Let the conditions of the statement be fulfilled and put $C_1' = C_1 - e^{-1}(D)$ and $C_2' = C_2 - e^{-1}(D)$. We clearly have $e(C_1) \cap e(C_2) = \emptyset$ (because, otherwise, there is a point $y \in e(C_1) \cap e(C_2)$, which means that $e^{-1}(\{z\}) \cap C_1 \neq \emptyset \neq e^{-1}(\{z\}) \cap C_2$ - this is a contradiction as $e^{-1}(\{z\})$ is connected). Therefore, $e(C_1') \cap e(C_2') = \emptyset$. This yields $C_i' = e^{-1}(e(C_i'))$ for $i = 1, 2$, hence $e(C_i')$ is connected for $i = 1, 2$ by Lemma 1. Suppose that $\mathbb{Z}^2 - D$ is connected. Then $e^{-1}(\mathbb{Z}^2 - D) = C_1' \cup C_2'$ is connected by Lemma 1. This is a contradiction because $\emptyset \neq C_i' \subseteq C_i$ for $i = 1, 2$, C_1 and C_2 are disjoint and $C_1 \cup C_2$ is not connected. Therefore, $\mathbb{Z}^2 - D = e(C_1') \cup e(C_2')$ is not connected and, consequently, $e(C_1')$ and $e(C_2')$ are components of $\mathbb{Z}^2 - D$.

The following result is proved in [15]:

Proposition 2. *Every cycle in the graph (with the vertex set $\mathbb{Z}^2$) a portion of which is shown in the following figure is a Jordan curve in $(\mathbb{Z}^2, w)$:*

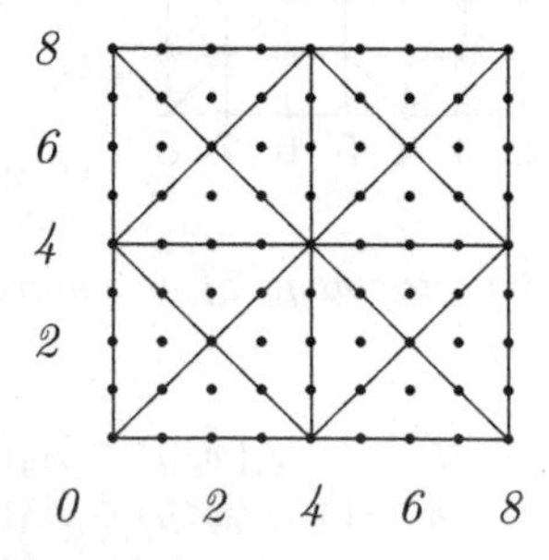

Using Propositions 1-2 and Theorems 1-4, we may identify Jordan curves among the simple closed curves in the Khalimsky and Marcus-Wyse planes, in $(\mathbb{Z}^2, v)$ and in $(\mathbb{Z}^2, u)$. This will be done in the following four Theorems which can be proved in much the same way and, therefore, we will present only the proof of the last of them.

Theorem 5. *Let D be a simple closed curve in the Khalimsky plane such that every point $z \in D$ with both coordinates odd satisfies $A_4(z) \cap D = \emptyset$. Then D is a Jordan curve in the Khalimsky plane.*

Theorem 6. *Let D be a cycle in the connectedness graph of the Marcus-Wyse topology such that there exists a point $(x, y) \in \mathbb{Z}^2$ with $(x + 2k, y + 2l) \notin D$ whenever $k, l \in \mathbb{Z}$. Then D is a Jordan curve in the Marcus-Wyse plane.*

Theorem 7. *Let D be a simple closed curve in $(\mathbb{Z}^2, v)$ such that every pair of different points $z_1, z_2 \in D$ with both coordinates even satisfies $A_4(z_1) \cap A_4(z_2) \subseteq D$. Then D is a Jordan curve in $(\mathbb{Z}^2, v)$.*

Corollary 1. *Every cycle in the graph a portion of which is shown in the following figure is a Jordan curve in $(\mathbb{Z}^2, v)$:*

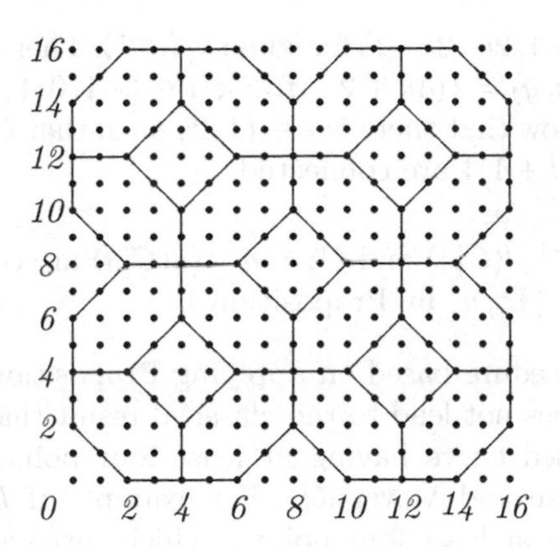

Theorem 8. *Every simple closed curve D in $(\mathbb{Z}^2, u)$ which is a cycle in the graph a portion of which is shown in the following figure is a Jordan curve in $(\mathbb{Z}^2, u)$:*

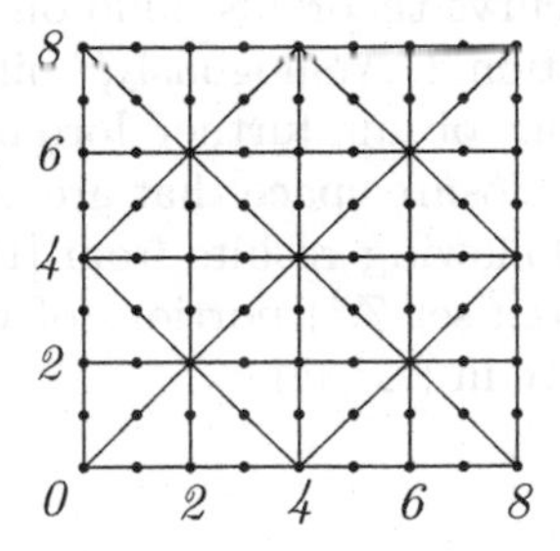

Proof. Let D be a simple closed curve in $(\mathbb{Z}^2, u)$ which is a cycle in the graph a portion of which is shown in the above figure. By Theorem 4, u is the quotient pretopology of w generated by d. It immediately follows from the definition of d that there exists a unique cycle C in the graph from Proposition 2 such that $d(C) = D$. By Proposition 2, C is a Jordan curve in $(\mathbb{Z}^2, w)$. Let C_1, C_2 be the two components of $\mathbb{Z}^2 - C$ and put $C_i' = C_i - d^{-1}(D)$ for $i = 1, 2$. Let $(x, y) \in D$ be a point and write $d^{-1}(x, y)$ briefly instead of $d^{-1}(\{(x, y)\})$. Clearly, $d^{-1}(x, y) \nsubseteq C$ if and only if both x and y are odd ($d^{-1}(x, y)$ is a singleton if x or y is even). Thus, let $(x, y) \in D$ be a point with both x and y odd. Then one of the following two cases occurs:

(1) $(x, y) = (2k + 2l + 1, 2k - 2l + 1)$ for some $k, l \in \mathbb{Z}$. Then $d^{-1}(x, y) = A_4(4k, 4l + 2)$, hence $C \cap d^{-1}(x, y) = \{(4k, 4l + 2 + i),\ i \in \{-1, 0, 1\}\}$. Therefore, there is $i \in \{1, 2\}$ such that $(4k - 1, 4l + 2) \in C_i - C_i'$ and $(4k + 1, 4l + 2) \in C_{3-i} - C_{3-i}'$ while $(4k - 2, 4l + 2) \in C_i'$ and $(4k + 2, 4l + 2) \in C_{3-i}'$, $i \in \{1, 2\}$. The points $(4k - 1, 4l + 1)$ and $(4k - 1, 4l + 3)$ are the only points that belong to C_i and are adjacent (in the connectedness graph of w) to $(4k - 1, 4l + 2)$. But both of these points are adjacent also to $(4k - 2, 4l + 2) \in C_i'$. Thus, $C_i - \{(4k - 1, 4l + 2)\}$ is connected and, by similar arguments, also $C_{3-i} - \{(4k + 1, 4l + 2)\}$ is connected.

(2) $(x, y) = (2k + 2l + 1, 2l - 2k - 1)$ for some $k, l \in \mathbb{Z}$. Then $d^{-1}(x, y) = A_4(4k + 2, 4l)$, hence $C \cap d^{-1}(x, y) = \{(4k + 2 + i, 4l),\ i \in \{-1, 0, 1\}\}$. Now, analogously to (1), we can easily show that there is $i \in \{1, 2\}$ such that $C_i - \{(4k + 2, 4l - 1)\}$ and $C_{3-i} - \{(4k + 2, 4l + 1)\}$ are connected.

It follows that $C_1' = d^{-1}(d(C_1'))$ and $C_2' = d^{-1}(d(C_2'))$ are connected. Therefore, D is a Jordan curve in $(\mathbb{Z}^2, u)$ by Proposition 1.

Remark 1. a) The procedure based on applying Propositions 1 and 2 and used to prove Theorem 8 does not lead to the classical result that, in the Khalimsky plane, any simple closed curve having at least four points is a Jordan curve (Khalimsky, Kopperman and Meyer [5]). For example, if D is a simple closed curve in $(\mathbb{Z}^2, t)$ (having at least four points) which turns at a point with both coordinates odd at the angle $\frac{3\pi}{4}$, then none of the Jordan curves C in $(\mathbb{Z}^2, w)$ identified in Theorem 5 fulfills $f(C) = D$. Theorem 5 is weaker than the classical result. As for the Marcus-Wyse topology, a Jordan curve theorem different from Theorem 6 was proved in [7].

b) The previous four Jordan curve theorems build on the Jordan curve theorem for $(\mathbb{Z}^2, w)$ given in Proposition 2. Analogously, building on further Jordan curve theorems for $(\mathbb{Z}^2, w)$, we may obtain further Jordan curve theorems in the three topological spaces and the closure space that are quotients of $(\mathbb{Z}^2, w)$. For example, we may build on the following results from [15]: Every cycle in each of the five graphs (with the vertex set $\mathbb{Z}^2$) portions of which are shown in the following figures is a Jordan curve in $(\mathbb{Z}^2, w)$.

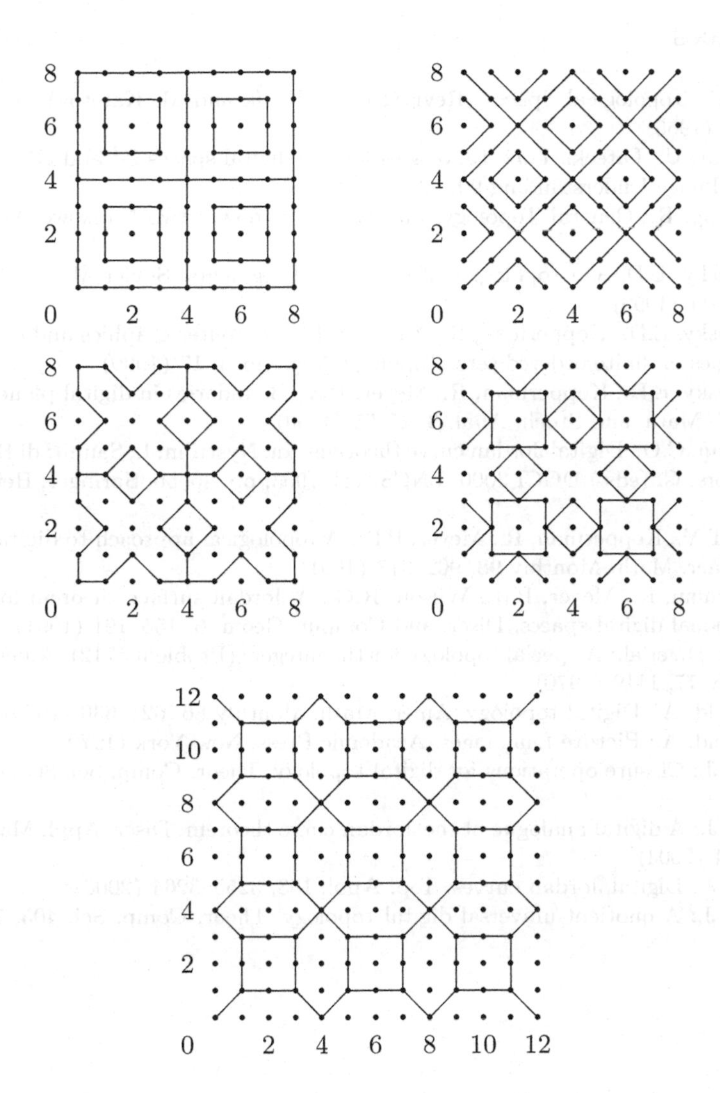

Conclusion. It is well known that, in computer imagery, Jordan curves play an important role because they represent boundaries of regions of digital images. It is therefore useful to work with a connectedness structure on $\mathbb{Z}^2$ possessing a considerable variety of Jordan curves. We discuss four such structures, namely topologies w, t, s, v and a pretopology u. The results obtained show that these may be used as background structures to solve problems of digital image processing, especially those closely related to boundaries (image data compression, pattern recognition, boundary detection and contour filling, etc.).

Acknowledgement. The author acknowledges partial support from Ministry of Education of the Czech Republic, research plan no. MSM0021630518.

References

1. Čech, E.: Topological Spaces (Revised by Z. Frolík and M. Katětov). Academia, Prague (1966)
2. Eckhardt, U., Latecki, L.J.: Topologies for the digital spaces $\mathbb{Z}^2$ and $\mathbb{Z}^3$. Comput. Vision Image Understanding 90, 295–312 (2003)
3. Engelking, R.: General Topology. Państwowe Wydawnictwo Naukowe, Warszawa (1977)
4. Khalimsky, E.D.: On topologies of generalized segments. Soviet Math. Dokl. 10, 1508–1511 (1999)
5. Khalimsky, E.D., Kopperman, R., Meyer, P.R.: Computer graphics and connected topologies on finite ordered sets. Topology Appl. 36, 1–17 (1990)
6. Khalimsky, E.D., Kopperman, R., Meyer, P.R.: Boundaries in digital planes. Jour. of Appl. Math. and Stoch. Anal. 3, 27–55 (1990)
7. Kiselman, C.O.: Digital Jordan curve theorems. In: Nyström, I., Sanniti di Baja, G., Borgefors, G. (eds.) DGCI 2000. LNCS, vol. 1953, pp. 46–56. Springer, Heidelberg (2000)
8. Kong, T.Y., Kopperman, R., Meyer, P.R.: A topological approach to digital topology. Amer. Math. Monthly 98, 902–917 (1991)
9. Kopperman, R., Meyer, P.R., Wilson, R.G.: A Jordan surface theorem for three-dimensional digital spaces. Discr. and Comput. Geom. 6, 155–161 (1991)
10. Marcus, D., et al.: A special topology for the integers (Problem 5712). Amer. Math. Monthly 77, 1119 (1970)
11. Rosenfeld, A.: Digital topology. Amer. Math. Monthly 86, 621–630 (1979)
12. Rosenfeld, A.: Picture Languages. Academic Press, New York (1979)
13. Šlapal, J.: Closure operations for digital topology. Theor. Comp. Sci. 305, 457–471 (2003)
14. Šlapal, J.: A digital analogue of the Jordan curve theorem. Discr. Appl. Math. 139, 231–251 (2004)
15. Šlapal, J.: Digital Jordan curves. Top. Appl. 153, 3255–3264 (2006)
16. Šlapal, J.: A quotient-universal digital topology. Theor. Comp. Sci. 405, 164–175 (2008)

Decomposing Cavities in Digital Volumes into Products of Cycles*

Ainhoa Berciano[1], Helena Molina-Abril[2], Ana Pacheco[2], Paweł Pilarczyk[3], and Pedro Real[2]

[1] Departamento de Didactica de la Matematica y de las CC. Experimentales, Universidad del Pais Vasco-Euskal Herriko Unibertsitatea, Ramon y Cajal, 72. 48014-Bilbao (Bizkaia), Spain
ainhoa.berciano@ehu.es
http://www.ehu.es/aba

[2] Departamento de Matematica Aplicada I, Universidad de Sevilla, Avenida Reina Mercedes, Spain
{habril,ampm,real}@us.es
http://ma1.eii.us.es/

[3] Centro de Matemática, Universidade do Minho, Campus de Gualtar, 4710-057 Braga, Portugal
http://www.pawelpilarczyk.com/

Abstract. The homology of binary 3–dimensional digital images (digital volumes) provides concise algebraic description of their topology in terms of connected components, tunnels and cavities. Homology generators corresponding to these features are represented by nontrivial 0–cycles, 1–cycles and 2–cycles, respectively. In the framework of cubical representation of digital volumes with the topology that corresponds to the 26–connectivity between voxels, we introduce a method for algorithmic computation of a coproduct operation that can be used to decompose 2–cycles into products of 1–cycles (possibly trivial). This coproduct provides means of classifying different kinds of cavities; in particular, it allows to distinguish certain homotopically non-equivalent spaces that have isomorphic homology. We define this coproduct at the level of a cubical complex built directly upon voxels of the digital image, and we construct it by means of the classical Alexander-Whitney map on a simplicial subdivision of faces of the voxels.

Keywords: homology, cubical homology, cubical set, cell complex, digital image, cavity, cycle, Alexander Whitney diagonal, chain homotopy, algebraic gradient vector field.

* This work has been partially supported by "Computational Topology and Applied Mathematics" PAICYT research project FQM-296, "Andalusian research project" PO6-TIC-02268, Spanish MEC project MTM2006-03722 and the Austrian Science Fund under grant P20134-N13.

S. Brlek, C. Reutenauer, and X. Provençal (Eds.): DGCI 2009, LNCS 5810, pp. 263–274, 2009.

1 Introduction

Over the recent decades, considerable progress has been made in the development of topological methods for the analysis of digital images. In particular, effective algorithms and efficient software for the computation of homology groups and their generators have been under heavy development; see [1,2,3,4,5,6,7,8,9,10] for some of this work. These tools have already proved their usefulness in applications, e.g. [11,12,13,14,15], and their potential in multi-dimensional digital image analysis is undeniable.

An n–dimensional binary digital image can be perceived as a union of closed n–dimensional hypercubes in $\mathbb{R}^n$ with respect to a uniform rectangular lattice with the topology induced from $\mathbb{R}^n$, which corresponds to the topology of $(3^n - 1)$–connectivity between the n-dimensional pixels (26–connectivity in the 3-dimensional case). The homology of such a set provides information on the number of connected components and holes of various dimensions. Computing homology generators allows to locate the connected components in the digital image and enclose the various holes geometrically.

A homology generator of dimension 1 is represented at chain level by a nontrivial 1–cycle and encloses some tunnel. A homology generator of dimension 2 is represented by a nontrivial 2–cycle and encloses some cavity. In this paper we are interested in decomposing 2–cycles into products of 1–cycles (which may turn out to be trivial in some cases), in order to classify the different types of cavities in digital volumes, and identify the related tunnels, if any (see Fig. 1). Such a method will provide a means of more thorough analysis of the geometry of the digital volumes, important e.g. for the purpose of structural pattern recognition. It will also allow to distinguish certain substantially different (homotopically non-equivalent) cases that give rise to isomorphic homology groups.

In order to address the problem in question, we consider the Alexander-Whitney diagonal, a canonical associative coproduct that is defined on the chain complex canonically associated to a simplicial complex K, and factorizes higher-dimensional cycles into lower-dimensional ones. We adapt this theoretical machinery into the context of digital imagery. We work directly at the level of the cubical complexes inferred from the cubical voxels, their faces, edges and vertices. However, our approach is different from the classical Serre diagonal [16].

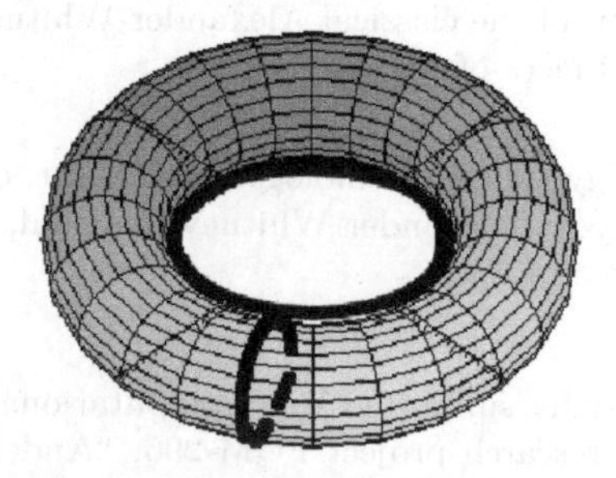

Fig. 1. Decomposing the cavity of a torus into cycles

In fact, we do not use here cocycles or cohomology notions, but we remain at the easier level of cycles and homology.

Although the methods developed in this paper are dimension-independent and are valid in a very generic context, for the sake of clarity of presentation we restrict our attention to 3–dimensional binary digital images (also referred to as digital volumes), and we consider coefficients in the ring $\mathbb{F}_2$.

Our approach to the computation of the cubical version of the Alexander-Whitney diagonal for cavities in a digital volume can be summarized as follows. First, given a digital volume K, we represent it by means of a cubical chain complex, as described in Section 2. We compute its homology along with homology generators and a homology gradient vector field [7]. The latter object captures the deformation process for obtaining the minimal homological expression for K, and allows to instantly determine the homology class of every cycle. We describe this construction in Section 3. Then we specify a simplicial-valued diagonal for each cubical 2–cell using two basic techniques: simplicial subdivision of the cartesian product of simplices and chain homotopy equivalence. We express the simplicial result in terms of cubical 1–chains, and in this way we derive an explicit formula for the cubical version of this diagonal in Section 4. Finally, we apply the Alexander-Whitney diagonal to each representative nontrivial 2–cycle (homology generator) by means of the natural linear extension of the diagonal defined for the 2–cells. The overall method combining all these mechanisms is described step by step in Section 5. Examples of the application of this algorithm are discussed in Section 6.

2 Simplicial and Cubical Complexes

In this section we introduce cubical cell complexes which we use to represent binary digital images, and we also mention simplicial complexes which will be used as an intermediate step in the construction of the Alexander-Whitney diagonal for cubical complexes in Section 4.

Throughout the paper, we consider the finite field $F_2 = \{0, 1\}$ as the ground ring of coefficients, which simplifies many formulas and makes the entire construction reflect better the combinatorial aspect of this approach. Note that the homology of geometric objects embedded in $\mathbb{R}^3$ is torsion free, so we do not lose any information in this context by this choice of coefficients.

Let K be an n–dimensional cell complex. Denote the set of its q–dimensional cells by $K^{(q)}$. The corresponding chain complex $(C_q(K), \partial_q)_{q\in\mathbb{Z}}$ consists of the fields of q-chains C_q over $\mathbb{F}_2$, whose generators correspond to the cells in $K^{(q)}$, and a family of homomorphisms $\partial_q\colon C_q \to C_{q-1}$ such that $\partial_{q-1} \circ \partial_q = 0$, which correspond to geometric boundaries of cells. Each q–chain $c \in C_q$ is then a formal sum $\sigma_1 + \cdots + \sigma_k$ of selected cells $\sigma_i \in K^{(q)}$. We denote $\sigma \in c$ if $\sigma \in K^{(q)}$ is a summand of c. Note that $C_q(K) = 0$ whenever $q < 0$ or $q > n$.

The case of a simplicial complex is well established [17], and the boundary map ∂_q on each q–simplex $\sigma = \langle v_0, \ldots, v_q\rangle$ is defined as follows: $\partial_q(\sigma) = \sum\langle v_0, \ldots, \hat{v}_i, \ldots, v_q\rangle$, where the hat over v_i means that v_i is omitted (see Fig. 2 for an illustration of the boundary of a 2–simplex).

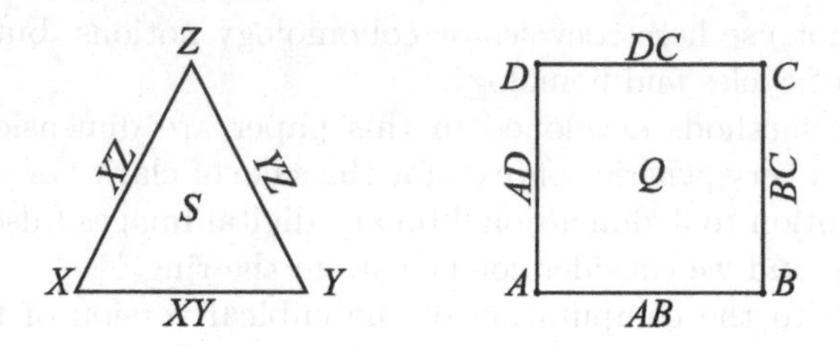

Fig. 2. The boundary of a 2–simplex S (left) and a 2–cube Q (right) over $\mathbb{F}_2$: $\partial_2 S = XY+YZ+XZ$ and $\partial_2 Q = AB+BC+DC+AD$. The boundary of each edge consists of the two endpoints, e.g., $\partial_1 AB = A+B$. It is easy to see that in both cases $\partial_1 \circ \partial_2 = 0$.

The case of a cubical complex is less typical, so we recall some definitions in order to avoid any ambiguity. The reader is referred to [5] for a comprehensive introduction. An *elementary interval* is an interval of the form $I = [k, k+1]$ (the *non-degenerate* case) or a set $I = \{k\}$ (the *degenerate* case), where $k \in \mathbb{Z}$. An *elementary cube* in $\mathbb{R}^n$ is the cartesian product of n elementary intervals, and the number of non-degenerate intervals in this product is its *dimension*.

In order to define the boundary of a q-cube $\sigma = I_1 \times \cdots \times I_n \in K_q^n$ (see Fig. 2 for an illustration of ∂_2), denote by $k_1, \ldots, k_q$ those indices that correspond to non-degenerate intervals $I_{k_j} = [a_j, b_j]$ in σ. Define $A_{k_j}\sigma := I_1 \times \cdots \times I_{j-1} \times \{a_j\} \times I_{j+1} \times \cdots \times I_n$ and $B_{k_j}\sigma := I_1 \times \cdots \times I_{j-1} \times \{b_j\} \times I_{j+1} \times \cdots \times I_n$. Then $\partial_q(\sigma) := \sum_{i=1}^{q}(A_{k_i}\sigma + B_{k_i}\sigma)$.

We identify an n-dimensional binary digital image with the set $K^{(n)}$ of n-dimensional elementary cubes in $\mathbb{R}^n$ corresponding to its black pixels (in 2D), or voxels (in $3D$), or n-dimensional picture elements (in general). We add all the lower-dimensional cubes contained in the cubes in $K^{(n)}$ in order to obtain a cell complex K that represents the digital image.

3 Homology of Cell Complexes

In this section we describe a homology computation procedure for finite cell complexes, in which a chain contraction is constructed in addition to the homology module and homology generators. As it will be made clear in Section 5, this construction is crucial for the successful method for the computation of the Alexander-Whitney diagonal.

Denote a cell complex under consideration by (K, ∂), where $\partial: C(K) \to C(K)$ is its boundary map. A chain $a \in C_q(K)$ is called a *q–cycle* if $\partial_q(a) = 0$. If $a = \partial_{q+1}(a')$ for some $a' \in C_{q+1}(K)$ then a is called a *q–boundary.* The *q-th homology group* $H_q(K)$ *of* (K, ∂) is the quotient group of q–cycles and q–boundaries. The homology class of a chain $a \in C_q(K)$ is denoted by $[a]$. If $\mathcal{C} = \{C_q, \partial_q\}$ and $\mathcal{C}' = \{C'_q, \partial'_q\}$ are chain complexes then a *chain map* $f: \mathcal{C} \to \mathcal{C}'$ is a family of homomorphisms $\{f_q: C_q \to \mathcal{C}'_q\}_{q\geq 0}$ such that $\partial'_q f_q = f_{q-1}\partial_q$.

It is clear that the problem of computing homology of a cell complex (K, ∂) can be reduced to solving the equation $\partial x = 0$ up to boundary. For the actual algorithms, we shall follow a computational algebraic approach for computing homology in terms of chain homotopy equivalences, developed first by Eilenberg and Mac Lane in the 1960s [18]. This approach was exhaustively used later in algebraic-topological theories like effective homology [19] and homological perturbation theory [20], as well as in discrete settings, like discrete Morse theory [21] and the AT-model [4,22].

In order to describe our approach to homology computation and to the construction of Alexander-Whitney diagonal, we are going to use the notion of gradient vector fields. Let (K, ∂) be a finite cell complex. A linear map $\phi: C_*(K) \to C_{*+1}(K)$ is called an *algebraic gradient vector field* (or an *algebraic gvf* for short) over K if $\phi\phi = 0$. An algebraic gvf over K is called a *combinatorial gradient vector field* (or a *combinatorial gvf* for short) over K (see [21]) if for any cell $a \in K^{(q)}$, either $\phi(a) = 0$, or $\phi(a) = b$ for some $b \in K^{(q+1)}$ (see Fig. 3). An algebraic gvf satisfying the condition $\phi\partial\phi = \phi$ is called an *algebraic integral operator*. An algebraic integral operator which is only non-null for a unique cell $a \in K^{(q)}$ is called a *(combinatorial) integral operator* (see [3]). An algebraic integral operator satisfying the additional condition $\partial\phi\partial = \partial$ is called a *homology gvf* (see [22]).

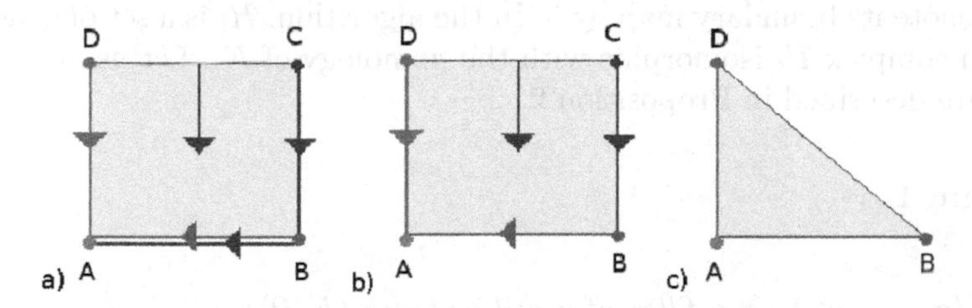

Fig. 3. A sample algebraic gradient vector field (a), a sample combinatorial gradient vector field (b), and the resultant image after applying $\phi(C) = CB$ (c). The difference between (a) and (b) is that $\phi(C) = CB + AB$ in (a), while $\phi(C) = BC$ in (b).

It turns out that a homology gvf over K determines a strong algebraic relationship connecting $C(K)$ and its homology vector space $H(K)$. In order to show this, we use the notion of a *chain contraction* $(f, g, \phi): (C, \partial) \to (C', \partial')$ between two chain complexes, which is a triple of linear maps $f: C_* \to C'_*$, $g: C'_* \to C_*$ and $\phi: C_* \to C_{*+1}$ that satisfy the following conditions: (a) $\mathrm{Id}_C - gf = \partial\phi + \phi\partial$; (b) $fg = \mathrm{Id}_{C'}$; (c) $f\phi = 0$; (d) $\phi g = 0$; (e) $\phi\phi = 0$. This is a classical notion in homological algebra and algebraic topology, but it is an exotic concept within the digital imagery setting [3,23,4]. The following proposition can be derived in a straightforward manner from [20], using the integral operator language.

Proposition 1. *Let (K, ∂) be a finite cell complex. An algebraic integral operator $\phi: C_*(K) \to C_{*+1}(K)$ over K gives rise to a chain contraction (π, ι, ϕ) from $C(K)$ onto its chain subcomplex $\mathrm{Im}\pi$, where π is a projection and ι is the inclusion map. Reciprocally, if (f, g, ϕ) is a chain contraction from $C(K)$ to another chain complex C', then ϕ is an algebraic integral operator.*

Note that given an algebraic integral operator, the projection π can be instantly determined by the formula $\pi = \mathrm{Id}_{C(K)} - \partial\phi - \phi\partial$.

The following result is a refinement of Proposition 1 and justifies the correctness of the homology computation algorithm that we use.

Proposition 2 (see [7]). *Let (K, ∂) be a finite cell complex. A homology gvf $\phi: C_*(K) \to C_{*+1}(K)$ over K gives rise to a chain contraction (π, ι, ϕ) from $C(K)$ onto its chain subcomplex isomorphic to the homology of K, where π is the projection $\pi = \mathrm{Id}_{C(K)} - \partial\phi - \phi\partial$, and ι is the inclusion map $\iota: H(K) \to C(K)$. Reciprocally, if (f, g, ϕ) is a chain contraction from $C(K)$ to its homology $H(K)$, then ϕ is a homology gvf.*

We apply the homology computation process given in [7,22], in which the incremental homology algorithm introduced in [1] was adapted for obtaining a homology gvf. We assume that the input cell complex (K, ∂) is given together with *filter* $\mathcal{K}$ for K, that is, an ordered set of cells $\mathcal{K} = \mathcal{K}_m = \langle c_1, \dots, c_m \rangle$ such that $\{c_1, \dots, c_m\} = \bigcup_{q \in \mathbb{Z}} K^{(q)}$, and for each $j = 1, \dots, m$, all the faces of c_j are contained in the subset $\{c_1, \dots, c_{j-1}\}$. For each $i = 0, \dots, m$, we represent the cell subcomplex K_i of K consisting of $\{c_0, \dots, c_i\}$ by the filter $\mathcal{K}_i := \langle c_0, \dots, c_i \rangle$, and we denote its boundary map by ∂. In the algorithm, $\mathcal{H}_i$ is a set of generators of a chain complex H_i isomorphic with the homology of K_i. Let us note that ϕ, ι and π are described in Proposition 2.

Algorithm 1

INPUT:

$\mathcal{K} = \langle c_1, \dots, c_m \rangle$ — *a filter of a cell complex* (K, ∂);
∂ — *the boundary operator on* K.

PSEUDOCODE:

$\mathcal{H}_0 := \{c_0\}$; $\phi_0(c_0) := 0$; $\pi_0(c_0) := c_0$;
`for` $i := 1$ `to` m `do`
  $\phi_i(c_i) := 0$; $\pi_i(c_i) := \bar{c}_i := c_i + \phi_{i-1}\partial(c_i)$;
  `if` $\partial\bar{c}_i = 0$ `then`
    `for` $j := 0$ `to` $i - 1$ `do`
      $\phi_i(c_j) := \phi_{i-1}(c_j)$;
    $\mathcal{H}_i := \mathcal{H}_{i-1} \cup \{\bar{c}_i\}$;
  `otherwise express` $\partial\bar{c}_i$ `as a sum of elements of` $\mathcal{H}_{i-1}$:
      $\partial\bar{c}_i = \sum_{j=1}^{r_i} u_j$, `where each` $u_j \in \mathcal{H}_{i-1}$;
    `choose any summand` u_k, `where` $k \in \{1, \dots, r\}$;
    `for` $j := 0$ `to` $i - 1$ `do`
      $\xi_j^i := c_j + (\phi_{i-1}\partial + \partial\phi_{i-1})(c_j)$;
      `express` ξ_j^i `as a sum of elements of` $\mathcal{H}_{i-1}$:
        $\xi_j^i = \sum_{l=1}^{t_{ij}} v_l$, `where each` $v_l \in \mathcal{H}_{i-1}$;
      `if` $u_k = v_l$ `for some` $l \in \{1, \dots, t_{ij}\}$ `then`
        $\phi_i(c_j) := \phi_{i-1}(c_j) + \bar{c}_i$;
      `else` $\phi_i(c_j) := (\phi_{i-1}(c_j)$;

$\pi_i(c_j) := c_j + (\phi_i \partial + \partial \phi_i)(c_j);$
$\mathcal{H}_i := \mathcal{H}_{i-1} \setminus \{u_k\}.$

OUTPUT:
$\mathcal{H}_m$ — *a set of homology generators of* K;
(π_m, ι, ϕ_m) — *a chain contraction in which* ϕ_m *is a homology gvf over* K.

The complexity of this algorithm is $O(m^3)$. Note that expressing $\partial \bar{c}_i$ and ξ^i_j as a sum of elements of $\mathcal{H}_{i-1}$ is straightforward, because π is compatible with the boundary operator. Therefore, solving the corresponding system of linear equations is trivial, and does not increase the complexity of the algorithm.

The application of the algorithm to a simple cubical set is shown in the Appendix.

4 Alexander-Whitney Diagonal of a Cubical 2–Cell

In this section we specify the cubical version of the Alexander-Whithey diagonal for an elementary 2–cube. For that purpose, we use the basic techniques of simplicial cartesian product subdivision and chain homotopy equivalence.

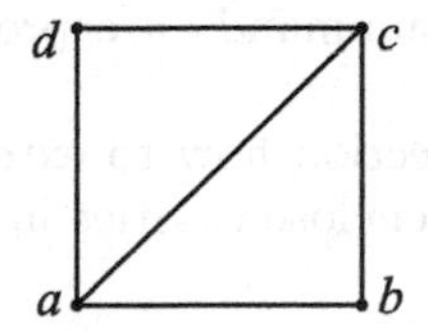

Fig. 4. Simplicial complex of a 2–cube (square): two 2–simplices (triangles), five 1–simplices (edges), and four 0–simplices (vertices)

Let Q be the cubical complex of an elementary 2–cube and let S be one simplicial subdivision, as illustrated in Fig. 4. Consider the combinatorial integral operator $\phi: C_*(S) \to C_{*+1}(S)$ such that $\phi(ac) = acd$. By Proposition 1, ϕ gives rise to a chain contraction (π, ι, ϕ) from $C(S)$ to its chain subcomplex $\mathrm{Im}(\pi)$. In this example $\pi(ac) = ab + cb$, $\pi(abc) = 0$, $\pi(acd) = acd + abc$ and $\pi(\sigma) = \sigma$ for the rest of elements.

We now use the chain contraction (π, ι, ϕ) to transfer the simplicial Alexander-Whitney diagonal into the cubical setting (from $C_*(S)$ to $C_*(Q)$). Recall from [17] that if (w_1, w_2, w_3) is a 2–simplex (a triangle) then

$$AW((w_1, w_2, w_3)) = (w_1) \otimes (w_1, w_2, w_3) + \\ +(w_1, w_2) \otimes (w_2, w_3) + (w_1, w_2, w_3) \otimes (w_3)$$

where $AW: C_*(S) \to C_*(S) \otimes C_*(S)$.

From the homological point of view, the only nontrivial term in this decomposition is the second one, so one can consider the Alexander-Whitney diagonal to decompose the 2–simplex (w_1, w_2, w_3) into the product of two 1–simplices $(w_1, w_2) \otimes (w_2, w_3)$, as illustrated in Fig. 5.

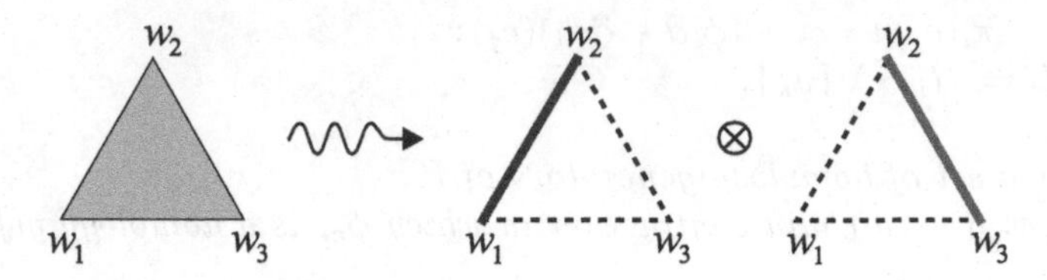

Fig. 5. Decomposition of a 2–simplex (a triangle) into 1–simplices (segments) via the Alexander-Whitney diagonal

A formula for the cubical Alexander-Whitney diagonal on the 2-cube *abcd* (see Fig. 4) transferred from the simplicial version of the Alexander-Whitney diagonal by means of the formula $AW_c(abcd) = \pi \otimes \pi\Big(AW\big(\iota(abcd)\big)\Big)$ is as follows:

$$AW_c(abcd) = b \otimes abcd + abcd \otimes a + bc \otimes cd + bc \otimes ad + cd \otimes ad + ad \otimes ad. \quad (1)$$

where $AW_c\colon C_*(Q) \to C_*(Q) \otimes C_*(Q)$.

Note that $\iota(abcd) = abc + acd$. In particular, the result is a combination in which the first two summands are trivial (as the product of the 2–cube and a vertex), and the next four summands are products of elementary cubes of dimension 1 (see Fig. 6).

We shall show in the next section how to extend this formula from a single 2–cube to entire 2–chains that enclose cavities in a cubical object.

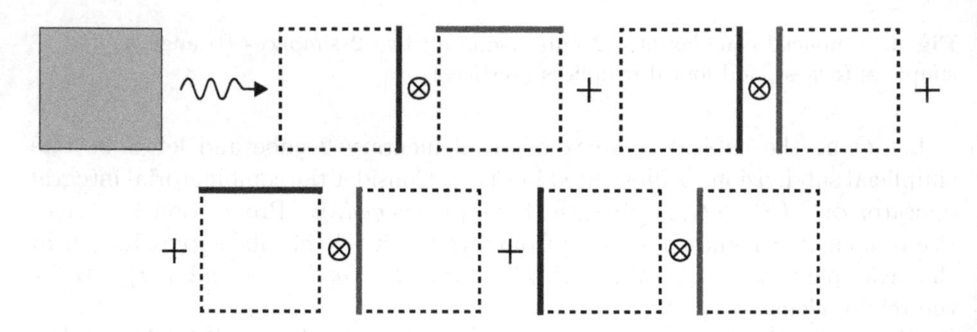

Fig. 6. Decomposition of a cubical 2–cell into a product of 1–cells via the cubical version of the Alexander-Whitney diagonal

5 Computing the Alexander-Whitney Diagonal for Cavities in a Digital Volume

In this section, we combine the results of the previous sections into an algorithmic method for determining 2–cycles (cavities) in a digital volume and decomposing them into products of 1–cycles.

Let (K, ∂) be a cubical complex of a binary digital volume, as described in Section 2. Let $\mathcal{K}$ be a filter for K. A straightforward filter for K can be constructed by first considering all the 0–cells in a certain order, then all the 1–cells, and so on.

Let $\mathcal{H} = \{\bar{c}_1, \ldots, \bar{c}_r\}$ and (π, ι, ϕ) be the output of Algorithm 1 applied to $\mathcal{K}$ and ∂. Let $\{\bar{c}_{s_1}, \ldots, \bar{c}_{s_t}\}$ be the 2–dimensional elements of $\mathcal{H}$. If this set is empty then there are no cavities to decompose. Otherwise, the set $\{\bar{c}_{s_1}, \ldots, \bar{c}_{s_t}\} = \mathcal{H} \cap C_2(K)$, where $t \in \{1, \ldots, r\}$, consists of representative 2–cycles in K that correspond to the generators of $H_2(K)$. In particular, each $\bar{c}_{s_i} = \sum_{j=1}^{t_i} c_{ij}$ is a sum of elementary 2–cubes enclosing at least one cavity in K.

We apply the formula (1) derived in Section 4 for the Alexander-Whitney diagonal to each c_{ij}, $i = 1, \ldots, t$, $j = 1, \ldots, t_i$, and we obtain a sum of tensor products $AW_c(\bar{c}_{s_i}) = \sum_{j=1}^{r_i} a_{ij} \otimes b_{ij}$, where each a_{ij} and b_{ij} is a 1–cube. Finally, we apply the map π to these pairs of chains, and in this way we obtain a description of the Alexander-Whitney diagonal in cubical homology: $AW_c \colon H_2(K) \to H_1(K) \otimes H_1(K)$.

6 Examples and Conclusions

In this paper, we introduced an algorithmic method for decomposing cavities in digital volumes into products of cycles. In this section, we show some specific examples that illustrate the usefulness of the Alexander-Whitney diagonal for the analysis of binary digital volumes.

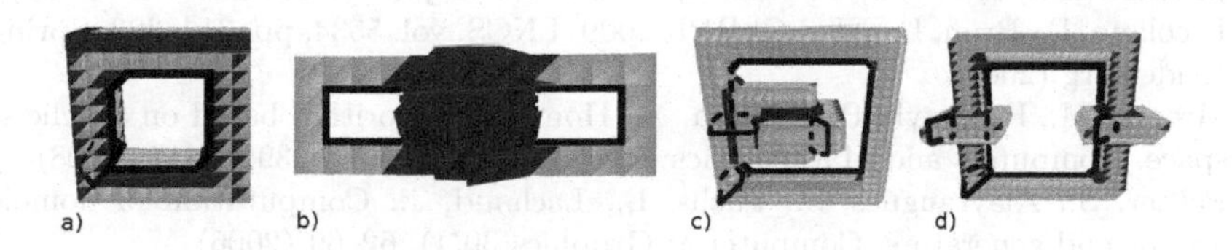

Fig. 7. Sample digital volumes with cavities: (a) a topological torus, (b) a topological sphere with two handles, (c) a configuration with three tori, (d) another configuration with three tori. 1–cycles are indicated in each example.

Let us consider the examples illustrated in Fig. 7. As shown in Table 1, the digital volumes (a) and (b) have the same homological information in terms of Betti numbers (number of connected components (β_0), number of 1-cycles (β_1) and number of cavities (β_2)). But computing the Alexander-Whitney diagonal, however, we can easily distinguish them.

The null result of the Alexander-Whitney diagonal in (b) is due to the fact that the cavity described by the sphere does not present non-trivial 1-cycles.

Similar situation occurs with examples (c) and (d). They have identic Betti numbers, but in order to distinghish them, we have to solve the problem of distinguishing the decompositions in terms of homology generators. This problem can be reduced to the problem of an isomorphism of groups in algebra, which is a problem of high computational complexity in general. We do not focus on this problem in this paper, which will be our aim in future work. Let us restrict to say that we can differenciate the respecting decomposition in (c) and (d) using matrix arguments.

References

1. Delfinado, C., Edelsbrunner, H.: An incremental algorithm for Betti numbers of simplicial complexes on the 3–sphere. Comput. Aided Geom. Design 12, 771–784 (1995)
2. Dey, T., Guha, S.: Computing homology groups of simplicial complexes in $\mathbb{R}^3$. Journal of the ACM 45(2), 266–287 (1998)
3. González-Diaz, R., Jiménez, M., Medrano, B., Molina-Abril, H., Real, P.: Integral operators for computing homology generators at any dimension. In: Ruiz-Shulcloper, J., Kropatsch, W.G. (eds.) CIARP 2008. LNCS, vol. 5197, pp. 356–363. Springer, Heidelberg (2008)
4. González-Diaz, R., Real, P.: On the cohomology of *3d* digital images. Discrete Applied Math. 147, 245–263 (2005)
5. Kaczynski, T., Mischaikow, K., Mrozek, M.: Computational homology. Applied Mathematical Sciences (2004)
6. Mischaikow, K., Mrozek, M., Pilarczyk, P.: Graph approach to the computation of the homology of continuous maps. Foundations of Computational Mathematics 5(2), 199–229 (2005)
7. Molina-Abril, H., Real, P.: Cell at-models for digital volumes. In: Torsello, A., Escolano, F., Brun, L. (eds.) GbRPR 2009. LNCS, vol. 5534, pp. 314–323. Springer, Heidelberg (2009)
8. Mrozek, M., Pilarczyk, P., Żelazna, N.: Homology algorithm based on acyclic subspace. Computers and Mathematics with Applications 55, 2395–2412 (2008)
9. Peltier, S., Alayrangues, S., Fuchs, L., Lachaud, J.: Computation of homology groups and generators. Computer & Graphics 30(1), 62–69 (2006)
10. Computational Homology Project, `http://chomp.rutgers.edu/`
11. Gameiro, M., Mischaikow, K., Kalies, W.: Topological Characterization of Spatial-Temporal Chaos. Physical Review E 70 3 (2004)
12. Gameiro, M., Pilarczyk, P.: Automatic homology computation with application to pattern classification. RIMS Kokyuroku Bessatsu B3, 1–10 (2007)
13. Krishan, K., Gameiro, M., Mischaikow, K., Schatz, M., Kurtuldu, H., Madruga, S.: Homology and symmetry breaking in Rayleigh-Bénard convection: Experiments and simulations. Physics of Fluids 19, 117105 (2007)
14. Niethammer, M., Stein, A., Kalies, W., Pilarczyk, P., Mischaikow, K., Tannenbaum, A.: Analysis of blood vessel topology by cubical homology. In: Proc. of the International Conference on Image Processing, vol. 2, pp. 969–972 (2002)
15. Żelawski, M.: Pattern recognition based on homology theory. Machine Graphics and Vision 14, 309–324 (2005)
16. Serre, J.: Homologie singulière des espaces fibrés, applications. Annals of Math. 54, 429–505 (1951)

17. Hatcher, A.: Algebraic Topology. Cambridge University Press, Cambridge (2001)
18. Eilenberg, S., Mac Lane, S.: On the groups $h(\pi, n)$, i, ii, iii. Annals of Math. 58, 60, 60, 55–106,48–139, 513–557 (1953, 1954)
19. Sergeraert, F.: The computability problem in algebraic topology. Advances in Mathematics 104, 1–29 (1994)
20. Barnes, D.W., Lambe, L.A.: A fixed point approach to homological perturbation theory. Proc. Amer. Math. Soc. 112, 881–892 (1991)
21. Forman, R.: A Discrete Morse Theory for Cell Complexes. In: Yau, S.T. (ed.) Topology and Physics for Raoul Bott. International Press (1995)
22. Molina-Abril, H., Real, P.: Advanced homological information on 3d digital volumes. In: da Vitoria Lobo, N., Kasparis, T., Roli, F., Kwok, J.T., Georgiopoulos, M., Anagnostopoulos, G.C., Loog, M. (eds.) S+SSPR 2008. LNCS, vol. 5342, pp. 361–371. Springer, Heidelberg (2008)
23. González-Diaz, R., Medrano, B., Real, P., Sanchez-Pelaez, J.: Algebraic topological analysis of time-sequence of digital images. In: Ganzha, V.G., Mayr, E.W., Vorozhtsov, E.V. (eds.) CASC 2005. LNCS, vol. 3718, pp. 208–219. Springer, Heidelberg (2005)

Appendix. Example for Algorithm 1

Sample application of the homology computation algorithm to a simple cell complex illustrated in Fig. 8 has been shown in Table 2.

Table 1. Betti numbers and the Alexander-Whitney diagonal computed for the examples shown in Fig. 7. Generators of H_1 are labeled α_i, generators of H_2 — γ_j.

Example	β_0	β_1	β_2	AW_c
(a)	1	2	1	$\gamma_1 = \alpha_1 \otimes \alpha_2$
(b)	1	2	1	0
(c)	1	4	3	$\gamma_1 = \alpha_1 \otimes \alpha_3$, $\gamma_2 = \alpha_2 \otimes \alpha_3$, $\gamma_3 = \alpha_4 \otimes \alpha_3$
(d)	1	4	3	$\gamma_1 = \alpha_1 \otimes \alpha_2$, $\gamma_2 = \alpha_2 \otimes \alpha_3$, $\gamma_3 = \alpha_2 \otimes \alpha_4$

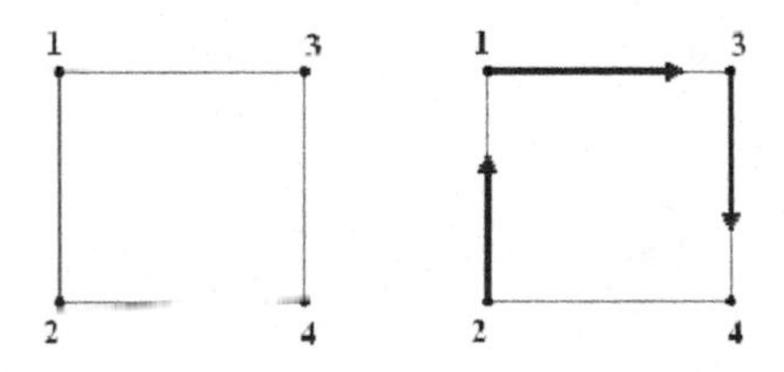

Fig. 8. A sample cell complex and a homology gvf for it expressed in a pictorial way

Table 2. Monitoring of algorithm 1 using the complex shown in Fig. 8

i	0	1	2	3	4	5	6	7
σ	$\langle 1\rangle$	$\langle 2\rangle$	$\langle 3\rangle$	$\langle 4\rangle$	$\langle 1,2\rangle$	$\langle 1,3\rangle$	$\langle 2,4\rangle$	$\langle 3,4\rangle$
π	$\langle 1\rangle$							
ϕ	0							
π	$\langle 1\rangle$	$\langle 2\rangle$						
ϕ	0	0						
π	$\langle 1\rangle$	$\langle 2\rangle$	$\langle 3\rangle$					
ϕ	0	0	0					
π	$\langle 1\rangle$	$\langle 2\rangle$	$\langle 3\rangle$	$\langle 4\rangle$				
ϕ	0	0	0	0				
π	$\langle 2\rangle$	$\langle 2\rangle$	$\langle 3\rangle$	$\langle 4\rangle$	$\langle 1,2\rangle$			
ϕ	$\langle 1,2\rangle$	0	0	0	0			
π	$\langle 3\rangle$	$\langle 3\rangle$	$\langle 3\rangle$	$\langle 4\rangle$	0	$\langle 1,3\rangle+\langle 1,2\rangle$		
ϕ	$\langle 1,3\rangle$	$\langle 1,2\rangle+\langle 1,3\rangle$	0	0	0	0		
π	$\langle 4\rangle$	$\langle 4\rangle$	$\langle 4\rangle$	$\langle 4\rangle$	0	0	$\langle 2,4\rangle+\langle 1,2\rangle+\langle 1,3\rangle$	
ϕ	$\langle 1,2\rangle+\langle 2,4\rangle$	$\langle 2,4\rangle$	$\langle 2,4\rangle+\langle 1,2\rangle+\langle 1,3\rangle$	0	0	0	0	
π	$\langle 4\rangle$	$\langle 4\rangle$	$\langle 4\rangle$	$\langle 4\rangle$	0	0	$\langle 2,4\rangle+\langle 1,2\rangle+\langle 1,3\rangle$	$\langle 3,4\rangle+\langle 2,4\rangle+\langle 1,2\rangle+\langle 1,3\rangle$
ϕ	$\langle 1,2\rangle+\langle 2,4\rangle$	$\langle 2,4\rangle$	$\langle 2,4\rangle+\langle 1,2\rangle+\langle 1,3\rangle$	0	0	0	0	0

The intermediary homology results are

$H_1=\langle 1\rangle$; $H_2=\langle 1\rangle,\langle 2\rangle$; $H_3=\langle 1\rangle,\langle 2\rangle,\langle 3\rangle$; $H_4=\langle 1\rangle,\langle 2\rangle,\langle 3\rangle,\langle 4\rangle$; $H_5=\langle 2\rangle,\langle 3\rangle,\langle 4\rangle$; $H_6=\langle 3\rangle,\langle 4\rangle$; $H_7=\langle 4\rangle$; $H_8=\langle 4\rangle,\ \langle 3,4\rangle+\langle 2,4\rangle+\langle 1,2\rangle+\langle 1,3\rangle$

Thinning Algorithms as Multivalued $\mathcal{N}$-Retractions

Carmen Escribano, Antonio Giraldo, and María Asunción Sastre*

Departamento de Matemática Aplicada, Facultad de Informática
Universidad Politécnica de Madrid, Campus de Montegancedo
Boadilla del Monte, 28660 Madrid, Spain

Abstract. In a recent paper we have introduced a notion of continuity in digital spaces which extends the usual notion of digital continuity. Our approach, which uses multivalued maps, provides a better framework to define topological notions, like retractions, in a far more realistic way than by using just single-valued digitally continuous functions. In particular, we characterized the deletion of simple points, one of the most important processing operations in digital topology, as a particular kind of retraction.

In this work we give a simpler algorithm to define the retraction associated to the deletion of a simple point and we use this algorithm to characterize some well known parallel thinning algorithm as a particular kind of multivalued retraction, with the property that each point is retracted to its neighbors.

Keywords: Digital images, digital topology, continuous multivalued function, simple point, retraction, thinning algorithm.

1 Introduction

The notion of continuous function is the fundamental concept in the study of topological spaces, therefore it should play an important role in Digital Topology.

There have been some attempts to define a reasonable notion of continuous function in digital spaces. The first one goes back to A. Rosenfeld [14] in 1986. He defined continuous function in a similar way as it is done for continuous maps in $\mathbb{R}^n$. It turned out that continuous functions agreed with functions taking 4-adjacent points into 4-adjacent points or, equivalently, with functions taking connected sets to connected sets.

More results related with this type of continuity were proved by L. Boxer in [1,2,3,4]. In these papers, he introduces such notions as homeomorphism, retracts and homotopies for digitally continuous functions, applying these notions to define a digital fundamental group, digital homotopies and to compute the fundamental group of sphere-like digital images. However, as he recognizes in [3],

* The authors have been supported by MICINN MTM2006-0825 (A.Giraldo) and UPM-CAM2009-Q061010133.

S. Brlek, C. Reutenauer, and X. Provençal (Eds.): DGCI 2009, LNCS 5810, pp. 275–287, 2009.

there are some limitations with the homotopy equivalences he get. For example, while all simple closed curves are homeomorphic and hence homotopically equivalent with respect to the Euclidean topology, in the digital case two simple closed curves can be homotopically equivalent only if they have the same cardinality.

A different approach was suggested by V. Kovalevsky in [12], using multivalued maps. He calls a multivalued function continuous if the pre-image of an open set is open. He considers, however, that another important class of multivalued functions is that of connectivity preserving mappings. The multivalued approach to continuity in digital spaces has also been used by R. Tsaur and M. Smyth in [15], where a notion of continuous multifunction for discrete spaces is introduced: A multifunction is continuous if and only if it is "strong" in the sense of taking neighbors into neighbors with respect to Hausdorff metric. See the introduction of [5] for a discussion of the limitations of all these multivalued approaches.

In a recent paper [5] the authors presented a theory of continuity in digital spaces which extends the one introduced by Rosenfeld. Our approach uses multivalued maps and provides a better framework to define topological notions, like retractions, in a far more realistic way than by using just single-valued digitally continuous functions. In particular, we characterized the deletion of simple points, one of the most important processing operations in digital topology, as a particular kind of retraction.

In this work we deepen into the properties of this family of continuous maps, now concentrating on parallel deletion of simple points and thinning algorithms.

In section 2 we revise the basic notions on digital topology required throughout the paper. In section 3 we recall the notion of continuity for multivalued functions and its basic properties, in particular those related with the notion of a digital retraction. In section 4 we give a new proof of our previous result characterizing the deletion of simple points in terms of digitally continuous multivalued functions, giving a new simpler algorithm to find the multivalued function associated to the deletion of a simpler point. Sections 5 is devoted to parallel deletion of simple points. We use our new algorithm to characterize some well known parallel thinning algorithms as digital multivalued retractions with the property that each point is retracted to its neighbors.

For information on Digital Topology we recommend the survey [10] and the books by Kong and Rosenfeld [11], and by Klette and Rosenfeld [8].

We are grateful to the referees for helpful comments and suggestions.

2 Single Valued Continuity in Digital Spaces

We consider $\mathbb{Z}^2$ as model for the digital plane.

Two points in the digital plane $\mathbb{Z}^2$ are 8-adjacent if they are different and their coordinates differ in at most a unit. They are said 4-adjacent if they are 8-adjacent and differ in at most a coordinate. Given $p \in \mathbb{Z}^2$ we define $\mathcal{N}(p)$ as the set of points 8-adjacent to p, i.e. $\mathcal{N}(p) = \{p_1, p_2, \ldots, p_8\}$. This is also denoted as $\mathcal{N}_8(p)$. Analogously, $\mathcal{N}_4(p)$ is the set of points 4-adjacent to p (with the above notation $\mathcal{N}_4(p) = \{p_2, p_4, p_6, p_8\}$).

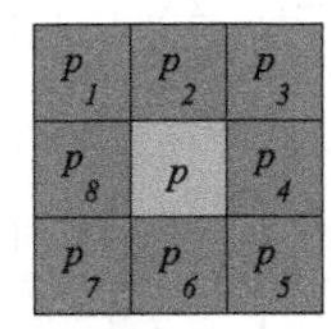

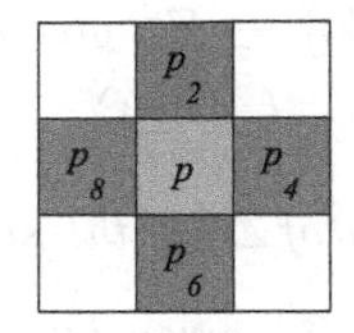

Fig. 1. $\mathcal{N}_8(p)$ and $\mathcal{N}_4(p)$ with their points labeled as used in the paper

A k-path P in $\mathbb{Z}^2$ ($k \in \{4, 8\}$) from the point q_0 to the point q_r is a sequence $P = \{q_0, q_1, q_2, \ldots, q_r\}$ of points such that q_i is k-adjacent to q_{i+1}, for every $i \in \{0, 1, 2, \ldots, r-1\}$. If $q_0 = q_r$ then it is called a closed path. A set $S \subset \mathbb{Z}^2$ is k-connected if for every pair of points of S there exists a k-path contained in S joining them. A k-connected component of S is a k-connected maximal set.

Jordan curve theorem for $\mathbb{R}^2$ states that any simple closed arc divides the plane in 2 connected components. A classical result in digital topology states that any simple closed k-arc divides the digital plane in two $\bar{k}$-connected components, where $\bar{k} = 4$ if $k = 8$, $\bar{k} = 8$ if $k = 4$ and a closed k-arc is a closed k-path $P = \{q_0, q_1, q_2, \ldots, q_r = q_0\}$ such that the only pairs of k-adjacent points of P are $(q_0, q_1), (q_1, q_2), \ldots, (q_{r-1}, q_r)$. Through the paper, given k, $\bar{k}$ will denote the complementary adjacency $\bar{k} = 12 - k$.

Let $f : X \subset \mathbb{Z}^2 \longrightarrow \mathbb{Z}^2$ be a function. According to [14], f is (k, k')-continuous if and only if f sends k-adjacent points to k'-adjacent points.

Examples of (k, k)-continuous functions are the identity, any constant map, translations $f(z) = z + r$, inversions $f(z_1, z_2) = (z_2, z_1)$,... On the other hand, a expansion like $f(z) = 2z$ is not continuous if X is connected with more than one point. Moreover, any $(k, 4)$-continuous function is $(k, 8)$-continuous and any $(8, k')$-continuous function is $(4, k')$-continuous.

In [14], Rosenfeld stated and proved several results about digitally continuous functions related to operations with continuous functions, intermediate values property, almost-fixed point theorem, Lipschitz conditions, one-to-oneness, ... Boxer [1,2,3] expanded this notion to digital homeomorphisms, retractions, extensions, homotopies, digital fundamental group, induced homomorphisms, ...(see also [7] and [9] for previous related results).

3 Digitally Continuous Multivalued Functions

In the following definition we recall the concept of subdivision of $\mathbb{Z}^2$. This notion and the ones that follow can be defined for $\mathbb{Z}^n$ (the general definition can be found in [5]).

Definition 1. *The first subdivision of $\mathbb{Z}^2$ is formed by the set*

$$\mathbb{Z}_1^2 = \left\{ \left(\frac{z_1}{3}, \frac{z_2}{3} \right) \mid (z_1, z_2) \in \mathbb{Z}^2 \right\}$$

and the $3:1$ *map* $i : \mathbb{Z}_1^2 \hookrightarrow \mathbb{Z}^2$ *given by* $i\left(\frac{z_1}{3}, \frac{z_2}{3}\right) = (z_1', z_2')$ *where* (z_1', z_2') *is the point in* $\mathbb{Z}^2$ *closer to* $\left(\frac{z_1}{3}, \frac{z_2}{3}\right)$.

The r*-th subdivision of* $\mathbb{Z}^2$ *is the set* $\mathbb{Z}_r^2 = \left\{\left(\frac{z_1}{3^r}, \frac{z_2}{3^r}\right) \mid (z_1, z_2) \in \mathbb{Z}^2\right\}$ *and the* $3^r : 1$ *map* $i_r : \mathbb{Z}_r^2 \hookrightarrow \mathbb{Z}^2$ *given by* $i_r\left(\frac{z_1}{3^r}, \frac{z_2}{3^r}\right) = (z_1', z_2')$ *where* (z_1', z_2') *is the point in* $\mathbb{Z}^2$ *closer to* $\left(\frac{z_1}{3^r}, \frac{z_2}{3^r}\right)$. *Observe that* $i_1 = i$ *and* $i_r = i \circ i \circ \overset{(r)}{\ldots} \circ i$.

Moreover, if we consider in $\mathbb{Z}^2$ *a* k*-adjacency relation, we can consider in* $\mathbb{Z}_r^2$, *in an immediate way, the same adjacency relation, i.e.,* $\left(\frac{z_1}{3^r}, \frac{z_2}{3^r}\right)$ *is* k*-adjacent to* $\left(\frac{z_1'}{3^r}, \frac{z_2'}{3^r}\right)$ *if and only if* (z_1, z_2) *is* k*-adjacent to* (z_1', z_2').

Proposition 1. i_r *is* k*-continuous as a function between digital spaces.*

Definition 2. *Given* $X \subset \mathbb{Z}^2$, *the* r*-th subdivision of* X *is the set* $X_r = i_r^{-1}(X)$.

Intuitively, if we consider X made of pixels, the r-th subdivision of X consists in replacing each pixel with 9^r pixels and the map i_r is the inclusion. The reason to divide a pixel in 3×3 pixels in each subdivision (and not, for example in 2×2) is due to the fact that the 3×3 mask is at the basis of most digital topology operations (see, for example, Remark 3 or Theorem 1).

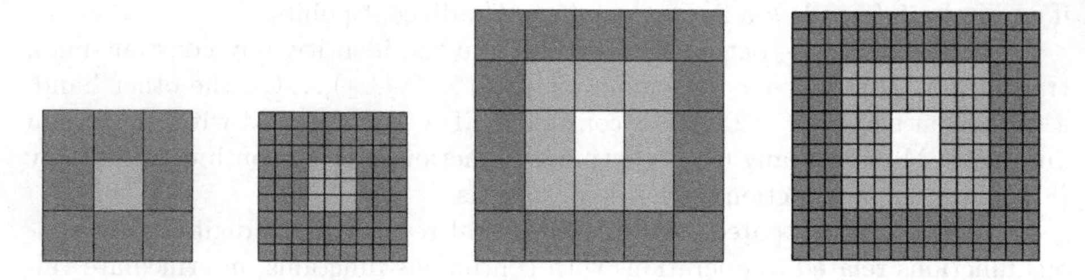

Fig. 2. Some sets and subsets and their first subdivisions (see Proposition 2)

Remark 1. *Given* $X, Y \subset \mathbb{Z}^2$, *any function* $f : X_r \longrightarrow Y$ *induces in an immediate way a multivalued function* $F : X \longrightarrow Y$ *where* $F(x) = \bigcup_{x' \in i_r^{-1}(x)} f(x')$.

Definition 3. *Consider* $X, Y \subset \mathbb{Z}^2$. *A multivalued function* $F : X \longrightarrow Y$ *is said to be a* (k, k')*-continuous multivalued function if it is induced by a* (k, k')*-continuous (single-valued) function from* X_r *to* Y *for some* $r \in \mathbb{N}$.

In the following remark we state some properties of digitally continuous multivalued functions. For more results and details the reader is referred to [5].

Remark 2. *Any single-valued digitally continuous function is continuous as a multivalued function. In particular, any constant map is continuous as a* *multivalued map. Moreover, if $F : X \longrightarrow Y$ $(X, Y \subset \mathbb{Z}^2)$ is a (k, k')-continuous multivalued function, then*

i) $F(x)$ is k'-connected, for every $x \in X$,
ii) if x and y are k-adjacent points of X, then $F(x)$ and $F(y)$ are k'-adjacent subsets of Y,
iii) F takes k-connected sets to k'-connected sets,
iv) if $X' \subset X$ then $F|_{X'} : X' \longrightarrow Y$ is a (k, k')-continuous multivalued function,
v) the composition of continuous multivalued function is a continuous multivalued function.

Remark 3. *In [6] the morphological operations of dilation, erosion and closing have also be modeled as digitally continuous multivalued maps.*

Definition 4. *Let $X \subset \mathbb{Z}^2$ and $Y \subset X$. We say that Y is a k-retract of X if there exists a k-continuous multivalued function $F : X \longrightarrow Y$ (a multivalued k-retraction) such that $F(y) = \{y\}$ if $y \in Y$.*

If moreover $F(x) \subset \mathcal{N}(x)$ for every $x \in X \setminus Y$, we say that F is a multivalued $(\mathcal{N}, k)$-retraction.

The following results are given in [5]. Their proofs are based on the subdivisions shown in Figure 2.

Proposition 2. *i) The boundary ∂X of a square X is not a k-retract of X.*
ii) The outer boundary ∂X of an annulus X is a k-retract of X.

These results improve those for single-valued maps. For them, the boundary of a filled square is not a retract of the whole square [1] but neither is the outer boundary of a squared annulus a digital retract of it.

4 Sequential Deletion of Simple Points as Retractions

It may seem that the family of continuous multivalued functions could be too wide, therefore not having good properties. In this section we show that this is not the case. We show, in particular, that the existence of a k-continuous multivalued function from a set X to $X \setminus \{p\}$ which leaves invariant $X \setminus \{p\}$ is closely related to p being a k-simple point of X.

If $X \subset \mathbb{Z}^2$, a point $p \in X$ is k-simple ($k = 4, 8$) in X (see [10]) if its deletion does not change the topology of X in the sense that after deleting it:

– no k-connected component of X vanishes or is split in several components,
– no $\bar{k}$-connected component of $\mathbb{Z}^2 \setminus X$ is created or is merged with the background or with another such component (remind that, as defined in section 2, $\bar{k} = 4$ if $k = 8$ and $\bar{k} = 8$ if $k = 4$).

A k-simple point can be locally detected by the following characterization. A point p is k-simple if the number of k-connected components of $\mathcal{N}(p) \cap X$ which are k-adjacent to p is equal to 1 and $\mathcal{N}_{\bar{k}}(p) \cap X^c \neq \emptyset$ (this last condition is equivalent to p being a k-boundary point of X).

The following theorem is a restatement of a result given in [5] and presents a new and simpler algorithm to define the map F associated to the deletion of a simple point. The differences with the algorithm in [5] are basically two: we only require to consider the first subdivision of X (and not the second as it happened in [5]), and we obtain smaller images for the deleted simple points which are more suitable for the purposes of this paper. Although we state the result for k-connected sets, this is not a loss of generality because for a general set X it would be applied to the connected component containing the simple point we want to delete.

Theorem 1. *Let $X \subset \mathbb{Z}^2$ be k-connected and consider $p \in X$. Suppose there exists a k-continuous multivalued function $F : X \longrightarrow X \setminus \{p\}$ such that $F(x) = \{x\}$ if $x \neq p$ and $F(p) \subset \mathcal{N}(p)$. Then p is a k-simple point.*
The converse is true under the following conditions:

a) for $k = 8$ it is always true and, moreover, we can impose that $F(p) \subset \mathcal{N}_4(p)$ whenever p is not 4-isolated,
b) for $k = 4$ it is true if and only p is not 8-interior to X.

Proof. The proof is similar to that of Theorem 2 in [5]. To define our new and simpler algorithm for the converse statement, consider a simple point p in the hypothesis of the theorem. Then we have two excluding possibilities

1. $\mathcal{N}(p) \cap X \subset \{p_1, p_3, p_5, p_7\}$, then $k = 8$ and $\mathcal{N}(p) \cap X$ consists on just one of those points, say p_i, and we define $F(p) = p_i$,
2. $\mathcal{N}_4(p) \cap X \neq \emptyset$, in which case there exist 2 points in $\mathcal{N}_4(p)$ in opposite sides of p such that one of them is in X and the other one is not in X (because p is a k-simple point and for $k = 4$ we exclude the case of p being a interior point).
   In this case, we rotate the neighborhood of p in order that, if we label $\mathcal{N}(p)$ and $i^{-1}(p)$ as in the following figure, then $p_2 \notin X$ and $p_6 \in X$.

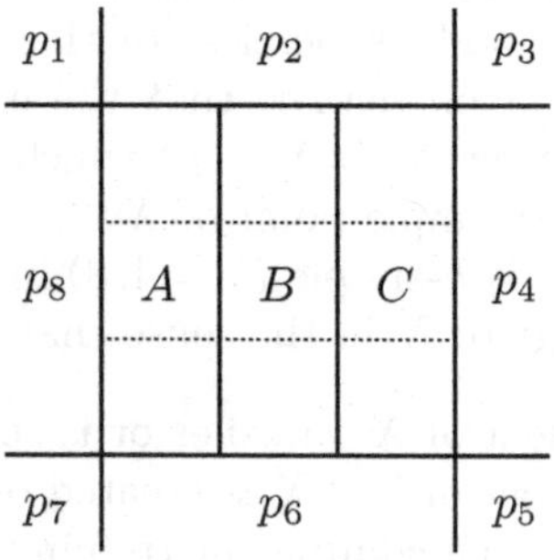

p_1	p_2	p_3
p_8	A B C	p_4
p_7	p_6	p_5

Fig. 3. Labels for $i^{-1}(p)$ and $\mathcal{N}(p)$

Then we define f according to the following cases:

a) $\underline{k=8}$. We define $f(B) = p_6$ and
  i) if $p_3 \in \mathcal{N}(p) \cap X$, then $f(C) = p_4$, otherwise $f(C) = p_6$,
  ii) if $p_1 \in \mathcal{N}(p) \cap X$, then $f(A) = p_8$, otherwise $f(A) = p_6$,

b) $\underline{k=4}$. We define $f(B) = p_6$ and
  i) if $p_4 \in \mathcal{N}(p) \cap X$, then $f(C) = p_5$, otherwise $f(C) = p_6$,
  ii) if $p_8 \in \mathcal{N}(p) \cap X$, then $f(A) = p_7$, otherwise $f(A) = p_6$.

Therefore, the multivalued map F induced by f is defined by $F(x) = \{x\}$ if $x \neq p$ and

a) for $k = 8$,

$$F(p) = \begin{cases} \{p_6\} & \text{if } \mathcal{N}(p) \cap X \subset \{p_4, p_5, p_6, p_7, p_8\} \\ \{p_4, p_6\} & \text{if } \mathcal{N}(p) \cap X = \{p_3, p_4, p_5, p_6, p_7, p_8\} \\ \{p_6, p_8\} & \text{if } \mathcal{N}(p) \cap X = \{p_4, p_5, p_6, p_7, p_8, p_1\} \\ \{p_4, p_6, p_8\} & \text{if } \mathcal{N}(p) \cap X = \{p_3, p_4, p_5, p_6, p_7, p_8, p_1\}, \end{cases}$$

b) for $k = 4$,

$$F(p) = \begin{cases} \{p_6\} & \text{if } \mathcal{N}(p) \cap X \subset \{p_1, p_3, p_5, p_6, p_7\} \\ \{p_5, p_6\} & \text{if } \{p_4, p_5, p_6, p_7\} \subset \mathcal{N}(p) \cap X \subset \{p_1, p_3, p_4, p_5, p_6, p_7\} \\ \{p_6, p_7\} & \text{if } \{p_5, p_6, p_7, p_8\} \subset \mathcal{N}(p) \cap X \subset \{p_1, p_3, p_5, p_6, p_7, p_8\} \\ \{p_5, p_6, p_7\} & \text{if } \{p_4, p_5, p_6, p_7, p_8\} \subset \mathcal{N}(p) \cap X. \end{cases}$$

Remark 4. *It is important to note that if p is a k-simple point, then $F(p)$ is contained in the component of $\mathcal{N}(p) \cap X$ which is k-adjacent to p. This is useful when defining F in a specific situation.*

Since the composition of k-continuous multivalued functions is a k-continuous multivalued function, we have the following result.

Corollary 1. *Let $X \subset \mathbb{Z}^2$ be k-connected and consider $Y \subset X$ such that Y is obtained from X by a sequential deletion of k-simple points such that none of them is 8-interior in the remainder. Then there exists a k-continuous multivalued k-retraction from X to Y.*

It is interesting to note that the ideas in Theorem 2 in [5] and in the previous theorem can be applied also to pairs of points whose simultaneous deletion does not change the k-topology. Such points, called k-simple pairs, are essential to verify the correctedness of parallel thinning algorithms and are locally characterized as follows: A pair $\{p, q\}$ is a k-simple pair if and only if at least one of them is not a k-interior point and the number of k-components of $\mathcal{N}(p, q) \cap X$ which are k-adjacent to $\{p, q\}$ is 1, where $\mathcal{N}(p, q) = ((\mathcal{N}(p) \cup \mathcal{N}(q)) \setminus \{p, q\}$.

The following result, proved in [6], characterizes the deletion of simple pairs in a similar way as Theorem 1 does it for simple points.

Theorem 2. *Consider $X \subset \mathbb{Z}^2$ k-connected and consider a pair $\{p, q\} \subset X$ of 4-adjacent points of X. Suppose that there exists a k-continuous multivalued function $F : X \longrightarrow X \setminus \{p, q\}$ such that $F(x) = \{x\}$ if $x \neq p, q$, and $F(p), F(q) \subset \mathcal{N}(p, q)$. Then $\{p, q\}$ is a k-simple pair.*

The converse is true under the following conditions:

a) for $k = 8$ it is always true and, moreover, we can impose that $F(\{p,q\}) \subset \mathcal{N}_4(p,q) = ((\mathcal{N}_4(p) \cup \mathcal{N}_4(q)) \setminus \{p,q\}$ whenever $\{p,q\}$ is not a 4-connected component of X,
b) for $k = 4$ it is true if and only p or q is not 8-interior to X.

Observe that in the above theorem we do not require that $F(p) \subset \mathcal{N}(p)$ and $F(q) \subset \mathcal{N}(q)$ as in $(\mathcal{N}, k)$-retractions. If we consider this additional requirement we obtain the following result, which deals with the case of a k-simple pair made of k-simple points, also proved in [6] (note that a pair of 4-adjacent k-simple points needs not be a k-simple pair as shown by the example in Figure 4).

Theorem 3. *Consider $X \subset \mathbb{Z}^2$ k-connected and consider a pair $\{p,q\}$ of 4-adjacent points of X. Suppose that there exists a k-continuous multivalued function $F : X \longrightarrow X \setminus \{p,q\}$ such that $F(x) = \{x\}$ if $x \neq p,q$, $F(p) \subset \mathcal{N}(p)$ and $F(q) \subset \mathcal{N}(q)$. Then $\{p,q\}$ is a k-simple pair of k-simple points.*
The converse is true under the following conditions:

a) for $k = 8$ it is always true and, moreover, we can impose that $F(p) \subset \mathcal{N}_4(p)$ and $F(q) \subset \mathcal{N}_4(q)$,
b) for $k = 4$ it is true if and only p or q is not 8-interior to X.

5 Thinning Algorithms as Multivalued $(\mathcal{N}, k)$-Retractions

It is well known that the parallel deletion of simple points needs not to preserve topology. The simplest example is given by the following figure.

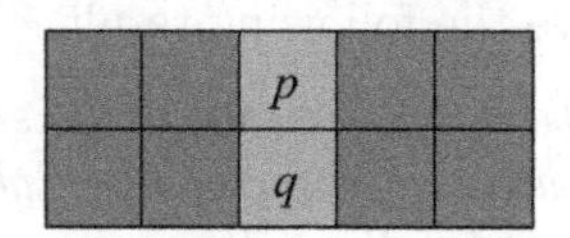

Fig. 4. A not k-deletable pair of k-simple points

Observe that p and q are both simple points but if we delete both then the connectedness (and hence the topology) of the set changes.

Two well known conditions that guarantee the preservation of topology in a parallel deletion of simple points are the following:

i) We can delete only north (or west or south or east) boundary simple points (a north boundary point p of X is any point of X such that $p_2 \notin X$ where p_2 is the point of $\mathcal{N}(p)$ just above p).
ii) We can delete only simple points of a subfield, for example, points whose integer coordinates add to an even number.

In both cases the connected components formed by only two points have to be considered independently.

In general thinning algorithms are divided in subiterations [8]. For example, in the first case, at odd subiterations a point is deleted if it is k-simple and north or west boundary while at even subiterations a point is deleted if it is south or east boundary (deleting simultaneously points with two different orientations, for example north and west, obliges for $k = 8$, to consider independently the connected components formed by two or three points mutually 8-adjacent). In the second case, at odd subiterations we delete all 8-simple points whose integer coordinates add to an even number, while in the even subiterations we delete all 8-simple points whose integer coordinates add to an odd number.

The basic notion behind thinning algorithms based on the parallel deletion of simple points is the following.

Definition 5. *Let $X \subset \mathbb{Z}^2$ and $D \subset X$. D is called k-deletable ($k = 4, 8$) in X if its deletion does not change the topology of X in the sense that after deleting D:*

- *no k-connected component of X vanishes or is split in several components,*
- *no $\bar{k}$-connected component of $\mathbb{Z}^2 \setminus X$ is created or is merged with the background or merged with another such component,*

Ronse [13] called these sets strongly k-deletable.

The following theorem relates deletable sets and multivalued retractions.

Theorem 4. *Let $X \subset \mathbb{Z}^2$ be k-connected and consider $D \subset X$. If D is deletable such that D does not contain 8-interior points, then there exists a multivalued k-retraction from X to $X \setminus D$.*

Proof. If D is k-deletable, then its points can be sequentially deleted in such a way that any of these points is a simple point of the remainder [13]. Therefore, by Corollary 1, there exists a k-continuous multivalued function $F : X \longrightarrow X \setminus D$ such that $F(x) = \{x\}$ if $x \in X \setminus D$.

Although the deletion of a k-simple point is a $(\mathcal{N}, k)$-retraction, when deleting simple points sequentially, the resulting composition of all these $(\mathcal{N}, k)$-retractions needs not to be a $(\mathcal{N}, k)$-retraction as the following example shows.

Example 1. *Consider X and $D = \{p_1, p_2, p_3\}$ are in Figure 5.*

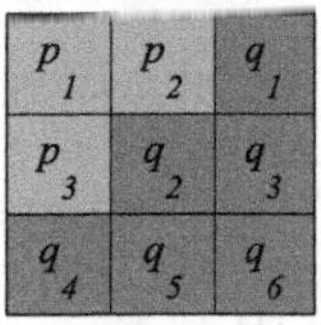

p_1	p_2	q_1
p_3	q_2	q_3
q_4	q_5	q_6

Fig. 5. A sequential deletion of simple points which is not a $\mathcal{N}$-retraction

Then, since p_1 is 4-simple we may delete it by a $(\mathcal{N},4)$-retraction taking it to $\{p_2\}$ or $\{p_3\}$. Suppose we take it to p_2 (if we choose p_3 the result is analogous). Since p_2 is simple in the remainder, we may delete it by a $(\mathcal{N},4)$-retraction taking p_2 to $\{q_1,q_3\}$ or $\{q_2,q_3\}$ (even if we choose to delete p_3 before). But then, the composition of these two retractions would take p_1 to $\{q_1,q_3\}$ or $\{q_2,q_3\}$ and, although it is a 4-retraction it is not a $(\mathcal{N},4)$-retraction.

On the other hand, if we started by deleting p_2 instead of p_1 (starting with p_3 would be analogous), we may delete it by a $(\mathcal{N},4)$-retraction taking it to $\{q_2\}$. Now, p_3 is not 4-simple in the remainder so we should delete p_1 by a $(\mathcal{N},4)$-retraction taking it to $\{p_3\}$. Finally, p_3 would be taken to $\{q_2,q_5\}$ or $\{q_4,q_5\}$. But then, the composition of these three retractions would take p_1 to $\{q_4,q_5\}$ and, although it is a 4-retraction it is not a $(\mathcal{N},4)$-retraction.

However, it is possible to define a $(\mathcal{N},4)$-retraction F which deletes p_1,p_2,p_3 by defining $F(p_1)=\{q_2\}$, $F(p_2)=\{q_3\}$ and $F(p_3)=\{q_5\}$.

In the next two results we show that the usual thinning algorithms based on conditions (i) and (ii) stated at the beginning of this section have subiterations which can be modeled by $(\mathcal{N},k)$-retractions which can be constructed explicitly, using only the first subdivision. As a consequence, being the composition of all their subiterations, each of these thinning algorithms can be modeled as a digitally continuous multivalued function.

Theorem 5. *Let $X\subset\mathbb{Z}^2$ be a k-connected set with more than 2 points and let D be a subset of north boundary k-simple points of X. Then there exists a multivalued $(\mathcal{N},k)$-retraction $F: X\longrightarrow X\setminus D$.*

Proof. We show first how to define the images of all the points of D in such a way that their images are contained in $X\setminus D$. To do that, consider $p\in D$ and label the points of $\mathcal{N}(p)$ as in figure 1.

k=4: If $p_6\in X$, if we define $F(p)$ as in Theorem 1 then it is easy to see that $F(p)$ does not include any north boundary point. On the other hand, if $p_6\notin X$, since p is 4-simple, $p_4\notin X$ or $p_8\notin X$. Suppose $p_4\notin X$ (the case $p_8\notin X$ is symmetric). Then $p_8\in X\setminus D$, because p_8 can not be simple, and we define, according to Theorem 1, $F(p)=p_8$.

k=8: If $p_6\in X$, if we define $F(p)$ as in Theorem 1 then it is easy to see that $F(p)$ does not include any north boundary point. On the other hand, if $p_6\notin X$, since p is 4-simple, $\{p_3,p_4,p_5\}\cap X=\emptyset$ or $\{p_7,p_8,p_1\}\cap X=\emptyset$. Suppose $\{p_3,p_4,p_5\}\cap X=\emptyset$ (the other case is symmetric). Then, we define $F(p)=p_8$ if $p_8\notin D$ and $F(p)=p_7$ if $p_8\in D$.

We see now that the function F defined as above is k-continuous. Suppose $D=\{d_1,d_2,\dots,d_n\}$. Consider, for every $i=1,2,\dots,n$, the multivalued k-continuous map $F_i: X\longrightarrow X\setminus\{d_i\}$ which leaves fixed all points in $X\setminus\{d_i\}$ and such that $F_i(d_i)$ is defined as above. Then the above defined $F: X\longrightarrow X\setminus D$ agrees with the composition $F=F_n\circ F_{n-1}\circ\cdots\circ F_2\circ F_1$ and hence is a k-continuous multivalued function. Therefore, F is a multivalued $(\mathcal{N},k)$-retraction from X to $X\setminus D$.

Theorem 6. *Let $X \subset \mathbb{Z}^2$ be a k-connected set with more than 2 points and let D be a subset of 8-simple points of X such that all the points of D have coordinates with even sum. Then there exists a multivalued $(\mathcal{N}, k)$-retraction $F : X \longrightarrow X \setminus D$.*

Proof. We have to show how to define, in a coherent way, the images of all the points of D. To do that, consider $p \in D$. By the algorithm in Theorem 1, we have to consider two cases:

- $\mathcal{N}(p) \cap X \subset \{p_1, p_3, p_5, p_7\}$, then and $\mathcal{N}(p) \cap X$ consists on just one of those points, say p_i, and we define $F(p) = p_i$. Moreover, since p_i is the only neighbor of p, then p_i can not be 8-simple and hence $p_i \notin D$.
- $\mathcal{N}_4(p) \cap X \neq \emptyset$, in which case $F(p) \subset \mathcal{N}_4(p)$ where $\mathcal{N}_4(p) \cap D = \emptyset$.

Therefore, we can define the images of all the points of D according the algorithm in Theorem 1.

The proof that the function F defined as above is k-continuous is similar to that of the previous theorem.

The proof of the continuity of F in the previous two results suggests the following result.

Proposition 3. *Let $X \subset \mathbb{Z}^2$ be a k-connected set and let $D = \{d_1, d_2, \ldots, d_n\} \subset X$ be a set of k-simple points of X. Suppose that, for every $i = 1, 2, \ldots, n$, there exists a multivalued k-continuous map $F_i : X \longrightarrow X \setminus \{d_i\}$ which leaves fixed all points in $X \setminus \{d_i\}$ and such that $F_i(d_i) \subset \mathcal{N}(p) \cap (X \setminus D)$. Then the composition $F = F_n \circ F_{n-1} \circ \cdots \circ F_2 \circ F_1$ is a multivalued $(\mathcal{N}, k)$-retraction from X to $X \setminus D$.*

Consider now the following result, proved in [6], which can be seen as the converse of some results presented in this paper.

Theorem 7. *Let $X \subset \mathbb{Z}^2$ be a k-connected set and let D be a set of k-simple points of X such that there exists a multivalued $(\mathcal{N}, k)$-retraction $F : X \longrightarrow X \setminus D$. Then D is k-deletable.*

The condition of D formed by k-simple points is necessary and, on the other hand, the condition of the points in D being k-simple can not be deduced from the rest of the hypothesis, as shown by the examples in Figure 6.

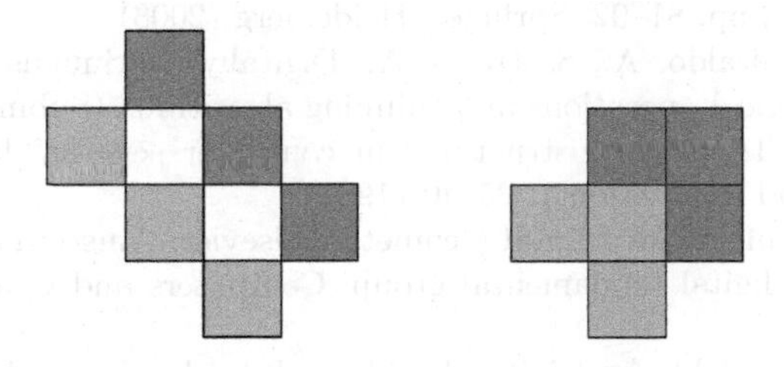

Fig. 6. Left: A $(\mathcal{N}, 8)$-retractable but not 8-deletable set of non 8-simple points. Right: A $(\mathcal{N}, 8)$-retractable and 8-deletable set of points not all of them 8-simple

The proof of this result is based on Ronse [8] sufficient conditions for the parallel deletion of the points of a subset D of a set X to preserve topology.

If we consider Proposition 3 and Theorem 7 together we would obtain the following local criterion for the parallel deletion of the points of a subset D of a set X to preserve topology.

Theorem 8. *Let $X \subset \mathbb{Z}^2$ be a k-connected set and let $D = \{d_1, d_2, \ldots, d_n\} \subset X$ be a set of k-simple points of X. Suppose that, for every $i = 1, 2, \ldots, n$, there exists a multivalued k-continuous map $F_i : X \longrightarrow X \setminus \{d_i\}$ which leaves fixed all points in $X \setminus \{d_i\}$ and such that $F_i(d_i) \subset \mathcal{N}(p) \cap (X \setminus D)$. Then D is k-deletable.*

6 Conclusion

In this paper we have continued the program started in [5] on digitally continuous multivalued maps, now focusing on retractions, and in particular, on the notion, introduced here, of $(\mathcal{N}, k)$-retraction. These types of retractions have the property that each point is retracted to its neighbors. We have modeled the deletion of simple points, one of the most important processing operations in digital topology, as a $(\mathcal{N}, k)$-retraction, and we have given a simple algorithm, requiring only the first subdivision, to define explicitly this retraction. Moreover, we have extended this algorithm to characterize some well known thinning algorithms.

References

1. Boxer, L.: Digitally continuous functions. Pattern Recognition Letters 15, 833–839 (1994)
2. Boxer, L.: A Classical Construction for the Digital Fundamental Group. Journal of Mathematical Imaging and Vision 10, 51–62 (1999)
3. Boxer, L.: Properties of Digital Homotopy. Journal of Mathematical Imaging and Vision 22, 19–26 (2005)
4. Boxer, L.: Homotopy properties of Sphere-Like Digital Images. Journal of Mathematical Imaging and Vision 24, 167–175 (2006)
5. Escribano, C., Giraldo, A., Sastre, M.A.: Digitally continuous multivalued functions. In: Coeurjolly, D., Sivignon, I., Tougne, L., Dupont, F. (eds.) DGCI 2008. LNCS, vol. 4992, pp. 81–92. Springer, Heidelberg (2008)
6. Escribano, C., Giraldo, A., Sastre, M.A.: Digitally continuous multivalued functions, morphological operations and thinning algorithms (Submitted)
7. Khalimsky, E.: Topological structures in computer science. Journal of Applied Mathematics and Simulation 1, 25–40 (1987)
8. Klette, R., Rosenfeld, A.: Digital Geometry. Elsevier, Amsterdam (2004)
9. Kong, T.Y.: A digital fundamental group. Computers and Graphics 13, 159–166 (1989)
10. Kong, T.Y., Rosenfeld, A.: Digital Topology: Introduction and survey. Computer Vision, Graphics and Image Processing 48, 357–393 (1989)
11. Kong, T.Y., Rosenfeld, A. (eds.): Topological algorithms for digital image processing. Elsevier, Amsterdam (1996)

12. Kovalevsky, V.: A new concept for digital geometry. In: Ying-Lie, O., et al. (eds.) Shape in Picture. Proc. of the NATO Advanced Research Workshop, Driebergen, The Netherlands, Computer and Systems Sciences (1992), vol. 126. Springer, Heidelberg (1994)
13. Ronse, C.: A topological characterization of thinning. Theoretical Computer Science 43, 31–41 (1988)
14. Rosenfeld, A.: Continuous functions in digital pictures. Pattern Recognition Letters 4, 177–184 (1986)
15. Tsaur, R., Smyth, M.B.: Continuous multifunctions in discrete spaces with applications to fixed point theory. In: Bertrand, G., Imiya, A., Klette, R. (eds.) Digital and Image Geometry. LNCS, vol. 2243, pp. 75–88. Springer, Heidelberg (2002)

Characterization of Simple Closed Surfaces in $\mathbb{Z}^3$: A New Proposition with a Graph-Theoretical Approach

Rémy Malgouyres and Jean-Luc Toutant

LAIC - Université d'Auvergne

Abstract. In the present paper, we propose a topological characterization of digital surfaces. We introduce simple local conditions on the neighborhood of a voxel. If each voxel of a 26-connected digital set satisfies them, we prove a Jordan theorem and ensure that this set is strong 6-separating in $\mathbb{Z}^3$. Thus, we consider it as a digital surface.

Introduction

The mathematical study of digital images is composed of digital topology and digital geometry. They aim at providing digital analogues for the usual continuous geometrical and topological notions and should fit the conception of integer only algorithms.

The main difficulty in finding appropriate digital analogues comes from the fact that we consider a discrete space ($\mathbb{Z}^3$), that is, a set of isolated points with integer only coordinates, usually called voxels, in which continuity is no more defined and Euclidean distance can not be used easily.

Intuitively, a simple closed surface may be seen as a closed surface which *does not fold upon itself*. The study of their digital analogues, namely digital simple closed surfaces, remains a key point for representation of digital objects. These shapes are indeed well fitted for representing borders of digital objects. Moreover, they are at the crossing of digital geometry and digital topology: They should be good digital approximation of continuous simple closed surfaces and also should satisfy topological properties such as separating the space or being without singularities.

In the continuous case, it is possible to detect locally the singularities of a surface, this leads to a local characterization of simple closed surfaces. In the digital space $\mathbb{Z}^3$, we look for a local, that is, in a bounded neighborhood of the considered voxel, characterization allowing the same detection.

In the present paper, we study digital surfaces with a graph theoretical approach. A surface is then defined as a thin set of voxels linked by adjacency relations. Our contribution consists in a new notion of 26-connected digital surfaces. We propose some specific properties. If each voxel of a 26-connected digital set satisfies them in its neighborhood, then this set separates the space in two 6-connected components. Moreover, it is a strong 6-separating set in $\mathbb{Z}^3$, i.e., the removal of any of its voxels leads to a non-separating set.

S. Brlek, C. Reutenauer, and X. Provençal (Eds.): DGCI 2009, LNCS 5810, pp. 288–299, 2009.

In the sequel, we first recall some basic notions on digital adjacencies, sets and paths. Then, we present the well accepted definition of closed digital curves and introduce some notions of closed digital surfaces before stating our local characterization and proving a Jordan theorem for it.

1 Basic Notions

1.1 Digital Adjacencies and Connectedness

In the digital plane and the digital space, we consider different adjacency relations:

- two pixels $\mathbf{p}$ et $\mathbf{q} \in \mathbb{Z}^2$ are *4-adjacent* (respectively *8-adjacent*) if $\|\mathbf{p}-\mathbf{q}\|_1 = 1$ (respectively $\|\mathbf{p}-\mathbf{q}\|_\infty = 1$),
- two voxels $\mathbf{v}$ et $\mathbf{w} \in \mathbb{Z}^3$ are *6-adjacent* (respectively *26-adjacent*) if $\|\mathbf{p}-\mathbf{q}\|_1 = 1$ (respectively $\|\mathbf{p}-\mathbf{q}\|_\infty = 1$).

They lead to different types of digital neighborhoods, called k-neighborhood ($k \in \{4,8\}$ in the plane and $k \in \{6,26\}$ in the space). The *k-neighborhood* of a digital set E, $\mathcal{N}_k(\mathrm{E})$ is the set of voxels k-adjacent to at least one voxel of E.

$\mathcal{N}_k^*(\mathrm{E})$ denotes $\mathcal{N}_k(\mathrm{E}) \setminus \mathrm{E}$ and, if F is also a digital set, $\mathcal{N}_k^{\mathrm{F}}(\mathrm{E})$ denotes $\mathcal{N}_k(\mathrm{E}) \cap \mathrm{F}$.

A sequence of voxels $\mathrm{P} = (\mathbf{p_i})_{1 \leqslant i \leqslant n_\mathrm{P}}$ is called a *k-connected digital path* if, for all $i \in \{1, \ldots, n_\mathrm{P}-1\}$, $\mathbf{p_i}$ and $\mathbf{p_{i+1}}$ are k-adjacent. In the sequel, a path X contains n_X elements denoted by $\mathbf{x_1}, \ldots, \mathbf{x_{n_X}}$. $\mathrm{P}^{-1}$is the reverse path of P and we have $(\mathrm{P}^{-1})^{-1} = \mathrm{P}$. If P and Q are two k-connected digital paths such that $\mathbf{p_{n_P}}$ and $\mathbf{q_1}$ are k-adjacent, $\mathrm{P} \cdot \mathrm{Q}$ means the concatenation of the paths P and Q, and we have $(\mathrm{P} \cdot \mathrm{Q})^{-1} = \mathrm{Q}^{-1} \cdot \mathrm{P}^{-1}$.

A digital set E is said to be k-connected if, for all $\mathbf{v}, \mathbf{w} \in \mathrm{E}$, there exists a k-connected path linking them. A digital set E is said to be *k-separating* in a superset F if the complement of E in F, $\overline{\mathrm{E}}$, admits exactly two distinct k-connected components. Moreover, if, for all $\mathbf{v} \in \mathrm{E}$, $\mathrm{E} \setminus \{\mathbf{v}\}$ is no more k-separating in F, E is a *strong k-separating* set in F.

1.2 Digital Curve

The notion of digital path is not sufficient to introduce topological properties and digital curves. A. Rosenfeld has proposed different improvements [1].

Definition 1 (Infinite digital curve [1]). *A* k-connected infinite digital curve *is a k-connected digital path* P *such that:*

1. $\mathbf{p_i} = \mathbf{p_j}$ *if and only if* $i = j$,
2. $\mathbf{p_i}$ *and* $\mathbf{p_j}$ *are k-adjacent if and only if* $j = i \pm 1$.

This definition is a discrete analogue to the notion of non closed simple curves if we consider infinite arc. It does not allow to reach closed simple curves without a slightly change.

Definition 2 (Closed digital curve [1]). *A k-connected closed digital curve is a k-connected digital path* P *such that:*

1. $\mathbf{p_i} = \mathbf{p_j}$ *if and only if* $i = j$,
2. $\mathbf{p_i}$ *and* $\mathbf{p_j}$ *are k-adjacent if and only if* $j = i \pm 1 \mod n_{\mathrm{P}}$.

This definition has the same properties as the latter, unless it ensures a k-adjacency relationship between the first voxel, $\mathbf{p_1}$, and the last one, $\mathbf{p_{n_P}}$.

They both can be expressed as strong separating digital set in the digital plane.

Proposition 1. *Let* E *be a digital subset of* $\mathbb{Z}^2$*. Then,* E *is an infinite or closed 4-connected (respectively 8-connected) digital curve if and only if* E *is a strong 8-separating (respectively strong 4-separating) digital set in* $\mathbb{Z}^2$*.*

Note that the adjacency relationship used for the curve and the adjacency relationship used for the complement are different. In the digital plane, good pair of adjacencies are $(4, 8)$ and $(8, 4)$, in the digital space, they are in $\{(26, 6), (18, 6), (6, 18), (6, 26)\}$, but we mainly focus on $(26, 6)$.

1.3 Digital Surface

The definition of digital curve as strong separating set is very simple but does not extend easily in higher dimension: R. Malgouyres has shown that there is no local characterization of strong 6-separating digital sets in $\mathbb{Z}^3$ [2]. D. G. Morgenthaler and A. Rosenfeld have early proposed a definition of digital surfaces more restrictive than the notions of strong separating set [3], the definition of digital (m, n)-surfaces. It involves two adjacency relations, the first, m, for the digital surface itself and the second, n, for its complement.

This definition is too restrictive, at least in the case of the $(26, 6)$-surfaces: even digital planes [4] are not included in such a class of digital surfaces. More general definitions, accepting digital planes, have been proposed, in particular, by R. Malgouyres and G. Bertrand [5,6,7]. Finally, the class of digital simplicity surfaces is the largest proposed [8,9] with the graph theoretical approach. It is based on the notion of simple points [10] and includes previously mentioned definitions. Nevertheless, this definition does not accept strong 6-separating sets which are good candidates to be digital surfaces. For example, the gray set drawn in Figure 1 does not satisfy the condition to be a simplicity surface.

2 A New Proposition of Digital Surfaces

2.1 Definition

Definition 3. *Let* E *be a 26-connected digital set such that, for all* $\mathbf{v} \in \mathrm{E}$*:*

1. $\mathcal{N}_{26}^{\overline{\mathrm{E}}}(\mathcal{N}_{26}^{\mathrm{E}}(\mathbf{v}))$ *admits exactly two 6-connected components,*
2. *for all* $\mathbf{w} \in \mathcal{N}_{26}^{\mathrm{E}}(\mathbf{v})$*,* $\mathbf{w}$ *is 6-connected to each of these 6-connected components.*

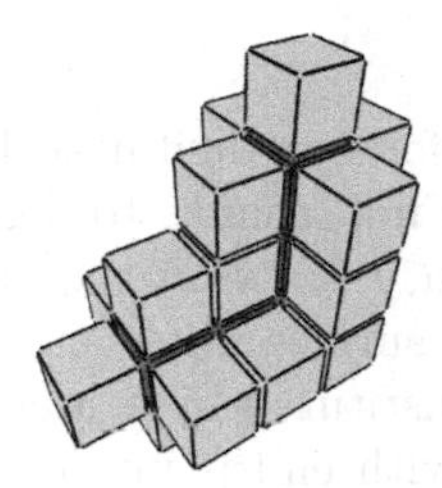

Fig. 1. Strong 6-separating digital set rejected by the definition of simplicity surfaces

Then E is called a digital surface.

The gray digital set drawn in Figure 1 is a digital surface according to definition 3.

Such a definition remains useless without a Jordan theorem. Throughout the present paper, S will denote a digital set satisfying Definition 3. In order to prove that the set S separates its complement $\overline{\mathrm{S}}$ in two distinct 6-connected components, we base our approach on the following assertion: considering two voxels not in S, the parity of the number of intersections with S remains the same for all paths between them. Then, by highlighting paths with odd and even number of intersections, we can state that S separates $\mathbb{Z}^3$. We need first to define what we mean by intersection:

Definition 4 (intersection). *Let* $\mathrm{P} = \mathrm{A} \cdot \mathrm{B} \cdot \mathrm{C}$ *be the concatenation of three 6-connected paths such that:*

1. $\mathbf{a_1}, \mathbf{a_{n_A}}, \mathbf{c_1}, \mathbf{c_{n_C}} \notin \mathrm{S}$
2. *for all* $i \in \{1, \ldots, n_\mathrm{B}\}$, $\mathbf{b_i} \in \mathrm{S}$.

Then, B *is called an* intersection *of* P *with* S, *if* $\mathbf{a_{n_A}}$, $\mathbf{c_1}$ *belong to distinct* 6-*connected components of* $\mathcal{N}_{26}(\mathrm{B}) \cap \overline{\mathrm{S}}$.

For the sake of clarity, we denote by $\mathbb{I}_S(\mathrm{P})$ the set of all intersections of P with S and by $\sharp\mathbb{I}_S(\mathrm{P})$ its cardinality, that is the number of intersections between P and S.

The property we look for can be formalized by: two 6-connected digital paths with same end points not in S, namely P and P′, should satisfy:

$$\sharp\mathbb{I}_S(\mathrm{P}) \equiv \sharp\mathbb{I}_S(\mathrm{P}') \pmod 2. \tag{1}$$

A 6-connected path P between voxels not in S can be viewed as the concatenation of n paths, $\mathrm{R_1}, \ldots, \mathrm{R_n}$ such that for all $i \in \{1, \ldots, n\}$, $\mathrm{R_i}$ is a path such that its end points are not in S. According to this decomposition:

$$\sharp\mathbb{I}_\mathrm{S}(\mathrm{P}) = \sum_{i=1}^{n} \sharp\mathbb{I}_\mathrm{S}(\mathrm{R_i}). \tag{2}$$

3 Jordan Theorem

To prove a Jordan's theorem for our digital surfaces, we reduce the problem according to different results. First, thanks to digital homotopy, we only study directly equivalent paths. Then, results on the neigborhood of an intersection between a path and a digital surface allow us to only consider two sorts of paths in our proofs: a path constituted by a unique voxel not in S and a path constituted of three elements with end points not in S and middle point in it. Endly, we highlight local deformations which preserve the parity of the number of intersections of paths with S.

3.1 Digital Homotopy

We first present some results on digital homotopy due to T. Y. Kong [11] which allow us to later restrict our study.

Definition 5 (Unit square [11]). *Let* $(\mathbf{e_1}, \mathbf{e_2}, \mathbf{e_3})$ *be the canonical basis of the Euclidean three-dimensional space. Then a* unit square K *is a set of four pixels such that:*

$$\mathrm{K} = \{\mathbf{v}, \mathbf{v} + \mathbf{e_i}, \mathbf{v} + \mathbf{e_j}, \mathbf{v} + \mathbf{e_i} + \mathbf{e_j}\},$$

with $(i, j) \in \{1, 2, 3\}^2$ *and* $i \neq j$.

Definition 6 (Reduced form of digital path [11]). *Let* P *be a 6-connected digital path. The* reduced form R *of* P *is obtained by removing all but one voxel from every set of consecutive equal voxels. Thus,* R *is such that for all* $i \in \{1, \ldots, r_{n_\mathrm{R}} - 1\}$, $\mathbf{r_i} \neq \mathbf{r_{i+1}}$ *and we write:*

$$\mathrm{P} \equiv \mathrm{R}. \tag{3}$$

The equivalence and the direct equivalence between digital paths are then define as follows:

Definition 7 (Directly equivalent digital paths [11]). *Let* P *and* P' *be two 6-connected digital paths in a digital set* E *such that* $\mathbf{p_1} = \mathbf{p'_1}$ *and* $\mathbf{p_{n_P}} = \mathbf{p'_{n_{P'}}}$. P *and* P' *are directly equivalent if one of the following properties is satisfied:*

- P *and* P' *have the same reduced form,*
- P *and* P' *differ only in a unit square* K.

Definition 8 (Equivalent digital paths [11]). *Let* P *and* P' *be two 6-connected digital paths with same end points in a digital set* E. P *and* P' *are said to be equivalent if it exists a sequence of 6-connected digital paths* $\mathrm{P_1}, \ldots, \mathrm{P_n}$ *such that:*

1. P *is directly equivalent to* $\mathrm{P_1}$,
2. P' *is directly equivalent to* $\mathrm{P_n}$,
3. *for all* $i \in \{1, \ldots, n-1\}$, $\mathrm{P_i}$ *is directly equivalent to* $\mathrm{P_{i+1}}$.

According to Definition 8, we limit our study to prove that two directly equivalent digital paths in $\mathbb{Z}^3$ have the same parity for their number of intersections with S. The first case of Definition 7 can be directly stated:

Proposition 2. *Let* P *and* P′ *be two* 6*-connected paths such that* $P \equiv P'$*. Then, we have:*

$$\sharp\mathbb{I}_S(P) = \sharp\mathbb{I}_S(P'). \tag{4}$$

Proof. The repetition of a voxel in a path cannot add or delete intersection since such a configuration involves at least three voxels. □

In the second case of Definition 7, the preservation of the parity of $\sharp\mathbb{I}_S(P)$ requires more investigation and we introduce some useful technical lemmas before stating it.

3.2 Preliminaries

We present results on the number of 6-connected components adjacent to a path included in S, we will use thoughout the next part.

Lemma 1. *Let* P *be a* 6*-connected digital path with all of its elements in a surface* S *and such that* $\mathcal{N}_{26}^{\overline{S}}(\mathcal{N}_{26}^{S}(P))$ *admits* c 6*-connected components. Let also* $\mathbf{v}$ *be a voxel belonging to* S *and* 6*-adjacent to* $\mathbf{p_{n_P}}$*. Then,* $\mathcal{N}_{26}^{\overline{S}}(\mathcal{N}_{26}^{S}(P \cdot \mathbf{v}))$ *admits at most* c 6*-connected components.*

Proof. By definition of S, $\mathcal{N}_{26}^{\overline{S}}(\mathcal{N}_{26}^{S}(\mathbf{v}))$ admits only two 6-connected components and $\mathbf{p_{n_P}}$ is 6-adjacent to both of them. The voxels included in $\mathcal{N}_{26}^{\overline{S}}(\mathcal{N}_{26}^{S}(\mathbf{v}))$ and not in $\mathcal{N}_{26}^{\overline{S}}(\mathcal{N}_{26}^{S}(P))$ are all 6-connected to voxels of $\mathcal{N}_{26}^{\overline{S}}(\mathcal{N}_{26}^{S}(P))$. Consequently, $\mathcal{N}_{26}^{\overline{S}}(\mathcal{N}_{26}^{S}(P \cdot \mathbf{v}))$ admits at most c 6-connected components. □

Proposition 3. *Let* P *be a* 6*-connected digital path with all of its elements in a surface* S*. Then* $\mathcal{N}_{26}^{\overline{S}}(\mathcal{N}_{26}^{S}(P))$ *admits at most two distinct* 6*-connected components.*

Proof. P is at least constituted of a voxel $\mathbf{v} \in S$. By definition of S, $\mathcal{N}_{26}^{\overline{S}}(\mathcal{N}_{26}^{S}(\mathbf{v}))$ admits exactly two 6-connected components. Then, by Lemma 1 and induction, $\mathcal{N}_{26}^{\overline{S}}(\mathcal{N}_{26}^{S}(P))$ admits at most two 6-connected components. □

A stronger result can be enounced in the case of intersections:

Corollary 1. *Let* I *be an intersection of a* 6*-connected digital path* P *with a surface* S*. Then* $\mathcal{N}_{26}^{\overline{S}}(\mathcal{N}_{26}^{S}(I))$ *admits exactly two distinct* 6*-connected components.*

Proof. By definition of an intersection, this set admits at least two 6-connected components. By Proposition 3, it admits at most two 6-connected components. □

A last result on connected components in the complement of a path in a digital surface S is:

Lemma 2. *Let* P *be a 6-connected digital path with all of its elements in a surface* S*. Then, for all* $i \in \{1, \ldots, n_{\mathrm{P}}\}$, $\mathbf{p_i}$ *is* 6*-adjacent to each* 6*-connected component of* $\mathcal{N}_{26}^{\overline{\mathrm{S}}}\left(\mathcal{N}_{26}^{\mathrm{S}}(\mathrm{P})\right)$.

Proof. First, consider that $\mathcal{N}_{26}^{\overline{\mathrm{S}}}\left(\mathcal{N}_{26}^{\mathrm{S}}(\mathrm{P})\right)$ admits only one 6-connected component. By definition of S, for all $i \in \{1, \ldots, n_{\mathrm{P}}\}$, $\mathbf{p_i}$ is 6-adjacent to at least two voxels of $\mathcal{N}_{26}^{\overline{\mathrm{S}}}\left(\mathcal{N}_{26}^{\mathrm{S}}(\mathbf{p_i})\right)$ Since $\mathcal{N}_{26}^{\overline{\mathrm{S}}}\left(\mathcal{N}_{26}^{\mathrm{S}}(\mathrm{P})\right) = \bigcup_{i=1}^{n_{\mathrm{P}}}\left(\mathcal{N}_{26}^{\overline{\mathrm{S}}}\left(\mathcal{N}_{26}^{\mathrm{S}}(\mathbf{p_i})\right)\right)$, $\mathbf{p_i}$ is 6-adjacent to the 6-connected set $\mathcal{N}_{26}^{\overline{\mathrm{S}}}\left(\mathcal{N}_{26}^{\mathrm{S}}(\mathrm{P})\right)$.

Then, consider that $\mathcal{N}_{26}^{\overline{\mathrm{S}}}\left(\mathcal{N}_{26}^{\mathrm{S}}(\mathrm{P})\right)$ admits exactly two 6-connected components:

1. By Definition 3, $\mathcal{N}_{26}^{\overline{\mathrm{S}}}\left(\mathcal{N}_{26}^{\mathrm{S}}(\mathbf{p_1})\right)$ admits two 6-connected components such that $\mathbf{p_1}$ is 6-adjacent to both of them.
2. Now suppose that $\mathcal{N}_{26}^{\overline{\mathrm{S}}}\left(\mathcal{N}_{26}^{\mathrm{S}}(\{\mathbf{p_1}, \ldots, \mathbf{p_k}\})\right)$ with $1 < k < n_{\mathrm{P}}$ admits two 6-connected components and for all $i \in \{1, \ldots, k\}$, $\mathbf{p_i}$ is 6-adjacent to both of them. By condition 2 of Definition 3, $\mathcal{N}_{26}^{\overline{\mathrm{S}}}\left(\mathcal{N}_{26}^{\mathrm{S}}(\mathbf{p_{k+1}})\right)$ admits two 6-connected components such that $\mathbf{p_k}$ and $\mathbf{p_{k+1}}$ are 6-adjacent to both of them. Since $\mathbf{p_k}$ is 6-adjacent to both 6-connected components of $\mathcal{N}_{26}^{\overline{\mathrm{S}}}\left(\mathcal{N}_{26}^{\mathrm{S}}(\{\mathbf{p_1}, \ldots, \mathbf{p_k}\})\right)$, so is $\mathbf{p_{k+1}}$.
3. By induction, for all $i \in \{1, \ldots, n_{\mathrm{P}}\}$, $\mathbf{p_i}$ is 6-adjacent to the two 6-connected components of $\mathcal{N}_{26}^{\overline{\mathrm{S}}}\left(\mathcal{N}_{26}^{\mathrm{S}}(\mathrm{P})\right)$.

□

Thanks to those results, we deduce some simplifications on the problem we consider and formalized by Equation 1. We have:

Lemma 3. *Let* $\mathrm{P} = \mathbf{w_1} \cdot \mathrm{Q} \cdot \mathbf{w_2}$ *be a* 6*-connected digital path such that* Q *is only composed of voxels of* S *and* $\mathbf{w_1}$ *and* $\mathbf{w_2} \notin \mathrm{S}$. *Then, for all* $i \in \{1, \ldots, n_{\mathrm{Q}}\}$, *it exists voxels* $\mathbf{w'_1}$ *and* $\mathbf{w'_2}$ 6*-adjacent to* $\mathrm{q_i}$ *such that:*

$$\sharp \mathbb{I}_{\mathrm{S}}(\mathrm{P}) = \sharp \mathbb{I}_{\mathrm{S}}(\mathbf{w'_1} \cdot \mathbf{q_i} \cdot \mathbf{w'_2}). \tag{5}$$

Proof. By Proposition 3 and Lemma 2, it exists $\mathbf{w'_1}$ and $\mathbf{w'_2}$ 6-adjacent to $\mathbf{q_i}$ such that it exists 6-connected paths in $\mathcal{N}_{26}(\mathrm{Q}) \cap \overline{\mathrm{S}}$ linking, respectively, $\mathbf{w_1}$ and $\mathbf{w'_1}$, and, $\mathbf{w_2}$ and $\mathbf{w'_2}$.

In the sequel, we will consider deformations which are the substitution in a path of a unique voxel $\mathbf{v}$ by a specific path with end points on $\mathbf{v}$. If the unique voxel is not in S, we can consider without loss of generality a path only constituted by $\mathbf{v}$. When it belongs to S, we consider, again, without loss of generality, a path $\mathbf{w_1} \cdot \mathbf{v} \cdot \mathbf{w_2}$ with $\mathbf{w_1}, \mathbf{w_2} \notin \mathrm{S}$.

3.3 Deformation in a Unit Square

Studying the number of intersections, with a surface S, of digital paths differing in K, requires to take in account all possible subpaths in K. We consider here different local configurations, namely U-turn, minimal loop and minimal path, to define equivalence classes and reduce the problem.

Adding or deleting a U-turn in a digital path. A U-turn is the following configuration:

Definition 9 (U-turn). *A* U-turn *is a 6-connected digital path* $U = (\mathbf{u_1}, \mathbf{u_2}, \mathbf{u_3})$ *such that* $\mathbf{u_1} = \mathbf{u_3}$. *A* pointed U-turn $U(\mathbf{v})$ *is a U-turn such that* $\mathbf{u_1} = \mathbf{v}$.

The deletion or the addition of a U-turn in a 6-connected digital path does not change the parity of its number of intersections with a digital surface:

Lemma 4. *Let* $\mathbf{v} \in \mathbb{Z}^3$. *Let also* Q *and* R *be two 6-connected digital paths such that* $P = Q \cdot \mathbf{v} \cdot R$ *and* $P' = Q \cdot U(\mathbf{v}) \cdot R$ *are 6-connected digital paths with extreme points not in* S. *Then, we have:*

$$\sharp\mathbb{I}_S(P) \equiv \sharp\mathbb{I}_S(P') \pmod 2. \tag{6}$$

Proof.

1. $\mathbf{v} \notin S$: without loss of generality, we consider $P = \mathbf{v}$. Thus, we have $\sharp\mathbb{I}_S(P) = 0$.
   (a) $\mathbf{u_2} \notin S$: $P' = U(\mathbf{v})$ contains no voxel of S, thus we have $\sharp\mathbb{I}_S(P') = 0$.
   (b) $\mathbf{u_2} \in S$: $\mathbf{u_2}$ is not an intersection of P′ with S since $\mathbf{u_1} = \mathbf{u_3}$. Thus, we have $\sharp\mathbb{I}_S(P') = 0$.
2. $\mathbf{v} \in S$: without loss of generality, we consider $P = \mathbf{w_1} \cdot \mathbf{v} \cdot \mathbf{w_2}$ with $\mathbf{w_1}$ and $\mathbf{w_2} \notin S$.
   (a) $\mathbf{u_2} \notin S$:
      i. $\mathbf{v} \in \mathbb{I}_S(P)$: it means that $\sharp\mathbb{I}_S(P) = 1$ and that $\mathbf{w_1}$ and $\mathbf{w_2}$ are in distinct 6-connected components of $\mathcal{N}_{26}(\mathbf{v}) \cap \overline{S}$. Either $\mathbf{u_2}$ is in the component of $\mathbf{w_1}$ or of $\mathbf{w_2}$. Without loss of generality, we focus on the first case (for the second one, consider P^{-1}). $\mathbf{u_1}$ is not an intersection of P′ with S while $\mathbf{u_3}$ is such an intersection. Then, we have $\sharp\mathbb{I}_S(P') = \sharp\mathbb{I}_S(P)$.
      ii. $\mathbf{v} \notin \mathbb{I}_S(P)$: it means that $\sharp\mathbb{I}_S(P) = 0$ and that $\mathbf{w_1}$ and $\mathbf{w_2}$ are in the same 6-connected components of $\mathcal{N}_{26}(\mathbf{v}) \cap \overline{S}$. Either $\mathbf{u_2}$ is in this component and both $\mathbf{u_1}$ and $\mathbf{u_3}$ are not intersections of P′ with S, or $\mathbf{u_2}$ is in the other component and both $\mathbf{u_1}$ and $\mathbf{u_3}$ are intersections of P′ with S. To sum up, we have $\sharp\mathbb{I}_S(P) = \sharp\mathbb{I}_S(P')$.
   (b) $\mathbf{u_2} \in S$: since $\mathbf{v} \in S$ and $U(\mathbf{v}) \subseteq S$, $\sharp\mathbb{I}_S(P) = \sharp\mathbb{I}_S(P')$.

□

A direct consequence concerns path containing the concatenation of a subpath and its reverse path:

Corollary 2. *Let* A, B *and* C *be three* 6*-connected paths such that* $\mathrm{P} = \mathrm{A} \cdot \mathrm{B} \cdot \mathrm{B}^{-1} \cdot \mathrm{C}$ *and* $\mathbf{a_1}$, $\mathbf{c_{n_C}} \notin \mathrm{S}$. *Then, we have:*

$$\sharp\mathbb{I}_S\,(\mathrm{P}) \equiv \sharp\mathbb{I}_S\,(\mathrm{A} \cdot \mathbf{b_1} \cdot \mathrm{C}) \pmod 2. \tag{7}$$

Proof. $\mathrm{B} \cdot \mathrm{B}^{-1}$ is such that $(\mathbf{b_1}, \dots, \mathbf{b_{n_B}}, \mathbf{b_{n_B}}, \dots, \mathbf{b_1})$. For all voxel $\mathbf{v}$ and $\mathbf{w_1}$, $\mathbf{w_2} \notin \mathrm{S}$, the path $\mathbf{w_1} \cdot \mathbf{v} \cdot \mathbf{v} \cdot \mathbf{w_2}$ is directly equivalent to the path $\mathbf{w_1} \cdot \mathbf{v} \cdot \mathbf{w_2}$ and such a deformation trivially does not modify the number of intersections with S. Then, we have:

$$\begin{aligned}
\sharp\mathbb{I}_S\,(\mathrm{P}) &\equiv \sharp\mathbb{I}_S\,(\mathrm{A} \cdot (\mathbf{b_1}, \dots, \mathbf{b_{n_B-1}}, \mathbf{b_{n_B}}, \mathbf{b_{n_B-1}}, \dots, \mathbf{b_1}) \cdot \mathrm{C}) \pmod 2,\\
&\equiv \sharp\mathbb{I}_S\,(\mathrm{A} \cdot (\mathbf{b_1}, \dots, \mathbf{b_{n_B-2}}) \cdot \mathrm{U}(\mathbf{b_{n_B-1}}) \cdot (\mathbf{b_{n_B-2}}, \dots, \mathbf{b_1}) \cdot \mathrm{C}) \pmod 2,\\
&\equiv \sharp\mathbb{I}_S\,(\mathrm{A} \cdot (\mathbf{b_1}, \dots, \mathbf{b_{n_B-2}}, \mathbf{b_{n_B-1}}, \mathbf{b_{n_B-2}}, \dots, \mathbf{b_1}) \cdot \mathrm{C}) \pmod 2,\\
&\equiv \dots,\\
&\equiv \sharp\mathbb{I}_S\,(\mathrm{A} \cdot (\mathbf{b_1}, \mathbf{b_2}, \mathbf{b_1}) \cdot \mathrm{C}) \pmod 2,\\
&\equiv \sharp\mathbb{I}_S\,(\mathrm{A} \cdot \mathrm{U}(\mathbf{b_1}) \cdot \mathrm{C}) \pmod 2,\\
&\equiv \sharp\mathbb{I}_S\,(\mathrm{A} \cdot \mathbf{b_1} \cdot \mathrm{C}) \pmod 2,
\end{aligned}$$

□

Adding or deleting a minimal loop in a digital path. The second local configuration we study is the minimal loop:

Definition 10 (Minimal loop). *A* minimal loop *is a* 6*-connected path* $\mathrm{L} = (\mathbf{l_i})_{1 \leqslant i \leqslant 5}$ *such that:*

1. L *does not contains any U-turn,*
2. $\mathbf{l_1} = \mathbf{l_5}$.

A pointed minimal loop $\mathrm{L}(\mathbf{v})$ *is a minimal loop such that* $\mathbf{l_1} = \mathbf{v}$.

Note that a loop is included in a unit square and is a parametrisation of it.

As in the case of a U-turn, the addition or deletion of a minimal loop in a path preserves the parity of its number of intersections with a digital surface.

Lemma 5. *Let* $\mathbf{v} \in \mathbb{Z}^3$ *be a voxel. Let also* Q *and* R *be* 6*-connected digital paths such that* $\mathrm{P} = \mathrm{Q} \cdot (\mathbf{v}) \cdot \mathrm{R}$ *and* $\mathrm{P}' = \mathrm{Q} \cdot \mathrm{L}(\mathbf{v}) \cdot \mathrm{R}$ *are* 6*-connected digital paths with extreme points not in* S. *Then, we have:*

$$\sharp\mathbb{I}_S\,(\mathrm{P}) \equiv \sharp\mathbb{I}_S\,(\mathrm{P}') \pmod 2. \tag{8}$$

Proof. This proof is similar to the one of Lemma 4. □

Changing the irreductible path of a unit square in a digital path The last specific configuration presented is the minimal path in a unit square.

Definition 11 (minimal path in a unit square). *Let* K *be a unit square. A path is said to be* minimal in K *if it is a strict subpath of a minimal loop included in* K. *Such a path between voxels* $\mathbf{v}$ *and* $\mathbf{w}$ *in* K *is denoted as* $\mathrm{M_K}(\mathbf{v}, \mathbf{w})$.

Considering two voxels in a unit square, there exists only two minimal path linking them.

Definition 12. *Let* K *be a unit square and* $M_K(\mathbf{v},\mathbf{w})$ *be a minimal path between voxels* $\mathbf{v}$ *and* $\mathbf{w}$ *in* K*. Then, the only other minimal path in* K *with same extreme points is denoted by* $\overline{M_K}(\mathbf{v},\mathbf{w})$*. Let* $L(\mathbf{v}) = A\cdot\mathbf{w}\cdot B$ *such that* $M_K(\mathbf{v},\mathbf{w}) = A\cdot\mathbf{w}$*. Then,* $\overline{M_K}(\mathbf{v},\mathbf{w})$ *is such that:*

$$M_K(\mathbf{v},\mathbf{w})\cdot\left(\overline{M_K}(\mathbf{v},\mathbf{w})\right)^{-1} = A\cdot\mathbf{w}\cdot\mathbf{w}\cdot B \equiv L(\mathbf{v}). \tag{9}$$

Substitute a minimal path by its complementary minimal path does not change the parity of the number of intersections with a digital surface.

Lemma 6. *Let* K *be a unit square and* $(\mathbf{v},\mathbf{w}) \in K^2$*. Let also* Q *and* R *be two* 6*-connected digital paths such that* $P = Q\cdot M_K(\mathbf{x},\mathbf{y})\cdot R$ *is a* 6*-connected digital path with extreme points not in* S*. Then, we have:*

$$\sharp\mathbb{I}_S\,(P) = \sharp\mathbb{I}_S\left(Q\cdot\overline{M_K}(\mathbf{v},\mathbf{w})\cdot R\right) \pmod 2. \tag{10}$$

Proof.

$$\begin{aligned}
\sharp\mathbb{I}_S\,(P) &= \sharp\mathbb{I}_S\left(Q\cdot M_K(\mathbf{v},\mathbf{w})\cdot R\right)\\
&\equiv \sharp\mathbb{I}_S\left(Q\cdot L(\mathbf{v})^{-1}\cdot M_K(\mathbf{v},\mathbf{w})\cdot R\right) \pmod 2\\
&\equiv \sharp\mathbb{I}_S\left(Q\cdot\overline{M_K}(\mathbf{v},\mathbf{w})\cdot M_K(\mathbf{v},\mathbf{w})^{-1}\cdot M_K(\mathbf{v},\mathbf{w})\cdot R\right) \pmod 2\\
&\equiv \sharp\mathbb{I}_S\left(Q\cdot\overline{M_K}(\mathbf{v},\mathbf{w})\cdot\mathbf{w}\cdot R\right) \pmod 2\\
&\equiv \sharp\mathbb{I}_S\left(Q\cdot\overline{M_K}(\mathbf{v},\mathbf{w})\cdot R\right) \pmod 2
\end{aligned}$$

□

According to results on U-turns, minimal loops and paths, we can conclude on the preservation of the parity of the number of intersections between a path and a digital surface in the case of the second condition of Definition 7.

Proposition 4. *Consider two* 6*-connected digital paths* P *and* P′ *with end points not in* S *such that they differ only in a unit square* K*. Then, we have:*

$$\sharp\mathbb{I}_S\,(P) \equiv \sharp\mathbb{I}_S\,(P') \pmod 2. \tag{11}$$

Proof. Deleting or adding U-turns and consecutive duplicates preserves the parity of the number of intersections with S. Without loss of generality, the study is reduced to subpaths in K constituted by the concatenation of several (0 to infinity) minimal loops and a minimal path in K. The same occurs with minimal loops. Again, without loss of generality, the study reduced and we only have to consider two subpaths between the input voxel $\mathbf{v}$ and the output voxel K of the paths P and P′, namely $M_K(\mathbf{v},\mathbf{w})$ and $\overline{M_K}(\mathbf{v},\mathbf{w})$. Lemma 6 allows to conclude and prove the proposition. □

3.4 Main Theorem

The following theorem is a direct consequence of Proposition 2 and Proposition 4:

Theorem 1. *Consider two directly equivalent 6-connected digital paths* P *and* P′ *with end points not in* S. *Then, we have:*

$$\sharp\mathbb{I}_S(\mathrm{P}) \equiv \sharp\mathbb{I}_S(\mathrm{P}') \pmod 2. \tag{12}$$

And, if we consider now Definition 8 about equivalent digital paths, we have:

Corollary 3. *Consider two* 6*-connected digital paths* P *and* P′ *linking two voxels* **a** *and* **b**. *Then, we have:*

$$\sharp\mathbb{I}_S(\mathrm{P}) \equiv \sharp\mathbb{I}_S(\mathrm{P}') \pmod 2. \tag{13}$$

Since all path in $\mathbb{Z}^3$ are equivalent, we can easily define path with no intersection or path with only one intersection with S. We can define a partition of $\overline{\mathrm{S}}$ into two sets according to a given voxel $\mathbf{v} \notin \mathrm{S}$: the voxels linked to $\mathbf{v}$ by a path having an even number of intersections with S and the ones linked to $\mathbf{v}$ by a path having an odd number of intersections with S. Hence, we prove that S separates the space in at least two 6-connected components. Moreover, we prove a stronger result:

Proposition 5. S *is a* 6*-separating set (in* $\mathbb{Z}^3$*).*

Proof. Suppose that $\overline{\mathrm{S}}$ admits more than two 6-connected components. Since S is 26-connected, either it exists a voxel $\mathbf{v} \in \mathrm{S}$ adjacent to more than two 6-connected components, or it exists two 26-adjacent voxels $\mathbf{v}, \mathbf{w} \in \mathrm{S}$ such that $\mathcal{N}_{26}^{\overline{\mathrm{S}}}(\{\mathbf{v}, \mathbf{w}\})$ admits three 6-connected components. By definition of S, neither the first case nor the second one can appear and $\overline{\mathrm{S}}$ admits exaclty two 6-connected components. □

We can now state the main result of the present paper:

Theorem 2. S *is a strong* 6*-separating set (in* $\mathbb{Z}^3$*).*

Proof. Proposition 5 ensures that $\overline{\mathrm{S}}$ admits exactly two distinct 6-connected components. Moreover, by definition of S, for all voxel $\mathbf{v} \in \mathrm{S}$, it exists two voxels $\mathbf{w_1}$ and $\mathbf{w_2} \notin \mathrm{S}$ 6-adjacent to $\mathbf{v}$ such that $\sharp\mathbb{I}_\mathrm{S}(\mathbf{w_1} \cdot \mathbf{v} \cdot \mathbf{w_2}) = 1$. It means that each voxel of S is 6-adjacent to both components of $\mathbb{Z}^3 \cap \overline{\mathrm{S}}$. □

Conclusion

In this paper, we have introduced a new notion of 26-connected digital surfaces which extends the one of digital curves. The local characterization stays very simple since it is only based on usual digital adjacency relations. A Jordan theorem is proved thanks to the fact that the neighborhood in $\overline{\mathrm{S}}$ of a crossing between a digital surface S and a 6-connected digital path admits at most two 6-connected components. Finally, we present digital surfaces consisting in 26-connected and strong 6-separating sets. On the same principle, it is possible to define digital surfaces consisting in 6-connected and strong 26-separating sets.

At this point, the main drawback of our characterization is the fact that we consider, for each voxel, a neighborhood larger than the 26-neighborhood. From a practical point of view, deeper investigations are required to limit the study to the 26-neighborhood and, thus, reduce the computation.

The main task to achieved is now to compare our notion of simple closed digital surfaces to previous ones. The class of simplicity surfaces is currently the largest one. The comparison with this class required some technical works. Indeed characterizations are not defined with the same tools (simplicity graph, geodesic neigborhood,...). The easiest way to proceed seems to be the study and the comparison of the sets of admissible local configurations for each definition.

References

1. Rosenfeld, A.: Arcs and curves in digital pictures. Journal of the ACM 20(1), 81–87 (1973)
2. Malgouyres, R.: There is no local characterization of separating and thin objects in z^3. Theoretical Computer Science 163(1&2), 303–308 (1996)
3. Morgenthaler, D.G., Rosenfeld, A.: Surfaces in three-dimensional digital images. Information and Control 51(3), 227–247 (1981)
4. Reveillès, J.P.: Géométrie discrète, calcul en nombres entiers et algorithmique. Thèse d'Etat, Université Louis Pasteur, Strasbourg (1991)
5. Bertrand, G., Malgouyres, R.: Some topological properties of discrete surfaces. In: Miguet, S., Ubéda, S., Montanvert, A. (eds.) DGCI 1996. LNCS, vol. 1176, pp. 325–336. Springer, Heidelberg (1996)
6. Malgouyres, R., Bertrand, G.: A new local property of strong n-surfaces. Pattern Recognition Letters 20(4), 417–428 (1999)
7. Malgouyres, R.: A definition of surfaces of z^3: a new 3d discrete jordan theorem. Theoretical Computer Science 186(1-2), 1–41 (1997)
8. Couprie, M., Bertrand, G.: Simplicity surfaces: a new definition of surfaces in z^3. In: SPIE Vision Geometry V, vol. 3454, pp. 40–51 (1998)
9. Ciria, J.C., Domínguez, E., Francés, A.R.: Separation theorems for simplicity 26-surfaces. In: Braquelaire, A., Lachaud, J.-O., Vialard, A. (eds.) DGCI 2002. LNCS, vol. 2301, pp. 45–56. Springer, Heidelberg (2002)
10. Bertrand, G.: On p-simple points. Compte-Rendus de l'Académie des Sciences, Série 1, Mathématique 321(8), 1077–1084 (1995)
11. Kong, T.Y.: A digital fundamental group. Computers & Graphics 13(2), 159–166 (1989)

Border Operator for Generalized Maps

Sylvie Alayrangues[1], Samuel Peltier[1], Guillaume Damiand[2], and Pascal Lienhardt[1]

[1] XLIM-SIC, Université de Poitiers, CNRS, Bât. SP2MI, Téléport 2, Bvd Marie et Pierre Curie, BP 30179 86962 Futuroscope Chasseneuil Cedex France
{alayrangues,peltier,lienhardt}@sic.univ-poitiers.fr
[2] Université de Lyon, CNRS, LIRIS, UMR5205, F-69622, France
guillaume.damiand@liris.cnrs.fr

Abstract. In this paper, we define a border operator for generalized maps, a data structure for representing cellular quasi-manifolds. The interest of this work lies in the optimization of homology computation, by using a model with less cells than models in which cells are regular ones as tetrahedra and cubes. For instance, generalized maps have been used for representing segmented images. We first define a face operator to retrieve the faces of any cell, then deduce the border operator and prove that it satisfies the required property : border of border is void. At last, we study the links between the cellular homology defined from our border operator and the classical simplicial homology.

1 Introduction

Computing topological properties onto subdivided objects is of interest in different communities of computer science (e.g. computational topology, geometric modeling, discrete geometry, image analysis). Among all topological properties (e.g. Euler characteristic, orientability, connectivity), many works show that homology is a powerful one: it can be computed in the same way in any dimension, it is directly linked with the structure of an object and the homological information can be represented in the object. Different approaches and optimizations have been proposed (e.g. computation of Betti numbers for torsion free objects [5], computation of homology generators [16]).

Most homological results are related to triangulated objects (abstract simplicial sets [15], Δ-complexes [10], semi-simplicial sets [11], simplicial sets [14]). Within these structures, simplices are oriented cells and for any dimension $d > 0$, $d+1$ face operators are defined for each $d-$simplex (each face operator maps a $d-$simplex onto a $(d-1)-$simplex; this application is trivial for $0-$simplices). Border operators can then be defined as the alternate sum of the face operators. Homology groups are defined from these border operators. As far as we know, such border operators are defined only for the "regular" structures of simplicial, cubical and simploidal sets [17].

In this paper, we address the problem of computing homology groups for cellular structures (e.g. cell-tuples [4], incidence graphs [18], orders [3]). Such

S. Brlek, C. Reutenauer, and X. Provençal (Eds.): DGCI 2009, LNCS 5810, pp. 300–312, 2009.

structures are used in both geometric modeling and image analysis. Any object, represented by any cellular structure, can always be canonically converted onto a simplicial one. Thus, homology of a cellular object can be computed through its associated triangulation, but in this way, the optimization of the cellular structure is lost as the simplicial object is made of more cells than the cellular one. The second drawback is that it is not straightforward to map a simplicial homology generator onto the cellular structure. In order to avoid these drawbacks, we chose to define a border operator for cellular structures.

More precisely, we define a border operator for the structure of generalized maps, which is well-suited for representing cellular quasi-manifolds[1] [13]. It can be noted that there exists several equivalent or derived structures which allow to represent more general cellular complexes or subclasses of quasi-manifolds and this work could be extended for these structures [8]. Moreover structures based on combinatorial maps have already been used in many fields, e.g. image analysis [7], discrete objects modeling [2]. Besides several links between generalized maps and other cellular structures (e.g. cell-tuples [4], orders [3]) have been exhibited and conversion operators between subclasses of such models have also been proposed [1,12].

In order to define a border operator for generalized maps, the idea is to match the classical simplicial approach: cells are given an orientation and for each couple (c_i, c'_{i-1}), where c_i is an $i-$cell and c'_{i-1} an $(i-1)-$cell, the incidence number between c_i and c'_{i-1} indicates how many times c_i is attached to c'_{i-1}, taking the orientation into account. This operator is actually defined on a restricted subclass of generalized maps which is sufficient to encode most useful subdivisions. We prove that the operator ∂ we define is a border operator, i.e. satisfies the property $\partial\partial(c_i) = 0$ for any cell $c_i, i > 0$. Thus, notions of chain, cycle, boundary and homology can be defined. We then study the links between the homology defined from ∂ and the simplicial homology (i.e. the homology computed on the associated simplicial structure).

The main advantage of defining a border operator on cellular structures through the explicit characterization of incidence numbers is that the usual methods for homology computation based on Smith normal form reductions of incidence matrices can be used while the optimization of the cellular representation is preserved, and all the optimizations defined for simplicial structures can directly be applied when computing the homology of cellular objects (e.g. computation of generators [16], optimization for incidence matrix reductions into their Smith normal form [6], optimization for sparse incidence matrices [9]). We are currently leading experimental studies to get an accurate estimation of the cellular optimization. In this article, we essentially focus on establishing a theoretical framework. More precisely, we introduce new notions on generalized maps: (un)signed incidence numbers (from section 3 to section 4). These notions are then used to define a new cellular border operator (section 5). All these definitions are based on classical notions related to generalized maps which are

[1] Informally, an n-dimensional cellular quasi-manifold is a collection of n-cells attached along $(n-1)$-cells such that at most two n-cells are incident to a given $(n-1)$-cell.

carefully recalled (from section 2) in order to ease the understanding and provide a self-contained paper.

In section 2, basic notions of generalized maps and cells are recalled. We also define the notions of compacted cells and face operators (adapted from [8]). This notion is based on an intrinsic property of cellular quasi-manifolds: each cell is a cellular quasi-manifold and can hence be represented with a generalized map of the same dimension. Faces operators associate with each i-dimensional cell, all $(i-1)$-cells belonging to its border. The notion of incidence numbers between cells of generalized maps (without taking into account their orientation) is presented in section 3. The notions of oriented cell and incidence numbers between oriented cells are defined in section 4. In section 5, the border operator of generalized maps is given and we study, in section 6, the links between the cellular homology we have defined and the simplicial homology. Finally, we conclude and give some hints for future works.

2 Generalized Maps and Corresponding Cell Subdivisions

A generalized map does not explicitly encode the cells of the associated cellular quasi-manifold [12]. It is constructed on more atomic elements called darts. Each dart is incident to exactly one cell of each dimension. The topological structure of the model is encoded through involutions linking the darts. An n-dimensional generalized map is equipped with $(n+1)$ involutions[2]. Informally, the involution numbered k relates two darts belonging to the two adjacent k-cells[3] incident to the same i-cells for all i different from k (Prop. 2 of Def. 1). Moreover cellular quasi-manifolds also have the interesting property to be locally everywhere a cellular quasi-manifold. In particular, the interior of each cell of such a subdivision is a cellular quasi-manifold (Prop. 3 of the Def. 1).

Definition 1. *(Generalized map [13]) Let $n \geq 0$. An n-dimensional generalized map (or nG-map) is $G = (D, \alpha_0, \dots, \alpha_n)$ where:*

1. *D is a finite set of darts;*
2. *$\forall i, 0 \leq i \leq n, \alpha_i$ is an involution on D;*
3. *$\forall i, j, 0 \leq i < i+2 \leq j \leq n, \alpha_i \alpha_j$ is an involution.*

Subdivided curves are encoded with a set of darts related by two involutions denoted by α_0 and α_1; α_0 links darts incident to a same edge and α_1 relates darts incident to a same vertex (Fig. 1(a)). Similarly, subdivided surfaces are encoded with a set of darts related by three involutions denoted by α_0, α_1 and α_2; α_0 (resp. α_1, α_2) links darts incident to a same edge and face (resp. vertex and face, vertex and edge) (Fig. 1(b)).

Defining the border operator associated with the subdivision requires first to construct the cells of the quasi-manifold from the definition of generalized maps. Actually, each cell is fully and implicitly defined by the set of darts incident to it. Defining a cell usually consists in extracting its corresponding set of darts.

[2] An involution f on S is a one-to-one mapping from S to S such that $f = f^{-1}$.
[3] not necessarily different: a cell can be attached to itself.

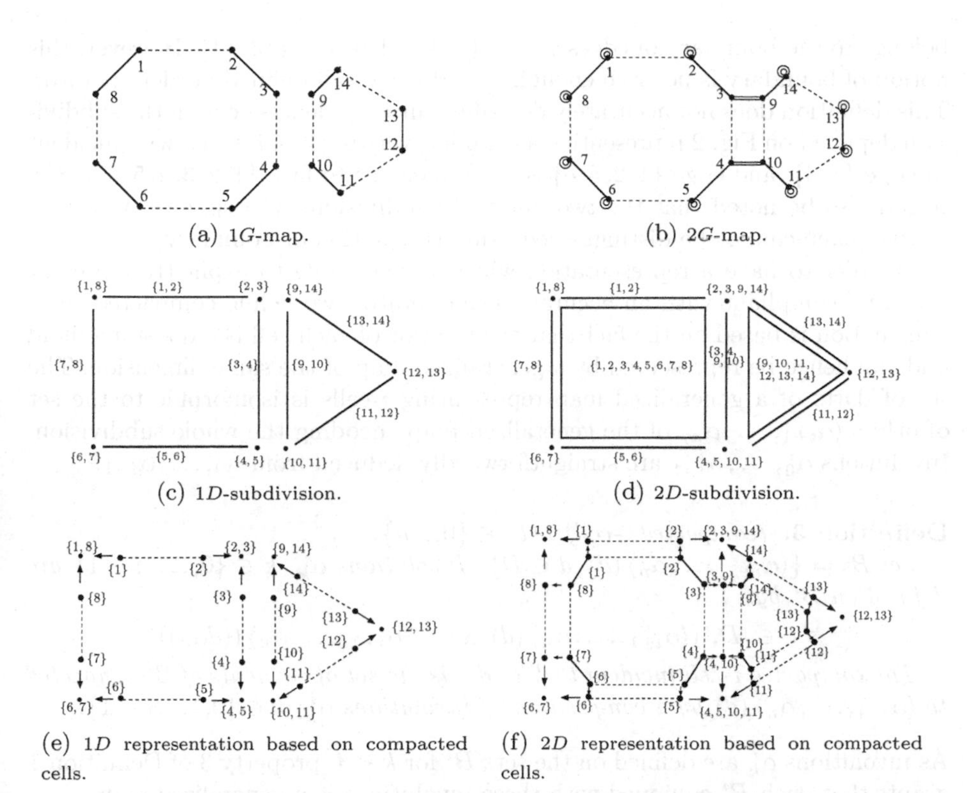

(a) 1G-map.
(b) 2G-map.
(c) 1D-subdivision.
(d) 2D-subdivision.
(e) 1D representation based on compacted cells.
(f) 2D representation based on compacted cells.

Fig. 1. Different representations of a $1D$-subdivision (left) and a $2D$-subdivision (right). In the generalized maps depicted on Fig. 1(a) and 1(b), darts are depicted by dots and numbered from 1, involution α_0 is represented by dotted lines, α_1 by plain lines and α_2 by double lines. On the corresponding subdivisions (Fig. 1(c) and 1(d)), cells are explictly depicted and are associated with their corresponding sets of darts. Face operators are depicted by arrows on the compacted cell decomposition of both subdivisions (Fig. 1(e) and 1(f)).

Definition 2. *(i-cell) Let d be a dart and $i \in N = \{0, ., n\}$. The i-cell incident to d, denoted by $C_i(d)$, is the orbit*[4]*:*

$$\langle\rangle_{N-\{i\}}(d) = \langle \alpha_0, \ldots, \alpha_{i-1}, \alpha_{i+1}, \ldots, \alpha_n \rangle (d)$$

The $1D$-subdivision (resp. $2D$-subdivision) encoded by the $1G$-map (resp. $2G$-map) of Fig.1(a) (resp. 1(b)) is depicted on Fig.1(c) (resp. 1(d)).

This definition of cell is well suited to determine which cells belong to the boundary of another one. More precisely, a cell is in the boundary of another one if they share at least one dart. In the $1D$ example, vertex $\{1, 8\}$ belongs to the boundary of edges $\{1, 2\}$ and $\{7, 8\}$ and in the $2D$ example, vertex $\{2, 3, 9, 14\}$

[4] Let $\{\Pi_0, ..., \Pi_n\}$ be a set of permutations on D. The orbit of an element d relatively to this set of permutations is $\langle \Pi_0, ..., \Pi_n \rangle (b) = \{\Phi(b), \Phi \in \langle \Pi_0, ..., \Pi_n \rangle\}$, where $\langle \Pi_0, \ldots, \Pi_n \rangle$ denotes the group of permutations generated by $\{\Pi_0, \ldots, \Pi_n\}$.

belongs to the boundary of edges $\{1,2\}$, $\{3,4,9,10\}$ and $\{13,14\}$. However, this notion of boundary is not rich enough to deduce a well-defined border operator. This definition does not accurately describe multi-incidences (e.g. in the subdivision depicted on Fig. 2 representing a cylinder, vertex $\{1,6,7,8\}$ is twice incident to edge $\{7,8\}$, and edge $\{1,2,5,6\}$ is twice incident to face $\{1,2,3,4,5,6,7,8\}$). It can also be noted that the two minimal subdivisions of a sphere and a projective plane cannot be distinguished using this notion of boundary[1].

In order to have a representation which is both able to explicitly represent cells and compliant with an accurate face operator, we define compacted cells. This notion is based on the fact that the interior of each cell is a quasi-manifold and can hence be represented by a generalized map of the same dimension. The set of darts of a generalized map representing i-cells is isomorphic to the set of orbits $\langle\alpha_{i+1},\dots,\alpha_n\rangle$ of the generalized map encoding the whole subdivision. Involutions $\alpha_0^i,\dots,\alpha_{i-1}^i$ are straightforwardly deduced from $\alpha_0,\dots,\alpha_{i-1}$.

Definition 3. *(compacted i-cell) Let $i \in \{0,.,n\}$.*

Let $B^i = \{\langle\alpha_{i+1},\dots,\alpha_n\rangle(d), d \in D\}$. Involutions α_k^i, $k \in \{0,\dots,i-1\}$ are defined on B^i by:

$$\forall d \in D,\ (\langle\alpha_{i+1},\dots,\alpha_n\rangle(d))\alpha_k^i = \langle\alpha_{i+1},\dots,\alpha_n\rangle((d\alpha_k))$$

The compacted i-cell incident to d, $c_i(d)$, is the set of elements of B^i connected to $\langle\alpha_{i+1},\dots,\alpha_n\rangle(d)$ by a composition of involutions α_k^i, $k \in \{0,\dots,i-1\}$.

As involutions α_k^i are defined on the sets B^i for $k < i$, property 3 of Definition 1 grants that each B^i equipped with these involutions is a generalized map.

The notion of compacted cells is illustrated on Fig. 1(e) ($1D$-subdivision) and 1(f) ($2D$-subdivision). For instance, edge $\{3,4,9,10\}$ of Fig. 1(d) is represented by the α_0^1-connected set $\{\langle\alpha_2\rangle(3) = \{3,9\}, \langle\alpha_2\rangle(4) = \{4,10\}\}$ (Fig. 1(f)).

A set of face operators describes the links between cells and cells of their boundaries. The ith face operator formalizes an incidence relation between a compacted i-cell and a compacted $(i-1)$-cell incident to it. More precisely, it associates darts of B^i and B^{i-1} in such a way that it is coherent upon the cells.

Definition 4. *(face operator) Let $i \in \{1,.,n\}$.*

The face operator ∂_i, is a mapping from B^i to B^{i-1} defined by:

$$\langle\alpha_{i+1},\dots,\alpha_n\rangle(d) \xrightarrow{\partial_i} \langle\alpha_i,\dots,\alpha_n\rangle(d)$$

It can be proven that $\partial_i\alpha_k^{i-1} = \alpha_k^i\partial_i$ for all $k \in [0,i-2]$ (Prop. 3 of Def. 1) . ∂_i^{-1} denotes the preimage of ∂_i. For instance, on Fig. 1(e), the image of the darts $\langle\rangle(2) = \{2\}$ and $\langle\rangle(3) = \{3\}$ (belonging to B^1) by the face operator, ∂_1, is the set $\{2,3\} = \langle\alpha_1\rangle(2) = \langle\alpha_1\rangle(3)$ (belonging to B^0). On Fig. 1(f), the image by ∂_1 of the singleton $\langle\alpha_2\rangle(1) = \{1\}$ belonging to B^1 is the set $\{1,8\} = \langle\alpha_1,\alpha_2\rangle(1)$ belonging to B^0. The image of the darts of B^2, $\langle\rangle(2) = \{2\}$ and $\langle\rangle(3) = \{3\}$ by ∂_2 are respectively the sets $\{2\} = \langle\alpha_2\rangle(2)$ and $\{3,9\} = \langle\alpha_2\rangle(3)$.

3 Unsigned Incidence Numbers

The unsigned incidence number is defined for any pair of cells whose dimensions differ by exactly 1. It counts how many times the cell of lesser dimension is present in the boundary of the cell of greater dimension. This number can be obtained from the face operator. Given a dart of the cell of lesser dimension, we only need to compare its preimage under the face operator and the cell of greater dimension. The unsigned incidence number is equal to the number of elements of the preimage whose darts also belong to the cell. Otherwise stated, for two cells, $C_i(d)$ and $C_{i-1}(d')$, the unsigned incidence number is equal to the number of distinct orbits $\langle\alpha_{i+1},\ldots,\alpha_n\rangle$ contained in $\langle\alpha_i,\ldots,\alpha_n\rangle(d')$ whose darts are also included in $C_i(d) = \langle\rangle_{N-\{i\}}(d)$.

Definition 5. *(unsigned incidence number) Let $i \in \{1,.,n\}$. Let $C_i(d)$ and $C_{i-1}(d')$ be two cells of the subdivision encoded by G. The* incidence number, *$(C_i(d) : C_{i-1}(d'))$, is:*

$$\mathrm{card}(\{\langle\alpha_{i+1},\ldots,\alpha_n\rangle(d'') \in \partial_i^{-1}(\langle\alpha_i,\ldots,\alpha_n\rangle(d')) \\ s.t.\ \langle\alpha_{i+1},\ldots,\alpha_n\rangle(d'') \subseteq C_i(d)\})$$

Property 1. $(C_i(d) : C_{i-1}(d'))$ does not depend on the chosen darts. This incidence number is hence denoted by $(C_i : C_{i-1})$.

The proof is a straightforward consequence of property 3 of Definition 1.

For instance, let us compute the incidence number between the edge incident to 14 and the face incident to 1 on Fig. 1(f). The edge is actually represented by the set $\{\langle\alpha_2\rangle(14) = \{14\}, \langle\alpha_2\rangle(13) = \{13\}\}$ connected by the involution α_0^1 and the face by the set $\{\{1\},\{2\},\{3\},\{4\},\{5\},\{6\},\{7\},\{8\}\}$ connected by α_0^2, α_1^2. The preimage of the singleton $\langle\alpha_2\rangle(14) = \{14\}$ (belonging to B^1) is the singleton $\langle\rangle(14) = \{14\}$ (belonging to B^2) which is not included into the face incident to dart 1. The unsigned incidence number of both cells is hence 0. Let us consider the edge incident to dart 9 relatively to the same face. This edge is the set $\{\langle\alpha_2\rangle(9), \langle\alpha_2\rangle(10)\} = \{\{3,9\},\{4,10\}\}$ connected via α_0^1. The preimage of $\langle\alpha_2\rangle(9)$ is the set made of darts $\{3\}$ and $\{9\}$. Its intersection with the face $\{\{1\},\{2\},\{3\},\{4\},\{5\},\{6\},\{7\},\{8\}\}$ contains exactly one element: $\{3\}$. The incidence number between both cells is thus equal to 1.

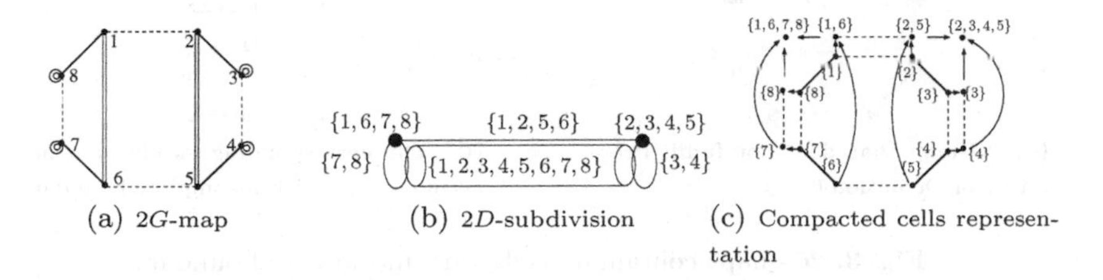

(a) 2G-map (b) 2D-subdivision (c) Compacted cells representation

Fig. 2. Representation of a cylinder

Let us consider, on Fig. 2, the vertex incident to dart 1, $\{\langle\alpha_1,\alpha_2\rangle(1) = \{1,6,7,8\}\}$, and the edge incident to the same dart, $\{\langle\alpha_2\rangle(1) = \{1,6\}, \langle\alpha_2\rangle(2) = \{2,5\}\}$. The preimage of $\{1,6,7,8\}$ is the set $\{\{1,6\},\{7\},\{8\}\}$. Only one subset of the preimage is included in the edge, the incidence number is thus equal to 1. The edge incident to dart 8 is $\{\langle\alpha_2\rangle(8) = \{8\}, \langle\alpha_2\rangle(7) = \{7\}\}$ connected by α_0^1. Two subsets of the preimage of the vertex incident to dart 1, $\{\{1,6,7,8\}\}$, are included in the edge incident to 8. The corresponding incidence number is hence 2. This subdivision contains a single face $\langle\rangle(1) = \{\{1\},\{2\},\{3\},\{4\},\{5\},\{6\},\{7\},\{8\}\}$. The preimage of $\{8\}$, used to compute the incidence number between the edge incident to dart 8 and the face of the subdivision, is the singleton $\{8\}$. It is included in the face, the incidence number associated to this pair of cells is hence 1. The preimage of $\alpha_2(1) = \{1,6\}$, corresponding to the edge incident to dart 1, is the set $\{\{1\},\{6\}\}$. Both elements belong to the face, the corresponding incidence number is thus 2. Note that considering the other element of the same edge, $\alpha_2(2) = \{2,5\}$ instead of dart $\{1,6\}$ leads to the same result. Its preimage is the set $\{\{2\},\{5\}\}$ whose both elements also belong to the face: this is consistent with Property 1.

The definitions and properties presented in the beginning of this section hold for any kind of generalized maps, e.g. maps whose involutions may have fixed points[5]. Such generalized maps are able to encode cellular quasi-manifolds whose cells may have incomplete boundaries [13], i.e. the boundary of an i-cell is not equal to the union of incident j-cells, $j < i$. In the following, we intend to define a border operator acting on cells and need hence the cells of the subdivision to have a well-defined boundary. We restrict thus the remaining of the study to the subclass of generalized maps defined as follows:

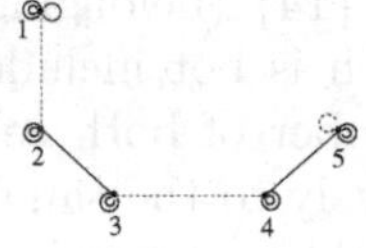

(a) $2G$-map that does not fulfill Property 1 of Definition 6: α_0 and α_1 have fixed points (resp. dart 5 and 1)

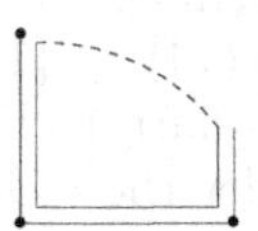

(b) An edge and a face of the corresponding subdivision do not have a complete boundary

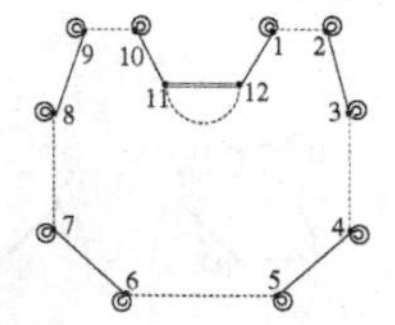

(c) $2G$-map that does not fulfill Property 2 of Definition 6

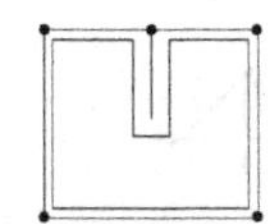

(d) The corresponding subdivision has an inner edge with incomplete boundary

Fig. 3. $2G$-maps containing cells with incomplete boundary

[5] b is a fixed point for some involution $\alpha \Leftrightarrow b\alpha = b$.

Definition 6. *An nG-map, G, is said to have cells with* complete boundary *if:*

1. α_i *is without fixed point,* $0 \leq i \leq n-1$
2. *G is without self-bending:*
   $\forall d \in D, \forall i \in \{0, \ldots, n\}, \langle \alpha_{i+1}, \ldots, \alpha_n \rangle (d) \cap \langle \alpha_0, \ldots, \alpha_{i-1} \rangle (d) = \{d\}$.

Counter-examples are displayed on Fig. 3.

Property 2. In such a generalized map, involutions defined on each compacted i-cell, $(\alpha_0^i, \ldots, \alpha_{i-1}^i)$, are without fixed points.

Note that this restriction to generalized maps without self-bending does not forbid subdivisions to have some multi-incidence (Fig. 2).

4 Cell Orientation and Signed Incidence Numbers

This section aims at characterizing more precisely the incidence between two cells not only by studying how many times a cell is incident to another one but also how these cells are connected, i.e. along which "directions". For instance, a cell which is incident twice to another one may be once incident along one "direction" and once along the opposite "direction". Such considerations are accurate only if it is possible to define the notion of "directions", that is to give an orientation to each cell of the subdivision. We define thus the notions of orientable and oriented (compacted) cells. As cells of a subdivision can be encoded by generalized maps, we recall first the notion of orientability for generalized maps.

Definition 7. *(orientable generalized map) A generalized map is* orientable *if its set of darts can be partitioned into two classes such that if d belongs to one class,* $d\alpha_i$ *belongs to the other for all* $i \in \{0, \ldots, n\}$ *when* $d\alpha_i \neq d$.

Definition 8. *(oriented generalized map) Let G be an orientable nG-map. The* oriented *generalized map,* $\overrightarrow{G}$, *is G in which one of the two classes is chosen*[6]. *In* $\overrightarrow{G}$, *the chosen class is denoted by* $\overrightarrow{G}^+$, *the other one by* $\overrightarrow{G}^-$.

The inverse orientation is denoted by $-\overrightarrow{G}$.

A single dart d may be chosen to represent the class $\overrightarrow{G}$, *the oriented generalized map is hence denoted by* $\overrightarrow{G}(d)$

In the remaining of the paper, we only deal with orientable or not orientable cellular subdivisions, whose cells have complete boundaries and are orientable (Fig. 5(a) and 5(b)). For instance, the Klein bottle is not orientable but all its (compacted) cells are (Fig. 4(b)). It can be easily proved that the compacted cells of such subdivisions are orientable. So an oriented compacted cell is denoted either by $\overrightarrow{c}_i$ or by $-\overrightarrow{c}_i$. In practice, each dart can be given a mark, + or −, for each dimension i. This mark points out the orientation class of the corresponding compacted i-cell to which the dart belongs. To obtain a consistent marking, all

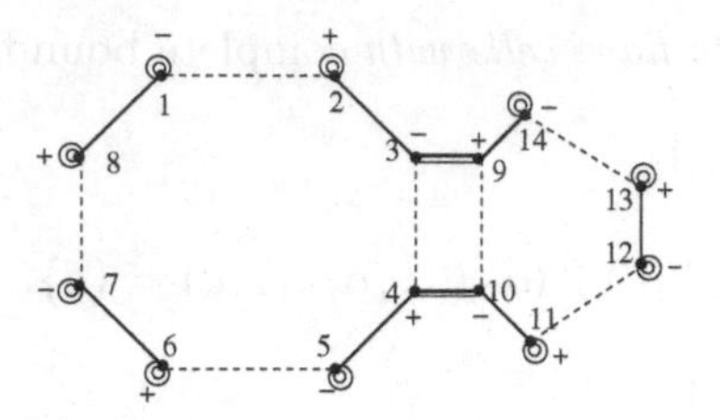

(a) Orientable 2G-map with an orientation.

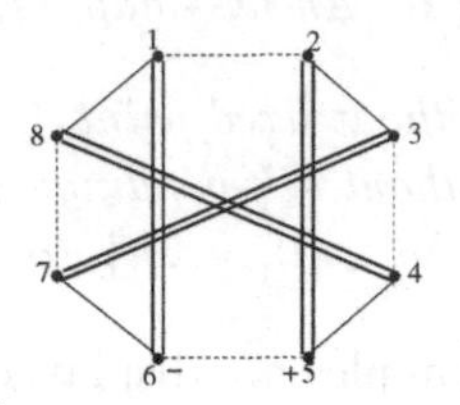

(b) Non orientable 2G-map encoding a Klein bottle.

Fig. 4. Orientability of 2G-maps

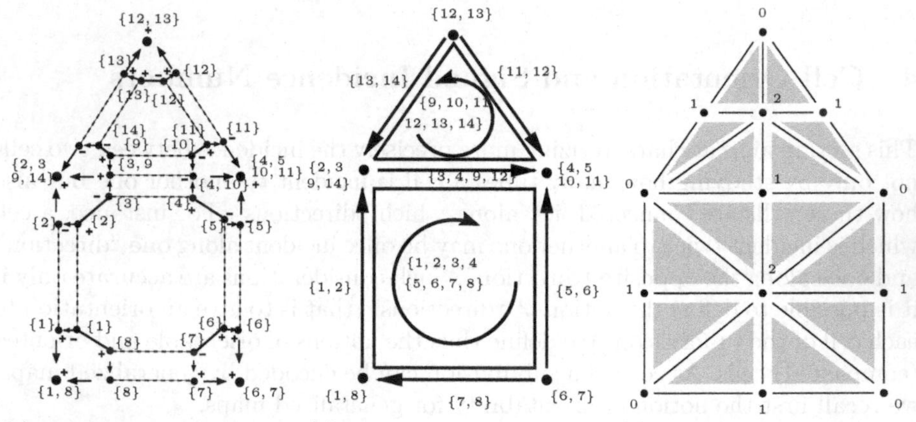

(a) Representation of the subdivision with oriented compacted cells.

(b) 2D-subdivision where each cell is oriented from + to −.

(c) Numbered simplicial object.

darts contained in an $\langle \alpha_{i+1}, \ldots, \alpha_n \rangle$-orbit must have the same mark for the ith dimension (Fig. 5(a)).

Defining the signed incidence number between an i-cell and an $(i-1)$-cell is quite similar to defining their associated unsigned incidence number. The preimage under ∂_i of some dart $\langle \alpha_{i+1}, \ldots, \alpha_n \rangle (d)$ of a compacted $(i-1)$-cell, $c_{i-1}(d)$, is examined. The elements of this preimage that are included in the i-cell, $c_i = c_i(d')$ are counted but a coefficient $+1$ or -1 is used to take their relative orientation into account.

Definition 9. *(signed incidence number) Let* $i \in \{1, \ldots, n\}$. *Let* $\overrightarrow{c_{i-1}}$ *and* $\overrightarrow{c_i}$ *be two compacted oriented cells of the subdivision, and* $\overrightarrow{C_{i-1}}$, $\overrightarrow{C_i}$ *the associated oriented cells. Let d be a dart incident to* C_{i-1}. *The signed incidence number of* $(\overrightarrow{C_i} : \overrightarrow{C_{i-1}}) = (\overrightarrow{c_i} : \overrightarrow{c_{i-1}})$ *is:*

[6] In practice, each dart can be given a mark, e.g. + or − if it belongs to $\overrightarrow{G}^+$ or to $\overrightarrow{G}^-$ (Fig. 4(a)).

$$\sum_{\substack{\langle\alpha_{i+1},\ldots,\alpha_n\rangle(d')\in\partial_i^{-1}(\langle\alpha_i,\ldots,\alpha_n\rangle(d)) \\ s.t.\langle\alpha_{i+1},\ldots,\alpha_n\rangle(d')\subseteq C_i}} sign(\overrightarrow{c_i}(d'), \overrightarrow{c_{i-1}}(d))$$

$$sign(\overrightarrow{c_i}(d'), \overrightarrow{c_{i-1}}(d)) = \begin{array}{l} +1 \ if d \in \overrightarrow{c_{i-1}}^+ \ and\ d' \in \overrightarrow{c_i}^+ \ or\ d \in \overrightarrow{c_{i-1}}^- \ and\ d' \in \overrightarrow{c_i}^-, \\ -1 \ otherwise. \end{array}$$

Property 3. The definition of the signed incidence number is consistent.

The proof is similar to the proof of Property 1 for unsigned incidence number but both possible orientations of c_{i-1} and c_i have to be taken into account.

5 Border Operator

Having defined the notion of signed incidence number between any pair $(\overrightarrow{C}_i, \overrightarrow{C}_{i-1})$, we can define the border of an i-cell by extending this notion linearly onto all $(i-1)$-cells belonging to the boundary of an i-cell.

Definition 10. *(Border Operator) Let $\mathfrak{C}_{i-1}$ be the set of $(i-1)-$cells. Let $\overrightarrow{C}_i$ be an oriented i-cell, the border of $\overrightarrow{C}_i$ is:*

$$\overrightarrow{C}_i\partial^C = \sum_{\overrightarrow{C}^j_{i-1}\in\mathfrak{C}_{i-1}} (\overrightarrow{C}_i : \overrightarrow{C}^j_{i-1})\overrightarrow{C}^j_{i-1}$$

This definition can be extended for any integer weighted sum of cells, classically called chains of cells. Moreover this notion of border is compliant with the definition of an homology because of property 4:

Property 4. $\partial^C\partial^C = 0$

The proof is based on the following idea. Let d be a dart incident to a given oriented compacted i-cell, $\overrightarrow{c_i}$ and let $\overrightarrow{c_{i-2}}$ be the oriented compacted $(i-2)$-cell incident to d, such that d belongs to $\overrightarrow{c_i}^+$ and $\overrightarrow{c_{i-2}}^+$. The preimage of $\langle\alpha_{i-1},\ldots,\alpha_n\rangle(d)$ under ∂_{i-1} contains $\langle\alpha_i,\ldots,\alpha_n\rangle(d)$, whose preimage under ∂_i contains $\langle\alpha_{i+1},\ldots,\alpha_n\rangle(d)$. This orbit is obviously included in the i-cell. Note that $d\alpha_{i-1}$ belongs to $\overrightarrow{c_i}^-$. The image of $\langle\alpha_{i+1},\ldots,\alpha_n\rangle(d\alpha_{i-1})$ under ∂_i, $\langle\alpha_i,\ldots,\alpha_n\rangle(d\alpha_{i-1})$ belongs to a compacted $(i-1)$-cell, and its image under ∂_{i-1} is still $c_{i-2}(d) = \overrightarrow{c_{i-2}}$. $\overrightarrow{c_{i-2}}$ is hence present in $(\overrightarrow{c_i})\partial^C\partial^C$, once with a sign for $\langle\alpha_{i+1},\ldots,\alpha_n\rangle(d)$ and once with the opposite sign for $\langle\alpha_{i+1},\ldots,\alpha_n\rangle(d\alpha_{i-1})$. This property is true whatever the orientations of incidente $(i-1)$-cells are. For instance, on Fig. 5(a), consider dart 2 and the 0-cell and the 2-cell to which it is incident. In the preimage of the vertex incident to 2, two elements are "incident" to the 2-cell with different orientations. Both elements are related by α_1^1.

The incidence matrix used to compute the homology groups of a generalized map, belonging to the subclass considered here, is straightforwardly obtained via the signed incidence number.

6 Towards the Equivalence between Cellular and Simplicial Homology

In this section, the relation between simplicial homology and the cellular homology we have defined from the cellular boundary operator ∂^c is studied. It is always possible to associate with any nG-map, a simplicial object such that:

- each dart corresponds to a $n-$simplex numbered $\{0, \cdots, n\}$,
- two $n-$simplices obtained from two darts linked by α_i share a $(n-1)-$face numbered $\{0, \cdots, \widehat{i}, \cdots, n\}$.

For example in Fig. 5(c), each dart corresponds to a triangle numbered $\{0, 1, 2\}$. The two triangles corresponding to the darts 4 and 5 are linked by α_1 and incident to an edge numbered $\{0, 2\}$.

Thus, a compacted i-cell is made of one vertex v numbered i, and a collection of j-simplices, $1 \leq j \leq i$, incident to v, numbered by integers lowers than i (denoted by $\{\ldots, i\}$).

Given a dart d and its associated oriented compacted cell $\vec{c_i}(d)$, we define $\vec{c_i}(d)\tau$ as the sum of the i-dimensional simplices of $c_i(d)$, taken positivey if the corresponding dart has the orientation of d, and negatively otherwise. The operator τ can directly be extended by linearity to sum of compacted cells.

If we denote the cellular boundary operator ∂^c and the simplicial one by ∂^s, we can show for any compacted i-cell c_i, that $c_i\partial^c\tau = c\tau\partial^s$. This is due to the fact that the interior of any compacted cell is a quasi manifold, so only one $(i-1)$-simplex σ belongs to the boundary of two i-simplices linked by α_j^i. Since they have inverse orientation, σ does not appear in the boundary of c_i.

We can thus deduce that for any cellular cycle[7] z (resp. boundary), $z\tau$ is a simplicial cycle (resp. boundary): for instance, let z be a cellular cycle, then $z\partial = 0 \Rightarrow z\partial^c\tau = 0 \Rightarrow z\tau\partial^s = 0 \Rightarrow z\tau$ is a cycle.

In order to prove the equivalence between cellular and simplicial homology, the converse has to be proved: for any simplicial cycle (resp. boundary) z, then there exists a cellular cycle (resp. boundary) z^c such that z and $z^c\tau$ are homologuous.

It can easily be shown that, if a simplicial i-cycle (resp. i-boundary) z is exclusively made of i-simplices associated with darts of compacted i-cells (i.e. simplices numbered $\{0, \cdots, i\}$), then there exists a cellular cycle (resp. boundary) z^c such that $z^c\tau = z$ (this is due to the fact that compacted i-cells are quasi-manifolds, so if z is exclusively made of i-simplices numbered $\{0, \cdots, i\}$, then all the simplices corresponding to a cell must appear in z, or else there is no chance that z is a cycle). When a simplicial i-cycle contains some i-simplices *internal* to a cell triangulation $c\tau$ (i.e. $dim(c) > i$ and these simplices are numbered $\{\cdots, dim(c)\}$), then we need to show that we can replace this chain of simplices by an homologuous chain such that each simplex is numbered $\{\cdots, dim(c) - 1\}$. So we could always replace a part of z by an homologuous chain, so that z remains a cycle and by recursion, we finally would obtain a simplicial cycle z' homologuous to z and exclusively made of simplices numbered $\{0, \cdots, i\}$.

[7] z is a cycle if $z\partial = 0$; z is a boundary, if c exists, such that $c\partial = z$

Thus, each cellular cycle (resp. boundary) could be associated with a simplicial cycle (resp. boundary) and reciprocally. Thus there will be an isomorphism between cellular and simplicial homology groups.

7 Conclusion

To conclude, we have defined a border operator for generalized maps having orientable cells with complete boundaries. It has been shown that this border operator defines a cellular homology and the proof of its equivalence with the simplicial homology is actually under study. Techniques of homology computation and their optimizations classically defined on simplicial structures can now directly be used for cellular structures.

For future works, we are studying some possible optimizations for particular cases of cellular structures (e.g. orientable generalized maps) and the extension of the border operator definition to more general cellular structures (complex maps, incidence graphs...).

In order to efficiently compute the homology of any cellular structure, another approach is needed: in the general case, cells may not be orientable. A track that should be followed is that any cellular structure can be interpreted as a simplicial one (structured into cells), and built as a succession of specific operations. So, we are working on the adaptation of incremental approaches defined for the computation of simplicial structures homology [19].

References

1. Alayrangues, S., Daragon, X., Lachaud, J.-O., Lienhardt, P.: Equivalence between closed connected n-g-maps without multi-incidence and n-surfaces. J. Math. Imaging Vis. 32(1), 1–22 (2008)
2. Andres, E., Breton, R., Lienhardt, P.: Spamod: design of a spatial modeler. In: Bertrand, G., Imiya, A., Klette, R. (eds.) Digital and Image Geometry. LNCS, vol. 2243, pp. 90–107. Springer, Heidelberg (2002)
3. Bertrand, G.: New notions for discrete topology. In: Bertrand, G., Couprie, M., Perroton, L. (eds.) DGCI 1999. LNCS, vol. 1568, pp. 218–228. Springer, Heidelberg (1999)
4. Brisson, E.: Representing geometric structures in d dimensions: Topology and order. Discrete & Computational Geometry 9, 387–426 (1993)
5. Delfinado, C.J.A., Edelsbrunner, H.: An incremental algorithm for betti numbers of simplicial complexes on the 3-sphere. Comput. Aided Geom. Design 12(7), 771–784 (1995)
6. Dumas, J.-G., Heckenbach, F., Saunders, B.D.: Computing simplicial homology based on efficient smith normal form algorithms. In: Algebra, Geometry, and Software Systems, pp. 177–206 (2003)
7. Dupas, A., Damiand, G.: First results for 3D image segmentation with topological map. In: Coeurjolly, D., Sivignon, I., Tougne, L., Dupont, F. (eds.) DGCI 2008. LNCS, vol. 4992, pp. 507–518. Springer, Heidelberg (2008)
8. Elter, H., Lienhardt, P.: Cellular complexes as structured semi-simplicial sets. International Journal of Shape Modeling 1(2), 191–217 (1995)

9. Giesbrecht, M.: Probabilistic computation of the smith normal form of a sparse integer matrix. In: Cohen, H. (ed.) ANTS 1996. LNCS, vol. 1122, pp. 173–186. Springer, Heidelberg (1996)
10. Hatcher, A.: Algebraic Topology. Cambridge University Press, Cambridge (2002)
11. Hu, S.T.: On the realizability of homotopy groups and their operations. Pacific J. of Math. 1(583-602) (1951)
12. Lienhardt, P.: Topological models for boundary representation: a comparison with n-dimensional generalized maps. Comput. Aided Design 23(1), 59–82 (1991)
13. Lienhardt, P.: N-dimensional generalized combinatorial maps and cellular quasi-manifolds. Int. J. on Comput. Geom. & App. 4(3), 275–324 (1994)
14. May, J.P.: Simplicial Objects in Algebraic Topology. Van Nostrand (1967)
15. Munkres, J.R.: Elements of algebraic topology. Perseus Books, Cambridge (1984)
16. Peltier, S., Alayrangues, S., Fuchs, L., Lachaud, J.-O.: Computation of homology groups and generators. Comput. & Graph. 30, 62–69 (2006)
17. Peltier, S., Fuchs, L., Lienhardt, P.: Simploidals sets: Definitions, operations and comparison with simplicial sets. Discrete App. Math. 157, 542–557 (2009)
18. Rossignac, J., O'Connor, M.: A dimension-independant model for pointsets with internal structures and incomplete boundaries. Geometric modeling for Product Engineering, 145–180 (1989)
19. Sergeraert, F.: Constructive algebraic topology. SIGSAM Bulletin 33(3), 13–13 (1999)

Computing Homology: A Global Reduction Approach

David Corriveau[1] and Madjid Allili[2]

[1] Dept. of Computer Science, Université de Sherbrooke, Sherbrooke, QC, J1K2R1
david.corriveau@usherbrooke.ca

[2] Dept. of Computer Science, Bishop's University, Sherbrooke, QC, J1M0C8
mallili@ubishops.ca

Abstract. A new algorithm to compute the homology of a finitely generated chain complex is proposed in this work. It is based on grouping algebraic reductions of the complex into structures that can be encoded as directed acyclic graphs. This leads to sequences of projection maps that reduce the number of generators in the complex while preserving its homology. This organization of reduction pairs allows to update the boundary information in a single step for a whole set of reductions which shows impressive gains in computational performance compared to existing methods. In addition, this method gives the homology generators for a small additional cost.

Keywords: Homology Computation, Reduction, Directed Acyclic Graphs, Generators.

1 Introduction

Computation of homology has become a very important tool in many applications and domains such as dynamical systems, image processing and recognition, and visualization. In dynamics, invariants such as the Conley index are computed using homology algorithms. In digital image analysis, topological invariants are useful in shape description, indexation, and classification. Among shape descriptors based on homology theory, there is the Morse shape descriptor [1,2], and the persistence barcodes for shape [3]. The necessity of improved algorithms appears evident as new applications of the homology computation arise in research for very large data sets. The classical approach to compute homology of a chain complex with integer coefficients reduces to the calculation of the Smith Normal Form (SNF) of the boundary matrices which are in general sparse [4,5]. Unfortunately, this approach has a very poor running-time and its direct implementation yields exponential bounds. Uses and improvements of this method can be found in works that appeared recently [6,7]. Another generic approach is the method of reduction proposed in [8,5]. Several algorithms based on the idea of reduction and that improve the time complexity for particular types of data sets have been designed. For cubical sets embedded in $\mathbb{R}^3$, Mrozek et al.

S. Brlek, C. Reutenauer, and X. Provençal (Eds.): DGCI 2009, LNCS 5810, pp. 313–324, 2009.

proposed a method [9] based on the computation of acyclic subspaces by using lookup tables. For problems dealing with higher dimensions, Mrozek and Batko developed an algorithm [10] based on the concept of coreduction which showed interesting performance results on cubical sets. However problems in which data is higher dimensional and not cubical are still difficult to handle. The classical methods spend most of their time in updating the boundary homomorphisms after each reduction step.

We propose a method that identifies reduction pairs and organizes them in a structure that allows to efficiently update the boundary information in a single step for a whole set of reductions. This reduces to the minimum the manipulations of the data structures that store the boundary information. Starting at an arbitrary cell A with a face a in its boundary, we identify successive adjacent admissible pairs of reduction and build the longest possible sequence originating at A. Several sequences can originate at the same cell, and each sequence can be seen as a path in a directed acyclic graph, called a *reduction DAG*, where nodes are reduction pairs and oriented edges denote an adjacency relation between the reduction pairs. Given a reduction DAG across a chain complex, we achieve a global simplification of the complex by performing all the reductions in the DAG at once. We will establish direct algebraic formulas that allow to update the boundary of remaining cells. The net advantage of our approach is that the boundaries are not explicitly updated after each reduction step. Instead, this information is always preserved implicitly within the data structures and processed globally for each reduction DAG allowing to reduce considerably the computational time.

2 Chain Complexes and Homology

A finitely generated free *chain complex* $(\mathcal{C}, \partial)$ with coefficients in a ring $\mathcal{R}$ is a sequence of finitely generated free abelian groups $\{C_q\}_{q\in\mathbb{Z}}$ together with a sequence of homomorphisms, called boundary maps or operators, $\{\partial_q : C_q \to C_{q-1}\}_{q\in\mathbb{Z}}$ satisfying $\partial_{q-1} \circ \partial_q = 0$ for all $q \in \mathbb{Z}$. Typically, $C_q = 0$ for $q < 0$. For each p, the elements of C_p are called p-chains and the kernel of $\partial_p : C_p \to C_{p-1}$ is called the group of p-cycles and denoted $Z_p = \{c \in C_p \mid \partial_p c = 0\}$. The image of ∂_{p+1}, called the group of boundaries, is denoted by $B_p = \{c \in C_p \mid \exists\, b \in C_{p+1} \text{ such that } \partial_{p+1} b = c\}$. B_p is a subgroup of Z_p because of the property $\partial_p \circ \partial_{p+1} = 0$. The quotient groups $H_p := Z_p/B_p$ are the homology groups of the chain complex $\mathcal{C}$. The reduction of a chain complex is a procedure that consists of removing successively pairs of generators from the bases of its chain groups while preserving the homology of the original complex. At the algebraic level, each removal of a pair of generators that form a reduction pair is equivalent to a collection of projection maps $\{\pi_d : C_d \to C_d\}_d$ that send each generator of the removed pair into 0. Moreover, it projects the other cells into the remaining generators taking into account the modifications of their boundaries caused by the removal of the pair. Let $C'_d = \pi_d(C_d)$ for each d. It is shown in [5] that C' is a chain complex and $H_*(C') \cong H_*(C)$, that is, the homologies of C' and C coincide. To calculate the homology of the original chain complex C, the idea is to define a sequence of projections associated to the removal of pairs in all dimensions and then compute the homology of the resulting chain complex that is the image of the successive projections. A sequence of projections is complete if no other projection can be added to the sequence. Such a sequence always exits when the chain complex is finitely generated. Let (C, ∂) be the original chain complex and (C^f, ∂^f) is the final chain complex obtained after a complete sequence of projections. We already know that the two complexes have the same homology, moreover it is easily observed that if $\partial^f = 0$ then $H_*(C) \cong H_*(C^f) \cong C^f$, that is, the Betti number β_d is given by the number of d-cells remaining in the complex C^f.

Formally, a reduction pair and its associated collection of projection maps are defined as follows.

Definition 1. *Let $\mathcal{C} = \{C_d, \partial_d\}$ be an abstract chain complex. A pair of generators (a, B) such that $a \in C_{m-1}$, $B \in C_m$ and $\langle \partial B,\ a\rangle = \pm 1$ is called a* reduction pair. *It induces a collection of group homomorphisms*

$$\pi_d c := \begin{cases} c - \frac{\langle c,\ a\rangle}{\langle \partial B,\ a\rangle} \partial B & \text{if } d = m-1, \\ c - \frac{\langle \partial c,\ a\rangle}{\langle \partial B,\ a\rangle} B & \text{if } d = m, \\ c & \text{otherwise,} \end{cases}$$

where $c \in C_d$ and $\langle\ \cdot\ ,\ \cdot\ \rangle$ is the canonical bilinear form on chains.

3 Computation of Homology by Grouping Reductions

Forming RDAGs: For a fixed dimension and starting from an arbitrary cell, we form the maximal possible directed acyclic graph (DAG) whose nodes are reduction pairs and a directed edge between two pairs (a, B) and (c, D) means that (a, B) is adjacent to (c, D) (see Section 3.1). A directed path in the DAG corresponds to a reduction sequence. That type of reduction sequence affects only adjacent cells to the path and leaves other cells unchanged. Information about each reduction is carried out by the reduced cells. After performing the whole set of reductions, we obtain the reduced complex by extracting the boundary information from the data structures associated to the cells visited by the reduction DAGs. We introduce the main concepts through an example. The example details how the projections of cells are calculated given a reduction sequence. This will show what information is exactly needed to be kept about the original complex in order to be able to rebuild the reduced complex.

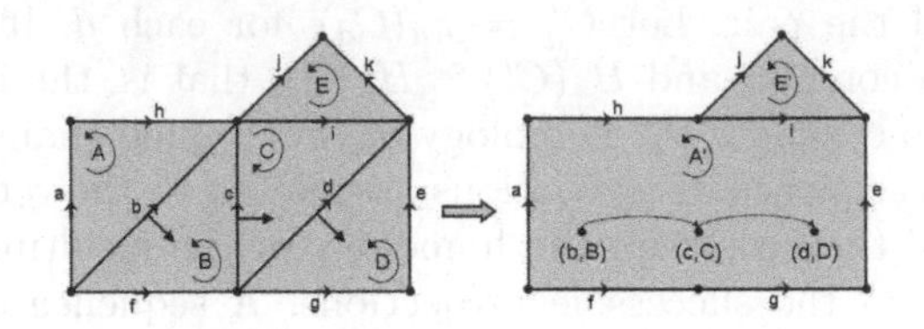

Fig. 1. Example of collapsing by using RDAGs

Example 1. Consider the complex given in Figure 1. The depicted sequence of reduction pairs originating at the cell A is (b, B), (c, C) and (d, D). The projection maps at the level of the 2-cells give the following

$$(b, B): \quad \pi_2^b \alpha = \alpha - \frac{\langle \partial\alpha,\ b\rangle}{\langle \partial B,\ b\rangle} B = \alpha - \lambda_b B$$

$$(c, C): \quad \pi_2^c \alpha = \alpha - \frac{\langle \partial\alpha,\ c\rangle}{\langle \partial C,\ c\rangle} C = \alpha - \lambda_c C$$

$$(d, D): \quad \pi_2^d \alpha = \alpha - \frac{\langle \partial\alpha,\ d\rangle}{\langle \partial D,\ d\rangle} D = \alpha - \lambda_d D$$

It follows that the projection map associated to the whole sequence is $\pi_2 = \pi_2^d \circ \pi_2^c \circ \pi_2^b$. The 2-cell A is the only cell adjacent to the sequence while the 2-cell E is not adjacent. Their respective projections with respect to the sequence are

$$\pi_2 A = \pi_2^d \circ \pi_2^c (A - \lambda_b B) = A - \lambda_b B + \lambda_b \lambda_c C - \lambda_b \lambda_c \lambda_d D.$$

$$\pi_2 E = E' = \pi_2^d \circ \pi_2^c \left(E - \frac{\langle \partial E,\ b\rangle}{\langle \partial B,\ b\rangle} B \right) = \pi_2^d \circ \pi_2^c (E - 0 \cdot B) = \ldots = E.$$

Each coefficient in the projection is called the coefficient of contribution of the corresponding cell to the projection of A.

3.1 Projection Formulas for Grouped Reductions

We define how to organize the reductions and the complex using reduction DAGs. A cell A_1 is *adjacent* to the pair (a_2, A_2) if the cells A_1 and A_2 are of the same dimension m and a_2 is a cell with dimension $m-1$ in the boundaries of both A_1 and A_2. A pair (a_1, A_1) is *adjacent* to a pair (a_2, A_2) if A_1 is adjacent to (a_2, A_2). A *reduction DAG* is a directed acyclic graph whose nodes are reduction pairs and a directed edge between two pairs (a_1, A_1) and (a_2, A_2) means that (a_1, A_1) is adjacent to (a_2, A_2).

A *path* P from (a_1, A_1) to (a_n, A_n) in a reduction DAG G is a *reduction sequence* $(a_1, A_1), \ldots, (a_n, A_n)$ whose elements are nodes in G and (a_i, A_i) is adjacent to (a_{i+1}, A_{i+1}), for $i = 1...n-1$. A cell A is said to be *adjacent* to a reduction sequence if it is adjacent to some pair of the sequence. A path from

(a_1, A_1) to (a_n, A_n) in a reduction DAG G is said to *originate at* A_0 if A_0 is adjacent to (a_1, A_1). A cell not appearing in any reduction DAG is called a *projected* cell. The cells that appear in a reduction DAG are called *reduced* cells.

In the following theorem, we give a formula to calculate the projection of a cell by considering the contribution of a single path and ignoring the input from other paths and sub-paths. This formula is called "path projection" and is proved using induction on the length of the sequence of reductions.

Theorem 1. *Let* $P : (a_1, A_1), \ldots, (a_n, A_n)$ *be a path originating at* A_0. *The* path projection of A_0 *by the path* P *is given by* $\pi_P(A_0) = A_0 + \sum_{i=1}^{n} \Gamma_i A_i$, *where* $\Gamma_i = (-1)^i \prod_{j=1}^{i} \lambda_j$, *and* $\lambda_j = \frac{\langle \partial A_{j-1},\, a_j \rangle}{\langle \partial A_j,\, a_j \rangle}$ *for* $1 \leq i, j \leq n$. Γ_i *is called the* coefficient of contribution *of* A_i *in the path projection of* A_0.

Proof: The proof is easily done by induction on the length of the path. □

Now, we can write $\pi_{P_n}(A_0) = A_0 + \underbrace{\sum_{i=1}^{n} \Gamma_i A_i}_{\Psi_{P_n}}$. We denote by Ψ_{P_n} the *projection chain* of the cell A_0 by the path P_n.

Corollary 1. *Let* $P_1, P_2, \ldots, P_k$ *be disjoint non overlapping paths originating at* A_0. *The total projection of* A_0 *by* $P_1, \ldots, P_k$ *denoted by* $\pi_{P_1, P_2, \ldots, P_k}(A_0) = A_0 + \sum_{i=1}^{k} \Psi_{P_i}$ *where* Ψ_{P_i} *is the* projection chain *of the cell* A_0 *by the path* P_i.

Corollary 2. *Let* T *be a reduction tree originating at* A_0, *then the projection of* A_0 *by* T *is equal to the sum of* A_0 *and the* projection chain *of* A_0 *by* T.

Proof: Typically, the trees originating at A_0 can occur as pure paths, in which case the associated projections are given previously. Otherwise, we can find paths that share a common ancestral branch that originates at A_0. This is seen as a bifurcation as shown in Figure 2. In this case, the ancestral branch is a path $B : (b_1, B_1), \ldots, (b_{n_B}, B_{n_B})$ which is extended by two paths $C : (c_1, C_1), \ldots, (c_{n_C}, C_{n_C})$ and $D : (d_1, D_1), \ldots, (d_{n_D}, D_{n_D})$. The combination of

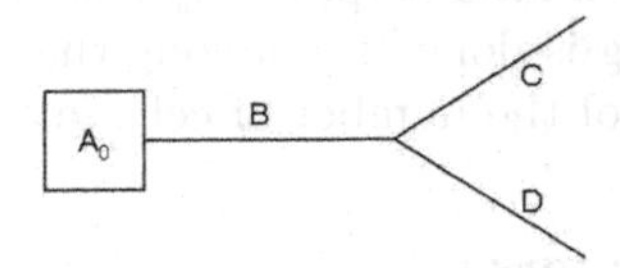

Fig. 2. A path B bifurcates into two paths C and D

the results in Theorem 1 and Corollary 1 can be used to show that the total projection of A_0 by $T = (BC), (BD)$ is given by

$$\pi_T(A_0) = A_0 + \sum_{i=1}^{n_B} \Gamma_i B_i + \Gamma_{n_B} \left(\sum_{j=1}^{n_C} \alpha_j C_j + \sum_{k=1}^{n_D} \beta_k D_k \right)$$

The term $\sum_{i=1}^{n_B} \Gamma_i B_i + \Gamma_{n_B} \left(\sum_{j=1}^{n_C} \alpha_j C_j + \sum_{k=1}^{n_D} \beta_k D_k \right) = \pi_T(A_0) - A_0 = \Psi_T(A_0)$ is called the *projection chain of the tree.* □

Theorem 2. *Let A_0 be a cell adjacent to a RDAG in a chain complex C. The projection of A_0 by the RDAG is equal to A_0 to which we add the projection chains of A_0 by all the trees in the RDAG that originate at A_0, that is*

$$\pi_{RDAG}(A_0) = A_0 + \sum_{T \in \mathcal{T}_{A_0}} \Psi_T(A_0)$$

where $\mathcal{T}_{A_0}$ is the collection of all reduction trees in the RDAG originating at A_0.

Proof: The results proved for paths in corollaries 1 and 2 can be easily extended for the case of trees to find the formula for the projection of A_0.

Building a Simplified Complex: Using reduction DAGs to compute the homology of a chain complex is a recursive process. At each recursion level, the algorithm simplifies the complex by constructing reduction DAGs on the complex and saving the associated projections into appropriate data structures. This is performed simultaneously for each dimension. This process eventually stops when it is impossible to add another reduction pair to any reduction DAG. In that case, the algorithm will build the associated simplified complex and continue the reduction process on the simplified complex. The simplified complex is rebuilt from the projected cells only (reduced cells are not considered). The boundaries of the cells are updated using global projection formulas that allow to calculate the incidence numbers between cells of contiguous dimensions. Note that the reduced cells are not completely removed from the structures since they may be needed to recover homology generators expressed in terms of cells of the original complex as we explain in subsection 3.2. Contrary to the classical case where the boundary updating is done at each reduction step and may concern cells that can be reduced at a later step, a major benefit of this new approach is that the boundary updating is done only among the projected cells which often constitute a small fraction of the number of cells in the original complex.

3.2 Calculating Generators

Let us consider the complex of a plane quotiented by its boundary as illustrated in Figure 3. This complex is homeomorphic to a 2-sphere and the 2-cell A

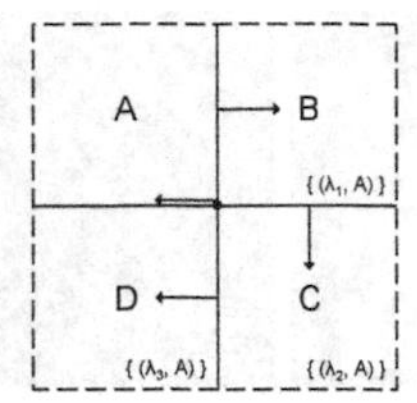

Fig. 3. A Complex homeomorphic to a 2-sphere

represents the 2-generator $A' = A + \lambda_1 B + \lambda_2 C + \lambda_3 D$. As illustrated in Figure 3, after each reduction, the projection coefficients λ_i are saved into the projection lists maintained in each reduced cell. Thus, to get the 2-generator associated to the 2-cell A, one has to scan through each reduced 2-cell and extract the projection coefficients associated to A. Projections correspond to generators but they can be expressed with cells of the simplified complex at any level of the recursion. However, in order for the generators to carry geometric meaning, it makes sense only to express them with cells from the original complex. Due to the recursive simplifications, a projected cell at a previous level of the recursion may become a reduced cell at a later level of the recursion. We illustrate this in Figure 4. In this example, there are two levels of recursion. The final simpli-

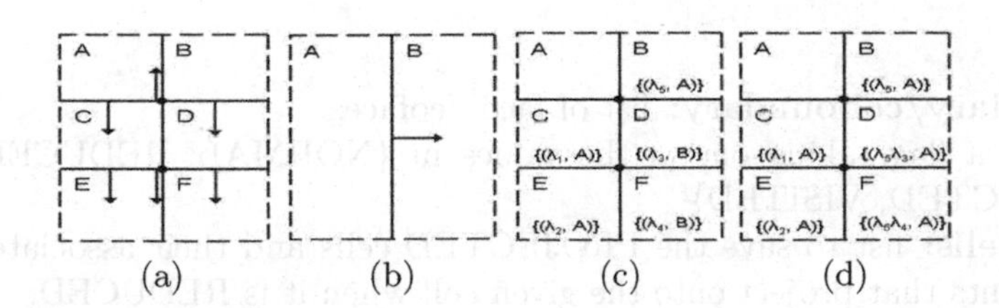

Fig. 4. (a) Reduction 1 of the initial complex. (b) Reduction 2 of the simplified complex. (c) Initial complex after returning from the second reduction. (d) Initial complex after returning from the first reduction.

fied complex is shown in 4(b). At this step, the cell A represents a 2-generator that is expressed as $A'' = A' + \lambda_5 B'$. Returning from the last recursion, we know that $A' = A + \lambda_1 C + \lambda_2 E$ and $B' = B + \lambda_3 D + \lambda_4 F$. The cell A represents a projected cell at both recursion levels (4(b) and 4(c)) and requires no special consideration. On the other hand, the cell B is a projected cell at the first level of recursion but becomes a reduced cell at the second recursion level. Consequently, returning from the second recursion level, we scan through each 2-cell and whenever B is encountered in one of the projection lists, it is replaced by $\lambda_5 A$. This is shown in Figure 4(d). Finally, the generator is expressed by $A'' = A + \lambda_5 B + \lambda_1 C + \lambda_5 \lambda_3 D + \lambda_2 E + \lambda_5 \lambda_4 F$. In Figure 5 we show different 1-generators that we extracted from 3D models.

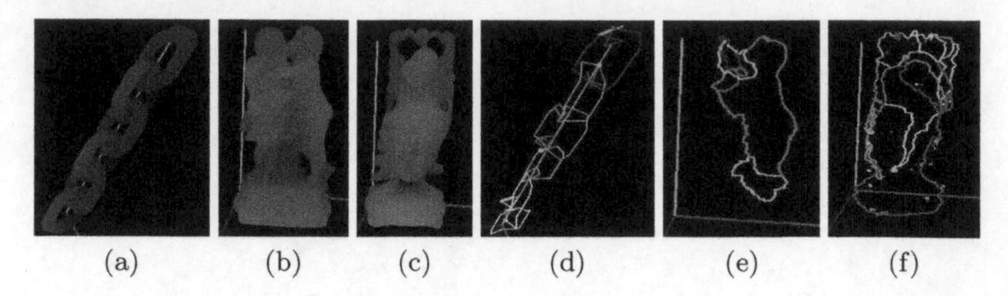

Fig. 5. (a,b,c) The 3D models of a chain, two kissing children and a Buddha. (d,e,f) The calculated holes (one dimensional generators).

4 Data Structures and Algorithms

The first data structure represents a chain complex.

ChainComplex:

- **E** : two dimensional array of cells organized by their respective dimension, that is E[d][i] denotes the i-th d-cell.

In our implementation, the pointer to a cell is used to identify the cell. These identifiers are saved in **E**. Only the pointers of the projected cells are copied into the simplified complex.

Cell:

- **boundary/coboundary:** list of faces/cofaces.
- **state:** a flag taking one of the values in {NORMAL, REDUCED, PROJECTED, VISITED}.
- **projCells:** list to save the PROJECTED cells and their associated coefficients that project onto the given cell when it is REDUCED.
- **nbUpdates:** approximates the number of updates to a **projCells** list when the given cell is reduced by one of its cofaces.

Initially, all cells are set to the NORMAL state. REDUCED is assigned to the cells in a reduction sequence and PROJECTED is assigned to the cells adjacent to a reduction sequence. The VISITED flag is assigned to the d-cells that don't have any $(d+1)$ coface that can form an admissible reduction pair. The flag helps to avoid testing more than once if the given d-cells have an admissible reduction pair. The **projCells** list is used by the REDUCED cells to keep track of the coefficients of contribution of every PROJECTED cell for which it contributes. To build a sequence of reduction pairs, **nbUpdates** is used to select reduction pairs that should minimize the amount of updates to the **projCells** list. We now give the principal steps of the algorithm.

◇ *HOM_RDAG*(**ChainComplex** K) returns the Betti numbers of a chain complex K by using reduction DAGs.

1. (*Initialization*) For all cells in K, set its **state** to NORMAL, empty **projCells**, set **nbUpdates** to the number of cofaces.
2. (*Reduce cells*) Proceed by decreasing order of dimension. Order the d-cells by increasing value of **nbUpdates**. Following this order, start a reduction DAG (call *BuildRDAG*()) from each cell whose state is NORMAL.
   For dimension 0, change remaining NORMAL 0-cells to PROJECTED.
3. (*Build the simplified complex*) Call *BuildSimplifiedComplex*(K). This method extracts the boundary information from the data structures to build a simplified complex K'.
   Continue to recursively simplify K' by calling *HOM_RDAG*(K') until there are no reduction pair left.
   Test if all boundaries are trivial. If not, then continue the reduction process with another method such as SNF.
4. (*Return the Betti numbers*) Assign the number of PROJECTED d-cells to β_d.

◇ *BuildRDAG*(**Cell** c) builds a reduction DAG from cell c.

1. (*Initialization*) Save c.**nbUpdates** into *nbUpdatesLimit* for later use.
2. (*Find a reduction pair*) Find a coface B of c such that B is NORMAL or VISITED. If a coface B is found, then call *PairCells*(c,B), otherwise set c to VISITED and exit *BuildRDAG*().
3. (*Expand the reduction DAG*) Expand the reduction DAG (repeat step 2) from all NORMAL faces of B that have **nbUpdates** $\leq$ *nbUpdatesLimit*. Proceed in a breadth first search approach.

◇ *PairCells*(**Cell** c, **Cell** B) adds the reduction pair (c, B) to the sequence and updates the data structures accordingly.

1. (*Update **projCells***) For all cofaces C of c different than B, if C is REDUCED, then add $\lambda_c * C$.**projCells** to B.**projCells**. Otherwise, set C to PROJECTED and add $\lambda_c * C$ to B.**projCells**, where $\lambda_c = -\frac{\langle \partial C,\ c\rangle}{\langle \partial B,\ c\rangle}$.
2. (*Update the **nbUpdates** variable*) For all faces a of B different than c, add $|B$.**projCells**$| - 1$ to c.**nbUpdates**. For all faces of c, remove one to **nbUpdates**.

◇ *BuildSimplifiedComplex*(**Complex** K) extracts the boundary and coboundary information from the data structures to build a simplified complex K'.

1. (*Update **boundary***) Proceed by increasing order of dimension. Let c be a PROJECTED d-cell of K. For all cofaces C of c, iterate through C.**projCells**. Let B be a $(d+1)$-cell in C.**projCells** and λ_B its associated coefficient. Add $\lambda_B \langle \partial C,\ c\rangle\, c$ to B.**boundary** and update c.**coboundary** accordingly.
   Finally, remove all REDUCED cells remaining in the **boundary** and **coboundary** of PROJECTED cells. Copy the pointers of all PROJECTED cells into $K'.E$.

5 Experimental Results and Discussion

We experimented with datasets from d-balls, d-spheres, tori, Bing's houses, 3D medical scans or randomly generated d-complexes. Each dataset contains many files of complexes of various data sizes. The size of a complex is measured as the number of cells plus the number of links (entries in boundary lists). A test is an execution of the algorithm on a file of a specified data set. We performed 30 tests for each file and computed the average, maximum and standard deviation of the time used for calculating the homology, the generators and sorting. We measured the sorting time because it has a time complexity of $\Theta(n \cdot \log n)$ and we wanted to verify that sorting would not monopolize the computation time. Also, we saved statistics on the number of recursions to see its influence on the global performance. We grouped and summarized the results per dataset. Table 1 shows the comparison between the performance of the RDAGs algorithm and the classical reduction method on different datasets. We can observe a significant improvement of the new method. The results suggest subquadratic time complexity for the RDAGs algorithm which is what we measured in Table 2.

Table 1. Reduction DAGs algorithm versus the classical reduction method

Dataset	Data Size	Times Faster
3-Ball	115905	195.4
4-Ball	27841	147.3
2-Sphere	517145	65.0
3-Sphere	134777	182.5
Torus	418608	36.0
Heart	545492	50.5
3-Complex	265322	410.0
5-Complex	431186	458.8
8-Complex	330285	185.9

The second column (**N**) gives the number of files within the dataset. The third and fourth columns show the parameters α and β that best-fit the equation "time $\approx \beta \cdot \text{data\_size}^{\alpha}$", where time is expressed in seconds. We obtained the best-fits from the times measured on the individual files of each dataset. In the last four columns, we present the times in milliseconds that we get from the best-fit equation for different data sizes. Those times approximate the real times measured in real experiments. In Table 2 we report the approximative time that our algorithm uses to calculate homology on various datasets.

In Table 2, we observe that $\alpha \approx 1.16$ and $\beta \approx 5E-8$ for d-balls and d-spheres without regard of the dimension. Except for few values, α is relatively constant while β gradually decreases as the dimension increases. We explain this by the fact that the ratio of exterior face reductions versus interior face reductions increases with the dimension. We observe that the approximation times and the

Table 2. Approximated performance with respect to dataset, data size and dimension

Dataset	N	Equation		Time (msec) Vs Data Size			
		β	α	10^4	10^5	10^6	10^7
2-Ball	15	9.05E-8	1.11	2.49	32.11	413.66	5329.04
3-Ball	11	4.63E-8	1.15	1.84	26.04	367.77	5194.95
4-Ball	11	4.26E-8	1.14	1.55	21.35	294.72	4068.27
5-Ball	9	5.66E-8	1.12	1.71	22.53	297.04	3915.76
6-Ball	5	3.56E-8	1.15	1.42	20.02	282.78	3994.39
2-Sphere	17	5.77E-8	1.18	3.03	45.83	693.71	10499.67
3-Sphere	16	2.79E-8	1.22	2.12	35.12	582.91	9673.96
4-Sphere	10	3.38E-8	1.17	1.62	23.93	353.93	5235.00
5-Sphere	9	5.52E-8	1.14	2.00	27.67	381.89	5271.56
6-Sphere	5	3.30E-8	1.17	1.58	23.36	345.55	5111.09
7-Sphere	3	1.31E-8	1.24	1.19	20.76	360.80	6270.05
Torus	5	1.48E-7	1.07	2.82	33.13	389.28	4573.64
Bing's House	5	2.18E-7	1.04	3.15	34.55	378.84	4153.90
Heart	6	6.84E-8	1.13	2.26	30.55	412.15	5559.76
Brain	6	3.13E-8	1.20	1.97	31.30	496.07	7862.20
2-Complex	9	6.78E-8	1.24	6.18	107.46	1867.37	32451.12
3-Complex	9	7.89E-9	1.36	2.17	49.78	1140.45	26126.25
4-Complex	9	2.82E-9	1.45	1.78	50.15	1413.35	39833.56
5-Complex	9	4.95E-9	1.38	1.64	39.32	943.20	22625.87
6-Complex	9	3.04E-9	1.41	1.33	34.11	876.75	22535.83
7-Complex	9	2.42E-8	1.23	2.01	34.18	580.52	9858.60
8-Complex	9	2.26E-9	1.43	1.19	31.92	859.23	23126.42

Table 3. (a) Betti numbers of the files in the heart and brain datasets. (b) Number of reduction calls per dataset.

File Name	Number of Generators
heart192	(11, 9, 0)
heart160	(4, 15, 2)
heart128	(39, 12, 6)
heart96	(30, 107, 19)
heart64	(38, 228, 104)
heart32	(111, 220, 208)
brain192	(11, 0, 0)
brain160	(109, 1, 0)
brain128	(230, 25, 0)
brain96	(255, 153, 4)
brain64	(134, 407, 21)
brain32	(41, 524, 252)

(a)

Dataset	N	Reduction Calls		
		Avg	Max	SD
Torus	5	1.71	2.20	0.46
Heart	6	3.62	4.33	0.33
Brain	6	3.18	3.33	0.20
2-Complex	9	2.76	3.33	0.34
3-Complex	9	2.68	3	0.27
4-Complex	9	2.94	3.56	0.35
5-Complex	9	2.51	2.78	0.19
6-Complex	9	2.55	2.89	0.21
7-Complex	9	2.89	3.22	0.24
8-Complex	9	2.61	3	0.28
All others	N/A	1	1	0

(b)

parameters α and β for the heart and brain datasets, remain in the same order as those of d-balls and d-spheres despite their high Betti numbers as shown in Table 3(a).

For these two datasets, the high Betti numbers were not a relevant factor for the performance of the algorithm. We think it is because they contain a high ratio of exterior face reductions, which cost near to nothing. If we compare with the random d-complexes then the topology is an important factor. Indeed, as the data size increased we observed bigger differences in the time performance. Another interesting aspect of the algorithm is the number of recursions which is expressed as the number of reduction calls in Table 3(b).

6 Conclusion

Our experimentations show that this algorithm performs significantly faster than the classical reduction method. In addition, for all the datasets that we tested, which cover a wide range of types of data, its global performance indicated a subquadratic time complexity. Moreover, it allows to calculate the homology generators at a small additional time cost. Interestingly enough, in all the datasets we dealt with, the algorithm needed only to construct few intermediate complexes to obtain the homology of the original complex.

References

1. Allili, M., Corriveau, D., Ziou, D.: Morse Homology Descriptor for Shape Characterization. Proceedings of the 17th ICPR 4, 27–30 (2004)
2. Allili, M., Corriveau, D.: Topological Analysis of Shapes using Morse Theory. Computer Vision and Image Understanding 105, 188–199 (2007)
3. Collins, A., Zomorodian, A., Carlsson, G., Guibas, L.: A Barcode Shape Descriptor for Curve Point Cloud Data. Computers and Graphics 28, 881–894 (2004)
4. Munkres, J.R.: Elements of Algebraic Topology. Addison-Wesley, Reading (1984)
5. Kaczynski, T., Mischaikow, K., Mrozek, M.: Computational Homology. Appl. Math. Sci. Series, vol. 157. Springer, New York (2004)
6. Storjohann, A.: Near Optimal Algorithms for Computing Smith Normal Forms of Integer Matrices. In: Proceedings of 1996 International Symposium on Symbolic and Algebraic Computation, ISSAC 1996, pp. 267–274 (1996)
7. Peltier, S., Alayrangues, S., Fuchs, L., Lachaud, J.: Computation of Homology Groups and Generators. Computers and Graphics 30, 62–69 (2006)
8. Kaczynski, T., Mrozek, M., Slusarek, M.: Homology Computation by Reduction of Chain Complexes. Computers and Mathematics with App. 35, 59–70 (1998)
9. Mrozek, M., Pilarczyk, P., Zelazna, N.: Homology Algorithm Based on Acyclic Subspace. Computers and Mathematics with App. 55(11), 2395–2412 (2008)
10. Mrozek, M., Batko, B.: Coreduction Homology Algorithm. Discrete and Computational Geometry (in press)
11. Allili, M., Corriveau, D., Derivière, S., Kaczynski, T., Trahan, A.: Discrete Dynamical System Framework for Construction of Connections between Critical Regions in Lattice Height Data. J. Math. Imaging Vis. 28(2), 99–111 (2007)

Surface Sketching with a Voxel-Based Skeleton

Jean-Luc Mari

Information and System Science Laboratory (LSIS)
Computer Graphics Team ("Image & Models")
University of Marseille 2
ESIL, Campus de Luminy, case 925, 13288 Marseille cedex 9, France
mari@univmed.fr
http://jean-luc.mari.perso.esil.univmed.fr/

Abstract. In this paper, we present a method to generate a first approximation surface from a volumic voxel-based skeleton. This approach preserves the topology described by the discrete skeleton in a 3D grid considering the 26-adjacency: if a cycle is sketched, then there is a hole in the resulting surface, and if a closed hull is designed, then the output has a cavity. We verify the same properties for connected components. This surrounding basic polyhedron is computed with simple geometrical rules, and can be a good starting point for 3D shape design from a discrete voxel skeleton. As an example application, we use this rough mesh as the *control polyhedron* of a subdivision surface in order to model multiresolution objects.

Keywords: Discrete geometry, geometrical modeling, voxel, automatic mesh generation, topological skeleton, sketch.

1 Introduction

We wish to compute a polyhedral surface that preserves the initial topology of a 3D discrete sketch designed interactively. To do this, we characterize a skeleton from a sketch made of voxels. This sketch defines the connection relations by creating edges and triangles, and represents the global shape of the object.

From this shape descriptor, our approach generates a polyhedron M whose topology is characterized by the skeleton according to the 26-adjacency. To do this, we first create points around vertices of the skeleton. The location of these points is the key problem in our method. Subsequently, they are linked to form a basic triangulation. No edges must cross, and no triangles must intersect. The closed resulting mesh is adequate to initialize, for instance, a subdivision surface process.

One asset of subdivision surfaces is that they can handle arbitrary topologies. Therefore, using the aforementioned generated mesh M as the *control polyhedron* for subdivision surfaces ensures that we produce a sequence of meshes of equivalent topology. If M is homotopic to the skeleton, then so are the refined surfaces.

S. Brlek, C. Reutenauer, and X. Provençal (Eds.): DGCI 2009, LNCS 5810, pp. 325–336, 2009.

Another advantage of subdivision surfaces is their natural support for *multiresolution*. Once the control polyhedron is generated according to our method, several *levels of detail* (LoDs) become available within a single model. Thus, depending on the context, a more or less refined mesh can be used. For example, low-resolution meshes are well-suited for preview, whereas high-resolution models of the same object are useful for final rendering.

To define the global appearance of an object, the designer is asked to draw a 3D sketch in a regular cubic grid. This means that creating a shape is fast and simple, as it uses *voxel* modeling to build an entity from "cubes". The method we develop has topological guarantees on the resulting mesh. The sketch, converted into a skeleton, gives a suitable shape descriptor for further finer geometrical modifications. In this paper, we use the word *skeleton* even if in discrete geometry, it is directly connected to the medial axis, to the set of the centers of maximal balls, or to an entity obtained by thinning. In geometrical modeling, the skeleton is a structure located *inside* the shape, that can give information on the shape or can be used to animate the surrounding shape. The use of the term *skeleton* is made to connect the two worlds (discrete objects and shape design) with due consideration.

We think that the most natural neighborhood to define a basic shape is the 26 one: as soon as two voxels touch, they are said to be adjacent. Thus, we characterize the surface to design with this adjacency: if two voxels are not connected this way, then the related surrounding surface does not have to be connected to preserve the topology of the 3D sketch.

This paper is divided into three main parts: we give a brief overview of related techniques in Section 2. Section 3 is dedicated to the description of our approach. Starting from a discrete 3D sketch, we create a topological skeleton and then create a basic surrounding polyhedron. In Section 4, we validate our method and present an application example in the frame of multiresolution design using subdivision surfaces. Finally, we give directions for future work in Section 5.

2 Related Work

The most prominent offset surfaces based on skeletons are undoubtedly *implicit surfaces* [5]. Implicit modeling provides an intuitive way to generate surrounding shapes, starting with a skeleton built from simple primitives: points, segments, curves, triangles or patches. The resulting surfaces are continuous and naturally smooth. Moreover, they can be expressed mathematically.

Usually, implicit surfaces are defined by a skeletal shape and a potential function. The choice of a scalar value associated with a function determines an isosurface around the skeleton. There are many approaches for design [6], reconstruction [3] or animation [21].

Among implicit techniques, convolution approaches are particularly adapted to interactive modeling [20]. Thanks to a different distance definition, such surfaces do not present bumps on junction areas, which is a problem with classical implicit surfaces [1].

In the same way, another intuitive way to model shapes interactively is implicit *virtual sculpture*. The object is manipulated like clay using specific tools instead of skeleton primitives [10].

Variational implicit surfaces are used in the *BlobMaker project* to model free-form shapes using sketches. The main operations are based on inflations to create 3D forms from a 2D stroke [2].

The approach based on *Skins* [17] deals with skeleton implicit modeling and subdivision surfaces. It is a surface representation that uses particles to sculpt an object interactively. Nevertheless, the computation time remains long and auto-intersection issues can appear when handling the shape (the topology of the skeleton is not connected with the surrounding subdivision mesh).

However, implicit modeling techniques suffer the same classical problem: it is difficult to control the topology of the final surface because *unwanted blendings* can appear when two yet unconnected skeletons elements are close. To sort out this point, *blending graphs* are often used to allow skeleton primitives to be joined or not. This process slows down interactive handling and makes it less intuitive. Moreover, the second drawback of implicit surfaces is the real time rendering that is hard to obtain. Indeed, the computation time of all the potential functions becomes rapidly prohibitive when the number of primitives increases.

To generate shapes interactively, designers usually use primitive objects like cubes, spheres or cones to obtain the rough appearance of the final form. Modifiers like *extrusion* or *spinning* (surfaces of revolution) allow modeling more complex entities, but the use of shape descriptors like skeletons is rare. Most of the time, the opposite order of steps occurs: the shape is designed first and a skeleton is added afterwards for animation.

Many works deal with medial axis and medial surfaces applied to the generation of polygonal meshes [11,19]. We do not develop this family of approaches because they are not related to sketching, but rather to finite element meshing. Besides, we do not explore further classical *marching type surfaces* [14] because cracks can appear on some configurations, and yet the generated mesh has to be a closed surface. On the other hand, KENMOCHI's et al. marching cubes type approach is topologically robust, and can in this way produce the boundary mesh of discrete objects [12]. This method is efficient and connected to our issue, but the resulting surface is not as wished: an offset of the initial voxel set. Indeed the voxels' centers are the mesh vertices themselves (for instance, a single voxel would not generate a surrounding hull).

The idea of using a voxel-based skeleton as both a shape descriptor and, above all, a topological structure was introduced in [15]. The authors presented further developments and validation [16] towards obtaining a model that comprises three entities: a structural and topological *inner skeleton*, a geometrical *external layer* and a *transition layer*. However, to obtain the vertices of the surrounding mesh, sample points are computed on an implicit surface and then connected to obtain a polyhedron. These steps can fail because if the cloud of points is not dense enough, then the triangulation can collapse.

Therefore, one asset of the above multilayer approach is the control of the topology by using the natural neighborhood of digital volumes: the topology of the surrounding layer is defined by the inner skeleton. To follow this idea, we propose defining the skeleton from a 3D sketch designed in a cubic grid on one hand, and meshing specific surrounding points according to topological rules on the other hand, regardless of the density of these points.

3 Surrounding Mesh Generation

In this section, we develop the mesh generation process. In a first step, a designer draws a 3D sketch in a cubic grid. Then this sketch is converted to a skeleton comprising vertices, edges and triangles that characterize the adjacency between voxels. Finally, a mesh is automatically generated around the skeleton structure. This polyhedron represents the global shape, and is a valid starting point for multiresolution by subdivision surfaces. All the operations on the generated mesh are done with respect to the topology of the sketch.

In the first subsection, we spell out the overview of the process, going from the sketch to the automatically generated polyhedron. In the second subsection, we explain the construction of the 3D skeleton. In the third subsection, we describe the computation of the surrounding polyhedron according to particular topological rules.

3.1 Overview of the Process: From the Sketch to the Polyhedron

The mesh generation process creates three objects: first a discrete sketch made of voxels, then the related skeleton that describes the topology of the shape, and finally the surrounding meshed surface (Figure 1).

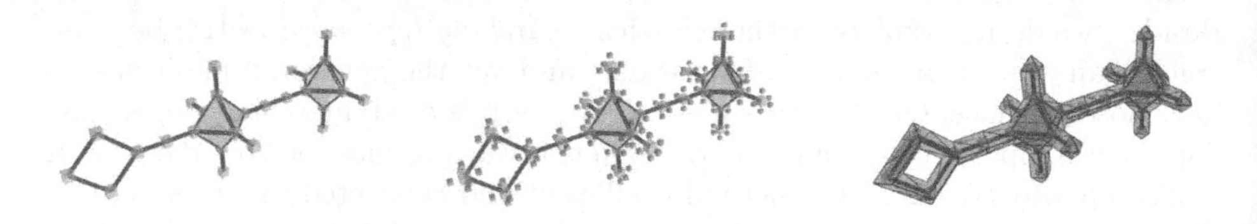

Fig. 1. Connected sketch (skeleton), surrounding points and basic polyhedron

We start from a set of voxels interactively set in a 3D grid by a designer. The designer first chooses the grid resolution (for example, a subset 32^3 of the discrete space $\mathbb{Z}^3$). Then, the designer draws voxels, knowing that the 26-adjacency will be used to determine the connected voxels.

During the voxel-edition step, the algorithm constructs a skeleton: segments and triangles are generated to emphasize the connection relations between voxels.

Once the skeleton is computed, based on the discrete sketch, a specific algorithm is used to generate a basic shape whose topology is defined by the skeleton.

This is done because the vertices of the surface are set up in safe locations, where further meshing cannot perturb the topology.

The resulting mesh is a relevant geometrical descriptor of the object. It characterizes the global shape, at low resolution level. This is why the mesh can be considered to be a *control polyhedron* for subdivision. Using an approximating scheme like LOOP's will produce a surface slightly *inside* the volume of the initial polyhedron. This ensures topological preservation, even at high resolution level.

3.2 Construction of the 3D Skeleton

We call the inner structure of the shape to be designed the *skeleton*. It is made of three kinds of elements: the vertices (edited interactively by a manipulator), the edges and the triangles (computed to code the adjacency relation between the vertices). In other words, the skeleton consists of the initial sketch plus segments and triangles.

Interactive Edition of the Sketch. The 3D sketch is edited interactively in the discrete space $\mathbb{Z}^3$: each voxel is located according to a regular grid (Figure 2).

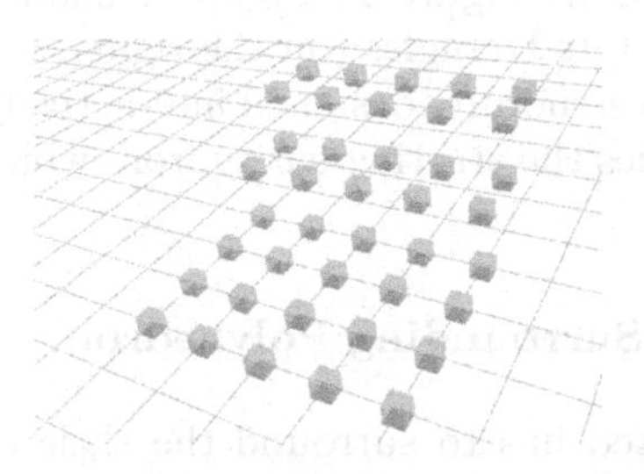

Fig. 2. Edition of the sketch in a 3D grid

Following this representation of the 3D sketch, an *adjacency* list is constructed for each voxel. The maximal number of neighbors is 26, due to the chosen connection relation. This set of voxels defines the points of the skeleton.

Adjacency Representation of the Voxels. For each voxel of all adjacency lists, an edge is added to the skeleton structure. The edges materialize pairs of connected voxels according to the 26-adjacency.

During the voxel edition phase, when an edge about to be added crosses another edge, its incorporation is canceled. This happens only for the 2×2 configuration (Figure 3).

A list of triangles is made from the list of edges, by scanning 3-cycles of edges. An elementary edition sequence is the following: three points form a triangle (cycle of 3 edges) and a fourth point is added to constitute a second triangle, without edge crossing (thanks to the middle edge removal routine seen previously).

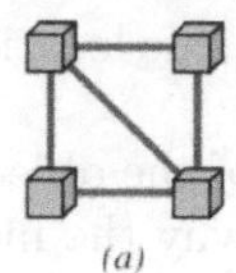

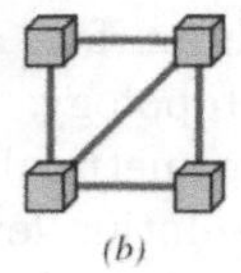

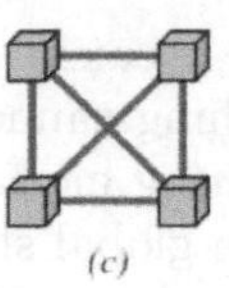

Fig. 3. Good edge construction *(a,b)*; bad construction *(c)*

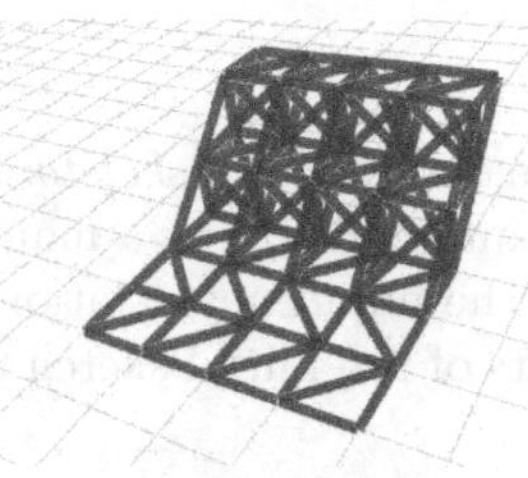

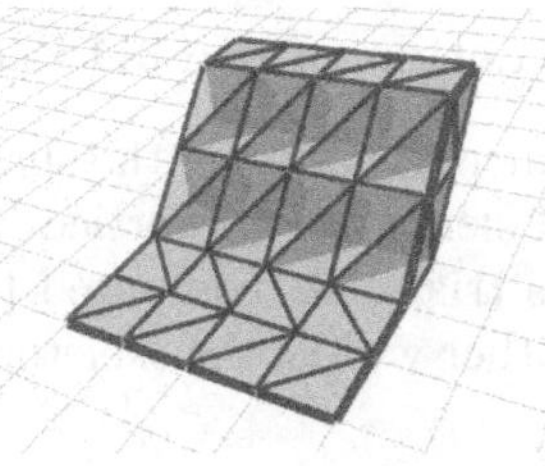

Fig. 4. Edges and triangles of the skeleton

Considering the voxels of Figure 2, Figure 4 shows the edges of the final skeleton (on the left) and its triangles (on the right).

The resulting skeleton structure, designed interactively in a three-dimensional environment, characterizes the starting entity from which the surrounding polyhedron is computed.

3.3 Computing the Surrounding Polyhedron

The mesh to be computed has to surround the skeleton as well as preserving its topology. In particular, no auto-intersection must appear. Moreover, if the skeleton has a cycle or a hull, then the generated polyhedron must include a hole or a cavity. To this end, specific rules are elaborated to create vertices and to mesh them.

Creating the Vertices. The vertices of the basic polyhedron are created as follows. For each voxel of the skeleton, up to six vertices of the mesh can be generated. Depending on adjacency rules, some points of the polyhedron are not created in order to get an actual surrounding mesh, with no points inside the wanted shape.

To simplify the problem, let us consider the size of a voxel as $10 \times 10 \times 10$. The distance between the centers of two 6-adjacent voxels is then 10 *units*.

Let $S(x, y, z)$ be a voxel of the skeleton. For a single voxel S, six points of the mesh are generated: $P_1(x-4, y, z)$, $P_2(x+4, y, z)$, $P_3(x, y-4, z)$, $P_4(x, y+4, z)$, $P_5(x, y, z-4)$ and $P_6(x, y, z+4)$. This prevents interaction between non-adjacent voxels. The points of the mesh associated to a voxel of the skeleton will never be located further than four units away, to prevent auto-intersection during the meshing step (Figure 5, on the left).

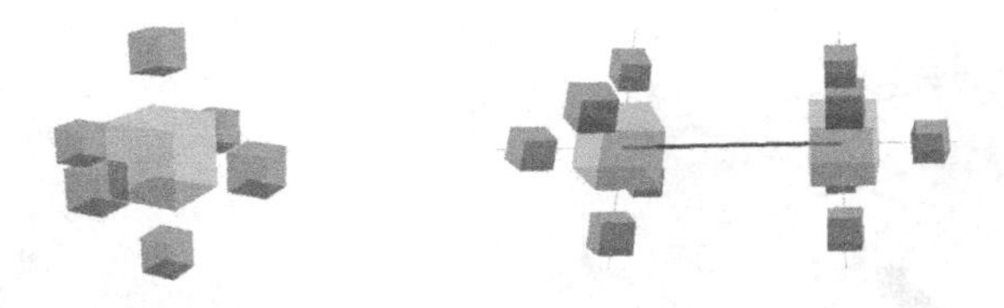

Fig. 5. Surrounding points of a single voxel (on the left) and of two 6-adjacent voxels (on the right)

However, when two voxels are 6-connected, only five points per voxel have to be generated, instead of six. This is done thanks to *face-connection flags*, constructed during the edition of the sketch. Six boolean values are allocated to each voxel (*up*, *down*, *left*, *right*, *front* and *back*): when the latter has face neighbors (6-adjacent connected voxels), the related values are set to *'true'*. Therefore, no points of the mesh are created in the specific directions of 6-connected voxels (Figure 5, on the right).

Naturally, for an inner voxel of the skeleton having six face-neighbors, no points of the mesh are generated at all.

Meshing the Vertices. Starting from the generated points, a critical step remains: meshing these vertices with the aim of getting a surrounding polyhedron that reflects the topology of the designed sketch.

The idea is first to generate all possible edges between the points P_i of the mesh, following specific linking rules, and then in a second phase to build the triangles of the mesh to get the final polyhedral crust of the object.

Before adding an edge E_i to the mesh, a series of tests is done.

Let P_1 and P_2 be the points that define E_i. The latter is added to the list of edges if the coordinates satisfy one of the following conditions (the coordinates x, y and z are equivalent; thus, the three axes have to be permuted to obtain all conditions):

- $P_1x = P_2x \pm 10$ and $P_1y = P_2y$ and $P_1z = P_2z$
- $P_1x = P_2x \pm 10$ and $P_1y = P_2y \pm 10$ and $P_1z = P_2z$
- $P_1x = P_2x \pm 10$ and $P_1y = P_2y \pm 6$ and $P_1z = P_2z \pm 6$
- $P_1x = P_2x \pm 10$ and $P_1y = P_2y \pm 4$ and $P_1z = P_2z \pm 4$ (to connect diagonal edges as shown in Figure 6 on the left).
- $P_1x = P_2x \pm 10$ and $P_1y = P_2y \pm 6$ and $P_1z = P_2z \pm 4$ (Figure 6 on the right).

These rules link all the points related to adjacent voxels that must be connected to form a surrounding mesh.

In addition, an edge E_i is not added to the list if:

- The new edge E_i intersects an existing edge in the list.
- E_i intersects an edge or a triangle of the skeleton.

For testing intersections of an edge with a triangle of the skeleton, we use the approach described in [8].

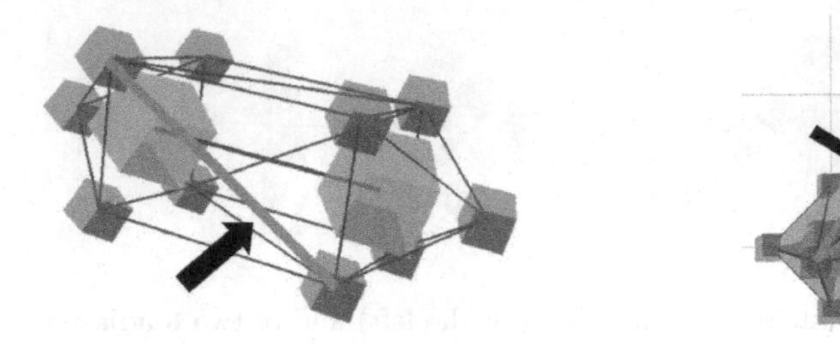

Fig. 6. Linking vertices of the mesh according to specific rules

The final step consists of constructing the triangles of the polyhedron: this computation is similar to the one detailed in Section 3.2: the list of edges is scanned and all 3-cycles produce triangles. However, one more test is done: if an edge of the skeleton intersects a triangle of the mesh, then the latter is not stored as part of the resulting polyhedron.

Figure 7 illustrates the mesh related to the simplest skeleton composed of one single vertex (on the left), and the surrounding polyhedra related to skeletons composed of two and three voxels (center and right).

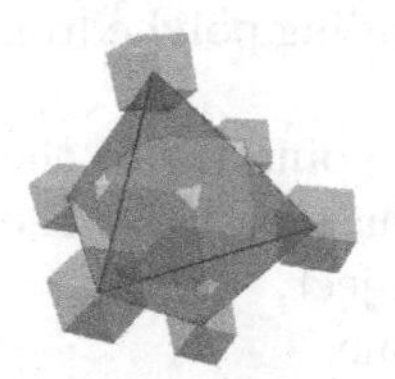

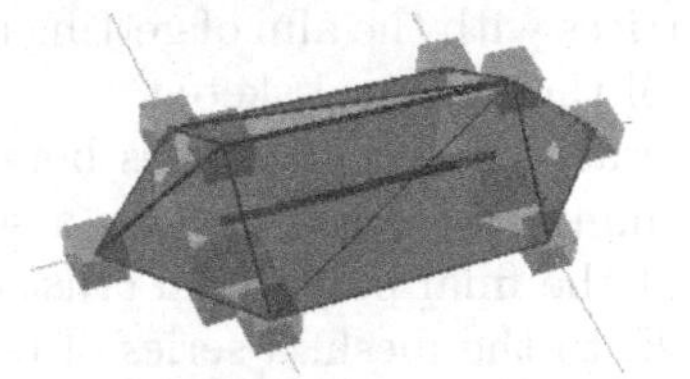

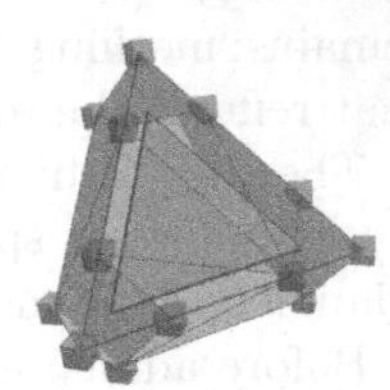

Fig. 7. A single voxel produces an octahedron (on the left); polyhedra related to simple skeletons (on the center and on the right)

4 Validation and Application to Multiresolution Design

The multiresolution feature is the ability of a model to enclose several *levels of detail* (called *LoD* in literature) to characterize an object. Depending on the context, the object can be represented either as a rough shape (a low resolution mesh within the LoD pyramid) or as a refined shape (a detailed and high model). Therefore, the choice of the LoD can save computation time when the model is rendered as a small entity in a 3D scene: only a few polygons need to be displayed to obtain a satisfying visual appearance.

4.1 Subdivision Surfaces

In this framework, *subdivision surfaces* for shape modeling emerged about 20 years ago, following works of Catmull and Clark [7] and Doo and Sabin [9]. This approach has been developed only recently in a large range of applications in computer graphics [18]. One reason of this development is the rise of multiresolution techniques, which focus on surfaces whose geometry is more and more complex.

The basic idea is that every polygon of a mesh can be subdivided according to a specific scheme. The result of successive refinements of this sequence is a smooth surface when the number of iterations approaches infinity. Yet subdivision surfaces are visually satisfactory after few iterations. The generic process starts with a given polygonal mesh, called the *control polyhedron*. A refinement scheme is then applied to this mesh, creating new vertices and new faces.

We settle on LOOP subdivision scheme [13] for three reasons: it is a multiresolution technique to get several levels of detail, it is an *approximating* scheme and it is dedicated to triangulations.

The approximation mesh ensures that the multiresolution surface is *inside* the volume of the control polyhedron computed in Section 3. In this way, no topological degenerations can appear on the different LoDs, because no surface auto-intersections can occur.

As the generated mesh is only made of triangles, LOOP scheme is the most appropriate subdivision technique.

Thus, on the one hand, we have a basic polyhedron that is automatically generated from a sketch. Its topology is characterized by the skeleton. On the other hand, we can get several levels of detail of a triangulated model by using subdivision surfaces thanks to the LOOP subdivision scheme. To obtain a multiresolution model of arbitrary topology, we combine our mesh construction approach with subdivision surfaces. At high resolution levels, it is then easy to edit details and local features on a sketched model.

4.2 Example

The "Strange Insect". We describe step by step the construction of a *strange insect*, to illustrate the topology characterization (it is a surface of genus 1, and the related object has two cavities).

Sketching the Insect — First, global features of the shape to design are drawn in a cubic grid, as seen in Figure 8 (a,b,c). This is done by using a 3D pointer and by displacing an *active edition plane*.

While editing the voxels of the sketch, the vertices, the edges and the triangles are constructed to characterize the topology of the object.

Surrounding Polyhedron — In a second step, a surrounding polyhedron is automatically generated following the rules defined in Section 3.3. Figure 8 (d) shows the created vertices after the meshing phase.

The resulting polyhedron surrounds the sketch. The hole and the two cavities of the insect, initially defined by a cycle and two hulls, are properly present.

Levels of Detail — Starting from the surrounding polyhedron of the sketch, levels of detail are generated using the LOOP subdivision scheme. From left to right, Figure 9 shows the initial control polyhedron M, two levels of subdivision

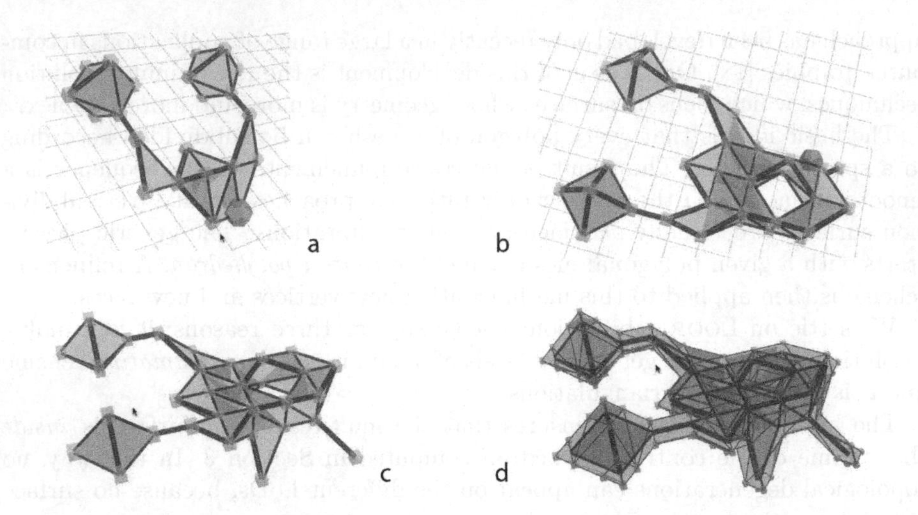

Fig. 8. Sketching in a 3D grid (a,b,c); generation of the mesh (d)

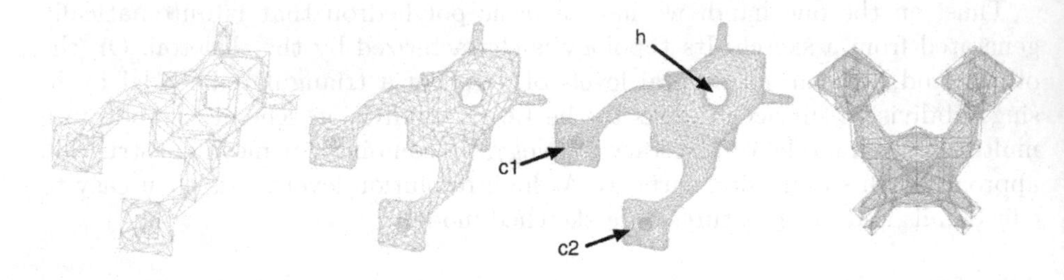

Fig. 9. Some subdivision levels

(the two cavities and the hole are indicated) and the superimposition of M and a high level of subdivision. M surrounds the refined mesh, thanks to the approximation scheme.

Topological Guarantees. At this stage, topology preservation is guaranteed because:

- the voxels of the sketch are connected according to the 26-adjacency to form the skeleton;
- the surrounding polyhedron preserves holes and cavities of the skeleton, without creating unwanted local connections between close areas of the mesh;
- the levels of detail obtained thanks to the approximating subdivision algorithm are slightly internal to the control polyhedron.

Further Deformations. In Figure 10, meshes are edited further (after two subdivision iterations) using the free software BLENDER [4]. The insect is deformed locally and globally with the proportional editing tool.

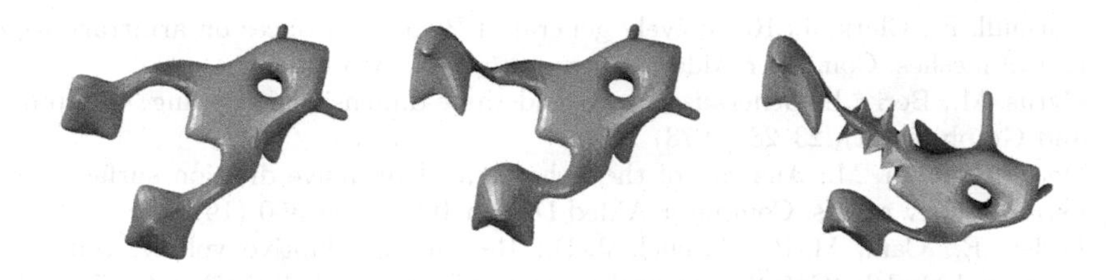

Fig. 10. Mesh editing using a 3D modeling software

5 Conclusion and Further Work

The approach we propose in this paper consists of generating a mesh automatically from a voxel-based shape descriptor. The latter is designed interactively in a 3D grid. It represents the global appearance of the object to model. In addition to that, we develop an example application by combining subdivision surfaces with the resulting control polyhedron in order to obtain a multiresolution feature.

Our main contribution is the elaboration of a new method to generate surfaces from a topological mesh, coupling it with subdivision surfaces to obtain several levels of detail of the model.

As future work, thanks to the multiresolution feature of the surrounding surface, the sketch could be designed in the same way, by editing large or small *cubes* (within an octree) depending on the size of the global morphological details to characterize. A second perspective could be to use the interactive sketch like a discrete skeleton for animation, as *bones* or *armatures*, as 3D modelers would do.

References

1. Alexe, A., Barthe, L., Cani, M.-P., Gaildrat, V.: Shape modelling by sketching using convolution surfaces. In: Pacific Graphics (Short Papers), Macau, China (October 2005)
2. De Araujo, B., Jorge, J.: Blobmaker: free-form modelling with variational implicit surfaces. In: Proceedings of "12 ° Encontro Português de Computação Grafica" (12 ° EPCG), Porto, Portugal, pp. 17–26 (2003)
3. Bittar, E., Tsingos, N., Gascuel, M.-P.: Automatic reconstruction of unstructured 3D data: combining a medial axis and implicit surfaces. In: Computer Graphics Forum (Eurographics 1995 Proc.), vol. 14, pp. 457–468 (1995)
4. Blender: Free open source 3d content creation suite, available for all major operating systems under the gnu general public license, `http://www.blender.org/`
5. Bloomenthal, J., Bajaj, C., Blinn, J., Cani-Gascuel, M.-P., Rockwood, A., Wyvill, B., Wyvill, G.: Introduction to implicit surfaces. Computer Graphics and Geometric Modeling series. Morgan Kaufmann Publisher, San Francisco (1997)
6. Bloomenthal, J., Wyvill, B.: Interactives techniques for implicit modeling. Computer Graphics 24(2), 109–116 (1990)

7. Catmull, E., Clark, J.: Recursively generated B-spline surfaces on arbitrary topological meshes. Computer Aided Design 10(6), 350–355 (1978)
8. Cyrus, M., Beck, J.: Generalized two- and three-dimensional clipping. Computers and Graphics 3(1), 23–28 (1978)
9. Doo, D., Sabin, M.: Analysis of the behaviour of recursive division surfaces near extraordinary points. Computer Aided Design 10(6), 356–360 (1978)
10. Ferley, E., Cani, M.-P., Gascuel, J.-D.: Resolution adaptive volume sculpting. Graphical Models (GMOD), Special Issue on Volume Modelling 63, 459–478 (2001)
11. Hoffmann, C.M.: Medial-axis approach to mesh generation. In: Lecture Notes: Princeton Conference, Computer Science Department. Purdue University (July 1996)
12. Kenmochi, Y., Imiya, A., Ichikawa, A.: Boundary extraction of discrete objects. Computer Vision and Image Understanding 71(3), 281–293 (1998)
13. Loop, C.: Smooth subdivision surfaces based on triangles. Master's thesis, University of Utah, Department of Mathematics (1987)
14. Lorensen, W.E., Cline, H.E.: Marching Cubes: a high resolution 3D surface construction algorithm. Computer Graphics 21(4) (July 1987)
15. Mari, J.-L., Sequeira, J.: A new modeling approach by global and local characterization. In: 11th International Conference on Computer Graphics, GraphiCon 2001, Nizhny Novgorod, Russia, September 2001, pp. 126–131 (2001)
16. Mari, J.-L., Sequeira, J.: Closed free-form surface geometrical modeling – a new approach with global and local characterization. International Journal of Image and Graphics (IJIG) 4(2), 241–262 (2004)
17. Markosian, L., Cohen, J.M., Crulli, T., Hugues, J.: Skin: a constructive approach to modeling free-form shapes. In: Computer Graphics Proceedings (SIGGRAPH 1999), pp. 393–400 (1999)
18. Schröder, P., Zorin, D.: Subdivision for modeling and animation. In: Course notes of Siggraph 1998, ACM SIGGRAPH (1998)
19. Sheehy, D.J., Armstrong, C.G., Robinson, D.J.: Shape description by medial surface construction. IEEE Transactions On Visualization & Computer Graphics 2, 62–72 (1996)
20. Sherstyuk, A.: Interactive shape design with convolution surfaces. In: Shape Modeling International 1999, International Conference on Shape Modeling and Applications, Aizu-Wakamatsu, Japan (March 1999)
21. Wyvill, B., McPheeters, C., Wyvill, G.: Animating soft objects. The Visual Computer 2(4), 235–242 (1986)

Minimal Offsets That Guarantee Maximal or Minimal Connectivity of Digital Curves in nD

Valentin E. Brimkov[1], Reneta P. Barneva[2], and Boris Brimkov[3]

[1] Mathematics Department, SUNY Buffalo State College, Buffalo, NY 14222, USA
brimkove@buffalostate.edu

[2] Department of Computer Science, SUNY Fredonia, NY 14063, USA
barneva@cs.fredonia.edu

[3] University at Buffalo, Buffalo, NY 14260-1000, USA
bbrimkov@buffalo.edu

Abstract. In this paper we investigate an approach of constructing a digital curve by taking the integer points within an offset of a certain radius of a continuous curve. Our considerations apply to digitizations of arbitrary curves in arbitrary dimension n. As main theoretical results, we first show that if the offset radius is greater than or equal to $\sqrt{n}/2$, then the obtained digital curve features maximal connectivity. We also demonstrate that the radius value $\sqrt{n}/2$ is the minimal possible that always guarantees such a connectivity. Moreover, we prove that a radius length greater than or equal to $\sqrt{n-1}/2$ guarantees 0-connectivity, and that this is the minimal possible value with this property. Thus, we answer the question about the minimal offset size that guarantees maximal or minimal connectivity of an offset digital curve.

Keywords: Digital geometry, digital curve, digital object connectivity, curve offset.

1 Introduction

Digital curves appear to be basic primitives in digital geometry. They have been used in volume modeling as well as in certain algorithms for image analysis, in particular in designing discrete multigrid convergent estimators [1–3].

A basic approach to modeling curves and surfaces in computer aided geometric design is the one based on computing the *offset* of a curve/surface Γ – the locus of points at a given distance d from Γ (see, e.g., [4–6] and the bibliography therein). As a rule, such sort of considerations resort to results from algebraic geometry, such as Grobner basis computation (see, e.g., [7, 8]). Note, however, that, usually, when looking for a curve offset defined by a distance d, one does not really care about the properties of the set of integer points enclosed by the offset. Exceptions (in 2D) are provided, e.g., by some recent works that define digital conics [9, 10].

In view of the above, it would indeed be quite natural to define a digital curve by the set of integer points within the offset of a continuous curve. With

S. Brlek, C. Reutenauer, and X. Provençal (Eds.): DGCI 2009, LNCS 5810, pp. 337–349, 2009.

no doubt it would be useful to know more about the pros and cons of digital curves obtained by an offset approach. In fact, this is the idea of the definition of a 2D digital line, which consists of the integer points that belong to a strip determined by two parallel lines at an appropriate distance from each other, so that the obtained digital set is connected. Assuring connectivity of a digital curve is a fundamental issue, as it can be seen, e.g., in [9, 11–17]. Offset approach has also been used to define digital circles and spheres, whose connectivity properties have been studied, as well [18, 19].

In the present paper we provide a complete answer (from a digital geometry point of view) to the most fundamental theoretical questions concerning digital curves defined by offsets. As main theoretical results, we first show that, in dimension n, if the offset radius is greater than or equal to $\sqrt{n}/2$, then the obtained digital curve features maximal connectivity. We also demonstrate that the radius value $\sqrt{n}/2$ is the minimal possible that always guarantees such a connectivity. Moreover, we prove that a radius length greater than or equal to $\sqrt{n-1}/2$ guarantees 0-connectivity, and that this is the minimal possible value with this property. All results apply to digitizations of arbitrary curves in an arbitrary dimension. With these two results we answer the question about the minimal offset size that guarantees maximal or minimal connectivity of an offset digital curve.

The paper is organized as follows. In the rest of the present section we recall some basic notions of digital geometry to be used in the sequel. Section 2, contains some subsidiary lemmas and constructions. In Section 3 we present the main theoretical results. We conclude with some final remarks in Section 4.

Some Definitions and Notations

Throughout we conform to terminology used in [20] (see also [21, 22]).

Our considerations take place in the *grid cell model* which consists of the grid cells of $\mathbb{Z}^n$, together with the related topology. In this model, a regular orthogonal grid subdivides $\mathbb{R}^n$ into n-dimensional hypercubes (e.g., unit squares for $n = 2$ or unit cubes for $n = 3$, also considered as *n-cells*) defining a class $\mathbb{C}_n^{(n)}$. These are usually called *hypervoxels*, or *voxels*, for short. Let $\mathbb{C}_n^{(k)}$ be the class of all k-dimensional cells of n-dimensional hypercubes, for $0 \leq k \leq n$. The grid-cell space $\mathbb{C}_n$ is the union of all classes $\mathbb{C}_n^{(k)}$, for $0 \leq k \leq n$. The $(n-1)$-cells, 1-cells, and 0-cells of a voxel are referred to as *facets*, *edges*, and *vertices*, respectively.

We say that two voxels v, v' are k-adjacent for $k = 0, 1, \ldots, n-1$, if they share a k-cell.

An n-dimensional (nD) digital object S is a finite set of voxels. A *k-path* (where $0 \leq k \leq n-1$) in S is a sequence of voxels from S such that every two consecutive voxels of the path are k-adjacent. Two voxels of S are *k-connected* (in S) iff there is a k-path in S between them. A subset G of S is *k-connected* iff there is a k-path connecting any two voxels of G.

A maximal (by inclusion) k-connected subset of a digital object S is called a *k-(connected) component* of S. Components are non-empty and distinct k-components are disjoint.

Given a set $M \subseteq \mathbb{R}^n$, by $|M|$ we denote its cardinality, by $M_{\mathbb{Z}}$ its *Gauss digitization* $M \cap \mathbb{Z}^n$, and by $conv(M)$ its convex hull.

Throughout the paper, the Euclidean norm is assumed. For a hyperball in $\mathbb{R}^n$ we will also use the term an *n-ball.* An n-ball of radius r and center c will be denoted by $B(c, r)$, and the corresponding sphere by $S(c, r)$. By $\overline{pq}$ we denote the segment with end-points p and q.

By $\{0, 1\}^n$ we denote the vertices of the unit hypercube. These are the points $(x_1, x_2, \ldots, x_n)$ where $x_i = 0$ or 1 for $1 \leq i \leq n$.

Finally, a *curve* $\gamma \in \mathbb{R}^n$ is a continuous mapping $\gamma : I \to \mathbb{R}^n$, where I is a closed interval of real numbers.

2 Subsidiary Lemmas and Constructions

2.1 Lemmas

The following two folklore facts will be used in the proofs that follow.

Lemma 1. *Any closed n-ball $B \subset \mathbb{R}^n$ with a radius greater than or equal to $\sqrt{n}/2$ contains at least one integer point.*

Follows from the fact that the longest diagonal in a grid hypercube of $\mathbb{Z}^n$ has length $\sqrt{n}$.

Lemma 2. *Let A and B be digital sets, each of which k-connected. If there are points $p \in A$ and $q \in B$ that are k-adjacent, then $A \cup B$ is k-connected.*

Corollary 1. *If A and B are digital sets, each of which is k-connected, and $A \cap B \neq \emptyset$, then $A \cup B$ is k-connected.*

Lemma 3. *If a closed n-ball $B \subset \mathbb{R}^n$ contains exactly two integer points, then these are $(n-1)$-adjacent.*

Proof. Assume that B contains integer points p and q that are not $(n-1)$-adjacent. We will show that then B contains at least a third point.

Let $p = (p_1, p_2, \ldots, p_n)$, $q = (q_1, q_2, \ldots, q_n)$. Since p and q are not $(n-1)$-adjacent, at least two of their coordinates differ. W.l.o.g., let $p_n < q_n$ (so, we also have $p_i \neq q_i$ for at least one i, $1 \leq i \leq n-1$).

Let B have center $c = (c_1, c_2, \ldots, c_n)$ and radius r. Then $(c_1 - p_1)^2 + (c_2 - p_2)^2 + \cdots + (c_n - p_n)^2 \leq r^2$ and $(c_1 - q_1)^2 + (c_2 - q_2)^2 + \cdots + (c_n - q_n)^2 \leq r^2$. Now we observe that: if $(c_n - p_n)^2 < (c_n - q_n)^2$ (which is equivalent to $p_n + q_n > 2c_n$), then $q' = (q_1, q_2, \ldots, q_{n-1}, p_n) \in B$; if $(c_n - p_n)^2 > (c_n - q_n)^2$ (i.e., iff $p_n + q_n < 2c_n$), then $p' = (p_1, p_2, \ldots, p_{n-1}, q_n) \in B$; and if $(c_n - p_n)^2 = (c_n - q_n)^2$ (i.e., iff $p_n + q_n = 2c_n$), then both $p', q' \in B$. □

We will also use the following fundamental lemma.

Lemma 4. *Given a closed n-ball $B \subset \mathbb{R}^n$ with $B_{\mathbb{Z}} \neq \emptyset$, $B_{\mathbb{Z}}$ is $(n-1)$-connected.*

Proof. We prove the statement by induction on $|B_{\mathbb{Z}}|$. If $|B_{\mathbb{Z}}| = 0$ or 1, we are done. The case $|B_{\mathbb{Z}}| = 2$ is possible only if the two elements of $B_{\mathbb{Z}}$ are $(n-1)$-adjacent (otherwise $B_{\mathbb{Z}}$ would contain a third element adjacent to one of the two, as follows from Lemma 3). Assume the statement is true for $|B_{\mathbb{Z}}| = k$, $k \geq 2$. Now let B be an n-ball with $|B_{\mathbb{Z}}| = k+1$. Let B^0 be an n-ball that contains exactly two points on its bounding sphere S^0 and $B^0_{\mathbb{Z}} = B_{\mathbb{Z}}$.

Such a ball can easily be constructed, e.g., by performing the following steps. If S (the bounding sphere of B) contains exactly two integer points, we are done and set $S^0 = S$. If S contains more than two integer points, go to Step 3. If S contains exactly one integer point, go to Step 2. If S contains no integer points, then do the following. *Step 1.* Decrease the radius of B with a sufficiently small $\varepsilon > 0$, until the corresponding sphere S includes integer point(s). If there are exactly two such points, then we are done and set $S^0 = S$. If there is exactly one such point, go to Step 2, otherwise go to Step 3. *Step 2.* Let p be the integer point on S and O the center of S. Decrease the radius of S with a sufficiently small $\varepsilon > 0$ by moving its center along the line $\overline{Op}$ towards p until other integer point(s) meet S. If exactly one integer point meets S, we are done and set $S^0 = S$. Otherwise, go to Step 3. *Step 3.* Let p and q be the end-points of an edge of $conv(B_{\mathbb{Z}})$ and t the midpoint of $\overline{pq}$. Increase the radius of the corresponding bounding sphere S with $\varepsilon > 0$ by moving the center of S along $\overline{tO}$ in direction opposite to t, so that the obtained sphere S^0 contains p and q and encloses the same integer points as S, but no other integer points. Since the set of integer points is discrete, this is obviously possible for sufficiently small ε. Then the only integer points on the bounding sphere S^0 of the obtained ball B^0 will be p and q.

Let p and q be the only integer points on S^0. Then, similar to the construction of S^0, one can find a ball B^1, such that $B^1_{\mathbb{Z}} = B^0_{\mathbb{Z}} \setminus p = B_{\mathbb{Z}} \setminus p$ (e.g., by decreasing the radius of B with a sufficiently small $\varepsilon > 0$ by moving its center along $\overline{Oq}$ towards q, while keeping q on S). The latter set is at least $(n-1)$-connected by the inductional hypothesis. Analogously, we obtain that there is a ball B^2, for which $B^2_{\mathbb{Z}} = B^0_{\mathbb{Z}} \setminus q = B_{\mathbb{Z}} \setminus q$, the latter set being $(n-1)$-connected by the inductional hypothesis. Then, by Corollary 1, $B_{\mathbb{Z}}$ is $(n-1)$-connected, as well. □

We complete this section with two more technical facts for future reference.

Lemma 5. *Let $S(c, r)$ be a sphere that contains the points $(0, 0, \ldots, 0)$, $(1, 0, \ldots, 0), \ldots, (0, \ldots, 0, 1)$. Then $c = (1/2, 1/2, \ldots, 1/2)$ and $r = \sqrt{n}/2$ (and so, S contains all vertices of the unit hypercube, that are $(x_1, x_2, \ldots, x_n)$, $x_i = 0$ or 1 for $1 \leq i \leq n$).*

Proof. Since $(0, 0, \ldots, 0) \in S$, we have $c_1^2 + c_2^2 + \cdots + c_n^2 = r^2$. Since $(1, 0, \ldots, 0) \in S$, we have $(c_1 - 1)^2 + c_2^2 + \cdots + c_n^2 = r^2$. Then $c_1^2 = (c_1 - 1)^2$, that yields $c_1 = 1/2$. Analogously, we get $c_2 = c_3 = \cdots = c_n = 1/2$ and thus $r^2 = n/4$, so $r = \sqrt{n}/2$. Obviously, every point of $\{0, 1\}^n$ satisfies the sphere equation. □

Lemma 6. *The n-ball* $B(c + \varepsilon, \sqrt{n}/2)$, *where* $c = (1/2, 1/2, \ldots, 1/2)$, $\varepsilon = (\varepsilon_1, \varepsilon_2, \ldots, \varepsilon_n)$, *and* $|\varepsilon_i| < 1/2$, $\varepsilon_i \neq 0$ *for at least one* i, $1 \leq i \leq n$, *contains at least one of the elements of* $\{0,1\}^n$ *in its interior.*

Proof. W.l.o.g., assume that $\varepsilon_i \geq 0$ for $1 \leq i \leq k$ and $\varepsilon_i \leq 0$ for $k+1 \leq i \leq n$, for some k, $0 \leq k \leq n$ (for $k = 0$ all ε_i's are nonpositive while for $k = n$ all of them are nonnegative). $B(c+\varepsilon, \sqrt{n}/2)$ is defined by the inequality $(1/2 + \varepsilon_1 - x_1)^2 + (1/2 + \varepsilon_2 - x_2)^2 + \cdots + (1/2 + \varepsilon_n - x_n)^2 \leq n/4$. The terms of the lemma also imply that

$$(1/2+\varepsilon_1-1)^2+\cdots+(1/2+\varepsilon_k-1)^2+(1/2+\varepsilon_{k+1})^2+\cdots+(1/2+\varepsilon_n)^2 < n/4, \quad (1)$$

since each additive does not exceed $1/4$ (equalities are possible if the corresponding $\varepsilon_i = 0$). Moreover, inequality (1) is strict, since at least one $\varepsilon_i \neq 0$. Hence, the point $(1, 1, \ldots, 1, 0, 0, \ldots, 0)$ belongs to the interior of B, as stated. □

Next, we provide a dynamic geometric interpretation of the concept of offset. It will be instrumental to the proof of the main results which are presented afterward.

2.2 Geometric Interpretation

Let γ be a curve in $\mathbb{R}^n$ with endpoints p and q. Place an n-ball B of radius r at point p. The space included in the n-ball is part of the offset.

Now move B so that its center continuously runs over γ until it reaches the other endpoint q. See Fig. 1, left. The space traced by B while its center is moving from p to q constitutes the *r-offset* of γ. It will be denoted by $\mathit{Off}(\gamma, r)$. We will identify ball's position by the position of its center.

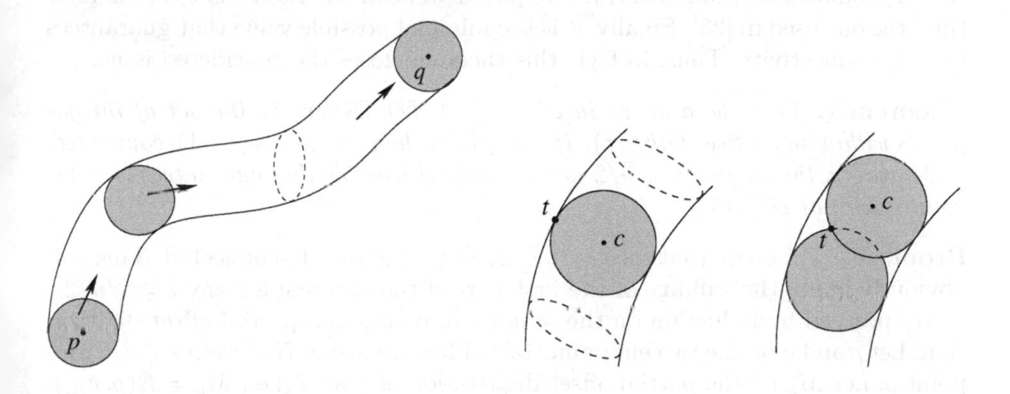

Fig. 1. Geometric interpretation. *Left:* The curve offset as a trace of a moving ball along the curve. *Middle:* An integer point t on the offset border is met by the moving ball. *Right:* An integer point t of the offset interior, when it is met by the moving ball, and when it leaves the ball.

When B is in its initial position (centered at p), it may or may not contain integer points. If the radius of B is sufficiently large, B always contains one or more integer points, as follows from Lemma 1.

During the motion of B along the curve, some integer points may leave it while others may enter it. Note that, depending on the curve specificity, it may be possible for an integer point to leave the ball, and later on to be met by it again. The following is a plain fact.

Fact 1. *During its motion from its initial position p to its final position q, the ball B meets all integer points that are contained in* $\mathrm{Off}(\gamma, r)$ *and constitute the offset digitization of γ.*

The subregion of *Off*(γ, r) traced by B when its center reaches a point $s \in \gamma$, $s \neq q$, will be called a *partial offset of γ at point s*. The set of integer points in such a partial offset will be called a *partial offset digitization* of γ at s.

3 Main Results

In this section we resolve the problem about the offset sizes that guarantee certain types of connectivity.

3.1 Minimal Offset Preserving Maximal Connectivity: First Main Theorem

It was shown in [23] that if $D(L)$ is a digital straight line defined by a cylindrical r-offset of a straight line in $\mathbb{R}^3$, then $D(L)$ is at least 1-connected provided that $r \geq \sqrt{3}$. The following theorem strengthens this last result in several ways. First, it applies to arbitrary curves, in arbitrary dimension, and proves maximal $(n-1)$-connectivity. Moreover, the required value of the radius is twice smaller than the one used in [23]. Finally, it is the minimal possible value that guarantees $(n-1)$-connectivity. Thus, in fact, this theorem closes the considered issue.

Theorem 1. *Let γ be a curve in $\mathbb{R}^n$, $n \geq 2$. Let $D(\gamma, r)$ be the set of integer points within an r-offset Off(γ, r). If $r \geq \sqrt{n}/2$, then $D(\gamma, r)$ is $(n-1)$-connected.*

Moreover, the value $r = \sqrt{n}/2$ is the minimal possible that guarantees $(n-1)$-connectivity of $D(\gamma, r)$.

Proof. We will prove that for $r = \sqrt{n}/2$, $D(\gamma, r)$ is $(n-1)$-connected. This will obviously imply the validity of the first part of the theorem for any $r \geq \sqrt{n}/2$.

We proceed by induction on the number of points in a partial offset digitization. Let p and q be the two endpoints of γ. Place an n-ball B of radius $\sqrt{n}/2$ at a point p. Let M_p be the partial offset digitization of γ at p (i.e., $M_p = B(p, r)_{\mathbb{Z}}$). By Lemma 1, $|M_p| \geq 1$. If we choose an arbitrary element of $B(p, r)_{\mathbb{Z}}$, it is $(n-1)$-connected. Note that by Lemma 4, $B(p, r)_{\mathbb{Z}}$ is $(n-1)$-connected, as well.

Let M_s be the partial offset digitization of γ at an arbitrary point $s \in \gamma$. By Lemma 4, $M_s \geq 1$. As an inductional hypothesis, suppose that M_s is $(n-1)$-connected. If $M_s = D(\gamma, r)$, we are done. Otherwise, move B along γ towards q.

During the motion, at a certain moment B will meet integer point(s) that are not currently in it. Note however that these may already be in the partial offset digitization. If this is the case, keep moving until reaching integer point(s) not yet included in $D(\gamma, r)$. Then these points are added to the current content of $D(\gamma, r)$. Assume that this happens when the center of B is at some point $c \in \gamma$. We will show that M_c is $(n-1)$-connected.

Let t be a new integer point met by B at position c, and $t \notin M_s$. Let $B(c, r)$ be the ball in that position. To simplify the calculations, assume without loss of generality that the center $c = (c_1, c_2, \ldots, c_n)$ belongs to the positive ortant and that t is at the origin of the coordinate system (i.e., $t = (0, \ldots, 0)$). Then the sphere $S(c, r)$ that bounds $B(c, r)$ has an equation

$$(c_1 - x_1)^2 + (c_2 - x_2)^2 + \cdots + (c_n - x_n)^2 = (\sqrt{n}/2)^2 = n/4, \tag{2}$$

and the distance from c to t equals the radius of the sphere, that is

$$\sqrt{c_1^2 + c_2^2 + \cdots + c_n^2} = \sqrt{n}/2. \tag{3}$$

Consider all $(n-1)$-neighbors of t in the nonnegative ortant, that are the points

$$(0, 0, \ldots, 0, 1),\ (0, 0, \ldots, 1, 0),\ \ldots,\ (1, 0, \ldots, 0, 0). \tag{4}$$

If some of these was (were) already in the partial offset when B was approaching t, we are done. Assume the opposite, i.e., none of these was in the partial offset before that moment. This is possible only if at the point when B meets t, the points from (4) are either outside or on $S(c, r)$.

Point $(1, 0, \ldots, 0)$ does not belong to the interior of $B(c, r)$ if and only if

$$(c_1 - 1)^2 + c_2^2 + \cdots + c_n^2 \geq n/4.$$

Keeping Eq. (2) in mind, the latter is equivalent to $c_1 \leq 1/2$. Analogously, we obtain

$$c_1 \leq 1/2,\ c_2 \leq 1/2,\ \ldots,\ c_n \leq 1/2. \tag{5}$$

The latter inequalities imply

$$c_1^2 + c_2^2 + \cdots + c_n^2 \leq n/4. \tag{6}$$

If at the meeting point of B with t some of the points (4) is outside $S(c, r)$, then the corresponding inequality in (5) would be strict. Then inequality (6) would be strict too, which contradicts Eq. (3).

Thus, it remains to consider the possibility when all $(n-1)$-neighbors of t are met by ball B at the same time (and thus all of them at the same time become elements of M_c). Lemma 5 implies that then all points $\{0, 1\}^n$ will belong to $S(c, r)$. By Lemma 6, there is a point $w \in \{0, 1\}^n$ that has been in the interior of B as B's center was approaching c, i.e., $w \in M_s$. Now we have that both M_s and $B(c, r)_{\mathbb{Z}} = \{0, 1\}^n$ are $(n-1)$-connected, and $w \in M_s \cap B(c, r)_{\mathbb{Z}}$. Then, $M_c = M_s \cup B(c, r)_{\mathbb{Z}}$ is $(n-1)$-connected, as well, by Corollary 1.

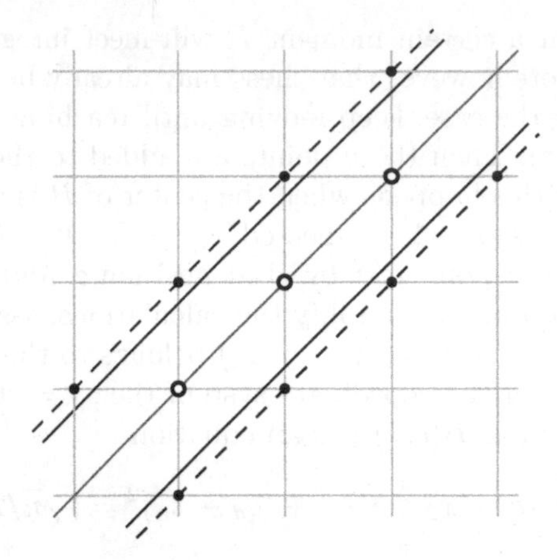

Fig. 2. The integer points within a straight line offset of radius $\sqrt{2}/2 - \varepsilon$ may not be 1-connected, even for an arbitrarily small $\varepsilon > 0$

Since the set $D(\gamma, r)$ is finite and in view of Fact 1, eventually all integer points within the offset will enter $D(\gamma, r)$, which will be $(n-1)$-connected. This completes the proof of the first part of the theorem.

Fig. 2 provides an example, which demonstrates that the value $r = \sqrt{n}/2$ is the minimal possible that guarantees $(n-1)$-connectivity of the digital line defined by an r-offset. This completes the proof of the theorem. □

Remark 1. The example of Fig. 2 makes also clear that Theorem 1 fails to hold if the cylindrical offset is not a closed set containing the whole cylindrical boundary.

Theorem 1 suggests that $D(\gamma, \sqrt{n}/2)$ can be considered as an $(n-1)$-connected digital curve that is a digitization of a given continuous curve γ. Note that every point of that digitization is at a distance from γ no greater than $\sqrt{n}/2$.

3.2 Minimal Offset Preserving Minimal Connectivity: Second Main Theorem

In this section we obtain a lower bound for an offset radius r that guarantees 0-connectivity of $D(\gamma, r)$. Moreover, we show that this radius value is the minimal possible with this property.

Theorem 2. *Let γ be a curve in $\mathbb{R}^n$, $n \geq 2$. Let $D(\gamma, r)$ be the set of integer points within an r-offset $Off(\gamma, r)$. If $r \geq \sqrt{n-1}/2$, then $D(\gamma, r)$ is at least 0-connected.*

Moreover, the value $r = \sqrt{n-1}/2$ is the minimal possible that guarantees 0-connectivity of the digital line defined by r-offset.

Proof. We will prove that for $r = \sqrt{n-1}/2$, $D(\gamma, r)$ is 0-connected. This will clearly imply the validity of the first part of the theorem for any $r \geq \sqrt{n-1}/2$.

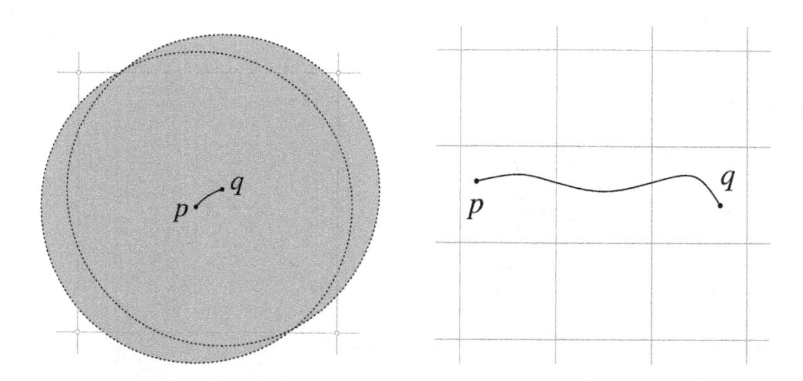

Fig. 3. *Left:* If the digitized curve is too short, its r-offset (where $r < \sqrt{n}/2$) may not contain any integer points. *Right:* A curve that fully crosses a pixel.

If $r < \sqrt{n}/2$, then obviously the curve offset may contain no integer points (see Fig. 3, left). If that is the case, i.e., if $D(\gamma, \sqrt{n-1}/2) = \emptyset$, we are done. In the rest of the proof we will assume the nontrivial case $D(\gamma, \sqrt{n-1}/2) \neq \emptyset$.

With a reference to the *Geometric Interpretation* from Section 2.2, consider a curve $\gamma \subset \mathbb{R}^n$ with endpoints p and q. Place an n-ball B of radius r at point p.

To prove the first part of the theorem, we proceed by induction on the number of $(n-1)$-facets of hypercubes intersected by γ. (Note that two consecutive intersections may occur at the same $(n-1)$-facet.) We exclude from further consideration non-interesting cases such as the one in Fig. 3, left, assuming that γ goes through (i.e., "enters" and "leaves") at least one grid hypercube (Fig. 3, right).

Let Q be a grid hypercube intersected by γ. It has $2n$ $(n-1)$-facets $Q_1^{(n-1)}, Q_2^{(n-1)}, \ldots, Q_{2n}^{(n-1)}$. Let $Q_i^{(n-1)}$ be one of these facets intersected by γ and let $u = \gamma \cap Q_i^{(n-1)}$. Denote by $L_i^{(n-1)}$ the affine hull of $Q_i^{(n-1)}$. It is a hyperplane in $\mathbb{R}^n$ that is a translated copy of a Euclidean subspace $\mathbb{R}^{n-1}$. Clearly, $L_i^{(n-1)}$ contains a subset of the grid-points $\mathbb{Z}^n$ which is a translated copy of a $\mathbb{Z}^{n-1}$ space. Denote it by $\mathbb{Z}_{L_i}^{n-1}$. Let $B_i^n(u, \sqrt{n-1}/2)$ be an n-ball centered at u. Then $B_i^{n-1}(u, \sqrt{n-1}/2) = B_i^n(u, \sqrt{n-1}/2) \cap L_i^{(n-1)}$ is an $(n-1)$-ball with the same center and radius as B_i^n. By Lemma 1 applied to $B_i^{n-1}(u, \sqrt{n-1}/2)$ and $L_i^{(n-1)}$, we have that $B_i^{n-1}(u, \sqrt{n-1}/2)$ contains an integer point from $\mathbb{Z}_{L_i}^{n-1}$.

We have already seen that before the first intersection, the partial offset digitization may be an empty set. If it is non-empty, then the partial offset will contain integer points that are vertices of the same hypercube (the one which contains the portion of the curve that determines the considered partial offset). Hence, the partial offset digitization will be at least 0-connected.

Assume that after k intersections, the partial offset digitization $S_k \subseteq D(\gamma, \sqrt{n-1}/2)$ of integer points within the partial $\sqrt{n-1}/2$-offset of γ traversed so far, is 0-connected. The last set of integer points added to the current partial offset digitization includes the integer points in an $(n-1)$-ball

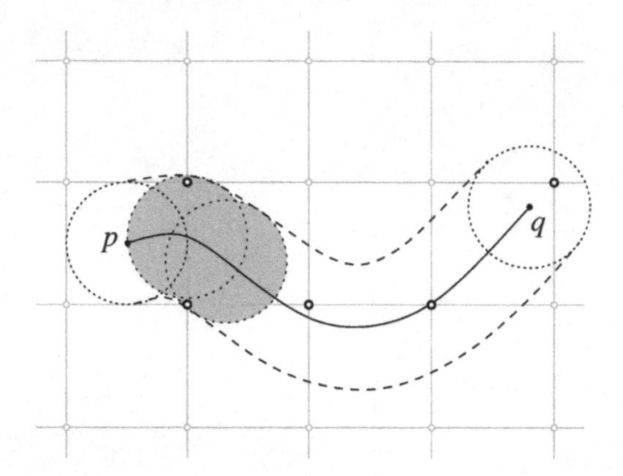

Fig. 4. A 2D example of a 0-connected digital curve obtained by an offset of radius $\sqrt{n-1}/2 = 1/2$. The points of the digital curve are marked by small circles.

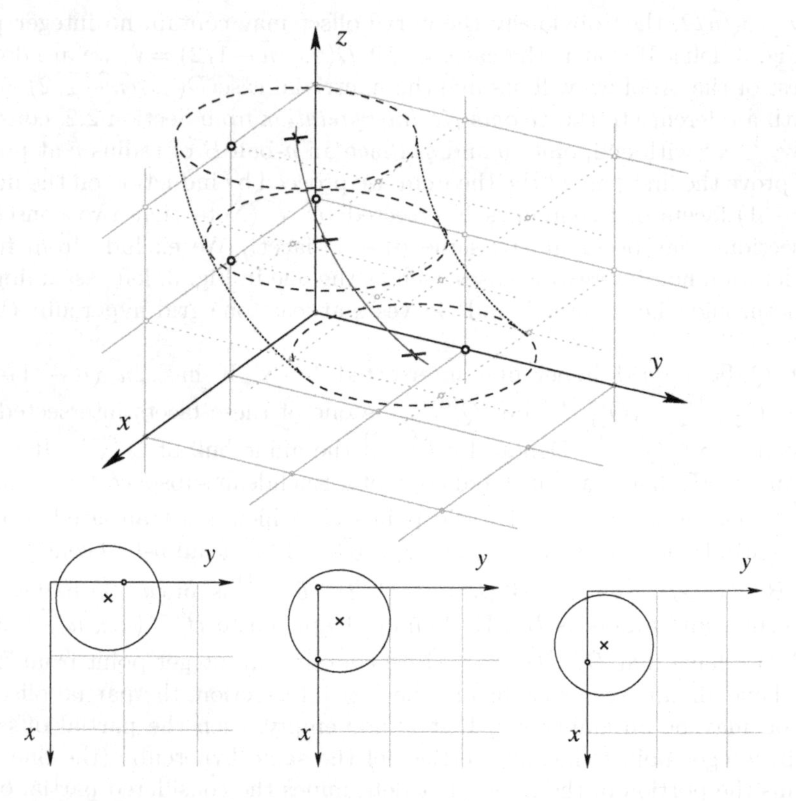

Fig. 5. *Top:* A 3D example of a 0-connected digital curve obtained by an offset of radius $\sqrt{n-1}/2 = \sqrt{2}/2$. The points of the digital curve are marked by small circles. *Bottom:* Offset sections at levels $z = 0$ (*left*), $z = 1$ (*middle*), and $z = 2$ (*right*).

$B_k^{n-1}(u, \sqrt{n-1}/2)$, which in turn includes the integer points on an $(n-1)$-facet $Q_k^{(n-1)}$ of some grid hypercube Q. (Note: Here k labels the last $(n-1)$-facet intersected by γ; these k facets may belong to different grid hypercubes, in general).

Moving along γ inside Q towards crossing another $(n-1)$-facet of Q, the n-ball B may involve different sets of integer points. By Lemma 4, each of them is $(n-1)$-connected. Moreover, every time when the ball contains some integer points, some of these will be among the cube vertices. The reason is that the integer point that is closest to the ball center will obviously belong to the n-cube that contains the ball center. Hence, all new integer points that are met by the ball B will be 0-connected to S_k.

Eventually, the ball center will meet an $(n-1)$-facet $Q_{k+1}^{(n-1)}$ of Q at a certain point $v = \gamma \cap Q_{k+1}^{(n-1)}$. At that moment, $D(\gamma, \sqrt{n-1}/2)$ will include some vertices of $Q_{k+1}^{(n-1)}$, which follows from Lemma 1. However, each of these vertices is at least 0-adjacent to those of $Q_k^{(n-1)}$. Thus the set of all integer points included

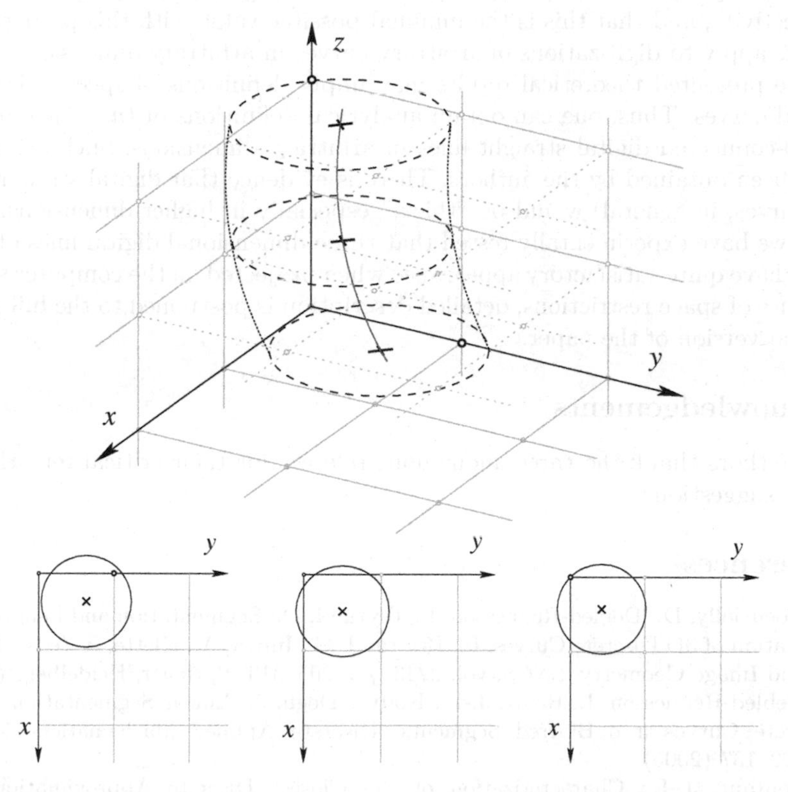

Fig. 6. *Top:* A 3D example of a disconnected digital set obtained by an offset of radius slightly less than $\sqrt{n-1}/2 = \sqrt{2}/2$. The points of the set are marked by small circles. *Bottom:* Offset sections at levels $z = 0$ (*left*), $z = 1$ (*middle*), and $z = 2$ (*right*).

in $D(\gamma, \sqrt{n-1}/2)$ when the center of B reaches point v, will be at least 0-connected. See Figures 4 and 5 for two- and three-dimensional examples, respectively.

The validity of the second part follows from the example presented in Fig. 6. □

Theorem 2 suggests to consider $D(\gamma, \sqrt{n-1}/2)$ as a 0-connected digital curve which is a digitization of a curve γ. Every point of that digitization is at a distance from γ no greater than $\sqrt{n-1}/2$.

4 Concluding Remarks

In this paper we investigated an approach of constructing a digital curve by taking the integer points within an offset of a certain radius of the curve. We showed that if the offset radius is greater than or equal to $\sqrt{n}/2$, then the obtained digital curve features maximal connectivity. We also demonstrated that the radius value $\sqrt{n}/2$ is the minimal possible that always guarantees such a connectivity. Moreover, we showed that radius length $\sqrt{n-1}/2$ guarantees 0-connectivity, and that this is the minimal possible value with this property. All results apply to digitizations of arbitrary curves in arbitrary dimensions.

The presented theoretical results may imply definitions of special classes of digital curves. Thus, one can obtain analytical definitions of $(n-1)$-connected and 0-connected digital straight lines in arbitrary dimensions. Such definitions have been obtained by the authors. There is evidence that digital straight lines (or curves, in general) would be "thick," especially in higher dimensions. However, we have experimentally tested that three-dimensional digital lines of these types have quite satisfactory appearance when projected on the computer screen. Because of space restrictions, detailed description is postponed to the full length journal version of the paper.

Acknowledgements

The authors thank the three anonymous referees for their critical remarks and useful suggestions.

References

1. Coeurjolly, D., Debled-Rennesson, I., Teytaud, O.: Segmentation and Length Estimation of 3D Discrete Curves. In: Bertrand, G., Imiya, A., Klette, R. (eds.) Digital and Image Geometry. LNCS, vol. 2243, pp. 295–313. Springer, Heidelberg (2002)
2. Debled-Rennesson, I., Rémy, J.-L., Rouyer-Degli, J.: Linear Segmentation of Discrete Curves into Blurred Segments. Discrete Applied Mathematics 151(1-3), 122–137 (2005)
3. Toutant, J.-L.: Characterization of the Closest Discrete Approximation of a Line in the 3-dimensional Space. In: Bebis, G., Boyle, R., Parvin, B., Koracin, D., Remagnino, P., Nefian, A., Meenakshisundaram, G., Pascucci, V., Zara, J., Molineros, J., Theisel, H., Malzbender, T. (eds.) ISVC 2006. LNCS, vol. 4291, pp. 618–627. Springer, Heidelberg (2006)

4. Anton, F., Emiris, I., Mourrain, B., Teillaud, M.: The Offset to an Algebraic Curve and an Application to Conics. In: Gervasi, O., Gavrilova, M.L., Kumar, V., Laganá, A., Lee, H.P., Mun, Y., Taniar, D., Tan, C.J.K. (eds.) ICCSA 2005. LNCS, vol. 3480, pp. 683–696. Springer, Heidelberg (2005)
5. Arrondo, E., Sendra, J., Sendra, J.R.: Genus Formula for Generalized Offset Curves. J. Pure and Applied Algebra 136(3), 199–209 (1999)
6. Hoffmann, C.M., Vermeer, P.J.: Eliminating Extraneous Solutions for the Sparse Resultant and the Mixed Volume. J. Symbolic Geom. Appl. 1(1), 47–66 (1991)
7. Anton, F.: Voronoi Diagrams of Semi-algebraic Sets. Ph.D. thesis, The University of British Columbia, Vancouver, British Columbia, Canada (2004)
8. Cox, D., Little, J., O'Shea, D.: Using Algebraic Geometry. Springer, New York (1998)
9. Debled-Rennesson, I., Domenjoud, E., Jamet, D.: Arithmetic Discrete Parabolas. In: Bebis, G., Boyle, R., Parvin, B., Koracin, D., Remagnino, P., Nefian, A., Meenakshisundaram, G., Pascucci, V., Zara, J., Molineros, J., Theisel, H., Malzbender, T. (eds.) ISVC 2006. LNCS, vol. 4292, pp. 480–489. Springer, Heidelberg (2006)
10. Fiorio, C., Jamet, D., Toutant, J.-L.: Discrete circles: an Arithmetical Approach Based on Norms. In: Internat. Conference "Vision-Geometry XIV" SPIE, San Jose, CA, vol. 6066, p. 60660 (2006)
11. Andres, E.: Discrete Linear Objects in Dimension n: The Standard Model. Graphical Models 65(1-3), 92–111 (2003)
12. Cohen-Or, D., Kaufman, A.: 3D Line Voxelization and Connectivity Control. IEEE Computer Graphics & Applications 17(6), 80–87 (1997)
13. Debled-Rennesson, I.: Etude et Reconnaissance des Droites et Plans Discrets. Ph.D. Thesis, Université Louis Pasteur, Strasbourg, France (1995)
14. Figueiredo, O., Reveillès, J.-P.: New Results About 3D Digital Lines. In: Melter, R.A., Wu, A.Y., Latecki, L. (eds.) Internat. Conference "Vision Geometry V." SPIE, vol. 2826, pp. 98–108 (1996)
15. Jonas, A., Kiryati, N.: Digital Representation Schemes for 3d Curves. Pattern Recognition 30(11), 1803–1816 (1997)
16. Rosenfeld, A.: Arcs and Curves in Digital Pictures. Journal of the ACM 20(1), 81–87 (1973)
17. Kim, C.E.: Three Dimensional Digital Line Segments. IEEE Transactions on Pattern Analysis and Machine Intelligence 5(2), 231–234 (1983)
18. Andres, E.: Discrete Circles, Rings and Spheres. Computers & Graphics 18(5), 695–706 (1994)
19. Andres, E., Jacob, M.-A.: The Discrete Analytical Hyperspheres. IEEE Trans. Vis. Comput. Graph. 3(1), 75–86 (1997)
20. Klette, R., Rosenfeld, A.: Digital Geometry – Geometric Methods for Digital Picture Analysis. Morgan Kaufmann, San Francisco (2004)
21. Kong, T.Y.: Digital Topology. In: Davis, L.S. (ed.) Foundations of Image Understanding, pp. 33–71. Kluwer, Boston (2001)
22. Rosenfeld, A.: Connectivity in Digital Pictures. Journal of the ACM 17(3), 146–160 (1970)
23. Brimkov, V.E., Barneva, R.P., Brimkov, B., de Vieilleville, F.: Offset Approach to Defining 3D Digital Lines. In: Bebis, G., Boyle, R., Parvin, B., Koracin, D., Remagnino, P., Porikli, F., Peters, J., Klosowski, J., Arns, L., Chun, Y.K., Rhyne, T.-M., Monroe, L. (eds.) ISVC 2008, Part I. LNCS, vol. 5358, pp. 678–687. Springer, Heidelberg (2008)

Arithmetization of a Circular Arc

Aurélie Richard[1], Guy Wallet[2], Laurent Fuchs[1], Eric Andres[1], and Gaëlle Largeteau-Skapin[1]

[1] Laboratoire XLIM-SIC, Université de Poitiers, BP 30179 86962 Futuroscope Chasseneuil cedex, France
{arichard,Laurent.Fuchs,andres,glargeteau}@sic.univ-poitiers.fr
[2] Laboratoire MIA, Université de La Rochelle, Avenue Michel Crépeau 17042 La Rochelle cedex, France
Guy.Wallet@univ-lr.fr

Abstract. In this paper, we present an arithmetization of the Euler's integration scheme based on infinitely large integers coming from the nonstandard analysis theory. Using the differential equation that defines circles allows us to draw two families of discrete arc circles using three parameters, the radius, the global scale and the drawing scale. These parameters determine the properties of the obtained arc circles. We give criteria to assure the 8-connectivity. A global error estimate for the arithmetization of the Euler's integration scheme is also given and a first attempt to define the approximation order of an arithmetized integration scheme is provided.

Keywords: Discrete circle, Discrete arc circle, Arithmetization, Numerical scheme, Error order, Connectedness.

1 Introduction

The discretization of a curve on a computer screen has been the focus of a lot of research in the late sixties and beginning of the seventies [1, 2]. After that, for a long time, nothing fundamentally new has been proposed as people considered that most of the research in this area had been covered. In 1988, a novel definition of the discrete straight line by J-P. Reveillès [3–5] sparked a new interest in basic discrete geometry and the digitization of discrete curves and surfaces. Many new results especially in the analytical description of discrete objects have been proposed since [6]. The theoretical roots behind Reveillès' straight line formula can be found in Non Standard Analysis (NSA). This theory [7, 8] provides a framework where infinitely large and infinitely small numbers can be explicitly manipulated. It can be shown that, given an infinitely large number ω (the global scale), it is possible to establish an equivalence between the set of limited real numbers and a subset $\mathcal{HR}_\omega$ of the set $\mathbb{Z}$ of integers. The set $\mathcal{HR}_\omega$, with an additional structure, is called the Harthong-Reeb line [9–12]. The intuitive idea behind this concept is to consider $\mathbb{R}$ like $\mathbb{Z}$ seen from far away.

S. Brlek, C. Reutenauer, and X. Provençal (Eds.): DGCI 2009, LNCS 5810, pp. 350–361, 2009.

Many curves can be defined as a solution of the integration of a vector field. With an integration scheme like, for instance, the Euler scheme with an infinitely small integration step, we obtain an infinitely close approximation of the curve. With the help of an arithmetization process consisting in mapping the scheme from the real line to the Harthong-Reeb line, we obtain a discrete curve which is an exact representation of the continuous one. The global scale ω does not however allow to draw the curve on a computer screen because the points of the curve are infinitely far away from each other. A strongly contracting rescaling in the discrete world allows then a way of getting a scale where the curve can be seen on a computer screen. The approximation that appears in the process of drawing a discrete curve on a screen is not described anymore as a sampling error but as the result of a strongly contracting rescaling in the discrete world. For more details, readers may refer to [13, 14].

At the last DGCI (and in an extended version in the Pattern Recognition journal) [13, 14], we took a new look at the ideas that led to Reveillès' discrete straight line. In this paper we decided to reconsider Holin's work [15–17] on discrete circles generated by the preceding approach. Many different circle generation algorithms exist. Here the purpose is not to propose a new one but rather to understand the continuous-discrete relations.

As a first step, we carefully study the arithmetization of the Euler integration scheme on the differential equation defining circles. The goal of this paper was to apply the insight gained previously on the arithmetization process to a new set of curves. For the discrete straight lines, only one drawing scale appears (labeled β where $\omega = \beta^2$) besides the global scale ω. In the circle case, two different drawing scales appear (labeled α and β with $\omega = \alpha\beta$). These two drawing scales define two families of discrete circles with different properties.

The connectedness of the discrete circles obtained by such arithmetization methods has never been studied before. In the case of the Euler integration scheme, we show that inside the square $[-\beta, \beta - 1]^2$ every circular arc of both families (α and β circles) is 8-connected. This allows a global control over the topological properties of the discrete circle obtained with this method. We also study the quality of the approximation we obtain by the arithmetization of the Euler scheme. We give a way to estimate the error between a circle and its discretization. Thus we define the approximation order of the arithmetized integration scheme that has never been studied before.

This paper is organized as follows: first, we recall the basis about the arithmetization method and we apply it on the circles. Section 3 deals with connectedness conditions while section 4 deals with the error study. A short conclusion and some perspectives are proposed in section 5.

2 The Arithmetization Method Applied to Circles

In this section we recall our arithmetization method [14] using it on the differential equations that describe a circle. We obtain two arithmetized versions of these differential equations that will be the starting point of our study.

In the Euclidean plane, the circle $C(0, R)$ with center 0 and radius $R \in \mathbb{R}_+^*$ can be described as the solution of the differential equations:

$$\begin{cases} (x(0), y(0)) &= (0, R) \\ (x'(t), y'(t)) &= (-y(t), x(t)). \end{cases} \tag{1}$$

which express the fact that the tangent vector is orthogonal to the radius vector. These differential equations can be numerically solved using the Euler method. This gives us the following Euler scheme that computes an accurate numerical approximation of $C(0, R)$ when the step h tends to 0 in $\mathbb{R}_+$:

$$\begin{cases} (x_0, y_0) &= (0, R) \\ (x_{n+1}, y_{n+1}) &= (x_n, y_n) + (-y_n h, x_n h). \end{cases} \tag{2}$$

The general idea of the arithmetization method [14, 15] is to transform this continuous numerical scheme into a discrete one with only integer variables. This discrete scheme generates a digital planar curve $C_d(0, R)$ which is a discrete analog of the circle $C(0, R)$.

This analogy is proved to be theoretically sound using nonstandard analysis [14]. The approach is to perform a rescaling on $\mathbb{Z}$ such that the new unity is an infinitely large integer $\omega \in \mathbb{N}$. The structure we obtain is the Harthong-Reeb line $\mathcal{HR}_\omega$ [14] which is a numerical system isomorphic to the system $\mathbb{R}_{lim}$ of limited real numbers (i.e. real numbers that are not infinitely large). The isomorphism is given by the two mappings:

$$\begin{Bmatrix} \varphi_\omega : \mathcal{HR}_\omega & \longrightarrow \mathbb{R}_{lim} \\ X & \longmapsto X/\omega \end{Bmatrix} \quad \text{and} \quad \begin{Bmatrix} \psi_\omega : \mathbb{R}_{lim} & \longrightarrow \mathcal{HR}_\omega \\ x & \longmapsto \lfloor \omega x \rfloor \end{Bmatrix}$$

where $\lfloor x \rfloor$ is the integer part of $x \in \mathbb{R}$ such that $x = \lfloor x \rfloor + \{x\}$ with $0 \leq \{x\} < 1$.

Using these mappings, it can be shown that the digital curve $C_d(0, R)$ is isomorphic to the initial Euclidean circle $C(0, R)$. However, in this article we are concentrating on the properties of the digital curve $C_d(0, R)$ (see [13, 14] for more details on such a proof).

The first step is to describe more precisely the arithmetization of the Euler scheme (2). We suppose that the radius R of the circle is limited ($R \not\simeq +\infty$) and non infinitesimal ($R \not\simeq 0$). We choose two natural numbers ω and β such that ω, β and ω/β are infinitely large. Let $h = \frac{1}{\beta}$ ($h \simeq 0$). Using the mapping ψ_ω, the Euler scheme is translated to its integer version, *the arithmetization of the Euler scheme (2) at the scale* ω, by:

$$\begin{cases} (X_0, Y_0) &= (0, \lfloor \omega R \rfloor) \\ (X_{n+1}, Y_{n+1}) &= (X_n, Y_n) + (-Y_n \div \beta, X_n \div \beta) \end{cases} \tag{3}$$

where $\div$ is the usual Euclidean division and X_i, Y_i, $i = 1, 2, \ldots$ are integer variables. The points (X_i, Y_i) define the digital curve $C_d(0, R)$.

Unfortunately, the obtained curve $C_d(0, R)$ is largely non connected: for instance, if we consider the abscissae X_0 and X_1 of the two first points, we have $|X_1 - X_0| = \lfloor \omega R \rfloor \div \beta$ which is an infinitely large number roughly equal to $\frac{\omega}{\beta} R$.

This means that the gap between these two points, and in general between two consecutive points of the curve, is infinitely large.

The idea to obtain points that are closer together, is to perform a rescaling from the scale ω to the intermediate scale $\alpha = \frac{\omega}{\beta}$. To do so, it is convenient to choose ω as a multiple of β: $\omega = \alpha\beta$ with $\alpha \simeq +\infty$. Hence, the rescaling is given by the composition $\psi_\alpha \circ \varphi_\omega : X \mapsto \lfloor \alpha \frac{X}{\omega} \rfloor$. This defines two drawing scales α and β because one can of course also consider a rescaling to the scale β.

Since $\lfloor \alpha \frac{X}{\omega} \rfloor = X \div \beta$, we can use the decomposition $X = \widetilde{X}\beta + \widehat{X}$ for any integer $X \in \mathcal{HR}_\omega$ so $\widetilde{X} = X \div \beta \in \mathcal{HR}_\alpha$ and $\widehat{X} = X \bmod \beta \in \{0, \ldots, \beta - 1\}$. The integer $\widetilde{X} \in \mathcal{HR}_\alpha$ is interpreted as the result of the rescaling on X.

With this notations, we can rewrite the scheme (3) and we get *the arithmetization of the Euler scheme (2) at the drawing scale* α by:

$$\begin{cases} (\widetilde{X}_0, \widetilde{Y}_0) &= (0, \lfloor \omega R \rfloor \div \beta) \\ (\widehat{X}_0, \widehat{Y}_0) &= (0, \lfloor \omega R \rfloor \bmod \beta) \\ (F_n^1, F_n^2) &= ((-\widetilde{Y}_n\beta - \widehat{Y}_n) \div \beta, (\widetilde{X}_n\beta + \widehat{X}_n) \div \beta) \\ (\widetilde{X}_{n+1}, \widetilde{Y}_{n+1}) &= (\widetilde{X}_n + (\widehat{X}_n + F_n^1) \div \beta, \widetilde{Y}_n + (\widehat{Y}_n + F_n^2) \div \beta) \\ (\widehat{X}_{n+1}, \widehat{Y}_{n+1}) &= ((\widehat{X}_n + F_n^1) \bmod \beta, (\widehat{Y}_n + F_n^2) \bmod \beta) \end{cases} \tag{4}$$

where the relevant variables are $\widetilde{X}_i$ and $\widetilde{Y}_i$.

The points $(\widetilde{X}_i, \widetilde{Y}_i)$ define a discrete circle which depends on the numbers R, α and β denoted by $C_d^\alpha(0, R)$. Symmetrically, we can also consider *the arithmetization of the Euler scheme (2) at the drawing scale* β, which is the analogue of scheme (4), and is given by:

$$\begin{cases} (\widetilde{X}_0, \widetilde{Y}_0) &= (0, \lfloor \omega R \rfloor \div \alpha) \\ (\widehat{X}_0, \widehat{Y}_0) &= (0, \lfloor \omega R \rfloor \bmod \alpha) \\ (F_n^1, F_n^2) &= ((-\widetilde{Y}_n\alpha - \widehat{Y}_n) \div \beta, (\widetilde{X}_n\alpha + \widehat{X}_n) \div \beta) \\ (\widetilde{X}_{n+1}, \widetilde{Y}_{n+1}) &= (\widetilde{X}_n + (\widehat{X}_n + F_n^1) \div \alpha, \widetilde{Y}_n + (\widehat{Y}_n + F_n^2) \div \alpha) \\ (\widehat{X}_{n+1}, \widehat{Y}_{n+1}) &= ((\widehat{X}_n + F_n^1) \bmod \alpha, (\widehat{Y}_n + F_n^2) \bmod \alpha) \end{cases} \tag{5}$$

where the relevant variables are $\widetilde{X}_i$ and $\widetilde{Y}_i$. We get another discrete circle denoted by $C_d^\beta(0, R)$. In every case, the numerical experimentations show that connectedness of the digital curves $C_d^\alpha(0, R)$ and $C_d^\beta(0, R)$ is directly linked to the relative values of α and β. In the following section, we will establish a relation between these scale parameters in order to obtain 8-connectivity.

Remark. The schemes (4) and (5) describe two algorithms that are defined for every integer values of α and β and not only for nonstandard values. Thus, despite the nonstandard framework used in the theoretical introduction of the arithmetization method, we finally get two standard families of discrete curves $C_d^\alpha(0, R)$ and $C_d^\beta(0, R)$ depending on two integer parameters α and β. The link with the nonstandard foundation is that, in some natural meaning, the discrete curves $C_d^\alpha(0, R)$ and $C_d^\beta(0, R)$ converge toward the continuous circle $C(0, R)$ when α and β become infinitely large.

3 The Connectedness of $C_d^\alpha(0,R)$ and $C_d^\beta(0,R)$ Circle Arcs

Graphical results (figures 1 and 2) show that the connectedness of the preceding discrete circles depends on the relative values of the drawing scales α and β. The section aim is to provide a condition on the scales that assure 8-connectivity for the discrete circle arcs of $C_d^\alpha(0,R)$ and $C_d^\beta(0,R)$. It must be noticed that these section results are independent of the nonstandard framework.
Let us consider the solution $((\widetilde{X}_n,\widetilde{Y}_n))_{0\leq n}$ of the arithmetization of the Euler scheme (5) at the drawing scale α which is the discrete circle $C_d^\alpha(0,R)$. Let us call *an arc of* $C_d^\alpha(0,R)$ every subsequence of the form $((\widetilde{X}_n,\widetilde{Y}_n))_{k\leq n\leq k+p}$ for $k,p\in\mathbb{N}$ such that $p\geq 1$. In order to state the theorem that follows, let us introduce the square in $\mathbb{Z}^2$ centered at 0 with side length 2β:

$$R_\beta=\{(X,Y)\in\mathbb{Z}^2\ ;\ -\beta\leq X<\beta \text{ and } -\beta\leq Y<\beta\}.$$

Theorem 1. *Every arc of $C_d^\alpha(0,R)$ in the square R_β is 8-connected.*

Proof. Let $\Gamma=((\widetilde{X}_n,\widetilde{Y}_n))_{k\leq n\leq k+p}$ be an arc of $C_d^\alpha(0,R)$ such that $(\widetilde{X}_n,\widetilde{Y}_n)\in R_\beta$ for each $n=k,\ldots,k+p$. Then, Γ is 8-connected if and only if, for each $n=k,\ldots,k+p-1$, we have

$$-1\leq\widetilde{X}_{n+1}-\widetilde{X}_n\leq 1 \text{ and } -1\leq\widetilde{Y}_{n+1}-\widetilde{Y}_n\leq 1 \tag{6}$$

The proof is in two parts: in part (a) we will give a necessary and sufficient condition for the connectedness of Γ, in part (b) we will show that the condition $\Gamma\subset R_\beta$ is sufficient.

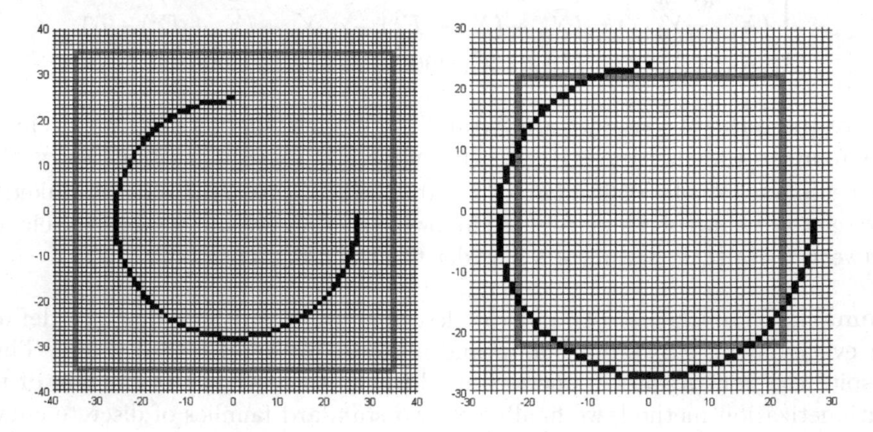

Fig. 1. Two examples of circle arcs at α scale (a): the one connected with $(R,\alpha,\beta)=(1,25,35)$ (b) the other one disconnected with $(R,\alpha,\beta)=(1,24,22)$. The gray square represents the region R_β.

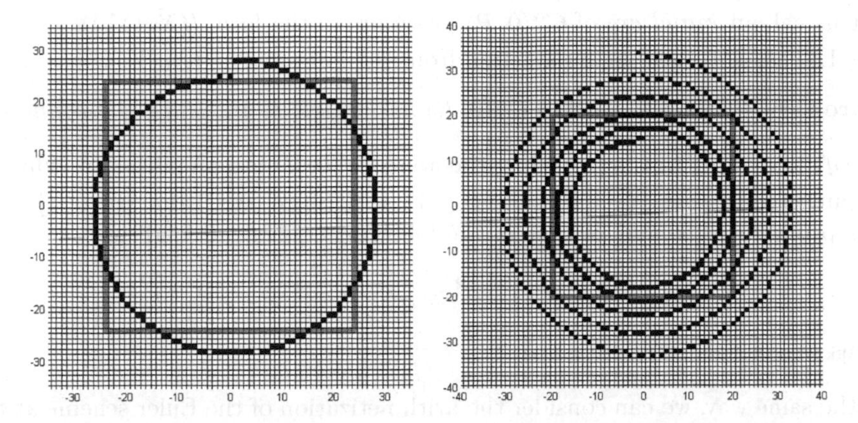

Fig. 2. The circle arcs of length (a) $\lfloor 2\pi\beta \rfloor$ with $(R, \alpha, \beta) = (1, 25, 40)$ (b) $\lfloor 10\pi\beta \rfloor$ with $(R, \alpha, \beta) = (1, 15, 20)$. The gray square represents the region R_β.

(a) *Equivalent conditions:* using the two schemes (3) and (4) and properties of the Euclidean division, we can see that the following conditions are equivalent:

$$
\begin{aligned}
-1 \le \widetilde{X}_{n+1} - \widetilde{X}_n \le 1 &\text{ and } -1 \le \widetilde{Y}_{n+1} - \widetilde{Y}_n \le 1 \\
-1 \le (\widehat{X}_n + F_n^1) \div \beta \le 1 &\text{ and } -1 \le (\widehat{Y}_n + F_n^2) \div \beta \le 1 \\
-\beta \le \widehat{X}_n + F_n^1 < 2\beta &\text{ and } -\beta \le \widehat{Y}_n + F_n^2 < 2\beta \\
-\beta - \widehat{X}_n \le (-\widetilde{Y}_n\beta - \widehat{Y}_n) \div \beta < 2\beta - \widehat{X}_n &\text{ and } -\beta - \widehat{Y}_n \le (\widetilde{X}_n\beta + \widehat{X}_n) \div \beta < 2\beta - \widehat{Y}_n \\
-\beta^2 - \widehat{X}_n\beta \le -\widetilde{Y}_n\beta - \widehat{Y}_n < 2\beta^2 - \widehat{X}_n\beta &\text{ and } -\beta^2 - \widehat{Y}_n\beta \le \widetilde{X}_n\beta + \widehat{X}_n < 2\beta^2 - \widehat{Y}_n\beta \\
-\beta^2 - \widehat{X}_n\beta \le -Y_n < 2\beta^2 - \widehat{X}_n\beta &\text{ and } -\beta^2 - \widehat{Y}_n\beta \le X_n < 2\beta^2 - \widehat{Y}_n\beta.
\end{aligned}
$$

Thus, Γ is 8-connected if and only if, for each $n = k, \dots, k + p - 1$, we have:

$$
-2\beta^2 + \widehat{X}_n\beta < Y_n \le \beta^2 + \widehat{X}_n\beta \text{ and } -\beta^2 - \widehat{Y}_n\beta \le X_n < 2\beta^2 - \widehat{Y}_n\beta. \tag{7}
$$

(b) *Sufficient condition:* Since $0 \le \widehat{X}_n \le \beta - 1$ and $0 \le \widehat{Y}_n \le \beta - 1$, we get the following two sequences of inequalities:

$$
\begin{aligned}
-2\beta^2 + \widehat{X}_n\beta \le -\beta^2 - \beta < -\beta^2 < \beta^2 \le \beta^2 + \widehat{X}_n\beta \\
-\beta^2 - \widehat{Y}_n\beta \le -\beta^2 < \beta^2 < \beta^2 + \beta \le 2\beta^2 - \widehat{Y}_n\beta.
\end{aligned}
$$

Thus, if $-\beta^2 \le X_n < \beta^2$ and $-\beta^2 \le Y_n < \beta^2$, then the condition (7) is verified. It is easy to see that, for $a \in \mathbb{Z}$ and $b \in \mathbb{N} \setminus \{0\}$, we have:

$$
-b^2 \le a < b^2 \iff -b \le a \div b < b.
$$

Since $\widetilde{X}_n = X_n \div \beta$ and $\widetilde{Y}_n = Y_n \div \beta$, the condition $-\beta^2 \le X_n < \beta^2$ is equivalent to $-\beta \le \widetilde{X}_n < \beta$ and $-\beta^2 \le Y_n < \beta^2$ to $-\beta \le \widetilde{Y}_n < \beta$. Hence, the 8-connectedness of Γ is a consequence of the condition:

$$
\forall n = k, \dots, k + p - 1 \quad -\beta \le \widetilde{X}_n < \beta \text{ and } -\beta \le \widetilde{Y}_n < \beta. \qquad \square
$$

Let us call an *initial arc* of $C_d^\alpha(0,R)$ every arc of the form $((\widetilde{X}_n,\widetilde{Y}_n))_{0\le n\le p}$ for $p\ge 1$. That is to say, an arc starting from the initial value $(\widetilde{X}_0,\widetilde{Y}_0)$.

Corollary 1. *If $\alpha R<\beta$, then $C_d^\alpha(0,R)$ has an initial arc which is 8-connected.*

Proof. According to the preceding theorem, $C_d^\alpha(0,R)$ has an initial arc which is 8-connected if $-\beta\le\widetilde{X}_0<\beta$ and $-\beta\le\widetilde{Y}_0<\beta$. Since $(\widetilde{X}_0,\widetilde{Y}_0)=(0,\lfloor\omega R\rfloor\div\beta)$, this property is reduced to each of the following equivalent condition:

$$-\beta\le\lfloor\omega R\rfloor\div\beta<\beta\Leftrightarrow-\beta^2\le\lfloor\omega R\rfloor<\beta^2\Leftrightarrow-\beta\le\alpha R<\beta.$$

Thus we get the result since $\alpha R\ge 0$. □

In the same way, we can consider the arithmetization of the Euler scheme at the drawing scale β. In this case, we have the following result:

Theorem 2.
- *Every arc of $C_d^\beta(0,R)$ in the square R_β is 8-connected.*
- *If $R<1$, then $C_d^\beta(0,R)$ has an initial arc which is 8-connected.*

Proof. The proof is similar to the previous one. □

Figures 1 and 2 provide a graphic illustration of these properties. Only graphic illustrations at drawing scale α are proposed here. The first one represents a discrete connected circle arc with $R=1$, $(\alpha,\beta)=(25,35)$ and a disconnected one with $R=1$, $(\alpha,\beta)=(24,22)$. The region R_β is also displayed (in gray). The second represents the discrete connected circle arc with length $\lfloor 2\pi\beta\rfloor$ with $R=1$, $(\alpha,\beta)=(25,40)$ and the initial arc of $C_d^\alpha(0,R)$ with length $\lfloor 10\pi\beta\rfloor$ with $R=1$, $(\alpha,\beta)=(25,30)$ which provides a spiral. This last one brings to light discontinuity outside of the square (each arc in the square is 8-connected while the ones which are outside can be disconnected).

If the arc is composed of $N=\lfloor 2\pi\beta\rfloor$ discrete points we should obtain a full circle but, in general, its first point is not connected to its last point (cf figure 2(a)). This arc is not a very good discretization of the underlying circle. The reason is that the Euler's scheme, as it is well known, accrues a small error at each integration step. When the curve is run one time the error is sufficiently large to "miss" the first point of the arc. This phenomenon depends of course on the integration scheme we use and also on the translation of the arithmetized scheme into the real domain. This is studied in the next section.

4 Global Error Estimate

The aim of this section is to study the error of our arithmetization method. To do that, let us describe a more general framework in which our result will be placed. Let $f:U\to\mathbb{R}$ be a function of class C^1 defined on an open set U of $\mathbb{R}\times\mathbb{R}^n$. We suppose that f is standard (that is to say independent of the nonstandard framework) and that f and its partial derivatives are bounded in U. Let us consider the continuous Cauchy problem:

$$\begin{cases} z(a) = b \\ z'(t) = f(t, z(t)) \end{cases} \tag{8}$$

with $a \in \mathbb{R}, b \in \mathbb{R}^n$, a and b limited. We consider the solution $z : [a, c] \rightarrow \mathbb{R}^n$ of (8) where $c \in \mathbb{R}$ is limited and such that $a \not\simeq c$ (i.e. a not infinitely close to c). This solution can be approximated by the Euler scheme:

$$\begin{cases} (t_0, z_0) &= (a, b) \\ (t_{n+1}, z_{n+1}) &= (t_n + \frac{1}{\beta}, z_n + \frac{1}{\beta} f(t_n, z_n)) \end{cases} \tag{9}$$

in which we choose an integration step $h = \frac{1}{\beta}$. For a infinitely large β (i.e. $\beta \simeq +\infty$), we have z_n infinitely close to $z(t_n)$ (i.e. $z_n \simeq z(t_n)$) for $0 \leq n \leq N$ with $N = \lfloor \beta c \rfloor$. Our arithmetization method allows us to introduce a similar scheme working only with integer variables:

$$\begin{cases} (T_0, Z_0) &= (\lfloor \omega a \rfloor, \lfloor \omega b \rfloor) \\ (T_{n+1}, Z_{n+1}) &= (T_n + \alpha, Z_n + F(T_n, Z_n) \div \beta) \end{cases} \tag{10}$$

where $F(T_n, Z_n) = \lfloor \omega f(\frac{T_n}{\omega}, \frac{Z_n}{\omega}) \rfloor$. This scheme results from the mapping ψ_ω and by substitution of the division by β by the Euclidean division by β.

Now using the reverse mapping φ_ω, we define a new scheme working with real variables:

$$\begin{cases} (t'_0, z'_0) &= (\frac{1}{\omega} \lfloor \omega a \rfloor, \frac{1}{\omega} \lfloor \omega b \rfloor) \\ (t'_{n+1}, z'_{n+1}) &= (t'_n + \frac{1}{\beta}, z'_n + \frac{1}{\beta} g(t'_n, z'_n)) \end{cases} \tag{11}$$

where $t'_n = \frac{T_n}{\omega}$, $z'_n = \frac{Z_n}{\omega}$ and $g(t'_n, z'_n) = \frac{1}{\alpha} \lfloor \frac{1}{\beta} \lfloor \omega f(t'_n, z'_n) \rfloor \rfloor$. We call this scheme the *trace* of the arithmetized scheme (10). We can remark that the two schemes (9) and (11) are very similar.

Our purpose is to estimate the distance between the value z'_n of the trace of the arithmetized scheme and the value $z(t_n)$ of the solution of the Cauchy problem (8). This distance defines the global error of our arithmetization method.

In order to fix our distance notion, we choose the infinity norm ($||x|| = ||(x_1, ..., x_n)|| = \max_{1 \leq i \leq n} |x_i|$), and then we have the following result:

Theorem 3. *There exists limited constants $\mathcal{L}, \mathcal{L}', \mathcal{L}'' \in \mathbb{R}$ such that*

$$\forall n \in [\![0, N]\!],\ ||z'_n - z(t_n)|| \leq \frac{\mathcal{L}}{\omega} + \frac{\mathcal{L}'}{\alpha} + \frac{\mathcal{L}''}{\beta}. \tag{12}$$

Proof. In order to compute an upper-bound of $||z'_n - z(t_n)||$, we split this expression in two parts:

$$||z'_n - z(t_n)|| \leq ||z'_n - z_n|| + ||z_n - z(t_n)|| \tag{13}$$

where z_n is the value the solution of the scheme (9) and then we find an upper bound separately for each part. The proof is thus split in two steps. By convention, to simplify the presentation of the following inequalities, all the indices n are in $[\![0, N]\!]$.

The first step is to compute an upper bound on $||z_n - z'_n||$. To do so, we consider $||z_{n+1} - z'_{n+1}||$. Using the usual triangular inequality, the schemes (9) and (11), we obtain:

$$||z_{n+1} - z'_{n+1}|| \leq ||z_n - z'_n|| + \frac{1}{\beta}||(f(t_n, z_n) - f(t'_n, z'_n))|| + \frac{1}{\beta}||(f(t'_n, z'_n) - g(t'_n, z'_n))||.$$

In order to bound the expression $||(f(t'_n, z'_n) - g(t'_n, z'_n))||$, we consider $g(t'_n, z'_n) = \frac{1}{\alpha}\left\lfloor \frac{1}{\beta}\lfloor \omega f(t'_n, z'_n)\rfloor \right\rfloor$. Using the integer part properties, this can be rewritten as

$$g(t'_n, z'_n) = f(t'_n, z'_n) - \frac{1}{\omega}\{\omega f(t'_n, z'_n)\} - \frac{1}{\alpha}\left\{\frac{1}{\beta}\lfloor \omega f(t'_n, z'_n)\rfloor\right\}$$

where $0 \leq \{\omega f(t'_n, z'_n)\} < 1$ and $0 \leq \{\frac{1}{\beta}\lfloor \omega f(t'_n, z'_n)\rfloor\} < 1$. Hence, we obtain $||g(t'_n, z'_n) - f(t'_n, z'_n)|| \leq \frac{2}{\alpha}$.

Now, let us consider the expression $||(f(t_n, z_n) - f(t'_n, z'_n))||$. Because f is a Lipschitz function (due to the class C^1 condition) we know that there exists $L_1 \in \mathbb{R}$, with L_1 limited, such that:

$$||f(t_n, z_n) - f(t'_n, z'_n)|| \leq L_1(|t_n - t'_n| + ||z_n - z'_n||). \tag{14}$$

In this last expression $|t_n - t'_n|$ is such that $\forall n$,
$|t_n - t'_n| = \left|t_{n-1} - \frac{1}{\beta} - t'_{n-1} + \frac{1}{\beta}\right| = |t_0 - t'_0| = \left|a - \frac{\omega a}{\omega} + \frac{\{\omega a\}}{\omega}\right| \leq \frac{1}{\omega}$.
Combining all the previous inequalities we obtain:

$$||z_{n+1} - z'_{n+1}|| \leq ||z_n - z'_n||\left(1 + \frac{L_1}{\beta}\right) + \frac{L_1}{\beta\omega} + \frac{2}{\omega} \leq ||z_n - z'_n||\left(1 + \frac{L_1}{\beta}\right) + \frac{L_1}{\omega} + \frac{2}{\omega}.$$

This is transformed as

$$\epsilon_{n+1} \leq \epsilon_n\left(1 + \frac{L_1}{\beta}\right) + \frac{L_2}{\omega} \tag{15}$$

where $L_2 = 2 + L_1$ and $\epsilon_{n+1} = ||z_{n+1} - z'_{n+1}||$. The following well-known lemma is now used.

Lemma 1 (Arithmetic-Geometric Sequence Properties). *Let e_n be a sequence of $\mathbb{R}^+$. If $e_{n+1} \leq Ae_n + B$ with $A, B \in \mathbb{R}^+$ then*

$$\forall n, e_n \leq A^n e_0 + B\frac{A^n - 1}{A - 1}.$$

Thanks to this lemma, we express ϵ_n according to ϵ_0 and obtain a bound for ϵ_n:

$$\epsilon_n \leq \epsilon_0\left(1 + \frac{L_1}{\beta}\right)^n + \frac{L_2}{\omega}\left(\frac{\left(1 + \frac{L_1}{\beta}\right)^n - 1}{\left(1 + \frac{L_1}{\beta}\right) - 1}\right).$$

Since $\frac{1}{\omega}$ is an upper-bound for $\epsilon_0 = ||z_0 - z'_0||$, we obtain:

$$\epsilon_n \leq \frac{1}{\omega}\left(1+\frac{L_1}{\beta}\right)^n + \frac{L_2}{L_1}\frac{1}{\alpha}\left(\left(1+\frac{L_1}{\beta}\right)^n - 1\right). \quad (16)$$

In order to find an upper bound of $\left(1+\frac{L_1}{\beta}\right)^n$, let us consider this other well-known lemma:

Lemma 2. $(1+u)^n \leq e^{nu}$, $\forall n \in \mathbb{N}$, $\forall u \geq 0$

The inequality $\frac{n}{\beta} \leq (c-a)$ expresses that the iteration number n is upper-bounded by the product of the length of the integration interval (that is a limited number) by β. Moreover, the exponential of a limited number is a limited number. Hence, using lemma 2, $\left(1+\frac{L_1}{\beta}\right)^n$ is bounded by a limited number. So, there exists limited constants $L_3, L_4 \in \mathbb{R}$ such that:

$$||z_n - z'_n|| \leq \frac{1}{\omega}L_3 + \frac{1}{\alpha}L_4. \quad (17)$$

The second step consists in computing an upper bound for $||z_n - z(t_n)||$ of equation (13). To do so, we consider $||z_{n+1} - z(t_{n+1})||$.
Let us use the Taylor expansion of $z(t_{n+1})$, there exists $\gamma_n \in]t_n, t_n + h[$ such that:

$$z(t_{n+1}) = z(t_n + h) = z(t_n) + hz'(t_n) + \frac{h^2}{2}z''(\gamma_n). \quad (18)$$

Since $]t_n, t_n + h[\subset [a,c]$, $z''(\gamma_n) \leq ||z''||_{[a,c]}$ where $||z''||_{[a,c]}$ is the norm of z'' in the interval $[a,c]$. So from (18) we have:

$$||z_{n+1} - z(t_{n+1})|| \leq ||z_n - z(t_n)|| + \frac{1}{\beta}||f(t_n, z_n) - f(t_n, z(t_n))|| + \frac{1}{2}\left(\frac{1}{\beta}\right)^2 ||z''||_{[a,c]}.$$

As previously, we set $\epsilon'_{n+1} = ||z_{n+1} - z(t_{n+1})||$ and we obtain the inequality:

$$\epsilon'_{n+1} \leq \epsilon_n(1 + \frac{1}{\beta}L_1) + \frac{1}{2}\left(\frac{1}{\beta}\right)^2 L_5,$$

where $L_5 \in \mathbb{R}$ is limited and such that $||z''||_{[a,c]} \leq L_5$. Now we can use again lemmas 1 and 2 and we obtain an upper bound for $||z_n - z(t_n)||$. There exists a limited constant $L_6 \in \mathbb{R}$ such that $||z_n - z(t_n)|| < \frac{1}{\beta}L_6$. Hence, from this last inequality and inequality (17) we can conclude that there exists limited constants $\mathcal{L}$, $\mathcal{L}'$ and $\mathcal{L}''$ in $\mathbb{R}$ such that: $||z'_n - z(t_n)|| \leq \frac{\mathcal{L}}{\omega} + \frac{\mathcal{L}'}{\alpha} + \frac{\mathcal{L}''}{\beta}$. □

In the particular case of the circles described in section 2, the exact solution is $z(t) = Re^{i(t+\frac{\pi}{2})}$, $t \in [0, 2\pi]$ and we obtain:

$$||z'_n - z(t_n)|| \leq \frac{e^{2\pi}}{\omega} + \frac{2(e^{2\pi}-1)}{\alpha} + \frac{\frac{R(e^{2\pi}-1)}{2}}{\beta}. \quad (19)$$

In the field of numerical analysis, integration schemes [18] come with an approximation order. This gives an idea of the numerical quality of the computed solutions. Until now, the order of arithmetized schemes had never been studied and the previous theorem allows us to define an approximation order of an arithmetized integration scheme which is the equivalent of the classical approximation order used in numerical analysis.

The classical approximation order is defined as the biggest integer such that there exists a constant $\mathcal{C} \in \mathbb{R}$ verifying $||z_n - z(t_n)|| \leq \mathcal{C}h^p$, where h is the integration step (small by definition). Each integration scheme has a precise order. For example, Euler method is of order 1. Other methods, such as the Heun scheme [18], is of order 2. From theorem 3, we propose now a definition for the approximation order of the Euler arithmetized scheme (10):

Definition 1 (Order). *The Euler arithmetized scheme (10) is of order p if there exists a limited number $\mathcal{L} \in \mathbb{R}$ such that:*

$$\forall n \in [\![0, N]\!],\ ||z'_n - z(t_n)|| \leq \frac{\mathcal{L}}{\beta^p}.$$

With this definition, the following corollary gives a sufficient condition to have order 1 for the arithmetized Euler scheme (10).

Corollary 2. *If $\frac{\beta}{\alpha}$ is limited then the arithmetized scheme (10) is of order 1.*

Proof. There is a limited number λ such that $\beta = \lambda\alpha$. Then, from theorem 3 $||z'_n - z(t_n)|| \leq \frac{1}{\beta}\left(\frac{\mathcal{L}}{\alpha} + \mathcal{L}'\lambda + \mathcal{L}\right)$ and $\frac{\mathcal{L}}{\alpha} + \mathcal{L}'\lambda + \mathcal{L}$ is clearly limited. □

The order of arithmetized integration schemes is currently under investigation by the authors and the previous definition can be extended to other schemes such as the Heun [18] integration scheme. However, the status of this approximation order is not so clear. For example, using a theorem similar to theorem 3, the arithmetized Heun scheme is of order at most 2 as expected from its corresponding original scheme. But, when studying connectedness, the condition found seems to impose that the obtained numerical solution is only of order 1. Numerical experiments show nonetheless that obtained circle arcs are "better" than those obtained by the arithmetized Euler scheme. This is still largely an open question.

5 Conclusion

In this paper we studied the arithmetization of the Euler integration scheme over the differential equation defining a circle. This work reconsiders a previous work done by Holin [15] and studies carefully the arithmetization process.

New results have been proposed; connectedness criteria for the two families of discrete circular arcs, a global error estimate between the arithmetized scheme trace and the initial Euler scheme, and a first definition of the approximation order of an arithmetized integration scheme. These results provide new insights in the way a curve can be approximated by the arithmetization process.

Our next goal is to generalize the preceding results to integration schemes with higher approximation orders. This will gives us a deeper understanding of the arithmetization process and improve the definition of the approximation order of an arithmetized integration scheme. Numerical experiments are under way. We have already observed that better circle arcs are obtained by Heun's method (as obtained by Holin) and that connectedness criteria still remain the same. This latter observation is surprising and is not yet fully understood.

References

1. Bresenham, J.: Algorithm for computer control of a digital plotter. ACM trans. Graphics 4, 25–30 (1965)
2. Bresenham, J.: A linear algorithm for incremental digital display of circular arcs. Comm. of ACM 20, 100–106 (1977)
3. Reveillès, J.P.: Mathématiques discrètes et analyse non standard. In: [19], pp. 382–390
4. Reveillès, J.P.: Géométrie discrète, Calcul en nombres entiers et algorithmique. PhD thesis, Université Louis Pasteur, Strasbourg, France (1991)
5. Reveillès, J.P., Richard, D.: Back and forth between continuous and discrete for the working computer scientist. Annals of Mathematics and Artificial Intelligence, Mathematics and Informatic 16, 89–152 (1996)
6. Andres, E.: Discrete circles, rings and spheres. Computer and Graphics 18, 695–706 (1994)
7. Nelson, E.: Internal set theory: A new approach to nonstandard analysis. Bulletin of the American Mathematical Society 83, 1165–1198 (1977)
8. Robinson, A.: Non-standard analysis. American Elsevier, New York (1974)
9. Harthong, J.: Une théorie du continu. In: Barreau, H., Harthong, J. (eds.) La mathématique non standard, Editions du CNRS, Paris, pp. 307–329 (1989)
10. Diener, M.: Application du calcul de Harthong-Reeb aux routines graphiques. In: [19], pp. 424–435
11. Harthong, J., Reeb, G.: Intuitionnisme 84. In: Barreau, H., Harthong, J. (eds.) La mathématique nonstandard, CNRS, pp. 213–252 (1989)
12. Harthong, J.: Éléments pour une théorie du continu. Astérisque 109/110, 235–244 (1983)
13. Fuchs, L., Largeteau-Skapin, G., Wallet, G., Andres, E., Chollet, A.: A first look into a formal and constructive approach for discrete geometry using nonstandard analysis. In: Coeurjolly, D., Sivignon, I., Tougne, L., Dupont, F. (eds.) DGCI 2008. LNCS, vol. 4992, pp. 21–32. Springer, Heidelberg (2008)
14. Chollet, A., Wallet, G., Fuchs, L., Largeteau-Skapin, G., Andres, E.: Insight in discrete geometry and computational content of a discrete model of the continuum. Pattern recognition (2009)
15. Holin, H.: Harthong-Reeb analysis and digital circles. The Visual Computer 8, 8–17 (1991)
16. Holin, H.: Harthong-Reeb circles. Séminaire non standard, Univ. de Paris 7 2, 1–30 (1989)
17. Holin, H.: Some artefacts of integer computer circles. Ann. Math. Artif. Intell. 16, 153–181 (1996)
18. Quarteroni, A., Sacco, R., Saleri, F.: Numerical mathematics. Springer, Heidelberg (2000)
19. Salanskis, J.M., Sinaceurs, H. (eds.): Le Labyrinthe du Continu. Springer, Heidelberg (1992)

On the Connecting Thickness of Arithmetical Discrete Planes

Eric Domenjoud[1], Damien Jamet[1], and Jean-Luc Toutant[2]

[1] Loria - Université Nancy 1 - CNRS, Campus Scientifique, BP 239, 54506 Vandœuvre-lès-Nancy, France
eric.domenjoud@loria.fr, damien.jamet@loria.fr
[2] LAIC, IUT Clermont-Ferrand, Campus des Cézeaux, 63172 Aubière Cedex
toutant@iut.u-clermont1.fr

Abstract. While connected rational arithmetical discrete lines and connected rational arithmetical discrete planes are entirely characterized, only partial results exist for the irrational arithmetical discrete planes. In the present paper, we focus on the connectedness of irrational arithmetical discrete planes, namely the arithmetical discrete planes with a normal vector of which the coordinates are not $\mathbb{Q}$-linear dependent. Given $\mathbf{v} \in \mathbb{R}^3$, we compute the lower bound of the thicknesses 2-connecting the arithmetical discrete planes with normal vector $\mathbf{v}$. In particular, we show how the translation parameter operates in the connectedness of the arithmetical discrete planes.

1 Introduction

In [1], J.-P. Reveillès introduced arithmetical discrete lines as sets of pairs of integers satisfying a double Diophantine inequality. The *arithmetical discrete line* with *normal vector* $\mathbf{v} \in \mathbb{R}^2 \setminus \{(0,0)\}$, *translation parameter* $\mu \in \mathbb{R}$ and *arithmetical thickness* $\omega \in \mathbb{R}$ is the set $\mathbb{L}(\mathbf{v}, \mu, \omega)$ defined by:

$$\mathbb{L}(\mathbf{v}, \mu, \omega) = \left\{\mathbf{x} = (x_1, x_2) \in \mathbb{Z}^2;\ 0 \leq \langle \mathbf{v}, \mathbf{x} \rangle + \mu < \omega\right\},$$

where $\langle \mathbf{v}, \mathbf{x} \rangle = v_1 x_1 + v_2 x_2$.

The definition of arithmetical discrete lines extends naturally to the definition of *arithmetical discrete planes* in the 3-dimensional discrete space $\mathbb{Z}^3$, and to the definition of *arithmetical discrete hyperplanes* in higher dimensions [2].

The problem of computing the minimal thickness connecting an arithmetical discrete plane has already been treated in several works [3,4,5,6]. While all these past works deal with rational arithmetical discrete planes (or hyperplanes), that is, with a normal vector $\mathbf{v}$ which can be expressed as an integer vector, the present work deals with any arithmetical discrete plane. More precisely, we deal with the following questions: given $\mathbf{v} = (v_1, v_2, v_3) \in \mathbb{R}^3 \setminus \{(0,0,0)\}$ with $\dim_{\mathbb{Q}}\{v_1, v_2, v_3\} \neq 1$, given $\mu \in \mathbb{R}$, how much is $\inf\{\omega \in \mathbb{R},\ \left\{\mathbf{x} = (x_1, x_2, x_3) \in \mathbb{Z}^3;\ 0 \leq v_1 x_1 + v_2 x_2 + v_3 x_3 + \mu < \omega\right\}$ is *connected*$\}$?

The present paper is organized as follows. After having introduced the basic notions and notations we use throughout the present paper in

S. Brlek, C. Reutenauer, and X. Provençal (Eds.): DGCI 2009, LNCS 5810, pp. 362–372, 2009.

Section 2, we deal with the connectedness of arithmetical discrete lines in Section 3. In Section 4, we deal with the connectedness of arithmetical discrete planes and show how to reduce the computation of $\Omega(\mathbf{v},\mu) = \inf\{\omega \in \mathbb{R},\ \{\mathbf{x} = (x_1, x_2, x_3) \in \mathbb{Z}^3;\ 0 \leq v_1x_1 + v_2x_2 + v_3x_3 + \mu < \omega\}$ is *connected*$\}$ to the one of a *smaller* vector (see Theorem 3 and Corollary 2). In Section 5, we are interested in computing explicitly $\Omega(\mathbf{v},\mu)$ and we provide formulas for two special cases.

2 Basic Notions and Notation

The aim of this section is to introduce the basic notions and notation we use throughout the present paper.

Let n be an integer equal or greater than 2. Let $\{\mathbf{e_1}, \mathbf{e_2}, \ldots, \mathbf{e_n}\}$ denote the canonical basis of the Euclidean vector space $\mathbb{R}^3$. Let us call *discrete set*, any subset of the *discrete space* $\mathbb{Z}^n$. In the following, for the sake of clarity, we denote by $(x_1, x_2, \ldots, x_n)$ the point $\mathbf{x} = \sum_{i=1}^{n} x_i\mathbf{e_i} \in \mathbb{R}^n$. An integer point $\mathbf{x} \in \mathbb{Z}^n$ is called a *voxel.*

Definition 1 (κ-Adjacency). *Let $\kappa \in \{0,1,,\ldots,n-1\}$. Two voxels $\mathbf{x},\mathbf{y} \in \mathbb{Z}^n$ are* κ-adjacent *if:*

$$\|\mathbf{x}-\mathbf{y}\|_\infty = 1 \text{ and } \|\mathbf{x}-\mathbf{y}\|_1 \leq n-\kappa.$$

REMARK — $\|\mathbf{x}\|_\infty = \max_{i\in\{1,2,\ldots,n\}} \{|x_i|\}$ and $\|\mathbf{x}\|_1 = \sum_{i=1}^{n} |x_i|$.

In other words, the voxel $\mathbf{x}$ and the voxel $\mathbf{y}$ are *κ-adjacent* if they are distinct, if the differences of their coordinates are at most 1 and $\mathbf{x}$ and $\mathbf{y}$ have at most $n-\kappa$ different coordinates. A *κ-path* is a (finite or infinite) sequence of consecutive κ-adjacent voxels. If $(\gamma_i)_{1\leq i\leq k}$ is a finite κ-path, then we say that γ *links* the voxel γ_1 to the voxel γ_k.

Definition 2 (κ-Connected Sets). *Let E be a discrete set and let $\kappa \in \{0,1,\ldots,n-1\}$. Then E is* κ-connected *if, for each pair of voxels $(\mathbf{x},\mathbf{y}) \in E^2$, there exists a κ-path in E linking $\mathbf{x}$ to $\mathbf{y}$.*

In [1], J.-P. Reveillès introduced the arithmetical discrete line as a set of integer points satisfying a double Diophantine inequality.

Definition 3 (Arithmetical Discrete Lines [1]). *Let $\mathbf{v} \in \mathbb{Z}^2\backslash\{(0,0)\}$, $\mu \in \mathbb{Z}$ and $\omega \in \mathbb{Z}$ with* $\gcd\{v_1, v_2\} = 1$. *The* arithmetical discrete line $\mathbb{L}(\mathbf{v},\mu,\omega)$ *with* normal vector $\mathbf{v}$, translation parameter μ *and* arithmetical thickness ω *is the discrete set defined by:*

$$\mathbb{L}(\mathbf{v},\mu,\omega) = \left\{\mathbf{x} \in \mathbb{Z}^2; 0 \leq \langle \mathbf{v},\mathbf{x}\rangle + \mu < \omega\right\},$$

where $\langle \mathbf{v},\mathbf{x}\rangle = v_1x_1 + v_2x_2$.

Let us notice that Definition 3 deals only with integer parameters $\mathbf{v}$, μ and ω. In fact, this definition extends in a natural way to any parameters (integer or not) and in higher dimensions as follows:

Definition 4 (Arithmetical Discrete Hyperplanes [1,2]). *Let $n \in \mathbb{N}$ greater than 2, let $\mathbf{v} \in \mathbb{R}^n \setminus \{(0,\dots,0)\}$, $\mu \in \mathbb{R}$ and $\omega \in \mathbb{R}$. The* arithmetical discrete hyperplane $\mathbb{H}(\mathbf{v},\mu,\omega)$ *with* normal vector $\mathbf{v}$, translation parameter μ *and* arithmetical thickness ω *is the discrete set defined by:*

$$\mathbb{H}(\mathbf{x},\mu,\omega) = \{\mathbf{v} \in \mathbb{Z}^n ; 0 \le \langle \mathbf{v},\mathbf{x}\rangle + \mu < \omega\},$$

where $\langle \mathbf{v},\mathbf{x}\rangle = v_1x_1 + \cdots + v_nx_n$.

If $n = 3$, the arithmetical discrete hyperplane $\mathbb{H}(\mathbf{v},\mu,\omega)$ is called an arithmetical discrete plane *and is denoted $\mathbb{P}(\mathbf{v},\mu,\omega)$.*

Let us first recall a useful lemma concerning a particular subclass of the arithmetical discrete planes:

Lemma 1 ([7]). *Let $n \in \mathbb{N}$ greater than 2, let $\mathbf{v} \in \mathbb{R}^n \setminus \{(0,\dots,0)\}$, $\mu \in \mathbb{R}$ and $\omega \in \mathbb{R}$.*

i) If $\dim_{\mathbb{Q}}\{v_1,\dots,v_n\} = 1$, then there exist $\mathbf{v}' \in \mathbb{Z}^n$, $\mu' \in \mathbb{Z}$ and $\omega' \in \mathbb{Z}$ with $\gcd\{v'_1,\dots,v'_n\} = 1$ such that $\mathbb{H}(\mathbf{v},\mu,\omega) = \mathbb{H}(\mathbf{v}',\mu',\omega')$.
ii) Let $\mathbf{v}' \in \mathbb{R}^n \setminus \{(0,\dots,0)\}$, $\mu' \in \mathbb{R}$ and $\omega' \in \mathbb{R}$. If $\mathbb{H}(\mathbf{v},\mu,\omega) = \mathbb{H}(\mathbf{v}',\mu',\omega')$ then the vectors $\mathbf{v}$ and $\mathbf{v}'$ are colinear.

Definition 5 (Rational Arithmetical Discrete Hyperplanes). *Let $n \in \mathbb{N}$ greater than 2, let $\mathbf{v} \in \mathbb{R}^n \setminus \{(0,\dots,0)\}$, $\mu \in \mathbb{R}$ and $\omega \in \mathbb{R}$. If $\dim_{\mathbb{Q}}\{v_1,\dots,v_n\} = 1$, then the arithmetical discrete hyperplane $\mathbb{H}(\mathbf{v},\mu,\omega)$ and its normal vector $\mathbf{v}$ are said to be* rational.

REMARK — Throughout the present paper, if $\mathbb{P}(\mathbf{v},\mu,\omega)$ is a rational arithmetical plane, then we assume, with no loss of generality, $\mathbf{v} \in \mathbb{Z}^3$ and $\mu \in \mathbb{Z}$.

3 1-Connectedness of (Rational or Irrational) Arithmetical Discrete Lines

Definition 6 (Minimal 1-Connecting Thickness). *Let $\mathbf{v} \in \mathbb{R}^2 \setminus \{(0,0)\}$ and $\mu \in \mathbb{R}$. The* minimal 1-connecting thickness *of $(\mathbf{v},\mu)$ is the number $\Omega_1(\mathbf{v},\mu)$ defined by:*

$$\Omega_1(\mathbf{v},\mu) = \inf\{\omega \in \mathbb{R} ; \mathbb{L}(\mathbf{v},\mu,\omega) \text{ is 1-connected}\}.$$

In [1], J.-P. Reveillès showed how the 1-connectedness of a rational arithmetical discrete line is entirely determined by its normal vector and its thickness. More precisely:

Theorem 1 ([1]). *Let $\mathbb{L}(\mathbf{v},\mu,\omega)$ be a rational arithmetical discrete line with $\omega \in \mathbb{Z}$ and $\gcd\{v_1,v_2\} = 1$. The arithmetical discrete line $\mathbb{L}(\mathbf{v},\mu,\omega)$ is 1-connected if and only if $\omega \ge \|\mathbf{v}\|_1$.*

Theorem 3 holds only for integer normal vectors with coprime coordinates. Moreover, if one does not suppose $\gcd\{v_1, v_2\} = 1$, then Theorem 3 does not hold any more. For instance let us consider $\mathbf{v} = 0\mathbf{e_1} + v_2\mathbf{e_2}$. For all $\mu \in \mathbb{R}$ and $\omega \in \mathbb{R}$, one checks:

$$\mathbb{L}(\mathbf{v}, \mu, \omega) \neq \emptyset \iff \mathbb{L}(\mathbf{v}, \mu, \omega) \text{ is 1-connected} \iff \omega \geq \mu - v_2 \left\lfloor \frac{\mu}{v_2} \right\rfloor .$$

A direct extension of Theorem 3 to any rational arithmetical discrete lines becomes:

Corollary 1. *Let* $\mathbf{v} \in \mathbb{Z}^2 \setminus \{(0,0)\}$ *and* $\mu \in \mathbb{Z}$.

$$\Omega_1(\mathbf{v}, \mu) = \mu - v_2 \left\lfloor \frac{\mu}{v_2} \right\rfloor, \qquad \textit{if } v_1 = 0 \textit{ and } v_2 \neq 0,$$

$$\Omega_1(\mathbf{v}, \mu) = \mu - v_1 \left\lfloor \frac{\mu}{v_1} \right\rfloor, \qquad \textit{if } v_1 \neq 0 \textit{ and } v_2 = 0,$$

$$\Omega_1(\mathbf{v}, \mu) = \|\mathbf{v}\|_1 - \gcd\{v_1, v_2\}, \qquad \textit{if } v_1 \neq 0 \textit{ and } v_2 \neq 0.$$

Remark — The third item of Corollary is not in contradiction with Theorem. Indeed, let $\mathbf{v} \in \mathbb{Z}^2 \setminus \{(0,0)\}$, $\mu \in \mathbb{Z}$ and $(x_1, x_2) \in \mathbb{Z}^2$, then one has:

$$v_1x_1 + v_2x_2 + \mu < \|\mathbf{v}\|_1 \iff v_1x_1 + v_2x_2 + \mu \leq \|\mathbf{v}\|_1 - \gcd\{v_1v_2\}.$$

Let us now extend Theorem 3 and Corollary 3 to irrational arithmetical discrete lines. One of our main ambition is to exhibit a unique formulation the properties we state. For that purpose, we introduce the following notation.
Notation — Let $\Phi : \mathbb{R}^2 \longmapsto \mathbb{R}^2$ the map defined by:

$$\Phi : \mathbb{R}^2 \mapsto \mathbb{R}^2$$
$$(x, y) \mapsto \begin{cases} \left(y - \left\lfloor \frac{y}{x} \right\rfloor x, x\right), & \text{if } x \neq 0, \\ (x, y) & \text{otherwise.} \end{cases}$$

Given $x \in \mathbb{R}$ and $y \in \mathbb{R}$, we set $x \wedge y = \lim_{n \to \infty} | (\Phi^n(x, y))_2 |$.

It is clear that, given $(x, y) \in \mathbb{Z}^2$, $x \wedge y = \gcd\{x, y\}$. Moreover, one states:

Lemma 2. *Let* $(x, y, z) \in \mathbb{R}^3$ *and* $\lambda \in \mathbb{R}$.

1. $\lambda x \wedge \lambda y = |\lambda|(x \wedge y)$.
2. *if* $\dim_{\mathbb{Q}}\{x, y\} = 2$, *then* $x \wedge y = 0$.
3. $x \wedge (y \wedge z) = (x \wedge y) \wedge z$.

Notation — Thanks to Lemma 2, for any $x_1, x_2, \ldots, x_k \in \mathbb{R}$, we denote $x_1 \wedge \cdots \wedge x_k$, or for short $\bigwedge_{i=1}^{k} x_i$ instead of $\underbrace{(\ldots((}_{k-1} x_1 \wedge x_2) \wedge x_3) \wedge \cdots \wedge x_k)$.

We now state the main result of the present section:

Theorem 2. *Let* $\mathbf{v} \in \mathbb{R}^2 \setminus \{(0,0)\}$ *and* $\mu \in \mathbb{R}$. *Then,*

$$\Omega_1(\mathbf{v},\mu) = \mu - v_2 \left\lfloor \frac{\mu}{v_2} \right\rfloor, \qquad \text{if } v_1 = 0 \text{ and } v_2 \neq 0,$$

$$\Omega_1(\mathbf{v},\mu) = \mu - v_1 \left\lfloor \frac{\mu}{v_1} \right\rfloor, \qquad \text{if } v_1 \neq 0 \text{ and } v_2 = 0,$$

$$\Omega_1(\mathbf{v},\mu) = \|\mathbf{v}\|_1 - v_1 \wedge v_2, \qquad \text{if } v_1 \neq 0 \text{ and } v_2 \neq 0.$$

Proof (Sketch). The case $\dim_{\mathbb{Q}}\{v_1, v_2\} = 1$ (in particular if $v_1 = 0$ or $v_2 = 0$) has already been treated (see Corollary 3 and Lemma 1). Assume that $\dim_{\mathbb{Q}}\{v_1, v_2\} = 2$. Then $v_1 \wedge v_2 = 0$. With no loss of generality, assume $0 \leq v_1 \leq v_2$ (one reduces to this assumption, up to an isomorphism of the unit cube $[-\frac{1}{2}, \frac{1}{2}]^3$).

i) The projection $\pi : \mathbb{L}(\mathbf{v}, \mu, \|\mathbf{v}\|_1) \longrightarrow \mathbb{Z}$ along the vector $\mathbf{e_1} + \mathbf{e_2}$ is one-to-one. Given $\mathbf{x}$ and $\mathbf{y}$ in $\mathbb{L}(\mathbf{v}, \mu, \|\mathbf{v}\|_1)$, one shows that $\mathbf{x}$ and $\mathbf{y}$ are 1-adjacent if, and only if, $|\pi(\mathbf{x}) - \pi(\mathbf{y})| = 1$. We deduce that the arithmetical discrete line $\mathbb{L}(\mathbf{v}, \mu, \|\mathbf{v}\|_1)$ is 1-connected.

ii) Let $\omega \geq \|\mathbf{v}\|_1$ and let $\mathbf{x} \in \mathbb{L}(\mathbf{v}, \mu, \omega)$. Then $\mathbf{x}$ is either in $\mathbb{L}(\mathbf{v}, \mu, \|\mathbf{v}\|_1)$ or vertically 1-linked with an element of $\mathbb{L}(\mathbf{v}, \mu, \|\mathbf{v}\|_1)$. Hence $\mathbb{L}(\mathbf{v}, \mu, \omega)$ is 1-connected.

iii) Let $\omega \in [0, \|\mathbf{v}\|_1[$. Since $\dim_{\mathbb{Q}}\{v_1, v_2\} = 2$, the set $\{\langle \mathbf{v}, \mathbf{x} \rangle + \mu, \mathbf{x} \in \mathbb{L}(\mathbf{v}, \mu, \|\mathbf{v}\|_1)\}$ is dense in $[0, \|\mathbf{v}\|_1[$. Let $\mathbf{x} \in \mathbb{L}(\mathbf{v}, \mu, \|\mathbf{v}\|_1)$ such that $\langle \mathbf{v}, \mathbf{x} \rangle + \mu \in [\omega, \|\mathbf{v}\|_1[$. Then, $\mathbb{L}(\mathbf{v}, \mu, \omega) \subseteq \mathbb{L}(\mathbf{v}, \mu, \|\mathbf{v}\|_1) \setminus \{x\}$. From i), we state that the set $\mathbb{L}(\mathbf{v}, \mu, \|\mathbf{v}\|_1) \setminus \{x\}$ is not 1-connected, and so is $\mathbb{L}(\mathbf{v}, \mu, \omega)$. □

4 2-Connectedness of Arithmetical Discrete Planes

The case of arithmetical discrete lines is somewhat confusing. Indeed, while an arithmetical discrete line is 1-connected if, and only if, it *separates* the discrete space $\mathbb{Z}^2$ into two 0-connected components, there exist 2-connected arithmetical discrete planes $\mathbb{P}(\mathbf{v}, \mu, \omega)$ with $\omega < \|\mathbf{v}\|_1$, that is, which do not *separate* the discrete space $\mathbb{Z}^3$ into two 0-connected components [2].

Definition 7 (Minimal 2-Connecting Thickness). *Let* $\mathbf{v} \in \mathbb{R}^3 \setminus \{(0,0,0\}$ *and* $\mu \in \mathbb{R}$. *The* minimal 2-connecting thickness *of* $(\mathbf{v}, \mu)$ *is the number* $\Omega_2(\mathbf{v}, \mu)$ *defined by:*

$$\Omega_2(\mathbf{v}, \mu) = \inf \{\omega \in \mathbb{R};\ \mathbb{P}(\mathbf{v}, \mu, \omega) \text{ is 2-connected}\}.$$

In [6], the authors showed how to reduce the computation of the minimal 2-connecting thickness of $(\mathbf{v}, 0)$, where $\mathbf{v}$ is a rational vector, to the one of a smaller vector $\mathbf{v}'$ (in terms of norm $\|\cdot\|_\infty$). Here we extend this reduction to the pairs $(\mathbf{v}, \mu)$ with $\mathbf{v}$ possibly irrational and $\mu \in \mathbb{R}$.

Theorem 3. *Let* $\mathbf{v} \in \mathbb{R}^3_+ \setminus \{(0,0,0\}$, $\mu \in \mathbb{R}$ *and* $\omega \in \mathbb{R}$. *The arithmetical discrete plane* $\mathbb{P}(\mathbf{v},\mu,\omega)$ *is 2-connected if and only if so is* $\mathbb{P}(\mathbf{v} + v_1(\mathbf{e_2}+\mathbf{e_3}), \mu, \omega + v_1)$.

In order to prove Theorem 3, we use a technical lemma providing a lower bound of $\Omega_2(\mathbf{v},\mu)$:

Lemma 3 (A Lower Bound [6]). *Let* $\mathbf{v} \in \mathbb{R}^3 \setminus \{(0,0,0\}$ *and* $\mu \in \mathbb{R}$. *We suppose that* $\mathbf{v}$ *does not have two coordinates equal to zero. Then* $\Omega_2(\mathbf{v},\mu) \geq \|\mathbf{v}\|_\infty$.

The case where $\mathbf{v}$ has at least one coordinate equal to zero has an explicit answer (see Lemma 4 in Section 5).

Proof (of Theorem 3). Let $\mathbf{v}' = \mathbf{v} + v_1(\mathbf{e_2}+\mathbf{e_3})$, let $\omega' = \omega + v_1$ and let us consider the linear bijection $\Psi_2 : \mathbb{R}^3 \longrightarrow \mathbb{R}^3$ and its inverse map $\Psi_2^{-1} : \mathbb{R}^3 \longrightarrow \mathbb{R}^3$ defined by:

$$\Psi_2 : \begin{array}{ccc} \mathbb{R}^3 & \longrightarrow & \mathbb{R}^3 \\ \begin{pmatrix} x_1 \\ x_2 \\ x_3 \end{pmatrix} & \longmapsto & \begin{pmatrix} x_1 - x_2 - x_3 \\ x_2 \\ x_3 \end{pmatrix} \end{array} \quad \text{and} \quad \Psi_2^{-1} : \begin{array}{ccc} \mathbb{R}^3 & \longrightarrow & \mathbb{R}^3 \\ \begin{pmatrix} x_1 \\ x_2 \\ x_3 \end{pmatrix} & \longmapsto & \begin{pmatrix} x_1 + x_2 + x_3 \\ x_2 \\ x_3 \end{pmatrix} \end{array}$$

One checks that Ψ_2 provides a bijection from $\mathbb{P}(\mathbf{v},\mu,\omega)$ to $\mathbb{P}(\mathbf{v}',\mu,\omega)$. Moreover, for each $\mathbf{x} \in \mathbb{Z}^3$, $\langle \mathbf{v}, \mathbf{x} \rangle = \langle \mathbf{v}', \Psi_2(\mathbf{x}) \rangle$.

1. Assume $\mathbb{P}(\mathbf{v},\mu,\omega)$ is 2-connected. Then, from Lemma 3, it follows that $\omega \geq \|\mathbf{v}\|_\infty$. Let us show that $\mathbb{P}(\mathbf{v}',\mu,\omega')$ is 2-connected too.
   Let $\mathbf{x} \in \mathbb{P}(\mathbf{v}',\mu,\omega')$. Either $\langle \mathbf{v}', \mathbf{x} \rangle + \mu \in [0,\omega[$ or $\langle \mathbf{v}', \mathbf{x} \rangle + \mu \in [\omega,\omega'[$.
   i) If $\langle \mathbf{v}', \mathbf{x} \rangle + \mu \in [0,\omega[$, then $\mathbf{x} \in \mathbb{P}(\mathbf{v}',\mu,\omega)$.
   ii) If $\langle \mathbf{v}', \mathbf{x} \rangle + \mu \in [\omega,\omega'[$, and since $\omega \geq \|\mathbf{v}\|_\infty \geq v_1$, then $\langle \mathbf{v}', \mathbf{x} - \mathbf{e_1} \rangle + \mu = \langle \mathbf{v}', \mathbf{x} \rangle + \mu - v_1 \in [0,\omega[$, and $\mathbf{x} - \mathbf{e_1} \in \mathbb{P}(\mathbf{v}',\mu,\omega)$.
   In other words, each $\mathbf{x} \in \mathbb{P}(\mathbf{v}',\mu,\omega')$ is either in $\mathbb{P}(\mathbf{v}',\mu,\omega)$ or 2-adjacent to an element of $\mathbb{P}(\mathbf{v}',\mu,\omega)$. Consequently, it remains to ensure that each pair of elements of $\mathbb{P}(\mathbf{v}',\mu,\omega)$ is 2-linked in $\mathbb{P}(\mathbf{v}',\mu,\omega')$. Moreover, since the restriction $\Psi_2 : \mathbb{P}(\mathbf{v},\mu,\omega) \longrightarrow \mathbb{P}(\mathbf{v}',\mu,\omega)$ of Ψ_2 is one-to-one, then it is sufficient to prove that the images, under Ψ_2, of two 2-adjacent elements of $\mathbb{P}(\mathbf{v},\mu,\omega)$ are 2-linked in $\mathbb{P}(\mathbf{v}',\mu,\omega')$ (see Figure 1).
   i) Let $\mathbf{x} \in \mathbb{P}(\mathbf{v},\mu,\omega)$ such that $\mathbf{x}+\mathbf{e_1} \in \mathbb{P}(\mathbf{v},\mu,\omega)$. Then $\Psi_2(\mathbf{x}+\mathbf{e_1}) - \Psi_2(\mathbf{x}) = \mathbf{e_1}$ and the images of $\mathbf{x}$ and $\mathbf{x}+\mathbf{e_1}$ are 2-adajcent in $\mathbb{P}(\mathbf{v}',\mu,\omega')$.
   ii) Let $\mathbf{x} \in \mathbb{P}(\mathbf{v},\mu,\omega)$ such that $\mathbf{x}+\mathbf{e_2} \in \mathbb{P}(\mathbf{v},\mu,\omega)$. Hence,

$$\begin{aligned} \langle \mathbf{v}', \Psi_2(\mathbf{x}+\mathbf{e_1}+\mathbf{e_2}) \rangle + \mu &= \langle \mathbf{v}, \mathbf{x}+\mathbf{e_1}+\mathbf{e_2} \rangle + \mu \\ &= \langle \mathbf{v}, \mathbf{x}+\mathbf{e_2} \rangle + \mu + v_1 \in [0, \omega + v_1[, \end{aligned}$$

   since $\mathbf{x}+\mathbf{e_2} \in \mathbb{P}(\mathbf{v},\mu,\omega)$. Hence $\Psi_2(\mathbf{x}+\mathbf{e_1}+\mathbf{e_2}) \in \mathbb{P}(\mathbf{v}',\mu,\omega')$. One checks that $\Psi_2(\mathbf{x}+\mathbf{e_1}+\mathbf{e_2}) - \Psi_2(\mathbf{x}) = \mathbf{e_2}$ and $\Psi_2(\mathbf{x}+\mathbf{e_2}) - \Psi_2(\mathbf{x}+\mathbf{e_1}+\mathbf{e_2}) = -\mathbf{e_1}$. It follows that $(\Psi_2(\mathbf{x}), \Psi_2(\mathbf{x}+\mathbf{e_1}+\mathbf{e_2}), \Psi_2(\mathbf{x}+\mathbf{e_2}))$ is a 2-path linking $\Psi_2(\mathbf{x})$ to $\Psi_2(\mathbf{x}+\mathbf{e_2})$ in $\mathbb{P}(\mathbf{v}',\mu,\omega')$.

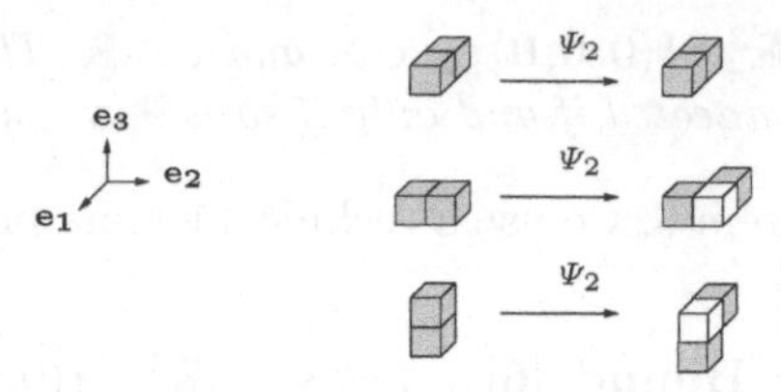

Fig. 1. The action of Ψ_2 on two 2-adjacent voxels of $\mathbb{P}(\mathbf{v}, \mu, \omega)$.: The right grey voxels are the images of the left grey ones, while the right white voxels are elements of $\mathbb{P}(\mathbf{v}', \mu, \omega')$ which allow us to 2-link the grey voxels. For ensuring the existence of such white voxels, we remind that the right grey ones belongs to $\mathbb{P}(\mathbf{v}', \mu, \omega)$ and, for any such voxels $\mathbf{v}$, the (white) voxel $\mathbf{v} + \mathbf{e_1}$ belongs to $\mathbb{P}(\mathbf{v}', \mu, \omega')$.

   iii) A similar reasoning shows that given $\mathbf{x} \in \mathbb{P}(\mathbf{v}, \mu, \omega)$ such that $\mathbf{x} + \mathbf{e_3} \in \mathbb{P}(\mathbf{v}, \mu, \omega)$, $\Psi_2(\mathbf{x})$ and $\Psi_2(\mathbf{x} + \mathbf{e_3})$ are 2-linked in $\mathbb{P}(\mathbf{v}', \mu, \omega')$.

2. Conversely, suppose $\mathbb{P}(\mathbf{v}', \mu, \omega')$ is 2-connected. From Lemma 3, one deduces that $\omega' \geq \|\mathbf{v}'\|_\infty$.
Let $\widetilde{\Psi}_2 : \mathbb{P}(\mathbf{v}', \mu, \omega') \longrightarrow \mathbb{P}(\mathbf{v}', \mu, \omega)$ be the surjective map defined by:

$$\widetilde{\Psi}_2 : \mathbb{P}(\mathbf{v}', \mu, \omega') \longrightarrow \mathbb{P}(\mathbf{v}, \mu, \omega)$$
$$\mathbf{v} \longmapsto \begin{cases} \Psi_2^{-1}(\mathbf{x}), & \text{if } \mathbf{x} \in \mathbb{P}(\mathbf{v}', \mu, \omega) \\ \Psi_2^{-1}(\mathbf{x} - \mathbf{e_1}), & \text{otherwise.} \end{cases}$$

Remind that each $\mathbf{x} \in \mathbb{P}(\mathbf{v}', \mu, \omega')$ is either in $\mathbb{P}(\mathbf{v}', \mu, \omega)$ or 2-adjacent to an element of $\mathbb{P}(\mathbf{v}', \mu, \omega)$. Since $\widetilde{\Psi}_2$ is surjective, then it remains to show that the image of two 2-adjacent voxels in $\mathbb{P}(\mathbf{v}', \mu, \omega')$ are either equal or 2-linked in $\mathbb{P}(\mathbf{v}, \mu, \omega)$ (see Figure 2).

   i) Let $\mathbf{x} \in \mathbb{P}(\mathbf{v}', \mu, \omega')$ such that $\mathbf{x} + \mathbf{e_1} \in \mathbb{P}(\mathbf{v}', \mu, \omega')$.
      - If $\langle \mathbf{v}', \mathbf{x} + \mathbf{e_1} \rangle + \mu \in [0, \omega[$, then $\widetilde{\Psi}_2(\mathbf{x} + \mathbf{e_1}) - \widetilde{\Psi}_2(\mathbf{x}) = \mathbf{e_1}$.
      - If $\langle \mathbf{v}', \mathbf{x} + \mathbf{e_1} \rangle + \mu \in [\omega, \omega'[$, then $\widetilde{\Psi}_2(\mathbf{x} + \mathbf{e_1}) = \widetilde{\Psi}_2(\mathbf{x})$.

   ii) Let $\mathbf{x} \in \mathbb{P}(\mathbf{v}', \mu, \omega')$ such that $\mathbf{x} + \mathbf{e_2} \in \mathbb{P}(\mathbf{v}', \mu, \omega')$.
      - If $\langle \mathbf{v}', \mathbf{x} + \mathbf{e_2} \rangle + \mu \in [0, \omega[$, then $0 \leq \langle \mathbf{v}', \mathbf{x} \rangle + \mu \leq \langle \mathbf{v}', \mathbf{x} \rangle + \mu + v_1 = \langle \mathbf{v}', \mathbf{x} + \mathbf{e_1} \rangle + \mu \leq \langle \mathbf{v}', \mathbf{x} \rangle + \mu + v_1 + v_2 = \langle \mathbf{v}', \mathbf{x} + \mathbf{e_2} \rangle + \mu \leq \omega$.
      Moreover, $\widetilde{\Psi}_2(\mathbf{x} + \mathbf{e_1}) - \widetilde{\Psi}_2(\mathbf{x}) = \mathbf{e_1}$ and $\widetilde{\Psi}_2(\mathbf{x} + \mathbf{e_2}) - \widetilde{\Psi}_2(\mathbf{x} + \mathbf{e_1}) = \mathbf{e_2}$ and we have shown that $(\widetilde{\Psi}_2(\mathbf{x}), \widetilde{\Psi}_2(\mathbf{x} + \mathbf{e_1}), \widetilde{\Psi}_2(\mathbf{x} + \mathbf{e_2}))$ is a 2-path linking $\widetilde{\Psi}_2(\mathbf{x})$ to $\widetilde{\Psi}_2(\mathbf{x} + \mathbf{e_2})$ in $\mathbb{P}(\mathbf{v}, \mu, \omega)$.
      - If $\langle \mathbf{v}', \mathbf{x} + \mathbf{e_2} \rangle + \mu \in [\omega, \omega'[$, then $\widetilde{\Psi}_2(\mathbf{x} + \mathbf{e_2}) = \widetilde{\Psi}_2(\mathbf{x} + \mathbf{e_2} - \mathbf{e_1}) = \widetilde{\Psi}_2(\mathbf{x}) + \mathbf{e_2}$ and we have shown that $\widetilde{\Psi}_2(\mathbf{x})$ and $\widetilde{\Psi}_2(\mathbf{x} + \mathbf{e_2})$ are 2-adjacent in $\mathbb{P}(\mathbf{v}, \mu, \omega)$.

□

Corollary 2. *Let* $\mathbf{v} \in \mathbb{R}^3_+ \setminus \{(0,0,0)\}$ *such that* $0 \leq v_1 \leq v_2, v_3$, *let* $\mu \in \mathbb{R}$ *and let* $\omega \in \mathbb{R}$. *The arithmetical discrete plane* $\mathbb{P}(\mathbf{v}, \mu, \omega)$ *is 2-connected if and only if so is* $\mathbb{P}(\mathbf{v} - v_1(\mathbf{e_2} + \mathbf{e_3}), \mu, \omega - v_1)$.

In terms of $\Omega_2(\mathbf{v}, \mu)$, Theorem 3 and Corollary 2 can be reformulated as follows

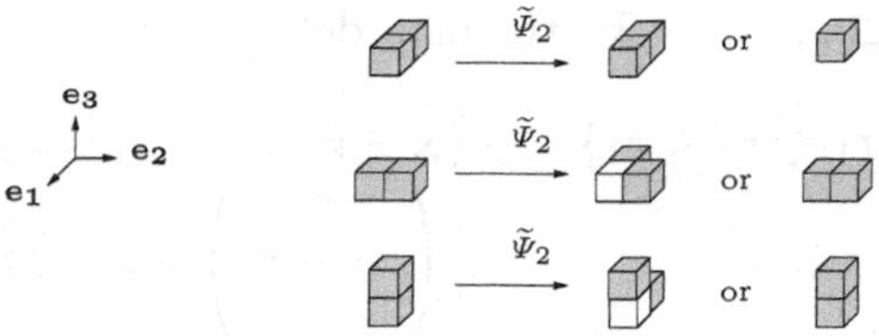

Fig. 2. The action of $\widetilde{\Psi}_2$ on two 2-adjacent voxels of $\mathbb{P}(\mathbf{v}', \mu, \omega')$

Corollary 3. *Let* $\mathbf{v} \in \mathbb{R}^3_+ \setminus \{(0,0,0)\}$ *such that* $0 \leq v_1 \leq v_2, v_3$ *and let* $\mu \in \mathbb{R}$. *Then*

$$\Omega_2(\mathbf{v}, \mu) = \Omega_2(\mathbf{v} - v_1(\mathbf{e_2} + \mathbf{e_3}), \mu) + v_1$$

5 Effective Computation of $\Omega_2(\mathbf{v}, \mu)$

Let $\mathbf{v} \in \mathbb{R}^3$ and $\mu \in \mathbb{R}$. In case of $v_1 = 0$, the reductions given in Theorem 3, Corollary 2 and Corollary 3 are useless. In fact, in that case we have a direct computation of $\Omega_2(\mathbf{v}, \mu)$ since for each $\omega \in \mathbb{R}$,

$$\mathbb{P}(\mathbf{v}, \mu, \omega) = \bigcup_{k \in \mathbb{Z}} \left(\{\mathbf{x} \in \mathbb{Z}^3;\, x_1 = 0 \text{ and } 0 \leq \langle \mathbf{v}, \mathbf{x} \rangle + \mu < \omega\} + k\mathbf{e_1}\right).$$

Roughly speaking, $\mathbb{P}(\mathbf{v}, \mu, \omega)$ is obtained by translating the arithmetical discrete line $\mathbb{L}((v_2, v_3), \mu, \omega)$ along the coordinate vector $\mathbf{e_1}$. Hence, in case of $v_1 = 0$, $\mathbb{P}(\mathbf{v}, \mu, \omega)$ is 2-connected if, and only if, the arithmetical discrete line $\mathbb{L}((v_2, v_3), \mu, \omega)$ is 1-connected. From Theorem 2, it follows:

Lemma 4. *Let* $\mathbf{v} \in \mathbb{R}^3 \setminus \{(0,0,0)\}$ *and let* $\mu \in \mathbb{R}$. *If* $v_1 = 0$, *then* $\Omega_2(\mathbf{v}, \mu) = \Omega_1((v_2, v_3), \mu)$. *More precisely,*

$$\Omega_2(\mathbf{v}, \mu) = \mu - v_3 \left\lfloor \frac{\mu}{v_3} \right\rfloor, \qquad \text{if } v_2 = 0 \text{ and } v_3 \neq 0,$$

$$\Omega_2(\mathbf{v}, \mu) = \mu - v_2 \left\lfloor \frac{\mu}{v_2} \right\rfloor, \qquad \text{if } v_2 \neq 0 \text{ and } v_3 = 0,$$

$$\Omega_2(\mathbf{v}, \mu) = \|\mathbf{v}\|_1 - v_2 \wedge v_3, \qquad \text{if } v_2 \neq 0 \text{ and } v_3 \neq 0.$$

Remark — Of course, the other cases ($v_2 = 0$ and $v_3 = 0$) are similar.

Let us now treat the general case. For clarity issues, and with no loss of generality, we reduce the computation of $\Omega_2(\mathbf{v}, \mu)$ to the vectors $\mathbf{v} \in \{\mathbf{x} \in \mathbb{R}^3_+, 0 \leq x_1 \leq x_2 \leq x_3\}^\star = \{\mathbf{x} \in \mathbb{R}^3_+, 0 \leq x_1 \leq x_2 \leq x_3\} \setminus \{(0,0,0)\}$. In fact, up to an isometry of the unit cube $[-0.5, 0.5]^3$, for each vector $\mathbf{v} \in \mathbb{R}^3$ and each $\mu \in \mathbb{R}$, there exists a vector $\mathbf{v}' \in \mathbb{R}^3$ such that $0 \leq v'_1 \leq v'_2 \leq v'_3$ and $\Omega_2(\mathbf{v}, \mu) = \Omega_2(\mathbf{v}', \mu)$.

Notation — Let $\Delta : \mathbb{R}^3 \longmapsto \mathbb{R}^3_+$ the map defined by:

$$\Delta : \left\{\mathbf{x} \in \mathbb{R}^3_+,\ 0 \le x_1 \le x_2 \le x_3\right\} \longmapsto \left\{\mathbf{x} \in \mathbb{R}^3_+,\ 0 \le x_1 \le x_2 \le x_3\right\}$$

$$\begin{pmatrix} x_1 \\ x_2 \\ x_3 \end{pmatrix} \mapsto \begin{cases} \begin{pmatrix} x_1 \\ x_2 - x_1 \\ x_3 - x_1 \end{pmatrix} & \text{if } x_1 \le x_2 - x_1 \le x_3 - x_1, \\[2ex] \begin{pmatrix} x_2 - x_1 \\ x_1 \\ x_3 - x_1 \end{pmatrix} & \text{if } x_2 - x_1 \le x_1 \le x_3 - x_1, \\[2ex] \begin{pmatrix} x_2 - x_1 \\ x_3 - x_1 \\ x_1 \end{pmatrix} & \text{otherwise.} \end{cases}$$

Roughly speaking, given a normal vector $\mathbf{v} \in \left\{\mathbf{x} \in \mathbb{R}^3_+,\ 0 \le x_1 \le x_2 \le x_3\right\}$, the operator Δ subtracts the lowest coordinate of $\mathbf{v}$ to the other ones and reorder the output vector. This operator is known as the fully subtractive algorithm, a two-dimensional continued fraction algorithm. For more informations, one can refer to [8].

Notation — Let $\mathbf{v} \in \left\{\mathbf{x} \in \mathbb{R}^3_+,\ 0 \le x_1 \le x_2 \le x_3\right\}$ and let $n \in \mathbb{N}$. We set $\mathbf{v}^{(n)} = \Delta^n(\mathbf{v}) \in \left\{\mathbf{x} \in \mathbb{R}^3_+,\ 0 \le x_1 \le x_2 \le x_3\right\}$.

Let $\mathbf{v} \in \left\{\mathbf{x} \in \mathbb{R}^3_+,\ 0 \le x_1 \le x_2 \le x_3\right\}$, $\mu \in \mathbb{R}$ and $n \in \mathbb{N}$. From Corollary 3, it follows

$$\|\mathbf{v}\|_1 - 2\Omega_2(\mathbf{v}, \mu) = \|\Delta^n(\mathbf{v})\|_1 - 2\Omega_2(\Delta^n(\mathbf{v}), \mu).$$

Given $\mathbf{v} \in \left\{\mathbf{x} \in \mathbb{R}^3_+,\ 0 \le x_1 \le x_2 \le x_3\right\}$, the sequences $(v_1^{(n)})_{n\in\mathbb{N}}$, $(v_2^{(n)})_{n\in\mathbb{N}}$ and $(v_3^{(n)})_{n\in\mathbb{N}}$ are non-increasing and minored by 0. Hence they are convergent. Let $\mathbf{v}' = \lim_{n\to\infty} \Delta^n(\mathbf{v})$, then $\mathbf{v}' = \Delta(\mathbf{v}')$ implies $v'_1 = 0$ and one deduces:

Theorem 4. *Let $\mathbf{v} \in \left\{\mathbf{x} \in \mathbb{R}^3_+,\ 0 \le x_1 \le x_2 \le x_3\right\}^\star$, let $\mu \in \mathbb{R}$ and let $\mathbf{v}' = \lim_{n\to\infty} \Delta^n(\mathbf{v})$. Then,*

$$\Omega_2(\mathbf{v}, \mu) = \frac{\|\mathbf{v}\|_1 - \|\mathbf{v}'\|_1 + 2\Omega_2(\mathbf{v}', \mu)}{2}.$$

Remark — Theorem 4 provides an explicit process to compute $\Omega_2(\mathbf{v}, \mu)$ whatever $\mathbf{v} \in \mathbb{R}^3$ and μ. Indeed, since $v'_1 = 0$, thanks to Lemma 4, we can explicitly compute $\Omega_2(\mathbf{v}', \mu)$.

Given $\mathbf{v} \in \left\{\mathbf{x} \in \mathbb{R}^3_+,\ 0 \le x_1 \le x_2 \le x_3\right\}$, two cases occur:

1. there exists $n_0 \in \mathbb{N}$ such that $v_1^{(n_0)} + v_2^{(n_0)} < v_3^{(n_0)}$,
2. for all $n_0 \in \mathbb{N}$, there exists $n \ge n_0$ such that $v_1^{(n)} + v_2^{(n)} \ge v_3^{(n)}$.

From Theorem 4, one deduces the following Corollaries.

Corollary 4 (The first case). *Let* $\mathbf{v} \in \left\{\mathbf{x} \in \mathbb{R}^3_+,\, 0 \le x_1 \le x_2 \le x_3\right\} \setminus \{(0,0,0)\}$, *let* $\mu \in \mathbb{R}$ *and let* $n_0 \in \mathbb{N}$ *such that* $v_1^{(n_0)} + v_2^{(n_0)} < v_3^{(n_0)}$.

i) If $v_1^{(n_0)} \wedge v_2^{(n_0)} = 0$*, then:*

$$\Omega_2(\mathbf{v}, \mu) = \frac{\|\mathbf{v}\|_1 - \gamma}{2} + \left(\mu - \gamma \left\lfloor \frac{\mu}{\gamma} \right\rfloor\right), \text{ with } \gamma = v_3^{(n_0)} - \left(v_1^{(n_0)} + v_2^{(n_0)}\right).$$

ii) If $v_1^{(n_0)} \wedge v_2^{(n_0)} \neq 0$*, then:*

$$\Omega_2(\mathbf{v}, \mu) = \frac{\|\mathbf{v}\|_1 - (v_1^{(n_0)} + v_2^{(n_0)}) + v_3^{(n_0)} + 2(v_1^{(n_0)} \wedge v_2^{(n_0)} - v_1^{(n_0)} \wedge v_2^{(n_0)} \wedge v_3^{(n_0)})}{2}.$$

Remark — The main advantage of Corollary 4 is to provide a possibly finite process to compute $\Omega_2(\mathbf{v}, \mu)$ for a *totally irrational* vector $\mathbf{v}$, that is, satisfying $\dim_{\mathbb{Q}}\{v_1, v_2, v_3\} = 3$.

Corollary 5 (The Second Case). *Let* $\mathbf{v} \in \left\{\mathbf{x} \in \mathbb{R}^3_+,\, 0 \le x_1 \le x_2 \le x_3\right\}$ *and assume:*

$$\forall n_0 \in \mathbb{N},\, \exists n \ge n_0,\, v_1^{(n)} + v_2^{(n)} > v_3^{(n)}. \tag{1}$$

Then

$$\Omega_2(\mathbf{v}, \mu) = \frac{v_1 + v_2 + v_3}{2}.$$

6 Conclusion

In the present paper we investigated, given $\mathbf{v} \in \mathbb{R}^3$ and $\mu \in \mathbb{R}$, the computation of $\inf\{\omega \in \mathbb{R},\, \mathbb{P}(\mathbf{v}, \mu, \omega) \text{ is } 2\text{-}connected\}$. The main improvement we made, compare to [6] is that we do not treat the rational case and the irrational case separately but in a common way and in a unique expression, thanks to the introduction of a notation inherited from the (multidimensional) continued fractions theory.

Moreover, we have shown the role of the translation parameter μ in the computation of $\Omega_2(\mathbf{v}, \mu)$. This result was already known for the rational arithmetical discrete planes but not for the irrational ones.

Due to the limitation of space, we do not have included the investigation of the 0-connectedness and the 1-connectedness of arithmetical discrete planes. In fact, this studies are similar to the one we did in the present paper. For explicit formulas see [6]. Reformulating in terms of possibly irrational arithmetical discrete planes is obvious.

In a further work, it will be interesting to extend all the present results to arithmetical discrete hyperplanes in any dimension. Another interesting way to investigate is the behavior of the class of arithmetical discrete (hyper)planes under the action of other multidimensional continued fractions algorithms.

References

1. Reveillès, J.P.: Géométrie discrète, calcul en nombres entiers et algorithmique. Thèse d'état, Université Louis Pasteur, Strasbourg (1991)
2. Andres, E., Acharya, R., Sibata, C.: Discrete analytical hyperplanes. CVGIP: Graphical Model and Image Processing 59(5), 302–309 (1997)
3. Gérard, Y.: Periodic graphs and connectivity of the rational digital hyperplanes. Theor. Comput. Sci. 283(1), 171–182 (2002)
4. Brimkov, V., Barneva, R.: Connectivity of discrete planes. Theor. Comput. Sci. 319(1-3), 203–227 (2004)
5. Jamet, D., Toutant, J.L.: On the connectedness of rational arithmetic discrete hyperplanes. In: Kuba, A., Nyúl, L.G., Palágyi, K. (eds.) DGCI 2006. LNCS, vol. 4245, pp. 223–234. Springer, Heidelberg (2006)
6. Jamet, D., Toutant, J.L.: Minimal arithmetic thickness connecting discrete planes. Discrete Applied Mathematics 157(3), 500–509 (2009)
7. Berthé, V., Fiorio, C., Jamet, D., Philippe, F.: On some applications of generalized functionality for arithmetic discrete planes. Image Vision Comput 25(10), 1671–1684 (2007)
8. Schweiger, F.: Multidimensional continued fractions. Oxford Science Publications. Oxford University Press, Oxford (2000)

Patterns in Discretized Parabolas and Length Estimation

Alain Daurat, Mohamed Tajine, and Mahdi Zouaoui

LSIIT CNRS UMR 7005, Université de Strasbourg,
Pôle API, Boulevard Sébastien Brant, 67400 Illkirch-Graffenstaden, France
{daurat,tajine,mzouaoui}@unistra.fr

Abstract. We estimate the frequency of the patterns in the discretization of parabolas, when the resolution tends to zero. We deduce that local estimators of length almost never converge to the length for the parabolas.

Keywords: Digital Curve, Pattern, Multigrid Convergence.

1 Introduction

Length estimation is an important domain of Image Analysis (see [1] for a review). In this paper, we will consider the problem of estimating the length of a curve from its discretizations at different resolutions. In particular we are interested in the comportment of estimators when the resolution tends to zero. We also restrict our study to special estimators called "local estimators" which consist in considering *patterns* which are pieces of fixed length of the discretized curve. Local estimators simply consist to fix a weight to each pattern and summing these weights to obtain the estimation of length (See Fig. 4 for illustration). So, if we want to study the estimated length by local estimators when the resolution tends to zero, we have at first to study the number of occurrences of a pattern of the discretization of digital curves. In fact an asymptotic result about the occurrence number of patterns for discretized general curves looks to be a quite hard problem, because the discretization process is not a continuous process (the integer part function is not continuous), so the estimation of the occurrence number of patterns cannot be deduced from Mathematical Analysis arguments, but by Number Theory arguments. The two first authors of this paper have already made this study for segments in [2]. In this paper we continue this work by considering another class of curves, the parabolas.

The paper is organized as follows: Section 2 describes the notations used in this paper, Section 3 will be devoted to the study of the frequency of patterns in parabolas, and finally Section 4 will apply the results of this study to the local estimators of length of parabolas. In this paper, some proofs are only sketched, the complete proofs are in the preprint [3].

2 Notations

In this section we precise the notations that will be used in all the paper.

S. Brlek, C. Reutenauer, and X. Provençal (Eds.): DGCI 2009, LNCS 5810, pp. 373–384, 2009.

- For $x \in \mathbb{R}$ we denote by $\lfloor x \rfloor$ (resp. $\lceil x \rceil$) the integer k such that $k \leq x < k+1$ (resp. $k-1 < x \leq k$).
- The fractional part of x is denoted $\langle x \rangle$ and is defined by $x = \lfloor x \rfloor + \langle x \rangle$.
- For $A, B \in \mathbb{Z}$, the discrete interval $\{A, A+1, \ldots, B-1, B\}$ is denoted $[\![A, B]\!]$.
- Let m be a positive integer. A pattern of size m is a function ω from $[\![0, m]\!]$ to $\mathbb{Z}$ such that $\omega(0) = 0$ and $\omega(k+1) \in \{\omega(k), \omega(k)+1\}$ (see Fig. 1). The set of patterns of size m is denoted $\mathcal{P}_m$.
- If X and Y are two real numbers such that $Y > 0$ then $X \bmod Y$ is the real number such that $0 \leq X \bmod Y < Y$ and $\frac{X - X \bmod Y}{Y} \in \mathbb{Z}$.
- For $r \in \mathbb{R}$ and $E \subset \mathbb{R}^2$, $rE = \{(rx, ry) | (x, y) \in E\}$.

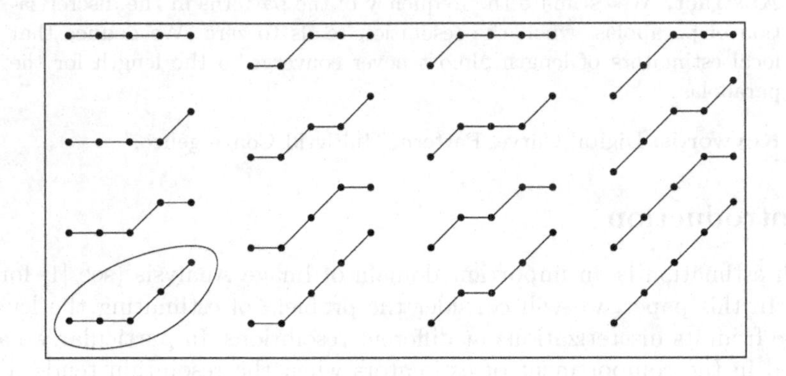

Fig. 1. The 16 patterns of size $m = 4$. Only the encircled one is not a digital segment.

3 Frequency of Patterns in Discrete Parabolas

Let $a, b \in \mathbb{R}$ such that $a < b$ and a derivable function $g : [a, b] \to \mathbb{R}$ which satisfies $0 \leq g'(x) \leq 1$ for all $x \in [a, b]$.

In all the following, for any $r > 0$ we use the notations:

- $A_r = \lceil \frac{a}{r} \rceil$, $B_r = \lfloor \frac{b}{r} \rfloor$, $N_r = B_r - A_r + 1$,
- $\mathcal{C}_r^g = r\{(X, Y) \in \mathbb{Z}^2 \mid A_r \leq X \leq B_r \text{ and } Y = \lfloor \frac{g(rX)}{r} \rfloor\}$.

The set $\mathcal{C}_r^g$ is the "naive" discretization of the graph of g at resolution r, and N_r is its number of points.

Let m be a positive integer. The pattern at position $X \in [\![A_r, B_r - m]\!]$ of size m of $\mathcal{C}_r^g$, denoted $\omega_{X,r,m}^g$ is defined by:

$$\omega_{X,r,m}^g(k) = \lfloor \frac{g(r(X+k))}{r} \rfloor - \lfloor \frac{g(rX)}{r} \rfloor.$$

The frequency of a pattern ω of size m in $\mathcal{C}_r^g$ is defined by:

$$F_r^g(\omega) = \frac{\text{card}\{X \in [\![A_r, B_r - m]\!] \mid \omega_{X,r,m}^g = \omega\}}{N_r - m}.$$

The aim of this section is the study of $F_r^g(\omega)$ for some functions g. For this we will approximate the curve by its tangents which will also be discretized, so we need some notions about digital straight lines:

For $u \in [0,1]$ and $v \in [0,1)$, let us denote $s_m^{u,v}$ the pattern of size m defined by:

$$s_m^{u,v}(k) = \lfloor uk + v \rfloor.$$

A pattern of this form is called a digital segment of size m.

For any pattern ω,

$$PI(\omega) = \{(u,v) \in [0,1]^2 \mid \lfloor uk + v \rfloor = \omega(k) \text{ for any } k \in [\![0,m]\!]\},$$

$$\operatorname{pinf}_u(\omega) = \inf\{v \mid (u,v) \in PI(\omega)\}, \tag{1}$$

$$\operatorname{psup}_u(\omega) = \sup\{v \mid (u,v) \in PI(\omega)\}, \tag{2}$$

$$\begin{aligned} FL_u(\omega) = 0 & \quad \text{if } \{v \mid (u,v) \in PI(\omega)\} = \emptyset \\ = \operatorname{psup}_u(\omega) - \operatorname{pinf}_u(\omega) & \quad \text{otherwise.} \end{aligned}$$

$PI(\omega)$ is called the preimage of ω, it is nonempty if and only if ω is a digital segment. $FL_u(\omega)$ is intuitively the frequency of the pattern ω in the discretized straight lines of slope u. (See [2,4] for more details and [5] for the generalization to slopes of planes).

In all the paper, the considered curves are parabolas corresponding to the function $g(x) = \alpha x^2$. We distinguish two cases: the case α irrational and the case α rational. In Subsection 3.1, we will see that for α irrational, the frequency $F_r^g(\omega)$ converges, when r is rational and tends to zero, to a quantity which can be expressed by using the function $x \mapsto FL_{g'(x)}(\omega)$. In Subsection 3.2, we study the case α rational, but we do not succeed to prove a similar result as in the case α irrational. Nevertheless we obtain a weaker result (The Tangent Lemma).

3.1 Parabolas of Equation $y = \alpha x^2$ with α Irrational

In this subsection we consider curves $\mathcal{C}_r^g$ for g defined on $[a,b]$ by $g(x) = \alpha x^2$ with α irrational, and $0 \le a < b \le \frac{1}{2\alpha}$. This last hypothesis is needed to have $0 \le g'(x) \le 1$ for $x \in [a,b]$.

The main result of this subsection is the following:

Theorem 1. *If $g(x) = \alpha x^2$ with $\alpha \notin \mathbb{Q}$ then for any pattern ω we have*

$$F_r^g(\omega) \xrightarrow[\substack{r \to 0 \\ r \in \mathbb{Q}}]{} \frac{1}{b-a} \int_a^b FL_{g'(x)}(\omega) \mathrm{d}x \tag{3}$$

The rest of this subsection is devoted to the proof of this Theorem.

The first needed lemma shows that the discretization of a curve and the discretization of its tangent are similar near the origin of the tangent and when the resolution tends to zero:

Lemma 1 (Tangent Lemma). *If g is defined on $[a, b]$ by $g(x) = \alpha x^2$ with α irrational, and $0 \leq a < b \leq \frac{1}{2\alpha}$ then*

$$\frac{\mathrm{card}\{X \in [\![A_r, B_r - m]\!] \mid \omega^g_{X,r,m} \neq s_m^{g'(rX),\langle \frac{g(rX)}{r} \rangle}\}}{N_r - m} \xrightarrow[r \in \mathbb{Q}]{r \to 0} 0.$$

Lemma 1 is illustrated by Fig. 2. In the last lemma we consider $r \in \mathbb{Q}$ because its proof needs the irrationality of αr.

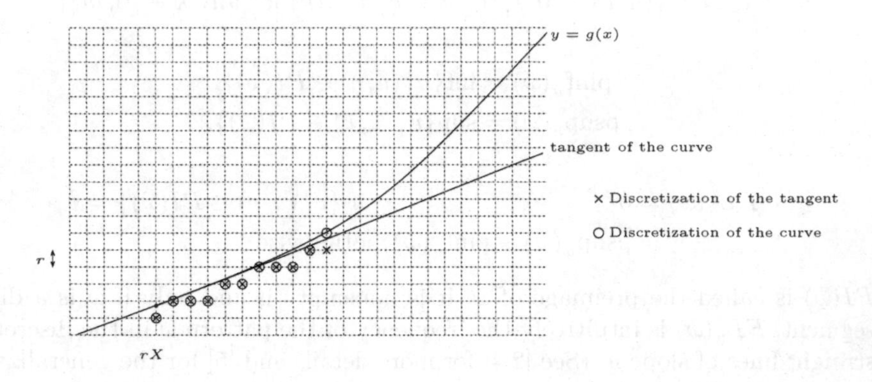

Fig. 2. Comparison of the discretization of the curve and the discretization of its tangent : here we have $\omega^g_{X,r,m} = s_m^{g'(rX),\langle \frac{g(rX)}{r} \rangle}$ for $m = 9$ but not for $m = 10$

Before starting the proof of Lemma 1 we need one more notation and two other lemmas. For $X \in [\![A_r, B_r]\!]$, we define $P_{r,k}(X)$ by:

$$P_{r,k}(X) = \frac{g(rX)}{r} + g'(rX)k = \alpha r X^2 + 2\alpha k r X = \alpha r((X + k)^2 - k^2).$$

This definition is motivated by the following lemma:

Lemma 2. *If $r < \frac{1}{\alpha m^2}$, then for any X we have $\omega^g_{X,r,m} \neq s_m^{g'(rX),\langle \frac{g(rX)}{r} \rangle}$ if and only if for all $k \in [\![0, m]\!]$ we have $\langle P_{r,k}(X) \rangle < 1 - \alpha r k^2$.*

Proof.

$$\begin{aligned}\omega^g_{X,r,m}(k) &= \lfloor \frac{g(r(X + k))}{r} \rfloor - \lfloor \frac{g(rX)}{r} \rfloor \\ &= \lfloor \frac{g(rX)}{r} + g'(rX)k + \alpha r k^2 \rfloor - \lfloor \frac{g(rX)}{r} \rfloor\end{aligned}$$

We know that $\lfloor u + v \rfloor = \lfloor u \rfloor + \lfloor v \rfloor$ iff $\langle u \rangle + \langle v \rangle < 1$.

So $\omega^g_{X,r,m}(k) = \lfloor \frac{g(rX)}{r} + g'(rX)k \rfloor - \lfloor \frac{g(rX)}{r} \rfloor$ iff $\langle \frac{g(rX)}{r} + g'(rX)k \rangle + \langle \alpha r k^2 \rangle < 1$.

With the hypothesis $r < \frac{1}{\alpha m^2}$ we have $\langle \alpha r k^2 \rangle = \alpha r k^2$. But:

$$\begin{aligned}\lfloor \frac{g(rX)}{r} + g'(rX)k \rfloor - \lfloor \frac{g(rX)}{r} \rfloor &= \lfloor \langle \frac{g(rX)}{r} \rangle + g'(rX)k \rfloor - \lfloor \langle \frac{g(rX)}{r} \rangle \rfloor \\ &= s_m^{g'(rX), \langle \frac{g(rX)}{r} \rangle}(k)\end{aligned}$$

So $\omega^g_{X,r,m} = s_m^{g'(rX), \langle \frac{g(rX)}{r} \rangle}$ iff for all $k \in [\![0, m]\!]$ we have $\langle P_{r,k}(X) \rangle < 1 - \alpha r k^2$. □

Lemma 3. *Let I be an interval of $[0,1]$. We have $T_{r,k}(I) \xrightarrow[r \in \mathbb{Q}]{r \to 0} \mu(I)$ where*

$$T_{r,k}(I) = \frac{1}{N_r} \mathrm{card}\{X \in [\![A_r, B_r]\!] \mid \langle P_{r,k}(X) \rangle \in I\}$$

and $\mu(I)$ is the usual length of I.

The proof of this lemma uses Weyl's argument, as in the proof of Theorem 1 of [2] or Appendix A.1 of [5], but extended to the quadratic case following the same ideas as in [6, p6-7]. It is given in Appendix A.1 of [3].

Proof of Lemma 1. We recall that $\omega^g_{X,r,m} = s_m^{g'(rX), \langle \frac{g(rX)}{r} \rangle}$ iff for all $k \in [\![0, m]\!]$ we have $\langle P_{r,k}(X) \rangle < 1 - \alpha r k^2$ so:

$$\frac{\mathrm{card}\{X \in [\![A_r, B_r - m]\!] \mid \omega^g_{X,r,m} \neq s_m^{g'(rX), \langle \frac{g(rX)}{r} \rangle}\}}{N_r - m} \leq \frac{N_r}{N_r - m} \max_{k=0}^{m} T_{r,k}(I_{r,k})$$

where $I_{r,k} = [1 - \alpha r k^2, 1)$, so it is sufficient to show that $T_{r,k}(I_{r,k}) \xrightarrow[r \in \mathbb{Q}]{r \to 0} 0$ for any $k \in [\![0, m]\!]$. Let $\varepsilon > 0$, there exists $R_1 > 0$ such that for any $r < R_1$ we have $\alpha r k^2 < \frac{\varepsilon}{2}$. We know by Lemma 3 that there exists $R_2 > 0$ such that for any $r < R_2$ we have $T_{r,k}(I_{R_0,k}) \leq \mu(I_{R_0,k}) + \frac{\varepsilon}{2}$. If r is such that $r < \min(R_1, R_2)$, we have:

$$\begin{aligned}S_{r,k}(I_{r,k}) &\leq S_{r,k}(I_{R_1,k}) \\ &\leq \mu(I_{R_1,k}) + \frac{\varepsilon}{2} \\ &= \alpha r k^2 + \frac{\varepsilon}{2} \\ &\leq \frac{\varepsilon}{2} + \frac{\varepsilon}{2} = \varepsilon\end{aligned}$$

This finishes the proof of Lemma 1. □

Sketch of Proof of Theorem 1. We present here the main ideas of the proof of Theorem 1. The detailed proof is in Appendix A.2 of [3].

By using Lemma 1 we know that $F_r^g(\omega)$ has the same limit as r tends to zero as:

$$G_r^g(\omega) = \frac{\operatorname{card}\{X \in [\![A_r, B_r - m]\!] \mid s_m^{g'(rX),\langle \frac{g(rX)}{r}\rangle} = \omega\}}{N_r - m}.$$

But $s_m^{x,y} = \omega$ is equivalent to $(x, \langle y\rangle) \in PI(\omega)$ so:

$$G_r^g(\omega) = \frac{\operatorname{card}\{X \in [\![A_r, B_r - m]\!] \mid (g'(rX), \langle \frac{g(rX)}{r}\rangle) \in PI(\omega)\}}{N_r - m}$$

which has the same limit as $H_r(PI(\omega))$ where

$$H_r(E) = \frac{\operatorname{card}\{X \in [\![A_r, B_r]\!] \mid (g'(rX), \langle \frac{g(rX)}{r}\rangle) \in E\}}{B_r - A_r + 1}.$$

By applying Lemma 3 with $k = 0$ to the piece of the curve $y = g(x)$ restricted to the domain $g'^{-1}(\alpha_1) \leq x \leq g'^{-1}(\alpha_2)$, we can prove:

$$H_r([\alpha_1, \alpha_2) \times I) \xrightarrow[r \in \mathbb{Q}]{r \to 0} \frac{g'^{-1}(\alpha_2) - g'^{-1}(\alpha_1)}{b - a} \mu(I)$$

So by approximating $PI(\omega)$ as the union of rectangles

$$\bigcup_{i=1}^{n} [y_{i-1}, y_i) \times [\operatorname{pinf}_{y_i}(\omega), \operatorname{psup}_{y_i}(\omega))$$

we approximate $H_r(PI(\omega))$ by:

$$\sum_{i=1}^{n} \frac{g'^{-1}(y_i) - g'^{-1}(y_{i-1})}{b - a} FL_{y_i}(\omega)$$

which is a Riemann sum for $\int_a^b FL_{g'(x)}(\omega) dx$. □

Corollary 1. *If $g(x) = \alpha x^2$ with $\alpha \notin \mathbb{Q}$, then for any pattern ω which is not a digital segment we have*

$$F_r^g(\omega) \xrightarrow[r \in \mathbb{Q}]{r \to 0} 0.$$

Numerical Application: We illustrate Theorem 1 with an example. Consider the curve $\mathcal{C}$ defined $y = g(x) = \frac{1}{\sqrt{2}} x^2$ for x between $a = 0$ and $b = \frac{1}{\sqrt{2}}$, and the pattern ω of size $m = 3$ defined by $(\omega(0), \omega(1), \omega(2), \omega(3)) = (0, 1, 2, 2)$. We will compute the limit of the frequency of ω when the resolution tends to zero.

First we can compute easily $FL_\alpha(\omega)$ because $\alpha \mapsto FL_\alpha(\omega)$ is a continuous function which is affine between two m-Farey numbers (see [4]). So we deduce:

$$\begin{aligned} FL_\alpha(\omega) &= 0 && \text{if } \alpha \in [0, \tfrac{1}{2}] \\ &= 2\alpha - 1 && \text{if } \alpha \in [\tfrac{1}{2}, \tfrac{2}{3}] \\ &= 1 - \alpha && \text{if } \alpha \in [\tfrac{2}{3}, 1] \end{aligned}$$

Theorem 1 proves that:

$$\begin{aligned} F_r^g(\omega) \xrightarrow[r\in\mathbb{Q}]{r\to 0} \frac{1}{b-a}\int_a^b FL_{g'(x)}(\omega)\mathrm{d}x \\ =\sqrt{2}\int_0^{\frac{1}{\sqrt{2}}} FL_{\sqrt{2}x}(\omega)\mathrm{d}x \\ =\sqrt{2}\left(\int_{\frac{1}{2\sqrt{2}}}^{\frac{2}{3\sqrt{2}}}(2\sqrt{2}x-1)\mathrm{d}x + \int_{\frac{2}{3\sqrt{2}}}^{\frac{1}{\sqrt{2}}}(1-\sqrt{2}x)\mathrm{d}x\right) = \frac{1}{12} \end{aligned}$$

3.2 Parabolas of Equation $y = \alpha x^2$ with α Rational

Now we are interested in the case where α is rational. Theorem 1 can be generalized to α rational and to the irrational resolutions r because only the irrationality of αr is used in the proof of this theorem.

On the contrary, in the case α rational and rational resolutions, the only result we will prove in this subsection is about the Tangent Lemma. Moreover, we must impose some restrictions about the resolution r and the interval $[a, b]$ of definition of the parabola: $r\alpha = \frac{1}{p}$ where p is a prime number ; $a = 0$ and $b = \frac{1}{2\alpha}$. In all this subsection we suppose that these conditions are satisfied. Actually, we do not succeed to prove the Tangent Lemma in the general case for α rational.

Let $P_{r,k}(X) = r\alpha((X+k)^2 - k^2)$. We will prove that for any interval $I \subset [0,1]$

$$\lim_{\substack{r\to 0\\ \frac{1}{r\alpha}\text{ is prime}}} \frac{\mathrm{card}\{X \in [\![A_r, B_r - k]\!] \mid \langle P_{r,k}(x)\rangle \in I\}}{N_r - k} = \mu(I)$$

Definition 1. *Let p be a prime number and a be an integer number, we define the Legendre symbol $(\frac{a}{p})$ of a relatively to p by*

$$(\frac{a}{p}) = \begin{cases} 0 & \text{if } p \text{ divides } a \\ 1 & \text{if } a \text{ there exists } t\in\mathbb{Z} \text{ such that } a \bmod p = t^2 \bmod p \\ -1 & \text{otherwise.} \end{cases}$$

Property 1. (Pólya-Vinogradov inequality [7], [8, Chap. 23]) Let p be a prime number and M and N two positive integers then

$$|\sum_{n=M}^{M+N}(\frac{n}{p})| < \sqrt{p}\log(p)$$

Corollary 2. *Let $J = [\![M, M+N]\!]$ be an integer interval. Then*

$$|\frac{\mathrm{card}(J)}{2} - \mathrm{card}\{y \in I \mid (\frac{y}{p}) = 1\}| < \sqrt{p}\log(p).$$

Lemma 4. *Let α be a rational number and assume that $a = 0$ and $b = \frac{1}{2\alpha}$. Then, for any interval $I \subset [0,1]$*

$$\lim_{r \to 0} \frac{\mathrm{card}\{X \in [\![A_r, B_r - k]\!] \mid \langle P_{r,k}(X) \rangle \in I\}}{N_r - k} = \mu(I)$$

where the limit is taken on the r such that $r > 0$ and $r\alpha = \frac{1}{p}$ where p is a prime number.

Proof. The function $X \mapsto X^2 \bmod p$ from $[\![1, \frac{p-1}{2}]\!]$ to $\{Y \mid (\frac{Y}{p}) = 1\}$ is a bijection. Put $H_r = \dfrac{\mathrm{card}\{X \in [\![A_r, B_r - k]\!] \mid \langle P_{r,k}(x) \rangle \in I\}}{N_r}$. Then

$$\begin{aligned} H_r &= \frac{\mathrm{card}\{X \in [\![k, \lfloor \frac{p}{2} \rfloor]\!] \mid \langle \frac{X^2 - k^2}{p} \rangle \in I\}}{N_r - k} \\ &= \frac{2\mathrm{card}\{X \in [\![k, \frac{p-1}{2}]\!] \mid \frac{X^2 \bmod p - k^2}{p} \in I\}}{p + 1 - 2k} \end{aligned} \tag{4}$$

Thus

$$\mathrm{card}\{X \in [\![1, \frac{p-1}{2}]\!] \mid X^2 \bmod p \in J\} = \mathrm{card}\{Y \in J \mid (\frac{Y}{p}) = 1\} \text{ where } J = pI + k^2.$$

So $|\mathrm{card}\{X \in [\![k, \frac{p-1}{2}]\!] \mid X^2 \bmod p \in J\} - \mathrm{card}\{Y \in J \mid (\frac{Y}{p}) = 1\}| < k$.
By using Pólya-Vinogradov inequality we have:

$$\left|\frac{\mathrm{card}(J)}{2} - \mathrm{card}\{X \in [\![1, \frac{p-1}{2}]\!] \mid X^2 \bmod p \in J\}\right| < \sqrt{p}\log(p)$$

So, $|\frac{\mathrm{card}(J)}{2} - \mathrm{card}\{X \in [\![k, \frac{p-1}{2}]\!] \mid X^2 \bmod p \in J\}| < \sqrt{p}\log(p) + k$.
By (4) we have

$$\left|\frac{2}{p+1-2k}\frac{\mathrm{card}(J)}{2} - \frac{2}{p+1-2k}\mathrm{card}\{X \in [\![k, \frac{p-1}{2}]\!] \mid X^2 \bmod p \in J\}\right| < \frac{2\sqrt{p}\log(p) + 2k}{p+1-2k}$$

Thus, $\lim_{r \to 0} \dfrac{\mathrm{card}\{X \in [\![A_r, B_r - k]\!] \mid \langle P_{r,k}(X) \rangle \in I\}}{N_r - k} = \mu(I)$ □

Theorem 2 (Tangent Lemma for rational slopes and special rational resolutions). *For any rational $\alpha > 0$ and any a, b such that $0 \leq a < b \leq \frac{1}{2\alpha}$ we have*

$$\frac{\mathrm{card}\{X \in [\![A_r, B_r - m]\!] \mid \omega^g_{X,r,m} \neq s_m^{g'(rX), \langle \frac{g(rX)}{r} \rangle}\}}{N_r - m} \xrightarrow[\frac{1}{r\alpha} \text{ is prime}]{r \to 0} 0$$

Proof. The case where $a = 0$, $b = \frac{1}{2\alpha}$ can be proved exactly in the same way as for Lemma 1 by using Lemma 4 instead of Lemma 3. Consider the general case $[a,b] \subset [0, \frac{1}{2\alpha}]$. Then we know that:

$$\frac{\mathrm{card}\{X \in [\![A_r, B_r - m]\!] \,|\, \omega^g_{X,r,m} \neq s_m^{g'(rX),\langle \frac{g(rX)}{r} \rangle}\}}{N_r - m}$$

$$\leq \frac{\lfloor \frac{1}{2\alpha r} \rfloor + 1 - m}{N_r - m} \cdot \frac{\mathrm{card}\{X \in [\![0, \lfloor \frac{b}{r} \rfloor - m]\!] \,|\, \omega^g_{X,r,m} \neq s_m^{g'(rX),\langle \frac{g(rX)}{r} \rangle}\}}{\lfloor \frac{1}{2\alpha r} \rfloor + 1 - m}$$

$$\xrightarrow[\frac{1}{r\alpha} \text{ is prime}]{r \to 0} \frac{\frac{1}{2\alpha}}{b-a} \cdot 0 = 0 \qquad \text{because of the case } a = 0,\ b = \frac{1}{2\alpha}. \qquad \square$$

Unfortunately we do not successfully generalize Theorem 1 to rational α and some rational resolutions even if experimentally Theorem 1 seems to be true in all the cases.

4 Application to Local Estimators of Length

A local estimator is given by a weight function p from the set $\mathcal{P}_m$ of patterns of size m to $\mathbb{R}$. The estimated length of the curve $y = g(x)$ where $g : [a,b] \to \mathbb{R}$ at resolution r is given by:

$$l(p,g,r) = r \sum_{k=0}^{\lfloor \frac{B_r - m - A_r}{m} \rfloor} p(\omega^g_{X,r,m}(A_r + km)).$$

Theorem 3. *Let a, b such that $0 \leq a < b$. The length estimated by a local estimator of a parabola $y = \alpha x^2$, $x \in [a,b]$, ($\alpha \leq \frac{1}{2b}$) converges when the resolution r is rational and tends to zero, to the length of the parabola* only *for a finite number of irrational numbers α.*

Sketch of proof. For this proof we use the notation $[f(x)]_a^b = f(b) - f(a)$.
It is easy to see that:

$$l(p,g,r) - \frac{(b-a)}{m} \sum_{\omega \in \mathcal{P}_m} p(\omega) F'^g_r(\omega) \xrightarrow[r \to 0]{} 0$$

where

$$F'^g_r(\omega) = \frac{\mathrm{card}\{X \in [\![A_r, B_r - m]\!] \cap (A_r + m\mathbb{Z}) \,|\, \omega^g_{X,r,m} = \omega\}}{\lfloor \frac{B_r - m - A_r}{m} \rfloor}.$$

Consider again the curve defined by $g(x) = \alpha x^2$ for $\alpha \notin \mathbb{Q}$. The proof of Theorem 1 can be extended to prove that:

$$F'^g_r(\omega) \xrightarrow[r \in \mathbb{Q}]{r \to 0} \frac{1}{b-a} \int_a^b FL_{g'(x)}(\omega) \mathrm{d}x.$$

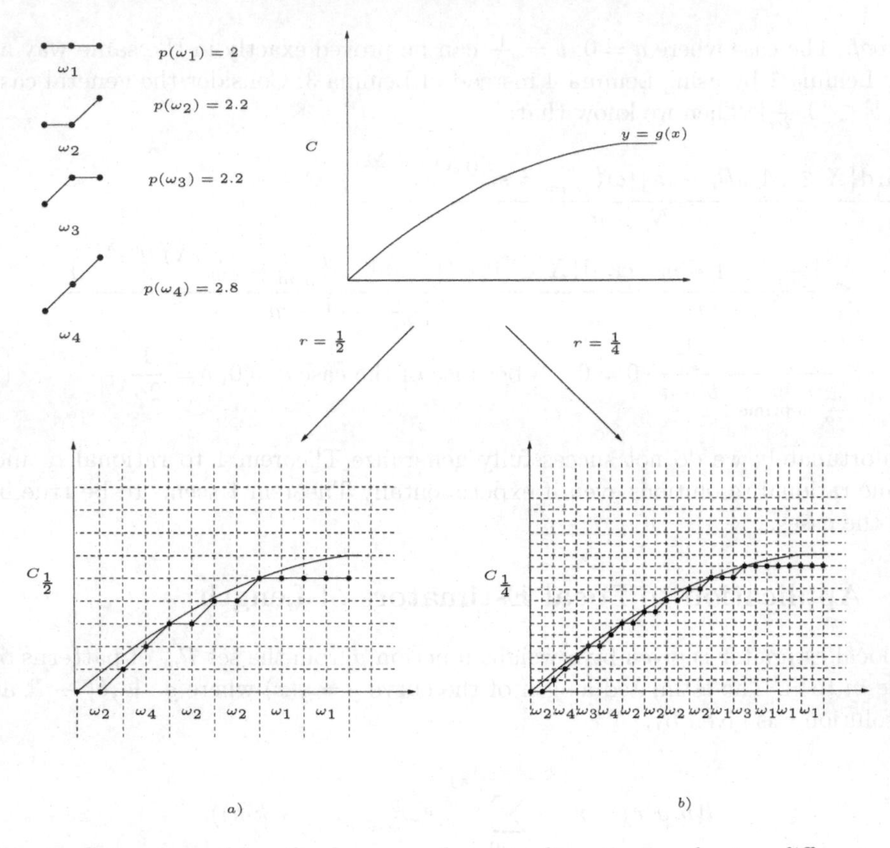

Fig. 3. Estimation of length of a curve from its discretization for two different resolutions: a) $l(p, g, \frac{1}{2}) = \frac{1}{2}(2p(\omega_1) + 3p(\omega_2) + 1p(\omega_4)) = 6.7$, b) $l(p, g, \frac{1}{4}) = \frac{1}{4}(4p(\omega_1) + 5p(\omega_2) + 2p(\omega_3) + 2p(\omega_4)) = 7.1$

So,

$$l(p, g, r) \xrightarrow[r \in \mathbb{Q}]{r \to 0} \frac{1}{m} \sum_{\omega \in \mathcal{P}_m} p(\omega) \int_a^b FL_{g'(x)}(\omega) \mathrm{d}x. \tag{5}$$

We know that $x \mapsto FL_x(\omega)$ is piecesewisely affine ([4]). From this property we deduce that we can partition the interval $[0, \frac{1}{2b}]$ in a finite number of intervals $(I_k)_{0 \le k \le n}$ such that $L_{est}(\alpha) = \lim_{\substack{r \to 0 \\ r \in \mathbb{Q}}} l(p, g, r)$ is of the form $\frac{A}{\alpha} + B\alpha + C$ on each interval I_k. (See Appendix A.3 of [3] for the details).

Let $L_{real}(\alpha)$ be the length of the parabola $\{(x, \alpha x^2) \mid x \in [a, b]\}$. We have:

$$L_{real}(\alpha) = \int_a^b \sqrt{1 + (2\alpha x)^2} \mathrm{d}x = \left[\frac{x\sqrt{1 + (2\alpha x)^2}}{2} + \frac{\arg\sinh(2\alpha x)}{4\alpha} \right]_a^b$$

Suppose that $L_{real}(\alpha) = L_{est}(\alpha)$ for an infinite number of irrational numbers α, then there exists an interval I_k of the previous partition of $[0, \frac{1}{2b}]$ such that

$L_{real}(\alpha) = L_{est}(\alpha)$ for an infinite number of irrational numbers $\alpha \in I_k$. On I_k we know that $L_{est}(\alpha)$ has the form $\frac{A}{\alpha} + B\alpha + C$. The functions $\alpha \mapsto \alpha L_{real}(\alpha)$ and $\alpha \mapsto \alpha(\frac{A}{\alpha} + B\alpha + C)$ are holomorphic in a open set of $\mathbb{C}$ containing $[0, \frac{1}{2b}]$ and are equal for an infinite number of $\alpha \in I_k \subset [0, \frac{1}{2b}]$. So by Theorem on the zeros of holomorphic functions [9, Cha. 10] they are equal on $[0, \frac{1}{2b}]$. So:

$$\alpha L_{real}(\alpha) = A + B\alpha^2 + C\alpha \qquad \text{for all } \alpha \in [0, \frac{1}{2b}]$$

We have:

$$\begin{aligned}\frac{\partial(\alpha L_{real}(\alpha))}{\partial \alpha} &= b\sqrt{1+(2\alpha b)^2} - a\sqrt{1+(2\alpha a)^2} \\ &= b - a + 2(b^3 - a^3)\alpha^2 + o(\alpha^2) \qquad \text{when } \alpha \to 0\end{aligned}$$

But $\frac{\partial(A+B\alpha^2+C\alpha)}{\partial\alpha} = 2B\alpha + C$, so $2(b^3 - a^3) = 0$ which is impossible if $b > a$. So the hypothesis that $L_{real}(\alpha) = L_{est}(\alpha)$ for an infinite number of irrational α is absurd. □

Corollary 3. *Let a, b such that $0 \leq a < b$. The length estimated by a local estimator of a parabola $y = \alpha x^2$, $x \in [a, b]$, does* not *converge when the resolution is rational and tends to zero, to the length of the curve for almost all $\alpha \in [0, \frac{1}{2b}]$.*

Numerical application: Again, we take the curve $y = g(x) = \frac{1}{\sqrt{2}}x^2$ for x between $a = 0$ and $b = \frac{1}{\sqrt{2}}$. Suppose that we consider the local estimator Chamfer 5-7-11 ([10]) with $m = 2$, $p(000) = 2$, $p(001) = p(011) = \frac{22}{10}$, $p(012) = \frac{28}{10}$. With Equation (5), we can prove that the estimation of the length of the parabola given by this local estimator converges to $L_{est} = \frac{23}{40}\sqrt{2} \approx 0.813172$, this limit is different from the length of the parabola which is $\frac{1}{2} + \frac{1}{4}\sqrt{2}\log(1+\sqrt{2}) \approx 0.811612$.

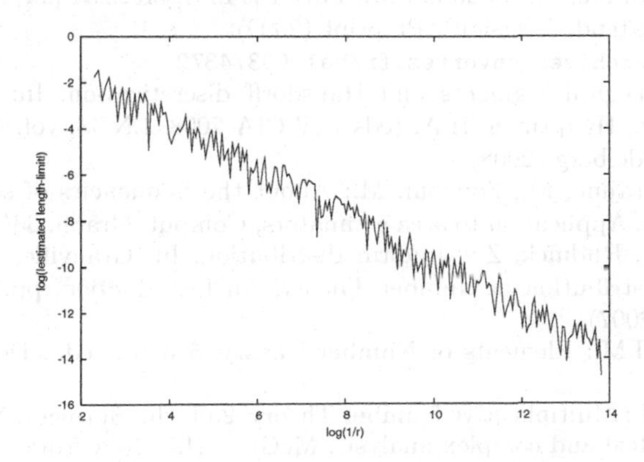

Fig. 4. Figure showing Convergence Speed of the Chamfer 5-7-11 to its limits for the parabola $y = \frac{1}{\sqrt{2}}x^2, \qquad 0 \leq x \leq \frac{1}{\sqrt{2}}$

Moreover Figure 4 shows how the length given by the estimator converges to its limit when the resolution tends to zero. It seems on this example that $l(p, g, r) - L_{est} = O(r)$.

5 Conclusion and Perspectives

In this paper, we have proved some local properties of discretizations of parabolas: First we show that locally discretization of parabola and discretization of its tangent often coincide (Tangent Lemma: Lemmas 1 and Theorem 2). In particular, asymptotically, the local patterns of a discretized parabola are digital segments. From this, we also give an explicit formula for the limit of frequency of a pattern of a parabola when the resolution tends to zero (Theorem 1). This has the important consequence that we can know to what tend local estimators of length for the parabolas, moreover it can be proved that this limit is often different from the length of the curve.

This work mainly brings two perspectives:

- The extension of Formula (3), which gives the limit of the frequency of pattern when the resolution tends to zero, to more general curves, in particular to the curves $y = P(x)$ when P is a polynomial of degree greater than 2.
- The application of this work for recognition of curve by just looking at patterns. For example, if the frequencies of patterns of a curve does not satisfy Theorem 1 then it is not a parabola of equation $y = \alpha x^2$.

References

1. Coeurjolly, D., Klette, R.: A comparative evaluation of length estimators. IEEE Trans. Pattern Anal. Mach. Intell. 26(2), 252–258 (2004)
2. Tajine, M., Daurat, A.: On local definitions of length of digital curves. In: Nyström, I., Sanniti di Baja, G., Svensson, S. (eds.) DGCI 2003. LNCS, vol. 2886, pp. 114–123. Springer, Heidelberg (2003)
3. Daurat, A., Tajine, M., Zouaoui, M.: Patterns in discretized parabolas and length estimation (extended version). Preprint (2009), `http://hal.archives-ouvertes.fr/hal-00374372`
4. Tajine, M.: Digital segments and Hausdorff discretization. In: Brimkov, V.E., Barneva, R.P., Hauptman, H.A. (eds.) IWCIA 2008. LNCS, vol. 4958, pp. 75–86. Springer, Heidelberg (2008)
5. Daurat, A., Tajine, M., Zouaoui, M.: About the frequencies of some patterns in digital planes. Application to area estimators, Comput. Graph. 33(1), 11–20 (2008)
6. Granville, A., Rudnick, Z.: Uniform distribution. In: Granville, A., Rudnick, Z. (eds.) Equidistribution in Number Theory, An Introduction, pp. 1–13. Springer, Heidelberg (2007)
7. Vinogradov, I.M.: Elements of Number Theory, 5th rev. edn. Dover, Cambridge (1954)
8. Davenport, H.: Multiplicative Number Theory, 2nd edn. Springer, New York (1980)
9. Rudin, W.: Real and complex analysis. McGraw-Hill, New York (1966)
10. Borgefors, G.: Distance transformations in digital images. Computer Vision, Graphics, and Image Processing 34(3), 344–371 (1986)

Universal Spaces for $(k, \overline{k})$–Surfaces*

J.C. Ciria[1], E. Domínguez[1], A.R. Francés[1], and A. Quintero[2]

[1] Dpto. de Informática e Ingeniería de Sistemas, Facultad de Ciencias, Universidad de Zaragoza, E-50009 – Zaragoza, Spain
[2] Dpto. de Geometría y Topología, Facultad de Matemáticas, Universidad de Sevilla, Apto. 1160, E-41080 – Sevilla, Spain

Introduction

In the graph–theoretical approach to Digital Topology, the search for a definition of digital surfaces as subsets of voxels is still a work in progress since it was started in the early 1980's. Despite the interest of the applications in which it is involved (ranging from visualization to image segmentation and graphics), there is not yet a well established general notion of digital surface that naturally extends to higher dimensions (see [5] for a proposal). The fact is that, after the first definition of surface, proposed by Morgenthaler [13] for $\mathbb{Z}^3$ with the usual adjacency pairs $(26, 6)$ and $(6, 26)$, each new contribution [10,4,9], either increasing the number of surfaces or extendeding the definition to other adjacencies, has still left out some objects considered as surfaces for practical purposes [12].

In this paper we find, for each adjacency pair $(k, \overline{k})$, $k, \overline{k} \in \{6, 18, 26\}$ and $(k, \overline{k}) \neq (6, 6)$, a homogeneous $(k, \overline{k})$-connected digital space whose set of digital surfaces is larger than any of those quoted above; moreover, it is the largest set of surfaces within that class of digital spaces as defined in [3]. This is an extension of a previous result for the $(26, 6)$-adjacency in [7].

1 A Framework for Digital Topology

This section summarizes the framework for Digital Topology we introduced in [3]. In this approach a *digital space* is a pair (K, f), where K is a polyhedral complex, representing the spatial layout of voxels, and f is a *lighting function* from which we associate to each digital image an Euclidean polyhedron, called its continuous analogue, that intends to be a continuous interpretation of the image.

In this paper we will only consider spaces of the form (R^3, f), where the complex R^3 is determined by the unit cubes in the Euclidean space $\mathbb{R}^3$ centered at points of integer coordinates. Each 3-cell in R^3 represents a voxel, so that a digital object displayed in an image is a subset of the set $\text{cell}_3(R^3)$ of 3-cells in R^3; while the lower dimensional cells in R^3 (actually, d-cubes, $0 \leq d < 3$) are used to describe how the voxels could be linked to each other. Notice that each d-cell $\sigma \in R^3$ can be associated to its center $c(\sigma)$. In particular, if $\dim \sigma = 3$

* This work has been partially supported by the project MTM2007–65726 MICINN, Spain.

S. Brlek, C. Reutenauer, and X. Provençal (Eds.): DGCI 2009, LNCS 5810, pp. 385–396, 2009.

then $c(\sigma) \in \mathbb{Z}^3$, so that every digital object O in R^3 can be naturally identified with a subset of the discrete space $\mathbb{Z}^3$. Henceforth we shall use this identification without further comment.

Lighting functions are maps of the form $f : \mathcal{P}(\text{cell}_3(R^3)) \times R^3 \to \{0,1\}$, where $\mathcal{P}(\text{cell}_3(R^3))$ stands for the family of all subsets of $\text{cell}_3(R^3)$; i.e, all digital objects. Each of these maps may be regarded as a "face membership rule", in the sense of Kovalevsky [11], that assigns to each digital object O the set of cells $f_O = \{\alpha \in R^3\, ;\, f(O,\alpha) = 1\}$. This set yields a continuous analogue as the counterpart of O in ordinary topology. Namely, the *continuous analogue* of O is the polyhedron $|\mathcal{A}^f_O| \subseteq \mathbb{R}^3$ triangulated by the subcomplex of the first derived subdivision of R^3, $\mathcal{A}^f_O$, consisting of all simplexes whose vertices are centers $c(\sigma)$ of cells $\sigma \in f_O$.[1] However, to avoid continuous analogues which are contrary to our usual topological intuition, lighting functions must satisfy the five properties below. We need some more notation to introduce them.

As usual, given two cells $\gamma, \sigma \in R^3$ we write $\gamma \leq \sigma$ if γ is a face of σ, and $\gamma < \sigma$ if in addition $\gamma \neq \sigma$. The interior of a cell σ is the set $\mathring{\sigma} = \sigma - \partial\sigma$, where $\partial\sigma = \cup\{\gamma\, ;\, \gamma < \sigma\}$ stands for the boundary of σ. We refer to [14] for further notions on polyhedral topology.

Next, we introduce two types of neighbourhoods of a cell $\alpha \in R^3$ in a given digital object $O \subseteq \text{cell}_3(R^3)$: the *star of α in O* which is the set $\text{st}_3(\alpha; O) = \{\sigma \in O\, ;\, \alpha \leq \sigma\}$ of voxels in O having α as a face, and the set $\text{st}^*_3(\alpha; O) = \{\sigma \in O\, ;\, \alpha \cap \sigma \neq \emptyset\}$ called the *extended star of α in O*. Finally, the *support* of O is the set $\text{supp}(O)$ of cells of R^3 (not necessarily voxels) that are the intersection of 3-cells in O; that is, $\alpha \in \text{supp}(O)$ if and only if $\alpha = \cap\{\sigma\, ;\, \sigma \in \text{st}_3(\alpha; O)\}$. To ease the writing, we use the following notation: $\text{st}_3(\alpha; R^3) = \text{st}_3(\alpha; \text{cell}_3(R^3))$ and $\text{st}^*_3(\alpha; R^3) = \text{st}^*_3(\alpha; \text{cell}_3(R^3))$.

A *lighting function* on R^3 is a map $f : \mathcal{P}(\text{cell}_3(R^3)) \times R^3 \to \{0,1\}$ for which the following five axioms hold for all $O \in \mathcal{P}(\text{cell}_3(R^3))$ and $\alpha \in R^3$; see [2,3].

(1) *object axiom*: if $\alpha \in O$ then $f(O,\alpha) = 1$;
(2) *support axiom*: if $\alpha \notin \text{supp}(O)$ then $f(O,\alpha) = 0$;
(3) *weak monotone axiom*: $f(O,\alpha) \leq f(\text{cell}_3(R^3),\alpha)$;
(4) *weak local axiom*: $f(O,\alpha) = f(\text{st}^*_3(\alpha; O),\alpha)$; and,
(5) *complement axiom*: if $O' \subseteq O \subseteq \text{cell}_3(R^3)$ and $\alpha \in R^3$ are such that $\text{st}_3(\alpha; O) = \text{st}_3(\alpha; O')$, $f(O',\alpha) = 0$ and $f(O,\alpha) = 1$, then the set $\alpha(O',O) = \cup\{\mathring{\omega}\, ;\, \omega < \alpha, f(O',\omega) = 0,\ f(O,\omega) = 1\} \subseteq \partial\alpha$ is non-empty and connected.

If $f(O,\alpha) = 1$ we say that f *lights* the cell α for the object O, otherwise f *vanishes* on α for O.

Example 1. The following are lighting functions on R^3: (a) $f_{\max}(O,\alpha) = 1$ if and only if $\alpha \in \text{supp}(O)$; (b) $g(O,\alpha) = 1$ if and only if $\text{st}_3(\alpha; R^3) \subseteq O$.

A digital space (R^3, f) is *Euclidean* if its continuous analogue is $\mathbb{R}^3$ (that is, if $f(\text{cell}_3(R^3),\alpha) = 1$ for each cell $\alpha \in R^3$) and, in addition, it is *homogeneous* in

[1] We often drop the "f" from the notation and also write $\mathcal{A}_{R^3}$ instead $\mathcal{A}^f_{\text{cell}_3(R^3)}$.

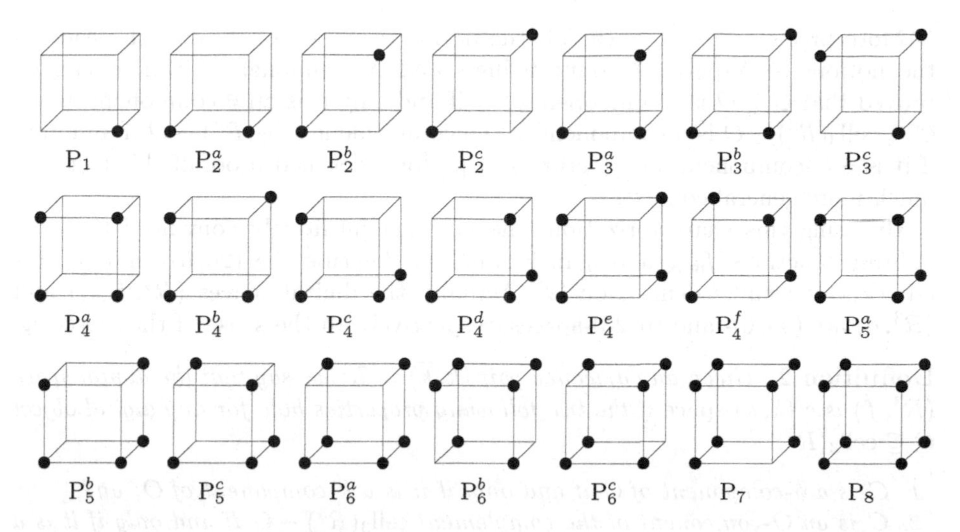

Fig. 1. Non-empty canonical 0-patterns around a vertex. For $d = 1, 2$, the list of d-patterns is longer because $\mathrm{st}_3^*(\alpha; O)$ may contain up to 12 (respectively, 18) voxels.

the sense that the continuous analogue of any object O does not change under isometries $\varphi : \mathbb{R}^3 \to \mathbb{R}^3$ preserving $\mathbb{Z}^3$ (i.e, $\varphi(|\mathcal{A}_O^f|) = |\mathcal{A}_{\varphi(O)}^f|$). It is obvious that Axiom 3 above is redundant for Euclidean spaces. Moreover, homogeneity allows us to rewrite Axiom 4 in terms of a minimal family of objects, called *(canonical) d-patterns*, consisting of all subsets of $\mathrm{st}_3^*(\alpha_d; R^3)$ which are distinct up to rotations or symmetries, where α_d is a fixed d-cell for $0 \leq d < 3$; see Fig. 1. More precisely, Axiom 4 becomes $f(O, \alpha) = f(\mathrm{P}(O, \alpha), \alpha_d)$, where $\mathrm{P}(O, \alpha) \subseteq \mathrm{st}_3^*(\alpha_d; R^3)$, called the *pattern of O in* α, is the unique canonical d-pattern for which there exists a isometry such that $\varphi(\mathrm{st}_3^*(\alpha; O)) = \mathrm{P}(O, \alpha)$.

2 $(k, \overline{k})$-Connected Digital Spaces

In this section we assume that (R^3, f) is Euclidean. Our aim is to find some necessary conditions on the lighting function f so that the space (R^3, f) provides us with the $(k, \overline{k})$-connectedness, as usually defined on $\mathbb{Z}^3$ by mean of adjacency pairs, $k, \overline{k} \in \{6, 18, 26\}$. For this, we first recall that a digital object O in (R^3, f) is said to be connected if its continuous analogue $|\mathcal{A}_O|$ is connected. On the other hand, the set of voxels $\mathrm{cell}_3(R^3) - O$, regarded as the complement of O, is declared to be O-connected if $|\mathcal{A}_{R^3}| - |\mathcal{A}_O|$ is connected. Moreover, we call $C \subseteq \mathrm{cell}_3(R^3)$ a component of O ($\mathrm{cell}_3(R^3) - O$) if it consists of all voxels σ whose centers $c(\sigma)$ belong to a component of $|\mathcal{A}_O|$ ($|\mathcal{A}_{R^3}| - |\mathcal{A}_O|$, respectively). These notions of connectedness are characterized by the following notions of adjacency

Definition 1. *Two cells* $\sigma, \tau \in O$ *are* $\emptyset$-adjacent *iff* $f(O, \alpha) = 1$ *for some common face* $\alpha \leq \sigma \cap \tau$. *Moreover,* $\sigma, \tau \in \mathrm{cell}_3(R^3) - O$ *are* O-adjacent *iff* $f(O, \alpha) = 0$ *for some* $\alpha \leq \sigma \cap \tau$.

More precisely, for $X \in \{\emptyset, O\}$, the notion of X-adjacency directly leads to the notions of X-path, X-connectedness and X-component. Then, it can be proved that $C \subseteq O$ is a component of O if and only it is an $\emptyset$-component, while $C \subseteq \text{cell}_3(R^3) - O$ is a component of the complement $\text{cell}_3(R^3) - O$ if and only if it is a O-component. See Section 4 in [2] for a detailed proof of this fact in a much more general context.

By using this characterization, one may get intuitively convinced that the lighting functions $f_{\max}$ and g in Example 1 describe the $(26, 6)$- and $(6, 26)$-adjacencies usually defined on $\mathbb{Z}^3$. Actually the digital spaces $(R^3, f_{\max})$ and (R^3, g) are $(26, 6)$- and $(6, 26)$-spaces, respectively, in the sense of the following

Definition 2. *Given an adjacency pair $(k, \overline{k})$ in $\mathbb{Z}^3$ we say that the digital space (R^3, f) is a $(k, \overline{k})$-*space *if the two following properties hold for any digital object $O \subseteq \text{cell}_3(R^3)$:*

1. *C is a $\emptyset$-component of O if and only if it is a k-component of O; and,*
2. *C is an O-component of the complement $\text{cell}_3(R^3) - O$ if and only if it is a $\overline{k}$-component.*

From now on, we assume that (R^3, f) is a Euclidean $(k, \overline{k})$-space, $k, \overline{k} \in \{6, 18, 26\}$ and $(k, \overline{k}) \neq (6, 6)$. Examples of these spaces can be found in [1,3,6].

Proposition 1. *Let $\text{P}(O, \alpha)$ the pattern of a digital object $O \subseteq \text{cell}_3(R^3)$ in a vertex $\alpha \in R^3$. The following properties hold:*

1. *If $\text{P}(O, \alpha) = \text{P}_8$ then $f(O, \alpha) = 1$.*
2. *$f(O, \alpha) = 0$ whenever $\text{P}(O, \alpha) \in \{\text{P}_0, \text{P}_1, \text{P}_2^a, \text{P}_2^b, \text{P}_3^a, \text{P}_4^a\}$.*
3. *If $\text{P}(O, \alpha) = \text{P}_2^c$, then $f(O, \alpha) = 1$ iff $k = 26$.*
4. *For $\overline{k} = 6$, $f(O, \alpha) = 1$ whenever $\text{P}(O, \alpha) \in X = \{\text{P}_3^c, \text{P}_4^b, \text{P}_4^e, \text{P}_4^f, \text{P}_5^b, \text{P}_5^c, \text{P}_6^b\}$.*
5. *If $\text{P}(O, \alpha) = \text{P}_6^c$, then $f(O, \alpha) = 1$ iff $\overline{k} \in \{6, 18\}$.*

Proof. Property (1) is an immediate consequence of the definition of Euclidean digital spaces and Axiom 4. Similarly, (2) follows from Axiom 2. To show (3) notice that $f(\text{st}_3(\alpha; O), \alpha) = 1$ iff $\text{st}_3(\alpha; O)$ is $\emptyset$-connected, and hence k-connected; however, it is a 26-connected set but is not 18-connected. For the proof of (4) and (5) let us consider the object $O_1 = (\text{cell}_3(R^3) - \text{st}_3^*(\alpha; R^3)) \cup \text{st}_3^*(\alpha; O)$. If $f(O, \alpha) = 0$, Axiom 4 implies that its complement $\text{cell}_3(R^3) - O_1$ would be O_1-connected, but it is not 6-connected if $\text{P}(O, \alpha) \in X$ and is not 18-connected if $\text{P}(O, \alpha) = \text{P}_6^c$. □

Proposition 2. *Let $O \subseteq \text{cell}_3(R^3)$ be a digital object and $\beta = \langle \alpha_1, \alpha_2 \rangle \in R^3$ be a 1-cell such that $f(O, \alpha_1) = f(O, \alpha_2)$. If $\text{st}_3(\beta; O) = \text{st}_3(\beta; R^3)$ then $f(O, \beta) = 1$. Moreover, if (R^3, f) is a $(k, 6)$-space and $\text{st}_3(\beta; O) = \{\sigma, \tau\}$, with $\beta = \sigma \cap \tau$, then $f(O, \beta) = 1$ as well.*

Proof. It suffices to find an object $O' \supseteq O$ such that $\text{st}_3(\beta; O') = \text{st}_3(\beta; O)$ and $f(O', \delta) = 1$ for each cell $\delta \in \{\beta, \alpha_1, \alpha_2\}$. In these conditions, it is readily checked that $\beta(O, O')$ is either empty or equal to the non-connected set $\{\alpha_1, \alpha_2\}$. Hence, $f(O, \beta) = 1$ by Axiom 5.

If $\mathrm{st}_3(\beta; O) = \mathrm{st}_3(\beta; R^3)$ we use the object $O_1 = \mathrm{cell}_3(R^3)$, while we take $O_2 = \mathrm{cell}_3(R^3) - \{\sigma', \tau'\}$, where $\{\sigma', \tau'\} = \mathrm{st}_3(\beta; R^3) - \mathrm{st}_3(\beta; O)$, if $\mathrm{st}_3(\beta; O) = \{\sigma, \tau\}$. In the first case, $f(O_1, \delta) = 1$, $\delta \in \{\beta, \alpha_1, \alpha_2\}$, since (R^3, f) is Euclidean. In the second, the equalities $f(O_2, \delta) = 1$ also hold since, in addition, the complement $\mathrm{cell}_3(R^3) - O_2$ is not 6-connected. □

Proposition 3. *Let $O \subseteq \mathrm{cell}_3(R^3)$ be a digital object and $\gamma \in \mathrm{supp}(O)$ a 2-cell. Then γ or some of its faces are lighted for O.*

Proof. Assume $f(O, \delta) = 0$ for each proper face $\delta < \gamma$, and let us consider the object $O_1 = (\mathrm{cell}_3(R^3) - \mathrm{st}_3(\beta_0; R^3) - \mathrm{st}_3(\beta_2; R^3)) \cup \mathrm{st}_3(\beta_0; O) \cup \mathrm{st}_3(\beta_2; O)$, where β_i, $0 \leq i \leq 3$, are the four 1-faces of γ, and $\alpha_i = \beta_i \cap \beta_{i+1 (\mathrm{mod}\ 4)}$ its vertices. Since (R^3, f) is homogeneous we know that $f(O_1, \alpha_0) = f(O_1, \alpha_1)$ and $f(O_1, \alpha_2) = f(O_1, \alpha_3)$. Thus, for $i = 0, 2$, the sets $\beta_i(O, O_1)$ are empty or non-connected and $f(O_1, \beta_i) = 0$ by Axiom 5. On the other hand, by Proposition 2 $f(O_1, \beta_1) = 1$ if f vanishes on both vertices α_1, α_2 for O_1. Hence, some cell in the set $\{\alpha_1, \beta_1, \alpha_2\}$, and similarly for $\{\alpha_3, \beta_3, \alpha_0\}$, is lighted for O_1. This way the sets $\gamma(O, O_1)$ and $\gamma(O_1, \mathrm{cell}_3(R^3))$ are non-connected, and $f(O_1, \gamma) = f(O, \gamma) = 1$ also by Axiom 5. □

3 $(k, \overline{k})$-Surfaces

Similarly to our previous definition of connectedness on a digital space (R^3, f), we use continuous analogues to introduce the notion of digital surface. Namely, a digital object S is a *digital surface* in the digital space (R^3, f), an *f-surface* for short, if $|\mathcal{A}_S|$ is a (combinatorial) surface without boundary; that is, if for each vertex $v \in \mathcal{A}_S$ its link $\mathrm{lk}(v; \mathcal{A}_S) = \{A \in \mathcal{A}_S \,;\, v, A < B \in \mathcal{A}_S \text{ and } v \notin A\}$ is a 1-sphere.

Along this section S will stand for an arbitrary f-surface S in a given Euclidean $(k, \overline{k})$-space (R^3, f), where $k, \overline{k} \in \{6, 18, 26\}$ and $(k, \overline{k}) \neq (6, 6)$. Our goal is to compute its continuous analogue; actually, we will determine the value $f(S, \delta)$ for almost each cell $\delta \in R^3$. A simple although crucial tool for this task is the following

Remark 1. By the definition of continuous analogues, each vertex in $\mathcal{A}_S$ is the center $c(\gamma)$ of a cell $\gamma \in R^3$ lighted for the surface S. Moreover, the cycle $\mathrm{lk}(c(\gamma); \mathcal{A}_S)$ is the complex determined by the set of cells $X_S^\gamma = \{\delta \in R^3 \,;\, \delta < \gamma \text{ or } \gamma < \delta \text{ and } f(S, \delta) = 1\}$, which is contained in $Y_S^\gamma = \{\delta \in R^3 \,;\, \delta < \gamma \text{ or } \gamma < \delta \text{ and } \delta \in \mathrm{supp}(S)\}$ by Axiom 2.

If δ is a 3-cell, Axioms 1 and 2 in the definition of lighting functions imply that $f(S, \delta) = 1$ if and only if $\delta \in S$. For 1-cells we will prove the following

Theorem 1. *Given a 1-cell $\beta \in R^3$, let $\alpha_1, \alpha_2 < \beta$ be its vertices and $\gamma_j \in R^3$ be the four 2-cells such that $\beta < \gamma_j$, $1 \leq j \leq 4$. Then $f(S, \beta) = 1$ if and only if one of the two following properties holds:*

1. $\mathrm{st}_3(\beta; S) = \{\sigma, \tau\}$, *with* $\beta = \sigma \cap \tau$ *and, moreover,* $f(S, \alpha_i) = 1$, $i = 1, 2$, *and* $f(S, \gamma_j) = 0$, $1 \leq j \leq 4$.
2. $\mathrm{st}_3(\beta; S) = \mathrm{st}_3(\beta; R^3)$, $f(S, \alpha_i) = 0$ *and* $f(S, \gamma_j) = 1$, $i = 1, 2$, $1 \leq j \leq 4$.

The proof of Theorem 1 needs the results in Sect. 2 as well as to know the lighting of some vertices of R^3 for the surface. However, the "only if" part is a consequence of the following

Proposition 4. *Let $\beta \in R^3$ be a 1-cell. Then $f(S,\beta) = 0$ if one of the following conditions holds:*

1. *$\mathrm{st}_3(\beta; S)$ consists of exactly three elements.*
2. *$\mathrm{st}_3(\beta; S) = \mathrm{st}_3(\beta; R^3)$ and $f(S,\alpha) = 1$ for some vertex $\alpha < \beta$.*

Proof. Assume on the contrary that $f(S,\beta) = 1$. We shall prove that the link $L = \mathrm{lk}(c(\beta); \mathcal{A}_S)$ is not a 1-sphere, which is a contradiction. If condition (1) holds, it is readily checked that the 2- and 3-cells of the set Y_S^β, as defined in Remark 1, do not determine a cycle in $\mathcal{A}_S$ unless $c(\alpha)$ also belongs to L for some vertex $\alpha < \beta$. But then $c(\alpha)$ is an end of all edges $\langle c(\alpha), c(\sigma_i)\rangle \in L$, where σ_i ranges over the star of β in S, and thus L is not a 1-sphere if $\mathrm{st}_3(\beta; S)$ has three or four elements. □

Proof. of Theorem 1 ("only if" part) If $f(S,\beta) = 1$ then $\beta \in \mathrm{supp}(S)$, and either $\mathrm{st}_3(\beta; S) = \{\sigma, \tau\}$, with $\beta = \sigma \cap \tau$, or $\mathrm{st}_3(\beta; S) = \mathrm{st}_3(\beta; R^3)$ by Proposition 4. In the first case only the two vertices of β may belong to Y_S^β, which must be lighted since $L = \mathrm{lk}(c(\beta); \mathcal{A}_S)$ is a cycle. In the second case, we already know, also by Proposition 4, that $f(S,\alpha_i) = 0$, $i = 1, 2$. Then $f(S,\gamma_j) = 1$ in order that L is a 1-sphere. □

Remark 2. If a vertex $\alpha \in R^3$ is lighted for the surface S the "only if" part of Theorem 1 allows us to be more precise than in Remark 1. Indeed, the set of cells X_S^α, determining the link $\mathrm{lk}(c(\alpha); \mathcal{A}_S)$, is contained in the disjoint union $\mathrm{st}_3(\alpha; S) \cup Z_S^\alpha$, where $Z_S^\alpha = \{\delta > \alpha\,;\, \mathrm{st}_3(\delta; S) = \{\sigma, \tau\}, \delta = \sigma \cap \tau\}$.

Using this remark we are able to describe locally the continuous analogue of an f-surface S around any vertex α which is lighted for it. Before, and in addition to those found in Proposition 1, we next identify some more patterns of an f-surface S around a vertex α for which $f(S,\alpha) = 0$.

Proposition 5. *Let $\mathrm{P}(S,\alpha)$ be the pattern of an f-surface S around a vertex $\alpha \in R^3$. If $\mathrm{P}(S,\alpha) \in \{\mathrm{P}_2^c, \mathrm{P}_3^b, \mathrm{P}_4^c, \mathrm{P}_4^d, \mathrm{P}_5^a, \mathrm{P}_5^c, \mathrm{P}_6^b, \mathrm{P}_7\}$ then $f(S,\alpha) = 0$.*

Proof. Assume that $f(S,\alpha) = 1$. For each voxel $\sigma \in \mathrm{st}_3(\alpha; S)$ let us consider the set $Z_\sigma = \{\delta \in Z_S^\alpha\,;\, \delta < \sigma\}$; see Remark 2. Since $L = \mathrm{lk}(c(\alpha); \mathcal{A}_S)$ is a cycle, $c(\sigma) \in L$ is in two edges of L; in other words, exactly two elements of Z_σ belong to X_S^α. However, if $\mathrm{P}(S,\alpha) \in \{\mathrm{P}_2^c, \mathrm{P}_3^b, \mathrm{P}_4^c, \mathrm{P}_4^d, \mathrm{P}_5^a\}$ it is easily found a voxel $\sigma \in \mathrm{st}_3(\alpha; S)$ for which the set Z_σ is a sigleton or the emptyset. On the other hand, for each of the patterns $\mathrm{P}_5^c, \mathrm{P}_6^b$ and P_7 there is a proper subset of voxels $A \subset \mathrm{st}_3(\alpha; S)$ such that, for each $\sigma \in A$, $Z_\sigma \subset X_S^\alpha$ since it consists of exactly two elements. But then one observes that the cells in $A \cup (\cup_{\sigma \in A} Z_\sigma)$ determines a cycle in L which leaves out the centers of the voxels in $\mathrm{st}_3(\alpha; S) - A$. □

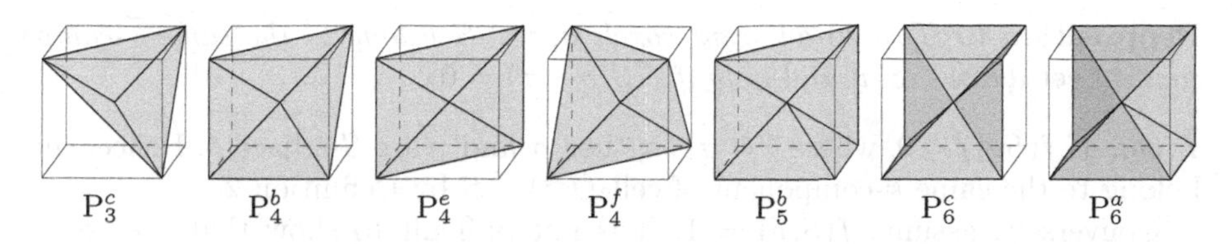

Fig. 2. Continuous analogue of a surface S around a vertex $\alpha \in R^3$

Remark 3. Given an f-surface S and a vertex $\alpha \in R^3$ such that $f(S,\alpha) = 1$ we know that $\mathrm{P}(S,\alpha) \in \{\mathrm{P}_3^c, \mathrm{P}_4^b, \mathrm{P}_4^e, \mathrm{P}_4^f, \mathrm{P}_5^b, \mathrm{P}_6^a, \mathrm{P}_6^c, \mathrm{P}_8\}$ from Propositions 1 and 5. For each of these patterns, except for P_8, we can determine the lighting of every cell in Z_S^α by using the same technique as in the proof above. That is, we completely know the continuous analogue of the surface inside the unit cube $C_\alpha \subset \mathbb{R}^3$ whose vertices are the centers of the eight voxels containing α; see Fig. 2.

Indeed, if $\mathrm{P}(S,\alpha) \in \{\mathrm{P}_3^c, \mathrm{P}_4^b, \mathrm{P}_4^e, \mathrm{P}_5^b, \mathrm{P}_6^c\}$ each set Z_σ, $\sigma \in \mathrm{st}_3(\alpha; S)$, consists of two cells. Therefore, all of them are lighted for the surface (i.e., $Z_S^\alpha \subset X_S^\alpha$).

For P_6^a, let σ, τ be the two voxels in $\mathrm{st}_3(\alpha; S)$ which are 6-adjacent to three other voxels in the star of α in S. One readily checks that for each voxel $\rho \in A = \mathrm{st}_3(\alpha; S) - \{\sigma, \tau\}$ the set Z_ρ has two elements and then it is contained in X_S^α; moreover, $X_S^\alpha = \mathrm{st}_3(\alpha; S) \cup (\cup_{\rho \in A} Z_\rho)$ since this set determines a cycle in $\mathrm{lk}(c(\alpha); \mathcal{A}_S)$. In other words, $\gamma = \sigma \cap \tau$ is the only 2-cell in $\mathrm{supp}(\mathrm{st}_3(\alpha; S))$ such that $f(S,\gamma) = 0$; moreover, also $f(S, \beta_i) = 0$ for the two edges $\alpha < \beta_1, \beta_2 < \gamma$.

Finally, at this point we can only determine the continuous analogue for P_4^f up to symmetries. We know that only two of the three 1-cells in each of the sets Z_{σ_i}, where $\mathrm{st}_3(\alpha; S) = \{\sigma_1, \sigma_2, \sigma_3, \sigma_4\}$, can be lighted for the surface. However the selection of two cells in Z_{σ_1} determines what cells in $Z_{\sigma_2}, Z_{\sigma_3}, Z_{\sigma_4}$ must be lighted in order that $\mathrm{lk}(c(\alpha); \mathcal{A}_S)$ is a cycle.

To complete our analysis of the lighting on vertices and edges of R^3 for S we use some separation properties of f-surfaces. Firstly, notice that, from our definition of $(k, \overline{k})$-space and the characterization of connectedness on a Euclidean space (R^3, f) in Sect. 2, the $\overline{k}$-components of $\mathrm{cell}_3(R^3) - S$ are characterizing the connected components of $\mathbb{R}^3 - |\mathcal{A}_S| = |\mathcal{A}_{R^3}| - |\mathcal{A}_S|$. Therefore, we obtain the following separation result as a corollary of the well-known Jordan-Brouwer Theorem; see Sect. 5 in [3].

Theorem 2. *Let S be a k-connected f-surface in a Euclidean $(k, \overline{k})$-space (R^3, f). Then S separates its complement* $\mathrm{cell}_3(R^3) - S$ *into two $\overline{k}$-components.*

Next proposition shows that the $\overline{k}$-componentes of $\mathrm{cell}_3(R^3) - S$ can be locally determined by using the notion of relative ball from polyhedral topology. Recall that given a closed topological surface $M \subset \mathbb{R}^3$ in the Euclidean space, a *relative ball* (B^3, B^2) in $(\mathbb{R}^3, M)$ is a pair of topological balls $B^2 \subset B^3$ such that $B^2 \cap \partial B^3 = \partial B^2$ and $B^2 = B^3 \cap M$. The key property of relative balls, which is also a consequence of the Jordan-Brouwer Theorem, states that the difference $B^3 - B^2$ has exactly two connected components, each contained in a distinct component of $\mathbb{R}^3 - M$; see [2] for details.

Proposition 6. *Two 26-adjacent voxels $\sigma, \tau \notin S$ belong to the same $\overline{k}$-component of* $\mathrm{cell}_3(R^3) - S$ *if and only if* $f(S, \sigma \cap \tau) = 0$.

Proof. If $f(S,\delta) = 0$, where $\delta = \sigma \cap \tau$, then σ and τ are S-adjacent. Hence, they belong to the same $\overline{k}$-component of $\mathrm{cell}_3(R^3) - S$ by Definition 2.

Conversely, assume $f(S,\delta) = 1$. It is not difficult to show that the pair of polyhedra $(|\mathrm{st}(c(\delta), \mathcal{A}_{R^3})|, |\mathrm{st}(c(\delta), \mathcal{A}_S)|)$ is a relative ball in $(\mathbb{R}^3, |\mathcal{A}_S|)$, where $\mathrm{st}(c(\delta), \mathcal{A}_X) = \{A \in \mathcal{A}_X \,;\, c(\alpha), A < B \in \mathcal{A}_X\}$ is the *star* of $c(\delta)$ in $\mathcal{A}_X$. Therefore, it will suffice to show that each $c(\sigma)$ and $c(\tau)$ belong to a distinct component of the difference $D = |\mathrm{st}(c(\delta), \mathcal{A}_{R^3})| - |\mathrm{st}(c(\delta), \mathcal{A}_S)|$.

Notice that $\dim \delta \leq 1$; otherwise, if δ is a 2-cell then $\delta \notin \mathrm{supp}(S)$, and $f(S,\delta) = 0$ by Axiom 2, since $\sigma, \tau \notin S$. If $\dim \delta = 1$ the "only if" part of Theorem 1 yields that $\mathrm{st}_3(\delta; S) = \{\sigma', \tau'\} = \mathrm{st}_3(\delta; R^3) - \{\sigma, \tau\}$. Then $|\mathrm{st}(c(\delta), \mathcal{A}_{R^3})|$ is the double cone from the vertices $v_1, v_2 < \delta$ over the unit square whose vertices are the centers of the voxels in $\mathrm{st}_3(\delta; R^3)$, while $|\mathrm{st}(c(\delta), \mathcal{A}_S)|$ is the double cone from v_1 and v_2 over the union of edges $\langle c(\delta), c(\sigma')\rangle \cup \langle c(\delta), c(\tau')\rangle$. After this description it is easily checked that $c(\sigma)$ and $c(\tau)$ belong to distinct components of the difference D.

If $\dim \delta = 0$, $|\mathrm{st}(c(\delta), \mathcal{A}_{R^3})|$ is the cube $C_\delta \subset \mathbb{R}^3$ with vertices at the centers of the eight voxels containing δ. Moreover, $\mathrm{P}(S,\delta) \in \{\mathrm{P}_3^c, \mathrm{P}_4^b, \mathrm{P}_4^e, \mathrm{P}_5^b, \mathrm{P}_6^c\}$ by Propositions 1 and 5. Then, $|\mathrm{st}(c(\delta), \mathcal{A}_S)|$ is the continuous analogue locally described in Remark 3 and the result follows; see Fig. 2. □

Corollary 1. *Assume $\overline{k} \in \{18, 26\}$. Then $f(S,\delta) = 0$ for each cell $\delta \in R^3$ satisfying one of the following conditions:*

1. *δ is an edge such that $\mathrm{st}_3(\delta; S) = \{\sigma, \tau\}$ with $\delta = \sigma \cap \tau$.*
2. *δ is a vertex and $\mathrm{P}(S,\delta) \in \{\mathrm{P}_3^c, \mathrm{P}_4^b, \mathrm{P}_4^e, \mathrm{P}_4^f, \mathrm{P}_5^b\}$.*

Proof. If δ is a vertex and $\mathrm{P}(S,\delta) = \mathrm{P}_4^f$ then each edge $\beta > \delta$ is the intersection of two 18-adjacent voxels in $\mathrm{st}_3(\delta; R^3) - S \subset \mathrm{cell}_3(R^3) - S$ and thus $f(S,\beta) = 0$. Therefore $f(S,\delta) = 0$ since, otherwise, the link $\mathrm{lk}(c(\delta); \mathcal{A}_S)$ is a discrete set of points consisting just of the centers of the four voxels in $\mathrm{st}_3(\delta; S)$. For all the remaining cases it is not difficult to find a pair of voxels σ', τ' in a 18-component of $\mathrm{st}_3(\delta; R^3) - S$ such that $\delta = \sigma' \cap \tau'$.

Propositions 1 and 5 as well as the corollary above provides us with all the information we need about the lighting of vertices for an f-surface S, which is summarized as follows. We are also ready to finish the proof of Theorem 1.

Theorem 3. *Let S be an f-surface in a Euclidean $(k, \overline{k})$-space (R^3, f), and let $\alpha \in R^3$ be a vertex such that $\mathrm{P}(S,\alpha) \notin \{\mathrm{P}_6^a, \mathrm{P}_8\}$. Then $\mathrm{P}(S,\alpha) \notin \mathbb{FP}_{k,\overline{k}}$ and, moreover, $f(S,\alpha) = 1$ if and only if $\mathrm{P}(S,\alpha) \in \mathbb{P}_{\overline{k}}$. The sets $\mathbb{FP}_{k,\overline{k}}$ and $\mathbb{P}_{\overline{k}}$, whose elements are respectively called $(k,\overline{k})$*-forbidden patterns *and $\overline{k}$*-plates, *are defined as follows: $\mathbb{FP}_{6,26}$, $\mathbb{FP}_{18,26}$, $\mathbb{FP}_{6,18}$ and $\mathbb{FP}_{18,18}$ are the empty set, $\mathbb{FP}_{26,26} = \mathbb{FP}_{26,18} = \{\mathrm{P}_2^c\}$, $\mathbb{FP}_{18,6} = \{\mathrm{P}_5^c, \mathrm{P}_6^b\}$ and $\mathbb{FP}_{26,6} = \{\mathrm{P}_2^c, \mathrm{P}_5^c, \mathrm{P}_6^b\}$; $\mathbb{P}_6 = \{\mathrm{P}_3^c, \mathrm{P}_4^b, \mathrm{P}_4^e, \mathrm{P}_4^f, \mathrm{P}_5^b, \mathrm{P}_6^c\}$, $\mathbb{P}_{18} = \{\mathrm{P}_6^c\}$ and $\mathbb{P}_{26} = \emptyset$.*

Proof. of Theorem1 ("if" part) If S is a $(k, \overline{k})$-f-surface for $\overline{k} \in \{18, 26\}$ and $\text{st}_3(\beta; S) = \{\sigma, \tau\}$ condition (1) does not hold and, so, there is nothing to prove. Indeed, for each vertex $\alpha < \beta$, $\text{P}(S, \alpha) \in \{\text{P}_2^b, \text{P}_3^b, \text{P}_3^c, \text{P}_4^b, \text{P}_4^e, \text{P}_4^f, \text{P}_5^b, \text{P}_5^c, \text{P}_6^b\}$; therefore $f(S, \alpha) = 0$ by Corollary 1 and Propositions 1 and 5. For the rest of cases the result is an immediate consequence of Proposition 2. □

Notice that in the theorem above we have excluded P_8 from our analysis. This pattern is usually considered as a small blob and it is widely rejected as part of a surface. Following this criterion, from now on we will only consider regular spaces according to the next definition.

Definition 3. *An Euclidean $(k, \overline{k})$-space is said to be* regular *if* $\text{P}(S, \alpha) \neq \text{P}_8$ *for any surface S and any vertex $\alpha \in R^3$.*

Remark 4. We have also eluded P_6^a in Theorem 3. Actually, a surface S containing this pattern can have two different but homeomorphic continuous analogues, depending on the value of $f(S, \alpha)$ for any vertex $\alpha \in R^3$ such that $\text{P}(S, \alpha) = \text{P}_6^a$ (recall that we are only considering homogeneous digital spaces). If $f(S, \alpha) = 1$, we have already described locally the continuous analogue of S around α in Remark 3. For describing the intersection $|\mathcal{A}_S| \cap C_\alpha$, where $C_\alpha \subset \mathbb{R}^3$ is the unit cube with vertices at the centers of the voxels in $\text{st}_3(\alpha; R^3)$, in case $f(S, \alpha) = 0$, let us consider the two edges $\beta_i = \langle \alpha, \alpha_i \rangle$ of R^3 such that $\text{st}_3(\beta_i; S) = \text{st}_3(\beta_i; R^3)$, $i = 1, 2$. If (R^3, f) is a regular space Theorem 3 yields that $f(S, \alpha_i) = 0$ for $i = 1, 2$ and, then, $f(S, \beta_i) = 1$ by Theorem 1; moreover, any 2-cell containing α is lighted for S. So that, the continuous analogue of S around α is the union of the two unit squares whose vertices are the centers of the voxels in $\text{st}_3(\alpha; S)$.

We finish this section showing that certain patterns around an edge are forbidden in a $(k, \overline{k})$-f-surface when $k, \overline{k} \in \{18, 26\}$.

Proposition 7. *Let S be a digital surface in a Euclidean $(k, \overline{k})$-space (R^3, f), $k, \overline{k} \in \{18, 26\}$. If $\beta \in R^3$ is an edge such that $\text{st}_3(\beta; S) = \{\sigma, \tau\}$, with $\beta = \sigma \cap \tau$, then σ and τ belong to a 6-component of $\text{st}_3^*(\beta; S)$.*

Proof. (sketch) Assume, on the contrary, that σ and τ are in distinct 6-components of $\text{st}_3^*(\beta; S)$. Then one checks that $\text{P}(S, \alpha_i) \in \{\text{P}_2^b, \text{P}_3^b, \text{P}_3^c, \text{P}_4^b, \text{P}_4^e, \text{P}_4^f, \text{P}_5^c\}$ for the two vertices $\alpha_1, \alpha_2 < \beta$, and thus $f(S, \delta) = 0$, for $\delta \in \{\beta, \alpha_1, \alpha_2\}$ by Corollary 1 and Theorem 3. From these facts we build a 18-connected object O_1, possibly infinite, such that $\text{P}(O_1, \beta') = \text{P}(S, \beta)$ for each edge $\beta' \in \text{supp}(O_1)$ which is a face of exactly two voxels in O_1. However, it can be proved that O_1 is not $\emptyset$-connected in (R^3, f), which is a contradiction. □

4 Universal $(k, \overline{k})$-Spaces

In this section we reach our goal: for each adjacency pair $(k, \overline{k}) \neq (6, 6)$, $k, \overline{k} \in \{6, 18, 26\}$, we define a lighting function on the complex R^3 which gives us a regular $(k, \overline{k})$-space $E_{k,\overline{k}} = (R^3, f_{k,\overline{k}})$ whose set of digital surfaces is the largest within that class of digital spaces. Namely, we will prove the following

Theorem 4. *Any digital surface S in a regular $(k, \overline{k})$-space (R^3, f) is also a digital surface in the space $E_{k,\overline{k}} = (R^3, f_{k,\overline{k}})$, which is called the* universal $(k, \overline{k})$-space.

In the definition of $E_{k,\overline{k}}$ we use the set of forbidden patterns introduced in Theorem 3 and the set of $\overline{k}$-plates, that are the elementary "bricks" from which f-surfaces of Euclidean $(k, \overline{k})$-spaces are built. Indeed, the lighting function $f_{k,\overline{k}}$ is defined as follows. Given a digital object $O \subseteq \text{cell}_3(R^3)$ and a cell $\delta \in R^3$, $f_{k,\overline{k}}(O, \delta) = 1$ if and only one of the following conditions holds:

1. $\dim \delta \geq 2$ and $\delta \in \text{supp}(O)$
2. $\dim \delta = 0$ and $\text{P}(O, \delta) \in \mathbb{P}_{\overline{k}} \cup \mathbb{FP}_{k,\overline{k}} \cup \{\text{P}_8\}$
3. $\dim \delta = 1$ and $\text{st}_3(\delta; O) = \text{st}_3(\delta; R^3)$ (*a square plate*), or
4. $\dim \delta = 1$ and $\text{st}_3(\delta; O) = \{\sigma, \tau\}$, with $\delta = \sigma \cap \tau$, and one of the next further conditions also holds: (a) for $\overline{k} = 6$, and $k \neq 6$, $f_{k,6}(O, \alpha_1) = f_{k,6}(O, \alpha_2)$, where α_1, α_2 are the two vertices of the 1-cell δ; or (b) if σ, τ belong to distinct 6-components of $\text{st}_3^*(\delta; O)$, for $k, \overline{k} \in \{18, 26\}$.

Notice that if $k = 6$, and $\overline{k} \in \{18, 26\}$, none of the conditions 4(a) and 4(b) hold. So, $f_{6,\overline{k}}(O, \delta) = 1$ for a 1-cell $\delta \in R^3$ if and only if $\text{st}_3(\delta; O) = \text{st}_3(\delta; R^3)$.

It is not difficult, but a tedious task, to check that each $f_{k,\overline{k}}$ is a lighting function, and to prove that $E_{k,\overline{k}}$ is actually a homogeneous $(k, \overline{k})$-space. To show that all of them are regular spaces, assume that O is a digital objet and $\alpha \in R^3$ is a vertex such that $\text{P}(O, \alpha) = \text{P}_8$; that is, $\text{st}_3(\alpha; O) = \text{st}_3(\alpha; R^3)$. Then, it is readily checked from the definition that $f_{k,\overline{k}}(O, \delta) = 1$ for each cell $\delta \geq \alpha$. Therefore, the unit cube $C_\alpha \subset \mathbb{R}^3$ centered at α, and whose vertices are the centers of all voxels in $\text{st}_3(\alpha; R^3)$, is contained in $|\mathcal{A}_O|$; thus no surface in $E_{k,\overline{k}}$ can contain the pattern P_8. This also proves that $E_{k,\overline{k}}$ is Euclidean since $f(\text{cell}_3(R^3), \delta) = 1$ for every cell $\delta \in R^3$.

In Remark 4 we suggested that the continuous analogue of a given surface in two different $(k, \overline{k})$-spaces may differ only if it contains the pattern P_6^a. Next results make more precise this in relation with the universal $(k, \overline{k})$-space.

Proposition 8. *Let S be a surface in a regular $(k, \overline{k})$-space (R^3, f). If $\delta \in R^3$ is a cell such that $\text{P}(S, \alpha) \neq \text{P}_6^a$ for all vertices $\alpha \leq \delta$, then $f(S, \delta) = f_{k,\overline{k}}(S, \delta)$.*

Proof. We may assume that $\delta \in \text{supp}(S)$, otherwise both f and $f_{k,\overline{k}}$ vanish on δ for S by Axiom 2 of lighting functions.

Assume firstly that δ is a vertex of R^3, and so $f(S, \delta) = 1$ iff $\text{P}(S, \delta) \in \mathbb{P}_{\overline{k}}$ by Theorem 3. Moreover, since (R^3, f) is regular the same theorem and the hypothesis ensures that $\text{P}(S, \delta) \notin \mathbb{FP}_{k,\overline{k}} \cup \{\text{P}_6^a, \text{P}_8\}$ and, under these conditions, the definition of $f_{k,\overline{k}}$ also yields that $f_{k,\overline{k}}(S, \delta) = 1$ iff $\text{P}(S, \delta) \in \mathbb{P}_{\overline{k}}$.

If δ is a 2-cell, $f_{k,\overline{k}}(S, \delta) = 1$ by definition. We reach a contradiction if $f(S, \delta) = 0$. Indeed, in that case Proposition 3 gives us a face $\alpha < \delta$ which is lighted for S. Since $\text{st}_3(\delta; S) \subseteq \text{st}_3(\alpha; S)$, condition (2) in Theorem 1 must hold if $\dim \alpha = 1$. Otherwise, if α is a vertex, then $\text{st}_3(\alpha; S)$ is a $\overline{k}$-plate by Theorem 3 and we showed that δ is then lighted in Remark 3.

Finally, if $\dim \delta = 1$ and $\delta \in \text{supp}(S)$ three cases are possible: (a) $\text{st}_3(\delta; S) = \{\sigma, \tau\}$, with $\delta = \sigma \cap \tau$; (b) $\text{st}_3(\delta; S) = \{\sigma, \tau, \rho\}$; and (c) $\text{st}_3(\delta; S) = \text{st}_3(\delta; R^3)$. In case (b) both $f_{k,\overline{k}}$ and f vanish on δ by definition and Theorem 1, respectively.

We claim that $f(S,\delta) = 1$ in case (c). Indeed, let α_1, α_2 be the vertices of the edge δ. Since $\mathrm{st}_3(\delta; S) \subseteq \mathrm{st}_3(\alpha_i; S)$ and (R^3, f) is regular we know that $\mathrm{P}(S,\alpha_i) \in \{\mathrm{P}_4^a, \mathrm{P}_5^a, \mathrm{P}_6^b, \mathrm{P}_7\}$; moreover, $\mathrm{P}(S,\alpha_i) \neq \mathrm{P}_6^b$ if $\overline{k} = 6$. In any case these patterns are not $\overline{k}$-plates. Then $f(S,\alpha_i) = 0$, $i = 1,2$, by Theorem 3, and Proposition 2 yields our claim. The case (a) requires two different arguments for $\overline{k} = 6$ and $\overline{k} \in \{18, 26\}$. If $\overline{k} = 6$, and so $k \neq 6$, we have already proved that $f(S,\alpha_i) = f_{k,\overline{k}}(S,\alpha_i)$ for the two vertices $\alpha_1, \alpha_2 < \delta$. Then also $f(S,\delta) = f_{k,\overline{k}}(S,\delta)$ by definition of $f_{k,\overline{k}}$ and Theorem 1. Finally, if $\overline{k} \in \{18, 26\}$ we know that $f(S,\delta) = 0$ by Corollary 1 and, moreover, $\mathrm{st}_3^*(\delta; S)$ is 6-connected if $k \neq 6$ by Proposition 7. Under these conditions we get $f_{k,\overline{k}}(S,\delta) = 0$ by definition. □

Proposition 9. *Let S be a surface in a regular $(k,\overline{k})$-space (R^3, f). Given a vertex $\alpha \in R^3$, let $C_\alpha \subset \mathbb{R}^3$ be the unit cube centered at α. If $\mathrm{P}(S,\alpha) = \mathrm{P}_6^a$ the two following properties hold:*

1. *$D_f^\alpha = |\mathcal{A}_S^f| \cap C_\alpha$ and $D_U^\alpha = |\mathcal{A}_S^{f_{k,\overline{k}}}| \cap C_\alpha$ are both 2-balls with common border; that is, $\partial D_f^\alpha = \partial D_U^\alpha$.*
2. *There exists a pl-homeomorphism $\varphi_\alpha : D_f^\alpha \to D_U^\alpha$ which extends the identity in the border; moreover, if $f(S,\alpha) = 0$ then $\varphi_\alpha = \mathrm{id}$.*

Proof. From the definition of the lighting function $f_{k,\overline{k}}$ it is readily checked that all the 2- and 3-cells $\delta > \alpha$, $\delta \in \mathrm{supp}(S)$, and also the two 1-cells $\beta_1, \beta_2 > \alpha$ such that $\mathrm{st}_3(\beta_i; S) = \mathrm{st}_3(\beta_i; R^3)$ are lighted for S, while $f_{k,\overline{k}}(S,\delta) = 0$ for any other cell $\delta \geq \alpha$. This way, the disk D_U^α is the union of the two unit squares defined by the centers of all voxels in $\mathrm{st}_3(\alpha; S)$. In particular, $D_U^\alpha = D_f^\alpha$ if $f(S,\alpha) = 0$ by Remark 4.

If $f(S,\alpha) = 1$ we know by Remark 3 that D_f^α is also a disk and from the above description of D_U^α it becomes clear that $\partial D_f^\alpha = \partial D_U^\alpha$. Moreover, φ_α can be defined as the conic extension of the identity that assigns the center $c(\alpha)$ to $c(\sigma \cap \tau)$, where $\sigma, \tau \in \mathrm{st}_3(\alpha; S)$ are the two only 3-cells which are 6-adjacent to three other 3-cells in that set.

Proof. (of Theorem 4) We claim that the polyhedra $|\mathcal{A}_S^f|$ and $|\mathcal{A}_S^{f_{k,\overline{k}}}|$ are *pl*-homeomorphic. Thus, the continuous analogues of S in both digital spaces (R^3, f) and $E_{k,\overline{k}}$ are combinatorial surfaces.

To ease the reading we will write $|\mathcal{A}_S^U|$ instead $|\mathcal{A}_S^{f_{k,\overline{k}}}|$, while keep $|\mathcal{A}_S^f|$ for the continuous analogue of S in the space (R^3, f). In order to define a *pl*-homeomorphism $\varphi : |\mathcal{A}_S^f| \to |\mathcal{A}_S^U|$ let us consider the sets of disks $\{D_f^\alpha = |\mathcal{A}_S^f| \cap C_\alpha\}_{\alpha \in A}$ and $\{D_U^\alpha = |\mathcal{A}_S^U| \cap C_\alpha\}_{\alpha \in A}$, where A stands for the set of vertices $\alpha \in R^3$ such that $\mathrm{P}(S,\alpha) = \mathrm{P}_6^a$ and $C_\alpha \subset \mathbb{R}^3$ is the unit cube centered at α. For each vertex $\alpha \in A$ we set $\varphi = \varphi_\alpha$, where $\varphi_\alpha : D_f^\alpha \to D_U^\alpha$ are the *pl*-homeomorphisms provided by Proposition 9, while from Proposition 8 we can define $\varphi : \overline{|\mathcal{A}_S^f| - \cup_{\alpha \in A} D_f^\alpha} \to \overline{|\mathcal{A}_S^U| - \cup_{\alpha \in A} D_U^\alpha}$ as the identity. Notice that $\varphi = \mathrm{id}$ if $f(S,\alpha) = 0$ for some vertex $\alpha \in A$, so we may assume that $f(S,\alpha) = 1$ for all of them. In order to check that φ and φ^{-1} are well defined it suffices to prove that $D_g^\alpha \cap D_g^{\alpha'} \subseteq$

$\partial D_g^\alpha \cap \partial D_g^{\alpha'}$ for each pair of distinct vertices $\alpha, \alpha' \in A$, where $g \in \{f, U\}$, and also $D_U^\alpha \cap \overline{(|\mathcal{A}_S^U| - \cup_{\lambda \in A} D_U^\lambda)} \subseteq \partial D_U^\alpha$ for each $\alpha \in A$.

Given $\alpha \in A$, let $\beta_1, \beta_2 > \alpha$ the two edges such that $\mathrm{st}_3(\beta_i; S) = \mathrm{st}_3(\beta_i; R^3)$ and $\gamma > \alpha$ the 2-cell having β_1 and β_2 as faces. From Remark 3 we know that $f(S, \delta) = 0$ for $\delta \in \{\gamma, \beta_1, \beta_2\}$ and then no other face of γ, but α, is lighted by f for S by Axiom 5. In particular, $\mathrm{P}(S, \alpha') \notin \mathbb{P}_{\overline{k}} \cup \{\mathrm{P}_6^\alpha\}$ for each vertex $\alpha \neq \alpha' < \gamma$ and the result follows. □

Theorem 4 suggests that, for each adjacency pair $(k, \overline{k}) \neq (6, 6)$, $k, \overline{k} \in \{6, 18, 26\}$, the $f_{k,\overline{k}}$-surfaces of the universal $(k, \overline{k})$-space $E_{k,\overline{k}}$ could be identified with $(k, \overline{k})$-surfaces of the discrete space $\mathbb{Z}^3$. This leads to the problem of characterizing the $f_{k,\overline{k}}$-surfaces just in terms of the adjacency pairs $(k, \overline{k})$, similarly to Morgenthaler's definition of $(26, 6)$-surfaces in [13].

References

1. Ayala, R., Domínguez, E., Francés, A.R., Quintero, A.: Digital lighting functions. In: Ahronovitz, E. (ed.) DGCI 1997. LNCS, vol. 1347, pp. 139–150. Springer, Heidelberg (1997)
2. Ayala, R., Domínguez, E., Francés, A.R., Quintero, A.: A digital index theorem. Int. J. Pattern Recog. Art. Intell. 15(7), 1–22 (2001)
3. Ayala, R., Domínguez, E., Francés, A.R., Quintero, A.: Weak lighting functions and strong 26−surfaces. Theoretical Computer Science 283, 29–66 (2002)
4. Bertrand, G., Malgouyres, R.: Some topological properties of surfaces in $\mathbb{Z}^3$. Jour. of Mathematical Imaging and Vision 11, 207–221 (1999)
5. Brimkov, V.E., Klette, R.: Curves, hypersurfaces, and good pairs of adjacency relations. In: Klette, R., Žunić, J. (eds.) IWCIA 2004. LNCS, vol. 3322, pp. 276–290. Springer, Heidelberg (2004)
6. Ciria, J.C., Domínguez, E., Francés, A.R.: Separation theorems for simplicity 26−surfaces. In: Braquelaire, A., Lachaud, J.-O., Vialard, A. (eds.) DGCI 2002. LNCS, vol. 2301, pp. 45–56. Springer, Heidelberg (2002)
7. Ciria, J.C., De Miguel, A., Domínguez, E., Francés, A.R., Quintero, A.: A maximum set of (26,6)-connected digital surfaces. In: Klette, R., Žunić, J. (eds.) IWCIA 2004. LNCS, vol. 3322, pp. 291–306. Springer, Heidelberg (2004)
8. Ciria, J.C., De Miguel, A., Domínguez, E., Francés, A.R., Quintero, A.: Local characterization of a maximum set of digital $(26, 6)$−surfaces. Image and Vision Computing 25, 1685–1697 (2007)
9. Couprie, M., Bertrand, G.: Simplicity surfaces: a new definition of surfaces in $\mathbb{Z}^3$. In: SPIE Vision Geometry V, vol. 3454, pp. 40–51 (1998)
10. Kong, T.Y., Roscoe, A.W.: Continuous analogs fo axiomatized digital surfaces. Comput. Vision Graph. Image Process. 29, 60–86 (1985)
11. Kovalevesky, V.A.: Finite topology as applied to image analysis. Comput. Vis. Graph. Imag. Process. 46, 141–161 (1989)
12. Malandain, G., Bertrand, G., Ayache, N.: Topological segmentation of discrete surfaces. Int. Jour. of Computer Vision 10(2), 183–197 (1993)
13. Morgenthaler, D.G., Rosenfeld, A.: Surfaces in three–dimensional digital images. Inform. Control. 51, 227–247 (1981)
14. Rourke, C.P., Sanderson, B.J.: Introduction to piecewise–linear topology. Ergebnisse der Math, vol. 69. Springer, Heidelberg (1972)

A Linear Time and Space Algorithm for Detecting Path Intersection*

Srečko Brlek[1], Michel Koskas[2], and Xavier Provençal[3]

[1] Laboratoire de Combinatoire et d'Informatique Mathématique, Université du Québec à Montréal, CP 8888 Succ. Centre-ville, Montréal (QC) Canada H3C 3P8
brlek.srecko@uqam.ca

[2] UMR AgroParisTech/INRA 518, 16 rue Claude Bernard, 75231 Paris Cedex 05
michel.koskas@agroparistech.fr

[3] LAMA, Université de Savoie, 73376 Le Bourget du Lac, France
LIRMM, Université Montpellier II, 161 rue Ada, 34392 Montpellier, France
xavier.provencal@lirmm.fr

Abstract. For discrete sets coded by the Freeman chain describing their contour, several linear algorithms have been designed for determining their shape properties. Most of them are based on the assumption that the boundary word forms a closed and non-intersecting discrete curve. In this article, we provide a linear time and space algorithm for deciding whether a path on a square lattice intersects itself. This work removes a drawback by determining efficiently whether a given path forms the contour of a discrete figure. This is achieved by using a radix tree structure over a quadtree, where nodes are the visited grid points, enriched with neighborhood links that are essential for obtaining linearity.

Keywords: Freeman code, lattice paths, self-intersection, radix tree, discrete figures, data structure.

1 Introduction

Many problems in discrete geometry involve the analysis of the contour of discrete sets. A convenient way to represent them is to use the well-known Freeman chain code [1,2] which encodes the contour by a word w on the four letter alphabet $\Sigma = \{a, b, \overline{a}, \overline{b}\}$, corresponding to the unit displacements in the four directions (right, up, left, down) on a square grid. Among the many problems that have been considered in the literature, we mention : computations of statistics such as area, moment of inertia [3,4], digital convexity [5,6,7], and tiling of the plane by translation [8,9]. All of these problems are solved by using algorithms shown to be linear in the length of the contour word, but often it is assumed that the path encoded by this word does not intersect itself. While it is easy to check that a word encodes a closed path (by checking that the word contains as many a as $\overline{a}$, and as many b as $\overline{b}$), checking that it does not intersect

* With the support of NSERC (Canada).

S. Brlek, C. Reutenauer, and X. Provençal (Eds.): DGCI 2009, LNCS 5810, pp. 397–408, 2009.

itself requires more work. The problem amounts to check if a grid point is visited twice. Of course, one might easily provide an $\mathcal{O}(n \log n)$ algorithm where sorting is involved, or use hash tables providing a linear time algorithm on average but not in worst case.

The goal of this paper is to remove this major drawback. Indeed, we provide a linear time and space algorithm in the worst case checking if a path encoded by a word visits any of the grid points twice. Section 2 provides the basic definitions and notation used in this paper. It also contains the description of the data structures used in our algorithm: it is based on a quadtree structure [10], used in a novel way for describing points in the plane, combined with a radix tree (see for instance [11]) structure for the labels. In Section 3 the algorithm is described in details. The time and space complexity of the algorithm is carried out in Section 4, followed by a discussion on complexity issues, with respect to the size of numbers and bit operations involved. Finally a list of possible applications is provided showing its usefulness.

2 Preliminaries

A word w is a finite sequence of letters $w_1 w_2 \cdots w_n$ on a finite alphabet Σ, that is a function $w : [1..n] \longrightarrow \Sigma$, and $|w| = n$ is its *length*. Therefore the ith letter of a word w is denoted w_i, and sometimes $w[i]$ when we emphasize the algorithmic point of view. The empty word is denoted ε. The set of words of length n is denoted Σ^n, that of length at most n is $\Sigma^{\leq n}$, and the set of all finite words is Σ^*, the free monoid on Σ. Similarly, the number of occurrences of the letter $\alpha \in \Sigma$, is denoted $|w|_\alpha$. From now on, the alphabet is fixed to $\Sigma = \{a, b, \overline{a}, \overline{b}\}$. To any word $w \in \Sigma^*$ is associated a vector $\overrightarrow{w}$ by the morphism $\overrightarrow{\cdot} : \Sigma^* \longrightarrow \mathbb{Z} \times \mathbb{Z}$ defined on the elementary translation $\overrightarrow{\epsilon} = (\epsilon_1, \epsilon_2)$ corresponding to each letter $\epsilon \in \Sigma$:

$$\overrightarrow{a} = (1, 0), \overrightarrow{\overline{a}} = (-1, 0), \overrightarrow{b} = (0, 1), \overrightarrow{\overline{b}} = (0, -1),$$

and such that $\overrightarrow{u \cdot v} = \overrightarrow{u} + \overrightarrow{v}$. For sake of simplicity we often use the notation $\boldsymbol{u}$ for vectors $\overrightarrow{u}$.

The set of elementary translations allows to draw each word as a 4-connected path in the plane starting from the origin, going *right* for a letter a, *left* for a letter $\overline{a}$, *up* for a letter b and *down* for a letter $\overline{b}$. This coding proved to be very convenient for encoding the boundary of discrete sets and is well known in discrete geometry as the *Freeman chain code* [1,2]. It has been extensively used in many applications and allowed the design of elegant and efficient algorithms for describing geometric properties.

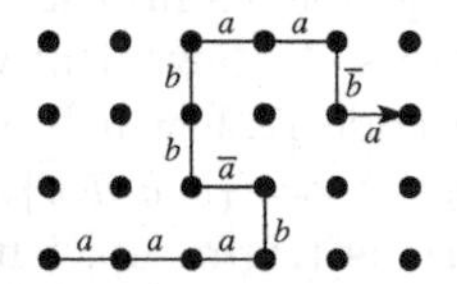

Fig. 1. Path encoded by the word $w = aaab\overline{a}\overline{b}bbaa\overline{b}a$

The underlying principle of our algorithm is to build a graph whose nodes represent points of the plane. For that purpose, the plane is partitioned as in Fig. 2, where the point $(2,1)$ is outlined with its four sons (solid arrows) and its four neighbors (dashed arrows). The sons of a node are grouped in grey zones, while dashed lines separate the levels of the tree. Each node has two possible states : *visited* or *not visited*. New nodes are created while reading the word $w = w_1 w_2 \cdots w_n$ from left to right. For each letter w_i, the node corresponding to the point $\overrightarrow{w[1..i]}$ is written as *visited* and of course, if at some point a node is visited twice then the path is not self-avoiding and the algorithm stops. During the process is built a graph $\mathcal{G} = (N, R, T)$ where N is a set of nodes associated to points of the plane, R and T are two distincts sets of oriented edges. The edges in R give a quadtree structure on the nodes while the edges in T are links from each node to its *neighbors*, for which we give a precise definition.

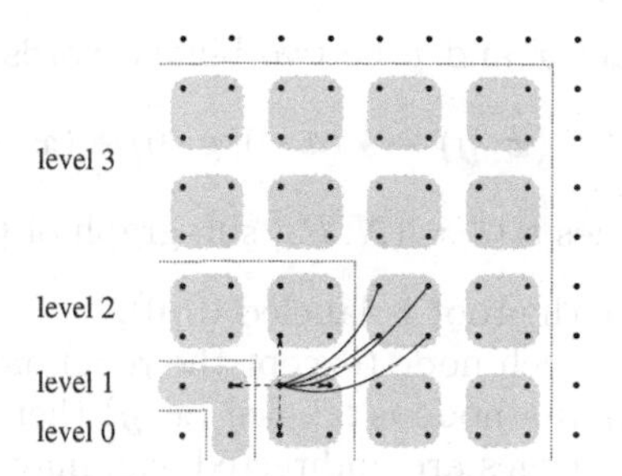

Fig. 2. Partition of $\mathbb{N} \times \mathbb{N}$

Definition 1. *Given a point* $(x, y) \in \mathbb{Z}^2$, *we say that* (x', y') *is a* neighbor *of* (x, y) *if there exists* $\epsilon \in \Sigma$ *such that* $(x', y') = (x, y) + \epsilon = (x + \epsilon_1, y + \epsilon_2)$.

When we want to discriminate the neighbors of a given point (x, y), for each $\epsilon \in \Sigma$, we say that (x', y') is an *ϵ-neighbor* of (x, y) if $(x', y') = (x, y) + \epsilon$.

2.1 Data Structure

First, we assume that the path is coded by a word w starting at the origin $(0, 0)$, and stays in the first quadrant $\mathbb{N} \times \mathbb{N}$. This means that the coordinates of all points are nonnegative. Subsequently, this solution is modified in order to remove this assumption. Note that in $\mathbb{N} \times \mathbb{N}$, each point has exactly four neighbors with the exception of the origin $(0, 0)$ which admits only two neighbors, namely $(0, 1)$ and $(1, 0)$, and the points on the half lines $(x, 0)$ and $(0, y)$ with $x, y \geq 1$ which admit only three neighbors (see Fig. 2).

Let $\mathbb{B} = \{0, 1\}$ be the base for writing integers. Words in $\mathbb{B}^*$ are conveniently represented in the *radix order* by a complete binary tree (see for instance [11,12]), where the level k contains all the binary words of length k, and the order is given by the breadth-first traversal of the tree. To distinguish a natural number $x \in \mathbb{N}$ from its representation we write $\boldsymbol{x} \in \mathbb{B}^*$. The edges are defined inductively by the rewriting rule $\boldsymbol{x} \longrightarrow \boldsymbol{x} \cdot 0 + \boldsymbol{x} \cdot 1$, with the convention that 0 and 1 are the labels of, respectively, the left and right edges of the node having value $\boldsymbol{x}$. This representation is extended to $\mathbb{B}^* \times \mathbb{B}^*$ as follows.

A quadtree with a radix tree structure for points in the integer plane. As usual, the concatenation is extended to the cartesian product of words by setting for $(\boldsymbol{x}, \boldsymbol{y}) \in \mathbb{B}^* \times \mathbb{B}^*$, and $(\alpha, \beta) \in \mathbb{B} \times \mathbb{B}$

$$(\boldsymbol{x}, \boldsymbol{y}) \cdot (\alpha, \beta) = (\boldsymbol{x} \cdot \alpha, \boldsymbol{y} \cdot \beta).$$

Let $\boldsymbol{x}$ and $\boldsymbol{y}$ be two binary words having same length. Then the rule

$$(\boldsymbol{x}, \boldsymbol{y}) \longrightarrow (\boldsymbol{x} \cdot 0, \boldsymbol{y} \cdot 0) + (\boldsymbol{x} \cdot 0, \boldsymbol{y} \cdot 1) + (\boldsymbol{x} \cdot 1, \boldsymbol{y} \cdot 0) + (\boldsymbol{x} \cdot 1, \boldsymbol{y} \cdot 1) \qquad (1)$$

defines a $\mathcal{G}' = (N, R)$, sub-graph of $\mathcal{G} = (N, R, T)$, such that :

(i) the root is labeled $(0, 0)$;
(ii) each node (except the root) has four sons;
(iii) if a node is labeled $(\boldsymbol{x}, \boldsymbol{y})$ then $|\boldsymbol{x}| = |\boldsymbol{y}|$;
(iv) edges are undirected, e.g. may be followed in both directions.

By convention, edges leading to the sons are labeled by pairs from the ordered set $\{(0,0), (0,1), (1,0), (1,1)\}$. These labels equip the quadtree with a *radix tree* structure for Equation (1) implies that (x', y') is a son of (x, y), if and only if

$$(x', y') = (2x + \alpha, 2y + \beta),$$

for some $(\alpha, \beta) \in \mathbb{B} \times \mathbb{B}$. Observe that any pair (x, y) of nonnegative integers is represented exactly once in this tree. Indeed, if $|\boldsymbol{x}| = |\boldsymbol{y}|$ (by filling with zeros at the left of the shortest one), the sequence of pairs of digits (the two digits in first place, the two digits in second place, and so on) gives the unique path in the tree leading to this pair. Of course the root may have up to three sons since no edge labeled $(0, 0)$ starts from the root.

Neighboring links. We superpose on G' the neighboring relation given by the edges of T (dashed lines). More precisely, for each elementary translation $\epsilon \in \Sigma$, each node Ⓩ $= (x, y)$ is linked to its ϵ-neighbor Ⓩ$+\epsilon$, when it exists. If a level k is fixed (see Fig. 2), it is easy to construct the graph

$$\mathcal{G}^{(k)} = (N^{(k)}, R^{(k)}, T^{(k)})$$

such that

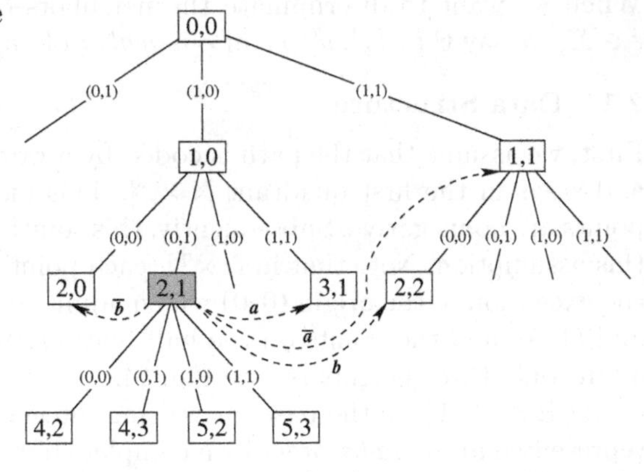

Fig. 3. The point $(2, 1)$ with its neighbors

(i) if $(\boldsymbol{x}, \boldsymbol{y}) \in N^{(k)}$, then $|\boldsymbol{x}| = |\boldsymbol{y}| = k$;
(ii) the functions $N^{(k)} \hookrightarrow \mathbb{N} \times \mathbb{N} \hookrightarrow \mathbb{B}^* \times \mathbb{B}^*$ are injective;
(iii) $R^{(k)}$ is the radix-tree representation : $(\mathbb{B}^{<k} \times \mathbb{B}^{<k}) \times (\mathbb{B} \times \mathbb{B}) \xrightarrow{\bullet} \mathbb{B}^{\leq k} \times \mathbb{B}^{\leq k}$;
(iv) the neighboring relation is $T^{(k)} \subseteq N \times (\mathbb{B} \times \mathbb{B}) \times N$.

Note that the labeling in Fig. 3 is superfluous: each node represents indeed an integer unambiguously determined by the path from the root using edges in R; similarly for the ordered edges. Moreover, if a given subset $M \subset \mathbb{N} \times \mathbb{N}$ has to be represented, then one may trim the unnecessary nodes so that the corresponding graph $\mathcal{G}_M$ is not necessarily complete.

3 The Algorithm

Adding 1 to an integer $\boldsymbol{x} \in \mathbb{B}^k$ is easily performed by a sequential function. Indeed, every positive integer can be written $\boldsymbol{x} = u1^i0^j$, where $i \geq 1$, $j \geq 0$, with $u \in \{\varepsilon\} \cup \{\mathbb{B}^{k-i-j-1} \cdot 0\}$. In other words, 1^j is the last run of 1's. The piece of code for adding 1 to an integer written in base 2 is

1 : **If** $j \neq 0$ **then** *Return* $u1^i0^{j-1}1$;
2 : **else If** $u = \varepsilon$ **then** *Return* $1 \cdot 0^i$;
3 : **else** *Return* $u \cdot 0^{-1} \cdot 1 \cdot 0^i$;
4 : **end if**
5 : **end if**

where 0^{-1} means to erase a 0. Clearly, the computation time of this algorithm is proportional to the length of the last run of 1's. Much better is achieved with the radix tree structure, where, given a node ⓩ, its *father* is denoted f(ⓩ), and we write $f(x, y)$ or $f(\boldsymbol{x}, \boldsymbol{y})$ if its label is (x, y). The following technical lemma is a direct adaptation to $\mathbb{B}^* \times \mathbb{B}^*$ of the addition above.

Lemma 1. *Let $G^{(k)}$ be the complete graph representing $\mathbb{B}^{\leq k} \times \mathbb{B}^{\leq k}$ for some $k \geq 1$, $\epsilon \in \Sigma$, and ⓩ $= (\boldsymbol{x}, \boldsymbol{y})$ be a node of N^k. If one of the four conditions holds:*

$$\text{(i) } \epsilon = a \text{ and } \boldsymbol{x}[\mathrm{k}] = 0, \quad \text{(ii) } \epsilon = \overline{a} \text{ and } \boldsymbol{x}[\mathrm{k}] = 1,$$
$$\text{(iii) } \epsilon = b \text{ and } \boldsymbol{y}[\mathrm{k}] = 0, \quad \text{(iv) } \epsilon = \overline{b} \text{ and } \boldsymbol{y}[\mathrm{k}] = 1,$$

then f(ⓩ) $= f$(ⓩ $+ \epsilon$). Otherwise, f(ⓩ) $+ \epsilon = f$(ⓩ $+ \epsilon$).

The process is illustrated for case (i) in the diagram on the right where the nodes $(10110, \bullet)$ and $(10111, \bullet)$ share the same father while fathers of neighboring nodes $(\bullet, 01011)$ and $(\bullet, 01011)$ are distinct but share the same neighboring relation.

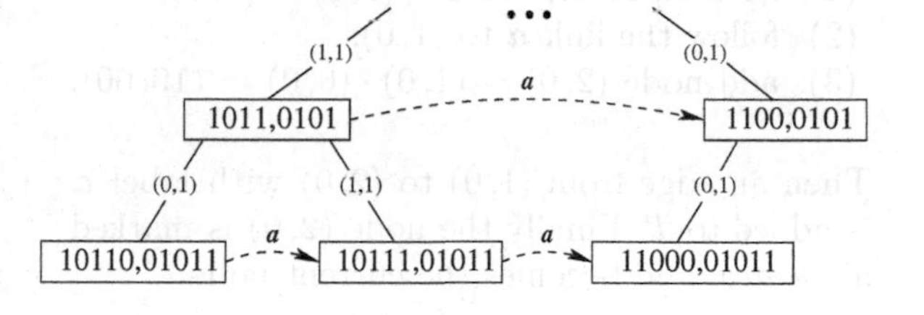

Now, assume that the node $(\boldsymbol{x}, \boldsymbol{y})$ exists and that its neighbor $(x+1, y+0)$ does not. If $|\boldsymbol{x}| = |\boldsymbol{y}| = k$, then the translation $(x, y) + (1, 0)$ is obtained in three steps by the following rules:

1. take the edge in R to $f(\boldsymbol{x}, \boldsymbol{y}) = (\boldsymbol{x}[1..k-1], \boldsymbol{y}[1..k-1])$;
2. take (or create) the edge in T from $f(\boldsymbol{x}, \boldsymbol{y})$ to ⓩ $= f(\boldsymbol{x}, \boldsymbol{y}) + (1, 0)$;
3. take (or create) the edge in R from ⓩ to ⓩ $\cdot (0, \boldsymbol{y}[k])$.

By Lemma 1, we have ⓩ $\cdot (0, \boldsymbol{y}[k]) = (x+1, y+0)$, so that it remains to add the neighboring link $(\boldsymbol{x}, \boldsymbol{y}) \overset{a}{\dashrightarrow} (x+1, y+0)$. Then, a nonempty word $w \in \Sigma^n$ is sequentially processed to build the graph $\mathcal{G}_w$, and we illustrate the algorithm on the input word $w = aabb$.

• *Initialization*: the algorithm starts with the graph containing only the node $(0,0)$ marked as visited. For convenience, the *non-visited* nodes $(0,1),(1,0)$, and the links from $(0,0)$ to its neighbors are also added. This is justified by the fact that the algorithm applies to nonempty words.

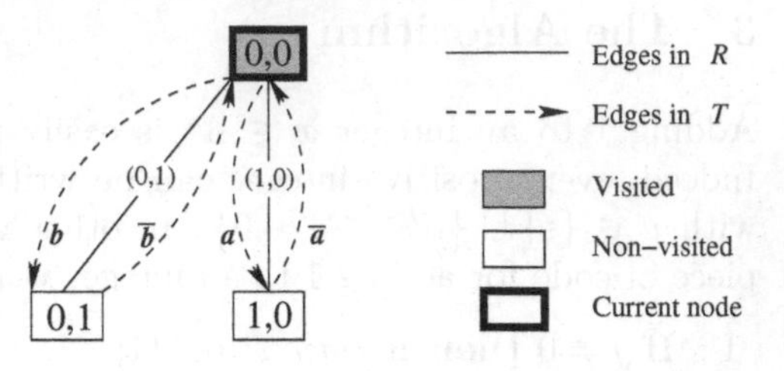

Fig. 4. Initial graph $\mathcal{G}_\varepsilon$

Since $(0,0)$ is an ancestor of all nodes, this ensures that every node has an ancestor linked with its neighbors. The *current node* is set to $(0,0)$ and this graph is called the *initial graph* $\mathcal{G}_\varepsilon$.

• *Read* $w_1 = a$: this corresponds to the translation $(0,0)+(1,0)$. A neighboring link labeled $\boldsymbol{a}$ starting from $(0,0)$ and leading to the node $(1,0)$ does exist, so the only thing to do is to follow this link and mark the node $(1,0)$ as visited. The current node is now set to $(1,0)$, and this new graph is called $\mathcal{G}_a$.

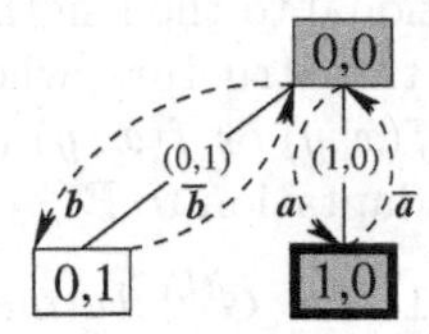

Fig. 5. Graph $\mathcal{G}_a$

• *Read* $w_2 = a$: this time, there is no edge in $\mathcal{G}_a$ labeled $\boldsymbol{a}$ starting from $(1,0)$. Using the translation rules above, we perform:

(1) go back to the father $f(1,0) = (0,0)$;
(2) follow the link $\boldsymbol{a}$ to $(1,0)$;
(3) add node $(2,0) \sim (1,0)\cdot(0,0) = (10,00)$.

Then an edge from $(1,0)$ to $(2,0)$ with label $\boldsymbol{a}$ is added to T. Finally the node $(2,0)$ is marked as *visited*, and becomes the current node.

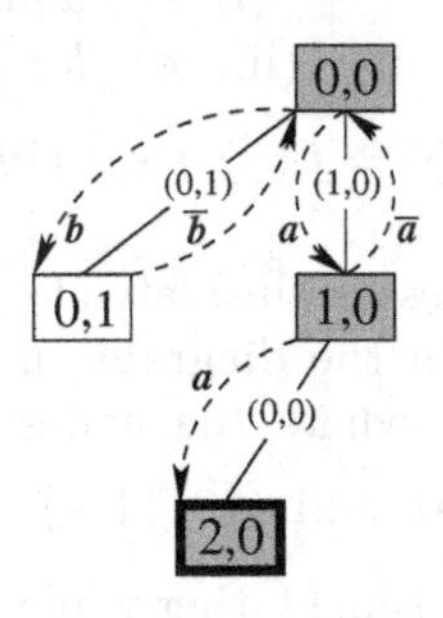

Fig. 6. Graph $\mathcal{G}_{aa}$

• *Read* $w_3 = b$: this amounts to perform the translation $(2,0)+(0,1)$. Since the edge to $f(2,0)$ is labeled by $(0,0)$, we know that the second coordinate of the current node $(2,0)$ is even. Therefore, $(2,1)$ and $(2,0)$ must be siblings, that is $f((2,0)+(0,1)) = f((2,0))$. What we need to do then is :

(1) go back to the father $f(2,0) = (1,0)$;
(2) follow the edge $\boldsymbol{b}$ if it exists;

Since it is does not exist, it must be created to reach the node $(2,1) \sim (10,01) = (1,0)\cdot(0,1)$. Again an edge from $(2,0)$ to $(2,1)$ with label $\boldsymbol{b}$ is added, $(2,1)$ is marked as *visited* and is now the current node.

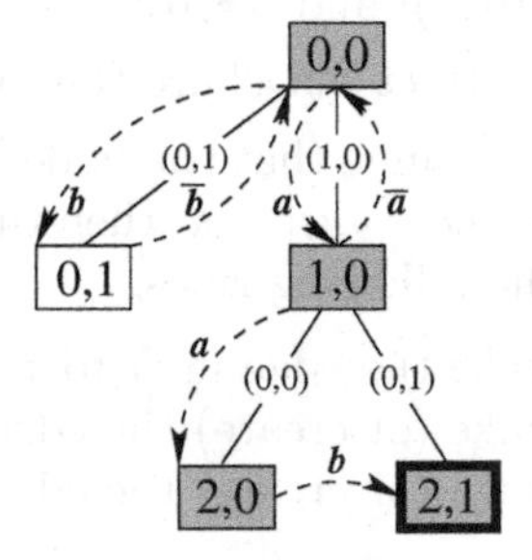

Fig. 7. Graph $\mathcal{G}_{aab}$

• *Read* $w_4 = b$: since $f((2,1))$ has no neighboring link labeled by $\boldsymbol{b}$, recursion is used to find (or build if necessary) the node corresponding to its translation by $\boldsymbol{b}$. This leads to the creation of the node $(1,1) \sim (0,0) \cdot (1,1)$ marked as *non-visited*. Then, the node $(2,2) \sim (1,1) \cdot (0,0)$ is added, marked as *visited*, and becomes the current node. Note that a neighboring links between $(1,0)$ and $(1,1)$, $(2,1)$ and $(2,2)$ are added to avoid eventual searches.

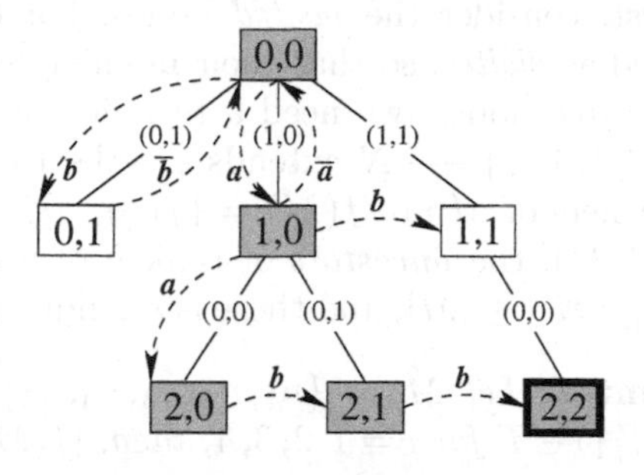

Fig. 8. Graph $\mathcal{G}_{aabb}$

The algorithm **readWord** sequentially reads $w \in \Sigma^*$, builds dynamically the graph G_w marking the corresponding node as visited, and determines if the path coded by w is self-intersecting, i.e. if some node is visited at least twice.

Algorithm 1 (readWord)
Input: $w \in \{a, b, \overline{a}, \overline{b}\}^*$

0 : $\mathcal{G} \leftarrow \mathcal{G}_\varepsilon$; ⓒ $\leftarrow$ root of $\mathcal{G}$;
1 : **For** i **from** 1 **to** $|w|$ **do**
2 : $\quad \epsilon \leftarrow w_i$;
3 : $\quad$ ⓩ $\leftarrow$ findNeighbor($\mathcal{G}$, ⓒ, ϵ);
4 : $\quad$ **If** ⓩ is *visited* **then**
5 : $\quad\quad w$ is self-intersecting.
6 : $\quad$ **end if**
7 : $\quad$ **Mark** ⓩ as *visited*;
8 : $\quad$ ⓒ $\leftarrow$ ⓩ;
9 : **end for**
10 : w is not self-intersecting.

Algorithm 2 (findNeighbor)
Input: $\mathcal{G} = (N, R, T)$; ⓒ $\in N$;
$\epsilon \in \{a, b, \overline{a}, \overline{b}\}$;

1 : **If** the link ⓒ $\overset{\epsilon}{\dashrightarrow}$ ⓩ does not exist **then**
2 : $\quad$ ⓟ $\leftarrow f$(ⓒ)
3 : $\quad$ **If** f(ⓒ $+ \epsilon$) $= f$(ⓒ) **then**
4 : $\quad\quad$ ⓡ $\leftarrow$ ⓟ;
5 : $\quad$ **else**
6 : $\quad\quad$ ⓡ $\leftarrow$ findNeighbor($\mathcal{G}$, ⓟ, ϵ);
7 : $\quad$ **end if**
8 : $\quad$ ⓩ $\leftarrow$ son of ⓡ corresponding to ⓒ$+\epsilon$;
9 : $\quad$ Add the neighboring link ⓒ $\overset{\epsilon}{\dashrightarrow}$ ⓩ.
10 : **end if**
11 : **return** ⓩ;

The algorithm **findNeighbor** finds, and creates if necessary, the ϵ-neighbor of a given node. Thanks to Lemma 1, testing the condition on line 3 is performed in constant time. At line 8, if the node ⓩ does not exist, it is created. Clearly, the time complexity of this algorithm is entirely determined by the recursive call on line 6 since all other operations are performed in constant time. Finally, note that after each call to **findNeighbor** on line 3 of **readWord**, there always exist a neighboring link ⓒ $\overset{\epsilon}{\dashrightarrow}$ ⓩ.

4 Complexity Analysis

The key for analyzing the complexity of this algorithm rests on the fact that each recursive call of Algorithm 2 requires the addition of a neighboring link. This implies that given a node ⓩ $\in N$, when all the neighboring links have been added, there will never be another recursive call on ⓩ. Since a node has at most 4 sons and each of these sons has at most 2 neighbors not sharing the same father, the number of recursive calls on a single node is bounded by 8. It remains to show that the number of nodes in the graph is proportional to $|w|$.

First, consider the *visited* nodes. For each letter read, exactly one node is marked as *visited*, so that their number is $|w|$. In order to bound the number of *non-visited* nodes, we need a technical lemma. Recall that the father function $f : N \setminus \{(0,0)\} \longrightarrow N$ extends to subset of nodes in the usual way: for $M \subseteq N$, the fathers of M are $f(M) = \{f(ⓢ) \mid ⓢ \in M\}$. Moreover, f can be iterated to get $f^h(M)$, the *ancestors* of rank h of a subset M. Clearly, f is a contraction since $|f(M)| \leq |M|$, and there is a unique ancestor of all nodes, namely the root.

Lemma 2. *Let* $M = \{n_1, n_2, n_3, n_4, n_5\} \subset N$ *a set of five nodes such that* $(n_i, n_{i+1}) \in T$ *for* $i = 1, 2, 3, 4$*, then,* $|f(M)| \leq 4$.

Proof. As shown in Fig. 2, the nodes sharing the same father split the plane in 2×2 squares. As a consequence, at least two of the nodes n_1, n_2, n_3, n_4, n_5 must share the same father, providing the bound $|f(M)| \leq 4$. ∎

This allows to bound the number of nodes using the fact that all non- visited nodes are ancestors of visited ones: the only exceptionis the initialisation step where the non-visited nodes $(0,1)$ and $(1,0)$ are created as leaves.

Lemma 3. *Given a word* $w \in \Sigma^n$ *and the graph* $\mathcal{G}_w = (N, R, T)$*, the number of nodes in* N *is in* $\mathcal{O}(n)$.

Proof. Let $N_v \subseteq N$ be the set of visited nodes, and h be the height of the tree (N, R). It is clear that $N = \bigcup_{0 \leq i \leq h} f^i(N_v)$, and so

$$|N| \leq \sum_{0 \leq i \leq h} |f^i(N_v)|. \tag{2}$$

By construction, the set N_v forms a sequence of nodes such that two consecutive ones are neighbors since they correspond to the path coded by w. Thus, by splitting this sequence of nodes in blocks of length 5, the previous lemma applies, and we have

$$|f(N_v)| \leq 4 \left\lceil \frac{|N_v|}{5} \right\rceil \leq \frac{4}{5}(|N_v| + 4). \tag{3}$$

By Lemma 1, two neighboring nodes either share the same father or have different fathers that are neighbors, so it is for the sets $f(N_v), f^2(N_v), \ldots, f^h(N_v)$. So, by combining inequations (2) and (3), the following bound is obtained

$$\begin{aligned} |N| &\leq \sum_{0 \leq i \leq h} |f^i(N_v)| \leq \sum_{0 \leq i \leq h} \left(\left(\frac{4}{5}\right)^i |N_v| + \sum_{0 \leq j \leq i} \left(\frac{4}{5}\right)^j 4 \right) \\ &\leq |N_v| \left(\frac{1}{1 - \frac{4}{5}} \right) + 4 \sum_{0 \leq i \leq h} \left(\frac{1}{1 - \frac{4}{5}} \right) \leq 5|N_v| + 20h. \end{aligned}$$

Since the height h of the tree (N, R) corresponds to the number of bits needed to write the coordinates of the nodes in N, $h \in \mathcal{O}(\log n)$ and thus $|N| \in \mathcal{O}(n)$. ∎

Note that the linearity constant obtained here is very large. Indeed, our goal here is to prove the linearity of the global algorithm, and not to provide a tight bound. With a more detailed analysis, the bound $|N| \leq 3|N_v| + 6h$ can be obtained for the number of nodes [13].

Theorem 1. *Given a word $w \in \Sigma^n$, determining if the path coded by w intersects itself is decidable in $\mathcal{O}(n)$.*

Proof. Lemma 3 implies that if the path w starts at $(0,0)$ and stays in the first quadrant, then determining whether w intersects itself or no is decidable in linear time. All that is needed to adapt this solution to any word $w \in \Sigma^*$ is to use four graphs $\mathcal{G}_{\mathrm{I}}, \mathcal{G}_{\mathrm{II}}, \mathcal{G}_{\mathrm{III}}, \mathcal{G}_{\mathrm{IV}}$ simultaneously.

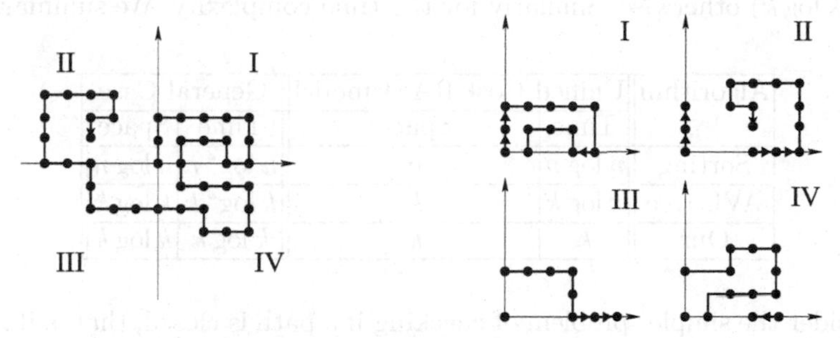

The cohesion between these graphs is ensured by *special nodes*, those representing points on axes. Since a point on an axis, distinct from the origin, is in exactly two quadrants, an additionnal link between the two nodes representing this point is added. These two nodes are *equivalent* since they represent the same point. The axis on which this point is located must also be stored in order to identify the quadrant in which the path enters. Switching from one graph to another is achieved with the following rules:

$$\text{(II) } a \leftrightarrow \overline{a}; \quad \text{(III) } a \leftrightarrow \overline{a}, b \leftrightarrow \overline{b}; \quad \text{(IV) } b \leftrightarrow \overline{b}.$$

This allows the processing of all coordinates as positive integers since their sign is determined by the quadrant. Consequently, each time a special node is created, its equivalent one is also created in the appropriate graph using the link between their fathers. ∎

Performance issues and comparison. Among the many ways of solving the intersection problem, the naive sparse matrix representation that requires an $O(n^2)$ space and initialization time is eliminated in the first round. When efficiency is concerned, there are two well-known approaches for solving it: one may store the coordinates of the visited points and sort them, then check if two consecutive sets of coordinates are equal or not (we call this *sorting algorithm*). One may also store the sets of coordinates in an AVL-tree and check for each new set of coordinates if it is already present or not (the *AVL algorithm*). Let us first assume that the path $w \in \Sigma^n$ is not self intersecting. Then the length of the largest coordinate is $O(\log n)$. But the largest coordinate is also $\Omega(\log n)$ because if the path is not self intersecting, the minimum coordinates are obtained when the points remain in a square centered on $(0, 0)$ with $\sqrt{n}$ side length. Since $\log(\sqrt{n}) = \frac{1}{2}\log n$ the largest coordinate is also $\Omega(\log n)$. Thus the storage of the largest coordinate is in $\Theta(\log n)$ and the whole storage costs $\Theta(n \log n)$.

Sorting n ordered pairs can be done in $\Theta(n \log n)$ swaps or comparisons. But each swap or comparison costs $\Theta(\log n)$ clock ticks. Then the whole computation time is $\Theta(n \log^2(n))$. In our algorithm, the storage cost and computation time are both $\Theta(k)$ where k is the index of the second occurrence of the point appearing at least twice in the path or the path length if it is not self intersecting. Unlike the sorting algorithm, there is no need to store the whole path: the computation is performed dynamically. With our algorithm, storing the necessary data costs, both on average and in worst case, $\mathcal{O}(k)$ if k can be stored in a machine word and $\mathcal{O}(k \log k)$ otherwise. Similarly for the time complexity. We summarize :

Algorithm	Unified Cost RAM model		General Case	
	Time	Space	Time	Space
Sorting	$n \log n$	n	$n \log^2 n$	$n \log n$
AVL tree	$k \log k$	k	$k \log^2 k$	$k \log k$
Our	k	k	$k \log k$	$k \log k$

Consider the simpler problem of checking if a path is closed, that is if for each $\epsilon \in \Sigma$ we have $|w|_\epsilon = |w|_{\bar{\epsilon}}$. The cost of storing the number $|w|_\epsilon$ of occurrences of each elementary step is in $\mathcal{O}(1)$ if each of these numbers can be stored in a single machine word, or in $\mathcal{O}(\log(n_\epsilon))$ otherwise. Increasing or decreasing the number $|w|_\epsilon$ by 1 costs $\mathcal{O}(1)$ on average in both cases, and at worst $\mathcal{O}(1)$ for the first case and $\mathcal{O}(\log n)$ otherwise. The total time is hence $\mathcal{O}(n)$ in the first case and $\mathcal{O}(n \log n)$ otherwise.

There are other ways to deal with paths having negative coordinates. Indeed, since the property of being self intersecting or not is invariant by translation, it suffices to translate the path conveniently. This can be achieved by making one pass on the word w to determine the starting node ⓢ as follows:

(a) ⓢ $\leftarrow (n, n)$, where $n = |w|$;

(b) ⓢ $\leftarrow (x, y)$ where x and y are determined from the extremal values.

In both cases it takes $\mathcal{O}(n)$ for reading the word, $\mathcal{O}(\log n)$ time and space to

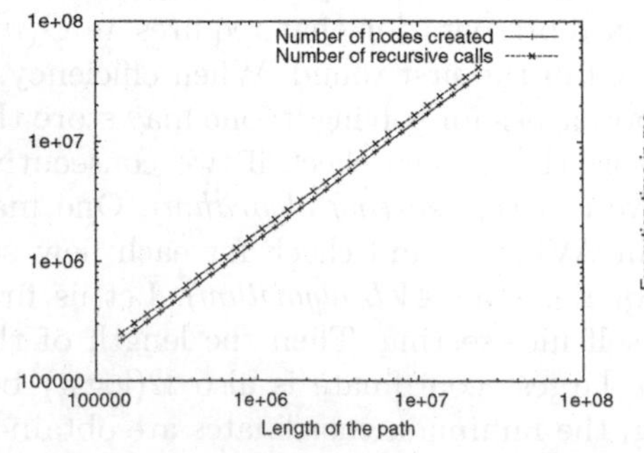

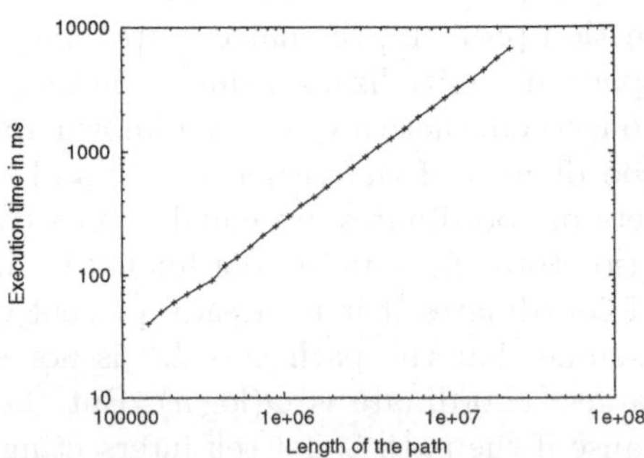

represent ⓢ. Then the path is encoded in the radix-tree starting from ⓢ. In our solution, we avoid the linear preprocessing for determining these values.

Numerical results. Our algorithms were implemented in C++ and tested on numerous examples. The results achieved for instance with $w_n = a^n b^n$ reveal a smaller linearity constant than the constant provided in the proof of Lemma 3, and confirmed their efficiency.

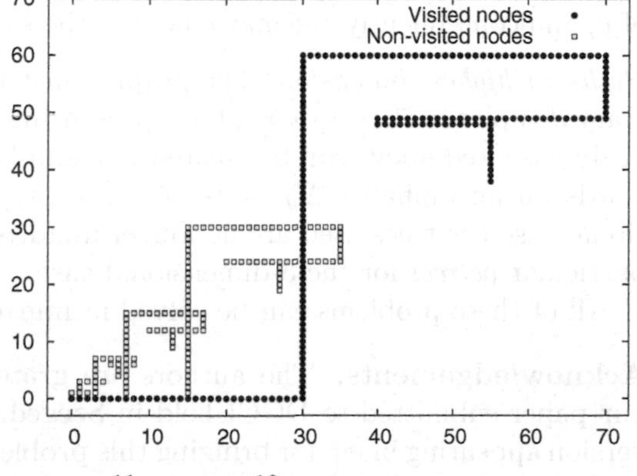

The radix-tree built for each word corresponds, as shown in Fig. 2, to points in the discrete plane $\mathbb{N} \times \mathbb{N}$. We provide on the right an illustration of the nodes involved in the radix-tree in the case of the word

$$w = a^{30} b^{60} a^{40} \overline{b}^{11} \overline{a}^{30} \overline{b} a^{15} \overline{b}^{10}.$$

5 Concluding Remarks

The first advantage of our algorithm is that ordering of edges can be used for avoiding labeling of both nodes and edges. Moreover, the neighboring relation T as presented is not implemented in its symmetric form. It could be easily done since each time a neighboring link ⓒ $\overset{\epsilon}{\dashrightarrow}$ ⓩ is added at line 9 of Algorithm 2, we can add its symmetric link ⓩ $\overset{\overline{\epsilon}}{\dashrightarrow}$ ⓒ at constant cost. This does not change the overall complexity, and further analysis is required for determining if it is worthwhile. On the other hand, our algorithm is useful for solving a series of related problems in discrete geometry.

Determining if a path w crosses itself. When a node is visited twice, deciding whether the path crosses itself or not amounts to check local conditions, describing all the possible configurations (See [8] Section 4.1).

Determining if $w \in \Sigma^n$ is the Freeman chain code of a discrete figure. It suffices to check that the last visited node is the starting one. This does not penalize the linear algorithms for determining, for instance, if a discrete figure is digitally convex [7], or if it tiles the plane by translation [8]. In the case of a self intersecting path, it also allows the decomposition of a discrete figure in elementary components, not necessarily disjoint.

Node multiplicity. By replacing the "visited/unvisited" labeling of nodes with a counter (set to 0 when a node is created), the number of times a node is visited is computed by replacing the lines 4, 5, 6 and 7 in Algorithm 1 by the incrementation of this counter. Then, the obsolete line 10 must be removed.

Intersection of distinct paths. Given two distinct paths u and v of length bounded by $n = \max\{|u|, |v|\}$, with starting nodes ⓢ and ⓡ respectively, their intersection is computed by constructing first the graph $\mathcal{G}_u$, inserting the node ⓡ in $\mathcal{G}_u$, and constructing $\mathcal{G}_{u,v} = \mathcal{G}_u + \mathcal{G}_v$. Again the overall algorithm remains in $\mathcal{O}(n)$. As a byproduct of this construction, given two nonintersecting closed paths u and v, it is decidable whether the interior of u is included in the interior of v; and consequently one may compute the exterior envelope of discrete figures.

Paths in higher dimension. The graph construction extends naturally to arbitrary d-tuples in $\mathbb{B}^* \times \cdots \times \mathbb{B}^*$, for representing numbers in $\mathbb{N}^d$. Therefore, all the problems cited above can be treated in a similar way, by processing sequentially words on an alphabet $\Sigma_d = \{\epsilon_1, \bar{\epsilon}_1, \ldots \epsilon_d, \bar{\epsilon}_d\}$, of size $2d$. In the multidimensional case the trees used are no longer quadtrees but higher order trees, and in particular *octrees* for the 3-dimensional case.

All of these problems can be solved in linear time and space complexity.

Acknowledgements. The authors are grateful to the anonymous referee of our paper submitted to DGCI held in Szeged, who also reviewed the extended version appearing in [8], for bringing this problem to our attention. A preliminary version (in French) of the results presented here appears in the doctoral thesis of Xavier Provençal [9] who is supported by a scholarship from FQRNT (Québec).

References

1. Freeman, H.: On the encoding of arbitrary geometric configurations. IRE Trans. Electronic Computer 10, 260–268 (1961)
2. Freeman, H.: Boundary encoding and processing. In: Lipkin, B., Rosenfeld, A. (eds.) Picture Processing and Psychopictorics, pp. 241–266. Academic Press, New York (1970)
3. Brlek, S., Labelle, G., Lacasse, A.: A note on a result of Daurat and Nivat. In: De Felice, C., Restivo, A. (eds.) DLT 2005. LNCS, vol. 3572, pp. 189–198. Springer, Heidelberg (2005)
4. Brlek, S., Labelle, G., Lacasse, A.: Properties of the contour path of discrete sets. Int. J. Found. Comput. Sci. 17(3), 543–556 (2006)
5. Debled-Rennesson, I., Rémy, J.L., Rouyer-Degli, J.: Detection of the discrete convexity of polyominoes. Discrete Appl. Math. 125(1), 115–133 (2003)
6. Brlek, S., Lachaud, J.O., Provençal, X.: Combinatorial view of digital convexity. In: Coeurjolly, D., Sivignon, I., Tougne, L., Dupont, F. (eds.) DGCI 2008. LNCS, vol. 4992, pp. 57–68. Springer, Heidelberg (2008)
7. Brlek, S., Lachaud, J.O., Provençal, X., Reutenauer, C.: Lyndon+Christoffel = digitally convex. Pattern Recognition 42, 2239–2246 (2009)
8. Brlek, S., Provençal, X., Fédou, J.M.: On the tiling by translation problem. Discr. Appl. Math. 157, 464–475 (2009)
9. Provençal, X.: Combinatoire des mots, géométrie discrète et pavages. PhD thesis, D1715, Université du Québec à Montréal (2008)
10. Finkel, R., Bentley, J.: Quad trees: A data structure for retrieval on composite keys. Acta Informatica 4(1), 1–9 (1974)
11. Knuth, D.E.: The Art of Computer Programming, Sorting and Searching, vol. 3. Addison-Wesley, Reading (1998)
12. Lothaire, M.: Applied Combinatorics on Words. Cambridge University Press, Cambridge (2005)
13. Labbé, S.: Personal communication (2009)

The <3,4,5> Curvilinear Skeleton

Carlo Arcelli, Gabriella Sanniti di Baja, and Luca Serino

Institute of Cybernetics "E. Caianiello", CNR, Via Campi Flegrei 34
80078 Pozzuoli (Naples), Italy
{c.arcelli,g.sannitidibaja,l.serino}@cib.na.cnr.it

Abstract. A new skeletonization algorithm is presented to compute the curvilinear skeleton of 3D objects. The algorithm is based on the use of the <3,4,5> distance transform, on the detection of suitable anchor points, and on iterated topology preserving voxel removal. The obtained skeleton is topologically correct, is symmetrically placed within the object and its structure reflects the morphology of the represented object.

1 Introduction

The curvilinear skeleton of a 3D object is a representation consisting of the union of curves that has received much attention in the literature; a comprehensive list of contributions can be found in two recent books [1,2]. Such a representation can be computed only for objects rid of cavities. In fact, one of the features of the skeleton is its topological equivalence with the object. Thus, for objects including cavities, a topologically equivalent linear representation cannot be obtained, since each cavity should be mapped into a cavity in the resulting set, which implies that a closed surface surrounding each cavity should be found in the representation.

Besides topological equivalence with the object (meaning that the curvilinear skeleton of a connected object has to be connected and should include as many closed curves, as many are the tunnels of the object), the curvilinear skeleton allows to reduce dimensionality (since a 3D object is represented by a union of curves), is symmetrically placed within the object, and each of the curves constituting it can be seen in correspondence with a part of the object. The latter property is particularly of interest in the framework of the structural approach to object analysis and recognition. In fact, the curvilinear skeleton can be split into the constituting curves and each curve can be used to analyze the part of the object it represents. The spatial relations among the curves of the skeleton can be used to derive information on the organization of the parts constituting the object. For example, this approach has been followed in the 2D case, when the skeleton is used in the framework of OCR systems. The character is thinned down to its skeleton, which is then decomposed into parts corresponding to the strokes forming the character. Also in the 3D case, decomposition of the curvilinear skeleton into its constituting curves can be useful for applications, e.g., for deformable object modeling.

One of the most appealing approaches to skeletonization is based on the use of the distance transform, e.g., [3,13]. As it is well known, all object elements in the distance

S. Brlek, C. Reutenauer, and X. Provençal (Eds.): DGCI 2009, LNCS 5810, pp. 409–420, 2009.

transform are assigned the value of their distance from a reference set. In the 2D case, the landscape paradigm can be followed according to which the 2D distance transform image is interpreted as a 3D elevation map, where distance values of pixels in the 2D image denote the elevations of those pixels with respect to the ground level, i.e., the reference set, in the 3D elevation map. The skeleton can be computed by detecting peaks and ridges in the 3D landscape and by suitably connecting them. Skeletons of 2D objects computed in this way are definitely centered within the objects and, if suitable criteria are used to connect ridges and peaks, are topologically correct. Another important feature of these skeletons is that they are reversible. In fact, the peaks and most of the ridges are centers of maximal discs in the distance transform and it is well known that the union of the maximal discs coincides in shape and size with the original object [14].

Centers of maximal balls in a 3D distance transform have the same property characterizing the centers of maximal discs in the 2D distance transform as concerns their envelope. However, except for the case of 3D objects with tubular shape, where the centers of maximal balls are mainly aligned along linear structures, the centers of maximal balls are often organized in patch structures. As a consequence, the inclusion of all centers of maximal balls in the curvilinear skeleton is not compatible with the requested linear structure of the skeleton. Thus, for 3D objects the curvilinear skeleton does not enjoy the reversibility property. However, the curvilinear skeleton is still of interest for applications if its morphological structure reflects sufficiently well the most salient aspects of the object's shape.

In the 3D case, the landscape paradigm could still be followed, by interpreting the 3D distance transform image as an elevation map in four dimensions. This process would be rather complex. In this paper, we prefer to resort to the classical approach, based on iterated topology preserving voxel removal, and to integrate it with criteria for the detection of suitable anchor points in the distance transform, so as to avoid unwanted shortening of skeleton branches.

We point out that also skeletonization methods that do not explicitly refer to the distance transform, e.g., those based on boundary evolution, do use information derived from the distance transform [15-19]. If the boundary of an object is interpreted as moving towards the inside of the object, the boundary motion can be modeled as a continuous wavefront by means of a partial differential equation and the singularities of the flow can be detected to compute the skeleton, by using the gradient of the Euclidean distance to the boundary (or the gradient of an approximation of the Euclidean distance).

Our skeletonization method consists of two phases. During the first phase, the 3D object is transformed into a topologically equivalent set consisting of both line- and patch-structures. During the second phase, such a set is furthermore compressed into the curvilinear skeleton. The distance transform is used both to reduce the computation time and to guarantee that the curvilinear skeleton is centered within the object and adequately represents it. The method is described in Section 2, together with some necessary notions and definitions. Experimental results are discussed in Section 3 and a brief conclusion is given in Section 4.

2 The Curvilinear Skeleton

We deal with binary voxel images in cubic grids, where the object F is the set of 1's and the background B is the set of 0's, and use the 26-connectedness and the 6-connectedness for the object and the background, respectively. The 3×3×3 neighborhood N(p) of a voxel p includes the 6 face-, the 12 edge- and the 8 vertex-neighbors of p (respectively denoted by n_f, n_e, n_v). The set N*(p) includes only the 6 face- and the 12 edge-neighbors of p. The generic neighbor of p is denoted by n_i, where i stands for f, e, or v.

We assume that F does not include any cavity otherwise the curvilinear skeleton could not be computed unless altering topology.

The number of 26-connected object components computed in N(p), and the number of 6-connected background components having p as face-neighbor and computed in N*(p) are respectively denoted by c_p and by c^*_p.

An object voxel p is *simple* if the two objects with and without p are topologically equivalent. In [20,21] it has been shown that p is simple iff c_p=1 and c^*_p=1.

The general scheme of the proposed skeletonization method is the following.

During the first phase, aimed at extracting from the 3D object a subset consisting of patches and lines, the distance transform of the input object from its complement is computed; a proper subset of the set of centers of maximal balls detected in the distance transform is taken as the set of anchor points; object voxels are checked in increasing distance value order and are removed (i.e., are assigned the background value 0) if they are not anchor points and their removal does not alter topology; the obtained set of patches and lines is simplified (i.e., is reduced to unit-thickness and some peripheral patches and lines regarded as non-meaningful are removed).

During the second phase, a classification of the voxels of the obtained set is done; the distance transform of the set is computed; a subset of the set of centers of maximal balls, as well as voxels classified as curve- branching- junction-voxels are taken as anchor points; voxels are checked in increasing distance value order and are removed if they are not anchor points and their removal does not alter topology; finally, the obtained set of lines is simplified (i.e., is reduced to the unit-wide curvilinear skeleton and peripheral branches regarded as non-meaningful are removed).

2.1 First Phase

The distance transform DT_F of F with respect to B is computed by using the <3,4,5> distance function [22], which provides a reasonably good approximation of the Euclidean distance and is particularly suited when dealing with digital objects.

The k-th layer of DT_F is the set including all voxels, whose distance value p satisfies the condition $3(k-1) < p \leq 3k$, see [23].

The border of the object at the k-th iteration of skeletonization consists of the voxels belonging to previous layers and that have not been removed at previous iterations, as well as of the voxels in the k-th layer. All other object voxels constitute the inside of the object at the k-th iteration.

In DT_F, a center of a maximal ball, CMB_F, is an object voxel p such that for each of its object neighbors n_i it results $n_i<p+w_i$, where w_i is 3, 4 or 5, depending on

whether i stands for f, e, or v. Actually, voxels with distance value 3 must be seen as having the *equivalent* distance value 1, in order the above rule always holds [23].

As pointed out in the Introduction, inclusion of all the $CMB_{F}s$ of the 3D object in its curvilinear skeleton is not generally possible. Thus, we perform a selection of the $CMB_{F}s$ to be considered as anchor points, so as to make the set resulting at the end of the first phase easier to be furthermore processed. We note that some $CMB_{F}s$ (termed $CMB1_{F}s$) either have no object neighbors in the successive more internal layer or have there at most vertex-neighbors, while the remaining $CMB_{F}s$ (termed $CMB2_{F}s$) have also face- or edge-neighbors in the successive more internal layer. While the $CMB1_{F}s$ are voxels that are equidistant from two opposite parts of the object border, the $CMB2_{F}s$, especially if sparse, are either due to discretization effects or are found in correspondence with two parts of the object border intersecting each other so as to form a 90 degree or larger dihedral angle. For example, if the original continuous object is a cylinder, some $CMB_{F}s$ can be created in DT_{F} of the discrete cylinder that do not carry relevant information, since they are not aligned along the main symmetry axis of the cylinder. In fact the circular border delimiting each section of the continuous cylinder is unavoidably transformed into a discrete polygon, which will be reflected by the presence of $CMB_{F}s$ in the distance transform due to the weak convexities along the polygonal border. If the original object is shaped like a brick, besides the $CMB_{F}s$ in the middle of the brick, which are due to opposite faces of the brick, other $CMB_{F}s$ will be found in correspondence with the 90 degree dihedral angles formed by two contiguous faces. The latter $CMB_{F}s$, if kept as anchor points, would generate in the resulting set some peripheral patches that would only make the second phase of the process more laborious. Thus, we select as anchor points only the $CMB1_{F}s$.

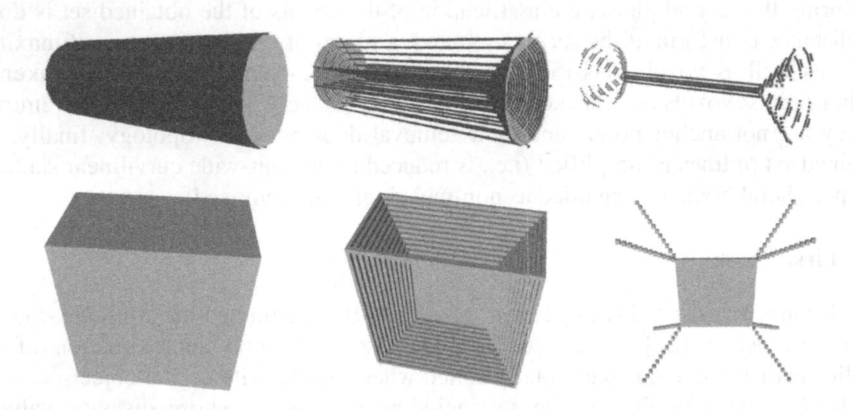

Fig. 1. A cylinder and a brick, left, all their $CMB_{F}s$, middle, and their $CMB1_{F}s$, right

Two simple examples are shown in Fig. 1. There, a cylinder and a brick, Fig. 1 left, all their $CMB_{F}s$, Fig. 1 middle, and the $CMB1_{F}s$, Fig. 1 right, are shown. Preserving as anchor points all the $CMB_{F}s$ would lead to sets including a number of peripheral patches that should be compressed anyway during the second phase of skeletonization. In turn, selecting as anchor points only the $CMB1_{F}s$ allows us to derive information from the most internal parts of the object and also to keep track of the 90 degree dihedral angles (see the linear subsets of the $CMB1_F$s for the brick, which are due to the triples of contiguous faces), while having a reduced set of patches and lines to handle during the second phase.

The voxels in DT_F are examined starting with those with the minimal distance value 3 in increasing distance order and, if they were not selected as anchor points, topology preserving removal is sequentially accomplished. Only simple voxels are removed. Thus, if p is the current distance value, a voxel p is removed if c_p=1 and c^*_p=1.

After all voxels with distance value p have been checked, before examining voxels with the successive distance value, we perform a process to maintain the superficial structure of those sets of non-removed voxels organized into patches. Any voxel p that has not been removed identifies and marks as non-removable its neighbors, if any, which are the most suitable for linking p to the inside. To this aim, the gradient from each non-removed voxel with distance value p towards each of its neighbors with distance value larger than p is computed. All neighbors for which the gradient results to be maximum are marked as non-removable. The gradient from p to its neighbor n_i is computed as $(n_i–p)/w_i$, where w_i is 3, 4 or 5, depending on whether i stands for f, e, or v.

Once all distance values have been taken into account, a set PL consisting of patches and lines is obtained. PL is nearly-thin, i.e., is at most 2-voxel thick.

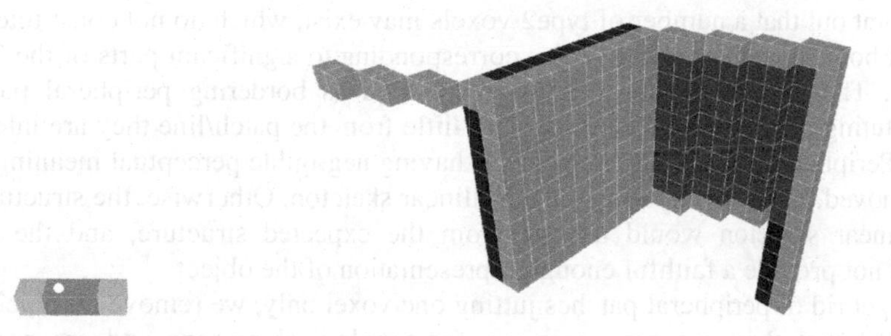

Fig. 2. One of the directional masks to check 2-voxel thickness, left; a set including 1- and 2-voxel thick patches and lines, right

The nearly-thin set PL is reduced to unit-thickness by means of final thinning. To this aim, object voxels are divided into two categories: type1 are the voxels characterized by $c^*_p \neq 1$, (see dark gray voxels in Fig. 2 right) and type2 are all other object voxels (see light gray and black voxels in Fig. 2 right). Final thinning is concerned only with type2 voxels that are placed in two voxel thick configurations. Thickening of PL can occur in face- or edge-direction. Thus, directional masks are successively used to identify thick sets of type2 voxels. In Fig. 2 left, one of the masks to detect thickening in face-direction is shown. In the mask, the two internal voxels are type2 voxels, while the external voxels are background voxels. The masks in the remaining face- and edge-directions have a similar structure. Any type2 voxel p in a thick configuration, according to the currently used directional mask, is sequentially removed provided that there exists exactly one 26-connected component of type2 voxels in N(*p*), and *p* has more than one neighboring object voxel. These two conditions are adequate to reduce thickness by removing only those voxels (e.g., the black voxels in Fig.2 right) placed in 2-voxel thick patches or lines, while preserving from removal voxels for which the directional mask is satisfied but they are placed along one-voxel thick patches or lines. For the running example shown in Fig. 3 left, the unit-wide set of patches and lines is given in Fig. 3 middle.

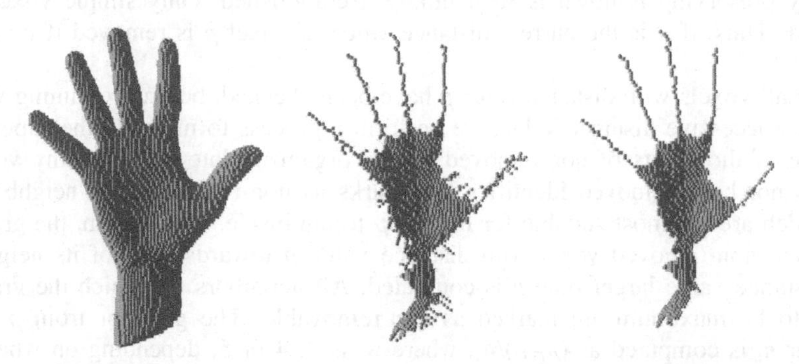

Fig. 3. An input 3D object, left, and the unit-wide set of patches and lines before pruning, middle, and after the 2 steps of pruning, right

Though PL has been reduced to unit-thickness, its structure may be still complex. We point out that a number of type2 voxels may exist, which do not constitute lines or do not border patches, perceived as corresponding to significant parts of the 3D input object. These type2 voxels can be understood as bordering peripheral patches or constituting peripheral lines jutting very little from the patch/line they are intersecting with. Peripheral patches and lines of PL having negligible perceptual meaning have to be removed, before computing the curvilinear skeleton. Otherwise, the structure of the curvilinear skeleton would deviate from the expected structure, and the skeleton would not provide a faithful enough representation of the object.

To get rid of peripheral patches jutting one voxel only, we remove all type2 voxels, provided that they are not necessary for topological reasons and are not tips of peripheral lines. Then, final thinning is applied again, since patches that were at most 4-voxel wide before removal of the type2 voxels, if any, might have been transformed into nearly-thin lines that have to be reduced to unit-thickness.

To remove non-significant peripheral lines, we resort to a pruning process, see e.g., [24,25]. A line of PL should be interpreted as significant if it corresponds to a region of the object that is perceived as important for shape analysis and recognition. In our opinion, since a strong relation exists between the centers of maximal balls along a line and size and shape of the object's region mapped into that line, the $CMB1_{F}s$ present on peripheral lines play a key role in deciding about pruning. Each peripheral line is traced starting from its end point, i.e., the voxel with only one object neighbor, until a branch point, i.e., a voxel with more than two object neighbors, is met. During tracing, the length L of the line, expressed in terms of the number of voxels constituting it, as well as the number N of $CMB1_{F}s$ along the line are computed. We use two pruning steps, performed with slightly different strategies. The first step is aimed at removing lines for which N is at most equal to a small threshold θ_1, to be fixed by the user (θ_1 has been set to 4 in this paper). As a result, lines with $L \leq \theta_1$, or lines for which, independently of their length L, is $N \leq \theta_1$ are removed. In both cases, these peripheral lines are interpreted as mapped into non-perceptually significant regions, due to the small number of $CMB1_{FS}$ that they include. In particular, removal of peripheral lines with $L \leq \theta_1$ is crucial to get rid of short lines, whose presence would inhibit removal of lines that were non-peripheral in the initial PL, but have still to be interpreted as mapped into non-perceptually significant regions. During the second pruning step, peripheral lines, possibly originating from voxels that have modified their status into that of end points due to the first pruning step, are removed provided that it is $N/L \leq \theta_2$, where θ_2 is a threshold to be fixed depending on problem domain. We have experimentally found that θ_2=0.25 is adequate as a default threshold. In Fig. 3 right, the pruned set PL resulting at the end of the first phase of the skeletonization process is shown.

2.2 Second Phase

The voxels of PL are classified during seven successive inspections of PL as follows:

Any voxel with at most two object neighbors that are not adjacent to each other is classified as *curve voxel.*

Any not yet classified voxel with a neighboring curve voxel is classified as *branching voxel.*

Any not yet classified voxel whose neighboring object voxels are all branching voxels is classified as *branching voxel.*

Any not yet classified voxel *p* with $c^*_p \neq 1$ is classified as *internal voxel.*

Any internal voxel *p* with more than two 6-connected components of background voxels in N(*p*) face- or edge-adjacent to *p*, or being any of the eight object voxels in a 2×2×2 configuration is re-classified *as junction voxel.*

Any internal voxel having a face- or an edge-neighbor classified as junction voxel, is re-classified as *extended junction voxel.*

Any remaining not yet classified voxel is classified as *edge voxel.*

The classification allows us also to distinguish patches into peripheral patches (those at least partially delimited by edge voxels) from internal patches (those completely delimited by junction voxels).

The distance transform DT_{PL} of the set of voxels classified as internal in PL is computed in two steps by using the <3,4,5> distance function. During the first step, all edge voxels are taken as reference set, all internal voxels constitute the set of voxels to be assigned a distance value, while all background, curve, branching, junction and extended junction voxels act as barriers to the propagation of distance information. Actually, extended junction voxels, whose role is to better separate intersecting patches, are permitted to receive distance information, since after all they are voxels internal in a patch. However, they do not propagate distance information, to prevent interaction between intersecting patches for which voxels in a patch are neighbors of voxels in another patch. During the first step, only voxels internal in peripheral patches are assigned their distance from the edge voxels delimiting the patches. In fact, if internal patches exist in PL, no distance information may reach their internal voxels during the first step, due to the barriers of junction and extended junction voxels delimiting the internal patches. Thus, the second step is accomplished. Extended junction voxels are re-classified as internal voxels; junction voxels delimiting the internal patches are re-classified as *delimiting junction voxels* and are assigned a sufficiently large value (e.g., the maximum distance value assigned during the first step plus 5). All delimiting junction voxels are taken as reference set, all internal voxels that have not yet an assigned distance value constitute the set of voxels that will receive distance information, while all other voxels in the image act as barriers to the propagation of distance information.

In DT_{PL} the centers of maximal balls, CMB_{PL}, are detected with the same criterion used to detect the CMB_F s in DT_F.

Analogously to the first phase of skeletonization, only the $CMB1_{PL}$s, i.e., the CMB_{PL}s having no neighbors or at most vertex-neighbors with a distance value characterizing a more internal layer, should be taken as anchor points. Actually, we filter out the $CMB1_{PL}$s such that the homologous voxels in DT_F are not $CMB1_F$. In this way, the anchor points are by all means symmetrically placed within the original 3D object. In turn, differently from the first phase, the selected $CMB1_{PL}$s are not the only anchor points. In fact, also the voxels of PL classified as curve, branching or junction voxels are all considered as anchor points, as these voxels are placed in key positions as regards shape information.

The voxels in DT_{PL} are examined in increasing distance order, starting with those with the minimum distance value and, if they are not anchor points, topology preserving removal is sequentially accomplished. Only simple voxels are removed.

Once all distance values have been taken into account, the nearly-thin curvilinear skeleton CS is obtained.

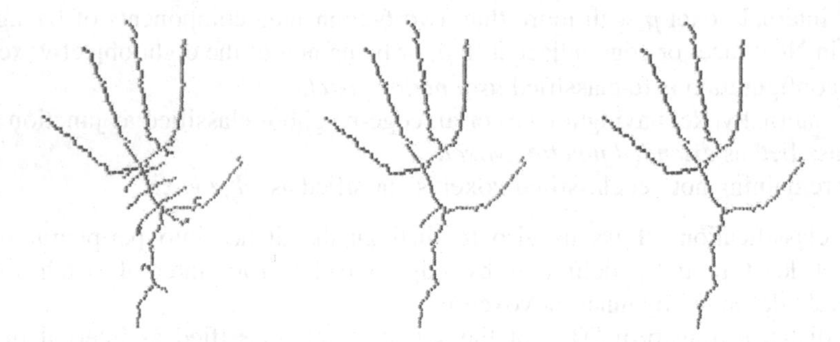

Fig. 4. The unit-wide curvilinear skeleton CS before pruning, left, and after the first step, middle, and the second step of pruning, right

The set CS is reduced to the unit-wide curvilinear skeleton by means of final thinning. Only voxels placed in 2-voxel thick configurations are possibly removed. Directional masks to detect thickening in face- or edge-direction are used as in the first phase. Any voxel *p* found to be placed in a thick configuration, according to the currently used directional mask, is sequentially removed provided that it is a simple voxel and has more than one neighboring object voxel. The result for the running example is shown in Fig. 4 left.

The set CS may still include some peripheral branches, which do not correspond to relevant parts of the 3D input object. Their removal is obtained by means of pruning, analogously to what has been done in the first phase. The only difference is that during the second pruning step the number N is now obtained by counting the selected $CMB1_{PL}$s as well as the remaining anchor points, i.e., the curve, branching and junction voxels along the peripheral branches. The values for the two thresholds θ_1 and θ_2 are here equal to those used in the first phase. In Fig. 4 middle and right, the curvilinear skeletons, as resulting after the first and the second pruning step, are shown.

3 Experimental Results

We have applied our skeletonization algorithm to a large collection of images taken from publicly available shape repositories [26,27]. A few examples of images are shown in Fig. 5 together with their curvilinear skeletons.

Fig. 5. Some 3D objects and their <3,4,5> curvilinear skeletons

Any skeletonization method should be evaluated in terms of the properties expected for the skeleton. The curvilinear skeleton is expected: 1) to be topologically equivalent to the object, 2) to be symmetrically placed within the object, and 3) to reflect the object's shape, by the presence of branches that can be seen as corresponding to parts of the object perceived as significant.

The first two properties are by all means satisfied by the <3,4,5> curvilinear skeleton. In fact, our voxel removal process has been accomplished (in both phases of skeletonization) sequentially, voxel after voxel, by using the notion of simple voxel. In turn, skeleton centrality is guaranteed by the criteria used in the two phases to select the anchor points, which are all symmetrically placed within the 3D object.

As for the third property, by looking at the examples in Fig. 5 we observe that the curvilinear skeletons have a number of branches corresponding to the main parts perceived as composing the objects. We remind that differently from the 2D case, the curvilinear skeleton does not generally enjoy the reversibility property. This means that by applying the reverse distance transformation to each single branch of the curvilinear skeleton we cannot completely recover the part of the 3D object mapped into that branch. However, the reverse distance transformation, computed by using for the voxels of the curvilinear skeleton the distance values pertaining to their homologous voxels in DT_F, can still be used to identify the main parts constituting the object, even if only in a rough manner.

As an example, the objects reconstructed from the curvilinear skeletons in Fig. 5 are given in Fig. 6. We note that the reconstructed objects may be regarded as sketched versions of the corresponding input objects. Thus, we can argue that the <3,4,5> curvilinear skeleton enjoys also the third property.

Fig. 6. Objects reconstructed form the curvilinear skeletons in Fig. 5

4 Concluding Remarks

In this paper, a method to compute the <3,4,5> curvilinear skeleton of 3D objects rid of cavities has been presented. Skeletonization is accomplished in two phases. The first phase is aimed at obtaining a subset of the 3D object consisting of 2D and 1D manifolds (patches and lines). The second phase is aimed at obtaining the curvilinear skeleton, exclusively consisting of 1D manifolds.

Both phases of the algorithm take advantage of the <3,4,5> distance transform. In the first phase, the distance transform DT_F of the object F is used to detect the anchor points (a suitable subset of the set of centers of maximal balls) and to guide iterated voxel removal (performed sequentially in increasing distance value order, and based on the notion of simple voxel). In the second phase, voxel classification of the set PL, resulting at the end of the first phase, is done to correctly identify the voxels constituting the reference set and the barriers. The distance transform DT_{PL} is computed and anchor points are detected as a subset of the set of center of the maximal balls as well as voxels classified as curve, branching and junction voxels. The distance transform DT_{PL} is also used to guide iterated voxel removal (performed sequentially in increasing distance value order, and based on the notion of simple voxel). Final thinning and pruning are also included in both phases of skeletonization.

The use of the <3,4,5> distance, which is a good approximation of the Euclidean distance, coupled with the pruning steps, makes the skeletonization algorithm rather stable under object rotation. Moreover, the use of the distance transform is advantageous from a computational point of view, since the voxels constituting the border of the object at any iteration of skeletonization are directly accessed by taking into account the distance values assigned to them.

References

1. Siddiqi, K., Pizer, S.M. (eds.): Medial Representations. Springer, Heidelberg (2008)
2. De Floriani, L., Spagnuolo, M. (eds.): Shape Analysis and Structuring. Springer, Heidelberg (2008)
3. Arcelli, C., Sanniti di Baja, G.: A width-independent fast thinning algorithm. IEEE Trans. on PAMI 7, 463–474 (1985)
4. Klein, F.: Euclidean skeletons. In: Proc. 5th Scandinavian Conf. Image Anal., pp. 443–450 (1987)
5. Arcelli, C., Sanniti di Baja, G.: A one-pass two-operations process to detect the skeletal pixels on the 4-distance transform. IEEE Trans. PAMI 11, 411–414 (1989)
6. Xia, Y.: Skeletonization via the realization of the fire front's propagation and extinction in digital binary shapes. IEEE Trans. PAMI 11(10), 1076–1086 (1989)
7. Arcelli, C., Sanniti di Baja, G.: Euclidean skeleton via center-of-maximal-disc extraction. Image and Vision Computing 11, 163–173 (1993)
8. Kimmel, R., Shaked, D., Kiryati, N.: Skeletonization via distance maps and level sets. Computer Vision and Image Understanding 62(3), 382–391 (1995)
9. Sanniti di Baja, G., Thiel, E.: Skeletonization algorithm running on path-based distance maps. Image and Vision Computing 14, 47–57 (1996)
10. Pudney, C.: Distance-ordered homotopic thinning: a skeletonization algorithm for 3D digital images. Computer Vision and Image Understanding 72(3), 404–413 (1998)
11. Zhou, Y., Kaufman, A., Toga, A.W.: Three-dimensional skeleton and centerline generation based on an approximate minimum distance field. The Visual Computer 14(7), 303–314 (1998)
12. Borgefors, G., Nystrom, I., Sanniti di Baja, G.: Computing skeletons in three dimensions. Pattern Recognition 32(7), 1225–1236 (1999)

13. Sanniti di Baja, G., Svensson, S.: Surface skeletons detected on the d6 distance transform. In: Amin, A., Pudil, P., Ferri, F., Iñesta, J.M. (eds.) SPR 2000 and SSPR 2000. LNCS, vol. 1876, pp. 387–396. Springer, Heidelberg (2000)
14. Blum, H.: Biological shape and visual science. Journal of Theoretical Biology 38, 205–287 (1973)
15. Leymarie, F., Levine, M.D.: Simulating the grassfire transform using an active contour model. IEEE Trans. PAMI 14(1), 56–75 (1992)
16. Kimia, B.B., Tannenbaum, A., Zucker, S.W.: Shape, shocks, and deformations I: the components of two-dimensional shape and the reaction-diffusion space. International Journal of Computer Vision 15, 189–224 (1995)
17. Siddiqi, K., Bouix, S., Tannenbaum, A., Zucker, S.W.: Hamilton-Jacobi skeletons. International Journal of Computer Vision 48(3), 215–231 (2002)
18. Dimitrov, P., Damon, J.N., Siddiqi, K.: Flux invariants for shape. In: Proc. IEEE Conf. CVPR 2003, Madison, WI, vol. 1, pp. 835–841 (2003)
19. Giblin, P.J., Kimia, B.B.: A formal classification of 3D medial axis points and their local geometry. IEEE Trans. PAMI 26(2), 238–251 (2004)
20. Saha, P.K., Chaudhuri, B.B.: Detection of 3D simple points for topology preserving transformations with application to thinning. IEEE Trans. PAMI 16(10), 1028–1032 (1994)
21. Bertrand, G., Malandain, G.: A new characterization of three-dimensional simple points. Pattern Recognition Letters 15(2), 169–175 (1994)
22. Borgefors, G.: Digital distance transforms in 2D, 3D, and 4D. In: Chen, C.H., Wang, P.P.S. (eds.) Handbook of Pattern Recognition and Computer Vision, pp. 157–176. World Scientific, Singapore (2005)
23. Svensson, S., Sanniti di Baja, G.: Using distance transforms to decompose 3D discrete objects. Image and Vision Computing 20, 529–540 (2002)
24. Shaked, D., Bruckstein, A.M.: Pruning medial axes. Computer Vision and Image Understanding 69(2), 156–169 (1998)
25. Svensson, S., Sanniti di Baja, G.: Simplifying curve skeletons in volume images. Computer Vision and Image Understanding 90(3), 242–257 (2003)
26. AIM@SHAPE Shape Repository, `http://shapes.aimatshape.net/viewmodels.php`
27. Shilane, P., Min, P., Kazhdan, M., Funkhouser, T.: The Princeton Shape Benchmark. In: Shape Modeling International, Genova, Italy (June 2004)

A Discrete λ-Medial Axis*

John Chaussard, Michel Couprie, and Hugues Talbot

Université Paris-Est, LIGM, Équipe A3SI, ESIEE Paris, France
{j.chaussard,m.couprie,h.talbot}@esiee.fr

Abstract. The λ-medial axis was introduced in 2005 by Chazal and Lieutier as a new concept for computing the medial axis of a shape subject to filtering with a single parameter. These authors proved the stability of the λ-medial axis under small shape perturbations. In this paper, we introduce the definition of a discrete λ-medial axis (DLMA). We evaluate its stability and rotation invariance experimentally. The DLMA may be computed by efficient algorithms, furthermore we introduce a variant of the DLMA, denoted by DL'MA, which may be computed in linear-time. We compare the DLMA and the DL'MA with the recently introduced integer medial axis and show that both DLMA and DL'MA provide measurably better results.

In the 60s, Blum [7,8] introduced the notion of medial axis or skeleton, which has since been the subject of numerous theoretical studies and has also proved its usefulness in practical applications. Although initially introduced as the outcome of a propagation process, the medial axis can also be defined in simple geometric terms. In the continuous Euclidean space, the two following definitions can be used to formalize this notion: let X be a bounded subset of $\mathbb{R}^n$;

a) The skeleton of X consists of the centers of the balls that are included in X but that are not included in any other ball included in X.

b) The medial axis of X consists of the points $x \in X$ that have several nearest points on the boundary of X.

The skeleton and the medial axis differ only by a negligible set of points (see [22]), in general the skeleton is a strict subset of the medial axis.

In this paper, we focus on medial axes in the discrete grid $\mathbb{Z}^2$ or $\mathbb{Z}^3$, which are centered in the shape with respect to the Euclidean distance.

A major difficulty when using the medial axis in applications (*e.g.*, shape recognition), is its sensitivity to small contour perturbations, in other words, its lack of stability. A recent survey [1] summarizes selected relevant studies dealing with this topic. This difficulty can be expressed mathematically: the transformation which associates a shape to its medial axis is only semi-continuous. This fact, among others, explains why it is usually necessary to add a filtering step (or pruning step) to any method that aims at computing the medial axis.

* This work has been partially supported by the "ANR BLAN07-2_184378 MicroFiss" project.

S. Brlek, C. Reutenauer, and X. Provençal (Eds.): DGCI 2009, LNCS 5810, pp. 421–434, 2009.

Hence, there is a rich literature devoted to skeleton or medial axis pruning, in which different criteria were proposed in order to discard "spurious" skeleton points or branches: see [4,24,3,21,2,27,17,5,13], to cite only a few. However, we lack theoretical justification, that is, a formalized argument that would help to understand why a filtering criterion is better than another.

In 2005, Chazal and Lieutier introduced the λ-medial axis and studied its properties, in particular those related to stability [9]. Consider a bounded subset X of $\mathbb{R}^n$, as for example, for $n = 2$, the region enclosed by the solid curve depicted in Fig. 1 (left). Let x be a point in X, we denote by $\Pi(x)$ the set of points of the boundary of X that are closest to x. For example in Fig. 1, we have $\Pi(x) = \{a, b\}$, $\Pi(x') = \{a', b'\}$ and $\Pi(x'') = \{a''\}$. Let λ be a non-negative real number, the λ-medial axis of X is the set of points x of X such that the smallest ball including $\Pi(x)$ has a radius greater than or equal to λ. Notice that the 0-medial axis of X is equal to X, and that any λ-medial axis with $\lambda > 0$ is included in the medial axis according to definition b). We show in Fig. 1 (right) two λ-medial axes with different values of λ.

A major outcome of [9] is the following property: informally, for "regular" values of λ, the λ-medial axis remains stable under perturbations of $\overline{X}$ that are small with regard to the Hausdorff distance. Typical non-regular values are radii of locally largest maximal balls.

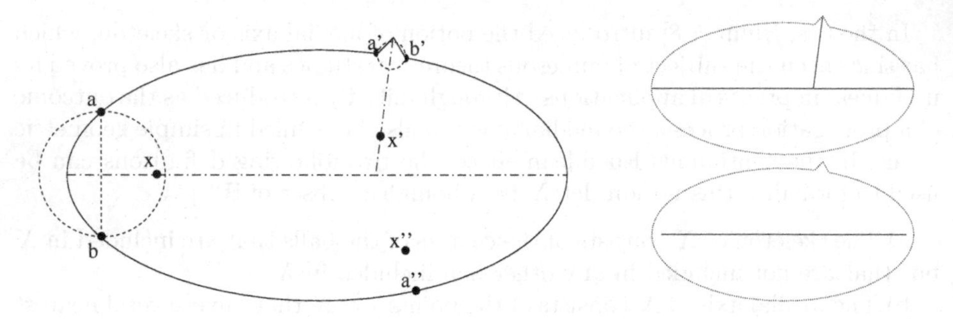

Fig. 1. Illustration of the λ-medial axis. Left: Points x, x' and x'' and their respective closest boundary points. Top right: λ-medial axis with $\lambda = \epsilon$, a very small positive real number. Bottom right: λ-medial axis with $\lambda = d(a', b') + \epsilon$.

This property is a strong argument in favor of the λ-medial axis, especially in the absence of such result for other proposed criteria.

In the discrete grids, namely $\mathbb{Z}^2$ and $\mathbb{Z}^3$, a similar filtering criterion has been considered in independent works [21,17]. It consists of selecting for each medial axis point only two closest boundary points, and using the distance between these two points as the filtering criterion. The work of Hesselink *et al.* [17] provides a linear-time algorithm to compute a filtered medial axis based on this criterion, which exhibits good noise robustness properties in practice.

In this paper, we introduce the definition of a discrete λ-medial axis (DLMA) in $\mathbb{Z}^n$. We evaluate experimentally its stability and rotation invariance in 2D. In this experimental study, we compare it with the previously introduced integer medial axis [18,17] and show that the DLMA provides measurably better results. Furthermore, we introduce a variant of the DLMA which may be computed in linear time, for which the results are very close to those of the DLMA, and which is only slightly slower than the one proposed in [17].

1 Discrete λ-Medial Axis

Let $x = (x_1, \ldots, x_n), y = (y_1, \ldots, y_n) \in \mathbb{R}^n$, we denote by $d(x, y)$ the Euclidean distance between x and y, that is, $d(x, y) = (\sum_{k=1}^{n}(y_k - x_k)^2)^{\frac{1}{2}}$. Let X be a finite subset of $\mathbb{R}^n$ or $\mathbb{Z}^n$. We set $d(y, X) = \min_{x \in X}\{d(y, x)\}$. We denote by $|X|$ the number of elements of X, and by $\overline{X}$ the complement of X.

Let E be either $\mathbb{R}^n$ or $\mathbb{Z}^n$. Let $x \in E, r \in \mathbb{R}^+$, we denote by $B_r(x)$ the *ball of radius r centered on x*, defined by $B_r(x) = \{y \in E \mid d(x, y) \leq r\}$. We also define $B_r^<(x) = \{y \in E \mid d(x, y) < r\}$.

For each point $x \in \mathbb{Z}^n$, we define the *direct neighborhood of x* as $N(x) = \{y \in \mathbb{Z}^n \mid d(x, y) \leq 1\}$. The direct neighborhood comprises $2n + 1$ points.

Let S be a nonempty subset of E, and let $x \in E$. The *projection of x on S*, denoted by $\Pi_S(x)$, is the set of points y of S which are at minimal distance from x ; more precisely,

$$\Pi_S(x) = \{y \in S \mid \forall z \in S, d(y, x) \leq d(z, x)\}.$$

If X is a subset of E, the *projection of X on S* is defined by $\Pi_S(X) = \bigcup_{x \in X} \Pi_S(x)$.

In the case of a subset X of $\mathbb{Z}^n$, it may be easily seen that $\Pi_{\overline{X}}(X)$ is composed of the points of $\overline{X}$ that are in the direct neighborhood of a point of X, that is, $\Pi_{\overline{X}}(X) = \{y \in \overline{X} \mid N(y) \cap X \neq \emptyset\}$. We call $\Pi_{\overline{X}}(X)$ the *(external) boundary of X*.

Let $S \subset \mathbb{R}^n$, we denote by $R(S)$ the radius of the smallest ball enclosing S, that is, $R(S) = \min\{r \in \mathbb{R} \mid \exists y \in \mathbb{R}^n, B_r(y) \supseteq S\}$.

The λ-medial axis may now be defined based on these notions.

Definition 1 ([9]). *Let X be an open bounded subset of $\mathbb{R}^n$, and let $\lambda \in \mathbb{R}^+$. The* λ-medial axis *of X is the set of points x in X such that $R(\Pi_{\overline{X}}(x)) \geq \lambda$.*

We cannot transpose this definition straightforwardly to $\mathbb{Z}^n$. To see why, consider a horizontal ribbon in $\mathbb{Z}^2$ with constant even width and infinite length. It can be seen that any point of this set has a projection that is reduced to a singleton. Hence, if we keep the same definition, any λ-medial axis of this object with $\lambda > 0$ would be empty.

It is the reason why we change the projection for the extended projection (see [13]).

Let $X \subseteq \mathbb{Z}^n$, and let $x \in X$. The *extended projection of x on $\overline{X}$*, denoted by $\Pi^e_{\overline{X}}(x)$, is the union of the sets $\Pi_{\overline{X}}(y)$, for all y in $N(x)$ such that $d(y, \overline{X}) \leq d(x, \overline{X})$.

The last condition ($d(y, \overline{X}) \leq d(x, \overline{X})$) is set in order to avoid getting several medial axis points when only one is sufficient. We are now ready to introduce the definition of our discrete λ-medial axis.

Definition 2. *Let X be a finite subset of $\mathbb{Z}^n$, and let $\lambda \in \mathbb{R}^+$. We define the function $\mathcal{F}_X$ which associates, to each point x of X, the value $\mathcal{F}_X(x) = R(\Pi^e_{\overline{X}}(x))$. The* discrete λ-medial axis (or DLMA) of X *is the set of points x in X such that $\mathcal{F}_X(x) \geq \lambda$.*

In Fig. 2 (right), we show two examples of DLMAs of a shape X. We also note that the function $\mathcal{F}_X$ (displayed on the left of the figure) can be computed once and stored as a grayscale image, and that any DLMA of X is a threshold of this function at a particular value λ. More examples are given in Fig. 3. Notice that DLMA has not, in general, the same topology as the original shape.

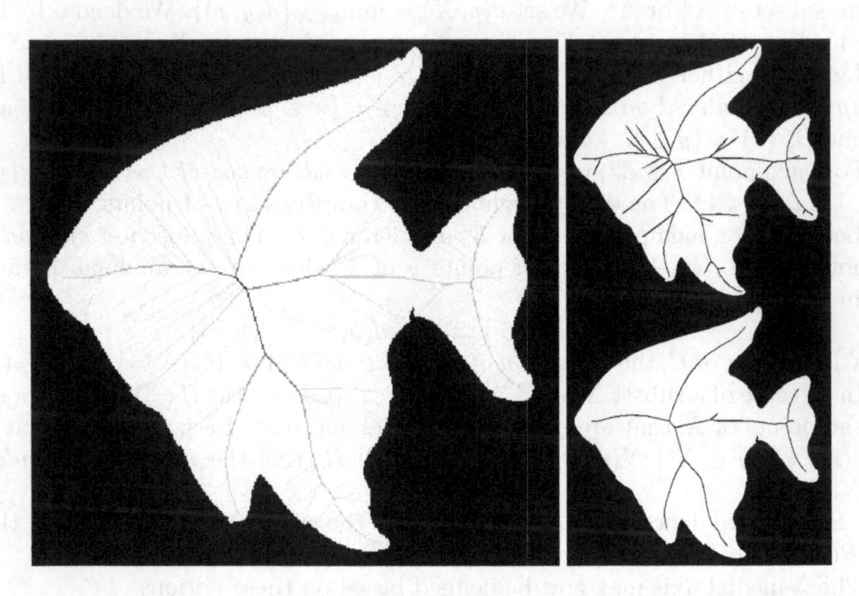

Fig. 2. Left: The function $\mathcal{F}_X$ superimposed to the shape X. Darkest colors represent highest values of $\mathcal{F}_X(x)$. Any DLMA of X is a threshold of this function at a particular value λ. Top right: discrete 10-medial axis. Bottom right: discrete 30-medial axis of X.

2 Integer Medial Axis

Let us first recall the definition of the integer medial axis [17], which is conceptually close to the DLMA, and defined in the same framework of discrete grids.

Let X be a subset of $\mathbb{Z}^n$, and let $x \in X$. We denote by $\Pi'_{\overline{X}}(x)$ the element of $\Pi_{\overline{X}}(x)$ that is smallest with regard to the lexicographic ordering of its coordinates.

Definition 3 ([17])**.** *Let X be a finite subset of $\mathbb{Z}^n$, and let $\gamma \in \mathbb{R}^+$. The* γ-integer medial axis (or GIMA) of X *is the set of points x in X such that*

i) $\exists y \in N(x)$, $d(\Pi'_{\overline{X}}(x), \Pi'_{\overline{X}}(y)) > \gamma$, and

ii) $d(m - \Pi'_{\overline{X}}(x)) \leq d(m - \Pi'_{\overline{X}}(y))$, *where* $m = \frac{x+y}{2}$.
The integer medial axis *of* X *is the* γ*-integer medial axis of* X *for* $\gamma = 1$.
Notice that the GIMA is not invariant under a permutation of the coordinates.

There are indeed some links between the GIMA and the DLMA. In 2D, in the case of a point x that is, conceptually, a "regular skeleton point" (neither a branch extremity nor a branch junction), condition i) is similar to the condition $R(\Pi_{\overline{X}}(x)) \geq \lambda$ in Def. 1.

Condition ii) plays a role analogous to condition $d(y, \overline{X}) \leq d(x, \overline{X})$ in the definition of the extended projection, that is to get a thinner axis.

However, these definitions differ, leading to sensible differences in the results of these two transformations (see Fig. 3). We analyse quantitatively these differences in Sec. 5.

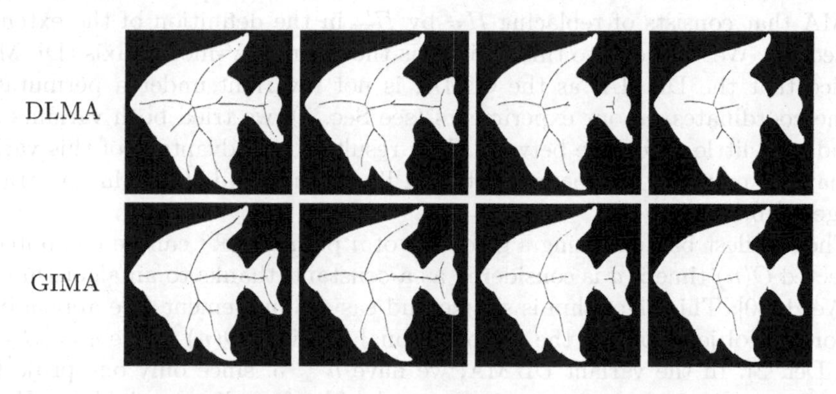

Fig. 3. Results of DLMA (first row) and GIMA (second row) for parameter values yielding similar reconstruction ratios (see Sec. 5)

3 Topology Preservation

It is easy to see that a DLMA or a GIMA of a given shape X does not exhibit, in general, the same homotopy type as X (see *e.g.* Fig. 3).

In order to guarantee topology preservation, a popular method [14,29,28,25] consists of performing a homotopic thinning of X with the constraint of retaining the points of its filtered medial axis M, or equivalently, of iteratively removing simple points [19,6,12] from X that do not belong to M. A priority function is needed in order to specify which points must be considered at each step of the thinning. In the general case, the choice of this priority function is not easy (see [28,13]). In the case of a DLMA, choosing the map $\mathcal{F}_X$ as priority function yields satisfying results (see, for example, Fig. 4).

4 Algorithms

The projection $\Pi_{\overline{X}}(x)$, for all points x in X, can be computed in optimal time and space (that is, in $O(N)$ where $N = \sum_{x \in X} |\Pi_{\overline{X}}(x)|$) thanks to an algorithm due to D. Coeurjolly [13].

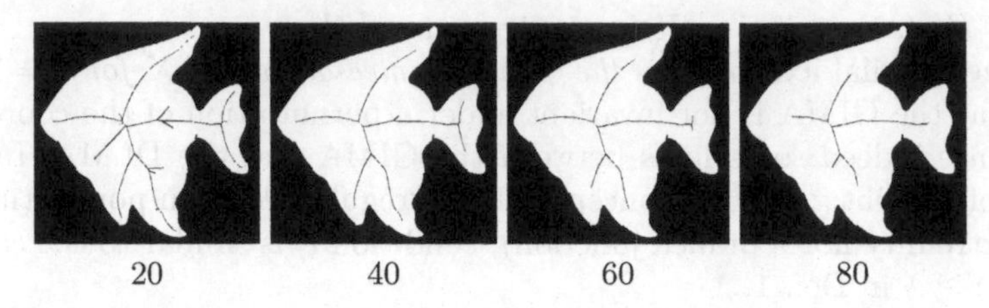

Fig. 4. Results of homotopic thinning of X constrained by the DLMA for parameter values 20, 40, 60 and 80, and with $\mathcal{F}_X$ as priority function

In order to avoid computing the map $\Pi_{\overline{X}}$, which requires a data structure like an array of lists to store all the sets $\Pi_{\overline{X}}(x)$, we propose a variant of the DLMA that consists of replacing $\Pi_{\overline{X}}$ by $\Pi'_{\overline{X}}$ in the definition of the extended projection. We will refer to this variant as the discrete λ'-medial axis (DL'MA). Notice that the DL'MA, as the GIMA, is not invariant under a permutation of the coordinates. In our experiments (see Sec. 5), we tried both variants and found very little difference between their results. One advantage of this variant is that the map $\Pi'_{\overline{X}}$ can be computed in linear time and stored in an array of integers [10,23,17].

The smallest ball enclosing a given set of n points in $\mathbb{R}^d$ can be computed in expected $O(n)$ time if d is considered as a constant, thanks to an algorithm due to Welzl [30]. This algorithm is simple and easy to implement (see appendix).

For each object point x, the computational cost is in $O(n)$ where $n = |\Pi^e_{\overline{X}}(x)|$ (see Def. 2). In the variant DL'MA, we have $n \leq 5$, since only one projection point is considered for x and, at worse, each of its four direct neighbors. Hence, computing the radius of the smallest enclosing ball can be performed in constant time.

Globally, the DL'MA procedure runs in linear time with regard to the number of object points.

Thanks to local characterizations, in 2D and 3D, checking whether a point is simple or not can be done in constant time. The homotopic thinning described in section 3 may be implemented to run in $O(n \log n)$ complexity, where n is the number of object points, using *e.g.* a balanced binary tree to store and retrieve the candidate points and their priority.

5 Results and Comparisons

First, let us give some definitions that are useful for describing our experiments and analyzing their results.

Let X be a finite subset of $\mathbb{Z}^n$. Let Y be a subset of X, we set $REDT_X(Y) = \bigcup_{y \in Y} B^<_{d(y,\overline{X})}(y)$. The transformation $REDT_X$ is sometimes called *reverse Euclidean distance transform* [11].

It is well known that any object can be fully reconstructed from its medial axis, more precisely, we have $X = REDT_X(M)$ whenever M is the (exact and non-filtered) medial axis of X. This property holds if M is the set of the centers of maximal balls of X, the DLMA of X with $\lambda = 1$, or the integer medial axis of X. However, it is no longer true if we consider filtered medial axes, *e.g.* DLMA or GIMA with arbitrary λ or γ.

Then, it is interesting to measure how much information about the original object is lost when we raise the filtering parameter. We set

$$\mathcal{L}_X(\lambda) = \frac{|X \setminus REDT_X(LM_X(\lambda))|}{|X|}, \quad \mathcal{I}_X(\gamma) = \frac{|X \setminus REDT_X(IM_X(\gamma))|}{|X|},$$

where $LM_X(\lambda)$ is the DLMA of X and $IM_X(\gamma)$ is the GIMA of X. In words, $\mathcal{L}_X(\lambda)$ (resp. $\mathcal{I}_X(\gamma)$) is the area of the difference between X and the set reconstructed from its DLMA (resp. GIMA), divided by the area of X. We call $\mathcal{L}_X(\lambda)$ and $\mathcal{I}_X(\gamma)$ the *(normalized) residuals* corresponding to filtering values λ and γ, respectively.

Fig. 5 shows the evolution of the residuals for the GIMA, DLMA and DL'MA. Since some differences may not be negligible, to ensure a fair evaluation we will compare the results of both methods for approximately equal values of their residuals, rather than for equal values of their parameters.

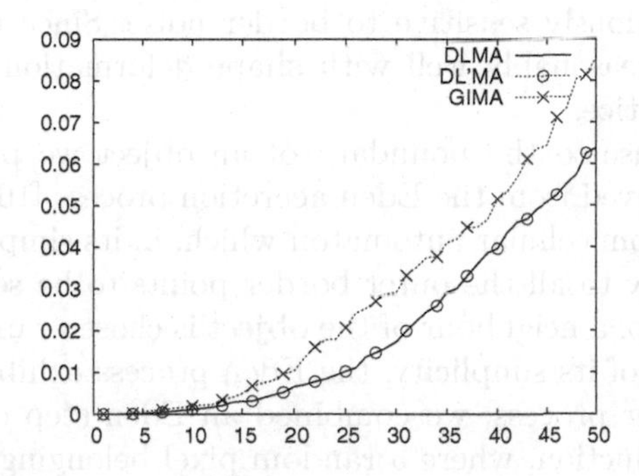

Fig. 5. Residuals $\mathcal{L}_X(\lambda)$ (2 variants) and $\mathcal{I}_X(\gamma)$, for the set X depicted in Fig. 2. Horizontal axis: the value of the parameter (λ or γ). Vertical axis: the value of the residual. Notice that curves corresponding to DLMA and DL'MA are superimposed.

For comparing shapes or medial axes, we use the Hausdorff distance (see below), and also a dissimilarity measure proposed by Dubuisson and Jain [15]. The drawback of Hausdorff distance for measuring shape dissimilarity is its extreme sensibility to outliers, the latter measure avoids this drawback.

Let X, Y be two subsets of $\mathbb{R}^n$. We set

$$H(X|Y) = \max_{x \in X}\{\min_{y \in Y}\{d(x,y)\}\},$$

and $d_H(X,Y) = \max\{H(X|Y), H(Y|X)\}$ is the *Hausdorff distance between X and Y*. We set

$$D(X|Y) = \frac{1}{|X|} \sum_{x \in X} \min_{y \in Y}\{d(x,y)\},$$

and $d_D(X,Y) = \max\{D(X|Y), D(Y|X)\}$ is the *Dubuisson and Jain's dissimilarity measure between X and Y* (called *dissimilarity* in the sequel for the sake of brevity).

We conducted our experiments on a database of 216 shapes provided by B.B. Kimia [26]. The 216 images are divided into 18 classes (birds, cars, etc.), Fig. 6 shows one (reduced) image of each class.

Fig. 6. A sample of the 216 shapes of Kimia's database

5.1 Stability

Skeletons are notoriously sensitive to border noise. Since the λ-medial axis is supposed to cope reasonably well with shape deformation, it is useful to test how it fares in practice.

To introduce noise to the boundary of an object we propose deforming it using a process derived from the Eden accretion process [16]. The Eden process is an iterative random cellular automaton which, in its simplest form, attributes an equal probability to all the outer border points to be set to 1 at each step. That is, at each step, a neighbour of the object is chosen randomly and added to the shape. In spite of its simplicity, the Eden process exhibits good asymptotic isotropy [20]. In our process, we combined an Eden step of accretion with an opposite step of reduction, where a random pixel belonging to the inner border is turned to zero (i.e. set to belong to the background). In addition, we required both accretion and diluvion steps to concern only simple points. This way at each step the object's area remains constant and retains the homotopy type of the original.

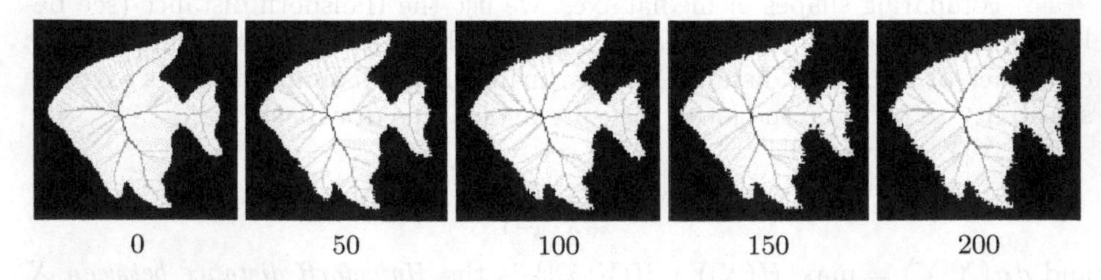

Fig. 7. Illustration of the border deformation process, after 0 (original image), 50, 100, 150 and 200 steps. The function $\mathcal{F}_X$ is superimposed to each shape X.

Table 1. Hausdorff distance

	50	100	150	200	all
GIMA	5.52	6.58	7.09	7.59	6.70
DLMA	5.37	6.31	6.81	7.32	6.45
DL'MA	5.35	6.32	6.79	7.32	6.45

Table 2. Dissimilarity

	50	100	150	200	all
GIMA	.551	.772	.889	1.016	.807
DLMA	.547	.763	.881	.998	.797
DL'MA	.551	.768	.885	1.004	.802

We denote by $E(X, n)$ the result of applying n steps of this process to the shape X. In this experiment, we compare the (filtered) medial axis of an original shape with the one of a deformed shape.

In Table 1 (resp. Table 2) we give the average Hausdorff distance (resp. dissimilarity) between $M(X)$ and $M(E(X, n))$ on the 216 shapes of Kimia's database, for parameter values $\gamma, \lambda, \lambda'$ varying from 1 to a value giving 30% residual. Results are given for values 50, 100, 150, 200 of n and all results averaged together. Results indicate a slight advantage of λ- and λ'-medial axes over the integer medial axis.

5.2 Rotation Invariance

Rotation invariance is an important property of the medial axis that holds in the continuous framework. If R_θ denotes the rotation of angle θ and center 0, and M denotes the medial axis transform, the rotation invariance property states that $M(R_\theta(X)) = R_\theta(M(X))$, whatever X and θ.

In a discrete framework, this property can only hold for particular cases (*e.g.*, when θ is a multiple of 90 degrees). Nevertheless, we can experimentally measure the dissimilarity between $M(R_\theta(X))$ and $R_\theta(M(X))$ for different instances, and different definitions of the medial axis. The lower this dissimilarity, the more stable under rotation the method is.

Fig. 8 and Fig. 9 show detailed results of such experiments. We used a rotation by interpolation algorithm, followed by a threshold. We adjusted the parameters λ and γ such that the residuals for each axis were approximately 1%, 3% and 5%.

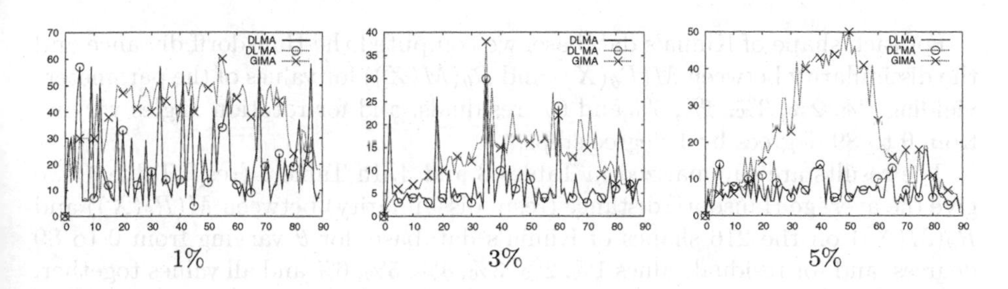

Fig. 8. Hausdorff distance between $M(R_\theta(X))$ and $R_\theta(M(X))$, for residual values 1%, 3%, 5%. Horizontal axis: θ. Vertical axis: Hausdorff distance.

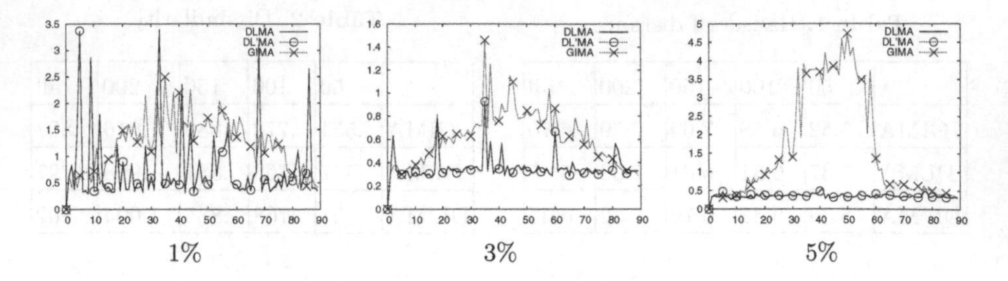

Fig. 9. Dissimilarity between $M(R_\theta(X))$ and $R_\theta(M(X))$, for residual values 1%, 3%, 5%. Horizontal axis: θ. Vertical axis: Dissimilarity.

Table 3. Hausdorff distance

	2%	4%	6%	all
GIMA	6.76	6.59	7.34	6.94
DLMA	6.17	5.48	5.71	6.03
DL'MA	6.17	5.43	5.71	5.99

Table 4. Dissimilarity

	2%	4%	6%	all
GIMA	.620	.628	.796	.677
DLMA	.563	.480	.580	.559
DL'MA	.566	.479	.602	.557

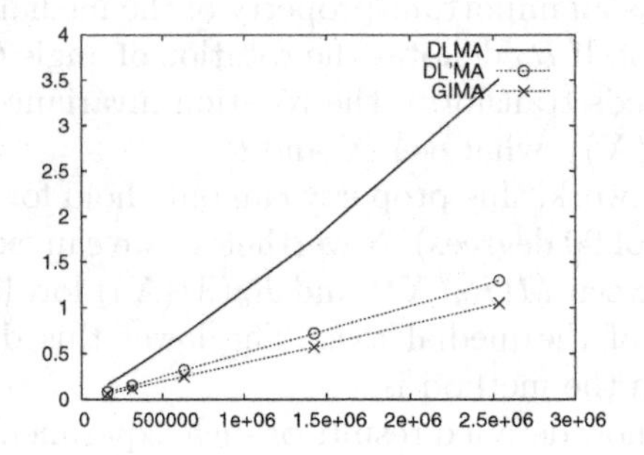

Fig. 10. Computing times (in seconds, vert. axis) versus image sizes (horiz. axis)

For each shape of Kimia's database, we computed the Hausdorff distance and the dissimilarity between $M(R_\theta(X))$ and $R_\theta(M(X))$ for values of the parameters yielding 1%, 2%, 3%, 4%, 5% and 6% residuals, and for rotation angles varying from 0 to 89 degrees by 1 degree steps.

The results are summarized in Tables 3 and 4. In Table 3 (resp. Table 4) we give the average Hausdorff distance (resp. dissimilarity) between $M(R_\theta(X))$ and $R_\theta(M(X))$ on the 216 shapes of Kimia's database, for θ varying from 0 to 89 degrees, and for residual values 1%, 2%, 3%, 4%, 5%, 6% and all values together. We see that the average dissimilarity or Hausdorff distance is almost always less for the DLMA (or DL'MA) than for the integer medial axis, around 15% less in the mean. The difference between both methods is more important (greater than 20%) for parameter values yielding highest residuals.

5.3 Computing Time

In Fig. 10, we show the results of computing time measurements that we performed on an Intel Core 2 Duo processor at 1.83 GHz. Computing times for the GIMA are the lowest, but DL'MA is only slightly slower (and also linear).

6 Conclusion

In this paper, we introduced the definition of a discrete λ-medial axis (DLMA) and compared it with the integer medial axis. We showed the DMLA provides good robustness to boundary noise and invariance by rotation in practice. We proposed an improved DM'LA algorithm with linear complexity. Future work will include a more thorough study of rotational invariance and border noise robustness, both in 2D and 3D, including comparison with traditional pruning. We will also study in more details the relationship of the DLMA to the discrete bisector function [28,13] and application of this work to shape classification and object splitting.

References

1. Attali, D., Boissonnat, J.-D., Edelsbrunner, H.: Stability and computation of the medial axis — a state-of-the-art report. In: Mathematical Foundations of Scientific Visualization, Computer Graphics, and Massive Data Exploration, pp. 1–19. Springer, Heidelberg (to appear, 2009)
2. Attali, D., Lachaud, J.O.: Delaunay conforming iso-surface, skeleton extraction and noise removal. Comp. Geom.: Theory and Applications 19, 175–189 (2001)
3. Attali, D., Montanvert, A.: Modelling noise for a better simplification of skeletons. In: ICIP, vol. 3, pp. 13–16 (1996)
4. Attali, D., Sanniti di Baja, G., Thiel, E.: Pruning discrete and semicontinuous skeletons. In: Braccini, C., Vernazza, G., DeFloriani, L. (eds.) ICIAP 1995. LNCS, vol. 974, pp. 488–493. Springer, Heidelberg (1995)
5. Bai, X., Latecki, L.J., Liu, W.Y.: Skeleton pruning by contour partitioning with discrete curve evolution. Trans. on PAMI 29(3), 449–462 (2007)
6. Bertrand, G.: Simple points, topological numbers and geodesic neighborhoods in cubic grids. Pattern Recognition Letters 15, 1003–1011 (1994)
7. Blum, H.: An associative machine for dealing with the visual field and some of its biological implications. Biol. prototypes and synthetic systems 1, 244–260 (1961)
8. Blum, H.: A transformation for extracting new descriptors of shape. In: Models for the Perception of Speech and Visual Form, pp. 362–380. MIT Press, Cambridge (1967)
9. Chazal, F., Lieutier, A.: The λ-medial axis. Graph. Models 67(4), 304–331 (2005)

10. Coeurjolly, D.: Algorithmique et géométrie discrète pour la caractérisation des courbes et des surfaces. PhD thesis, Université Lyon II, France (2002)
11. Coeurjolly, D.: d-dimensional reverse Euclidean distance transformation and Euclidean medial axis extraction in optimal time. In: Nyström, I., Sanniti di Baja, G., Svensson, S. (eds.) DGCI 2003. LNCS, vol. 2886, pp. 327–337. Springer, Heidelberg (2003)
12. Couprie, M., Bertrand, G.: New characterizations of simple points in 2D, 3D and 4D discrete spaces. Trans. on PAMI 31(4), 637–648 (2009)
13. Couprie, M., Coeurjolly, D., Zrour, R.: Discrete bisector function and Euclidean skeleton in 2D and 3D. Image and Vision Computing 25(10), 1543–1556 (2007)
14. Davies, E.R., Plummer, A.P.N.: Thinning algorithms: a critique and a new methodology. Pattern Recognition 14, 53–63 (1981)
15. Dubuisson, M.-P., Jain, A.K.: A modified hausdorff distance for object matching. In: 12th ICPR, vol. 1, pp. 566–568 (1994)
16. Eden, M.: A two-dimensional growth process. In: Fourth Berkeley Symp. on Math. Statist. and Prob., vol. 4, pp. 223–239. Univ. of Calif. Press (1961)
17. Hesselink, W.H., Roerdink, J.B.T.M.: Euclidean skeletons of digital image and volume data in linear time by the integer medial axis transform. Trans. on PAMI 30(12), 2204–2217 (2008)
18. Hesselink, W.H., Visser, M., Roerdink, J.B.T.M.: Euclidean skeletons of 3d data sets in linear time by the integer medial axis transform. In: 7th ISMM. Computational Imaging and Vision, vol. 30, pp. 259–268. Springer, Heidelberg (2005)
19. Yung Kong, T., Rosenfeld, A.: Digital topology: introduction and survey. Comp. Vision, Graphics and Image Proc. 48, 357–393 (1989)
20. Lee, T.M.C., Cowan, R.: A stochastic tesselation of digital space. In: 2nd ISMM, Fontainebleau, pp. 218–224. Kluwer, Dordrecht (1994)
21. Malandain, G., Fernández-Vidal, S.: Euclidean skeletons. Image and Vision Computing 16, 317–327 (1998)
22. Matheron, G.: Examples of topological properties of skeletons, vol. 2, pp. 217–238. Academic Press, London (1988)
23. Maurer, C.R., Qi, R., Raghavan, V.: A linear time algorithm for computing exact euclidean distance transforms of binary images in arbitrary dimensions. Trans. on PAMI 25(2), 265–270 (2003)
24. Ogniewicz, R.L., Kübler, O.: Hierarchic voronoi skeletons. Pattern Recognition 28(33), 343–359 (1995)
25. Pudney, C.: Distance-ordered homotopic thinning: a skeletonization algorithm for 3D digital images. Comp. Vision and Im. Understanding 72(3), 404–413 (1998)
26. Sharvit, D., Chan, J., Tek, H., Kimia, B.B.: Symmetry-based indexing of image databases. J. of Visual Comm. and Im. Representation 9(4), 366–380 (1998)
27. Svensson, S., Sanniti di Baja, G.: Simplifying curve skeletons in volume images. Comp. Vision and Im. Understanding 90(3), 242–257 (2003)
28. Talbot, H., Vincent, L.: Euclidean skeletons and conditional bisectors. In: VCIP 1992, SPIE, vol. 1818, pp. 862–876 (1992)
29. Vincent, L.: Efficient computation of various types of skeletons. In: Medical Imaging V, SPIE, vol. 1445, pp. 297–311 (1991)
30. Welzl, E.: Smallest enclosing disks (balls and ellipsoids). In: Maurer, H.A. (ed.) New Results and New Trends in Computer Science. LNCS, vol. 555, pp. 359–370. Springer, Heidelberg (1991)

Appendix

The following recursive function, introduced and analyzed in [30], computes the smallest ball enclosing the set of points P, and having the points of the set R on its boundary. The initial call is done with $R = \emptyset$. The result B is made of the center and the radius of the enclosing ball.

Algorithm 1. Function SmallestEnclosingBall(P, R)

Data : $P \subseteq \mathbb{R}^d, R \subseteq \mathbb{R}^d$
if $P = \emptyset$ *[or* $|R| = d+1$*]* **then** compute B directly;
else
  choose $p \in P$ at random;
  B = SmallestEnclosingBall($P \setminus \{p\}, R$);
  if *[B defined and]* $p \notin B$ **then** B = SmallestEnclosingBall($P \setminus \{p\}, R \cup \{p\}$);
return B;

Appearance Radii in Medial Axis Test Mask for Small Planar Chamfer Norms*

Jérôme Hulin and Édouard Thiel

Laboratoire d'Informatique Fondamentale de Marseille (LIF, UMR 6166),
Aix-Marseille Université,
163 Avenue de Luminy, Case 901, 13288 Marseille cedex 9, France
{Jerome.Hulin,Edouard.Thiel}@lif.univ-mrs.fr

Abstract. The test mask $\mathcal{T}_\mathcal{M}$ is the minimum neighbourhood sufficient to extract the medial axis of any discrete shape, for a given chamfer distance mask $\mathcal{M}$. We propose an arithmetical framework to study $\mathcal{T}_\mathcal{M}$ in the case of chamfer norms. We characterize $\mathcal{T}_\mathcal{M}$ for 3×3 and 5×5 chamfer norm masks, and we give an algorithm to compute the appearance radius of the vector $(2,1)$ in $\mathcal{T}_\mathcal{M}$.

Keywords: medial axis, chamfer norm, distance transform.

1 Introduction

The medial axis MA of a shape $\mathcal{S}$ is a representation having numerous applications in image analysis, computer vision, and several other fields. Given a distance d, we consider the distance balls which are included in $\mathcal{S}$. A ball B is *maximal* in $\mathcal{S}$ if B is not included in any single other ball included in $\mathcal{S}$. The MA of $\mathcal{S}$ (for d) is the set of centres and radii of all maximal balls in $\mathcal{S}$ [1]. Since MA is a covering, it is a reversible coding. In general, MA is not a minimal representation; we know that finding a minimum covering subset of MA is an NP-hard problem [2]. Variations on MA include for example the reduced MA [3] or the MA in higher resolution [4].

The MA can be efficiently detected by local tests on a distance transform DT. A binary image I is composed of shape points and background points; the DT of I is a copy of I, where each shape point is labelled with its distance to the background. The main discrete distances used are the chamfer (or weighted) distances [5] and the Squared Euclidean Distance (SED) [6].

The characterization of MA is simple for distances d_4 and d_8 (the simplest chamfer distances, also known as ℓ_1 and ℓ_∞ norms). In [1], Pfaltz and Rosenfeld have shown that MA is the set of local maxima in DT, using the 4- and the 8-neighbourhood, respectively. Arcelli and Sanniti di Baja [7] have studied the $\langle 3,4\rangle$ chamfer norm: after lowering some DT labels, the local maxima criterion over the 8-neighbourhood still holds. A general method, named the LUT (look-up table) method, has then been proposed by Borgefors in [8] for the $\langle 5,7,11\rangle$ chamfer norm, with a local test on the 5×5 neighbourhood.

* Work supported in part by ANR grant BLAN06-1-138894 (projet OPTICOMB).

S. Brlek, C. Reutenauer, and X. Provençal (Eds.): DGCI 2009, LNCS 5810, pp. 434–445, 2009.

In the LUT method, the LUT gives for each DT value of a point p and each neighbour q of p, the minimum DT value of q that would forbid p to be in MA. Naturally, the size of a sufficient neighbourhood depends on the width of the shape. The *test mask* $\mathcal{T}(R)$ is the minimum test neighbourhood sufficient to extract the MA of all shapes whose inner radii are no greater than R, the inner radius of $\mathcal{S}$ being the radius of a larger ball included in $\mathcal{S}$.

Algorithms to compute both LUT and $\mathcal{T}(R)$ are given by Rémy and Thiel in arbitrary dimension for chamfer norms and SED [9,10], with code available in dimensions 2 to 6 in [11]. In recent papers, Normand and Évenou have proposed a faster method for chamfer norms in 2D and 3D based on a polytope representation of chamfer balls [12,13].

We now aim to understand the properties of $\mathcal{T}(R)$. We have recently shown for SED in $\mathbb{Z}^n$ that $\mathcal{T}(R)$ tends to the set of visible vectors when R tends to infinity [14], a vector being visible if its coordinates are coprime.

In this paper we focus on 2-dimensional chamfer norms, which have been studied in [15]. We examine the test masks for small chamfer norms, namely the 3×3 chamfer $\langle a, b\rangle$ and 5×5 chamfer $\langle a, b, c\rangle$ masks. We show that for these norms, the chamfer mask itself is a sufficient test neighbourhood for MA. Moreover, the neighbour $\overrightarrow{c}(2, 1)$ may not be in $\mathcal{T}$ for 5×5 norms; we give an arithmetical criterion to know whether $\overrightarrow{c}$ belongs to $\mathcal{T}$. The *appearance radius* of a vector $\overrightarrow{v}$ is the smallest R for which $\overrightarrow{v} \in \mathcal{T}(R)$. We link our problem to the Frobenius problem [16], so as to provide a simple algorithm which computes the appearance radius of $\overrightarrow{c}$ in time $\mathcal{O}(bc)$. After some preliminary definitions in Section 2, we present the test mask in Section 3 and discuss the case of 3×3 chamfer norms. Section 4 deals with 5×5 chamfer norms, while Section 5 is devoted to the appearance radius of $\overrightarrow{c}$. We finally conclude in Section 6.

Due to the lack of space, the proofs are not in the paper; they are available in an electronic annex [17].

2 Preliminaries

2.1 The Discrete Space $\mathbb{Z}^n$

In the following, we consider $\mathbb{Z}^n$ both as a n-dimensional $\mathbb{Z}$-module (i.e., a discrete vector space) and as its associated affine space; we set $\mathbb{Z}^n_* = \mathbb{Z}^n \setminus \{0\}$. The Cartesian coordinates of a vector $\overrightarrow{v}$ are denoted by $(v_1, \ldots, v_n)$. A point p is called *visible* if there is no lattice point between O and p on the line segment $[Op]$, i.e., if the coordinates of p are coprime. An n dimensional *shape* $\mathcal{S}$ is a subset of $\mathbb{Z}^n$.

We call Σ^n the group of axial and diagonal symmetries in $\mathbb{Z}^n$ about centre O. The cardinal of the group is $\#\Sigma^n = 2^n\, n!$ (which is 8, 48 and 384 for $n = 2$, 3 and 4). A shape $\mathcal{S}$ is said to be *G-symmetrical* if for every $\sigma \in \Sigma^n$ we have $\sigma(\mathcal{S}) = \mathcal{S}$. The *generator* of a set $\mathcal{S} \subseteq \mathbb{Z}^n$ (or $\mathbb{R}^n$) is $\mathrm{G}(\mathcal{S}) = \{\, (p_1, \ldots, p_n) \in \mathcal{S} : 0 \leqslant p_n \leqslant p_{n-1} \leqslant \ldots \leqslant p_1 \,\}$.

Given two vectors $\overrightarrow{u}, \overrightarrow{v} \in \mathbb{Z}^n$ and a point $p \in \mathbb{Z}^n$, we define the *vector cone* $\mathcal{C}(\overrightarrow{u}, \overrightarrow{v}) = \overrightarrow{u}\mathbb{N} + \overrightarrow{v}\mathbb{N} = \{\alpha\overrightarrow{u} + \beta\overrightarrow{v} : \alpha, \beta \in \mathbb{N}\}$. Also, we define the *affine cone* $\mathcal{C}(p, \overrightarrow{u}, \overrightarrow{v}) = p + \mathcal{C}(\overrightarrow{u}, \overrightarrow{v})$.

2.2 The Frobenius Problem and Representable Integers

Let $x, a_1, \ldots, a_n$ be natural numbers, x is said $(a_1, \ldots, a_n)$*-representable* if $x \in a_1\mathbb{N} + \cdots + a_n\mathbb{N}$, i.e., if there are n non-negative integers $\lambda_1, \ldots, \lambda_n$ s.t. $x = \lambda_1 a_1 + \cdots + \lambda_n a_n$. The largest natural number that is not $(a_1, \ldots, a_n)$-representable is called the Frobenius number, denoted by $g(a_1, \ldots, a_n)$; it exists iff $a_1, \ldots, a_n$ are coprime. Sylvester proved that if a and b are coprime, then $g(a, b) = ab - a - b$ (see [16] for a complete study). For example, the set of all $(4, 7)$-representable integers are $4\mathbb{N} + 7\mathbb{N} = \{0, 4, 7, 8, 11, 12, 14, 15, 16, 18, 19, 20, 21, \ldots\}$; all integers greater than $g(4, 7) = 17$ are $(4, 7)$-representable.

Let x, a, b be three natural numbers; we define $[x]_{a,b}$ to be the largest integer (a, b)-representable and no greater than x:

$$[x]_{a,b} = \max\{\, y \in a\mathbb{N} + b\mathbb{N} \,:\, y \leqslant x \,\}. \tag{1}$$

By definition, an integer x is (a, b)-representable iff $[x]_{a,b} = x$. Following the above example, we have $[10]_{4,7} = 8$ and $[15]_{4,7} = 15$. The set of integers $\{[x]_{a,b}\}_{0 \leqslant x \leqslant k}$ can be efficiently computed by Alg. 1, in time $\mathcal{O}(k)$. It consists in a single scan of an array of size k, using the property that if an integer x is (a, b)-representable, then so are $x + a$ and $x + b$.

Algorithm 1. Compute_RI

```
input  : two natural numbers a and b (a ⩽ b), an integer k.
output: an array t verifying t[x] = [x]_{a,b} for all 0 ⩽ x ⩽ k.
allocate an array t of k + b + 1 integers, all initialized to 0 ;
for x ← 0 to k do
    if t[x] = x then                          /* x is (a,b)-representable */
        t[x + a] ← x + a ; t[x + b] ← x + b ;
        rep ← x ;        /* updating the largest representable integer */
    else t[x] ← rep ;
return t ;
```

2.3 Balls and Medial Axis

Given a distance d, the *ball* of centre $p \in \mathbb{Z}^n$ and radius $r \in \mathbb{R}$ is $\mathcal{B}(p, r) = \{\, q \in \mathbb{Z}^n : d(p, q) \leqslant r \,\}$. Since we consider discrete closed balls, any ball B may have an infinite number of real radii in a left-closed interval $[r_1, r_2[$. We define the *representable radius* of a given ball B, to be the radius of B which belongs to $\mathrm{Im}(d)$, namely r_1 with the above notation.

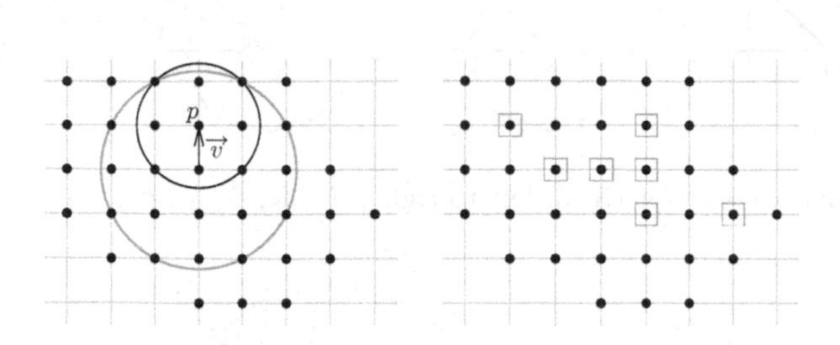

Fig. 1. A shape $\mathcal{S}$ drawn with bullets. Left: $\overrightarrow{v}$ forbids p to be a MA point of $\mathcal{S}$ since $I_p(\mathcal{S})$ is included in $I_{p-\overrightarrow{v}}(\mathcal{S})$ (equivalently, $H_{p-\overrightarrow{v}}(I_p(\mathcal{S})) \subseteq \mathcal{S}$). Right: the medial axis of $\mathcal{S}$, i.e., the centres of maximal balls of $\mathcal{S}$. In this figure we use the Euclidean distance.

Fig. 2. $H_q(B)$ is the smallest ball of centre q which contains B. The representable radius of $H_q(B)$ is $\mathcal{R}_q(B) = d(q, z)$.

Let $\mathcal{S}$ be a shape and $p \in \mathcal{S}$. We define $I_p(\mathcal{S})$ to be the largest ball of centre p and included in $\mathcal{S}$. The *inner radius* of $\mathcal{S}$ is the representable radius of a largest ball included in $\mathcal{S}$. We denote by $\mathrm{CS}^n(R)$ the class of all n-dimensional shapes whose inner radius are less than or equal to R. A ball included in $\mathcal{S}$ is called a *maximal ball* of $\mathcal{S}$ if it is not included in any other ball included in $\mathcal{S}$. The *Medial Axis* (MA) of a shape $\mathcal{S}$ is the set of centres (and radii) of all maximal balls of $\mathcal{S}$:

$$p \in \mathrm{MA}(\mathcal{S}) \Leftrightarrow p \in \mathcal{S} \text{ and } \forall q \in \mathcal{S} \setminus \{p\},\ I_p(\mathcal{S}) \nsubseteq I_q(\mathcal{S}). \tag{2}$$

If there exists $\overrightarrow{v} \in \mathbb{Z}^n_*$ such that $I_p(\mathcal{S}) \subseteq I_{p-\overrightarrow{v}}(\mathcal{S})$, we say that $\overrightarrow{v}$ *forbids* p to be a medial axis point of $\mathcal{S}$. See Fig. 1 for an example.

We denote by $H_q(B)$ the smallest ball of centre q which contains the ball B; and we define $\mathcal{R}_q(B)$ to be the representable radius of $H_q(B)$, see an example in Fig. 2. Accordingly, the medial axis of $\mathcal{S}$ can be expressed as:

$$p \in \mathrm{MA}(\mathcal{S}) \Leftrightarrow p \in \mathcal{S} \text{ and } \forall q \in \mathcal{S} \setminus \{p\},\ H_q\big(I_p(\mathcal{S})\big) \nsubseteq \mathcal{S}. \tag{3}$$

2.4 Chamfer Distances and Norms

Here we recall some results from [15], Sec. 4.2 and 4.3. A *chamfer mask* $\mathcal{M}$ in $\mathbb{Z}^n$ is a central-symmetric set $\mathcal{M} = \{(\overrightarrow{v_i}, w_i) \in \mathbb{Z}^n \times \mathbb{Z}_{+*}\}_{1 \leqslant i \leqslant m}$ containing at least a basis of $\mathbb{Z}^n$, where $(\overrightarrow{v_i}, w_i)$ are called *weightings*, $\overrightarrow{v_i}$ *vectors* and w_i *weights*. The *chamfer distance* $d_{\mathcal{M}}$ between two points $p, q \in \mathbb{Z}^n$ is

$$d_{\mathcal{M}}(p, q) = \min\left\{ \sum \lambda_i w_i : \sum \lambda_i \overrightarrow{v_i} = \overrightarrow{pq}\ ,\ 1 \leqslant i \leqslant m,\ \lambda_i \in \mathbb{Z}_+ \right\}. \tag{4}$$

We write it simply pq when no confusion can arise. By definition every chamfer distance is a translation invariant metric. Here we point out a link with representable integers: if $w_1, \ldots, w_m$ are the weights of a chamfer mask $\mathcal{M}$, then all representable radii belong to $w_1\mathbb{N} + \cdots + w_m\mathbb{N}$.

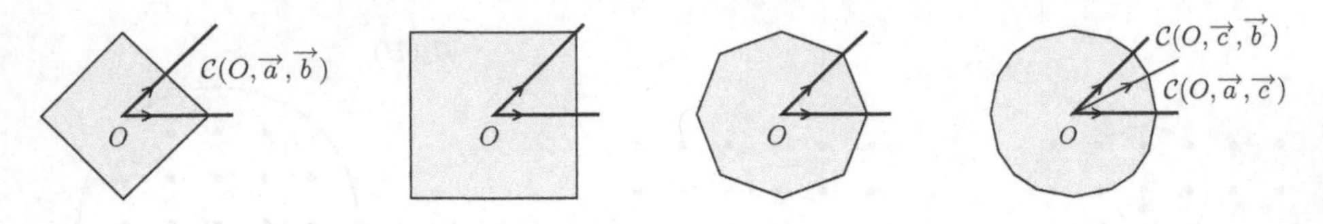

Fig. 3. Chamfer balls and influence cones for (from left to right) d_4, d_8, $d_{\langle 3,4 \rangle}$, $d_{\langle 5,7,11 \rangle}$

2b	a+b	2a	a+b	2b
a+b	b	a	b	a+b
2a	a	O	a	2a
a+b	b	a	b	a+b
2b	a+b	2a	a+b	2b

2b	c	2a	c	2b
c	b	a	b	c
2a	a	O	a	2a
c	b	a	b	c
2b	c	2a	c	2b

Fig. 4. Balls of radii a (shaded) and b (delimited by the thick line) for any minimal norm mask $\langle a, b \rangle$ (left) and $\langle a, b, c \rangle$ (right). Values indicate the chamfer distance to O.

Fig. 5. A ball B of radius $Op = 50$ for the norm mask $\langle 8, 11, 18 \rangle$, The generator of B is drawn with bullets. 50 is $(8, 18)$-representable: $50 = 4{*}8{+}18$. The dashed line represents the axis $O + \mathbb{R}\vec{c}$.

Consider $\mathcal{M}' = \{ O + \vec{v_i}/w_i \}_{1 \leqslant i \leqslant m} \in \mathbb{R}^n$ and let $B'_{\mathcal{M}} = \text{conv}\,(\mathcal{M}')$, then $B'_{\mathcal{M}}$ is a central-symmetric and convex polyhedron whose facets separate $\mathbb{R}^n$ in cones from O. A facet $\mathcal{F}$ of $B'_{\mathcal{M}}$ is generated by a subset $\mathcal{M}|_{\mathcal{F}} = \{ (\vec{v_j}, w_j) \}_{1 \leqslant j \leqslant n}$ of $\mathcal{M}$; if $\Delta_{\mathcal{F}} = \det \{ \vec{v_j} \}_{1 \leqslant j \leqslant n}$ is such that $|\Delta_{\mathcal{F}}| = 1$, then $\mathcal{F}$ is said *unimodular.* If each facet of $B'_{\mathcal{M}}$ is unimodular, then $d_{\mathcal{M}}$ is a norm in $\mathbb{Z}^n$ ([15], p. 53). We denote by $\| \vec{v} \|$ the chamfer norm of a vector $\vec{v}$.

Now let $d_{\mathcal{M}}$ be a chamfer norm, $\mathcal{F}$ a facet of $B'_{\mathcal{M}}$ and $\mathcal{M}|_{\mathcal{F}} = \{ (\vec{v_j}, w_j) \}_{1 \leqslant j \leqslant n}$; then for each point $p = (p_1, \ldots, p_n)$ in the cone $(O, \mathcal{F})$, called *influence cone* of $\mathcal{M}|_{\mathcal{F}}$, we have $d_{\mathcal{M}}(O, p) = p_1\, \delta_1 + \cdots + p_n\, \delta_n$, where $(\delta_1, \ldots, \delta_n) = \overrightarrow{\delta_{\mathcal{F}}}$ is a normal vector of $\mathcal{F}$, and δ_k is the *elementary displacement* for the kth coordinate:

$$\delta_k = \frac{(-1)^{n+k}}{\Delta_{\mathcal{F}}} \cdot \begin{vmatrix} v_{1,1} & \cdots & v_{1,k-1} & v_{1,k+1} & \cdots & v_{1,n} & w_1 \\ \vdots & & \vdots & \vdots & & \vdots & \\ v_{n,1} & \cdots & v_{n,k-1} & v_{n,k+1} & \cdots & v_{n,n} & w_n \end{vmatrix}^T . \tag{5}$$

In other words, if a vector $\vec{v}$ belongs to some influence cone $\mathcal{C}(\vec{v_1}, \ldots, \vec{v_n})$, then there is a minimal path from O to $O + \vec{v}$ which is only composed of the vectors $\vec{v_1}, \ldots, \vec{v_n}$, and so $\| \vec{v} \|$ is $(w_1, \ldots, w_n)$-representable.

From now on, we restrict the discussion to G-symmetrical masks. As a consequence, the chamfer balls are G-symmetrical. In $\mathbb{Z}^2$, let $\vec{a} = (1, 0)$, $\vec{b} = (1, 1)$, $\vec{c} = (2, 1)$; a common way to denote 3×3 masks is $\langle a, b \rangle = \{ (\vec{a}, a), (\vec{b}, b) \}$ (plus the G-symmetrical weightings) and $\langle a, b, c \rangle = \{ (\vec{a}, a), (\vec{b}, b), (\vec{c}, c) \}$

(ditto) for 5×5 masks. Note that $\mathrm{G}(\mathbb{Z}^2) = \mathcal{C}(\overrightarrow{a}, \overrightarrow{b})$. Widely used chamfer norms are $d_4 = \ell_1 = d_{\langle 1,2\rangle}$, $d_8 = \ell_\infty = d_{\langle 1,1\rangle}$, $d_{\langle 3,4\rangle}$ and $d_{\langle 5,7,11\rangle}$, see Fig. 3. Chamfer distances to O are indicated in Fig. 4 for $\langle a,b\rangle$ and $\langle a,b,c\rangle$ norm masks. Also, Fig. 5 illustrates a ball of radius 50 for the norm mask $\langle 8,11,18\rangle$. The conditions for a chamfer mask $\langle a,b\rangle$ to generate a norm are $a \leqslant b \leqslant 2a$; and according to (5), the elementary displacements in $\mathrm{G}(\mathbb{Z}^2)$ are $\delta_x = a$ and $\delta_y = b - a$. For $\langle a,b,c\rangle$ norm masks the constraints are $\{2a \leqslant c\,;\, c \leqslant a + b\,;\, 3b \leqslant 2c\}$, while the displacements are $\{\delta_x = a,\ \delta_y = c - 2a\}$ in the influence cone $\mathcal{C}(\overrightarrow{a}, \overrightarrow{c})$ and $\{\delta_x = c - b,\ \delta_y = 2b - c\}$ in the influence cone $\mathcal{C}(\overrightarrow{c}, \overrightarrow{b})$.

Without loss of generality, we will only consider *minimal* chamfer masks. A mask $\mathcal{M} = \{(\overrightarrow{v_i}, w_i)\}_{1\leqslant i\leqslant m}$ is said *minimal* if the removal of any single weighting of $\mathcal{M}$ modifies the distance, i.e, if $\forall\, 1 \leqslant i \leqslant m,\ d_{\mathcal{M}} \neq d_{\mathcal{M}\setminus\{(\overrightarrow{v_i}, w_i)\}}$. Considering Fig. 4, it is easily seen that a norm mask $\langle a,b\rangle$ is minimal iff $b < 2a$. Otherwise we would have $b = 2a$, so $(1,0) + (0,1)$ would be a minimal chamfer path from O to $(1,1)$, hence $\overrightarrow{b}$ would be redundant in the chamfer mask. In a similar way, we check at once that a norm mask $\langle a,b,c\rangle$ is minimal iff $c < a + b$.

Here are some elementary properties of covering relations between chamfer balls, illustrated in Fig. 6. These properties are extensively used in the proofs.

Lemma 1 (Representable radius). *Let $\mathcal{C}(\overrightarrow{v_1}, \overrightarrow{v_2})$ be an influence cone of a given 2D chamfer norm, and assume the vectors $\overrightarrow{v_1}$ and $\overrightarrow{v_2}$ have respective weights w_1 and w_2. If r is (w_1, w_2)-representable then for any vector $\overrightarrow{v}$ in $\mathcal{C}(\overrightarrow{v_1}, \overrightarrow{v_2})$, we have $\mathcal{R}_{O-\overrightarrow{v}}(\mathcal{B}(O,r)) = r + \|\overrightarrow{v}\|$.*

Lemma 2 (Covering a cone). *Let $\mathcal{C}(\overrightarrow{v_1}, \overrightarrow{v_2})$ be an influence cone of a given 2D chamfer norm, and B be a ball of centre O. For any vector $\overrightarrow{v}$ in the cone $\mathcal{C}(\overrightarrow{v_1}, \overrightarrow{v_2})$, we have $B \cap \mathcal{C}(O, \overrightarrow{v_1}, \overrightarrow{v_2}) = H_{O-\overrightarrow{v}}(B) \cap \mathcal{C}(O, \overrightarrow{v_1}, \overrightarrow{v_2})$. In other words, the balls B and $H_{O-\overrightarrow{v}}(B)$ coincide in the cone $\mathcal{C}(O, \overrightarrow{v_1}, \overrightarrow{v_2})$.*

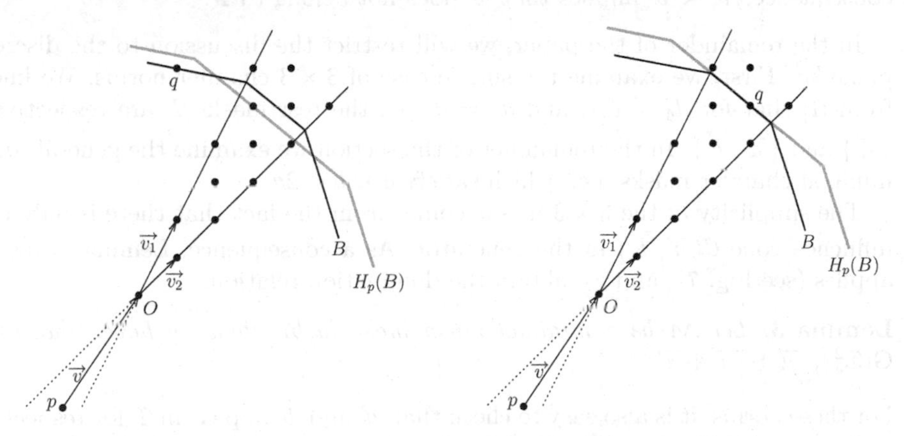

Fig. 6. Covering a ball B of center O and representable radius $r = d(O,q)$ in direction $\overrightarrow{v}$, with $\overrightarrow{v}$ being a vector in the influence cone $\mathcal{C}(\overrightarrow{v_1}, \overrightarrow{v_2})$. Left: r is not (w_1, w_2)-representable so $\mathcal{R}_p(B) < r + \|\overrightarrow{v}\|$. Right: r is (w_1, w_2)-representable, so there is a minimal chamfer path from p to q passing through O, hence $\mathcal{R}_p(B) = r + \|\overrightarrow{v}\|$.

3 The Medial Axis Test Mask $\mathcal{T}_\mathcal{M}$

Given a chamfer mask $\mathcal{M}$ in $\mathbb{Z}^n$, we define the *test mask* $\mathcal{T}_\mathcal{M}(R)$ to be the minimum neighbourhood sufficient to detect locally, for any $\mathcal{S}$ in $\mathrm{CS}^n(R)$ and any $p \in \mathcal{S}$, if p is a point of $\mathrm{MA}(\mathcal{S})$:

$$\begin{cases} \forall \mathcal{S} \in \mathrm{CS}^n(R), \forall p \in \mathcal{S}, \left(p \notin \mathrm{MA}(\mathcal{S}) \Rightarrow \exists \overrightarrow{v} \in \mathcal{T}_\mathcal{M}(R),\, I_p(\mathcal{S}) \subseteq I_{p-\overrightarrow{v}}(\mathcal{S})\right) \\ \mathcal{T}_\mathcal{M}(R) \text{ has minimum cardinality.} \end{cases} \tag{6}$$

If $p \in \mathrm{MA}(\mathcal{S})$, then no $\overrightarrow{v} \in \mathbb{Z}^n$ satisfies $I_p(\mathcal{S}) \subseteq I_{p-\overrightarrow{v}}(\mathcal{S})$. Otherwise, if $p \notin \mathrm{MA}(\mathcal{S})$, then $\mathcal{T}_\mathcal{M}(R)$ contains at least one vector $\overrightarrow{v}$ which forbids p from $\mathrm{MA}(\mathcal{S})$. We have shown the unicity of $\mathcal{T}_\mathcal{M}(R)$ for all $R \geqslant 0$ in [14] (the proof remains the same for any norm). As a corollary, if the considered norm is G-symmetrical, then so is $\mathcal{T}_\mathcal{M}(R)$.

Finally, we set $\mathcal{T}_\mathcal{M} = \lim_{R\to+\infty} \mathcal{T}_\mathcal{M}(R)$. We write it simply $\mathcal{T}$ when no confusion can arise, and by abuse of notation we often write $\mathcal{T}$ instead of $\mathrm{G}(\mathcal{T})$. The *appearance radius* $R_{app}(\overrightarrow{v})$ of a given vector $\overrightarrow{v}$ is the smallest radius R for which $\overrightarrow{v} \in \mathcal{T}_\mathcal{M}(R)$.

Definition 1 (Domination). *We write B_r for $\mathcal{B}(O,r)$. Given a distance d, a vector $\overrightarrow{v} \in \mathbb{Z}^n_*$ is dominated by a vector $\overrightarrow{u} \in \mathbb{Z}^n_*$ if for all radii $r \in \mathbb{R}_+$, $H_{O-\overrightarrow{u}}(B_r)$ is strictly included in $H_{O-\overrightarrow{v}}(B_r)$:*

$$\overrightarrow{v} \prec \overrightarrow{u} \Leftrightarrow \forall r \geqslant 0,\, H_{O-\overrightarrow{u}}(B_r) \subsetneq H_{O-\overrightarrow{v}}(B_r). \tag{7}$$

Note that by definition, $\prec$ is transitive. The basic idea of the domination relation is the following: if $\overrightarrow{v} \prec \overrightarrow{u}$, it is useless to test if a given point p belongs to some $\mathrm{MA}(\mathcal{S})$ in both directions $\overrightarrow{u}$ and $\overrightarrow{v}$, testing in direction $\overrightarrow{u}$ is sufficient. As a consequence, $\overrightarrow{v} \prec \overrightarrow{u}$ implies that $\overrightarrow{v}$ does not belong to $\mathcal{T}$.

In the remainder of the paper, we will restrict the discussion to the discrete plane $\mathbb{Z}^2$. First, we examine the simple case of 3×3 chamfer norms. We know from [1] that for $d_4 = d_{\langle 1 \rangle}$ and $d_8 = d_{\langle 1,1 \rangle}$, the test masks $\mathcal{T}$ are respectively $\{\overrightarrow{a}\}$ and $\{\overrightarrow{a}, \overrightarrow{b}\}$. In the remainder of the section we examine the general $\langle a, b \rangle$ minimal chamfer masks, i.e., which satisfy $a < b < 2a$.

The simplicity of the 3×3 masks comes from the fact that there is only one influence cone $\mathcal{C}(\overrightarrow{a}, \overrightarrow{b})$ in the generator. As a consequence, Lemma 1 always applies (see Fig. 7), and we obtain the domination relation:

Lemma 3. *Let $\mathcal{M}$ be a minimal norm mask $\langle a, b \rangle$, then we have: $\forall \overrightarrow{u}, \overrightarrow{v} \in \mathrm{G}(\mathbb{Z}^2_*)$, $\overrightarrow{u} \succ \overrightarrow{u} + \overrightarrow{v}$.*

For these norms, it is also easy to check that $\overrightarrow{a}$ and $\overrightarrow{b}$ appear in $\mathcal{T}$ for respective radius a and b: the fact that $a < b$ implies that the balls B_a of radii a and centre O are constant for all these norms (Fig. 4, left). B_a is the smallest ball having positive radius, and $p(1,0)$ does not belong to $\mathrm{MA}(B_a)$ so $\mathcal{T}(a) = \{\overrightarrow{a}\}$. Same reasonning applies to $B_b = \mathcal{B}(O, b)$: these balls are constant for all these norms

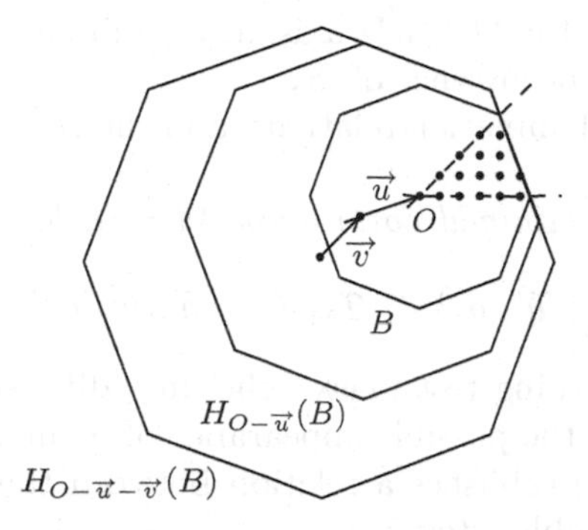

Fig. 7. For any $\langle a, b\rangle$ norm mask and $\overrightarrow{u}, \overrightarrow{v} \in \mathrm{G}(\mathbb{Z}^2)$, $\overrightarrow{u}$ dominates $\overrightarrow{u} + \overrightarrow{v}$

$(b < 2a)$. Since $\overrightarrow{b}$ is the only vector which forbids $q(1,1)$ from $\mathrm{MA}(B_b)$, we conclude that $\mathcal{T}(b) = \{\overrightarrow{a}, \overrightarrow{b}\}$.

Lemma 3 asserts that any vector $\overrightarrow{v} \in \mathrm{G}(\mathbb{Z}^2_*)$ is dominated by $\overrightarrow{a}$ or by $\overrightarrow{b}$ (or both). Adding the fact that $\mathcal{T}(b) = \{\overrightarrow{a}, \overrightarrow{b}\}$, it follows:

Theorem 1. *For any minimal norm mask $\mathcal{M} = \langle a, b\rangle$, we have $\mathcal{T}_{\mathcal{M}} = \{\overrightarrow{a}, \overrightarrow{b}\}$.*

4 Test Masks for 5 × 5 Chamfer Norms

4.1 The Test Masks $\mathcal{T}_{\langle a,b,c\rangle}$ Are Bounded

Compared to the case of 3 × 3 minimal norm masks, same reasonning holds here concerning the appearance of $\overrightarrow{a}$ and $\overrightarrow{b}$. Since $b \leqslant \frac{2}{3}c < c$ for $\langle a, b, c\rangle$ norms, balls of radii a and b are the same as for $\langle a, b\rangle$ masks (see Fig. 4, right). Thus, we have $\mathcal{T}(b) = \{\overrightarrow{a}, \overrightarrow{b}\}$. On the other hand, the domination relations are described in the following two lemmas:

Lemma 4 (Domination along $\overrightarrow{a}\mathbb{N}$ and $\overrightarrow{b}\mathbb{N}$). *For any minimal norm mask $\langle a, b, c\rangle$ and any $k \in \mathbb{N}_*$, we have $k\overrightarrow{a} \succ (k+1)\overrightarrow{a}$ and $k\overrightarrow{b} \succ (k+1)\overrightarrow{b}$.*

Lemma 5 (Domination by addition of $\overrightarrow{c}$). *For any minimal norm mask $\langle a, b, c\rangle$ and any $\overrightarrow{v} \in \mathrm{G}(\mathbb{Z}^2_*)$, we have $\overrightarrow{v} \succ \overrightarrow{v} + \overrightarrow{c}$.*

Let us mention that in the Euclidean case, the Lemma 4 holds for any vector [14]. However, this is generally false for chamfer norms. Geometrically speaking, this lemma holds for $\overrightarrow{a}$ and $\overrightarrow{b}$ because we consider G-symmetrical masks and each of these two vectors has the same direction as a G-symmetry axis of $\mathbb{Z}^2$.

Lemma 5 comes from the following observation: since the generator only contains two influence cones $\mathcal{C}(\overrightarrow{a}, \overrightarrow{c})$ and $\mathcal{C}(\overrightarrow{c}, \overrightarrow{b})$, the vector $\overrightarrow{c}$ belongs to all influence cones of $\mathrm{G}(\mathbb{Z}^2)$. Accordingly, for any point $p \in \mathrm{G}(\mathbb{Z}^2)$ which does not lie on the $O + \overrightarrow{a}\mathbb{N}$ nor the $O + \overrightarrow{b}\mathbb{N}$ axis, any minimal chamfer path from O to p contains at least one occurrence of $\overrightarrow{c}$.

According to these domination relations, and the transitivity of $\prec$, we obtain:

Theorem 2. *For any minimal norm mask* $\mathcal{M} = \langle a, b, c \rangle$*, we have*

$$\{\overrightarrow{a}, \overrightarrow{b}\} \subseteq \mathcal{T}_{\mathcal{M}} \subseteq \{\overrightarrow{a}, \overrightarrow{b}, \overrightarrow{c}\}.$$

We observe that concerning test masks, the only difference between 3×3 and 5×5 chamfer norms is the possible appearance of $\overrightarrow{c}$ in the case of 5×5 norms. The following section establishes a relation between the appearance of $\overrightarrow{c}$ and the notion of representable integers.

4.2 Appearance of $\overrightarrow{c}$ in the Test Masks $\mathcal{T}_{\langle a,b,c \rangle}$

The task is now to find all the chamfer masks $\mathcal{M} = \langle a, b, c \rangle$ for which $\overrightarrow{c} \in \mathcal{T}_{\mathcal{M}}$. We first illustrate the appearance of $\overrightarrow{c}$ for $\langle 5, 7, 11 \rangle$ in Fig. 8. B is the ball of centre O and radius 35. On the left is shown $H_{O-\overrightarrow{a}}(B)$: there is a point of B in the cone $\mathcal{C}(O, \overrightarrow{a}, \overrightarrow{c})$ which is at distance 35 from O (35 is (5, 11)-representable). Hence Lemma 1 yields $\mathcal{R}_{O-\overrightarrow{a}}(B) = 35 + 5 = 40$. However, there is a point p in the cone $\mathcal{C}(O, \overrightarrow{c}, \overrightarrow{b})$ at distance 40 from O, so $q = p - \overrightarrow{a}$ (circled point) belongs to $H_{O-\overrightarrow{a}}(B)$. On the other hand, q is at a distance 36 from O, so $q \notin B$. Given that B and $H_{O-\overrightarrow{c}}(B)$ are equal in $\mathcal{C}(O, \overrightarrow{c}, \overrightarrow{b})$ (using $\overrightarrow{v} = \overrightarrow{c}$ in Lemma 2), the point q does not belong to $H_{O-\overrightarrow{c}}(B)$.

The same conclusion can be drawn with $H_{O-\overrightarrow{b}}(B)$ on the right of the figure: the circled point q belongs to $H_{O-\overrightarrow{b}}(B)$, but not to B.

The configuration of Fig.8 is the smallest one for which neither $H_{O-\overrightarrow{a}}(B)$ nor $H_{O-\overrightarrow{b}}(B)$ are included in $H_{O-\overrightarrow{c}}(B)$, implying that the appearance radius of $\overrightarrow{c}$ is $\mathcal{R}_{O-\overrightarrow{c}}(B) = 35 + c = 46$. This phenomenon can be expressed in an arithmetical way:

Lemma 6 (Arithmetical expression of $R_{app}(\overrightarrow{c})$). *For any minimal norm mask* $\langle a, b, c \rangle$*, we have*

$$R_{app}(\overrightarrow{c}) = \min\left\{ r \in \mathbb{N} : \begin{cases} r + c - b < \Big[[r]_{a,c} + a \Big]_{b,c} \\ r + c - a < \Big[[r]_{b,c} + b \Big]_{a,c} \end{cases} \right\} + c. \qquad (8)$$

This lemma is the starting point of all remaining results of this paper. In the next section, this lemma will be used to deduce an upper bound of the appearance radius of $\overrightarrow{c}$, as well as a simple algorithm for computing $R_{app}(\overrightarrow{c})$. First, we provide a criterion to determine whether the system in (8) has a solution:

Theorem 3 (Appearance of $\overrightarrow{c}$). *For any minimal norm mask* $\mathcal{M} = \langle a, b, c \rangle$*,*

$$\overrightarrow{c} \in \mathcal{T}_{\mathcal{M}} \Leftrightarrow \gcd(a, c) + \gcd(b, c) \leqslant 2\,(a + b - c).$$

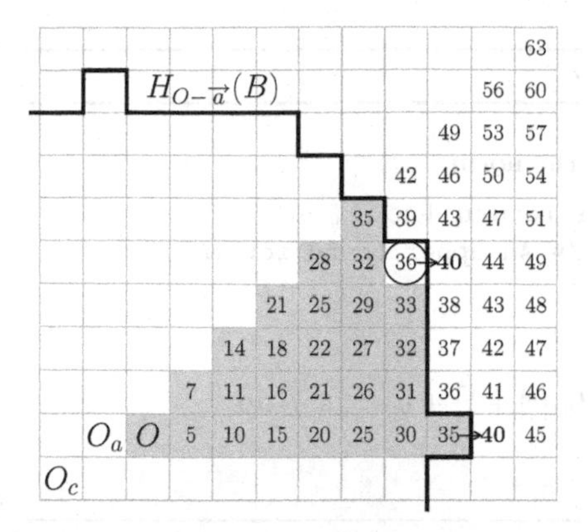

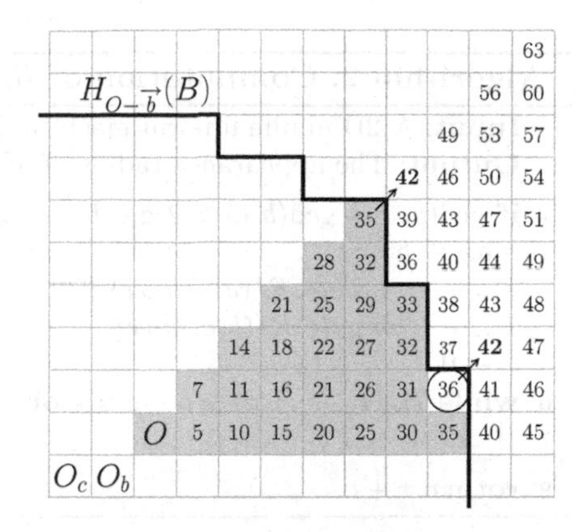

Fig. 8. Appearance of $\overrightarrow{c}$ for the mask $\langle 5, 7, 11\rangle$. A ball B of centre O and radius 35 (whose generator is shaded), the points $O_a = O - \overrightarrow{a}$, $O_b = O - \overrightarrow{b}$ and $O_c = O - \overrightarrow{c}$. Left : the circled point belongs to $H_{O-\overrightarrow{a}}(B)$ but not to B. Right : the circled point belongs to $H_{O-\overrightarrow{b}}(B)$ but not to B.

5 Appearance Radius of $\overrightarrow{c}$

5.1 A Bound for the Appearance Radius of $\overrightarrow{c}$

We now combine Lemma 6 and Sylvester's result on the Frobenius problem to obtain an upper bound of $R_{app}(\overrightarrow{c})$.

In the case were $\gcd(a, c) = \gcd(b, c) = 1$, it is easy to check that the system in (8) is satisfied for $r = g(b, c) + 1$, since for all $r > g(b, c)$, $\left[[r]_{a,c} + a\right]_{b,c} = [r + a]_{b,c} = r + a$, and similarly $\left[[r]_{b,c} + b\right]_{a,c} = r + b$. In this case, the system in (8) is therefore equivalent to $c < a + b$, which is a necessary condition for the minimality of $\langle a, b, c\rangle$. It follows from Lemma 6 that $R_{app}(\overrightarrow{c}) \leqslant g(b, c) + 1 + c$. Then, applying Sylvester's result $g(b, c) = bc - b - c$ gives $R_{app}(\overrightarrow{c}) < bc$.

We can generalize this property to the remaining masks:

Theorem 4. *Let $\mathcal{M} = \langle a, b, c\rangle$ be a minimal norm mask. If $\overrightarrow{c} \in \mathcal{T}_{\mathcal{M}}$ then $R_{app}(\overrightarrow{c}) < bc$.*

5.2 Computation of $R_{app}(\overrightarrow{c})$

Algorithm 2 computes the appearance radius of $\overrightarrow{c}$ in the MA test mask for any minimal norm mask $\mathcal{M} = \langle a, b, c\rangle$. The presence of $\overrightarrow{c}$ in the test mask (line 1) is determined according to Thm. 3. The calls *Compute_RI*(a, c, max) and *Compute_RI*(b, c, max) are computed in time $\mathcal{O}(max) = \mathcal{O}(bc)$, see Alg. 1. The main loop (lines 5–7) makes the arithmetical test provided by Lemma 6, for each radius r, from $r = 0$ to $r = R_{app}(\overrightarrow{c}) - c < bc - c$ (the bound comes from Thm. 4). Therefore, the time complexity of Alg. 2 is $\mathcal{O}(bc)$. As an example, this algorithm is run for $\langle 5, 7, 11\rangle$ in Fig. 9.

Algorithm 2. Computation of $R_{app}(\overrightarrow{c})$

Input: A 2D minimal norm mask $\langle a, b, c \rangle$.
Output: The appearance radius of $\overrightarrow{c}$ for the norm.

1 **if** $\gcd(a,c) + \gcd(b,c) > 2(a+b-c)$ **then return** $+\infty$;
2 $max \leftarrow b * c$; `/* An upper bound for` $R_{app}(\overrightarrow{c})$ `*/`
3 $T_{ac} \leftarrow Compute\_RI(a, c, max)$;
4 $T_{bc} \leftarrow Compute\_RI(b, c, max)$;
5 $r \leftarrow 0$;
6 **while** $T_{bc}\big[T_{ac}[r] + a\big] + b - c \leqslant r$ **or** $T_{ac}\big[T_{bc}[r] + b\big] + a - c \leqslant r$ **do**
7 $\quad r \leftarrow r + 1$;
8 **return** $r + c$

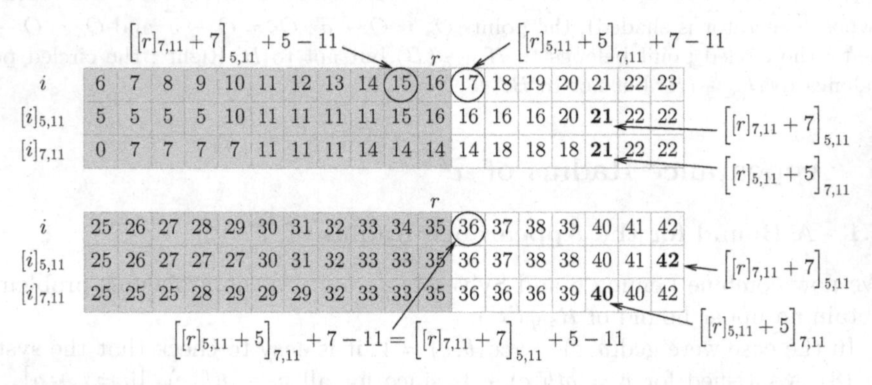

Fig. 9. Execution of Alg. 2 for the mask $\langle 5, 7, 11 \rangle$. The arrays $T_{ac}[.]$ et $T_{bc}[.]$ respectively contain the values of $[.]_{5,11}$ and $[.]_{7,11}$. Top: when $r = 16$ (configuration illustrated in Fig. 10). Bottom: when it detects the configuration of Fig. 8 (using same graphical rules), that is to say, for the appearance of $\overrightarrow{c}$ in $\mathcal{T}_{\langle 5,7,11 \rangle}$.

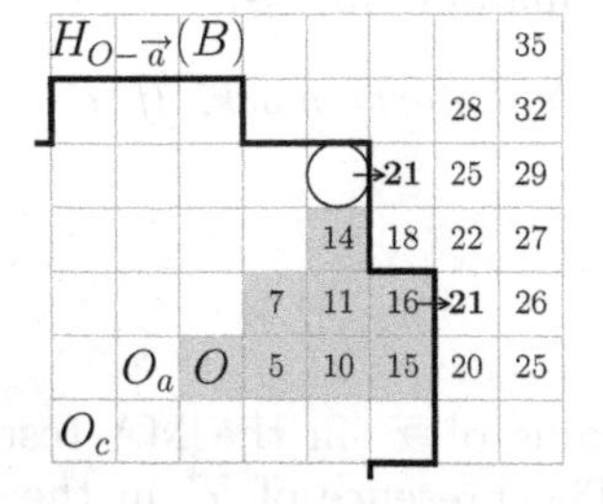

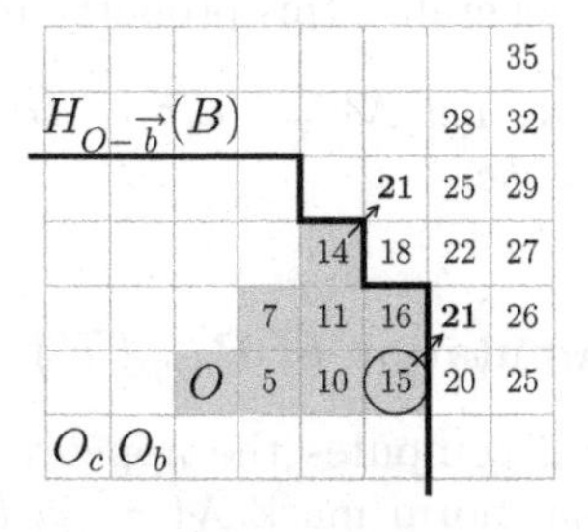

Fig. 10. Configuration for $\langle 5, 7, 11 \rangle$ and $r = 16$. B is the ball of centre O and radius r, its generator is shaded. Left: the circled point shows the result of $x = \big[[r]_{a,c} + a\big]_{b,c} + b - c = 17$. Since $x > r$, this point belongs to $H_{O-\overrightarrow{a}}(B) \setminus H_{O-\overrightarrow{c}}(B)$. Right: the circled point shows the result of $x = \big[[r]_{b,c} + b\big]_{a,c} + a - c = 15$. Since $x \leqslant r$, there is no point in $H_{O-\overrightarrow{b}}(B) \setminus H_{O-\overrightarrow{c}}(B)$. So this configuration does not lead to the appearance of $\overrightarrow{c}$.

6 Conclusion

We have characterized the MA test masks of all 2D chamfer norms $\langle a, b\rangle$ and $\langle a, b, c\rangle$. The next step is devoted to larger 2D masks. We have noticed that simple 7×7 chamfer masks (created by adding a single weighting $((3,1), d)$ or $((3,2), e)$ to a 5×5 mask) generally yield test masks with new properties: they can be larger than the chamfer mask itself, and can also contain non-visible vectors. We intend to use our arithmetical framework to describe these test masks.

References

1. Pfaltz, J.L., Rosenfeld, A.: Computer Representation of Planar Regions by their Skeletons. Comm. of ACM 10, 119–125 (1967)
2. Coeurjolly, D., Hulin, J., Sivignon, I.: Finding a Minimum Medial Axis of a Discrete Shape is NP-hard. Theoretical Computer Science 206(1-2), 72–79 (2008)
3. Coeurjolly, D., Montanvert, A.: Optimable Separable Algorithms to compute the Reverse Euclidean Distance Transformation and Discrete Medial Axis in Arbitrary Dimension. IEEE Trans. on PAMI 29(3), 437–448 (2007)
4. Saúde, A.V., Couprie, M., Lotufo, R.A.: Discrete 2D and 3D euclidean medial axis in higher resolution. Image Vision Comput. 27(4), 354–363 (2009)
5. Borgefors, G.: Distance Transformations in Arbitrary Dimensions. Computer Vision, Graphics and Image Processing 27, 321–345 (1984)
6. Hirata, T.: A Unified Linear-time Algorithm for Computing Distance Maps. Information Processing letters 58(3), 12–133 (1996)
7. Arcelli, C., Sanniti di Baja, G.: Finding Local Maxima in a Pseudo-Euclidean Distance Transform. Comp. Vision, Graphics and Image Proc. 43, 361–367 (1988)
8. Borgefors, G.: Centres of Maximal Disks in the 5-7-11 Distance Transform. In: 8th Scandinavian Conf. on Image Analysis, Tromsø, Norway, pp. 105–111 (1993)
9. Rémy, E., Thiel, E.: Medial Axis for Chamfer Distances: Computing Look-up Tables and Neighbourhoods in 2D or 3D. Pattern Rec. Letters 23(6), 649–661 (2002)
10. Rémy, E., Thiel, E.: Exact Medial Axis with Euclidean Distance. Image and Vision Computing 23(2), 167–175 (2005)
11. The Npic library and tools, http://www.lif.univ-mrs.fr/~thiel/npic/
12. Normand, N., Évenou, P.: Medial Axis LUT Computation for Chamfer Norms using H-polytopes. In: Coeurjolly, D., Sivignon, I., Tougne, L., Dupont, F. (eds.) DGCI 2008. LNCS, vol. 4992, pp. 189–200. Springer, Heidelberg (2008)
13. Normand, N., Évenou, P.: Medial Axis Lookup Table and Test Neighborhood Computation for 3D Chamfer Norms. Pattern Rec. 42(10), 2288–2296 (2009)
14. Hulin, J., Thiel, E.: Visible Vectors and Discrete Euclidean Medial Axis. Discrete and Computational Geometry (in press) (Available online December 10, 2008)
15. Thiel, E.: Géométrie des Distances de Chanfrein. Habilitation à Diriger les Rech., Univ. de la Méditerranée, (déc 2001), http://www.lif.univ-mrs.fr/~thiel/hdr/
16. Ramírez Alfonsín, J.L.: The Diophantine Frobenius Problem. Oxford Lectures Series in Mathematics and its Applications 30 (2005)
17. Annex with proofs: http://pageperso.lif.univ-mrs.fr/~jerome.hulin/dgci09/

Exact, Scaled Image Rotation Using the Finite Radon Transform

Imants Svalbe

School of Physics,
Monash University,
Australia

Abstract. In traditional tomography, a close approximation of an object can be reconstructed from its sinogram. The orientation (or zero angle) of the reconstructed image can be chosen to be any one of the many projected view angles. The Finite Radon Transform (FRT) is a discrete analogue of classical tomography. It permits exact reconstruction of an object from its discrete projections. Reordering the discrete FRT projections is equivalent to an exact digital image rotation. Each FRT-based rotation preserves the intensity of all original image pixels and allocates new pixel values through use of an area-preserving, angle-specific interpolation filter. This approach may find application in image rotation for feature matching, and to improve the display of zoomed and rotated images.

1 Introduction

A comparison of the artefacts produced by classic image rotation algorithms appears in [1]. The lack of a unique prescription to interpolate data on a 2D or higher dimension grid means image rotations, at some angles more than others, can lead to significant errors in the image intensities of the rotated image, especially after multiple, sequential rotations. The mapping of patterns onto grids is a problem well understood by those working in crafts such as tiling and tapestry.

We present here a method for the exact rotation of digital image data based on the Finite Radon Transform (FRT) [2]. The FRT generates a sinogram-like set of discrete projections, $R(t,m)$, from a discrete image, $I(x,y)$. The original data $I(x,y)$ can always be reconstructed exactly from the FRT projections $R(t,m)$. Translating an image, by offset (x_d, y_d), in FRT space is almost as simple as translation in image space. $R(t,m)$ gets re-mapped to $R(t',m)$, where $t' = x_d + my_d$. However to perform an image rotation in FRT space, $R(t,m)$ must be mapped to a new projection set, $R(t',m')$. This paper shows how to use the t and m values from a known set of FRT projections of an image to find the appropriate new t' and m' values that produce an exact, rotated version of the original image data. As the FRT space is an exact mapping of an image space, the same exact rotations can be performed in either image or FRT coordinates. Rotated pixels are treated as lattice points except for up-scaled rotation where, for interpolation, the pixels are treated as squares of unit area.

S. Brlek, C. Reutenauer, and X. Provençal (Eds.): DGCI 2009, LNCS 5810, pp. 446–456, 2009.

The FRT has been used as an alternative approach to the traditional back-projection or Fourier inversion methods to reconstruct real tomographic x-ray projection data [3]. An overview of the properties of FRT projections, relevant to the present work, appears in [4]. Early work on skewing and primitive rotation of images in FRT space, for images on square and hexagonal lattices, appeared in [5]. The use of the "natural angles" that occur on a discrete grid is central to this topic. The distribution and frequency of occurrence of the natural angles on discrete grids for the discrete Hough transform, a close relative of discrete Radon projections, was derived in [6]. An extended review of the FRT and its relation to other discrete projection methods is given in [7]. Asymmetric sets of projection data occur frequently in tomography, often through practical restrictions on the range of available acquisition angles. The FRT provides a means to symmetrise these projections and hence reconstruct images from asymmetric data sets [8,9]. The discrete Mojette projective transform [10] is closely related to the FRT. A recent, comprehensive review of the Mojette transform, its relationship to

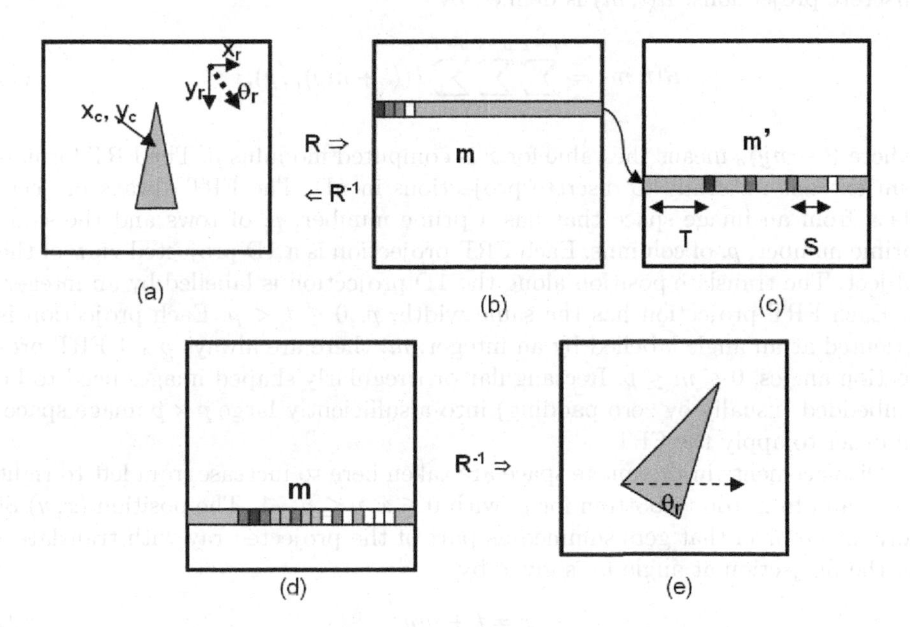

Fig. 1. Algorithm summary: a) Original image embedded into a larger (pxp) space, $I(x, y)$, with rotation centre (x_c, y_c), and rotation vector, (x_r, y_r). b) Applying the FRT forward projection operator, R, to $I(x, y)$ produces $R(t, m)$, comprised of $p + 1$ projection angles, and with each projection containing p translates. Only one row (labelled m) of $R(t, m)$ is shown. c) The rotation is performed by re-ordering all rows of the FRT (mapping each m to a new m') and then shifting (by operator T) and shuffling (by operator S) the content of each row to effect the rotation and scale (S and T map t to t'). d) The re-mapped rows are convolved with a linear filter, $f(t)$, to interpolate values for the empty pixels on the new, rotated lattice. e) Applying the inverse FRT operator, R^{-1}, reconstructs the original data as a rotated and up-scaled image, $I'(x, y)$.

the FRT and a comprehensive survey of applications for discrete projected data is given in [11]. Kingston, in chapter 6 of [11], showed how to generate multi-scale versions of a discrete image from its Mojette-projected data. That work prompted this application to scaled rotations using the FRT formalism.

Figure 1 shows an overview of the FRT-based rotation algorithm presented here. Section 2 gives a short review of the FRT. Section 3 details the step-by-step operation of the FRT rotation algorithm, including a comparison to a real space implementation of the same process. Section 4 provides some image rotation examples. Section 5 outlines possible extensions to this approach and points to areas where it might find application. A summary and conclusions are given in Section 6.

2 Finite Radon Transform (FRT)

The FRT, $R(t,m)$, maps discrete 2D image intensities, $I(x,y)$, into a set of 1D discrete projections. $R(t,m)$ is defined by

$$R(t,m) = \sum_{m=0}^{p-1}\sum_{t=0}^{p-1}\sum_{y=0}^{p-1} I(\langle t+my\rangle_p, y), \tag{1}$$

where $\langle t+my\rangle_p$ means the value for x is computed modulus p. The FRT formalism extends naturally to discrete projections in nD. The FRT always projects data from an image space that has a prime number, p, of rows and the same prime number, p, of columns. Each FRT projection is a 1D projected view of the object. The translate position along the 1D projection is labelled by an integer, t. Each FRT projection has the same width, p, $0 \leq t < p$. Each projection is oriented at an angle labelled by an integer, m, there are always $p+1$ FRT projection angles, $0 \leq m \leq p$. Rectangular or irregularly shaped images need to be embedded (usually by zero-padding) into a sufficiently large $p \times p$ image space, in order to apply the FRT.

Displacements in the image space are taken here to increase from left to right for x and from top to bottom for y, with $0 \leq x, y \leq p-1$. The position (x,y) of any image pixel that gets summed as part of the projected ray with translate t in the projection at angle m is given by

$$x = t + my. \tag{2}$$

For $m = p$, the summations are taken along the rows, $R(t,p) = \sum_{x=0}^{p-1} I(x,t)$. For $m = 0$, the sums are taken along columns of the image, $R(t,0) = \sum_{y=0}^{p-1} I(t,y)$.

Following (2), for pixels on a uniform, square lattice, we write, where α is an integer,

$$m = (\alpha p + x_m)/y_m. \tag{3}$$

The values of x_m and y_m are chosen to minimise $x_m^2 + y_m^2$. This choice is equivalent [4] to projecting across the image space at the angle θ_m that connects pixels that are nearest neighbours for any given value of t.

The FRT projections span the angle range from 0 to π through the four-fold symmetry based on $x_m : y_m$, $y_m : x_m$, $-y_m : x_m$ and $-x_m : y_m$, corresponding to θ_m, $\pi/2 - \theta_m$, $\pi/2 + \theta_m$ and $\pi - \theta_m$ respectively. The projection angle for $m' = p - m$, θ_{p-m}, corresponds to $180 - \theta_m$. Similarly, $\pi/2 + \theta_m$ and $\pi/2 - \theta_m$ have complemented m values. A list of m values and their associated discrete projection angles for $p = 17$ appears in Table 1. The algorithm described in the next section always chooses one of the $p + 1$ "natural" FRT angles from a $p \times p$ grid as the input rotation angle, $x_r : y_r$. The integer values of x_m and y_m, for a square lattice, range from 0 up to $\sqrt{2p/\sqrt{3}}$ [7]. There is almost complete overlap in the discrete angle sets that arise from different primes. This very useful property of the FRT is discussed in greater detail in [4].

Table 1. Discrete FRT projection angles for $p \times p$ images with $p = 17$. Only half of the $p + 1$ values of m are shown.

m	0	1	2	3	4	5	6	7	8	9
$x_m : y_m$	0:1	1:1	2:1	3:1	4:1	-2:3	1:3	-3:2	-1:2	1:2

Table 2. $x_f : y_f$ is the result of a discrete rotation, $x_r : y_r = 3 : 2$ applied to the FRT angles $x_m : y_m$ for $p = 17$, as given in Table 1

m	0	1	2	3	4	5	6	7	8	9
$x_f : y_f$	-2:3	1:5	4:7	7:9	10:11	-12:5	3:11	0:1	1:0	-1:8

3 Rotation Algorithm

If a discrete angle, $x_m : y_m$, is rotated by a discrete angle, $x_r : y_r$, then the result is a new discrete angle, $x_f : y_f$, given by

$$x_f : y_f = (x_m x_r - y_m y_r) : (x_m y_r + x_r y_m). \tag{4}$$

This follows from the definition of θ_i as $\tan^{-1}(y_i/x_i)$ and from $\tan(\theta_m + \theta_r) = (\tan\theta_m + \tan\theta_r)/(1 - \tan\theta_m \tan\theta_r)$. Here we take 1:0 to be 0°, pointing along the $+x$ direction, and 0:1 to be 90°, pointing along the $+y$ direction. We take $ax_i : ay_i$ to be the same angle as $x_i : y_i$. For real projections, $\pi - \theta_r$ is equivalent to $+\theta_r$, provided that effects such as the hardening of a polychromatic beam can be neglected. For rotations, these angles are distinct. Table 2 shows the angles, $x_f : y_f$, that result, using (4), when applying a rotation of $x_r : y_r = 3 : 2$ to the original set of $x_m : y_m$ FRT angles, as given in Table 1, for $p = 17$.

3.1 Mapping FRT Rotation m to New Value m'

After a rotation by $x_r : y_r$, some of the resulting rotated FRT discrete angles, as given by (4), do not appear in the original list of nearest neighbour projections for that p. However, each m value has a multiplicity of potential projection angles. All members of the rotated angle set can be mapped back into one of the known existing angles for that p. This cyclic mapping of FRT angle m into the corresponding rotated angle m' can be done directly, by re-writing (3) as

$$m' = (\alpha' p + x_f)/y_f, \tag{5}$$

where α' is some integer. Table 3 gives the mapping of m to m' for a 3 : 2 rotation of the FRT projection set for $p = 17$.

Table 3. Mapping of m to m' under a 3 : 2 rotation for $p = 17$

m	0	1	2	3	4	5	6	7	8	9	10	11	12	13	14	15	16	17
m'	5	7	3	14	4	1	9	17	11	2	16	6	0	13	15	8	12	10

3.2 Re-centring FRT Projection m'

The position of the pixel (x_c, y_c) about which the image rotates determines the translate index t_c for the ray from projection m that passes through the image centre. Let the centre point of the original image space be the point (x_0, y_0), with translate label t_0. Usually this point is the geometric centre of the image space, $x_0 = y_0 = p/2$. Following (2),

$$t_0 = x_0 - my_0 \qquad \text{and} \qquad t_c = x_c - m'y_c. \tag{6}$$

Then the shift, T, required to adjust a projection for a change from FRT row m to FRT row m', as a consequence of the rotation, is given by

$$T = x_c - x_0 + my_0 - m'y_c. \tag{7}$$

If $(x_0, y_0) = (x_c, y_c)$, then

$$T = y_0(m - m'). \tag{8}$$

Note that the centring operation depends only on the difference between m and m' and the location of the rotation centre.

3.3 Shuffling the m' Data to Preserve Translation Order

The pixel (we will label it as N) that was located immediately to the right of the centre pixel in the original image need to be located immediately to the right of the centre pixel in the new, rotated frame of reference. A unit increase in translate index for FRT row m requires that the same pixel N be found shifted by x_r pixels in the x direction and y_r pixels in the y direction. Then the required index, S, to correctly shuffle the data is given by

$$t_{N'} = x_c + x_r - m'(y_c + y_r) = t_N + x_r - m'y_r,$$
$$\text{so that} \qquad t_{N'} - t_N = x_r - m'y_r. \tag{9}$$

Because the pixel N has to be the translate next to the origin (a translate increment of one), then $S(x_r - m'y_r) = 1$, giving

$$S = \frac{1}{(x_r - m'y_r)}. \tag{10}$$

Note that the shuffling of the row data depends only on the new row index m' and the rotation angle, $x_r : y_r$.

3.4 Convolution of the FRT Row Data

The expansion of the pixel size upon rotation is compensated for by convolving each row of $R(t, m')$ with a linear filter, $f(t)$. This filter fills the gaps between the re-sized pixels. The original pixels are zoomed from a square of unit area, to a new pixel of length $\sqrt{(x_r^2 + y_r^2)}$ on each side, with a new pixel area, A, where $A = x_r^2 + y_r^2$. The pattern of interpolation weights is shown in Figure 2 for a rotation of 3 : 2. The area of the enlarged pixel is subdivided uniquely into proportional contributions that come from the original, unit area pixels. The only assumption made here is that the original image intensity can be distributed on the basis of the area that each original pixel intersects with the rotated lattice pixels (a uniform intensity pixel model). Integer weights are used to avoid rounding or truncation of intensities during the scaling process. Discrete 2D convolutions in the image space $I(x, y)$ can be implemented as 1D convolutions in the equivalent $R(t, m)$ FRT space [2].

Table 4 shows some example 1D FRT filter coefficients derived from the 2D weights given in Figure 2, for a 3 : 2 rotation at the new, rotated row index m' for $m' = 0$, 1 and 2. These filters are easy to implement as 1D cyclic convolutions of length p. The 1D FRT filter coefficients vary for each angle m. Because the coefficients have reflective symmetry, FRT angles $p - m$ and m share the same 1D coefficient set. Alternatively, sequential, shifted convolutions can be applied directly, using the coefficients on each individual row of the 2D filter, one row at a time. When they become long enough to wrap, the summed 1D filter coefficients add linearly, modulus p.

3.5 Implementing Rotations in FRT Space vs. Image Space

The aim of this paper was to show the "discrete sinogram" of the FRT can be reformatted to re-define the orientation of discrete reconstructed image data. The same discrete rotations can, of course, be implemented directly in image or Fourier space, without recourse to the FRT, because the image space and its FFT and FRT are one to one mappings. A real space algorithm similar to the present work, but with de-scaling of the rotated image, was presented in [12].

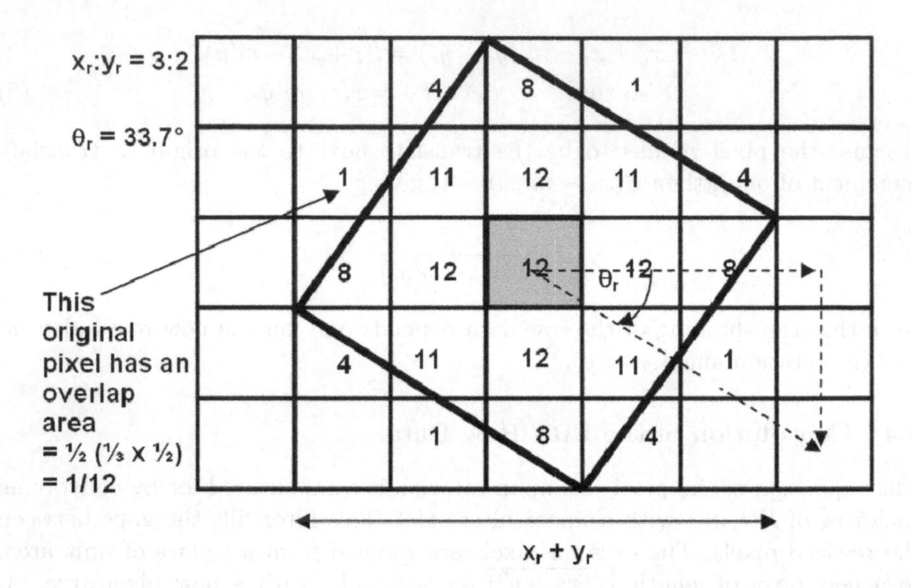

Fig. 2. A unit square (shown shaded) is rotated by the discrete angle 3 : 2 to create an enlarged pixel (with a border shown by the bold lines) which has an area 13 times larger ($2^2 + 3^2 = 13$) than the original pixel. The intersection of the zoomed and rotated pixel with the original grid uniquely defines the pixel interpolation weights (given by the integer entry in each original unit square). The FRT projection of these interpolation weights onto the row containing the centre pixel gives the equivalent 1D linear interpolation filter coefficients to apply on each row m of the FRT space. The sum of the 2D weights shown here is 156.

Table 4. Interpolation filter weights for a rotation of 3 : 2. The 1D circular convolution of FRT row data has a different set of weights for each value of m. The 1D filters for FRT angle $p - m$ is the same as that for m. The weights for row m are discrete FRT projections, at FRT angle m, of the 2D interpolation weights given in Figure 2. The centre pixel of the 1D FRT interpolation filter is shown in bold font. The 1D weights shown here each sum to 156.

m	1D FRT interpolation weights										
0				13	39	**52**	39	13			
1			5	27	29	**34**	29	27	5		
2	1	12	15	20	23	**14**	23	20	15	12	1

Indeed, the specific 1D convolution applied to each FRT row by the pixel interpolation weights (as shown in Table 5) may be easier to perform directly as a single 2D convolution in the image space, after applying the inverse FRT operation to reconstruct the rotated original image pixels.

On the other hand, the 2D rotation of the image pixels in image space is more easily performed as a simple re-assignment of m values and shuffling of each row in FRT space. The cost of performing the FRT (and its inverse) scales as $O(p^3)$, or as $O(p^2 \log_2 p)$ if a Fourier based method is used [7]. The overhead of performing the forward FRT on the original, zero-padded image data can be reduced substantially by summing only over the non-zero parts of the over-sized pxp image space. Similarly, the FRT space can be sampled to reconstruct only non-zero pixels in the output image, although the scaled, rotated image is usually chosen to nearly fill the pxp space. The FRT rotation method has a clear advantage if the application, such as feature matching, is done using projection rather than real space data.

4 Examples

The input image used here is a 64×64 subset of the Lena image. Figure 3 shows the result after example rotations of $2:3$ and $8:1$ respectively.

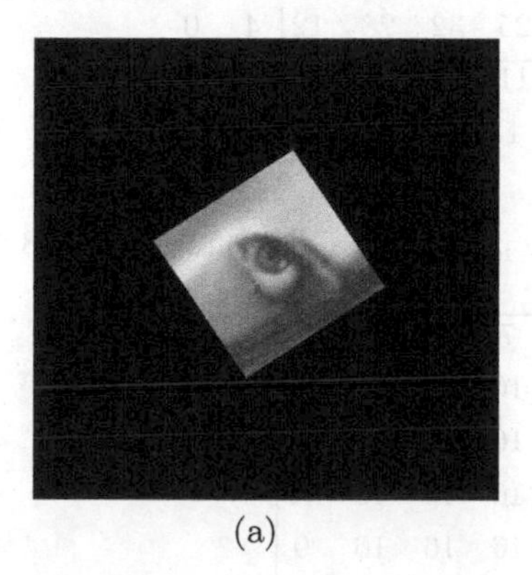

(a)

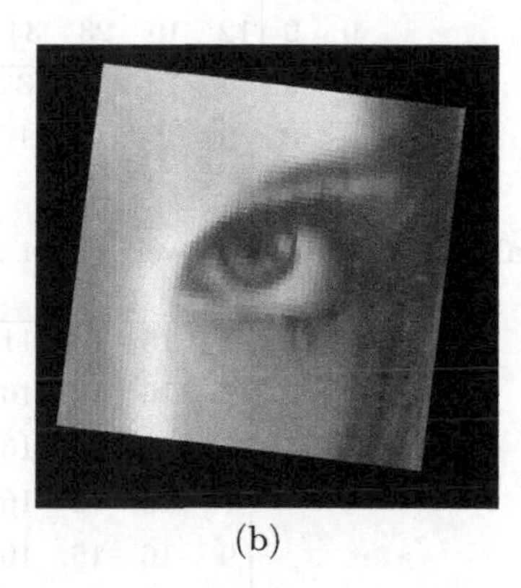

(b)

Fig. 3. Rotation by a) $2:3$ ($\theta_r = 56.3°$) and b) by $8:1$ ($\theta_r = 7.13°$). The periodicity of the FRT means the image will wrap if the rotated image exceeds the size of the initial $p \times p$ array. The final image size here is 647×647.

5 Applications and Extensions

The combination of tying image rotation to zoom or scale changes has advantages and disadvantages. The integer scaling factors $x_r : y_r$ do not need to be large to achieve fine angle rotations. Choosing x_r and $y_r \leq 10$ defines 32 distinct angles, with a mean spacing of 1.4° over the range 0 to $\pi/4$. The image size (p) must be able to be up-scaled by the factor $x_r + y_r$ for a rotation of $x_r : y_r$. This scale factor ranges from 2 for $1:1$, through to 19 for $9:10$. The image pixel area increases by $x_r^2 + y_r^2$, ranging from 2 for $1:1$, through to 181 for $9:10$. Using signed integers less than 11 for x_r and y_r provides a choice of 256 distinct rotation angles over 0 to 2π.

Table 5. Interpolation of a single pixel after successive rotations. Here a rotation of $3:2$ is followed by a rotation of $2:-1$. These sequential rotations should be equivalent to a direct rotation of 8:1. The original centre pixel is shown in bold. Note the asymmetry in interpolation values. The dotted box encloses the coefficients that should correspond to the direct rotation.

0	0	0	0	0	1	3	1	0	0	0	0	0
0	0	8	2	4	11	4	3	4	12	4	0	0
0	4	12	28	3	25	14	34	23	16	12	0	0
0	12	16	20	36	44	45	44	37	20	28	8	0
0	0	23	37	47	48	47	45	47	36	32	24	0
1	3	34	44	45	48	48	48	48	44	25	11	1
3	4	14	45	47	48	**48**	48	47	45	14	4	3
1	11	25	44	48	48	48	48	45	44	34	3	1
0	24	32	36	47	45	47	48	47	37	23	4	0
0	8	28	20	37	14	45	44	36	20	16	12	0
0	0	12	16	23	34	14	25	32	28	12	4	0
0	0	4	12	4	3	4	11	4	2	8	0	0
0	0	0	0	0	1	3	1	0	0	0	0	0

Table 6. Interpolation weights for a single pixel after a direct rotation of $8:1$

1	15	13	11	9	7	5	3	1
3	16	16	16	16	16	16	16	15
5	16	16	16	16	16	16	16	13
7	16	16	16	16	16	16	16	11
9	16	16	16	**16**	16	16	16	9
11	16	16	16	16	16	16	16	7
13	16	16	16	16	16	16	16	5
15	16	16	16	16	16	16	16	3
1	3	5	7	9	11	13	15	1

For particular rotations, the same image scale factors arise from different rotation angles. For example, $1:8$ and $4:7$ have a common image area zoom of 65 ($65 = 12 + 82 = 42 + 72$). This property occurs [4] only for integers that can be written as the product of primes p where $p = 4n + 1$ for some integer n, for example, 5, 13 and 17. Here $65 = 5 \times 13$.

A second rotation treats the newly interpolated values and the original pixels both as "original data" and remaps them (correctly) for the new angle. Performing a rotation of θ_r then $-\theta_r$ preserves the original data, but the final interpolated image values depend on the intermediate value of θ_r. The interpolation

Table 7. Interpolation weights for a single pixel after a rotation of 2 : 1, followed by a de-rotation of 2 : 1. These sequential rotations correspond to up-scaling the image pixel area by a factor of 5 in x and y. The original centre pixel is shown in bold. Note the asymmetry in interpolation values. The 5 × 5 dotted box encloses the coefficients that mimic an up-scaling of an image by a factor of 5.

0	0	0	0	1	3	1	0	0
0	0	3	9	6	4	3	0	0
1	3	6	10	12	6	10	3	0
3	4	6	13	15	13	12	9	0
1	6	10	15	**16**	15	10	6	1
0	9	12	13	15	13	6	4	3
0	3	10	6	12	10	6	3	1
0	0	3	4	6	9	3	0	0
0	0	1	3	1	0	0	0	0

filter for θ_r is the symmetric reflect of the filter for $-\theta_r$. The small differences in the images are best shown by the effects on interpolation for a single pixel.

Table 5 shows the interpolated pixel weights from up-scaling each pixel in the input image after a 3 : 2 rotation, followed by a 2 : −1 rotation, corresponding to a one-step rotation of 8 : 1. Table 6 shows the "correct" interpolation weights for a single step 8 : 1 rotation. Table 7 shows the effect of interpolation on a single pixel for a rotation of 2:1 followed by de-rotation of 2:-1.

6 Conclusions

This work demonstrated exact digital image rotation through a scheme that re-orders the discrete sinogram of projections defined under the FRT formalism. The synchronised scaling and rotation of image pixels preserves the exact, original pixel values, even after multiple successive rotations. Sub-sampling an image back to the original size recovers the original data exactly. The filter coefficients to interpolate the pixel intensities that lie between the zoomed and rotated original values are fixed by the selected discrete rotation angle.

Practical real and FRT space algorithms are given to implement the digital rotation scheme. The FRT implementation of discrete rotations will be an optimal approach for projection-based image processing operations. Manipulation of discrete rotations on grids is an important tool in the reconstruction of images from real projection data.

The algorithm, described here for square image grids, can be adapted for discrete projections on hexagonal lattices. On hexagonal lattices [4,5], the $x_m : y_m$ FRT projection m values are symmetric about 0, $\pi/3$, $2\pi/3$ and π. Another application of simultaneous scale and rotations may be to model the physical flow of vectors from discrete vortices on real lattices.

References

1. Danielsson, P.E., Hammerin, M.: High-accuracy rotation of images. Graphical Models and Image Processing (CVGIP) 54(4), 340–344 (1991)
2. Matúš, F., Flusser, J.: Image representation via a finite Radon transform. IEEE Transactions on Pattern Analysis and Machine Intelligence 15(10), 996–1006 (1993)
3. Svalbe, I., van der Spek, D.: Reconstruction of tomographic images using analog projections and the digital Radon transform. Linear Algebra and Its Applications 339, 125–145 (2001)
4. Svalbe, I.: Sampling properties of the discrete Radon transform. Discrete Applied Mathematics 139(1-3), 265–281 (2004)
5. Svalbe, I.: Image operations in discrete Radon space. In: Proc. of the Sixth Digital Image Computing Techniques and Applications, Dicta 2002, pp. 285–290 (2002)
6. Svalbe, I.: Natural representations for straight lines and the hough transform on discrete arrays. IEEE Transactions on Pattern Analysis and Machine Intelligence 11(9), 941–950 (1989)
7. Kingston, A., Svalbe, I.: Projective transforms on periodic discrete image arrays. Advances in Imaging and Electron Physics 139, 75–177 (2006)
8. Chandra, S., Svalbe, I.: A method for removing cyclic artefacts in discrete tomography using Latin squares. In: 19th International Conference on Pattern Recognition, December 2008, pp. 1–4 (2008)
9. Chandra, S., Svalbe, I., Guédon, J.P.: An exact, non-iterative Mojette inversion technique utilising ghosts. In: Coeurjolly, D., Sivignon, I., Tougne, L., Dupont, F. (eds.) DGCI 2008. LNCS, vol. 4992, pp. 401–412. Springer, Heidelberg (2008)
10. Servières, M., Normand, N., Guédon, J.P., Bizais, Y.: The Mojette transform: Discrete angles for tomography. In: Herman, G., Kuba, A. (eds.) Proceedings of the Workshop on Discrete Tomography and its Applications. Electronic Notes in Discrete Mathematics, vol. 20, pp. 587–606 (2005)
11. Guédon, J., Normand, N., Kingston, A., Parrein, B., Serviéres, M., Evenou, P., Svalbe, I., Autrusseau, F., Hamon, T., Bizais, Y., Coeurjolly, D., Boulos, F., Grail, E.: The Mojette Transform: Theory and Applications. ISTE-Wiley (2009)
12. Chen, X., Lu, S., Yuan, X.: Midpoint line algorithm for high-speed high-accuracy rotation of images. In: IEEE Conf. on Systems, Man and Cybernetics, Beijing, China, pp. 2739–2744 (1996)

Lower and Upper Bounds for Scaling Factors Used for Integer Approximation of 3D Anisotropic Chamfer Distance Operator

Didier Coquin and Philippe Bolon

LISTIC, Domaine Universitaire, BP 80439, 74944 Annecy le Vieux Cedex, France
didier.coquin@univ-savoie.fr

Abstract. For 3D images composed of successive scanner slices (e.g. medical imaging, confocal microscopy or computed tomography), the sampling step may vary according to the axes, and specially according to the depth which can take values lower or higher than 1. Hence, the sampling grid turns out to be parallelepipedic. In this paper, 3D anisotropic local distance operators are introduced. The problem of coefficient optimization is addressed for arbitrary mask size. Lower and upper bounds of scaling factors used for integer approximation are given. This allows, first, to derive analytically the maximal normalized error with respect to Euclidean distance, in any 3D anisotropic lattice, and second, to compute optimal chamfer coefficients. As far as large images or volumes are concerned, 3D anisotropic operators are adapted to the measurement of distances between objects sampled on non-cubic grids as well as for quantitative comparison between grey level images.

Keywords: Distance transformation, Chamfer distance, Anisotropic lattice.

1 Introduction

In image analysis, measuring distances between objects is often essential. The notion of distance is very useful to describe a pattern in a digital image or to characterize the location of objects inside a volume. This is useful in many different shape representations and shape recognition tasks. An overview of various applications can be found in [1,2,3]. The aim of a distance transformation is to compute the distance from a point to an object, i.e. to a set of points. The distance from point p to the object is the smallest distance from p to any point of the object. In other words, it is the distance from p to the nearest point q belonging to the object. The two main approaches to the computation of distance maps are Chamfer Distance operators [1] and Euclidean Distance operators [4,5]. In the first case, the distance between two pixels is defined as the length of the minimal path between them. This path is composed of a finite sequence of elementary steps to which weights are assigned. Distance maps are obtained by propagating local distances. In most cases, they do not yield the exact Euclidean distance between pixels. In the latter case, quasi-exact or exact Euclidean distances are obtained by taking the relative location between pixels into account. Interesting analysis are presented in [6,7,8].

S. Brlek, C. Reutenauer, and X. Provençal (Eds.): DGCI 2009, LNCS 5810, pp. 457–468, 2009.

To improve Chamfer Distance Transforms, efforts have been made in four directions:

(i) decreasing the sensitivity to rotation by means of a better approximation of the Euclidean distance. This is achieved by assigning weights to elementary displacements allowed by the local distance operator. These weights are optimized according to an error criterion which generally consists in minimizing the maximum difference between the computed distance and the Euclidean distance along a reference trajectory, which could be rectilinear [3] or circular [10,11].

(ii) increasing the dimension of the image space. Weighted distance transforms in 3D were introduced as early as 1984 [2]. Different approaches are possible to set the local coefficients, based either on a discrete model [12,13] or a continuous one [10,11,14]. Reference trajectories may be rectilinear [14] or spherical [10]. Coefficients for 5x5x5 local operators were proposed in [6,10,12,18]. Higher dimension distance transforms were used for skeletonisation in 4D space [15]. However, some practical difficulties may arise because of memory size requirements.

(iii) studying generic properties. Whether distances computed by means of distance transforms are metric or not may be of great interest for optimization purposes. Semi-regularity conditions state that any discrete path composed of repetitions of elementary displacements is optimal. In [16], Kiselman showed that a distance transform in $\mathbf{Z}^n$ producing a discrete norm is semi-regular. Moreover, such a semi-regularity distance is metric in $\mathbf{Z}^2$. Conditions for obtaining a discrete norm with a 5x5x5 operator are given in [12]. Metricity is considered in [13].

(iv) adapting the local operator to non cubic sampling grids. Imaging systems often produce images having different sampling steps according to the different axes. In most cases, the resulting image is composed of parallelepipedic voxels having two sides equal and the third different. For computed tomography, or confocal microscopy, the ratio between the largest to the shortest voxel dimension typically ranges from 1 to 10 [17,19]. Another way to decrease the maximum error between the Euclidean distance and the local distance is to increase the size of the mask operator. Since voxels may not be cubic, anisotropic masks have to be considered.

In this paper, the general case of MxMxV distance operators adapted to parallelepipedic grids composed of voxels having a WxWxP size is addressed. These so-called anisotropic operators yield the same performances as those of isotropic ones (MxMxM) with reduced complexity and computation time. In Section 2, optimal isotropic operators adapted to anisotropic grids are introduced. The influence of voxel depth P and operator size M is studied. Section 3 addresses a new issue of integer approximation implementation, by choosing a scaling factor between lower and upper bounds. Examples of integer chamfer masks are given.

2 3D Anisotropic Chamfer Distance Operator

The objective is to approximate the real Euclidean distance d_E. The optimization criterion consists in minimizing the maximum error between the local distance d_L and the Euclidean distance d_E. The maximum error is sometimes called the maximum absolute error. In both cases, it is the absolute value of the maximum difference between the chamfer local distance d_L and the Euclidean distance d_E. In [20] we have proposed a generalization to parallelepipedic grids of the approach developed by Verwer [10]. But unlike the approach followed by Verwer which directly gives the position and the amplitude of the maximum absolute error, our method allows error analysis at any point of the sphere. Thus we have a better control of optimization, especially in the case of the integer approximation which generally involves a displacement of the maximum absolute error location.

Given a grid, what is the optimal mask size ? In this paper, we consider the case of voxels having two sides equal (width W and height H) and the third different (depth P). This is the situation with most of 3D images such as those obtained by confocal microscopy or computed tomography [7]. It is also the case for grey level images where geometrical distortions (spatial domain) and radiometric distortions (grey level domain) are assigned different weights.

The directions in which voxels have the same size may be referred to as "horizontal" dimensions whereas the other directions may be referred to as "vertical" dimensions.

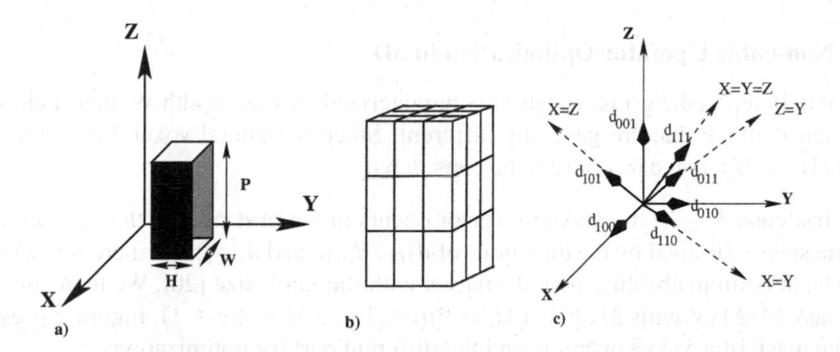

Fig. 1. a) Voxel's characteristic, b) cubic 3x3x3 operator, c) elementary displacements d_{ijk} for 3x3x3 mask

Local distance operators are characterized by the coefficients assigned to the elementary displacements. For a cubic MxMxM operator, with M = 2m+1, one needs to set coefficients d_{ijk} with $i, j, k \in \{0, 1, 2, ..., m\}$ and GCD(i,j,k)=1, i and j denote the number of steps in horizontal dimensions, k denotes the number of steps in the vertical dimension. The directions of the elementary displacements d_{ijk} are shown in Figure 1c, for a cubic 3x3x3 operator (M=3, and m=1). They define parts in the image space. We assume that $W \leq P$.

Let voxel O be the origin of a binary digitized 3D image. Let x, y and z be the coordinates in the image referential, and (OX,OY,OZ) the actual space referential. Let voxel Q(x,y,z) describe a sphere having fixed radius R. The equation of the trajectory is $d_E = OQ = \sqrt{(Wx)^2 + (Wy)^2 + (Pz)^2} = R$. We assume that the fixed value of R is large with respect to W and P, so that displacements between two adjacent voxels can be regarded as continuous.

Verwer [10] showed that the maximum error between d_E and d_L occurs in the cone having the greatest angle. For a cubic MxMxM operator (M=2m+1), the maximum error between the local distance d_L and the Euclidean distance d_E occurs in the part limited by the directions of d_{100}, d_{m10} and d_{m11}.

The values of the maximum error e_{max} (normalized by radius R) produced by using a 3D cubic operator of different sizes, when $W = H = P = 1$, are summarized in Table 1. It should be noticed that e_{max} decreases with the mask size [2,10].

Table 1. Maximal normalized error produced with isotropic operator in 3D ($W = H = P = 1$)

mask size	3x3x3	5x5x5	7x7x7	9x9x9	11x11x11
e_{max}%	6.019	2.411	1.223	0.725	0.476

The study of the error evolution e_{max} as a function of depth P shows that the larger P is, the larger the maximum error occurring in the part limited by the directions of d_{100}, d_{m10} and d_{m11}. By using non-cubic masks (mask of size MxMxV) it is possible to reduce the computation time and still to achieve the accuracy [20].

2.1 Non-cubic Operator Optimization in 3D

With parallelepipedic grids, voxels are characterized by their width W, their height H, and their depth P that are generally different. Since horizontal voxel dimensions are equal ($W = H$), two cases have to be considered.

- **First case** $P \geq 1$: the maximum error occurs in the **first part** of the first octant of the sphere (limited by the directions of d_{100}, d_{m10} and d_{m11}) and increases with P. The maximum absolute error decreases with the mask size [20]. We then choose a mask MxMxV with $M \geq V$ ($M = 2m + 1$ and $V = 2v + 1$). Figure 2 presents the mask of a 5x5x3 operator, and the different part for optimizations.
- **Second case** $0 < P \leq 1$: the maximum error occurs in the **last part** of the first octant of the sphere (limited by the directions of d_{001}, d_{01v} and d_{11v}) and decreases with P. The maximum absolute error decreases with the mask size. We then choose a mask MxMxV with $M \leq V$ ($M = 2m + 1$ and $V = 2v + 1$). Figure 3 presents the mask of a 3x3x5 operator, and the different parts for the optimizations.

Coefficient optimization: Two procedures are available:

(1) Minimizing the error in the **first part** limited by the directions of d_{100}, d_{m10} and d_{m11} (Fig. 2b). The local distance in this part is given by:

$$d_L(O, Q) = d_{100}.x + (d_{m10} - m.d_{100}).y + (d_{m11} - d_{m10}).z \tag{1}$$

It can be expressed as a function of y and z:

$$d_L(y, z) = \frac{d_{100}}{W}.\sqrt{(R)^2 - (Wy)^2 - (Pz)^2} + (d_{m10} - m.d_{100}).y + (d_{m11} - d_{m10}).z \tag{2}$$

Hence, the error $E(y, z) = d_L - d_E$ is:

$$E(y, z) = \frac{d_{100}}{W}.\sqrt{(R)^2 - (Wy)^2 - (Pz)^2} + (d_{m10} - m.d_{100}).y + (d_{m11} - d_{m10}).z - R \tag{3}$$

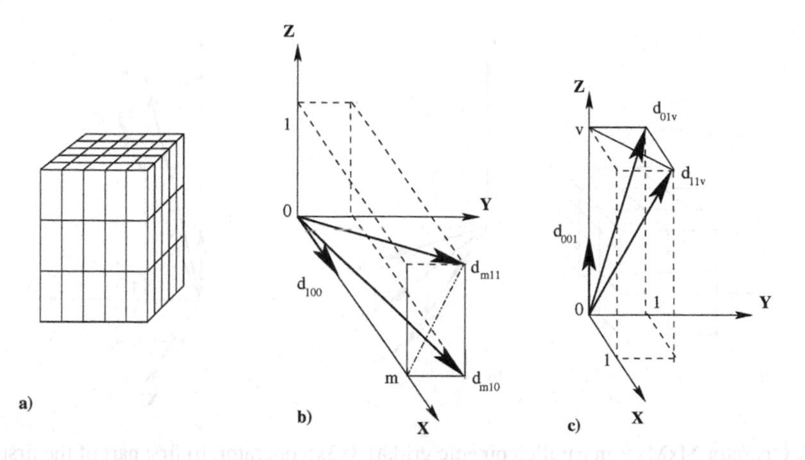

Fig. 2. Operator MxMxV in parallelepipedic grid, a) 5x5x3 operator, b) first part of the first octant of the sphere, c) last part of the first octant of the sphere.

$E(y, z)$ is extremal at the interval borders and when the partial first derivatives are zero. The maximum absolute normalized error $e_1^{'}$ is:

$$e_1^{'} = \frac{E_{max}}{R} = \left| 1 - \frac{d_{100}}{W} \right| \tag{4}$$

with

$$d_{100} = \frac{-2W + 2W\sqrt{1 + \lambda_1^{'}}}{\lambda_1^{'}} \tag{5}$$

and

$$\lambda_1^{'} = \frac{1}{W^2}(T_{m10} - mW)^2 + \frac{1}{P^2}(T_{m11} - T_{m10})^2 \tag{6}$$

Since $W = H = 1$, the error depends on P and m. Coefficients d_{ijk} are given by:

$$d_{ijk} = T_{ijk}\frac{d_{100}}{W} \tag{7}$$

with

$$T_{ijk} = \sqrt{(iW)^2 + (jW)^2 + (kP)^2} \tag{8}$$

(2) Minimizing the error in the **last part** limited by the directions of d_{001}, d_{01v} and d_{11v} (Fig. 2c) yields:

$$d_{001} = \frac{-2P + 2P\sqrt{1 + \lambda_2^{'}}}{\lambda_2^{'}} \tag{9}$$

with

$$\lambda_2^{'} = \frac{1}{W^2}(T_{11v} - T_{01v})^2 + \frac{1}{W^2}(T_{01v} - vP)^2 \tag{10}$$

Coefficients d_{ijk} are given by:

$$d_{ijk} = T_{ijk}\frac{d_{001}}{P} \tag{11}$$

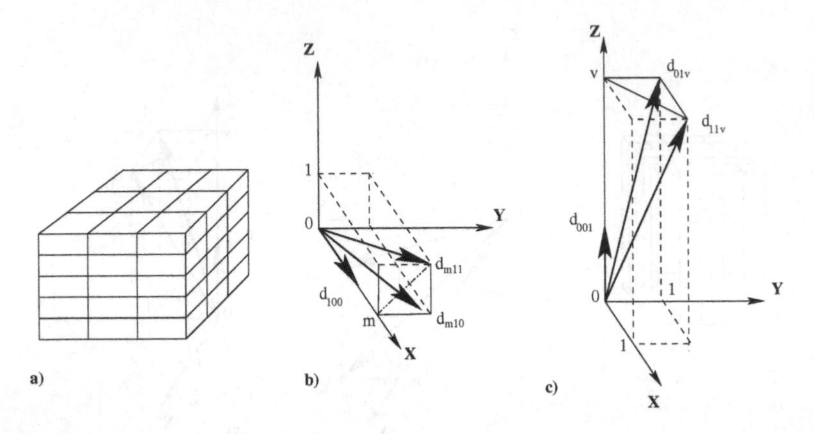

Fig. 3. Operator MxMxV in parallelepipedic grid,a) 3x3x5 operator, b) first part of the first octant of the sphere, c) last part of the first octant of the sphere

with

$$T_{ijk} = \sqrt{(iW)^2 + (jW)^2 + (kP)^2} \tag{12}$$

the maximum absolute normalized error $e_2^{'}$ is:

$$e_2^{'} = 1 - \frac{d_{001}}{P} \tag{13}$$

Since $W = H = 1$, the error depends on P and v.

For $P \geq 1$, the study of $e_1^{'}$ and $e_2^{'}$ as functions of voxel depth P yields that $e_1^{'}$ is steadily increasing function of P whereas $e_2^{'}$ is a steadily decreasing function. We have $e_1^{'} = e_2^{'}$ for $P = P_0$. Let P_0 be the solution to the equation $\lambda_1^{'} = \lambda_2^{'}$.

- if $P \leq P_0$: the second optimization procedure must be performed. Coefficients are given by eq.11. The maximum absolute normalized error is $e_{max} = e_2^{'}$.
- if $P \geq P_0$: the first optimization procedure must be performed. Coefficients are given by eq.7. The maximum absolute normalized error is $e_{max} = e_1^{'}$.

For $0 < P \leq 1$, the study of $e_1^{'}$ and $e_2^{'}$ as functions of voxel depth P yields that $e_1^{'}$ is steadily increasing function of P whereas $e_2^{'}$ is a steadily decreasing function. We have $e_1^{'} = e_2^{'}$ for $P = P_0^{'}$. Let $P_0^{'}$ be the solution to the equation $\lambda_1^{'} = \lambda_2^{'}$.

- if $P \leq P_0^{'} \leq 1$: the second optimization procedure must be performed. Coefficients are given by eq.11. The maximum absolute normalized error is $e_{max} = e_2^{'}$.
- if $1 \geq P \geq P_0^{'}$: the first optimization procedure must be performed. Coefficients are given by eq.7. The maximum absolute normalized error is $e_{max} = e_1^{'}$.

2.2 Performance Study

In this subsection, we study the performances of 3D anisotropic operators. Some examples of 3D operators of different dimensions are given.

The values of the maximum normalized error produced by using 3D operators of different sizes, when voxel dimensions are $W = H = 1$ and $P = 2$, are summarized in Table 2. It can be noticed that isotropic operators can be replaced by anisotropic ones (smaller in size) without reducing the performances (value of e_{max}).

Table 2. Maximal normalized error produced with 3D anisotropic operators for $W = H = 1, P = 2$

mask size	3x3x3	5x5x5	7x7x7	9x9x9	5x5x3
$e_{max}\%$	9.08	4.59	2.61	1.64	4.59
mask size	7x7x3	9x9x3	11x11x3	9x9x5	11x11x5
$e_{max}\%$	2.61	2.41	2.41	1.64	1.11

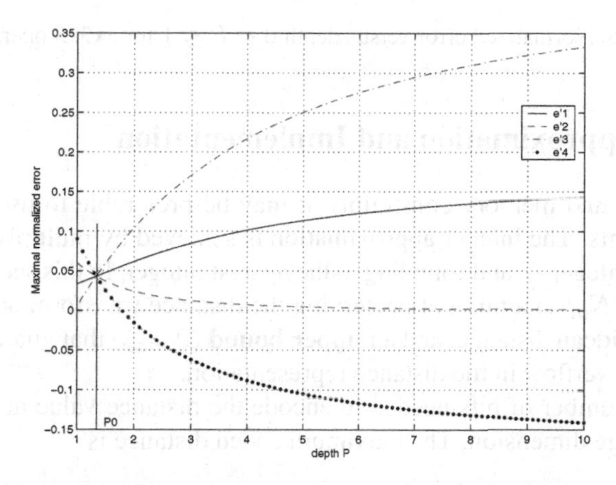

Fig. 4. Normalized error versus depth $P \geq 1$ for 5x5x3 operator

The error functions e_1' and e_2' produced by a 5x5x3 operator with the two optimization procedures given in Section 2.1, are shown in Figure 4. In this case, $P_0 = 1.54$. It is the solution to the equation $\lambda_1' = \lambda_2'$. The first optimization is better for $P \geq P_0$, and the second one for $P \leq P_0$. Curve e_3' represents the maximum error in the last part of the sphere with the first optimization, and curve e_4' gives the maximum error in the first part of the sphere with the second optimization. Curve e_2' in dashed green line for $P \leq P_0$ and curve e_1' in solid blue line for $P \geq P_0$ represent the maximum absolute normalized error obtained.

The error functions e_1' and e_2' produced by a 3x3x5 operator with the two optimization procedures given in 3.1 are shown in Figure 5. In this case $P_0' = 0.61$. It is the solution to the equation $\lambda_1' = \lambda_2'$. The first optimization is better for $P \geq P_0'$, and the second one for $P \leq P_0'$. Curve e_2' in a dashed green line for $P \leq P_0'$ and curve e_1' in a solid blue line for $P \geq P_0'$ represent the maximum absolute normalized errors obtained.

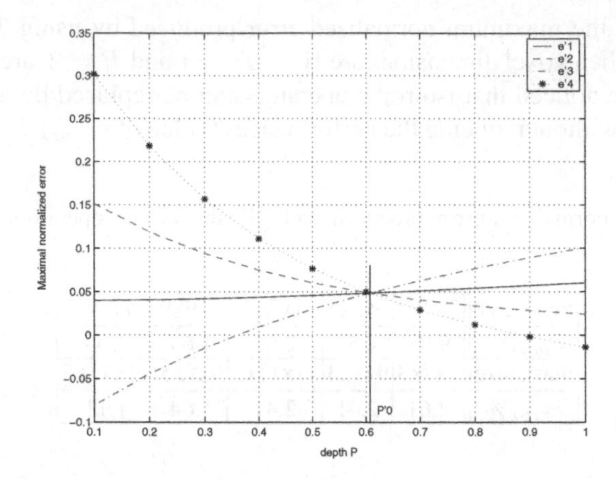

Fig. 5. Normalized error versus depth $0 < P \leq 1$ for 3x3x5 operator

3 Integer Approximation and Implementation

Because of time and memory constraints, it may be preferable to use operators with integer coefficients. The integer approximation is achieved by multiplying the real coefficients by an integer N and rounding to the nearest integer. In this section we propose a **lower bound** (N_{min}) aiming at preserving the distance transform accuracy with respect to the Euclidean distance, and an **upper bound** (N_{max}) that guarantees that there is no numerical overflow in the distance representation.

Let i be the number of bits needed to encode the distance value at each voxel. Let Dim be the image dimension. The maximum coded distance is

$$D_{max} = Dim.round(N.d_{111}) = Dim.(N.d_{111} + q) \tag{14}$$

where q is the rounding error such that $|q| \leq \frac{1}{2}$

For large N, the rounding error can be neglected, so that we have:

$$N_{max} < \frac{2^i}{Dim.d_{111}} \tag{15}$$

where

$$d_{111} = (\sqrt{2 + P^2})[\frac{-2 + 2\sqrt{1 + \lambda}}{\lambda}] \tag{16}$$

and

$$\lambda = (\sqrt{(m^2 + 1)} - m)^2 + \frac{1}{P^2}[\sqrt{(m^2 + 1 + P^2)} - \sqrt{(m^2 + 1)}]^2 \tag{17}$$

where $M = 2m + 1$ is the operator mask size.

The minimum value is chosen so that the rounding error is of the same order of magnitude as that of the distance transform.

Let E_{max} be the maximum error of the distance transform. As shown in [10], this maximum error is proportional to the distance R. It is obtained in the center of the first cone and in direction d_{100}. Let D_{100} be the integer representation of elementary displacement d_{100}. We have

$$D_{100} = round(N.d_{100}) \tag{18}$$

and the relative error induced by the integer approximation is

$$|\epsilon| = \frac{|round(N.d_{100}) - N.d_{100}|}{N.d_{100}} = \frac{|q|}{N.d_{100}} = \frac{1}{2.N.d_{100}} \tag{19}$$

with $|q| \leq \frac{1}{2}$

For $W = H = 1$, the maximum absolute normalized error induced by using a local distance operator is:

$$|e_{max}| = \frac{|E_{max}|}{R} = |1 - d_{100}| \tag{20}$$

By taking eq.20 and eq.5 into account, the minimum scaling factor value must satisfy

$$N_{min} > \frac{1}{2.d_{100}.e_{max}} \tag{21}$$

where

$$d_{100} = \frac{-2 + 2\sqrt{1+\lambda}}{\lambda} \tag{22}$$

and λ is given by eq.17. It can be seen that N_{min} depends on voxel depth P.

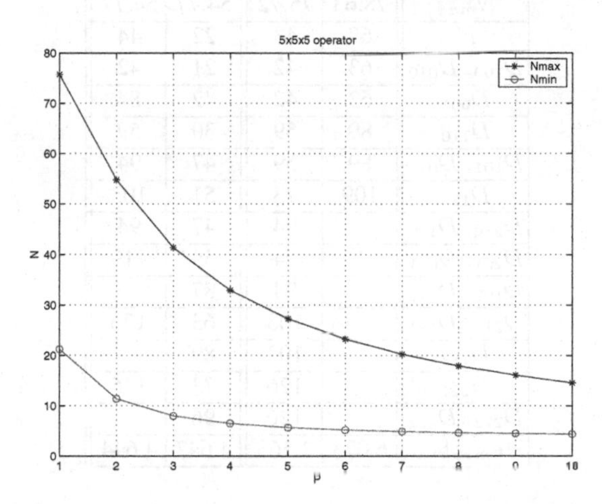

Fig. 6. Scaling factor bounds N_{max} and N_{min} for a 5x5x5 operator versus voxel depth P

It should be noticed that eq.15 is valid if the volume dimensions are the same in rows, columns and layers. For eq.18, it is assumed that the real coefficients are set to their optimum value.

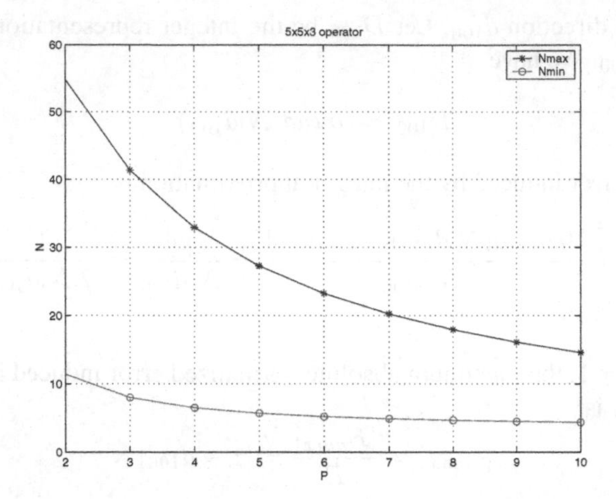

Fig. 7. Scaling factor bounds N_{max} and N_{min} for a 5x5x3 operator versus voxel depth P

Table 3. Best integer approximation of some isotropic $W = H = P = 1$ and anisotropic operators in 3D with and $W = H = 1, P = 2, i = 16$ bits, $Dim = 512$

W,H,P	(1,1,1)	(1,1,1)	(1,1,2)	(1,1,2)
size	3x3x3	5x5x5	5x5x5	5x5x3
N_{min}	8.84	21.25	11.42	11.42
N_{max}	78.63	75.72	54.77	54.77
N	67	43	22	44
D_{100}, D_{010}	63	42	21	42
D_{001}	63	42	42	84
D_{110}	89	59	30	59
D_{101}, D_{011}	89	59	47	94
D_{111}	109	73	51	103
D_{210}, D_{120}		94	47	94
D_{201}, D_{021}		94	59	119
D_{102}, D_{012}		94	87	
D_{211}, D_{121}		103	63	126
D_{112}		103	89	
D_{221}		126	73	145
D_{212}, D_{122}		126	96	
$e_{max}\%$	6.073	2.563	4.647	4.644

The variations of scaling factor bounds N_{max} and N_{min} as functions of parameter P, for 5x5x5 and 5x5x3 distance operators are presented in Figures 6 and 7.

In Table 3 we give the best integer approximation of some 3D cubic operators for $W = H = P = 1$ and non cubic operators for $W = H = 1$ and $P = 2$. We consider $Dim = 512$ and $i = 16$ bits. Scaling factors N and maximum absolute normalized errors e_{max} are given for each operator.

From Table 3, it can be seen that the maximum error obtained by these integer operators is close to the theoretical one (obtained with real operators of identical sizes). We notice that we have the same maximum error with an operator of size 5x5x3 as that obtained with a 5x5x5 operator, in the case $W = H = 1$ and $P = 2$.

The Distance Transform should be semi-regular [6]. Operators given in Table 3 satisfy these conditions.

4 Conclusion

In this paper, anisotropic local distance operators in parallelepipedic grids have been studied. These operators are useful for the analysis of 3D images such as those obtained by means of confocal microscopy or computed tomography where the sampling steps differ according to the axes.

According to the Mathematical Morphology approach, grey level images can be regarded as 3D objects in a 3D volume. Hence, 3D local distance operators turn out to be useful tools for various purposes such as image comparison.

Though exact Euclidean Distance operators are now available, Chamfer Distance operators are still of interest, especially for shape analysis, shape coding applications, and in 3D spaces.

By means of an appropriate optimization procedure, the same performances as those obtained by isotropic MxMxM operators can be achieved with lower computational complexity. Formulas providing optimal coefficients have been given for any mask size M and voxel elongation ratio P/W. Local distance operator performances have been studied for various mask sizes and voxel elongation ratio. The paper proposes a method to calculate lower and upper bounds for scaling factors in order to obtain integer approximation for the coefficients. This approach helps the algorithm perform in scenarios where the memory is limited.

References

1. Borgefors, G.: Applications using distance transforms. In: Arcelli, C., Cordella, L.P., Sanniti di Baja., G. (eds.) Aspects of Visual Form Processing, pp. 83–108. World Scientific, Singapore (1994)
2. Borgefors, G.: Distance transformations in arbitrary dimensions. Computer Vision, Graphics and Image Processing 27, 312–345 (1984)
3. Borgefors, G.: Distance transformations in digital images. Computer Vision, Graphics and Image Processing 34, 344–371 (1986)
4. Cuisenaire, O., Macq, B.: Fast Euclidean distance transformation by propagation using multiple neighborhood. Computer Vision and Image Understanding 76, 163–172 (1999)
5. Maurer Jr., C.R., Qi, R., Raghavan, V.: A linear time algorithm for computing exact Euclidean distance transforms of binary images in arbitrary dimensions. IEEE Transactions on Pattern Analysis and Machine Intelligence 25(2), 265–270 (2003)
6. Svensson, S., Borgefors, G.: Digital distance transforms in 3D images using information from neighbourhoods up to 5x5x5. Computer Vision and Image Understanding 88, 24–53 (2002)

7. Fouard, C., Malandain, G.: 3-D chamfer distances and norms in anisotropic grids. Image and Vision Computing 23, 143–158 (2005)
8. Fouard, C., Strand, R., Borgefors, G.: Weighted distance transforms generalize to modules and their computation on point lattices. Pattern Recognition 40, 2453–2474 (2007)
9. Hulin, J., Thiel, E.: Chordal axis on weighted distance transforms. In: Kuba, A., Nyúl, L.G., Palágyi, K. (eds.) DGCI 2006. LNCS, vol. 4245, pp. 271–282. Springer, Heidelberg (2006)
10. Verwer, B.: Local distances for distance transformations in two and three dimensions. Pattern Recognition Letters 12, 671–682 (1991)
11. Coquin, D., Bolon, P.: Discrete distance operator on rectangular grids. Pattern Recognition Letters 16, 911–923 (1995)
12. Remy, E., Thiel, E.: Optimizing 3D chamfer mask with norm constraints. In: Proceedings of International Workshop on Combinatorial Image Analysis, Caen, France, pp. 39–56 (2000)
13. Strand, R.: Weighted distances based on neighbourhood sequences. Pattern Recognition Letters 28(15) (2007)
14. Borgefors, G.: On digital distance transforms in three dimensions. Computer Vision and Image Understanding 64, 368–376 (1996)
15. Borgefors, G.: Weighted digital distance transforms in four dimensions. Discrete Applied Mathematics 125, 161–176 (2003)
16. Kiselman, C.: Regularity properties of distance transformations in image analysis. Computer Vision and Image Understanding 64, 390–398 (1996)
17. Sintorn, I.M., Borgefors, G.: Weighted distance transforms in rectangular grids. In: 11th International Conference on Image Analysis and Processing, Palermo, Italy, pp. 322–326 (2001)
18. Svensson, S., Borgefors, G.: Distance transforms in 3D using four different weights. Pattern Recognition Letters 23, 1407–1418 (2002)
19. Sintorn, I.M., Borgefors, G.: Weighted distance transforms for images using elongated voxel grids. In: Proc. 10th Discret Geometry for Computer Imagery, Bordeaux, France, pp. 244–254 (2002)
20. Chehadeh, Y., Coquin, D., Bolon, P.: A generalization to cubic and non cubic local distance operators on parallelepipedic grids. In: Proc. 5th Discret Geometry for Computer Imagery, pp. 27–36 (1995)

A Novel Algorithm for Distance Transformation on Irregular Isothetic Grids

Antoine Vacavant[1,2], David Coeurjolly[1,3], and Laure Tougne[1,2]

[1] Université de Lyon, CNRS
[2] Université Lyon 2, LIRIS, UMR5205, F-69676, France
[3] Université Lyon 1, LIRIS, UMR5205, F-69622, France
{antoine.vacavant,david.coeurjolly,laure.tougne}@liris.cnrs.fr

Abstract. In this article, we propose a new definition of the E^2DT (Squared Euclidean Distance Transformation) on irregular isothetic grids such as quadtree/octree or run-length encoded d-dimensional images. We describe a new separable algorithm to compute this transformation on every grids, which is independent of the background representation. We show that our proposal is able to efficiently handle various kind of classical irregular two-dimensional grids in imagery, and that it can be easily extended to higher dimensions.

1 Introduction

The representation and the manipulation of large-scale data sets by irregular grids is a today scientific challenge for many applications [9,12]. Here, we are interested in discrete distance definition and distance transformation [8,13] on this kind of multi-grids to develop fast and adaptive shape analysis tools. The Distance Transformation (DT) of a binary image consists in labeling each point of a discrete object $\mathcal{E}$ (*i.e.* foreground) with its shortest distance to the complement of $\mathcal{E}$ (*i.e.* background). This process is widely studied and developed on regular grids (see for example [4]). Some specific extensions of the DT to non-regular grids also exist, such as rectangular grids [15], quadtrees [18], *etc.* This article deals with generalizing the DT computation on *irregular isothetic grids* (or $\mathbb{I}$-grid for short) in two dimensions (2-D). The proposed method is easily extensible to higher dimensions. In 2-D, this kind of grids is defined by rectangular cells whose edges are aligned along the two axis. The size and position of those non-overlapping cells are defined without constraint (see next section for a formal definition). The quadtree decomposition and the RLE (Run Length Encoding) grouping schemes are examples of classical techniques in imagery which induce an $\mathbb{I}$-grid. Here, we focus our interest on generalizing techniques that compute the E^2DT (squared Euclidean DT) of a d-dimensional (d-D) binary image [2,10,14]. Many of those methodologies can be linked to the computation of a discrete Voronoi diagram of the background pixels [6,2]. In a previous work [16], we introduced a definition of E^2DT on $\mathbb{I}$-grids based on the background cells centers. After showing the link between this process and the computation of

S. Brlek, C. Reutenauer, and X. Provençal (Eds.): DGCI 2009, LNCS 5810, pp. 469–480, 2009.

the Voronoi diagram of those background points, we proposed two completing algorithms to handle various configurations of $\mathbb{I}$-grids. Unfortunately, we noticed that they suffer from non-optimal time complexity. Moreover, the result of this transformation strongly depends on the representation of the background.

In this article, we aim to provide a faster version of our previous separable algorithm for distance transformation on $\mathbb{I}$-grids such as quadtree or run-coded d-D images. Instead of a center based DT algorithm we now focus on a faster border based DT algorithm, by re-sampling the irregular border part of objects with a partly regular grid. This paper is organized as follows: (1) we present a new way to compute the E^2DT on $\mathbb{I}$-grids, based on the border of the background cells (Section 2). We also illustrate its relation with the computation of a Voronoi diagram of segments; (2) in Section 3, we describe an algorithm based on the work of R. Maurer *et al.* [10] to efficiently compute this new transformation in 2-D, but extensible to higher dimensions; (3) we finally show in Section 4 experimental results to compare our new contribution with the techniques we developed in [16] in terms of speed and time complexity.

2 Distance Transformations on $\mathbb{I}$-grids and Voronoi Diagrams

In this section, we first recall the concept of irregular isothetic grids ($\mathbb{I}$-grids), with the following definition [1]:

Definition 1 (2-D $\mathbb{I}$-grid). *Let $I \subset \mathbb{R}^2$ be a closed rectangular support. A 2-D $\mathbb{I}$-grid $\mathbb{I}$ is a tiling of I with non overlapping rectangular cells which edges are parallel to the X and Y axis. The position (x_R, y_R) and the size (l_R^x, l_R^y) of a cell R in $\mathbb{I}$ are given without constraint: $(x_R, y_R) \in \mathbb{R}^2$, $(l_R^x, l_R^y) \in \mathbb{R}^2$.*

We consider this definition of 2-D $\mathbb{I}$-grids for the rest of the paper, and we will shortly show that our contribution is easily extensible to the d-D case. In our framework, we consider *labeled* $\mathbb{I}$-grids, *i.e.* each cell of the grid has a *foreground* or *background* label (its value is respectively "0" or "1" for example). For an $\mathbb{I}$-grid $\mathbb{I}$, we denote by $\mathbb{I}_F$ and $\mathbb{I}_B$ the sets of foreground and background cells.

We can first consider that the distance between two cells R and R' is the distance between their centers. If we denote $p = (x_R, y_R)$ and $p' = (x_{R'}, y_{R'})$ these points, and $d_e^2(p, p')$ the squared Euclidean distance between them, the $\mathbb{I}$-CDT (Center-based DT on $\mathbb{I}$-grids) of a cell R is defined as follows [16]:

$$\mathbb{I}\text{–CDT}(R) = \min_{R'} \{d_e^2(p, p');\ R' \in \mathbb{I}_B\}, \tag{1}$$

and resumes to the E^2DT if we consider $\mathbb{I}$ as a regular (square or rectangular) discrete grid. However, this distance is strongly dependent of the background representation. In Figure 1, we present an example of the computation of the $\mathbb{I}$-CDT of two $\mathbb{I}$-grids where only the background cells differ. Since this definition is based on the cells centers position, the $\mathbb{I}$-CDT do not lead to the same distance map depending on the background coding.

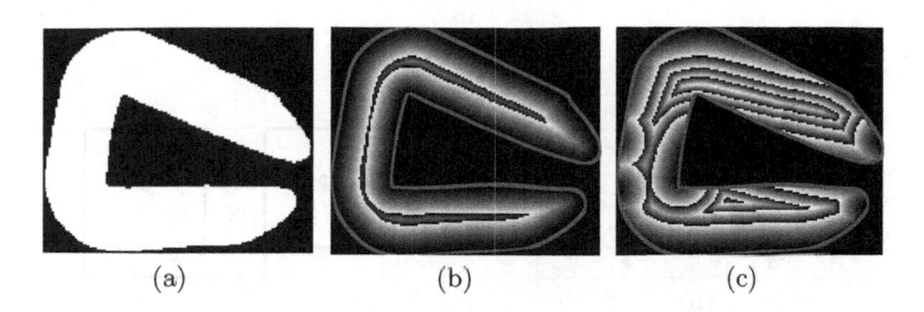

(a) (b) (c)

Fig. 1. The result of the $\mathbb{I}$-CDT of the complete regular grid (b) computed from the binary image (a) and an $\mathbb{I}$-grid where the foreground is regular and the background is encoded with a RLE scheme along Y (c). The distance value d of a cell is represented with a grey level $c = d \mod 255$. The contour of the object (background/foreground frontier) is recalled in each distance map (b) and (c) with a smooth curve.

We now introduce an alternative definition of the E^2DT by considering the shortest distance between a foreground cell center and the background/foreground boundary of the $\mathbb{I}$-grid. We suppose here that the intersection between two adjacent cells is a segment of dimension one (*i.e.* they respect the *e*-adjacency relation [1]). Let $\mathcal{S}$ be the set of segments at the interface between two adjacent cells $R \in \mathbb{I}_F$ and $R' \in \mathbb{I}_B$.The $\mathbb{I}$-BDT (Border-based DT on $\mathbb{I}$-grids) is then defined as follows:

$$\mathbb{I}\text{--BDT}(R) = \min_{s} \left\{ d_e^2(p, s);\ s \in \mathcal{S} \right\}. \quad (2)$$

Contrary to the $\mathbb{I}$-CDT given in Equation 1, the result of this process does not depend on the representation of the background. We can draw a parallel between those extensions of the E^2DT to $\mathbb{I}$-grids and the computation of a Voronoi diagram (VD). More precisely, as in the regular case, the $\mathbb{I}$-CDT can be linked with the VD of the background cells centers (*i.e. Voronoi sites* or *VD sites*) [16]. The VD of a set of points $\mathcal{P} = \{p_i\}$ is a tiling of the plane into *Voronoi cells* (or *VD cells*) $\{C_{p_i}\}$ [3]. If we now consider the background/foreground frontier to compute the $\mathbb{I}$-BDT, our definition implies that we compute a VD of segments, and not a classical VD of points (see Figure 2 for an example of these diagrams computed on a simple $\mathbb{I}$-grid). Hence, a simple approach to compute the $\mathbb{I}$-CDT is to pre-compute the complete VD of the background points [16], and to locate foreground points in the VD. In this case, the $\mathbb{I}$-CDT computation has a $\mathcal{O}(n \log n_B)$ time complexity, where n is the total number of cells, and n_B is the number of background cells. This technique is obviously not computationally efficient for every grids, and not adapted to dense grids [16]. To compute the $\mathbb{I}$-BDT, a similar synopsis can be drawn, where the computation of the VD of segments can be handled in $\mathcal{O}(n_S \log^2 n_S)$, where $n_S = 4n_B$ is the total number of segments belonging to the background/foreground frontier [7]. The extension of those transformation to d-D $\mathbb{I}$-grids is a hard work. A VD can be computed in d-D with a $\mathcal{O}(n_B \log n_B + n_B^{\lceil d/2 \rceil})$ time complexity (thanks to a gift-wrapping approach [3] for example). However, localizing a point in the VD is an arduous

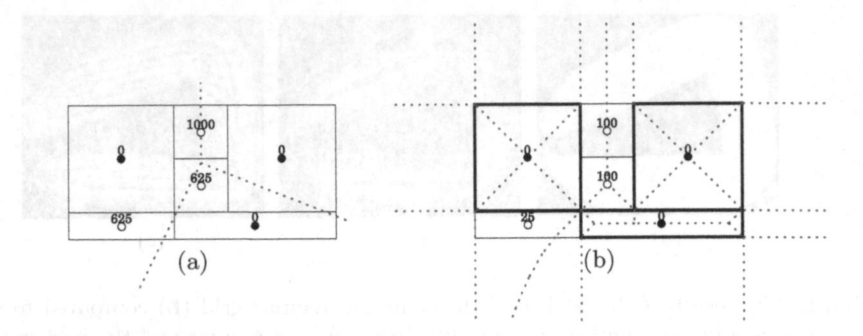

Fig. 2. Example of the VD of the background points to obtain the $\mathbb{I}$-CDT of a simple $\mathbb{I}$-grid (a). Background cell centers are depicted in black, and foreground ones in white. Distance values are presented above the cells centers. From the same grid, the computation of the $\mathbb{I}$-BDT (b) implies that we consider the VD of the segments belonging to the background/foreground frontier.

task, and an additional structure like subdivision grids [11] should be constructed to handle this operation.

Hence, we have shown that building the entire VD to obtain the $\mathbb{I}$-BDT is neither computationally efficient for every $\mathbb{I}$-grids, nor easily extensible to higher dimensions. We now propose a separable algorithm to compute the $\mathbb{I}$-BDT that can be extended to the d-D case.

3 A New Separable Algorithm to Compute the $\mathbb{I}$-BDT

3.1 R. Maurer *et al.* E²DT Algorithm on Regular Grids

The main idea of this separable (one phase by dimension) method [4,10] is that the intersection between the complete VD of background pixels (*i.e.* sites) and a line of the grid can be easily computed, then simplified. Indeed, for a row j, the VD sites can be "deleted" by respecting three remarks mainly based on the monotonicity of d_e distance [10]: (1) if we consider the line $l : y = j,\ j \in \mathbb{Z}$, then we only have to keep the nearest VD sites from l (Figure 3-b), (2) those sites can be ordered along the X axis, which means that it is not necessary to keep the complete VD, but only its intersection with the line l, (3) a VD site may be *hidden by* two neighbour sites, and thus not considered anymore (Figure 3-c). In this article, we show the extension of this technique on $\mathbb{I}$-grids by adapting these properties on these grids.

3.2 Separable Computation of the $\mathbb{I}$-BDT

To develop a separable process on $\mathbb{I}$-grids, we use a similar structure as the *irregular matrix* **A** associated to a labeled $\mathbb{I}$-grid $\mathbb{I}$ introduced in [16]. This data structure aims to organize the cells of the grid along the X and Y axis by adding virtual cells centers (see Figure 4 for an example). We used it in our

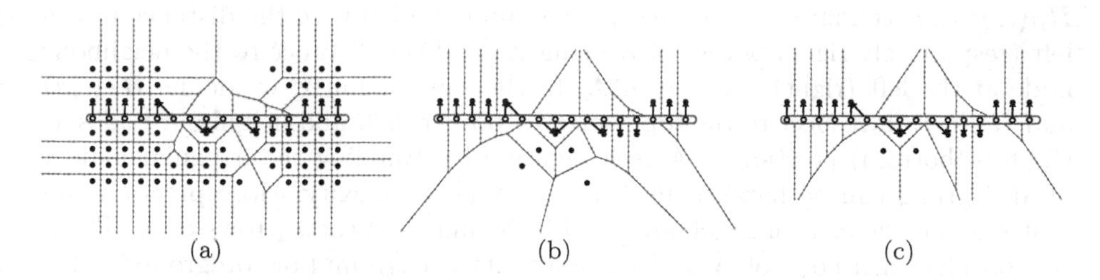

Fig. 3. In a regular grid, when we treat a row l with R. Maurer *et al.* algorithm, we consider the VD of background nodes like in (a). We only keep the nearest VD sites of l (b). We obtain (c) by deleting sites which associated VD cells intersect the l row. Arrows indicate the associated VD site of each foreground pixel of this row.

previous work to adapt the separable E^2DT algorithm of T. Saito *et al.* [14], devoted to compute the $\mathbb{I}$-CDT on $\mathbb{I}$-grids. The irregular matrix contains as many columns (respectively rows) as X-coordinates (Y-coordinates) in the grid. These coordinates are stored in two tables T_X and T_Y (and we denote $n_1 = |T_X|$ and $n_2 = |T_Y|$). At the intersection of two X and Y coordinates, a node in $\mathbf{A}$ has the same value as the cell containing it, and may represent the cell center or not (*i.e.* this is an *extra node*, see Figure 4-a). The extra nodes are used to propagate the distance values through the irregular matrix and then compute a correct distance value for each cell center. To apply the $\mathbb{I}$-BDT on this data structure, we also have to take into account the border and the size of the treated cells. We thus add for each node $\mathbf{A}(i,j)$ *border attributes* along X and Y axis that permit to propagate the distance values to cell borders through $\mathbf{A}$. For a node $\mathbf{A}(i,j)$ contained by the cell R in $\mathbb{I}$, we denote respectively by $H_L(i,j)$ and

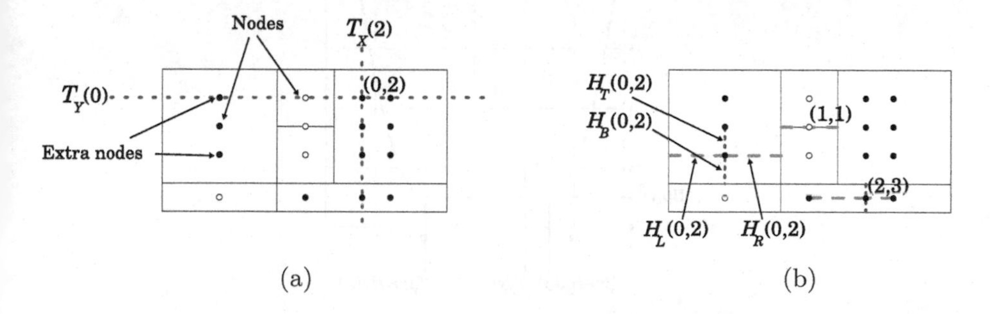

Fig. 4. Construction of the irregular matrix associated to the simple $\mathbb{I}$-grid illustrated in Figure 2. In (a), the extra node $\mathbf{A}(0,2)$ of the matrix is depicted at the intersection of the dotted lines. New nodes have the same value as the cell containing them. For the $\mathbb{I}$-BDT, we also need the shortest distance between a node in respect to its neighbour nodes and the border of the cell containing it (b). Examples of values for border attributes are also given along X (dashed lines) and Y (dotted lines). For instance, we have $H_T(1,1) = H_B(1,1) = 0$ since this node coincides with a cell horizontal border. We can also notice that $H_L(2,3) \neq H_R(2,3)$.

$H_R(i,j)$ the attributes that represent the minimum between the distance to the left (respectively right) border of R along X, and the distance to the neighbour node at the left (right) position in $\mathbf{A}$. In the same manner, we define $H_T(i,j)$ and $H_B(i,j)$ in respect to the top (bottom) border of R and neighbour nodes at the top (bottom) position in $\mathbf{A}$ (see Figure 4-b). Building the irregular matrix of an $\mathbb{I}$-grid $\mathbb{I}$ can be handled in $\mathcal{O}(n_1 n_2)$ time complexity. More precisely, we first scan all the cells of $\mathbb{I}$ to know the n_1 columns and the n_2 rows of $\mathbf{A}$. Then, we consider each node of $\mathbf{A}$ and we assign its background or foreground value and its border attributes.

Equation 2 can now be adapted on the irregular matrix with a minimization process along the two axis (see [17] for the proof):

Proposition 1 (Separable $\mathbb{I}$-BDT). *Let $\mathbf{A}$ be the associated irregular matrix of the 2-D $\mathbb{I}$-grid $\mathbb{I}$. Then the $\mathbb{I}$-BDT of $\mathbb{I}$ can be decomposed in two separable processes, and consists in computing the matrix $\mathbf{B}$ and $\mathbf{C}$ as follows:*

$$\mathbf{B}(i,j) = \min_x \left\{ \min\left(|T_X(i) - T_X(x) - H_R(x,j)|, |T_X(x) - T_X(i) - H_L(x,j)| \right) ;\; x \in \{0,...,n_1-1\},\; \mathbf{A}(x,j) = 0 \right\},$$
$$\mathbf{C}(i,j) = \min_y \left\{ \mathcal{G}_y(j);\; y \in \{0,\ldots,n_2-1\} \right\},$$

where $\mathcal{G}_y(j)$ is a flattened parabola given by:

$$\mathcal{G}_y(j) = \begin{cases} \mathbf{B}(i,y)^2 + (T_Y(j) - T_Y(y) - H_T(i,y))^2 & \text{if } T_Y(j) - T_Y(y) > H_T(i,y) \\ \mathbf{B}(i,y)^2 + (T_Y(y) - T_Y(j) - H_B(i,y))^2 & \text{if } T_Y(y) - T_Y(j) > H_B(i,y) \\ \mathbf{B}(i,y)^2 & \text{otherwise.} \end{cases}$$

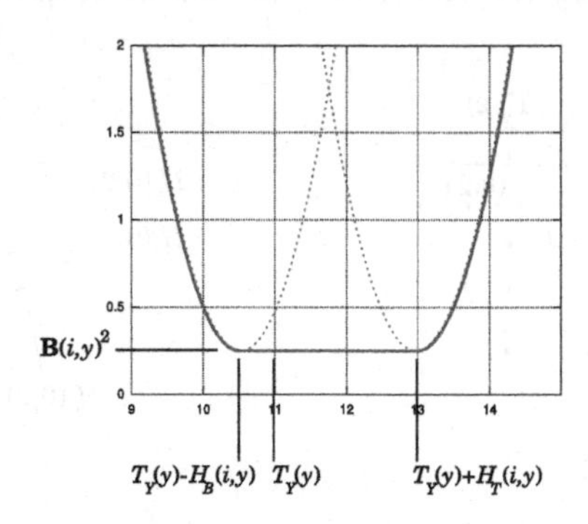

Fig. 5. Important features of a flattened parabola, composition of two half-parabolas (in dotted lines) and a constant function

We prove in [17] that this proposition is correct, *i.e.* this two-dimensional scheme consists in computing the $\mathbb{I}$-BDT given in Equation 2. This separable minimization process also permits to compute the $\mathbb{I}$-CDT (Equation 1), if we assign all border attributes to zero. In this case, this transformation is equivalent to a minimization process of a 2-D quadratic form [16], and we consider a set of classical parabolas along the Y axis, as in the regular case [14]. Here, a flattened parabola (see also Figure 5) is an *ad-hoc* function composed by two half-parabolas and a constant function. It represents a constant distance value from a node $\mathbf{A}(i,j)$ to its neighbour nodes and cell borders, and then increases as a classical quadratic function beyond those limits. The computation of matrix $\mathbf{C}$ in Proposition 1 thus consists in computing the lower envelope of a set of flattened parabolas. This operation is possible since these functions are monotone (composition of monotone functions), and there exist a single intersection or an infinity of intersections between two flattened parabolas. We give in [17] the proof of this property, and the relation between Equation 2 and Proposition 1.

3.3 Adaptation of R. Maurer *et al.* E^{2}DT Algorithm on $\mathbb{I}$-grids

The first stage of our method (first step of Algorithm 1) consists in scanning along X and in initializing the distance of each node of the irregular matrix. This is indeed a double linear scan for each row. Notice the use of H_R and H_L attributes to propagate the distance to cells borders. At the end of this stage, each node stores a squared distance (line 1). In the second part of our algorithm, we call the `Voronoi_IBDT` function (Algorithm 2) to build a partial VD intersected with each column i. As in the original algorithm [10], we use two stacks storing real successive coordinates of treated sites (h), and their squared distance (g). The first loop of this function (line 2) corresponds to the deletion of hidden sites, thanks to the `hidden_by()` predicate (Algorithm 3). In Algorithm 2, we also use two additional stacks, denoted by f_T and f_B to store the border attributes and update them (line 2). Thanks to these stacks, the second stage of our algorithm is achieved in linear time. By testing the value of l (line 2),

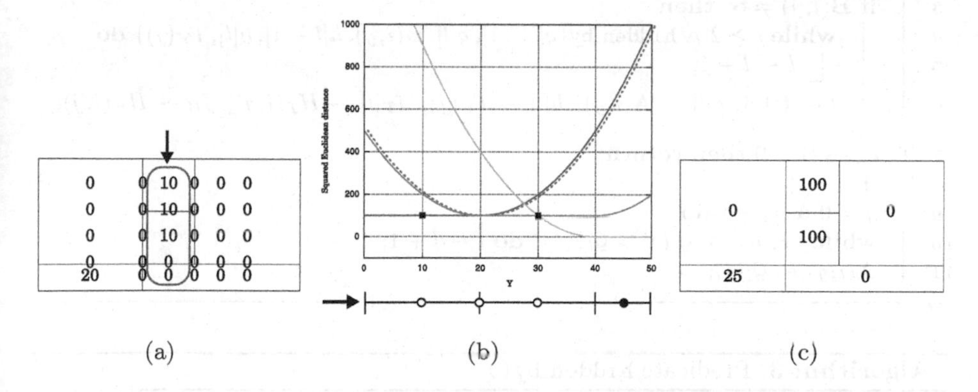

Fig. 6. For the column chosen in (a), the last phase of Algorithm 2 consists in considering the lower envelope of a set of flattened parabolas (b). At the bottom of this plot, background and foreground nodes are represented by black and white circles at the corresponding Y-coordinate. Cell borders are also represented (vertical dashes). Black squares represent where the cell centers are located along Y axis, and the associated $\mathbb{I}$-BDT value. We give in (c) the obtained $\mathbb{I}$-BDT for this $\mathbb{I}$-grid.

Algorithm 1. Separable computation of the $\mathbb{I}$-BDT inspired from [10]

input : the labeled $\mathbb{I}$-grid $\mathbb{I}$.
output: the $\mathbb{I}$-BDT of $\mathbb{I}$, stored in the irregular matrix $\mathbf{C}$.

1 build the irregular matrix $\mathbf{A}$ associated to $\mathbb{I}$;
2 **for** $j = 0$ **to** $n_2 - 1$ **do** {*First stage along X*}
3   **if** $\mathbf{A}(0, j) = 0$ **then** $\mathbf{B}(0, j) \leftarrow 0$;
4   **else** $\mathbf{B}(0, j) \leftarrow \infty$;
5   **for** $i = 1$ **to** $n_1 - 1$ **do**
6   **if** $\mathbf{A}(i, j) = 0$ **then** $\mathbf{B}(i, j) \leftarrow 0$;
7   **else**
8   **if** $\mathbf{B}(i-1, j) = 0$ **then** $\mathbf{B}(i, j) \leftarrow T_X(i) - T_X(i-1) - H_R(i-1, j)$;
9   **else** $\mathbf{B}(i, j) \leftarrow T_X(i) - T_X(i-1) + \mathbf{B}(i-1, j)$;
10   **for** $i = n_1 - 2$ **to** 0 **do**
11   **if** $\mathbf{B}(i+1, j) < \mathbf{B}(i, j)$ **then**
12   **if** $\mathbf{B}(i+1, j) = 0$ **then** $\mathbf{B}(i, j) \leftarrow T_X(i+1) - T_X(i) - H_L(i, j)$;
13   **else** $\mathbf{B}(i, j) \leftarrow T_X(i+1) - T_X(i) + \mathbf{B}(i-1, j)$;
14       $\mathbf{B}(i, j) \leftarrow \mathbf{B}(i, j)^2$;
15 **for** $i = 0$ **to** $n_1 - 1$ **do** {*Second stage along Y*}
16   `Voronoi_IBDT`(i);

Algorithm 2. Function `Voronoi_IBDT()` to build a partial VD along Y

input : the column i of the irregular matrix $\mathbf{B}$.
output: the $\mathbb{I}$-BDT of each node of the column i stored in the matrix $\mathbf{C}$.

1 $l \leftarrow 0$, $g \leftarrow \emptyset$, $h \leftarrow \emptyset$, $f_T \leftarrow \emptyset$, $f_B \leftarrow \emptyset$;
2 **for** $j = 0$ **to** $n_2 - 1$ **do**
3   **if** $\mathbf{B}(i, j) \neq \infty$ **then**
4   **while** $l \geq 2 \wedge$ `hidden_by`$\big(g[l-1], g[l], \mathbf{B}(i, j), h[l-1], h[l], T_Y(j)\big)$ **do**
5          $l \leftarrow l - 1$;
6       $l \leftarrow l + 1$, $g[l] \leftarrow \mathbf{A}(i, j)$, $h[l] \leftarrow T_Y(j)$, $f_T[l] \leftarrow H_T(i, j)$, $f_B \leftarrow H_B(i, j)$;
7 **if** $(n_s \leftarrow l) = 0$ **then return**;
8 $l \leftarrow 1$;
9 **for** $j = 0$ **à** $n_2 - 1$ **do**
10   **while** $l < n_s \;\wedge\; \mathcal{G}_l(j) > \mathcal{G}_{l+1}(j)$ **do** $l \leftarrow l + 1$;
11    $\mathbf{A}(i, j) \leftarrow \mathcal{G}_l(j)$;

Algorithm 3. Predicate `hidden_by()`

input : Y-coordinates of three points in $\mathbb{R}^2$ denoted by u_y, v_y, w_y, and their squared distance to the line $L : y = r$ denoted by $d_e^2(u, L)$, $d_e^2(v, L)$, $d_e^2(w, L)$.
output: is v hidden by u and w ?

1 $a \leftarrow v_y - u_y$, $b \leftarrow w_y - v_y$, $c \leftarrow a + b$;
2 **return** $c \times d_e^2(v, L) - b \times d_e^2(u, L) - a \times d_e^2(w, L) - abc > 0$;

we know if we have to scan again the stacks and to update the distance values of the nodes. Finally, we linearly scan the stacks to find the nearest border of the $\mathbf{A}(i,j)$ current node (line 2), and this step indeed consists in considering the lower envelope of a set of flattened parabolas $\{\mathcal{G}_l\}$, given by (see also Figure 6):

$$\mathcal{G}_l(j) = \begin{cases} g[l] + (T_Y(j) - h[l] - f_T[l])^2 & if\ T_Y(j) - h[l] > f_T[l] \\ g[l] + (h[l] - T_Y(j) - f_B[l])^2 & if\ T_Y(j) - h[l] > f_B[l] \\ \mathbf{B}(i,y)^2 & else, \end{cases} \quad (3)$$

which is the adaptation of Proposition 1 with the stacks g, h, f_T and f_B. We prove in [17] that this adaptation is correct, *i.e.* it computes a correct distance map for $\mathbb{I}$-BDT.

As a consequence, we have described a linear $\mathbb{I}$-BDT algorithm in respect to the associated irregular matrix size, *i.e.* in $\mathcal{O}(n_1 n_2)$ time complexity. Our new contribution is easily extensible to higher dimensions: we still realize the first step as an initialization phase, and for each dimension $d > 1$, we combine results obtained in dimension d-1. If we consider a labeled d-D $\mathbb{I}$-grid, which associated irregular matrix size is $n_1 \times \cdots \times n_d$, the time complexity of our algorithm is thus in $\mathcal{O}(n_1 \times \cdots \times n_d)$. In the next section, we present experiments to show the interest of the $\mathbb{I}$-BDT, and to point out that our algorithm is a very efficient approach to compute both $\mathbb{I}$-CDT and $\mathbb{I}$-BDT of various classical $\mathbb{I}$-grids.

4 Experimental Results

We first propose to present the result of our $\mathbb{I}$-BDT algorithm for the binary image depicted at the beginning of this article in Figure 1, digitized in various classical $\mathbb{I}$-grids used in imagery. Table 1 illustrates $\mathbb{I}$-BDT elevation maps where the background and the foreground of the original image are independently represented with a regular square grid ($\mathbb{D}$), a quadtree ($q\mathbb{T}$) and a RLE along Y ($\mathbb{L}$). We can notice in this figure that the result of the $\mathbb{I}$-BDT is independent of the representation of the background. The distance values in the foreground region are thus the same in the three elevation maps of a given column.

We now focus our interest on the execution time of our new algorithm, in respect to our previous work [16]. We consider the three algorithms presented in Table 2. The first algorithm represents the simple approach we discussed in Section 2, which is hardly extensible to d-D treatments, and $\mathbb{I}$-BDT computation. Algorithm 2 is described in [16] and is inspired from the quadratic form minimization scheme of T. Saito *et al.* [14]. Thanks to the flattened parabola definition, we can extend this algorithm to $\mathbb{I}$-BDT, but with a non-optimal time complexity, in respect to the irregular matrix size. Algorithm 3 is our new separable transformation inspired from [10]. In Figure 7, we present the three chosen images for our experiments, and in Table 3, we show the execution times for these three algorithms, for $\mathbb{I}$-grids built from three binary images. We have performed those experiments on a mobile workstation with a 2.2 Ghz Intel Centrino Duo processor, and 2 Gb RAM. We can first notice that our new contribution gives good results for $\mathbb{I}$-CDT. Indeed, Algorithm 3 is the fastest one for dense

(a) *noise* (b) *lena* (c) *canon*

Fig. 7. For the images *noise* (100x100), *lena* (208x220), *canon* (512x512) we consider in our experiments the associated $\mathbb{I}$-grids (regular grid, quadtree decomposition and RLE along Y scheme)

Table 1. Elevation maps representing the $\mathbb{I}$-BDT of each $\mathbb{I}$-grid. X and Y are the axis of the image, and Z corresponds to the distance value. The foreground (columns) and the background (rows) of the image are digitized independently. The color palette is depicted on the right of the figure.

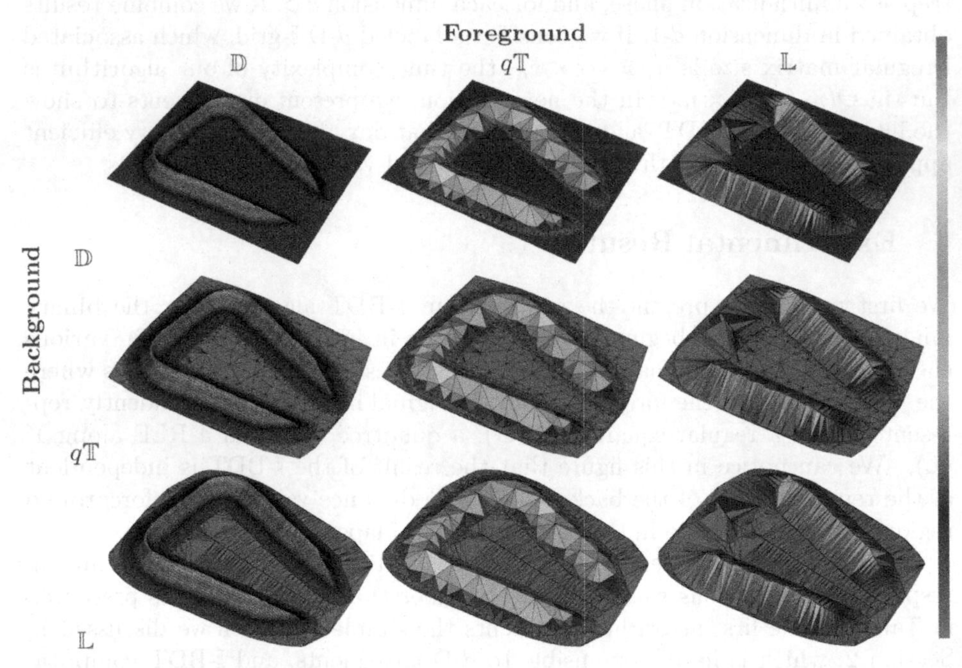

Table 2. The three compared algorithms, and their associated time and space complexities. We also check if an algorithm is extensible to d-D $\mathbb{I}$-grids and what kind of transformation it can perform ($\mathbb{I}$-CDT, $\mathbb{I}$-BDT).

Id.	Algorithm	Time	Space	d-D	$\mathbb{I}$-CDT	$\mathbb{I}$-BDT
1	Complete VD [16]	$\mathcal{O}(n \log n_B)$	$\mathcal{O}(n)$		✓	
2	From T. Saito *et al.* [14,16]	$\mathcal{O}(n_1 n_2 \log n_2)$	$\mathcal{O}(n_1 n_2)$	✓	✓	✓
3	From R. Maurer *et al.* [10]	$\mathcal{O}(n_1 n_2)$	$\mathcal{O}(n_1 n_2)$	✓	✓	✓

Table 3. We present execution times (in seconds) for each algorithm for the $\mathbb{I}$-CDT (a) and the $\mathbb{I}$-BDT (b) and for each $\mathbb{I}$-grid. Number inside parenthesis in (b) are the increasing rate in % between $\mathbb{I}$-CDT and $\mathbb{I}$-BDT execution times.

(a) $\mathbb{I}$-CDT

Image	Algorithm 1			Image	Algorithm 2			Image	Algorithm 3		
	$\mathbb{D}$	$q\mathbb{T}$	$\mathbb{L}$		$\mathbb{D}$	$q\mathbb{T}$	$\mathbb{L}$		$\mathbb{D}$	$q\mathbb{T}$	$\mathbb{L}$
noise	0.255	0.104	0.053	*noise*	0.037	0.077	0.044	*noise*	0.046	0.065	0.038
lena	1.413	0.145	0.081	*lena*	0.192	0.376	0.245	*lena*	0.185	0.166	0.135
canon	36.236	0.273	0.234	*canon*	1.678	1.322	1.316	*canon*	1.134	0.485	0.585

(b) $\mathbb{I}$-BDT

Image	Algorithm 2			Image	Algorithm 3		
	$\mathbb{D}$	$q\mathbb{T}$	$\mathbb{L}$		$\mathbb{D}$	$q\mathbb{T}$	$\mathbb{L}$
noise	0.047 (27)	0.100 (29)	0.054 (23)	*noise*	0.065 (42)	0.085 (31)	0.049 (29)
lena	0.256 (33)	0.491 (31)	0.320 (31)	*lena*	0.258 (40)	0.209 (26)	0.170 (26)
canon	2.248 (34)	2.107 (59)	2.020 (53)	*canon*	1.507 (33)	0.563 (16)	0.718 (23)

$\mathbb{I}$-grids (regular grids or $\mathbb{L}$ for *noise*), and is very competitive for sparse $\mathbb{I}$-grids (*e.g.* near one half second for $q\mathbb{T}$ and $\mathbb{L}$ based on image *canon*). In the latter case (*canon*), Algorithm 1 is slightly faster than our approach, but we recall that it is hardly extensible to higher dimensions. Algorithm 2, which we developed in our previous work [16], was interesting for dense grids, and is now overtaken by Algorithm 3. For the $\mathbb{I}$-BDT, Algorithm 2 and 3 suffer from an execution time increase, mainly due to the integration of the complex flattened parabola. But our contribution remains as the fastest algorithm, and the time increasing rate is moderate for the tested grids. Large sparse $\mathbb{I}$-grids like $q\mathbb{T}$ and $\mathbb{L}$ based on *canon* are still handled in less than one second.

5 Conclusion and Future Works

In this article, we have proposed a new extension of the E^2DT on $\mathbb{I}$-grids based on the background/foreground frontier, and independent of the background representation. We have also developed a new separable algorithm inspired from [10] that is able to efficiently compute the $\mathbb{I}$-BDT thanks to the irregular matrix structure. We have finally shown that our new contribution is adaptable to various configurations of $\mathbb{I}$-grids, with competitive execution time and complexities, in comparison with our previous proposals.

As a future work, we would like to develop the extension of the discrete medial axis transform [2] on $\mathbb{I}$-grids with our new definition. This would permit to propose a centred simple form of a binary object, independently of the representation of the background, and surely extensible to higher dimensions. As in this paper, we aim to propose a linear algorithm to construct an irregular medial axis, in respect to the irregular matrix size.

References

1. Coeurjolly, D., Zerarga, L.: Supercover Model, Digital Straight Line Recognition and Curve Reconstruction on the Irregular Isothetic Grids. Computer and Graphics 30(1), 46–53 (2006)
2. Coeurjolly, D., Montanvert, A.: Optimal Separable Algorithms to Compute the Reverse Euclidean Distance Transformation and Discrete Medial Axis in Arbitrary Dimension. IEEE Transactions on Pattern Analysis and Machine Intelligence 29(3), 437–448 (2007)
3. de Berg, M., van Kreveld, M., Overmars, M., Schwarzkopf, O.: Computational Geometry: Algorithms and Applications. Springer, Heidelberg (2000)
4. Fabbri, R., Costa, L.D.F., Torelli, J.C., Bruno, O.M.: 2D Euclidean Distance Transform Algorithms: A Comparative Survey. ACM Computing surveys 40(1), 1–44 (2008)
5. Fouard, C., Malandain, G.: 3-D Chamfer Distances and Norms in Anisotropic Grids. Image and Vision Computing 23(2), 143–158 (2005)
6. Hesselink, W.H., Visser, M., Roerdink, J.B.T.M.: Euclidean Skeletons of 3D Data Sets in Linear Time by the Integer Medial Axis Transform. In: 7th International Symposium on Mathematical Morphology, pp. 259–268 (2005)
7. Karavelas, M.I.: Voronoi diagrams in CGAL. In: 22nd European Workshop on Computational Geometry (EWCG 2006), pp. 229–232 (2006)
8. Klette, R., Rosenfeld, A.: Digital Geometry: Geometric Methods for Digital Picture Analysis. Morgan Kaufmann Publishers Inc., San Francisco (2004)
9. Knoll, A.: A Survey of Octree Volume Rendering Techniques. In: 1st International Research Training Group Workshop (2006)
10. Maurer, C.R., Qi, R., Raghavan, V.: A Linear Time Algorithm for Computing Exact Euclidean Distance Transforms of Binary Images in Arbitrary Dimensions. IEEE Transactions on Pattern Analysis and Machine Intelligence 25(2), 265–270 (2003)
11. Park, S.H., Lee, S.S., Kim, J.H.: The Delaunay Triangulation by Grid Subdivision. Computational Science and Its Applications, 1033–1042 (2005)
12. Plewa, T., Linde, T., Weirs, V. (eds.): Adaptive Mesh Refinement - Theory and Applications. In: Proceedings of the Chicago Workshop on Adaptive Mesh Refinement in Computational Science and Engineering, vol. 41 (2005)
13. Rosenfeld, A., Pfalz, J.L.: Distance Functions on Digital Pictures. Pattern Recognition 1, 33–61 (1968)
14. Saito, T., Toriwaki, J.: New Algorithms for n-dimensional Euclidean Distance Transformation. Pattern Recognition 27(11), 1551–1565 (1994)
15. Sintorn, I.M., Borgefors, G.: Weighted Distance Transforms for Volume Images Digitized in Elongated Voxel Grids. Pattern Recognition Letters 25(5), 571–580 (2004)
16. Vacavant, A., Coeurjolly, D., Tougne, L.: Distance Transformation on Two-Dimensional Irregular Isothetic Grids. In: Coeurjolly, D., Sivignon, I., Tougne, L., Dupont, F. (eds.) DGCI 2008. LNCS, vol. 4992, pp. 238–249. Springer, Heidelberg (2008)
17. Vacavant, A., Coeurjolly, D., Tougne, L.: A Novel Algorithm for Distance Transformation on Irregular Isothetic Grids. Technical report RR-LIRIS-2009-009 (2009), `http://liris.cnrs.fr/publis/?id=3849`
18. Vörös, J.: Low-Cost Implementation of Distance Maps for Path Planning Using Matrix Quadtrees and Octrees. Robotics and Computer-Integrated Manufacturing 17(6), 447–459 (2001)

Fully Parallel 3D Thinning Algorithms Based on Sufficient Conditions for Topology Preservation

Kálmán Palágyi and Gábor Németh

Department of Image Processing and Computer Graphics,
University of Szeged, Hungary
{palagyi,gnemeth}@inf.u-szeged.hu

Abstract. This paper presents a family of parallel thinning algorithms for extracting medial surfaces from 3D binary pictures. The proposed algorithms are based on sufficient conditions for 3D parallel reduction operators to preserve topology for $(26, 6)$ pictures. Hence it is self-evident that our algorithms are topology preserving. Their efficient implementation on conventional sequential computers is also presented.

1 Introduction

Thinning algorithms extract "skeletons" from binary pictures by using topology preserving reduction operations [8]. A 3D reduction operation does *not* preserve topology [7] if any object is split or is completely deleted, any cavity is merged with the background or another cavity, a new cavity is created, a hole (which doughnuts have) is eliminated or created.

A *3D binary picture* [8] is a mapping that assigns a value of 0 or 1 to each point with integer coordinates in the 3D digital space denoted by $\mathbb{Z}^3$. Points having the value of 1 are called *black* points, and those with a zero value are called *white* ones. A *simple* point is a black point whose deletion does not alter the topology of the picture [8].

Parallel thinning algorithms delete a set of black points. Topological correctness of parallel 3D thinning algorithms can be verified by the help of sufficient conditions for 3D parallel reduction operators to preserve topology [10,7,14].

Thinning algorithms use operators that delete some points which are not end-points, since preserving end-points provides important geometrical information relative to the shape of the objects. *Curve-thinning* algorithms are used to extract medial lines or centerlines, while *surface-thinning* ones produce medial surfaces. Curve-thinning preserves line-end points while surface-thinning does not delete surface end points [2,3,14].

This paper presents a family of 3D surface-thinning algorithms that are based on sufficient conditions for topology preservation proposed by Palágyi and Kuba [14]. The strategy which is used is called fully parallel: the same parallel reduction operator is applied at each iteration. Our algorithms use different types of surface-end points.

This paper is organized as follows. Section 2 gives the basic notions of 3D digital topology and the applied sufficient conditions for topology preservation.

S. Brlek, C. Reutenauer, and X. Provençal (Eds.): DGCI 2009, LNCS 5810, pp. 481–492, 2009.

Then in Section 3 we present three new fully parallel surface-thinning algorithms and their results on some test pictures. Finally, Section 4 proposes an efficient implementation of our algorithms, and we conclude in Section 5.

2 Basic Notions and Results

Let p be a point in the 3D digital space $\mathbb{Z}^3$. Let us denote $N_j(p)$ (for $j = 6, 18, 26$) the set of points that are *j-adjacent* to point p (see Fig. 1a).

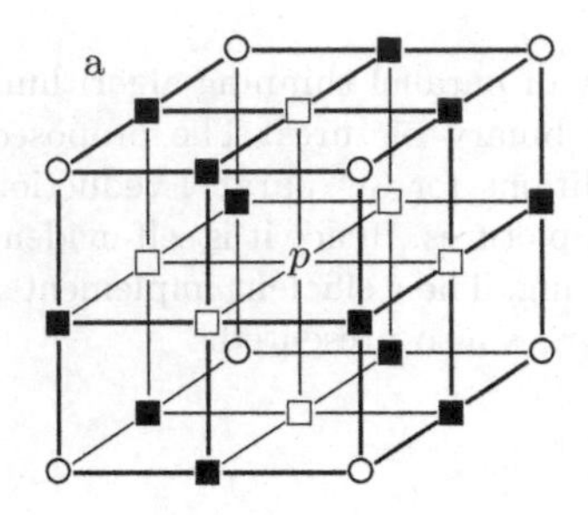

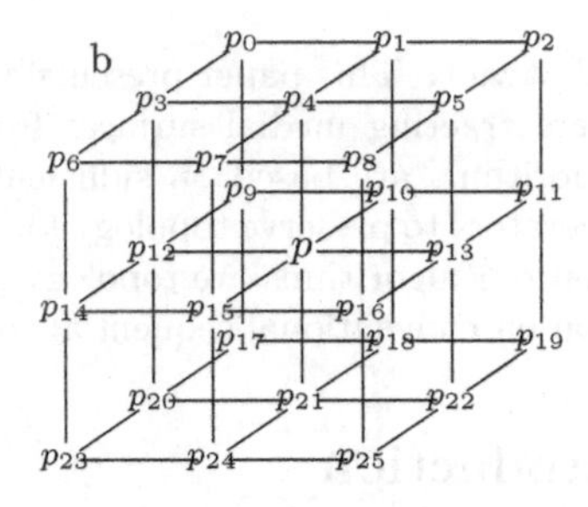

Fig. 1. Frequently used adjacencies in $\mathbb{Z}^3$ (a) and an indexing scheme to encode all possible $3 \times 3 \times 3$ configurations (b). The set $N_6(p)$ of the central point $p \in \mathbb{Z}^3$ contains point p and the 6 points marked "□". The set $N_{18}(p)$ contains $N_6(p)$ and the 12 points marked "■". The set $N_{26}(p)$ contains $N_{18}(p)$ and the 8 points marked "○".

The sequence of distinct points $\langle x_0, x_1, \ldots, x_n \rangle$ is called a *j-path* (for $j = 6, 18, 26$) of length n from point x_0 to point x_n in a non-empty set of points X if each point of the sequence is in X and x_i is j-adjacent to x_{i-1} for each $1 \leq i \leq n$ (see Fig. 1a). Note that a single point is a j-path of length 0. Two points are said to be *j-connected* in the set X if there is a j-path in X between them.

The *3D binary* (26, 6) *digital picture* $\mathcal{P}$ is a quadruple $\mathcal{P} = (\mathbb{Z}^3, 26, 6, B)$ [8]. Each element of $\mathbb{Z}^3$ is called a *point* of $\mathcal{P}$. Each point in $B \subseteq \mathbb{Z}^3$ is called a *black point* and has a value of 1 assigned to it. Each point in $\mathbb{Z}^3 \backslash B$ is called a *white point* and has a value of 0 assigned to it. Adjacency 26 is associated with the black points and adjacency 6 is assigned to the white points. A *black component* is a maximal 26-connected set of points in B, while a *white component* is a maximal 6-connected set of points in $\mathbb{Z}^3 \backslash B$. Here it is assumed that a picture contains finitely many black points.

A black point is called a *border point* in (26, 6) pictures if it is 6-adjacent to at least one white point. A black point is called an *interior point* if it is not a border point. A *simple* point is a black point whose deletion is a topology preserving reduction [8].

Parallel reduction operators delete a set of black points and not just a single simple point. Hence we need to consider what is meant by topology preservation when a number of black points are deleted simultaneously. The following theorem provides *sufficient conditions* for 3D parallel reduction operators to preserve topology.

Theorem 1. [14] *Let $\mathcal{O}$ be a parallel reduction operator. Let p be any black point in any picture $\mathcal{P} = (\mathbb{Z}^3, 26, 6, B)$ such that p is deleted by $\mathcal{O}$. Let $\mathcal{Q}$ be the family of all the sets of $Q \subseteq (N_{18}(p)\backslash\{p\}) \cap B$ such that $q_1 \in N_{18}(q_2)$, for any $q_1 \in Q$ and $q_2 \in Q$. The operator $\mathcal{O}$ is* topology preserving *if all of the following conditions hold:*

1. *p is simple in the picture $(\mathbb{Z}^3, 26, 6, B\backslash Q)$ for any Q in $\mathcal{Q}$.*
2. *No black component contained in a $2 \times 2 \times 2$ cube can be deleted completely by $\mathcal{O}$.*

Note that there is an alternative method for verifying the topological correctness of 3D parallel thinning algorithms. It is based on Ma's sufficient conditions for 3D parallel reduction operators to preserve topology for $(26, 6)$ pictures [10]. It is proved in [14] that if a parallel reduction operator satisfies the conditions of Theorem 1, then Ma's conditions are satisfied as well.

3 The New Thinning Algorithms

In this section, a family of thinning algorithms are presented for extracting medial surfaces from 3D $(26, 6)$ binary pictures. These algorithms use fully parallel strategy; the same parallel reduction operator is applied at each phase of the thinning process [5].

Deletion conditions of the proposed surface-thinning algorithms are based on the conditions of Theorem 1. Let us define their deletable points:

Definition 1. *Let us consider an arbitrary characterization of surface-end points that is called as type $\mathcal{T}$. A black point p in picture $(\mathbb{Z}^3, 26, 6, B)$ is $\mathcal{T}$-*deletable *if all of the following condition hold:*

1. *p is not a surface-end point of type $\mathcal{T}$.*
2. *p is a simple point.*
3. *p is simple in the picture $(\mathbb{Z}^3, 26, 6, B\backslash Q)$ for any $Q \subseteq (N_{18}(p)\backslash\{p\}) \cap B$, where set $Q \neq \emptyset$ and it contains mutually 18-adjacent points and all points in Q satisfy the first two conditions.*
4. *p does not come first in the lexicographic ordering of a black component $C \subseteq B$ which is contained in a $2 \times 2 \times 2$ cube and all points in C satisfy the first three conditions.*

Note that various characterizations of surface-end points yield different types of deletable points.

It can readily be seen that the simultaneous deletion of all $\mathcal{T}$-deletable points from any $(26, 6)$ picture is a topology preserving reduction (i.e., it satisfies both conditions of Theorem 1). Hence topology preservation by the following algorithm is guaranteed:

Fully parallel surface-thinning algorithm based on $\mathcal{T}$-deletable points

Input: picture $(\mathbb{Z}^3, 26, 6, B)$
Output: picture $(\mathbb{Z}^3, 26, 6, B')$

$B' = B$
repeat
$D = \{ p \mid p \text{ is } \mathcal{T}\text{-deletable in } (\mathbb{Z}^3, 26, 6, B') \}$
$B' = B' \setminus D$
until $D = \emptyset$

A surface-thinning algorithm does not delete surface-end points that are defined by the algorithm-specific characterization of end-points. There are numerous types of surface-end points that are used by the existing surface-thinning algorithms [1,2,3,4,6,9,11,13,14,18,20]. Hence we can get a family of fully parallel surface-thinning algorithms by applying various characterizations of surface-end points.

For simplicity, let us consider the following three types of surface-end points.

Definition 2. *A border point p in picture* $(\mathbb{Z}^3, 26, 6, B)$ *is a* surface-end point of type i *if the i-th condition holds:*

1. *There is no interior point in the set* $N_6(p) \cap B$.
2. *There is no interior point in the set* $N_{18}(p) \cap B$.
3. *There is no interior point in the set* $N_{26}(p) \cap B$.

Note that surface-end points of type 1 are considered by some existing 3D surface-thinning algorithms [1,15,16,17] and all the three types are used by the algorithms proposed by Manzanera et al. [13].

The proposed three algorithms are called as **Alg-**i that are determined by surface-end points of type i ($i = 1, 2, 3$). In experiments these fully parallel surface-thinning algorithms were tested on objects of different shapes. Here we present some illustrative examples below (Figs. 2-5).

Note that due to the considered types of surface end-points, our medial surfaces may contain 2-point thick surface patches [1,15,16,17]. Fortunately, it is not hard to overcome this problem here (e.g. by applying a final thinning step [1]). Note that the proposed three algorithms use symmetric deletion conditions, hence they are invariant by reflections and $k \cdot \pi/2$ rotations. We should mention that the skeleton (as a shape feature) is sensitive to coarse object boundaries. As a result, the "skeleton" produced generally includes unwanted segments that must be removed by a pruning step [19].

4 Implementation

One may think that the proposed algorithms are time consuming and it is rather difficult to implement them. That is why this section will present an efficient way for implementing algorithms **Alg-1**, **Alg-2**, and **Alg-3** on a conventional sequential computer.

Our method uses a pre-calculated look-up-table to encode the simple points in $(26, 6)$ pictures and an array to encode all possible sets that are considered in condition 3 of Definition 1. In addition, two lists are used to speed up the process: one for storing the border-points in the current picture; the other list is to store all deletable points in the current phase. At each iteration, the deletable points are deleted, and the list of border points is updated accordingly. The pseudocode of the proposed algorithms is given by the following:

```
procedure FULLY_PARALLEL_SURFACE_THINNING ( A, i )
  /* collect border points */
    border_list = < empty list >
    for each p = (x, y, z) in A do
      if p is border point then
        if p is surface-end point of type i then
          A[x, y, z] = 2
        else
          border_list = border_list + < p >
          A[x, y, z] = 3
  /* thinning process */
    repeat
      deleted = ITERATION_STEP ( border_list, A, i )
    until deleted = 0
```

The two parameters of the procedure are the array A (which stores the input picture to be thinned) and index i (that is to give the type of the considered surface-end points ($i = 1, 2, 3$)). In input array A, the value "1" corresponds to black points and the value "0" is assigned to white ones.

First, the input picture is scanned and all the border points that are not surface-end points of type i are inserted into the list *border_list*. Note that it is the only time consuming scanning of the entire array A. In order to avoid storing more than one copy of a border point in *border_list* and verifying the deletability of surface-end points again and again, the array A represents a five-color picture during the thinning process: the value of "0" corresponds to the white points, the value of "1" corresponds to (black) interior points, the value of "2" is assigned to all (black) surface-end points of type i, the value of "3" is assigned to all (black) border points in the actual picture that are not end-points (added to *border_list*), and the value of "4" corresponds to (black) points that satisfy the first two conditions of Definition 1.

The kernel of the **repeat** cycle corresponds to one iteration step of the process. Function ITERATION_STEP returns the number of deleted points by the actual iteration step. This number is then stored in the variable called *deleted*. The entire process terminates when no more points can be deleted. After the thinning, all points having a nonzero value belong to the medial surface. Note that array A contains the resultant "skeleton", hence the input and output pictures can be stored in the same array. The key part of the proposed method is the organization of function called ITERATION_STEP. Let us see its pseudocode:

```
function ITERATION_STEP ( border_list, A, i )
  /* collect i-deletable points */
    deleted = COLLECT_DELETABLE ( A, i, border_list, deletable_list )
  /* deletion */
    for each point p = (x, y, z) in deletable_list do
        A[x, y, z] = 0
        border_list = border_list − < p >
        for each point q = (x', y', z') being 6-adjacent to p do
            if A[x', y', z'] = 1 then
                border_list = border_list + < q >
    for each point p = (x, y, z) in border_list do
        if p is surface-end point of type i then
            A[x, y, z] = 2
            border_list = border_list − < p >
        else
            A[x, y, z] = 3
  return deleted
```

Function ITERATION_STEP uses an additional auxiliary data structure called *deletable_list*. Its processing task is composed of two basic parts. First, i-deletable points (see Definition 1) in the actual picture are transferred from *border_list* to *deletable_list*. Second, all the points in *deletable_list* are deleted from picture A and *border_list* is updated. If a border point is deleted, all interior points that are 6-adjacent to it become border points. These brand new border points of the actual picture are added to the *border_list*.

The pseudocode of the function COLLECT_DELETABLE is given by the following:

```
function COLLECT_DELETABLE ( A, i, border_list, deletable_list )
  deleted = 0
  deletable_list = < empty list >
  /* condition 2 */
    for each point p in border_list do
        index_26_nonzero = COLLECT_26_NONZERO ( p, A )
        if LUT_simple [index_26_nonzero] = 1  then
            A[x, y, z] = 4
            deletable_list = deletable_list + < p >
            deleted = deleted + 1
  /* condition 3 */
    for each point p in deletable_list do
        index_26_nonzero = COLLECT_26_NONZERO ( p, A )
        index_18_4 = COLLECT_18_4 ( p, A )
        for j = 1 to 66 do
            if ( index_18_4 AND Q[j] ) = Q[j]  then
                if LUT_simple [index_26_nonzero AND NOT Q[j]] = 0  then
                    deletable_list = deletable_list − < p >
                    deleted = deleted − 1
  return deleted
```

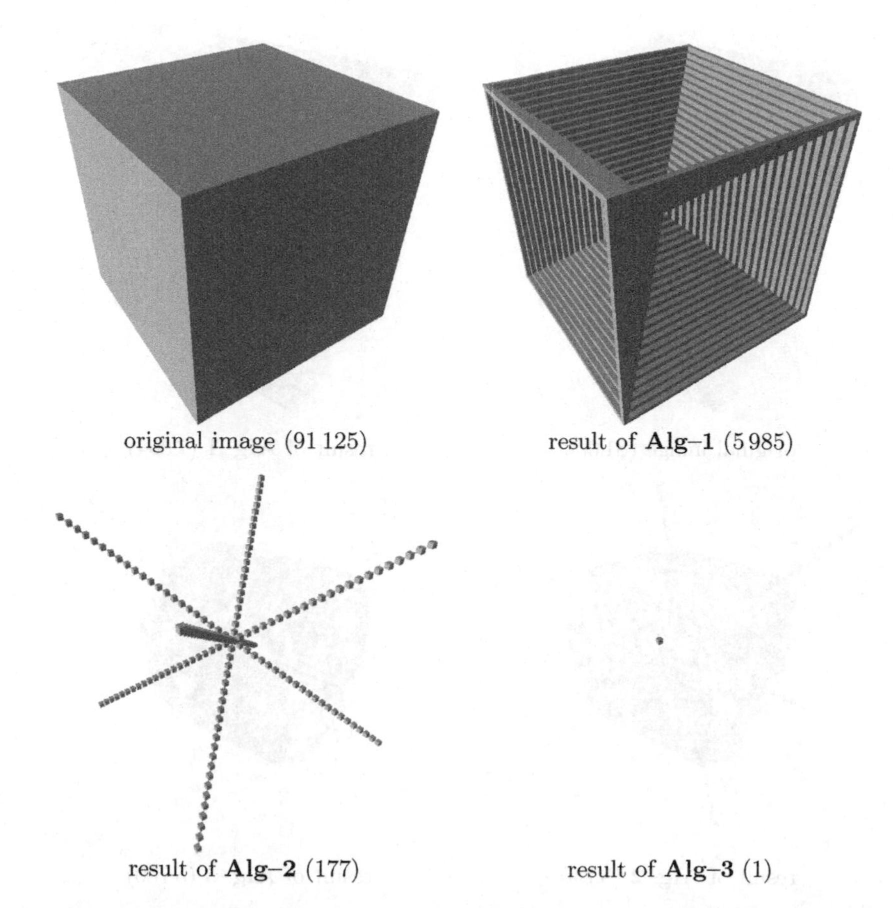

Fig. 2. The 3D image of a $45 \times 45 \times 45$ cube and its medial surfaces produced by our algorithms. Numbers in parentheses mean the count of black points.

The basic task of the function COLLECT_DELETABLE is comprised of two testing parts according to the conditions 2 and 3 of Definition 1. Since *border_list* contains all border points that are not surface-end point of type i in the actual picture, condition 1 of Definition 1 is satisfied by any points in *border_list*. Let us notice that the proposed algorithms **Alg-1**, **Alg-2**, and **Alg-3** can ignore condition 4 of Definition 1 (that is corresponds to condition 2 of Theorem 1), since a point that is 6-, 18-, or 26-adjacent to an interior point may not be an element of any black component contained in a $2 \times 2 \times 2$ cube.

This function uses two pre-calculated arrays as global variables. They are *LUT_simple* to encode the simple points and Q to encode all possible sets that are considered in condition 3 of Definition 1. We denote by "NOT" and "AND" the bitwise negation and the bitwise conjunction, respectively.

Simple points in (26, 6) pictures can be locally characterized; the simplicity of a point p can be decided by examining the set $N_{26}(p)$ [12]. There are 2^{26}

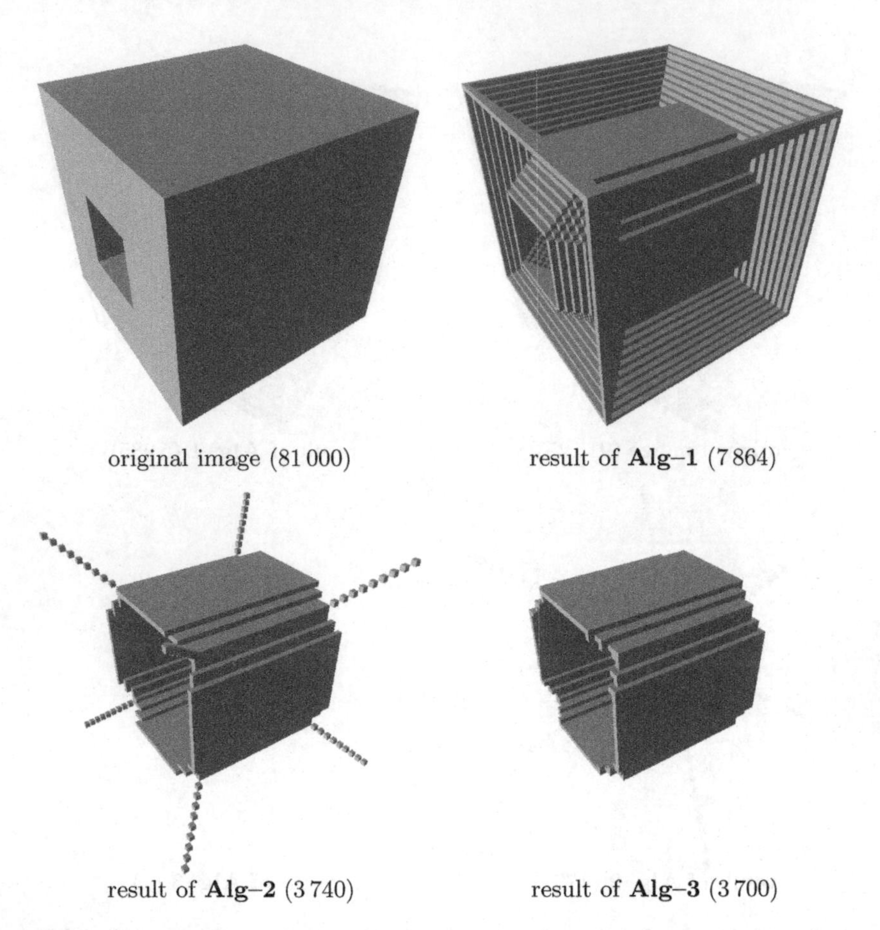

Fig. 3. The 3D image of a 45×45×45 cube with a hole and its medial surfaces produced by our algorithms. Numbers in parentheses mean the count of black points.

possible configurations in the $3 \times 3 \times 3$ neighborhood if the central point is not considered. Hence we can assign an index (i.e., a non-negative integer code) for each possible configuration and address a pre-calculated (unit time access) look-up-table *LUT_simple* having 2^{26} entries of 1 bit in size, therefore, it requires only 8 megabytes storage space in memory.

It is readily confirmed that there are 66 possible sets to be verified in condition 3 of Definition 1, since a set Q may contain only one, two, or three mutually 18-adjacent points, where any point in Q is 18-adjacent to the central point p. Note that we have to investigate at most 53 of these sets since p is a border point (i.e., it is 6-adjacent to at least one white point).

Function COLLECT_26_NONZERO finds the index of the given configuration (i.e., the $3 \times 3 \times 3$ neighborhood of the point p in question excluding p itself). Its result (i.e., *index_26_nonzero*) is calculated as

$$\sum_{k=0}^{25} 2^k \cdot p_k,$$

where $p_k = 1$ if the corresponding neighbor of point p (see Fig. 1b) is an object point in the actual image A, $p_k = 0$ otherwise ($k = 0, 1, \ldots, 25$). Function COLLECT_18_4 calculates the index *index_18_4* similarly: $p_k = 1$ if the corresponding neighbor is 18-adjacent to p and it has the value of "4" in the actual image A, $p_k = 0$ otherwise ($k = 0, 1, \ldots, 25$). Element $Q[j]$ of the pre-calculated array Q is encoded in a similar way: $p_k = 1$ if the corresponding neighbor of point p is in the j-th set of mutually 18-adjacent points, $p_k = 0$ otherwise ($j = 1, \ldots, 66$, $k = 0, 1, \ldots, 25$).

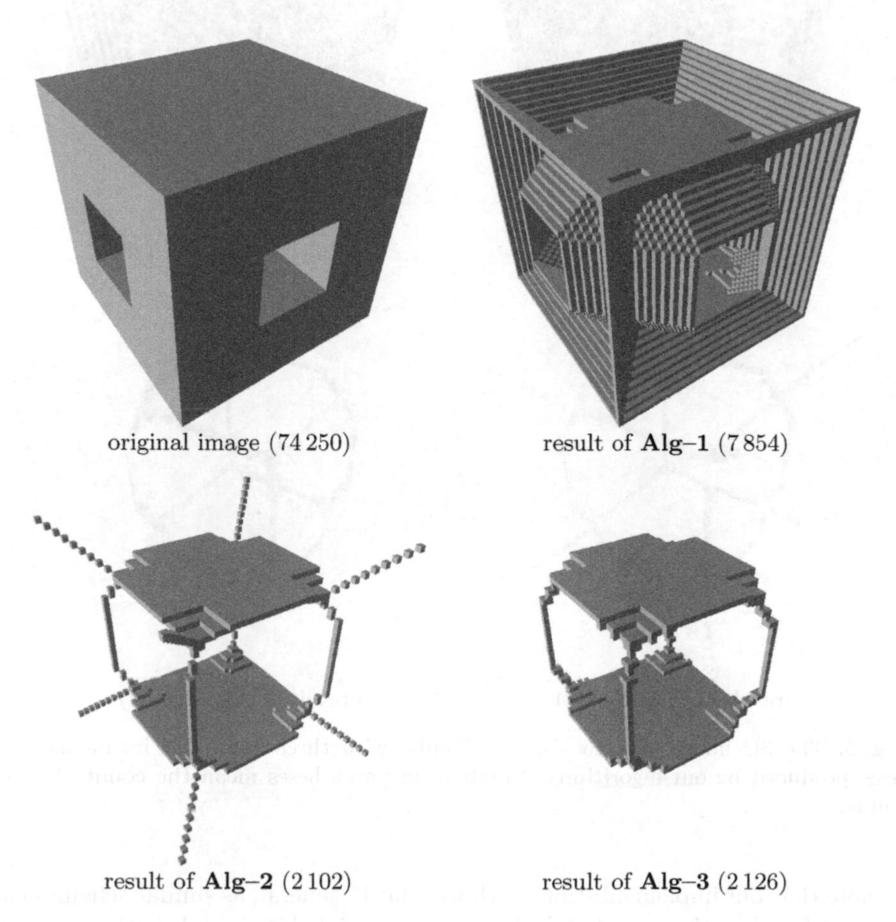

Fig. 4. The 3D image of a 45 × 45 × 45 cube with two holes and its medial surfaces produced by our algorithms. Numbers in parentheses mean the count of black points.

It is important that the support of all tests that are used is $3 \times 3 \times 3$. The proposed implementation method is fairly straightforward and computationally efficient (see Table 4). Here we find that the time complexity depends only on the number of object points and the compactness of the objects (i.e., volume to area ratio); but it does not depend on the size of the volume which contains the objects to be thinned.

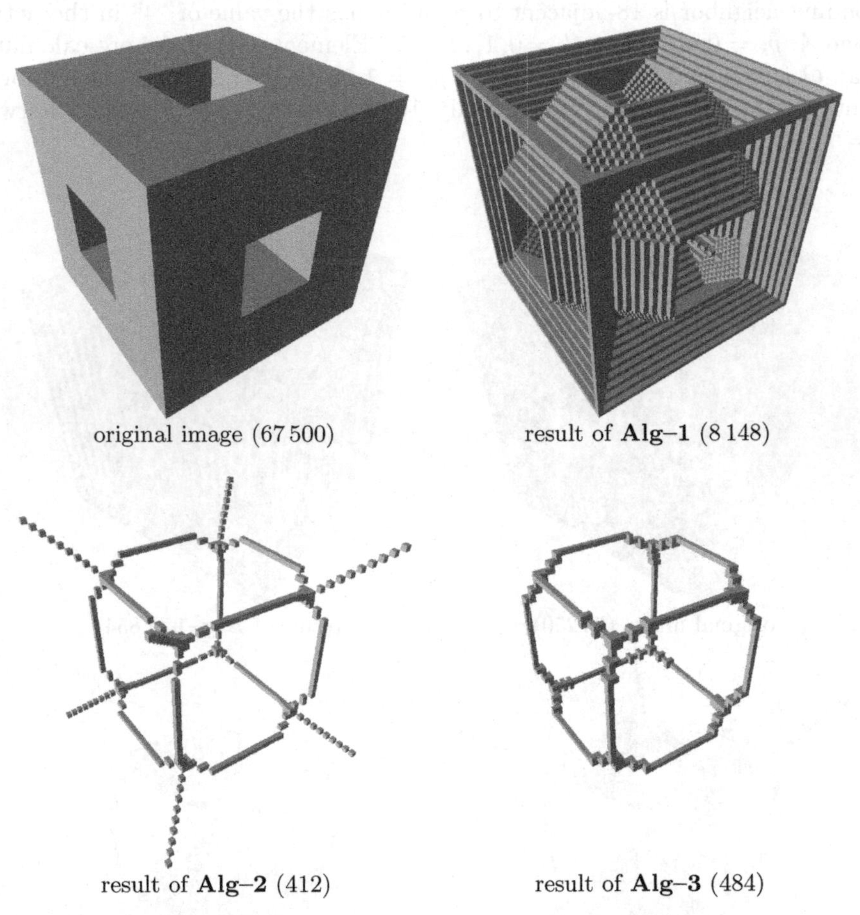

Fig. 5. The 3D image of a $45 \times 45 \times 45$ cube with three holes and its medial surfaces produced by our algorithms. Numbers in parentheses mean the count of black points.

Note that our implementation method is fairly general, as similar schemes can be used for the other parallel and some sequential thinning algorithms as well [17].

Table 1. Computation times for the considered four kinds of test objects. Our algorithms were run under Linux on an Intel Pentium 4 CPU 2.80 GHz PC. (Note, that just the thinning itself was considered here; reading the input volume, the 8 MB look-up-table, and writing the output image were not taken into account.)

Test object	Size	No. object points	No. iterations	No. skeletal points			Running time (sec.)		
				Alg–1	**Alg–2**	**Alg–3**	**Alg–1**	**Alg–2**	**Alg–3**
	45^3	91 125	23	5 985	177	1	0.06	0.06	0.05
	93^3	804 357	47	25 761	369	1	0.55	0.47	0.48
	141^3	2 803 221	71	59 361	561	1	2.03	1.71	1.69
	45^3	81 000	10	7 864	3 740	3 700	0.05	0.05	0.05
	93^3	714 984	21	33 384	15 104	15 024	0.46	0.43	0.44
	141^3	2 491 752	32	76 916	34 432	34 312	1.64	1.54	1.52
	45^3	74 250	10	7 854	2 102	2 126	0.05	0.04	0.04
	93^3	655 402	21	34 062	8 262	8 342	0.42	0.40	0.39
	141^3	2 284 106	32	78 718	18 542	18 678	1.56	1.38	1.40
	45^3	67 500	10	8 148	412	484	0.05	0.04	0.04
	93^3	595 820	21	35 660	780	980	0.40	0.36	0.36
	141^3	2 076 460	32	82 636	1 180	1 508	1.46	1.26	1.35

5 Conclusion and Future Work

In the present work, we proposed a new family of parallel 3D thinning algorithms. Our attention was focused on an efficient implementation of the proposed algorithms since fast skeleton extraction is extremely important for the large 3D shapes. Our future works concern subiteration and subfield based 3D thinning algorithms that are derived from sufficient conditions for parallel reduction operators to preserve the topology as well.

Acknowledgements

This research was partially supported by the TÁMOP-4.2.2/08/1/2008-0008 program of the Hungarian National Development Agency.

References

1. Arcelli, C., Sanniti di Baja, G., Serino, L.: New removal operators for surface skeletonization. In: Kuba, A., Nyúl, L.G., Palágyi, K. (eds.) DGCI 2006. LNCS, vol. 4245, pp. 555–566. Springer, Heidelberg (2006)
2. Bertrand, G., Aktouf, Z.: A 3D thinning algorithm using subfields. In: Proc. SPIE Conf. on Vision Geometry III, vol. 2356, pp. 113–124 (1994)

3. Bertrand, G.: A parallel thinning algorithm for medial surfaces. Pattern Recognition Letters 16, 979–986 (1995)
4. Gong, W.X., Bertrand, G.: A simple parallel 3D thinning algorithm. In: Proc. 10th IEEE Internat. Conf. on Pattern Recognition, ICPR 1990, pp. 188–190 (1990)
5. Hall, R.W.: Parallel connectivity-preserving thinning algorithms. In: Kong, T.Y., Rosenfeld, A. (eds.) Topological algorithms for digital image processing, pp. 145–179. Elsevier Science, Amsterdam (1996)
6. Jonker, P.: Skeletons in N dimensions using shape primitives. Pattern Recognition Letters 23, 677–686 (2002)
7. Kong, T.Y.: On topology preservation in 2-d and 3-d thinning. Int. Journal of Pattern Recognition and Artificial Intelligence 9, 813–844 (1995)
8. Kong, T.Y., Rosenfeld, A.: Digital topology: Introduction and survey. Computer Vision, Graphics, and Image Processing 48, 357–393 (1989)
9. Lee, T., Kashyap, R.L., Chu, C.: Building skeleton models via 3-D medial surface/axis thinning algorithms. CVGIP: Graphical Models and Image Processing 56, 462–478 (1994)
10. Ma, C.M.: On topology preservation in 3D thinning. CVGIP: Image Understanding 59, 328–339 (1994)
11. Ma, C.M., Wan, S.-Y.: A medial-surface oriented 3-d two-subfield thinning algorithm. Pattern Recognition Letters 22, 1439–1446 (2001)
12. Malandain, G., Bertrand, G.: Fast characterization of 3D simple points. In: Proc. 11th IEEE Internat. Conf. on Pattern Recognition, pp. 232–235 (1992)
13. Manzanera, A., Bernard, T.M., Pretêux, F., Longuet, B.: Medial faces from a concise 3D thinning algorithm. In: Proc. 7th IEEE Internat. Conf. Computer Vision, ICCV 1999, pp. 337–343 (1999)
14. Palágyi, K., Kuba, A.: A parallel 3D 12-subiteration thinning algorithm. Graphical Models and Image Processing 61, 199–221 (1999)
15. Palágyi, K.: A 3-Subiteration Surface-Thinning Algorithm. In: Kropatsch, W.G., Kampel, M., Hanbury, A. (eds.) CAIP 2007. LNCS, vol. 4673, pp. 628–635. Springer, Heidelberg (2007)
16. Palágyi, K.: A Subiteration-Based Surface-Thinning Algorithm with a Period of Three. In: Hamprecht, F.A., Schnörr, C., Jähne, B. (eds.) DAGM 2007. LNCS, vol. 4713, pp. 294–303. Springer, Heidelberg (2007)
17. Palágyi, K.: A 3D fully parallel surface-thinning algorithm. Theoretical Computer Science 406, 119–135 (2008)
18. Saha, P.K., Chaudhuri, B.B., Majumder, D.D.: A new shape-preserving parallel thinning algorithm for 3D digital images. Pattern Recognition 30, 1939–1955 (1997)
19. Shaked, D., Bruckstein, A.: Pruning medial axes. Computer Vision Image Understanding 69, 156–169 (1998)
20. Tsao, Y.F., Fu, K.S.: A parallel thinning algorithm for 3-D pictures. Computer Graphics and Image Processing 17, 315–331 (1981)

Quasi-Affine Transformation in Higher Dimension

Valentin Blot and David Coeurjolly

Université de Lyon, CNRS
Ecole Normale Supérieure de Lyon
Université Lyon 1, LIRIS, UMR5205, F-69622, France
valentin.blot@ens-lyon.fr, david.coeurjolly@liris.cnrs.fr

Abstract. In many applications and in many fields, algorithms can considerably be speed up if the underlying arithmetical computations are considered carefully. In this article, we present a theoretical analysis of discrete affine transformations in higher dimension. More precisely, we investigate the arithmetical paving structure induced by the transformation to design fast algorithms.

1 Introduction

In many computer vision and image processing applications, we are facing new constraints due to the image sizes both in dimension with 3-D and 3-D+t medical acquisition devices, and in resolution with VHR (Very High Resolution) satellite images. This article deals with high performance image transformation using quasi-affine transforms (QATs for short), which can be viewed as a discrete version of general affine transformations. QAT can approximate rotations and scalings, and in some specific cases, QAT may also be one-to-one and onto mappings from $\mathbb{Z}^n$ to $\mathbb{Z}^n$, leading to exact computations. A similar approach has been proposed in [1] with the notion of *Generalized Affine Transform.* The author demonstrated that any isometric transform (equivolume affine transform) in dimension n can be decomposed into $(n^2 - 1)$ fundamental skew transformations which parameters can be optimized to obtain a reversible transformation. Note that the bound has been improved to $(3n - 3)$ in [2]. This approach is a generalization of the decomposition of rotations in dimension 2 into three shears (see for example [3]). In this paper, we focus on QATs since they provide a wider class of transformation (contracting and dilating transformations are allowed). Furthermore, we investigate direct transformation algorithms without the need of intermediate computations.

In dimension 2, the QAT appeared in several articles [4,5,6,7,8]. To summarize the main results, the authors have proved several arithmetical results on QAT in 2-D leading to efficient transformation algorithms. More precisely, thanks to periodic properties of pavings induced by the reciprocal map, the image transformation can be obtained using a set of precomputed canonical pavings. In this paper, we focus on a theoretical analysis of n-dimensional QAT.

S. Brlek, C. Reutenauer, and X. Provençal (Eds.): DGCI 2009, LNCS 5810, pp. 493–504, 2009.

The idea is to investigate fundamental results in order to be able to design efficient transformation algorithms in dimension 2 or 3 as detailed in [9]. More precisely, we demonstrate the arithmetical and periodic structures embedded in $n-$dimensional QAT. Due to the space limitation, we only details the proofs of the main theorems (other proofs are available in the technical report [9]).

During the reviewing process of the paper, the article [10] was brought to our attention. Even if the framework differs, some results presented here can also be found in [10]. However, the closed formulas for the paving periodicity presented in Section 3.2 make our contribution complementary.

In Section 2, we first detail preliminary notations and properties. Then, Section 3 contains the main theoretical results leading to a generic n-D transformation algorithm sketched in Section 4.

2 Preliminaries

2.1 Notations

Before we introduce arithmetical properties of QAT in higher dimension, we first detail the notations considered in this paper. Let n denote the dimension of the considered space, V_i denote the ith coordinate of vector V, and $M_{i,j}$ denote the $(i,j)^{th}$ coefficient of matrix M. We use the notation $\gcd(a,b,\ldots)$ for the greatest common divisor of an arbitrary number of arguments, and $\operatorname{lcm}(a,b,\ldots)$ for their least common multiple.

Let $\left[\frac{a}{b}\right]$ denote the quotient of the euclidean division of a by b, that is the integer $q \in \mathbb{Z}$ such that $a = bq + r$ satisfying $0 \leq r < |b|$ regardless of the sign of b[1]. We consider the following generalization to $n-$dimensional vectors:

$$\left[\frac{V}{b}\right] = \begin{pmatrix} \left[\frac{V_0}{b}\right] \\ \vdots \\ \left[\frac{V_{n-1}}{b}\right] \end{pmatrix} \text{ and } \left\{\frac{V}{b}\right\} = \begin{pmatrix} \left\{\frac{V_0}{b}\right\} \\ \vdots \\ \left\{\frac{V_{n-1}}{b}\right\} \end{pmatrix} . \quad (1)$$

2.2 Quasi-Affine Transformation Definitions

Defined in dimension 2 in [4,5,6,7,8], we consider a straightforward generalization to $\mathbb{Z}^n$ spaces.

Definition 1 (QAT). *A quasi-affine transformation is a triple* $(\omega, M, V) \in \mathbb{Z} \times M_n(\mathbb{Z}) \times \mathbb{Z}^n$ *(we assume that* $\det(M) \neq 0$*). The associated application is:*

$$\begin{aligned} \mathbb{Z}^n &\longrightarrow \mathbb{Z}^n \\ X &\longmapsto \left[\frac{MX+V}{\omega}\right] . \end{aligned}$$

[1] $\left\{\frac{a}{b}\right\}$ denotes the corresponding remainder $\left\{\frac{a}{b}\right\} = a - b\left[\frac{a}{b}\right]$.

And the associated affine application is:

$$\begin{aligned} \mathbb{R}^n &\longrightarrow \mathbb{R}^n \\ X &\longmapsto \frac{MX+V}{\omega}\,. \end{aligned}$$

In other words, a QAT is the composition of the associated affine application and the integer part floor function.

Definition 2. *A QAT is said to be contracting if $\omega^n > |\det(M)|$, otherwise it is said to be dilating.*

In other words, a QAT is contracting if and only if the associated affine application is contracting. Note that if $\omega^n = |\det(M)|$, the QAT is dilating, even if the associated affine application is an isometry.

Definition 3. *The inverse of a QAT (ω, M, V) is the QAT:*

$$(\det(M), \omega\,\mathrm{com}(M)^t, -\,\mathrm{com}(M)^t V)\,, \tag{2}$$

where ^{t} denotes the transposed matrix and com(M) *the co-factor matrix of M*[2]*.*

The associated affine application of the inverse of a QAT is therefore the inverse of the affine application associated to the QAT. However, due to the nested floor function, the composition $f \cdot f^{-1}$ is not the identity function in the general case.

3 QAT Properties in Higher Dimensions

3.1 Pavings of a QAT

Without loss of generality, we suppose that the QAT is contracting. A key feature of a QAT in dimension 2 is the paving induced by the reciprocal map of a discrete point. In the following, we adapt the definitions in higher dimensions and prove that a QAT in $\mathbb{Z}^n$ also carries a periodic paving.

Definition 4 (Paving). *Let f be a QAT. For $Y \in \mathbb{Z}^n$, we denote:*

$$P_Y = \{X \in \mathbb{Z}^n \mid f(X) = Y\} = f^{-1}(Y)\,, \tag{3}$$

P_Y is called order 1 paving of index Y of f.

P_Y can be interpreted as a subset of $\mathbb{Z}^n$ (maybe empty) that corresponds to the reciprocal map of Y by f. We easily show that the set of pavings of a QAT forms a paving of the considered space (see Fig. 1). In dimension 2, this definition exactly coincides with previous ones [4,6,7,8,5].

---

[2] Remind that $M\,\mathrm{com}(M)^t = \mathrm{com}(M)^t M = \det(M) I_n$.

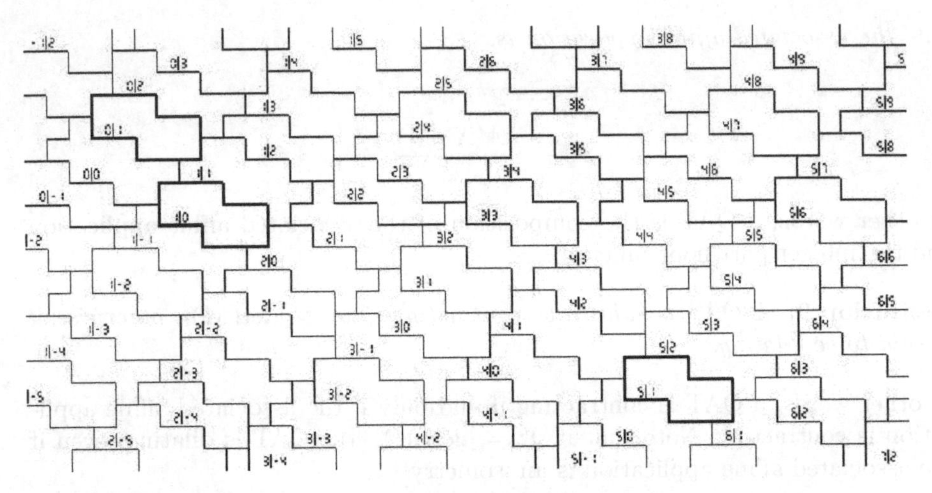

Fig. 1. Pavings of the QAT $\left(84, \begin{pmatrix} 12 & -11 \\ 18 & 36 \end{pmatrix}, \begin{pmatrix} 150 \\ -500 \end{pmatrix}\right)$ with their indexes (in 2D, these are the couples: $Y = (x, y)$)

Definition 5. *P_Y is said arithmetically equivalent to P_Z (denoted $P_Y \equiv P_Z$) if:*

$$\forall X \in P_Y, \exists X' \in P_Z, \left\{\frac{MX + V}{\omega}\right\} = \left\{\frac{MX' + V}{\omega}\right\} . \tag{4}$$

Again, this definition is equivalent (as shown below) to those given in the literature.

Theorem 1. *The equivalence relationship is symmetric,* i.e.*:*

$$P_Y \equiv P_Z \Leftrightarrow P_Z \equiv P_Y . \tag{5}$$

Proof. The proof is given [9].

Figure 1 illustrates arithmetically equivalent pavings: the pavings of index $(0, 1)$ and $(5, 1)$ are arithmetically equivalent (see Table 1).

Definition 6. *P_Y and P_Z are said geometrically equivalent if:*

$$\exists \boldsymbol{v} \in \mathbb{Z}^n, P_Y = T_{\boldsymbol{v}} P_Z , \tag{6}$$

where $T_{\boldsymbol{v}}$ denotes the translation of vector $\boldsymbol{v}$.

In Figure 1, the pavings of indexes $(0, 1)$ and $(1, 0)$ are geometrically equivalent. In image processing purposes, when we want to transform a $n-$dimensional image by a QAT, geometrically equivalent pavings will allow us to design fast transformation algorithms.

Table 1. Pavings of index $(0,1)$, $(5,1)$ and $(1,0)$ of the QAT $\left(84, \begin{pmatrix} 12 & -11 \\ 18 & 36 \end{pmatrix}, \begin{pmatrix} 150 \\ -500 \end{pmatrix}\right)$

$P_{0,1}$		$P_{5,1}$		$P_{1,0}$	
X	$\left\{\frac{MX+V}{\omega}\right\}$	X	$\left\{\frac{MX+V}{\omega}\right\}$	X	$\left\{\frac{MX+V}{\omega}\right\}$
$\begin{pmatrix} 3 \\ 15 \end{pmatrix}$	$\begin{pmatrix} 21 \\ 10 \end{pmatrix}$	$\begin{pmatrix} 27 \\ 3 \end{pmatrix}$	$\begin{pmatrix} 21 \\ 10 \end{pmatrix}$	$\begin{pmatrix} 6 \\ 11 \end{pmatrix}$	$\begin{pmatrix} 17 \\ 4 \end{pmatrix}$
$\begin{pmatrix} 5 \\ 14 \end{pmatrix}$	$\begin{pmatrix} 56 \\ 10 \end{pmatrix}$	$\begin{pmatrix} 29 \\ 2 \end{pmatrix}$	$\begin{pmatrix} 56 \\ 10 \end{pmatrix}$	$\begin{pmatrix} 8 \\ 10 \end{pmatrix}$	$\begin{pmatrix} 52 \\ 4 \end{pmatrix}$
$\begin{pmatrix} 4 \\ 15 \end{pmatrix}$	$\begin{pmatrix} 33 \\ 28 \end{pmatrix}$	$\begin{pmatrix} 28 \\ 3 \end{pmatrix}$	$\begin{pmatrix} 33 \\ 28 \end{pmatrix}$	$\begin{pmatrix} 7 \\ 11 \end{pmatrix}$	$\begin{pmatrix} 29 \\ 22 \end{pmatrix}$
$\begin{pmatrix} 6 \\ 14 \end{pmatrix}$	$\begin{pmatrix} 68 \\ 28 \end{pmatrix}$	$\begin{pmatrix} 30 \\ 2 \end{pmatrix}$	$\begin{pmatrix} 68 \\ 28 \end{pmatrix}$	$\begin{pmatrix} 9 \\ 10 \end{pmatrix}$	$\begin{pmatrix} 64 \\ 22 \end{pmatrix}$
$\begin{pmatrix} 3 \\ 16 \end{pmatrix}$	$\begin{pmatrix} 10 \\ 46 \end{pmatrix}$	$\begin{pmatrix} 27 \\ 4 \end{pmatrix}$	$\begin{pmatrix} 10 \\ 46 \end{pmatrix}$	$\begin{pmatrix} 6 \\ 12 \end{pmatrix}$	$\begin{pmatrix} 6 \\ 40 \end{pmatrix}$
$\begin{pmatrix} 5 \\ 15 \end{pmatrix}$	$\begin{pmatrix} 45 \\ 46 \end{pmatrix}$	$\begin{pmatrix} 29 \\ 3 \end{pmatrix}$	$\begin{pmatrix} 45 \\ 46 \end{pmatrix}$	$\begin{pmatrix} 8 \\ 11 \end{pmatrix}$	$\begin{pmatrix} 41 \\ 40 \end{pmatrix}$
$\begin{pmatrix} 7 \\ 14 \end{pmatrix}$	$\begin{pmatrix} 80 \\ 46 \end{pmatrix}$	$\begin{pmatrix} 31 \\ 2 \end{pmatrix}$	$\begin{pmatrix} 80 \\ 46 \end{pmatrix}$	$\begin{pmatrix} 10 \\ 10 \end{pmatrix}$	$\begin{pmatrix} 76 \\ 40 \end{pmatrix}$
$\begin{pmatrix} 4 \\ 16 \end{pmatrix}$	$\begin{pmatrix} 22 \\ 64 \end{pmatrix}$	$\begin{pmatrix} 28 \\ 4 \end{pmatrix}$	$\begin{pmatrix} 22 \\ 64 \end{pmatrix}$	$\begin{pmatrix} 7 \\ 12 \end{pmatrix}$	$\begin{pmatrix} 18 \\ 58 \end{pmatrix}$
$\begin{pmatrix} 6 \\ 15 \end{pmatrix}$	$\begin{pmatrix} 57 \\ 64 \end{pmatrix}$	$\begin{pmatrix} 30 \\ 3 \end{pmatrix}$	$\begin{pmatrix} 57 \\ 64 \end{pmatrix}$	$\begin{pmatrix} 9 \\ 11 \end{pmatrix}$	$\begin{pmatrix} 53 \\ 58 \end{pmatrix}$
$\begin{pmatrix} 5 \\ 16 \end{pmatrix}$	$\begin{pmatrix} 34 \\ 82 \end{pmatrix}$	$\begin{pmatrix} 29 \\ 4 \end{pmatrix}$	$\begin{pmatrix} 34 \\ 82 \end{pmatrix}$	$\begin{pmatrix} 8 \\ 12 \end{pmatrix}$	$\begin{pmatrix} 30 \\ 76 \end{pmatrix}$
$\begin{pmatrix} 7 \\ 15 \end{pmatrix}$	$\begin{pmatrix} 69 \\ 82 \end{pmatrix}$	$\begin{pmatrix} 31 \\ 3 \end{pmatrix}$	$\begin{pmatrix} 69 \\ 82 \end{pmatrix}$	$\begin{pmatrix} 10 \\ 11 \end{pmatrix}$	$\begin{pmatrix} 65 \\ 76 \end{pmatrix}$

Theorem 2. *If $P_Y \equiv P_Z$, then P_Y and P_Z are geometrically equivalent. Since $P_Y \equiv P_Z$, there exists $X \in P_Y$ and $X' \in P_Z$ such that:*

$$\left\{\frac{MX+V}{\omega}\right\} = \left\{\frac{MX'+V}{\omega}\right\}.$$

Then $\boldsymbol{v} = X - X'$ is the translation vector:

$$P_Y = T_{\boldsymbol{v}} P_Z\,.$$

In dimension 2, this theorem is also proved in [7].

Proof. The proof is given in [9].

In a computational point of view, if a paving P_Y has been already computed, and if we know that $P_Y \equiv P_Z$, then P_Z can be obtained by translation of P_Y. In Figure 1, the pavings of index $(0,1)$ and $(5,1)$ are arithmetically equivalent (see Table 1), therefore they are geometrically equivalent (as we can check on the figure). Note that the inverse implication is false: in Figure 1, the pavings of index $(0,1)$ and $(1,0)$ are geometrically equivalent but they are not arithmetically equivalent (see Table 1).

3.2 Paving Periodicity

Definition 7. *$\forall 0 \leq i < n$, We define the set $\mathcal{A}_i$ as follows:*

$$\mathcal{A}_i = \{\alpha \in \mathbb{N}^* \mid \exists(\beta_j)_{0\leq j<i} \in \mathbb{Z}^i, \forall (y_0,\dots,y_{n-1}) \in \mathbb{Z}^n, P_{y_0,\dots,y_i+\alpha,\dots,y_{n-1}} \equiv P_{y_0+\beta_0,\dots,y_{i-1}+\beta_{i-1},y_i,\dots,y_{n-1}}\}$$

Theorem 3 (Perdiodicity). *The set of QAT pavings is $n-$periodic, in other words*

$$\forall 0 \leq i < n, \mathcal{A}_i \neq \emptyset$$

Proof. The proof is given in Sect. A.1.

If we consider $\alpha = |det(M)|$ as in Sect. A.1, we have demonstrated the periodic structure of QAT pavings since $P_Y \equiv P_{Y+\alpha e_i}$ for each i. We investigate now the quantities α_i which are minimal for each dimension i,

Definition 8. *$\forall 0 \leq i < n$, let us consider $\alpha_i = \min(\mathcal{A}_i)$. We define $\{\beta^i_j\}_{0\leq j<i} \in \mathbb{Z}^i$ and $U_i \in \mathbb{Z}^n$ such that*

$$\forall (y_0,\dots,y_{n-1}) \in \mathbb{Z}^n, P_{y_0,\dots,y_i+\alpha_i,\dots,y_{n-1}} = T_{U_i} P_{y_0+\beta^i_0,\dots,y_{i-1}+\beta^i_{i-1},y_i,\dots,y_{n-1}}.$$

Thanks to Theorem 2 and using notations of Def. 8, let $X \in P_{y_0,\dots,y_i+\alpha_i,\dots,y_{n-1}}$ and $X' \in P_{y_0+\beta^i_0,\dots,y_{i-1}+\beta^i_{i-1},y_i,\dots,y_{n-1}}$, such that $\left\{\frac{MX+V}{\omega}\right\} = \left\{\frac{MX'+V}{\omega}\right\}$. Then, we have $U_i = X - X'$.

The quantities α_i, β^i_j and U_i can be computed in dimension 2 and 3 (see [9]) by using greatest common divisors and Euclide's algorithm. The computation of α_2 in 3D already involves a consequent number of intermediate variables, that is why these computations seem to be hard to generalize in arbitrary dimension. Therefore, we will suppose here that quantities α_i, β^i_j and U_i are given.

To paraphrase above results, α_i and its associated U_i and $\{\beta^i_j\}$ allows us to *reduce* the ith component of Y while preserving the geometrically equivalence relationship. If we repeat this *reduction* process to each component from $n-1$ down-to 0, we construct a point Y^0 such that P_Y and P_{Y^0} are geometrically equivalent. The following theorem formalizes this principle and defines the initial period paving P_{Y^0}.

Theorem 4. *$\forall (y_0,\dots,y_{n-1}) \in \mathbb{Z}^n$, we have $P_{y_0,\dots,y_{n-1}} = T_W P_{y^0_0,\dots,y^0_{n-1}}$ with*

$$W = \sum_{i=0}^{n-1} w_i U_i$$

$$\text{and } \forall n > i \geq 0, \begin{cases} w_i = \left\lceil \frac{y_i + \sum_{j=i+1}^{n-1} w_j \beta^j_i}{\alpha_i} \right\rceil \\ y^0_i = \left\{ \frac{y_i + \sum_{j=i+1}^{n-1} w_j \beta^j_i}{\alpha_i} \right\} \end{cases}.$$

Proof. The proof is given in Sect. A.2.

Therefore, if we already computed the pavings $P_{y^0_0,\dots,y^0_{n-1}}$ for $0 \leq y^0_i < \alpha_i$, we can obtain any paving by translation of one of these pavings.

3.3 Super-Paving of a QAT

We now describe how to compute these initial period pavings based on the notion of super-paving (see Fig. 2).

Definition 9. *A super-paving of a QAT is the set $\mathcal{P}$ such that*

$$\mathcal{P} = \bigcup_{0 \leq Y^0 < (\alpha_0, \ldots, \alpha_{n-1})} P_{Y^0}$$

In other words, the super-paving is the union of all pavings of the initial period. In dimension 2, this definition coincides with definitions given in [7,6,8].

Theorem 5. *$\mathcal{P}$ is the paving $P_{(0,\ldots,0)}$ of the QAT defined by:*

$$\left(\omega \operatorname{lcm}_{0 \leq i < n}(\alpha_i), \begin{pmatrix} \theta_0 & \cdots & 0 \\ \vdots & \ddots & \vdots \\ 0 & \cdots & \theta_{n-1} \end{pmatrix} M, \begin{pmatrix} \theta_0 & \cdots & 0 \\ \vdots & \ddots & \vdots \\ 0 & \cdots & \theta_{n-1} \end{pmatrix} V \right), \tag{7}$$

with $\forall 0 \leq i < n$,

$$\theta_i = \frac{\operatorname{lcm}_{0 \leq j < n}(\alpha_j)}{\alpha_i}.$$

Proof. The proof is given in [9].

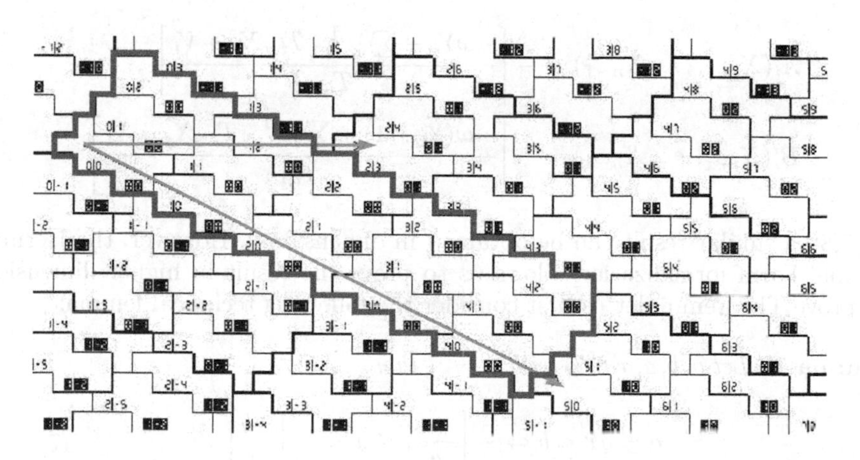

Fig. 2. Super-paving decomposition of the QAT defined in Fig. 1. Arrows illustrate a basis of the periodic structure, numbers on white background are the y_i, and numbers on black background are the w_i.

Hence, we can associate a canonical paving to each point of the super-paving. More precisely, the super-paving allows us to compute the equivalence classes for the arithmetical equivalence relationship between two pavings.

3.4 Paving Construction

In this section, we focus on an arithmetic paving construction algorithm. Hence, using the results of the previous section, such a construction algorithm will be used to compute canonical pavings in the super-paving.

Definition 10. *The matrix T is the Hermite Normal Form of the QAT matrix M if:*

- *T is upper triangular, with coefficients $\{T_{ij}\}$ such that $T_{ii} > 0$;*
- *$\exists H \in GL_n(\mathbb{Z}), MH = T$.*

If M is nonsingular integer matrix, the Hermite Normal Form exists. Note also that if $H \in M_n(\mathbb{Z})$, then $H \in GL_n(\mathbb{Z}) \Leftrightarrow |\det(H)| = 1$.

For example, given $\begin{pmatrix} 12 & -11 \\ 18 & 36 \end{pmatrix}$, we have: $\begin{pmatrix} 12 & -11 \\ 18 & 36 \end{pmatrix} \begin{pmatrix} 2 & 1 \\ -1 & 0 \end{pmatrix} = \begin{pmatrix} 35 & 12 \\ 0 & 18 \end{pmatrix}$.

Using the Hermite Normal Form, we can design a fast paving computation algorithm formalized in the following theorem:

Theorem 6. *$\forall Y \in \mathbb{Z}^n$, let $MH = T$ be the Hermite Normal Form of the QAT matrix M, then*

$$P_Y = \{HX \mid \forall n > i \geq 0, A_i(X_{i+1}, \ldots, X_{n-1}) \leq X_i < B_i(X_{i+1}, \ldots, X_{n-1})\} \tag{8}$$

With

$$A_i(X_{i+1}, \ldots, X_{n-1}) = -\left\lceil \frac{-\omega Y_i + \sum_{j=i+1}^{n-1} T_{i,j} X_j + V_i}{T_{i,i}} \right\rceil,$$

$$B_i(X_{i+1}, \ldots, X_{n-1}) = -\left\lceil \frac{-\omega (Y_i + 1) + \sum_{j=i+1}^{n-1} T_{i,j} X_j + V_i}{T_{i,i}} \right\rceil.$$

In [7,8], a similar result can be obtained in dimension 2. However, the Hermite Normal Form formalization allows us to prove the result in higher dimension. To prove Theorem 6, let us first consider the following technical lemma:

Lemma 1. *Let $a, b, q, x \in \mathbb{Z}$ with $q > 0$, then*

$$a \leq qx < b \Leftrightarrow -\left\lceil \frac{-a}{q} \right\rceil \leq x < -\left\lceil \frac{-b}{q} \right\rceil.$$

Proof. The proof is detailed in [9].

We can now prove the Theorem 6 (cf Sect. A.3). The implementation of the construction algorithm is straightforward: we just have to consider n nested loop such that the loop with level i goes from A_i to B_i quantities. See [9] for details in dimension 2 and 3.

Algorithm 1. Generic QAT algorithm for a contracting QAT

```
Input: a contracting QAT f := (ω, M, V), an image A : Z^n → Z
Output: a transformed image B : Z^n → Z
Compute the Hermite Normal Form of the matrix M;
Determine the minimal periodicities {α_i} and vectors {U_i};
Use Theorems 5 and 6 to compute the canonical pavings in the super-paving P;
foreach Y ∈ B do
    Find Y^0 and W such that P_Y = T_W P_{Y^0} by using Theorem 4;
    sum ← 0;
    foreach Z ∈ P_{Y^0} do
        c ← A(T_W Z);              // we read the color in the initial image
        sum ← sum + c;
    B(Y) ← sum/|P_{Y^0}|;                                  // we set the color
```

4 A Generic QAT Algorithm

In Algorithm 1, we give the generic algorithm applying a contracting QAT f to an image $\mathcal{A}$ (see Fig. 3). The principle is that we give to each pixel Y of image $\mathcal{B}$ the average color of the paving P_Y in image $\mathcal{A}$. If f is a dilating QAT, we obtain the very similar Algorithm 2 which principle is that firstly we replace f with f^{-1}, and then we give the color of each pixel Y of image $\mathcal{A}$ to each pixel of P_Y in image $\mathcal{B}$. In both algorithms, some elements cannot be computed in

Algorithm 2. Generic QAT algorithm for a dilating QAT

```
Input: a dilating QAT f := (ω, M, V), an image A : Z^n → Z
Output: a transformed image B : Z^n → Z
Replace f with f^{-1};
Compute the Hermite Normal Form of the matrix M;
Determine the minimal periodicities {α_i} and vectors {U_i};
Use Theorems 5 and 6 to compute the canonical pavings in the super-paving P;
foreach Y ∈ A do
    Find Y^0 and W such that P_Y = T_W P_{Y^0} by using Theorem 4;
    c ← A(Y);                      // we read the color in the initial image
    foreach Z ∈ P_{Y^0} do
        B(T_W Z) ← c;                                      // we set the color
```

arbitrary dimension n. Indeed, even if there exist algorithms to compute the Hermite Normal Form of an arbitrary square integer matrix [11], there is no generic algorithm to obtain the minimal periodicities $\{\alpha_i\}$.

In [9], we detail the computation of the minimal periodicities in dimension 2 and 3. We also demonstrate with a complete experimental analysis that algorithms 1 and 2 outperform classical techniques to transform an image by affine functions.

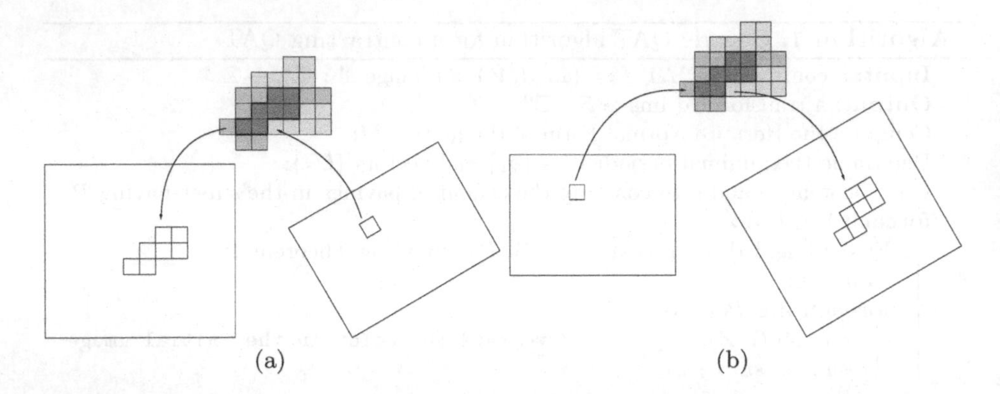

Fig. 3. Illustration in dimension 2 of the QAT algorithm when f is contracting (a) and dilating (b). In both cases, we use the canonical pavings contained in the super-paving to speed-up the transformation.

5 Conclusion and Future Works

In this paper, we have demonstrated that in higher dimension, Quasi-Affine Transformations contain arithmetical properties leading to the fact that the induced pavings are $n-$periodic. Furthermore, thanks to the Hermite Normal Form of the QAT matrix, we have presented efficient algorithms to construct a given paving and to compute a set of canonical pavings. From all these theoretical results, fast transformation algorithms have been designed in [9]. However, several future works exist. First, as detailed in Sections 3.1 and 3.3, the super-paving of a QAT contains a set of arithmetically distinct pavings. However, two arithmetical distinct pavings may be geometrically equivalent. Hence, a subset of the super-paving may be enough to design a fast algorithm. In dimension 2, in [4,6,7,8], the authors have investigated another structure, so-called generative strip, which removes some arithmetical distinct pavings whose geometry are identical. Even if the generalization in higher dimension of this object is not trivial, it may be interesting to investigate theoretical techniques to reduce the canonical paving set. Finally, a generic algorithm to compute the minimal periodicities is challenging.

References

1. Shizawa, M.: Discrete invertible affine transformations. In: 10th International Conference on Pattern Recognition, Atlantic City, vol. II, pp. 134–139. IEEE, Los Alamitos (1990)
2. Condat, L., Van De Ville, D., Forster-Heinlein, B.: Reversible, fast, and high-quality grid conversions. IEEE Transactions on Image Processing 17(5), 679–693 (2008)
3. Andres, E.: The quasi-shear rotation. In: Miguet, S., Ubéda, S., Montanvert, A. (eds.) DGCI 1996. LNCS, vol. 1176, pp. 307–314. Springer, Heidelberg (1996)
4. Nehlig, P., Ghazanfarpour, D.: Affine Texture Mapping and Antialiasing Using Integer Arithmetic. Computer Graphics Forum 11(3), 227–236 (1992)

5. Jacob, M.: Transformation of digital images by discrete affine applications. Computers & Graphics 19(3), 373–389 (1995)
6. Nehlig, P.: Applications quasi affines: pavages par images réciproques. Theoretical Computer Science 156(1-2), 1–38 (1996)
7. Jacob, M.: Applications quasi-affines. PhD thesis, Université Louis Pasteur, Strasbourg, France (1993)
8. qJacob-Da Col, M.: Applications quasi-affines et pavages du plan discret. Theoretical Computer Science 259(1-2), 245–269 (2001), english version, http://dpt-info.u-strasbg.fr/~jacob/articles/paving.pdf
9. Blot, V., Coeurjolly, D.: Quasi-affine transform in higher dimension. Technical Report RR–LIRIS-2009-010, Laboratoire LIRIS (April 2009), http://liris.cnrs.fr/publis?id=3853
10. Col, M.A.J.D., Tellier, P.: Quasi-linear transformations and discrete tilings. Theor. Comput. Sci 410(21-23), 2126–2134 (2009)
11. Storjohann, A., Labahn, G.: Asymptotically fast computation of hermite normal forms of integer matrices. In: ISSAC 1996: Proceedings of the 1996 international symposium on Symbolic and algebraic computation, pp. 259–266. ACM, New York (1996)

A Appendix: Proofs

A.1 Theorem 3

Proof. Given $0 \leq i < n$, let us suppose that $\forall 0 \leq j < i, \beta_j = 0$ and $\alpha = |\det(M)|$. Let $Y \in \mathbb{Z}^n$, $X \in P_Y$, and

$$X' = X + \frac{\det(M)}{|\det(M)|}\omega\,\mathrm{com}(M)^t e_i$$

with e_i being the $i-$th vector of the canonical basis of $\mathbb{R}^n$. We prove that $\mathcal{A}_i \neq \emptyset$ since $P_Y \equiv P_{Y+\alpha e_i}$:

$$\begin{aligned}
MX' + V &= MX + V + M\frac{\det(M)}{|\det(M)|}\omega\,\mathrm{com}(M)^t e_i \\
&= \omega Y + \left\{\frac{MX+V}{\omega}\right\} + \omega\frac{\det(M)}{|\det(M)|}M\,\mathrm{com}(M)^t e_i \\
&= \omega Y + \left\{\frac{MX+V}{\omega}\right\} + \omega\frac{\det(M)}{|\det(M)|}det(M)e_i \\
&= \omega Y + \left\{\frac{MX+V}{\omega}\right\} + \omega|\det(M)|e_i = \omega(Y+\alpha e_i) + \left\{\frac{MX+V}{\omega}\right\}
\end{aligned}$$

Hence, $\left\{\frac{MX'+V}{\omega}\right\} = \left\{\frac{MX+V}{\omega}\right\}$ and thus $X' \in P_{Y+\alpha e_i}$. Finally, $P_{Y+\alpha e_i} \equiv P_Y$ which proves that $\alpha \in \mathcal{A}_i$. □

A.2 Theorem 4

Proof. Let us denote $\mathcal{T}(j)$ the proposition

$$P_{y_0,\dots,y_{n-1}} = T_{\sum_{i=j}^{n-1} w_i U_i} P_{y_0+\sum_{i=j}^{n-1} w_i\beta_0^i,\dots,y_{j-1}+\sum_{i=j}^{n-1} w_i\beta_{j-1}^i,y_j^0,\dots,y_{n-1}^0}.$$

We consider the following induction: given $n > p \geq 0$, we suppose $\mathcal{T}(p+1)$ and prove $\mathcal{T}(p)$. As a consequence of Def. 8,

$$\forall(z_1,\dots,z_{n-1}) \in \mathbb{Z}^n, P_{z_0,\dots,z_p+\alpha_p,\dots,z_{n-1}} = T_{U_p} P_{z_0+\beta_0^p,\dots,z_{p-1}+\beta_{p-1}^p,z_p,\dots,z_{n-1}}.$$

Hence, $\forall k \in \mathbb{Z}$, we have

$$P_{z_0,\dots,z_p+k\alpha_p,\dots,z_{n-1}} = T_{kU_p} P_{z_0+k\beta_0^p,\dots,z_{p-1}+k\beta_{p-1}^p,z_p,\dots,z_{n-1}}.$$

With $\begin{cases} k = w_p \\ \forall 0 \leq j < p, z_j = y_j + \sum_{i=p+1}^{n-1} w_i\beta_j^i \\ \forall p \leq j < n, z_j = y_j^0 \end{cases}$, we obtain

$$\begin{aligned} &P_{y_0+\sum_{i=p+1}^{n-1} w_i\beta_0^i,\dots,y_{p-1}+\sum_{i=p+1}^{n-1} w_i\beta_{p-1}^i,y_p^0+w_p\alpha_p,y_{p+1}^0,\dots,y_{n-1}^0} \\ &\qquad = T_{w_pU_p} P_{y_0+\sum_{i=p}^{n-1} w_i\beta_0^i,\dots,y_{p-1}+\sum_{i=p}^{n-1} w_i\beta_{p-1}^i,y_p^0,\dots,y_{n-1}^0}. \end{aligned} \tag{9}$$

Since $\mathcal{T}(p+1)$ is true, and since $y_p + \sum_{j=p+1}^{n-1} w_j\beta_p^j = y_p^0 + \alpha_p w_p$, we have

$$P_{y_0,\dots,y_{n-1}} = T_{\sum_{i=p+1}^{n-1} w_iU_i} P_{y_0+\sum_{i=p+1}^{n-1} w_i\beta_0^i,\dots,y_{p-1}+\sum_{i=p+1}^{n-1} w_i\beta_{p-1}^i,y_p^0+\alpha_pw_p,y_{p+1}^0,\dots,y_{n-1}^0} \tag{10}$$

We can identify the left side of Eq. (A.2) to the right part of the right side of (10), summing up the translation vectors leads to $\mathcal{T}(p)$. Since $\mathcal{T}(n) : P_{y_0,\dots,y_{n-1}} = T_0 P_{y_0,\dots,y_{n-1}}$ is trivial, we prove $\mathcal{T}(0)$ and thus the theorem. □

A.3 Theorem 6

Proof. Let $X, Y, Z \in \mathbb{Z}^n$ such that $X = H^{-1}Z$, then $Z \in P_Y$ is equivalent to

$$\left\lfloor \frac{MZ+V}{\omega} \right\rfloor = Y \Leftrightarrow \left\lfloor \frac{TX+V}{\omega} \right\rfloor = Y \Leftrightarrow \forall 0 \leq i < n, \omega y_i \leq \sum_{j=i}^{n-1} T_{i,j}X_j + V_i < \omega(y_i+1)$$

$$\Leftrightarrow \forall 0 \leq i < n, \omega y_i - \sum_{j=i+1}^{n-1} T_{i,j}X_j - V_i \leq T_{i,i}X_i < \omega(y_i+1) - \sum_{j=i+1}^{n-1} T_{i,j}X_j - V_i.$$

Thanks to Lemma 1, $Z \in P_Y$ is equivalent to

$$\forall 0 \leq i < n, A_i(X_{i+1},\dots,X_{n-1}) \leq X_i < B_i(X_{i+1},\dots,X_{n-1}).$$ □

Solving Some Instances of the 2-Color Problem

S. Brocchi[1], A. Frosini[1], and S. Rinaldi[2]

[1] Dipartimento di Sistemi e Informatica
Università di Firenze, Firenze, Italy
brocchi@dsi.unifi.it, andrea.frosini@unifi.it
[2] Dipartimento di Scienze Matematiche ed Informatiche
Università di Siena, Siena, Italy
rinaldi@unisi.it

Abstract. In the field of Discrete Tomography, the *2-color problem* consists in determining a matrix whose elements are of two different types, starting from its horizontal and vertical projections. It is known that the one color problem has a polynomial time reconstruction algorithm, while, with $k \geq 2$, the k-color problem is NP-complete. Thus, the 2-color problem constitutes an interesting example of a problem just in the frontier between hard and easy problems.

In this paper we define a linear time algorithm to solve a set of its instances, where some values of the horizontal and vertical projections are constant, while the others are upper bounded by a positive number proportional to the dimension of the problem. Our algorithm relies on classical studies for the solution of the one color problem.

Keywords: Discrete tomography, polynomial time algorithm, k-color problem.

1 Introduction

Recently, a series of valuable improvements in transmission electron microscopy have finally allowed the resolution of an inspected object till reaching the atomic scale (HRTEM). More precisely, since these last years it is possible to collect a huge quantity of data about the density of a material by focusing beams of electrons across it, and then measuring the decreasing of their energies: denser areas of the material will absorb more energy from the beams. The HRTEM has been mainly applied to biological specimens' studies to find structural defects in materials, but it has also furnished the bases for a new technique, called QUANTITEM ([9], [11]), that performs a *quantitative* analysis of highly structured materials like crystals. In particular QUANTITEM allows to measure the exact number of atoms (or other primary constituents) of an object along a set of lines parallel to given directions. Such a measurements, called *projections*, are usually arranged as vectors or matrices having integer entries, and they are used to recover some geometrical properties of the considered object. Usually, the final goal is its faithful reconstruction. This problem is a typical example of *inverse problem*, and fits in the area of computer science called Discrete Tomography,

S. Brlek, C. Reutenauer, and X. Provençal (Eds.): DGCI 2009, LNCS 5810, pp. 505–516, 2009.

a branch of the more general Computerized Tomography where only discrete structures are considered. Other classical problems of interest for Discrete Tomography are to determine when one or more projections are consistent with at least one discrete object belonging to a given class, say *consistency problem*, and to search for different objects which share the projections, say *uniqueness problem*.

The choice of representing each physical planar object by means of a discrete set of points in the lattice $\mathbb{Z} \times \mathbb{Z}$ is commonly accepted, and so is also the highly simplified but relevant model, which uses matrices of elements on a finite set $\Sigma = \{x_1, \dots x_c\} \cup \{0\}$, where each entry $x_i \in \Sigma$ or 0 stands for the presence of an atom of the material x_i or the absence of atoms in the correspondent point of $\mathbb{Z} \times \mathbb{Z}$.

So, a fundamental parameter is the number c of different atoms which compose the object, and that may be detected by the crossing beams' decreasing energies. Actually, the QUANTITEM allows the resolution of polyatomic structures, collecting the obtained measurements in multidimensional vectors.

The simplest case is in presence of an homogeneous material, i.e. $c = 1$: this scenery has been deeply studied in the past years, starting from the classical result by Ryser who showed, in [10], how to reconstruct homogeneous planar sets of points from two projections in polynomial time. Successively, in [6] the authors extended Ryser's result to the lattice $\mathbb{Z}^d$, with $d \geq 2$, and proved the NP-hardness of the same reconstruction problem in presence of three or more projections.

Obviously, polyatomic discrete sets inherit this last result, and the successive researches focused only on the case of two projections, so it is commonly indicated as *c-color problem* the problem of reconstructing a planar object having c different types of atoms from two projections, w.l.g. the horizontal and the vertical ones.

In [7], the authors furnished a proof of the NP-hardness of the c-color problem, with $c \geq 6$. Later and with different techniques, in [2], this result was improved, by showing that the presence of three different types of atoms is sufficient to maintain the c-color problem NP-hard. Recently, this result has been definitely extended to $c = 2$ in [5].

The present paper fits into this research line: we consider a significative class of instances the 2-color problem where some constraints on the elements of the two vectors are imposed, and we provide an algorithm that finds a solution to each element in polynomial time. In a few words, the algorithm uses the imposed constraints to reduce a given instance I of 2-color into four solvable 1-color ones, and then acts on their solutions by adding few further elements in order to reconstruct one of the solutions of I.

The paper is organized as follows: in section 2 we recall some basic notions of discrete tomography, and we define the main problem. Then we give some results about the 1-color problem that will be necessary to prove the correctness of the solving algorithm. In section 3 we describe two procedures, one regarding the insertion of elements of a given color in a matrix with some particular restrictions, and the other concerning the splitting of a multiset of integers into two parts, satisfying a certain balance property. Finally, in section 4, we define a special set of instances of the 2-color problem, and then describe the algorithm which solves such instances. Conclusions are in section 5.

2 Notations and Preliminary Results

Following the standard model, we represent a planar discrete object S made up with c different types of atoms using a $m \times n$ matrix $A = (a_{i,j})$, whose elements belong to the set $\Sigma \cup \{0\}$, with $\Sigma = \{x_1, \dots x_c\}$. We say that A is a *c-colored matrix*, and its dimensions are given by the dimensions of the minimal bounding rectangle of S. The elements of A having value 0 can be considered as the background of the object.
For each $1 \leq i \leq m$, let

$$h_i^{x_k} = |\{a_{i,j} \,:\, a_{i,j} = x_k\}| \text{ with } 1 \leq j \leq n, \text{ and } x_k \in \Sigma,$$

and analogously, for each $1 \leq j \leq n$, let

$$v_i^{x_k} = |\{a_{i,j} \,:\, a_{i,j} = x_k\}| \text{ with } 1 \leq i \leq m, \text{ and } x_k \in \Sigma.$$

We define

$$H = ((h_1^{x_1}, \dots, h_1^{x_c}), \dots, (h_m^{x_1}, \dots, h_m^{x_c})) \text{ and}$$
$$V = ((v_1^{x_1}, \dots, v_1^{x_c}), \dots, (v_n^{x_1}, \dots, v_n^{x_c}))$$

as the multi-vectors of *horizontal* and *vertical projections* of A, respectively. The matrix A is said to be *consistent* with H and V, and by extension, H is said to be consistent with V. Obviously, when we deal with homogeneous objects, we have $|\Sigma| = 1$, and H and V are integer vectors.

These simple few notions allow us to define the general reconstruction problem:

c-color

Instance: two multi-vectors

$$H = ((h_1^{x_1}, \dots, h_1^{x_c}), \dots, (h_m^{x_1}, \dots, h_m^{x_c})) \text{ and}$$
$$V = ((v_1^{x_1}, \dots, v_1^{x_c}), \dots, (v_n^{x_1}, \dots, v_n^{x_c})).$$

Task: reconstruct a $m \times n$ $c-$colored matrix A having H and V as horizontal and vertical projections, if it exists, otherwise give FAILURE.

As a matter of fact, we observe that, by definition, every consistent instance $I = (H, V)$ of $c-$color, must satisfy the following simple properties: for each $x \in \Sigma$, $1 \leq i \leq m$ and $1 \leq j \leq n$

(i) $0 \leq h_i^x \leq n$;
(ii) $0 \leq v_j^x \leq m$;
(iii) $\sum_{i=1}^{m} h_i^x = \sum_{j=1}^{n} v_j^x$.

From now on we consider only instances of c-color that respect these three conditions, and such that the matrices obtained as solutions, if any, have dimension $m \times n$.

On the 1-Color Problem

Now we will start by considering the case of the reconstruction of a single homogeneous object, hence we will deal with matrices having two different values, which means that $\Sigma \cup \{0\}$, with $\Sigma = \{a\}$. This problem has been already extensively studied by Ryser in [10]. In the following we point out some properties common to all couples (H, V) of consistent instances, which will be useful in the rest of the paper. In such a context the superscript a of the elements of the projections H and V of a 1-colored matrix will be omitted, since no ambiguities can occur.

Theorem 1. *Let H be an integer vector. H is consistent with all the integer vectors V which satisfy conditions (ii), with $m = |H|$, and (iii), if and only if all the elements of H belong to a set $\{k, k-1\}$, with $k > 0$.*

Proof. ($\Leftarrow$) we will prove that, in this case, the application of the greedy algorithm defined by Ryser always provides a valid solution. We proceed by induction on the number of columns of the matrix A:

Base. If the matrix is formed by a single column and the projections respect our assumptions, the result trivially holds.

Inductive step. If the matrix A has more than one column, then set v_1 elements in the first column in priority in the lines where $h_i = k$. This operation is possible as for *(iii)* it stands that $|h_i : h_i > 0| \geq v_1$.

At this point the matrix formed by the columns $2, \ldots, m$ of A and with projections H' and $v_2, \ldots, v_m$, where H' contains the projections of H minus the projections of column 1, can be correctly filled by inductive hypothesis, since its values belong to the set $\{k, k-1\}$ or $\{k-1, k-2\}$.

($\Rightarrow$) let us proceed by contradiction assuming that there exists a vector H, which is compatible with each vector V satisfying *(ii)* and *(iii)*, and having two elements that differ at least by two. Let us consider the array $\hat{V} = \{n, \ldots, n, j, 0, \ldots, 0\}$, that satisfies *(iii)*. Moreover it is easy to verify that any matrix having $\hat{V}$ as vector of vertical projections must have any couple of horizontal projections differing of at most by 1, leading to a contradiction. □

3 Two Useful Procedures

In this paragraph we show two procedures that will be used in the main algorithm for solving particular instances of the 2-color problem. The first one, called *Insertion*, acts on a 2-colored matrix, and changes the positions of some elements inside it. The second, called *Balance* splits an integer vector into two parts, maintaining balanced the sums of their elements.

3.1 Procedure 1: *Insertion*

Let A be a 2-colored matrix of dimension $m \times n$ on the alphabet $\Sigma = \{a, b\} \cup \{0\}$. The matrix A is said to be $c-sparse$ if each of its rows contains only elements 0 except for at most c positions, which are filled with elements a, with $0 < c \leq n$. The procedure *Insertion* we are going to define performs the following:

Task. *given a set R of at most $min\{m, n - c\}$ rows of a c-sparse matrix A, replace, inside each row of R, an element 0 with an element b so that there exists at most one element b for each column of A.*

Procedure. *Insertion*

```
for each i ∈ R do
    set ok = false
    for each j such as 1 ≤ j ≤ n do
        if a_{i,j} = 0 and no element b is present in column j then
            set a_{i,j} = b
            set ok = true
            exit j loop
        end if
    next j
    if ok = false then give FAILURE;
next i
give matrix A as OUTPUT.
```

Theorem 2. *The procedure Insertion having as input a set R of rows of a c-sparse 2-colored matrix M never gives FAILURE.*

The proof directly follows from the observation that, that if a row i contains no elements 0 to be changed into a b, then at least $n - c$ elements b have already been placed in the rows of R, and this is a contradiction.

Lemma 1. *The procedure Insertion acts in $O(m\,n)$ computational time.*

The proof is immediate, since the elements of R are $O(m)$, and for each of them, at most $O(n)$ elements of A are scanned.

3.2 Procedure 2: *Balance*

Now, let us consider an integer vector $S = (s_1, \ldots, s_d)$, and two integers min and Max, such that $0 < min \leq s_k \leq Max$, with $k \in 1, \ldots, d$.

We define the procedure *Balance* which performs the following:

Task. *determine a rearrangement $\widetilde{S}$ of the elements of S such that*

$$0 \leq \sum_{k=1}^{d'} \widetilde{s}_k - \sum_{k=d'+1}^{d} \widetilde{s}_k \leq Max \ , \quad with\ d' = \lceil \tfrac{d}{2} \rceil. \tag{1}$$

Procedure. *Balance*

Step 1. Let $\hat{S}$ be the vector S increasingly ordered, and set $d' = \lceil \frac{d}{2} \rceil$;
Step 2. for each $1 \leq k \leq d'$, set $\widetilde{s}_k = \hat{s}_{2k-1}$;
Step 3. for each $d' + 1 \leq k \leq d$, set $\widetilde{s}_k = \hat{s}_{2(k-d')}$;

Step 4. give vector $\widetilde{S}$ as OUTPUT.

It is easy to see that the vector $\widetilde{S}$ satisfies the required task, in fact, if d is odd it holds

$$0 \leq \sum_{k=1}^{d'} \widetilde{s}_k - \sum_{k=d'+1}^{d} \widetilde{s}_k = \sum_{k=1}^{(d-1)/2} (\hat{s}_{2k-1} - \hat{s}_{2k}) + \hat{s}_d \leq \sum_{k=1}^{(d-1)/2} (\hat{s}_{2k-1} - \hat{s}_{2k+1}) + \hat{s}_d =$$

$$= \hat{s}_1 - \hat{s}_d + \hat{s}_d = \hat{s}_1 \leq Max,$$

otherwise, if d is even we can similarly check that

$$0 \leq \sum_{k=1}^{d'} \widetilde{s}_k - \sum_{k=d'+1}^{d} \widetilde{s}_k \leq Max - min.$$

Lemma 2. *The procedure Balance acts in $O(d + Max)$ computational time.*

The result is immediate, since $O(d + Max)$ is required to order the vector S using counting sort, and $O(d)$ to scan S and create $\widetilde{S}$.

Finally, we define $\sigma_{Balance}$ the permutation of the elements of S performed by *Balance* to obtain $\widetilde{S}$.

4 Solving Some Instances of the 2-Color Problem

In this paragraph we define a subset of instances of 2-color, and then we present a polynomial algorithm to solve the problem. These instances are defined by some constraints on their values: in a word, some elements of H and V are constant, while the remaining elements are smaller than a given number which is somehow proportional to the minimum between m and n, the dimensions of the reconstructed matrix. Our approach to the problem can be naturally generalized to the whole set of instances of 2-color, unfortunately without being able to polynomially bound the computational complexity. So, let us define the following problem:

2c-restricted

Instance. two consistent vectors of projections

$$H = ((h_1^a, h_1^b), \ldots, (h_m^a, h_m^b)) \text{ and } V = ((v_1^a, v_1^b), \ldots, (v_n^a, v_n^b))$$

such that

(1) all the elements of H and V are bounded by an integer M such that
   - $M = \lfloor min\{m, n\}/3 \rfloor$ if m and n are both even,
   - $M = \lfloor min\{m, n\}/4 \rfloor$ otherwise;

(2) there are two positive integers c_1 and c_2 such that for all $1 \leq i \leq m$, $1 \leq j \leq n$ we have $h_i^a = c_1$ and $v_j^b = c_2$.

Task. reconstruct a $m \times n$ 2−colored matrix A having H and V as horizontal and vertical projections, if it exists, otherwise give FAILURE.

The imposed constraints on the projections of $2c$ − *restricted* allow a fast reconstruction algorithm. The algorithm is mainly based on the fact that, for each of its instances, there exists a solution which can be divided into four disjoint parts, say NW, NE, SW, and SE, each of them containing only one color, say the dominant one, except for a fixed number of elements.

More precisely, given an instance I we split the non constant part H^b of H [resp. V^a of V], into two parts which differ at most by M, using the procedure *Balance*. Successively we arrange these parts into four instances of 1-color which are used to set the dominant color in the four parts NW, NE, SW, and SE. At this step it is not guaranteed that all these instances satisfy (iii), so we have to slightly modify them by setting aside some elements, at most M for each one, to make them consistent and, furthermore, solvable by Lemma 1.

Once we have reconstructed the four submatrices NW, NE, SW, and SE, we obtain a partial solution which turns into the final one, which we call A, after placing the previously removed elements by procedure *Insertion*. Theorem 2 assures that this process successfully ends.

Now we define the four submatrices NW, NE, SW, and SE of a $m \times n$ matrix A

NW is the submatrix formed by the intersection of the first $\lfloor m/2 \rfloor$ rows with the first $\lceil n/2 \rceil$ columns;
NE is the submatrix formed by the intersection of the first $\lfloor m/2 \rfloor$ rows with the last $\lfloor n/2 \rfloor$ columns;
SW is the submatrix formed by the intersection of the last $\lceil m/2 \rceil$ rows with the first $\lceil n/2 \rceil$ columns;
SE is the submatrix formed by the intersection of the last $\lceil m/2 \rceil$ rows with the last $\lfloor n/2 \rfloor$ columns.

So, let us define the algorithm.

Algorithm. *Reconstruct*

Step 1. let $H^b = (h_1^b, \ldots, h_m^b)$ [resp. $V^a = (v_1^a, \ldots, v_n^a)$].
Apply procedure *Balance* to H^b [resp. V^a], and let $\widetilde{H}^b$ [resp. $\widetilde{V}^a$] be its output. In this way the sums of the elements in the first and second half of $\widetilde{H}^b$ [resp. $\widetilde{V}^a$] have a bounded difference.
Let $\sigma^b_{Balance}$ [resp. $\sigma^a_{Balance}$] be one of the permutations of the elements of H^b [resp. V^a] that leads to $\widetilde{H}^b$ [resp. $\widetilde{V}^a$].
Compute the vector $\widetilde{H} = ((c_1, \widetilde{h}_1^b), \ldots, (c_1, \widetilde{h}_m^b))$ [resp. $\widetilde{V} = ((\widetilde{v}_1^a, c_2), \ldots, (\widetilde{v}_n^a, c_2))$].
Let $m' = \lceil m/2 \rceil$ and $n' = \lceil n/2 \rceil$.

Step 2.
now we compute and store in D^a_{SW} or D^a_{NE} the (number of) elements a that lie in the submatrices SW or NE of the final solution, where the dominant color is b.

The symmetrical computation of D^b_{SE} or D^b_{NW} is carried with respect to the color b and the submatrices SE and NW.

These numbers are computed in such a manner to allow the instances of the 1-color problems, which partially reconstruct the four submatrices of the final solution, to satisfy condition *(iii)*.

if $\sum_{k=1}^{n'} \widetilde{v}^a_k > m'\, c_1$

then set $D^a_{SW} = \sum_{k=1}^{n'} \widetilde{v}^a_k - m'\, c_1$, and $D^a_{NE} = 0$

else set $D^a_{NE} = m'\, c_1 - \sum_{k=1}^{n'} \widetilde{v}^a_k$, and $D^a_{SW} = 0$.

If $\sum_{k=1}^{m'} \widetilde{h}^b_k > (n-n')c_2$ then set $D^b_{NW} = \sum_{k=1}^{m'} \widetilde{h}^b_k - (n-n')c_2$, and $D^b_{SE} = 0$

else set $D^b_{SE} = (n-n')\, c_2 - \sum_{k=1}^{m'} \widetilde{h}^b_k$, and $D^b_{NW} = 0$.

W.l.g. assume that $D^a_{SW} = D^b_{SE} = 0$, *and , as a consequence,* $D^b_{NW} \geq 0$ *and* $D^a_{NE} \geq 0$.

Step 3. (where the submatrix NW is reconstructed)

Step 3.1: create the instance $I^a_{NW} = (H^{NW}, V^{NW})$ of 1-color such that
- $h^{NW}_i = c_1 - 1$, with $1 \leq i \leq D^a_{NE}$;
- $h^{NW}_i = c_1$, with $D^a_{NE} < i \leq m'$;
- $v^{NW}_j = \widetilde{v}^a_j$ with $1 \leq j \leq n'$.

Step 3.2: solve the instance I^a_{NW} of 1-color (by means of Ryser's algorithm, as described in [10]) and insert the values a in the correspondent submatrix NW of A.

Step 3.3: run the procedure *Insertion* with $R = \{1, \ldots, D^b_{NW}\}$ (notice that the submatrix NW is M-sparse). Store the indexes of the columns where the D^b_{NW} elements b are placed in R_{NW}.

Step 4. (where the submatrix NE is reconstructed);

Step 4.1: create the instance $I^b_{NE} = (H^{NE}, V^{NE})$ of 1-color such that
- $h^{NE}_i = \widetilde{h}^b_i - 1$, with $1 \leq i \leq D^b_{NW}$;
- $h^{NE}_i = \widetilde{h}^b_i$ with $D^b_{NW} < j \leq m'$;
- $v^{NE}_j = c_2$ with $1 \leq j \leq n - n'$.

Step 4.2: solve the instance I^b_{NE} of 1-color and insert the values b in the correspondent submatrix NE of A.

Step 4.3: run the procedure *Insertion* with $R = \{1, \ldots, D^a_{NE}\}$. Store the indexes of the columns where the D^a_{NE} elements a are placed in R_{NE}.

Step 5,6: (where the submatrix SW [resp. SE] is reconstructed) now act similarly as in Step 3.2 [resp. Step 4.2] to reconstruct the two remaining submatrices, with the following remark: in Step 5.1 [resp. Step 6.1], the two projections of the instance $I_{SW} = (H_{SW}, V_{SW})$ [resp. $I_{SE} = (H_{SE}, V_{SE})$], can be computed by simply subtracting from $\widetilde{H}$ and $\widetilde{V}$ the elements a [resp. b] stored in R_{NW} [resp. R_{NE}], i.e. those already placed in NW [resp. NE].

Step 7: permute the rows of A according to $(\sigma^b_{Balance})^{-1}$, then permute the columns of A according to $(\sigma^a_{Balance})^{-1}$, and finally return the updated matrix A as OUTPUT.

4.1 An Example of the Execution of the Algorithm

To clarify the steps of the reconstruction procedure we furnish the following example:

Let us represent a $m \times n$, 2-colored matrix by means of a $m \times n$ set of cells on a squared surface having two different colors, say red and blue; red, blue and void cells correspond to the elements a, b, and 0 of the matrix, respectively. Let us consider the instance $I = (H, V)$ such that

$$H = ((2,3),(2,3),(2,4),(2,3),(2,1),(2,2),(2,4),(2,4),(2,1),(2,4),(2,4),(2,3))$$
$$V = ((2,3),(3,3),(1,3),(3,3),(3,3),(1,3),(2,3),(2,3),(3,3),(2,3),(1,3),(1,3))$$

It is easy to check that H and V respect conditions (i), (ii) and (iii) for each color, and that I is an instance of $2c-restricted$.Since H^{blue} and V^{red} are already balanced, i.e. they satisfy equation (1), there is no need to perform procedure *Balance*, and so for sake of simplicity we skip Step 1 and consider $H = \widetilde{H}$, and $V = \widetilde{V}$. The steps of the procedure *Reconstruction* are sketched in Fig. 1, starting from a), where the instance I is depicted; furthermore, we have $m' = n' = 6$. In Step 2 we compute $D^{red}_{SW} = 1$, and $D^{red}_{NE} = 0$, then we compute $D^{blue}_{NW} = 0$ $D^{blue}_{SE} = 2$. These numbers represent the red and blue elements that have to be added to each of the four submatrices.

Since $D^{red}_{SW} \neq 0$ and $D^{blue}_{SE} \neq 0$, we start the reconstruction process from one of these two submatrices, say the SW submatrix, then we proceed with the other. So, two blue cells are moved from the submatrix SW to SE, and we create the instance of 1-color $I^{blue}_{SW} = (H^{SW}, V^{SW})$, with $H^{SW} = (3,3,1,4,4,3)$, and $V^{SW} = (3,3,3,3,3,3)$. Finally we solve I^{blue}_{SW}, as in Fig. 1, b), and we add, by means of the procedure *Insertion*, one red cell in the first free position, as required by D^{red}_{SW}; this concludes Step 3.

Similarly to Step 3, we perform Step 4, where we reconstruct the submatrix SE, and, successively we add two blue cells as required by D^{blue}_{SE}, as in Fig. 1, d).

Now, we perform Steps 5 and 6 after creating the instances $I^{NW} = (H^{NW}, V^{NW})$ and $I^{NE} = (H^{NE}, V^{NE})$, respectively, where $H^{NW} = (2,2,2,2,2,2)$, $V^{NW} = (1,3,1,3,3,1)$, $H^{NE} = (3,3,4,3,1,2)$, and $V^{NE} = (2,2,3,3,3,3)$.

The reconstruction of the two submatrices NW and NE, together with the whole solution of I, are depicted in Fig. 1, e).

4.2 Correctness

Now we prove that the algorithm *Reconstruct* having as input a generic instance I of $2c - restricted$ always leads to a solution by showing that:

i) each instance of 1-color related to the four submatrices NW, NE, SW and SE admits a solution;
ii) each reconstructed submatrix satisfies the conditions required by the procedure *Insertion*.

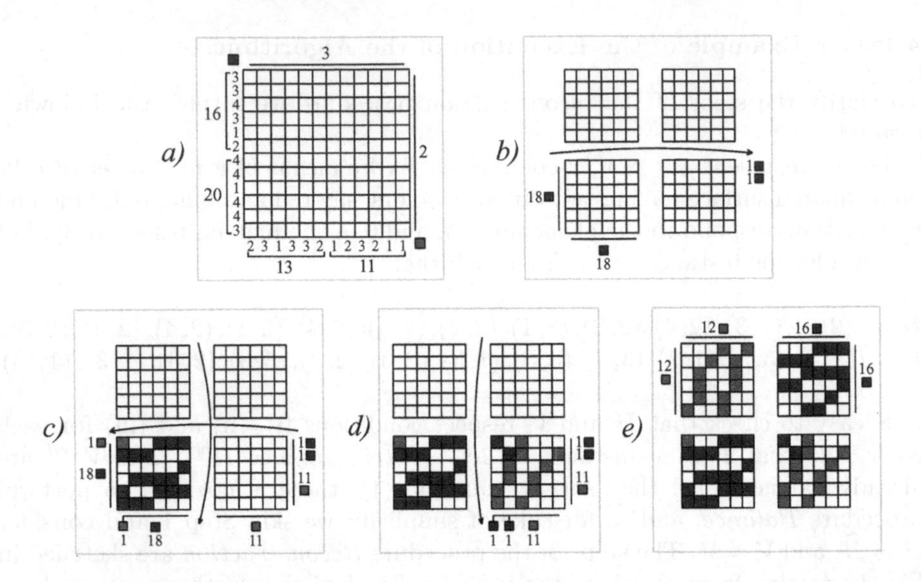

Fig. 1. The steps of *Reconstruction* on the instance I

Assertion i) is straightforward since each instance of 1-color related to the submatrices NW, NE, SW and SE created in Steps 3, 4, 5 and 6 of *Reconstruction* has at least one among the horizontal or the vertical projections whose values are in $\{k-1, k\}$, where $k = c_1$ or $k = c_2$. Further, the number of cells in the horizontal and vertical projections are consistent thanks to the use of the variables D. So, the hypothesis of Theorem 1 are satisfied (with respect to each instance), and consequently each instance admits a solution.

Assertion ii) follows after observing that

- each reconstructed submatrix is M-sparse;
- if m or n are odd, the values D^b_{NW}, D^a_{NE}, D^a_{SW}, and D^b_{SE} are positive and each of them bounded by $\lfloor n/2 \rfloor - M$. As an example, let us consider D^b_{NW}: its value is either zero or it holds $D^b_{NW} \leq \lceil (M-1)/2 + c_2/2 \rceil$ (remind that the application of the procedure *Balance* to H^b produces $\sum_{k=1}^{m'} \widetilde{h}^b_k - \sum_{k=m'+1}^{m} \widetilde{h}^b_k \leq M-1$). Let $min = \min\{m, n\}$, and observe that $c_2 \leq min/4$. Consequently

$$D^b_{NW} \leq \lceil (M-1)/2 + c_2/2 \rceil \leq (min-4)/8 + min/8 + 1 = min/4 + 1/2$$

and assuming that $min \geq 3$, we finally reach

$$D^b_{NW} \leq min/4 + 1/2 \leq min/2 \leq \lceil n/2 \rceil - M.$$

The same argument can be used to prove the analogous result for D^a_{NE}, D^a_{SW}, and D^b_{SE}.

With similar arguments, we can prove that if the dimensions m and n of the solution matrix are both even, the maximum value M of the elements of H^b and V^a becomes $\lfloor \min\{m,n\}/3 \rfloor$ as, in such a setting, the four values D^b_{NW}, D^a_{NE}, D^a_{SW}, and D^b_{SE} are upper bounded by $\lceil (M-1)/2 \rceil$.
- each submatrix has more than D^b_{NW} [resp. D^a_{NE}, D^a_{SW}, and D^b_{SE}] rows.

So, all the hypotheses to apply the procedure *Insertion* to each reconstructed submatrix NW, NE, SW and SE are satisfied, as desired.

4.3 Complexity

Theorem 3. *The algorithm Reconstruct finds a solution to a generic instance I of $2c-$ restricted in $O(mn)$ computational time.*

Proof We will analyze the complexity of each step of *Reconstruct*

Step 1: The procedure *Balance* acts in $O(m+n)$ (see Lemma 2) computational time, and the same for the computation of the vectors $\widetilde{H}$ and $\widetilde{V}$.
Step 2: All the computations are clearly performed in $O(n+m)$.
Steps $3-6$: Each step takes $O(mn)$ computational time, since Ryser's algorithm for solving the 1-color problem takes $O(mn)$, and the same holds for procedure *Insertion* (see Lemma 1).
Steps 7: The output matrix A is computed in $O(nm)$ computational time.

So, *Reconstruct* takes $O(m\,n)$ computational time as desired. □

5 Further Research

In this paper we show a class of instances of the 2-color problem which can be solved in linear (computational) time with respect to the dimensions of the solution matrix. The algorithm we propose is mainly based on the existence of a family of 2-colored matrices which, after an appropriate organization of its rows and columns, can be divided into disjoint submatrices each with only one color, so that we are allowed to use, for each of these cases, the standard Ryser's reconstruction algorithm for 1-colored matrices.

The proposed result gives birth to a new way of challenging subclasses of instances of the $2-$ *color* problem, searching for all those solutions where the two colors are somehow disjoint in each row and column. As a consequence, the proposed algorithm *Reconstruct* seems to allow a further series of refinements by using techniques that decompose each instance into smaller ones in the *divide et impera* fashion, and which may lead to the definition of the non polynomial set of instances of $2-$ *color*, to the limit; obviously the hardest task will be to keep polynomially bounded the amount of time needed for the whole process.

Our result suggests that the hard part of 2-color has to contain only those instances having non uniform projections for both colors, and it furnishes hints and inspiration for future researches.

References

1. Brocchi, S., Frosini, A., Picouleau, C.: Reconstruction of binary matrices under fixed size neighborhood constraints. Theoretical Computer Science 406, 1-2, 43-54 (2008)
2. Chrobak, M., Durr, C.: Reconstructing polyatomic structures from discrete X-rays: NP-completeness proof for three atoms. Theoretical computer science 259, 81–98 (2001)
3. Costa, M.C., de Werra, D., Picouleau, C., Schindl, D.: A solvable case of image reconstruction in discrete tomography. Discrete Applied Mathematics 148(3), 240–245 (2005)
4. Costa, M.C., de Werra, D., Picouleau, C.: Using graphs for some discrete tomography problems, CEDRIC report (2004), http://cedric.cnam/fr/
5. Dũrr, C., Guiñez, F., Matamala, M.: Reconstructing 3-colored grids from horizontal and vertical projections is NP-hard. In: Fiat, A., Sanders, P. (eds.) ESA 2009. LNCS, vol. 5757, pp. 776–788. Springer, Heidelberg (2009)
6. Gardner, R.J., Gritzmann, P., Pranenberg, D.: On the computational complexity of reconstructing lattice sets from their X-rays. Discrete Mathematics 202(1-3), 45–71 (1999)
7. Gardner, R.J., Gritzmann, P., Pranenberg, D.: On the computational complexity of determining polyatomic structures by X-rays. Theoretical computer science 233, 91–106 (2000)
8. Herman, G., Kuba, A.: Discrete Tomography: Foundations, Algorithms and Applications. Birkhauser, Basel (1999)
9. Kisielowski, C., Schwander, P., Baumann, F.H., Seibt, M., Kim, Y., Ourmazd, A.: An approach to quantitative high-resolution transmission electron microscopy of crystalline materials. Ultramicroscopy 58, 131–155 (1995)
10. Ryser, H.J.: Combinatorial properties of matrices of zeros and ones. Canad. J. Math. 9, 371–377 (1957)
11. Schwander, P., Kisielowski, C., Seibt, M., Baumann, F.H., Kim, Y., Ourmazd, A.: Mapping projected potential interfacial roughness, and composition in general crystalline solids by quantitative transmission electron microscopy. Phys. Rev. Lett. 71, 4150–4153 (1993)

Grey Level Estimation for Discrete Tomography

K.J. Batenburg, W. van Aarle, and J. Sijbers

IBBT - Vision Lab*
University of Antwerp, Belgium
joost.batenburg@ua.ac.be, wim.vanaarle@ua.ac.be
jan.sijbers@ua.ac.be

Abstract. Discrete tomography is a powerful approach for reconstructing images that contain only a few grey levels from their projections. Most theory and reconstruction algorithms for discrete tomography assume that the values of these grey levels are known in advance. In many practical applications, however, the grey levels are unknown and difficult to estimate. In this paper, we propose a semi-automatic approach for grey level estimation that can be used as a preprocessing step before applying discrete tomography algorithms. We present experimental results on its accuracy in simulation experiments.

1 Introduction

The field of *tomography* deals with the reconstruction of images (known as *tomograms*) from their projections, taken along a range of angles. Tomography has a wide range of applications in medicine, industry, and science. According to [1,2], the field of *discrete tomography* is concerned with the reconstruction of images from a small number of projections, where the set of pixel values is known to have only a few discrete values. It must be noted that the term "*discrete*" is often used to indicate the *discrete domain* of the image, i.e., when reconstructing lattice sets. In this paper, we adhere to the former definition, and focus on the reconstruction of images that consist of a small number of grey levels from their (non-lattice) X-ray projections.

A variety of reconstruction algorithms have been proposed for discrete tomography problems. In [3], an algorithm is presented for reconstructing binary images from a small number of projections. This primal-dual subgradient algorithm is applied to a suitable decomposition of the objective functional, yielding provable convergence to a binary solution. In [4], a similar reconstruction problem is modeled as a series of network flow problems in graphs, that are solved iteratively. Both [5] and [6] consider reconstruction problems that may involve more than two grey levels, employing statistical models based on Gibbs priors for their solution.

Besides the assumption on the grey levels, many of these algorithms incorporate additional prior knowledge of the original object that needs to be reconstructed. A preference for locally smooth regions is typically incorporated, as it

* This work was financially supported by the IBBT, Flanders and FWO, Flanders.

S. Brlek, C. Reutenauer, and X. Provençal (Eds.): DGCI 2009, LNCS 5810, pp. 517–529, 2009.

corresponds to a range of practical images. By combining such prior knowledge with knowledge of the grey levels, it is often possible to compute accurate reconstructions from just a small number of projections. In some cases, as few as three projections is already sufficient to compute a high-quality reconstruction.

A common assumption in all these reconstruction algorithms, is that the set of admissible grey levels is known *a priori*. Although this assumption is easy to satisfy in simulation experiments, obtaining prior knowledge of the set of possible grey levels is often not straightforward in practical applications, for several reasons:

- Even when the number of materials that constitute the scanned object is known, the densities of these materials may be unknown.
- When the materials and their densities *are* known in advance, calibration is required to translate the material properties into a reconstructed grey level. The calibration parameters depend not only on the scanned materials, but also on various properties of the scanner system. In many experimental settings, such information is not available.
- Calibration parameters of the scanning system may change over time. For example, the X-ray source of a micro-CT scanner heats up while scanning, changing the spectrum of emitted X-rays. While this may have a negligible effect on the calibration parameters during a single scan, batch processing of scans may change the parameters, and consequently change the grey levels in the reconstruction.

When several similar objects are scanned as a single batch, it may be possible to obtain a high quality reconstruction of one of those objects, based on a large number of projections. This reconstruction can then be used to estimate the admissible grey levels for the remaining objects, which can then be reconstructed from few projections. However, many important applications of discrete tomography deal with acquisition systems where it is not possible at all to obtain an accurate reconstruction without resorting to discrete tomography. In Materials Science, for example, discrete tomography is used to reconstruct 3D images of nanomaterials from a series of projection images acquired by an electron microscope [7]. As the sample is flat and its thickness increases with the incidence angle of the electron beam, projections can only be obtained for a limited range of angles, resulting in severe reconstruction artifacts for conventional algorithms.

Even in such cases, it may still be possible for an expert user to delineate certain areas in the reconstruction that are likely to have a constant grey level in the reconstruction. In this paper, we therefore consider a simpler version of the grey level estimation problem, where the user first selects an image region that can be expected to contain a constant grey level, based on an initial reconstruction. This reconstruction can be obtained by classical, non-discrete reconstruction algorithms. In certain cases, knowledge of such a constant region allows for reliable estimation of the grey level corresponding to the selected region.

The outline of this paper is as follows. In Section 2, the problem of grey level estimation is introduced, along with formal notation. In Section 3, we present a semi-automatic approach for estimating the grey levels when for each grey level a subset of corresponding pixels is given. Experimental results for a range of simulation experiments are described in Section 4. Section 5 discusses the possibilities for obtaining more accurate estimates, compared to the approach of this paper. Section 6 concludes the paper.

2 Problem and Model

The unknown physical object from which projection data has been acquired is represented by a grey value image $f : \mathbb{R}^2 \to \mathbb{R}$. We denote the set of all such functions f by $\mathcal{F}$, also called the set of *images*.

Projections are measured along lines $l_{\theta,t} = \{(x,y) \in \mathbb{R}^2 : x\cos\theta + y\sin\theta = t\}$, where θ represents the angle between the line and the y-axis and t represents the coordinate along the projection axis; see Fig. 1.

Denote the set of all functions $s : \mathbb{R} \times [0, 2\pi) \to \mathbb{R}$ by $\mathcal{S}$. The *Radon transform* $\mathcal{F} \to \mathcal{S}$ is defined by

$$R(f)(t,\theta) = \iint_{-\infty}^{\infty} f(x,y)\delta(x\cos\theta + y\sin\theta - t)\,\mathrm{d}x\,\mathrm{d}y.$$

with $\delta(.)$ denoting the Dirac delta function. The function $R(f) \in \mathcal{S}$ is also called the *sinogram* of f.

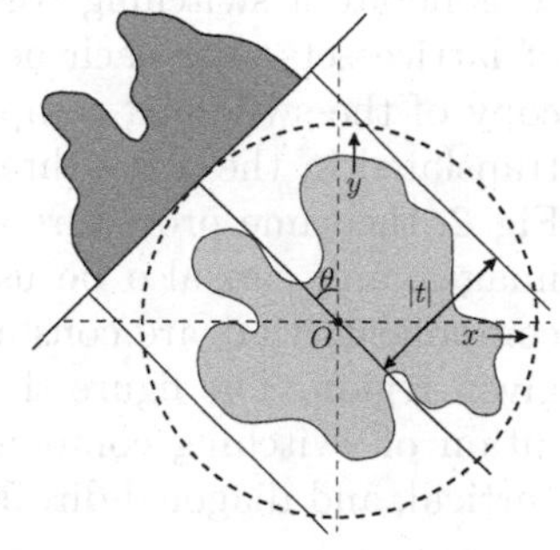

Fig. 1. Basic setting of transmission tomography

2.1 Grey Level Estimation Problem

The *reconstruction problem* in tomography concerns the recovery of f from its Radon transform $R(f)$. Here, we focus on images f for which the set of grey levels is a small, discrete set $G = \{g_1, \ldots, g_k\}$. We denote the set of all such images $f : \mathbb{R}^2 \to G$ by $\mathcal{G}$. The main problem of this paper consists of estimating the grey levels G from given projection data along a finite set of angles:

Problem 1. Let $D = \{\theta_1, \ldots, \theta_d\}$ be a given set of projection angles. Let $G = \{g_1, \ldots, g_k\}$ be an unknown set of *grey levels* and let $f \in \mathcal{G}$ be an unknown image. Suppose that k and $R(f)(t, \theta_i)$ are given for $i = 1, \ldots, d$, $t \in \mathbb{R}$. Reconstruct G.

It is clear that Problem 1 does not have a unique solution in general. A first requirement for a grey level to be recoverable is that it occurs in f as a measurable subset of $\mathbb{R}^2$. Also, when a small number of projections is available and k is large, it is straightforward to find examples where the grey values cannot be determined uniquely. In this paper, we therefore assume that additional prior knowledge is available, prescribing a set A on which the image is known to have a constant grey level. This leads to the following estimation problem:

Problem 2. Let $D = \{\theta_1, \ldots, \theta_d\}$ be a given set of projection angles. Let $G = \{g_1, \ldots, g_k\}$ be an unknown set of *grey levels* and let $f \in \mathcal{G}$ be an unknown image. Suppose that k and $R(f)(t, \theta_i)$ are given for $i = 1, \ldots, d$, $t \in \mathbb{R}$, and that for $i = 1, \ldots, k$, a set $A_i \subset \mathbb{R}^2$ is given such that $f(x, y) = g_i$ for all $(x, y) \in A_i$. Reconstruct G.

However, even specifying an entire region where the image is known to be constant is not sufficient to guarantee uniqueness of the solution of Problem 2. In particular, this problem occurs when the number of projections is small. Fig. 2 shows a well-known procedure to generate a so-called *switching component*: a non-zero image that has a zero projection in a given set of projection angles. The procedure is described in [1] (Section 4.3, p. 88), where it is used to generate a switching component in the context of lattice sets. For each new direction, a negated copy of the switching component is added, that is translated in the given direction. As shown in the Fig. 2, the same procedure applies directly to pixel images, and can also be used to create switching components that are constant, and non-zero, on a given region. The figure shows the subsequent addition of switching components in the horizontal, vertical, and diagonal direction, respectively.

Fig. 2. Example of a switching component for three directions, which is constant and nonzero on each of the black and white regions

Proposition 1. *Let* $D = \{\theta_1, \ldots, \theta_d\}$ *be a given set of projection angles. Let* $G = \{g_1, \ldots, g_k\}$, $A \subset \mathbb{R}^2$ *and* $g \in G$. *Let* $f \in \mathcal{G}$ *such that* $f(x, y) = g$ *for all* $(x, y) \in A$. *Then for each grey level* $\tilde{g} \in \mathbb{R}$, *there is an image* $\tilde{f} \in \mathcal{F}$ *such that* $\tilde{f}(x, y) = \tilde{g}$ *for all* $(x, y) \in A$ *and* $R(f)(t, \theta_i) = R(\tilde{f})(t, \theta_i)$ *for* $i = 1, \ldots, d$, $t \in \mathbb{R}$. *Moreover, there is such an image* $\tilde{f}$ *that has at most* $3k$ *grey levels.*

Proof. (sketch) By the construction depicted in Fig. 2, an image can be created that has a constant value of 1 on A, has constant zero projections in all given directions and only contains grey levels from $\{-1, 0, 1\}$. Let $\rho = \tilde{g} - g$. By adding a multiple of ρ times the switching component to f, an image $\tilde{f}$ that conforms to the proposition can be created. This image will have at most $3k$ grey levels, included in the set $\{g_1 - \rho, g_1, g_1 + \rho, g_2 - \rho, g_2, g_2 + \rho, \ldots, g_k - \rho, g_k, g_k + \rho\}$. Note that some of these grey levels may be negative, even if the image f has only nonnegative grey levels.

2.2 Discretization

In practice, a projection is measured at a finite set of detector cells, each measuring the integral of the object density along a line. Let m denote the total number of measured detector values (for all angles) and let $\boldsymbol{p} \in \mathbb{R}^m$ denote the measured projection data. We now discretize the images $f \in \mathcal{F}$ as well, represented on a square grid of width w. Let $n = w^2$, the number of pixels in the image. We assume that the image is zero outside this rectangle. Let $\boldsymbol{v} \in \mathbb{R}^n$ denote the discretized image of the object. The Radon transform for a finite set of angles can now be modeled as a linear operator $\boldsymbol{W}$, called the *projection operator*, that maps the image $\boldsymbol{v}$ to the projection data $\boldsymbol{p}$:

$$\boldsymbol{W}\boldsymbol{v} = \boldsymbol{p}. \tag{1}$$

We represent $\boldsymbol{W}$ by an $m \times n$ matrix $\boldsymbol{W} = (w_{ij})$. The vector $\boldsymbol{p}$ is called the *forward projection* or *sinogram* of $\boldsymbol{v}$.

The notation introduced above allows us to define the key problem of this paper:

Problem 3. Let $\boldsymbol{v} \in \mathbb{R}^n$ be an unknown image and let $g \in \mathbb{R}$ be an unknown grey level. Suppose that $\boldsymbol{p} = \boldsymbol{W}\boldsymbol{v}$ is given, and that a set $A \subset \{1, \ldots, n\}$ is given such that $v_i = g$ for all $i \in A$. Determine g.

2.3 Grey Level Penalty Function

Let $A \subset \{1, \ldots, n\}$ and $g \in \mathbb{R}$, as in Problem 3. Reorder the pixels $\{1, \ldots, n\}$ and the corresponding columns of $\boldsymbol{W}$, such that

$$\boldsymbol{W} = (\boldsymbol{W}_A \boldsymbol{W}_B), \tag{2}$$

where $\boldsymbol{W}_A$ contains the columns of $\boldsymbol{W}$ corresponding to the pixels in A. We then have

$$\boldsymbol{W}\boldsymbol{x} = (\boldsymbol{W}_A \boldsymbol{W}_B) \begin{pmatrix} \boldsymbol{g} \\ \boldsymbol{v}_B \end{pmatrix} = \boldsymbol{p}, \tag{3}$$

where $\boldsymbol{g}$ denotes a constant vector for which all entries are g. This leads to

$$\boldsymbol{W}_B \boldsymbol{v}_B = \boldsymbol{p} - \boldsymbol{W}_A \boldsymbol{g}, \tag{4}$$

which provides a necessary condition for a grey level estimate to be correct: Eq. (4) must have a solution. Clearly, this condition is not always sufficient, as emphasized by Prop. 1. Yet, if $|A|$ is a relatively large fraction of n, it can be expected that the condition is also sufficient, at least in many cases. Note that in practice, it may not be possible to solve Eq. (4) exactly, due to noise and discretization errors. Given A, one can measure the consistency of a grey level $\tilde{g}$ with the projection data $\boldsymbol{p}$ using the following *grey level penalty function*

$$P(\tilde{g}) = \min_{\boldsymbol{v}_B} ||\boldsymbol{p} - \boldsymbol{W}_A \tilde{\boldsymbol{g}} - \boldsymbol{W}_B \boldsymbol{v}_B||, \tag{5}$$

where $||\cdot||$ denotes a certain norm.

To minimize the grey level penalty, the penalty function must typically be evaluated in several points. A range of iterative methods are available for solving the minimization problem in Eq. (5). In this paper, we use the SIRT algorithm; see [8,9]. Let $\boldsymbol{v}^{(0)} = \boldsymbol{0}$. For $q = 1, 2, \ldots$, let $\boldsymbol{r}^{(q)} = \boldsymbol{p} - \boldsymbol{W}\boldsymbol{v}^{(q-1)}$. In each iteration q, the current reconstruction $\boldsymbol{v}^{(q-1)}$ is updated, yielding a new reconstruction $\boldsymbol{v}^{(q)}$, as follows:

$$v_j^{(q)} = v_j^{(q-1)} + \frac{1}{\sum_{i=1}^{n} w_{ij}} \sum_{i=1}^{m} \frac{w_{ij} r_i^{(q)}}{\sum_{j=1}^{m} w_{ij}}. \tag{6}$$

The SIRT algorithm converges to a solution of Eq. (5) where the norm to be minimized is a weighted sum of squares [9].

An important physical constraint in tomography problems, is that the attenuation coefficients of the scanned materials, corresponding to the grey levels in the reconstructed image, cannot be negative. Therefore, it seems that solving Eq. (5) for *nonnegative* $\boldsymbol{v}_B$ would yield a more powerful penalty function. However, this problem typically requires significantly more computation effort compared to the unconstrained variant (depending on the norm). As an alternative, we use a heuristic adaptation of the SIRT algorithm, where nonnegative entries of $\boldsymbol{v}_B$ are set to 0 after each iteration. This approach was already suggested in [8], and often results in more accurate reconstructions than the unconstrained version. We denote this version of SIRT by *SIRT-P*.

3 Estimation Approach

In this section, we propose the *Discrete Gray Level Selection* approach (DGLS) for estimating grey levels from projection data.

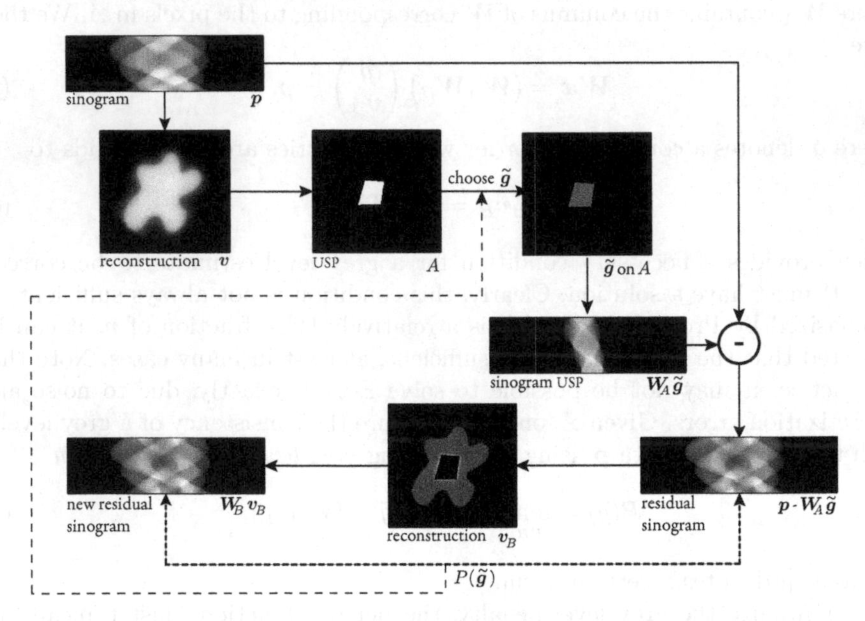

Fig. 3. Schematic overview of the Discrete Grey Level Selection procedure

Fig. 3 shows a schematic overview of the steps involved in estimating one or more grey levels. The first part of the procedure requires user interaction. Based on an initial reconstruction, obtained by a conventional (i.e., non-discrete) reconstruction algorithm, the user selects a region, the *user-selected part* (USP), for which it seems clear that it belongs to one of the grey levels. In this phase, a significant amount of implicit prior knowledge can be input by the user, such as the fact that the scanned object contains no holes, or knowledge about its topology. However, it can be difficult to select a proper region if the number of projections is small. The selected region should be as large as possible, but should certainly not contain any pixels that correspond to a different grey level. If the original image contains several grey levels, a region must be selected for each grey level that needs to be estimated. The regions A_i, along with the projection data, now form the input of the estimation algorithm.

The proposed algorithm estimates a single grey level at a time. After selecting the USP, an optimization algorithm is used to minimize the grey level penalty function. To evaluate the penalty function for a given grey level g, the sinogram $\boldsymbol{W}_A \boldsymbol{g}$ of the USP is first subtracted from the complete sinogram $\boldsymbol{p}$, forming the right-hand side of Eq. (4). Subsequently, consistency of the remaining part of the image with the remaining projection data is determined by evaluating the penalty function (e.g., using SIRT or SIRT-P). Based on the value of the penalty function, the grey level estimate is updated iteratively, until the penalty function is minimized. Evaluating the grey level penalty function can be computationally expensive. To find the minimum of this function using a small number of evaluations, Brent's method is used, as described in Chapter 5 of [10]).

4 Experiments and Results

Simulation experiments were performed based on four 256×256 phantom images: two binary images (Fig. 4a and b), one image with 3 grey levels (Fig. 4c) and one with 7 grey levels (Fig. 4d). From each phantom image, a parallel beam sinogram was computed with a varying number of projections, angular range, or noise level. Then, the phantom image was reconstructed using both the SIRT and SIRT-P algorithm. Finally, the grey levels of the reconstructed phantom were estimated, either with a simple median value (MED) of the user-selected part or with the proposed DGLS approach based on the same USP, using either SIRT or SIRT-P to compute the grey level penalty. For all images, the grey level of the background was assumed to be zero and was therefore not estimated. In all experiments, the USP for each grey level was computed by iteratively applying a binary erosion operation on the phantom region for that grey level, until a certain fraction of the pixels was left. Although in practice, the USP will be selected by the user, based on the reconstruction, using the phantom ensures that the USP is not affected by other parameters of the experiment. For both grey level estimation procedures, the absolute difference with respect to the true

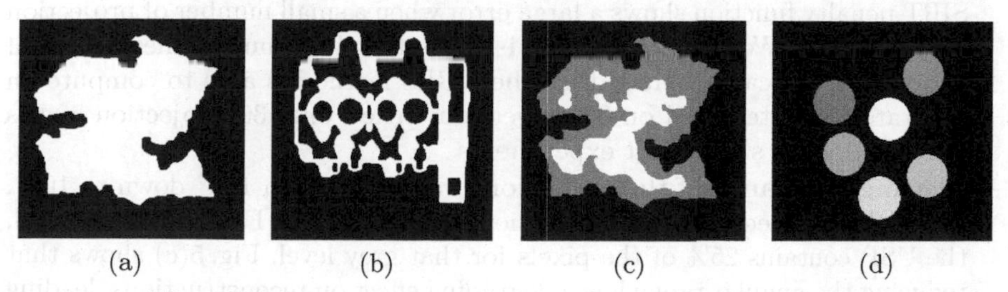

Fig. 4. Phantom images used for our simulation experiments

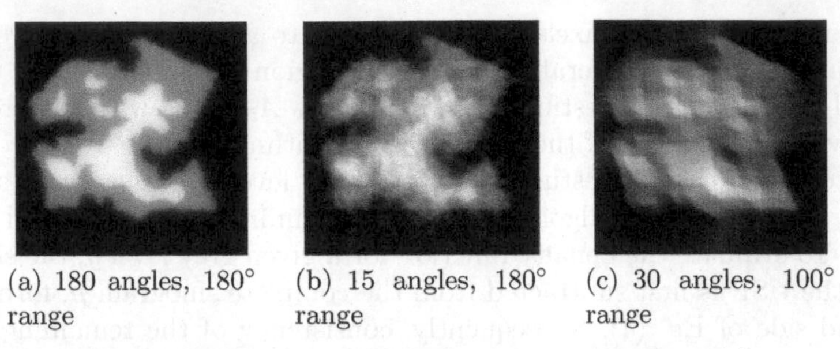

(a) 180 angles, 180° range (b) 15 angles, 180° range (c) 30 angles, 100° range

Fig. 5. SIRT reconstructions of phantom image 4(c)

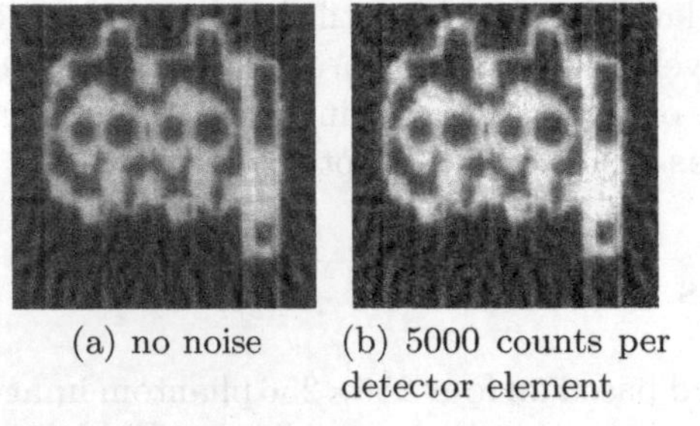

(a) no noise (b) 5000 counts per detector element

Fig. 6. SIRT reconstructions of phantom image 4(b)

Fig. 7. A USP of Phantom image 4(a)

grey level was computed. In case multiple grey levels had to be estimated (such as for phantom image 4(c) and 4(d)), the absolute differences were summed. The following series of experiments were run, varying only one parameter at a time:

- The **number of projections** from which the image was reconstructed was varied from 15 up to 180. In this experiment, the projections were equiangularly selected in the interval $[0°, 180°]$. For each grey level, the USP contains 25% of the pixels for that grey level. Fig. 5(a) and 5(b) show reconstructions of phantom image 4(c) for 180 and 15 projection angles, respectively.
  The results, plotted in Fig. 8, show that DGLS for both SIRT and SIRT-P generally yields much more accurate estimations than the MED estimations. Only when there are very few projection angles (e.g. 15), a significant error is visible. In phantom image 4(c) and 4(d), DGLS based on the unconstrained SIRT penalty function shows a large error when a small number of projection angles is used. We believe that this is related to the nonuniqueness issues of Prop. 1. As we can conclude that the DGLS method is able to compute an accurate estimate based on 30 projection angles, only 30 projection angles were used in all subsequent experiments.
- The **angular range** of the projections was varied from 180° down to 100°, from which 30 equiangular projections were selected. For each grey level, the USP contains 25% of the pixels for that grey level. Fig. 5(c) shows that reducing the angular range has a degrading effect on reconstructions, leading to artifacts. Fig. 9 shows that although the estimation error in absolute value

is larger than in the previous experiment, the DGLS error is significantly lower than the error of the median value.
- The **noise level** of the projections was varied. In this experiment, Poisson noise was applied to the projection data, for a varying source intensity. By *number of counts per detector element*, we refer to the measured detector count when there is no object blocking the path from source to detector. The higher this count, the higher the Signal-to-Noise ratio. For the reconstruction, 30 projection angles were equally spaced in $[0°, 180°]$. For each grey level, the USP again contains 25% of the pixels for that grey level. The effect of poisson noise on a reconstruction is visible in Fig. 6. Fig. 10 shows that for the investigated noise range, the noise level does not have a major impact on the estimation error, for both the MED and DGLS estimates. This can be explained by the fact that only 30 projections are used, such that the reconstruction errors due to the small number of projections are much more significant than those related to the noise in the projection data.
- The **size of the USP**, as a percentage of the total region for each grey level, was varied by iteratively applying an erosion filter as discussed above. Fig. 7 shows the user-selected part for phantom 4(a), 20% the size of the original phantom. Fig. 11 shows that accurate grey level estimation is possible even if the USP is relatively small, down to 10% of the object size.

The results suggest that the DGLS method is robust with respect to the number of projection angles, the angular range, the level of noise and the size of the user-selected part. Moreover, DGLS typically yields a significantly more accurate estimate compared to computing the median value of the user-selected part. The results also demonstrate that more accurate estimation can be achieved by using the heuristic SIRT-P method as a scoring function instead of SIRT.

On our test PC, running at 2.4GHz, the running time of a single grey level estimation in all experiments but the first one was around 4 minutes. This is mostly attributed to the SIRT or SIRT-P algorithm that has to be performed every time while minimizing the grey level penalty function.

5 Discussion

The grey level estimation problem, when posed in its general form, does not guarantee a unique solution. The experimental results show that even for a moderately large number of 15 projections, additional constraints may be necessary to obtain an accurate estimate of the grey levels using our proposed approach. Several techniques can be used to improve the accuracy of the estimate:

- **Use a More Accurate Scoring Function.** The projection distance, as computed using SIRT), does not incorporate nonnegativity constraints. Adding a nonnegativity heuristic to the reconstruction algorithm seems to result in improved accuracy of the grey level estimate, but its theoretical properties are hard to verify. A scoring function that incorporates both minimization of the projection distance and nonnegativity constraints (based on the nonnegative least squares problem) could result in more accurate scoring, at the expense of long running times.

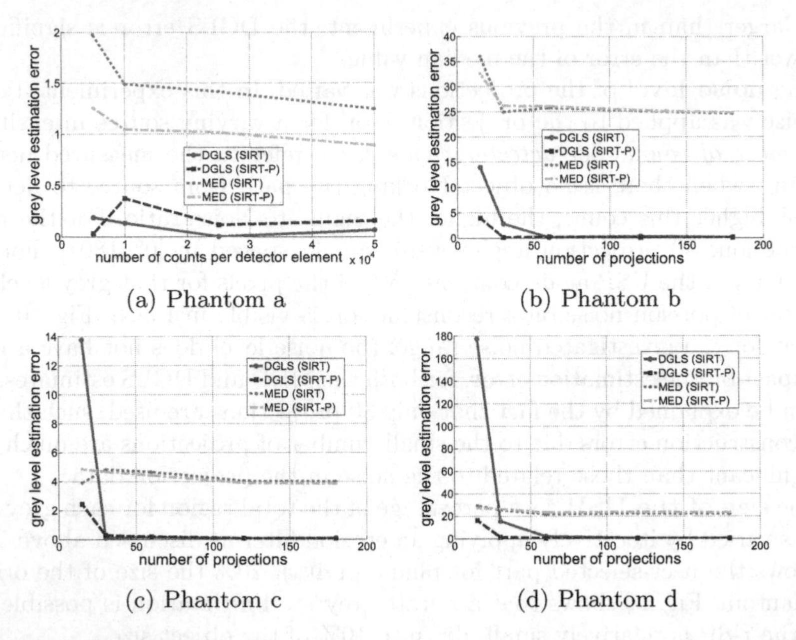

(a) Phantom a (b) Phantom b

(c) Phantom c (d) Phantom d

Fig. 8. Error of grey level estimation with varying numbers of projection angles

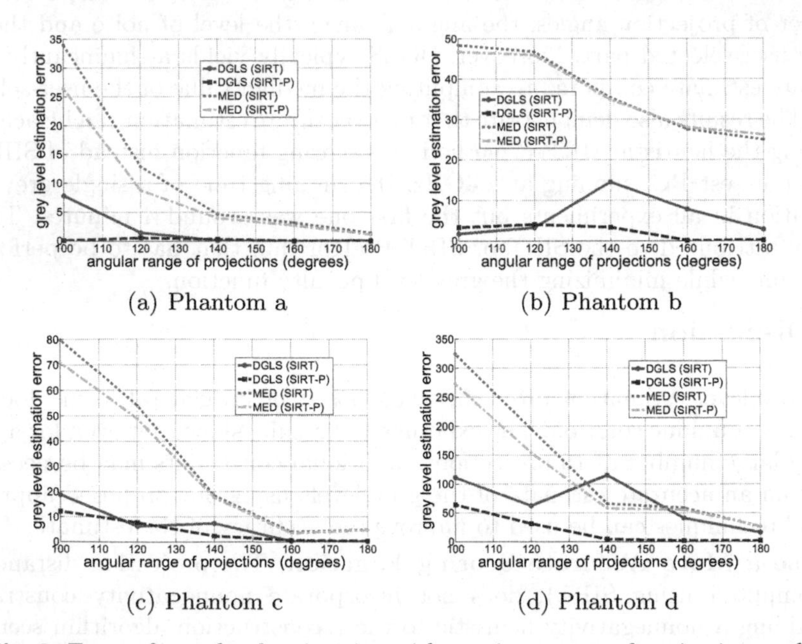

(a) Phantom a (b) Phantom b

(c) Phantom c (d) Phantom d

Fig. 9. Error of grey level estimation with varying ranges of projection angles

– **Simultaneous Estimation of Several Grey Levels.** In the approach presented in this paper, each of the grey levels is estimated independently. Simultaneous estimation of all grey levels, where the grey level in all user-selected

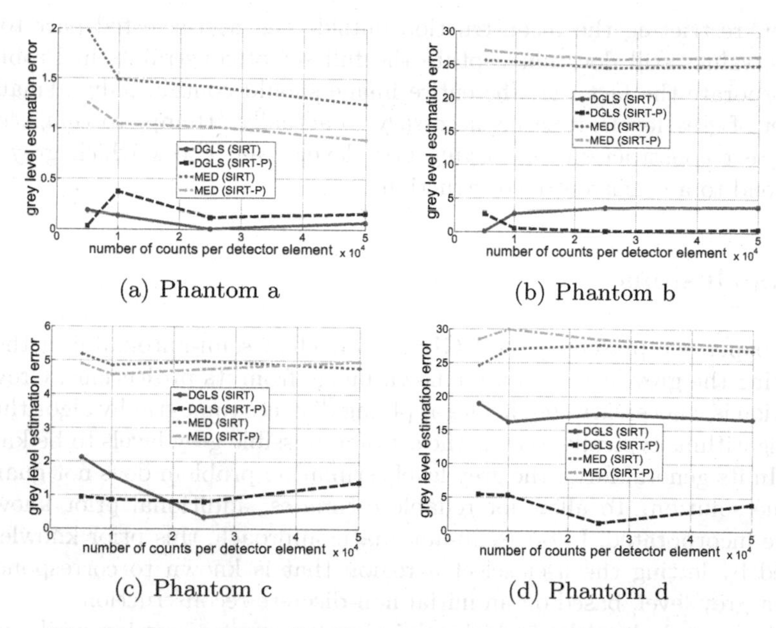

(a) Phantom a (b) Phantom b

(c) Phantom c (d) Phantom d

Fig. 10. Error of grey level estimation with different noise levels

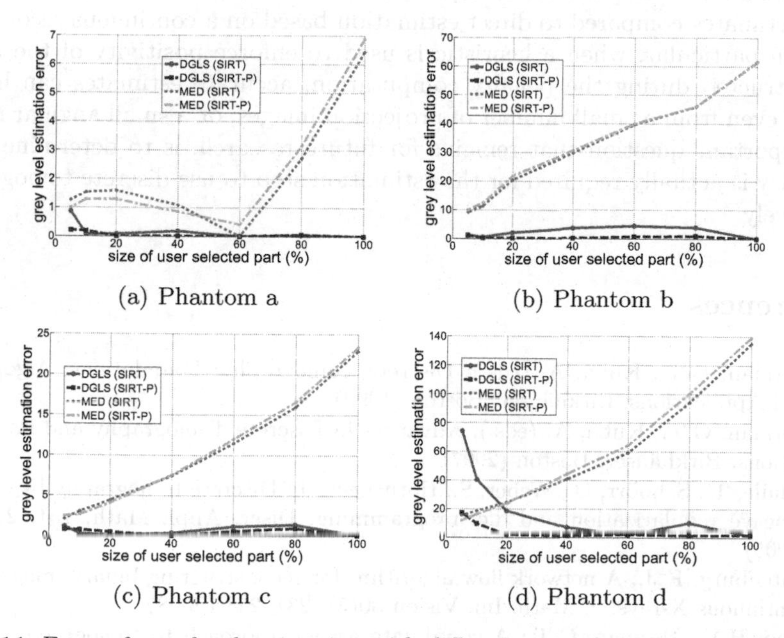

(a) Phantom a (b) Phantom b

(c) Phantom c (d) Phantom d

Fig. 11. Error of grey level estimation with different sizes of the user-selected part

parts is required to be constant, would provide more constraints for the estimation problem. However, this also leads to a higher dimensional search space (one dimension for each grey level), leading to longer running times.

However, restricting the reconstruction outside the user-selected part to non-negative values still does not capture the full set of constraints in Problem 1. To incorporate the fact that the entire image should contain only a small, discrete set of grey levels, it seems necessary to actually attempt to *compute* such a discrete reconstruction for varying grey levels, and check which grey levels correspond to a consistent reconstruction.

6 Conclusions

In this paper, we presented the DGLS approach: a semi-automatic method for estimating the grey levels of an unknown image from its projections. Grey level estimation is a necessary step before applying discrete tomography algorithms, as these algorithms typically assume the set of admissible grey levels to be known a priori. In its general form, the grey level estimation problem does not guarantee a unique solution. To allow for reliable estimates, additional prior knowledge must be incorporated. In our semi-automatic approach, this prior knowledge is included by letting the user select a region that is known to correspond to a constant grey level, based on an initial non-discrete reconstruction.

The proposed algorithm, which minimizes a penalty function while varying the grey level of the user-selected part, was shown to yield more accurate grey level estimates compared to direct estimation based on a continuous reconstruction. In particular, when a heuristic is used to enforce positivity of the image reconstructed during the penalty computation, accurate estimates can be obtained even from a small number of projection images, or a small angular range. An important question that remains for future research, is to determine what accuracy is actually required for the estimation step to use discrete tomography effectively.

References

1. Herman, G.T., Kuba, A. (eds.): Discrete Tomography: Foundations, Algorithms and Applications. Birkhäuser, Boston (1999)
2. Herman, G.T., Kuba, A. (eds.): Advances in Discrete Tomography and its Applications. Birkhäuser, Boston (2007)
3. Schüle, T., Schnörr, C., Weber, S., Hornegger, J.: Discrete tomography by convex-concave regularization and d.c. programming. Discr. Appl. Math. 151, 229–243 (2005)
4. Batenburg, K.J.: A network flow algorithm for reconstructing binary images from continuous X-rays. J. Math. Im. Vision 30(3), 231–248 (2008)
5. Liao, H.Y., Herman, G.T.: A coordinate ascent approach to tomographic reconstruction of label images from a few projections. Discr. Appl. Math. 151, 184–197 (2005)
6. Alpers, A., Poulsen, H.F., Knudsen, E., Herman, G.T.: A discrete tomography algorithm for improving the quality of 3DXRD grain maps. J. Appl. Crystall. 39, 582–588 (2006)

7. Batenburg, K., Bals, S., Sijbers, J., et al.: 3D imaging of nanomaterials by discrete tomography. Ultramicroscopy 109, 730–740 (2009)
8. Gilbert, P.: Iterative methods for the three-dimensional reconstruction of an object from projections. J. Theoret. Biol. 36, 105–117 (1972)
9. Gregor, J., Benson, T.: Computational analysis and improvement of SIRT. IEEE Trans. Medical Imaging 27(7), 918–924 (2008)
10. Brent, R.P.: Algorithms for minimization without derivatives. Prentice-Hall, Englewood Cliffs (1973)

The 1-Color Problem and the Brylawski Model

S. Brocchi[1], A. Frosini[1], and S. Rinaldi[2]

[1] Dipartimento di Sistemi e Informatica
Università di Firenze, Firenze, Italy
brocchi@dsi.unifi.it, andrea.frosini@unifi.it
[2] Dipartimento di Scienze Matematiche ed Informatiche
Università di Siena, Siena, Italy
rinaldi@unisi.it

Abstract. In discrete tomography, the *1-color* problem consists in determining the existence of a binary matrix with row and column sums equal to some given input values arranged in two vectors. These two vectors are said to be compatible if the associated *1-color* problem has at least a solution. Here, we start from a vector of projections, and we define an algorithm to compute all the vectors compatible with it, then we show how to arrange them in a partial order structure, and we point out some of its combinatorial properties. Finally, we prove that this poset is a sublattice of the Brylawski lattice too, and we check some common properties.

1 Introduction

The *1-color* problem is a decision problem which asks for the existence of a binary matrix compatible with two integer vectors, called horizontal and vertical projections, whose entries count the number of elements 1 in each row and column of the matrix, respectively. This problem has been approached by Ryser in [9] as a study on combinatorial properties of matrices of zeros and ones. His results are considered to be the milestones in discrete tomography, a discipline that studies geometrical properties of discrete sets of points from projections or from partial information on them.

One of the principal motivations of this discipline is the study of crystalline structures from information obtained through electron beams that cross the material, as described in [8] and [10]; many problems in discrete tomography are also related to network flow, timetabling or data security (see [6], [7] for a survey). Theoretical results on the *1-color* problem are often used in these more complex models.

If there exists a binary matrix whose vectors of projections are H and V, then we say that H and V are compatible. In this work we define a simple procedure to list all the vertical projections which are compatible with a given vector of horizontal projections. The obtained elements can be easily arranged in a partial order structure that can be proved to be a lattice. In order to do this, we go deeper into the strong connections between *1-color* and the Brylawski model (see

S. Brlek, C. Reutenauer, and X. Provençal (Eds.): DGCI 2009, LNCS 5810, pp. 530–538, 2009.

[2]), a structure which is related to a very different context: defined to study the integer partitions, it is often considered the ancestor of many dynamical models used for physical simulations of realistic environments (see i.e. [4]). In fact, the Brylawski model can be considered as a special case of the Ice Pile Model, described in [5].

In [2], the author uses some properties of the model to enumerate the matrices of 0s and 1s with prescribed row and column sums, and he furnishes a lower bound for their number which turns out to be exact in some cases; these studies take into account all the switching components inside a binary matrix, i.e. those elements that can be modified (switched) without altering both the vectors of projections.

Here different aspects of the *1-color* problem have been considered, in particular we study how a vector of projections can be changed in order to maintain its compatibility with a given second one in its orthogonal direction. Finally we show that all the computed vectors still form a lattice that inherits many properties of that of Brylawski.

This approach reveals its usefulness each time the vectors of projections are not given as an input, but they have to be computed through the decomposition of a more complex problem, as in the *n-color* case. For example, in [1] the authors decompose some instances of *2-color* into four *1-color*'s ones that allow a solution to exist.

Exploiting properties of this model, we hope to be able in future works to furnish solving algorithms for special instances of the *n-color* problem by using a *divide et impera* approach. The general idea is to decompose a problem in many *1-color* subproblems, then solve each of them, and finally merge together all the solutions, eventually taking care of some still unplaced elements, maintaining the consistency of the final solution.

This strategy relies on the property of the described lattice of Brylawski to detect elements to move from a subproblem to another without precluding the existence of a solution, in the intent of creating void positions where elements of different colors may find their placement.

It is remarkable that such a simple model can be applied as an invaluable tool to investigate problems in different research fields ranging from discrete tomography to dynamic models and combinatorics, allowing one to consider them in a unified perspective.

2 The *1-Color* Problem

Let us consider a $m \times n$ binary matrix $M = (m_{i,j})$, with $1 \leq i \leq m$ and $1 \leq j \leq n$; we define its horizontal and vertical projections as the vectors $H = (h_1, \dots, h_m)$ and $V = (v_1, \dots, v_n)$ of the sums of its elements for each row and each column, respectively, i.e.

$$h_i = \sum_{j=1 \dots n} m_{i,j} \quad \text{and} \quad v_j = \sum_{i=1 \dots m} m_{i,j}.$$

Let us consider the following problem:

1-Color (H, V)

Input: two vectors of integers H and V of length m and n respectively, such that for each $1 \leq i \leq m$ and $1 \leq j \leq n$, it holds
- $0 \leq h_i \leq n$;
- $0 \leq v_j \leq m$;
- $\sum h_i = \sum v_j$.

Question: does there exist a binary matrix such that H and V are its horizontal and vertical projections, respectively?

The *1-color* problem is known as *Consistency* problem, as well. Two vectors H and V are *feasible*, if they meet the requirements of the *1-color* problem; furthermore, they are *compatible* if *1-color* (H, V) has a positive answer.

We observe that from each matrix M having projections H and V we can compute a matrix M' having projections H' and V', where H' and V' are obtained from H and V by sorting their elements in decreasing order, respectively, by switching the rows and the columns of M. So, without loss of generality, from now on we will consider only vectors of projections sorted in decreasing order.

Starting from the vector $H = (h_1, \dots, h_m)$, let us define the vector $\overline{H} = (\overline{h}_1, \dots, \overline{h}_n)$ as follows:

$$\overline{h}_j = |\{\, h_i \in H \,:\, h_i \geq j \,\}| \,.$$

In [9], Ryser gives a necessary and sufficient condition for a couple of vectors to be compatible, i.e.

Theorem 1. *(Ryser, [9]) Let H and V be a couple of compatible vectors. The problem* 1-color (H, V) *has a positive answer if and only if it holds*

$$\sum_{j=l}^{n} v_j \geq \sum_{j=l}^{n} \overline{h}_j, \qquad \textit{with } 2 \leq l \leq n.$$

Obviously the vectors H and $\overline{H}$ are compatible; as a matter of fact the $m \times n$ matrix whose i-th row has h_i elements 1 followed by $n - h_i$ elements 0 has H and $\overline{H}$ as projections, and it is called *maximal.* Figure 1 shows an example of a matrix which is compatible with a couple of maximal vectors.

In the following we give an alternative characterization of the couples of consistent vectors that allows the construction of a poset with ease.

An immediate check reveals that the class of the matrices whose horizontal and vertical projections are H and V respectively, say $\mathcal{U}(H, V)$, reduces to a singleton if and only if $V = \overline{H}$ by showing that this element admits no switching components. Such a property of the vector $\overline{H}$ allows us to choose it as the minimum of the poset we define in Paragraph 3.

Now we consider the following procedure *Sup* that starts from an integer vector V and gives as output a vector V' equal to V except in two of its elements; we remark that *Sup* acts non deterministically in the choice of a couple of column indexes where the elements of V are modified:

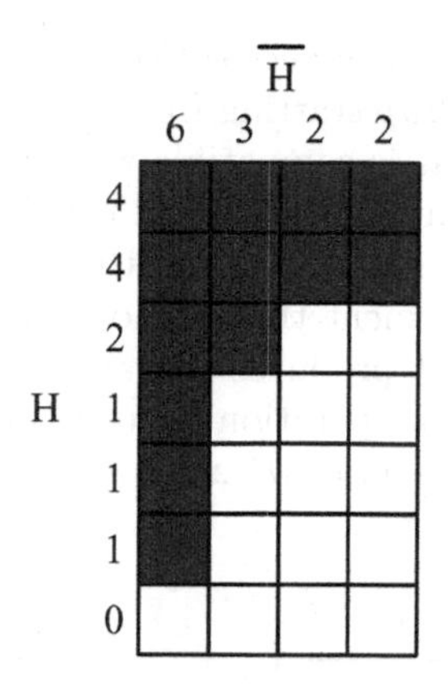

Fig. 1. The computation of the vector $\overline{H}$ and the maximal matrix associated to H and $\overline{H}$

Sup (V)

Step 1: Initialize $V' = V$;
Step 2: *If* there exist two column indexes j, j' such that $v_j - v_{j'} > 1$, and $v_j > v_{j+1}$ and $v_{j'-1} > v_{j'}$
then compute

$$v'_{j'} = v_{j'} + 1 \quad \text{and} \quad v'_j = v_j - 1;$$

and give V' as output
else give V' as output.

The hypothesis $v_j > v_{j+1}$ and $v_{j'-1} > v_{j'}$ maintain the array V' sorted. They impose that if in V there are many consecutive elements with the same value, then *Sup* chooses only one of them. Since we consider only ordered vectors, such a choice can be done without loss of generality.

We will write $Sup(V) \longrightarrow V'$ if V' can be obtained from V through an application of *Sup*.

Theorem 2. *Consider two compatible vectors H and V, and let $Sup(V) \longrightarrow V'$. The vectors H and V' are compatible.*

Proof. If $V' = V$ then the result is trivial, otherwise starting from a matrix M which is compatible with H and V, we construct a new matrix M' compatible with H and V' as follows: let j and j' be the column indexes modified by the procedure *Sup*. Since $v_j > v_{j'} + 1$, then there exist two elements $m_{i,j} = 1$, and $m_{i,j'} = 0$. The matrix M' is equal to M except for the two elements $m'_{i,j} = 0$, and $m'_{i,j'} = 1$. In this way the horizontal projections are the same in the two matrices M and M', while this latter has vertical projections equal to V'. □

Corollary 1. *Let V and V' be two vectors compatible with H. If $Sup(V) \longrightarrow V'$, then $|\mathcal{U}(H, V)| \leq |\mathcal{U}(H, V')|$.*

We start by showing that each vector V compatible with H can be obtained by iterating the procedure *Sup* starting from $\overline{H}$.

Consider a matrix M that is compatible with H and V as in Figure 2 (left). The sliding of all the elements of the matrix till the leftmost available positions in the same row, produces a matrix M' with projections H and $\overline{H}$, as shown in Figure 2 (right). Each element that is moved from a column j to a column $j' < j$ to obtain M' from M produces a change in V which can be considered as the application of the *Sup* operation to it, consequently all the moves of the elements which transform M into M' are a sequence of applications of *Sup* to derive $\overline{H}$ from V, and viceversa.

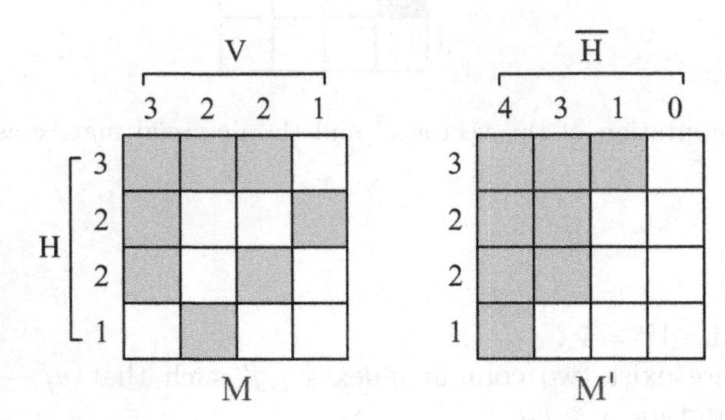

Fig. 2. Sliding all of the elements of a matrix M with projections H and V (left) to the leftmost possible positions, we obtain a matrix M' with projections H and $\overline{H}$ (right). Analyzing the differences between the two matrices, one can obtain all the applications of *Sup* needed to obtain V from $\overline{H}$.

Let us define $\mathcal{U}_n(H, \star)$ the class of the vectors which have dimension n, and which are compatible with H. Given two elements V and V' of $\mathcal{U}_n(H, \star)$, we say that V' is greater than or equal to V, say $V \leq_S V'$, if V' can be reached from V with a series (eventually void) of successive applications of *Sup*. From Theorem 2, it holds that V' is greater than V if the vectors which are compatible with V are a subset of those compatible with V'.

We observe that $\overline{H}$ is the minimum of the class $\mathcal{U}_n(H, \star)$; similarly, we can also define the maximum element of $\mathcal{U}_n(H, \star)$ as follows:

Theorem 3. *Let H be an integer vector. If*

$$V_M = (k, k, \ldots, k, k-1, \ldots, k-1)$$

is a vector feasible with H, then it is also compatible with H. Furthermore, if V is a vector compatible with H of the same dimension as V_M, then it holds $V \leq_S V_M$.

The theorem can be easily verified by observing that V_M can be obtained from any vector feasible with H through a finite number of applications of *Sup*: an array in this form is the only fixed point of the *Sup* operation. This can be easily understood acting symmetrically to what already done in the proof of Corollary 1.

This result was also proved by induction in [1] with the purpose to furnish a property that allows the decomposition of a two color problem of a certain family in several one color problems that always admit a solution.

3 The Brylawski Model

Let us recall the following two standard definitions: a *partially ordered set*, say *poset*, is a set P together with a binary relation $\leq$ on it which is reflexive, transitive and anti-symmetric. A poset in which any two elements have a unique *sup* and a unique *inf* element is said to be a *lattice*. An immediate check reveals that the couple $(\mathcal{U}_n(H,\star), \leq_S)$ is a poset.

Let H be a given integer vector, and let V be an integer vector whose elements sum to s, and which is compatible with H. Since by hypothesis the elements of V are ordered in decreasing order, we may regard V as an integer partition.

So, we relate the above defined order on $\mathcal{U}(H,\star)$ to the *dominance order* on the integer partition of s, defined by Brylawski in [2] as follows: given two integer partitions V and V'

$$V \leq_D V' \quad \text{if and only if} \quad \sum_{j=1}^{k} v_j \geq \sum_{j=1}^{k} v'_j$$

for each $1 \leq k \leq n$, with n being the common dimension of V and V', as usual.

The Brylawski model is very popular in combinatorics, and it can be viewed as an extension of the well-known sandpile model (see [3],[4]), where, under some assumptions the grains of sand are allowed to slide.

We immediately realize that the dominance order bases on the condition given by Ryser in Theorem 1; hence we can express the consistency of two vectors H and V using it:

Lemma 1. *H is compatible with V if and only if $\overline{H} \leq_D V$.*

It is immediate to verify that if $V \leq_S V'$, then it holds $V \leq_D V'$; such a correspondence allows us to use some results of Brylawski in the discrete tomography environment.

In particular, we notice that the procedure *Sup* applied to one element V in $\mathcal{U}_n(H,\star)$ does not compute the elements of the same class which lie immediately below it in the poset $(\mathcal{U}_n(H,\star), \leq_S)$, as shown in Fig. 3: let us consider the vector $\overline{H} = (3,1,0)$ which is the maximum element of the class $\mathcal{U}_n(H,\star)$, with $H = (2,1,1)$ (Fig. 3,(a)). The computation of *Sup* on $\overline{H}$ produces $(3,1,0) \longrightarrow (2,2,0)$ and $(3,1,0) \longrightarrow (2,1,1)$ (Fig. 3,(b)). Unfortunately, the obtained poset structure needs to be refined since $(2,1,1)$ is not an immediate successor of $\overline{H}$, as pointed out by the application of *Sup* to $(2,2,0)$ (Fig. 3,(c)). This makes the partial order quite difficult to analyze.

Brylawski defines two simple rules to avoid the problem, and he uses them to directly arrange the integer partition of s in a structure that turns out to be a lattice: let V be an integer partition of s, we construct an integer partition V' as follows

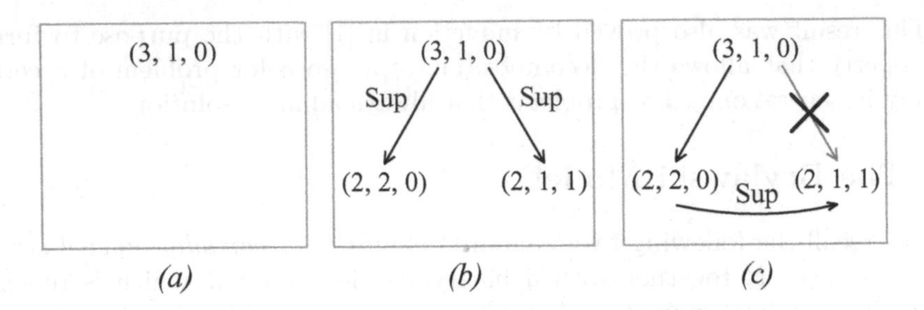

Fig. 3. The steps to determine the poset structure of $\mathcal{U}(H,\star)$, with $H=(2,1,1)$. The minimum and the maximum elements of the poset are $\overline{H}=(3,1,0)$ and $H_M=(2,1,1)$, respectively.

Vertical rule: if $v_i \geq v_{i+1}+2$, then $v'_i = v_i - 1, v'_{i+1} = v_{i+1}+1$, with $1 \leq i \leq n$. The other elements of V and V' coincide.

Horizontal rule: if there exist two indexes i, k such that

- $v_i - v_{i+k} = 2$;
- $v_{i+j} = v_i - 1$ for each $0 < j < k$;
- $v_{i+k} = v_i - 2$,

then $v'_i = v_i - 1, v'_{i+k} = v_{i+k}+1$. The other elements of V and V' coincide.

Brylawski proves that the order $\leq_B$ such that $V \leq_B V'$ if and only if V' can be derived from V by applying one of these two rules, is equivalent to the order $\leq_D$. Consequently it also holds that if $V \leq_S V'$, then $V \leq_B V'$.

This can be nicely rephrased saying that the procedure *Sup* intrinsically contains both these rules, but sometimes they act together in a single appliance of *Sup*.

Remark 1:

The procedure *Sup* preserves the length of its input vector, while if V and V' are respectively the input and the output partitions of the two rules of Brylawski, then it holds $0 \leq |V'| - |V| \leq 1$, with $|V|$ and $|V'|$ being the number of elements of V and V', respectively. This depends from the fact that *Sup* acts on two elements of an array by modifying only their values, and not their number, while the rules of Brylawsky are defined on integer partitions, without any constraints on the number of their elements except those due to the cardinality of the partition sum, i.e. a partition of n elements could generate a partition with $n+1$ elements by adding a 1 in the $(n+1)$th position; as an example the two element partition $(4,3)$ could generate the three element one $(4,2,1)$.

The following result holds:

Theorem 4. *The structure $(\mathcal{U}_n(H,\star), \leq_S)$ is a lattice.*

Proof. Since $\mathcal{U}_n(H,\star)$ has a maximum element $\overline{H}$, then each of its subsets has at least a *sup*. Now we prove that the *sup* of two of its elements, say V and V', is unique. Let assume $\hat{V} = sup\{V, V'\}$ in the Brylawski lattice. Since $|\overline{H}| = |\hat{V}|$ and $|\overline{H}| \leq |\hat{V}| \leq |V|$, then $|\hat{V}| = |V|$ by Remark 1. So there exists a sequence of applications of the Brylawski rules starting from $\overline{H}$ and leading to $\hat{V}$. Each of this application can be performed by an application of the procedure Sup, so $\hat{V} \in \mathcal{U}_n(H, \star)$ as desired. A similar argument holds to define the minimum of two elements of $\mathcal{U}_n(H, \star)$. □

So, just to make a comparison between the Brylawski lattice of the partition of an integer s, and the correspondent lattice $\mathcal{U}(H, \star)$, we may note that the latter contains only the elements having with limited dimensions. Obviously the two lattices collapse if we assume $n \geq s$.

Final Remarks:

Many new questions arise from the study of the lattice $(\mathcal{U}(H, \star), \leq_S)$, and interesting results on the Brylawski model naturally reflects as new properties on it. As an example, since it holds that $\overline{\overline{H}} = H$, then there exists a strong duality between the classes $(\mathcal{U}(H, \star), \leq_S)$ and $(\mathcal{U}(\star, \overline{H}), \leq_S)$, i.e. the class of all the horizontal projections consistent with $\overline{H}$.

In particular we have that if $H \leq H'$, then $\overline{H'} \leq \overline{H}$, so the procedure Sup inverts the order $\leq_S$. As an example, in Fig. 4 there are depicted two lattices: the one on the left is $(\mathcal{U}(H, \star), \leq_S)$, with $H = (2, 2, 1, 1, 1)$, while the one on the right is the upside-down version of its dual. The elements are linked by the *overline* operator.

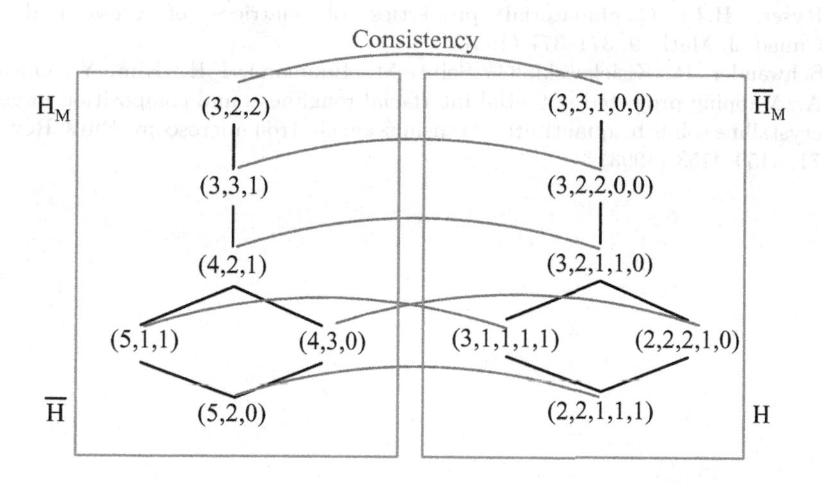

Fig. 4. Integer partitions of the number 7. To the left: poset of partitions with 3 elements, and no element greater than 5. To the right: poset of partitions with 5 elements, and no element greater than 3.

Obviously both the two lattices can be found embedded in the Brylawski lattice of the partition of the number seven. This implies that each Brylawski lattice is composed by many reverse symmetrical parts.

Finally, as an immediate consequence of the above observation, it holds the following combinatorial result:

Theorem 5. *The number of integer partitions of s having at most length m, and whose maximum value is n is equal to the number of integer partitions of s having at most length n, and whose maximum value is m.*

References

1. Brocchi, S., Frosini, A., Rinaldi, S.: Solving some instances of the two color problem (submitted)
2. Brylawski, T.: The lattice of integer partitions. Discrete Math. 6, 210–219 (1973)
3. Goles, E., Kiwi, M.A.: Games on line graphs and sand piles. Theoret. Comput. Sci. 115, 321–349 (1993)
4. Goles, E., Latapy, M., Magnien, C., Morvan, M., Phan, H.D.: Sandpile Models and Lattices: A Comprehensive Survey. Theoret. Comput. Sci. 322(2), 383–407 (2004)
5. Goles, E., Morvan, M., Phan, H.D.: Sand piles and order structure of integer partitions. Discrete Appl. Math. 117, 51–64 (2002)
6. Herman, G., Kuba, A. (eds.): Discrete Tomography: Foundations, Algorithms and Applications. Birkhauser, Basel (1999)
7. Herman, G., Kuba, A. (eds.): Advances in Discrete Tomography and Its Applications. Birkhauser, Basel (2007)
8. Kisielowski, C., Schwander, P., Baumann, F.H., Seibt, M., Kim, Y., Ourmazd, A.: An approach to quantitative high-resolution transmission electron microscopy of crystalline materials. Ultramicroscopy 58, 131–155 (1995)
9. Ryser, H.J.: Combinatorial properties of matrices of zeros and ones. Canad. J. Math. 9, 371–377 (1957)
10. Schwander, P., Kisielowski, C., Seibt, M., Baumann, F.H., Kim, Y., Ourmazd, A.: Mapping projected potential interfacial roughness, and composition in general crystalline solids by quantitative transmission electron microscopy. Phys. Rev. Lett. 71, 4150–4153 (1993)

Author Index